Hydrology

Water in its different forms has always been a source of wonder, curiosity and practical concern for humans everywhere. This textbook presents a coherent introduction to many of the concepts and relationships needed to describe the distribution and transport of water in the natural environment.

Continental water transport processes take place above, on and below the Earth's surface, and consequently the book is split into four main parts. Part I deals with water in the atmosphere. Part II introduces the transport of water on the surface. Water below the surface is the subject of Part III. Part IV is devoted to flow phenomena at the basin scale and statistical concepts useful in the analysis of hydrologic data. Finally, the book closes with a brief history of ideas concerning the hydrologic cycle. Hydrologic phenomena are dealt with at spatial and temporal scales at which they occur in nature. The physics and mathematics necessary to describe these phenomena are introduced and developed, and readers will require a working knowledge of calculus and basic fluid mechanics.

Hydrology – An Introduction is a textbook that covers the fundamental principles of hydrology, based on the course that Wilfried Brutsaert has taught at Cornell University for the past 30 years. The book will be invaluable as a textbook for entry-level courses in hydrology directed at advanced seniors and graduate students in physical science and engineering. In addition, the book will be more broadly of interest to professional scientists and engineers in hydrology, environmental science, meteorology, agronomy, geology, climatology, oceanology, glaciology and other Earth sciences.

WILFRIED BRUTSAERT is William L. Lewis Professor of Engineering at Cornell University. In a long and prestigious career in the research and teaching of hydrology, Professor Brutsaert has received many awards and honors, including: the Hydrology Award and Robert E. Horton Medal, American Geophysical Union; President, Hydrology Section, American Geophysical Union, from 1992 to 1994, Fellow of the American Geophysical Union and American Meteorological Society; the Ray K. Linsley Award, American Institute of Hydrology; Walter B. Langbein Lecturer, American Geophysical Union; International Award, Japan Society of Hydrology & Water Resources; Jule G. Charney Award, American Meteorological Society. He is a member of the National Academy of Engineering and has published two previous books, *Evaporation into the Atmosphere: Theory, History and Applications* (D. Reidel Publishing Company, 1982), and *Gas Transfer at Water Surfaces* (with G. H. Jirka, D. Reidel Publishing Company, 1984). He has authored and co-authored more than 170 journal articles.

HYDROLOGY
AN INTRODUCTION

WILFRIED BRUTSAERT

Cornell University

CAMBRIDGE UNIVERSITY PRESS
Cambridge, New York, Melbourne, Madrid, Cape Town, Singapore, São Paulo

Cambridge University Press
The Edinburgh Building, Cambridge CB2 2RU, UK

Published in the United States of America by Cambridge University Press, New York

www.cambridge.org
Information on this title: www.cambridge.org/9780521824798

First published 2005

Printed in the United Kingdom at the University Press, Cambridge

A catalog record for this book is available from the British Library

Library of Congress Cataloging in Publication data

ISBN-13 978-0-521-82479-8 hardback
ISBN-10 0-521-82479-6 hardback

CONTENTS

FOREWORD

Water in its different forms has always been a source of wonder, curiosity and practical concern for humans everywhere. The goal of this book is to present a coherent introduction to some of the concepts and relationships needed to describe the distribution and transport of water in the natural environment. Thus it is an attempt to provide a more thorough understanding, and to connect the major paradigms that bear upon the hydrologic cycle, that is the never-ending circulation of water over the continents of the Earth.

Continental water transport processes take place above, on and below the Earth's landsurfaces. Accordingly, in Part I, water is considered as it passes through the lower atmosphere; this part consists of a general description of atmospheric transport in Chapter 2, followed by the application of these concepts to precipitation and evaporation in Chapters 3 and 4, respectively. In Part II, water transport on the Earth's surface is dealt with; this part consists of a general description of the hydraulics of free surface flow in Chapter 5, which is then applied to overland runoff and streamflow routing in rivers in Chapters 6 and 7, respectively. Water below the surface is the subject of Part III; again, a general introduction to flow in porous materials in Chapter 8 is followed by applications to phenomena involving infiltration and capillary rise in Chapter 9, and groundwater drainage and baseflow in Chapter 10. Part IV is devoted to flow phenomena, mostly fluvial runoff, in response to precipitation at the catchment and river basin scales, which result from the combination of flows both above and below the Earth's surface, already treated at smaller scales separately in Parts II and III. Various interactions of these flow phenomena and the major paradigms regarding the subscale mechanisms are described in Chapter 11. This is followed by a treatment of the available parameterizations in Chapter 12. In Chapter 13 the fourth part of the book concludes with a brief description of some of the more common statistical concepts that are useful in the analysis of hydrologic data. Finally, as an afterword, Chapter 14 closes the book with a brief history of the ideas on the water cycle, which over the centuries evolved to our present understanding; Santayana's dictum may be a bit worn by now, but several recent reinventions of the hydrologic wheel could have been avoided, if the past had been better remembered.

These transport phenomena in the hydrologic cycle on land are treated at spatial and temporal scales, at which they are commonly encountered in everyday life and at which they are tractable with presently available data. Hydrology is a physical science, and the language of physics is mathematics. Accordingly, plausible assumptions are introduced and the mathematical formulations and parameterizations are derived, which describe the more relevant mechanisms involved in the different phases of the continental hydrologic cycle. The resulting equations are then examined and, if possible, solved for certain prototype situations and boundary conditions. The motivation for this is, first, to gain a better understanding of their structure and underlying assumptions, and of the physics they are intended to represent; and second, to provide the basis and background for more complex modeling exercises, simulations and predictions in practical applications.

The subject material covered in this book grew out of the lecture notes for my courses in hydrology and related topics in the School of Civil and Environmental Engineering, at Cornell University. I have not tried to cover all possible angles and points of view of the subject matter. Rather, I have followed a line of thought, which over the years I have come to find effective in conveying a broad understanding of the more important phenomena, and in stimulating further inquiry in the subject. Similarly, no attempt has been made to compile a complete bibliography. But the references that are listed refer to other works, so that it should be possible to trace back the more important developments.

As its subtitle indicates, this book is intended as an introduction; as such, it should be suitable as a textbook for an entry-level course in hydrology directed at advanced seniors and beginning graduate students in engineering and physical science, who have a working knowledge of calculus and basic fluid mechanics. The book contains much more material than can reasonably be covered in a first course. Thus it will depend on the objectives of the course, and on the orientation and level of the students, which specific topics should be selected for coverage. Naturally, the instructor should be the ultimate judge in this. However, to facilitate this selection, the text is printed in two different type formats. The main subject matter, which in the experience of the author can be suggested for inclusion in a first course, is presented in regular type. An effort has been made to lay out this part of the text in such a way that the student should be able to grasp the material with little or no reliance on the more advanced sections. For certain topics, clarification by an experienced instructor in the lectures will undoubtedly be helpful. Subject matter of a more advanced or specialized nature, is printed as indented text in a slightly smaller type and with a grey rule on the left-hand side of the page. This material is intended either as optional or explanatory reading for the first course, or as subject matter to be covered in a second and more advanced course. Sections of this second type of material have also been used as major portions in more specialized courses, namely in Groundwater Hydrology (Chapters 8, 9 and 10) and in Boundary Layer Meteorology (Chapters 2, 3 and 4) at Cornell.

The book is intended mainly for students of hydrology; it should, however, also be more broadly of interest to professional scientists and engineers, who are active in environmental matters, meteorology, agronomy, geology, climatology, oceanology, glaciology and other Earth sciences, and who wish to study some of the underlying concepts of hydrology, relevant to their discipline. In addition, it is hoped that the book will be of use to workers in fluid dynamics, who want to become acquainted with applications to some intriguing and fascinating phenomena in nature.

Wilfried Brutsaert

Ter nagedachtenis van mijn ouders Godelieve S. G. Bostijn (-B.) en Daniel P. C. B.
妻トヨに捧げる
And to the life of Siska, Hendrik, Erika and Karl.

NOTE ON THE TEXT

Readers should note that more advanced material in this book is printed in smaller type than the main subject matter, with a grey rule in the left-hand margin. A fuller explanation may be found in the Foreword.

1 INTRODUCTION

1.1 DEFINITION AND SCOPE

Hydrology is literally the science of water. Etymologically, the word has its roots in ancient Greek, and is a composite, made up of ὕδωρ, water, and λόγος, word. Obviously, defined this way, the term is much too broad to be very useful, as it would have ramifications in all scientific disciplines.

Actually, the word hydrology has not always been well defined and even as recently as the 1960s it was not very clear exactly what hydrology was supposed to cover and encompass. Price and Heindl (1968), in a survey of many of the definitions that had appeared in the literature over the previous 100 years, were compelled to conclude that the question "What is Hydrology?" had not been resolved by their review. Still, they felt that, in general, there seemed to be a consensus that hydrology is a physical science, which is concerned mainly with the water cycle of land and near-shore areas; moreover, there had been a tendency to broaden the term rather than to narrow it, even to the point of including socio-economic aspects.

Over the past few decades, however, with the growing activity level and the increasing maturity of this field of endeavor, a more precise definition has emerged. Hydrology is now widely (see, for example, Eagleson, 1991) accepted to be the science that deals with those aspects of the cycling of water in the natural environment that relate specifically with

- *the continental water processes*, namely the physical and chemical processes along the various pathways of continental water (solid, liquid and vapor) at all scales, including those biological processes that influence this water cycle directly; and with
- *the global water balance*, namely the spatial and temporal features of the water transfers (solid, liquid and vapor) between all compartments of the global system, i.e. atmosphere, oceans and continents, in addition to stored water quantities and residence times in these compartments.

Because it is defined as being concerned specifically with continental water processes, hydrology is a discipline distinct from meteorology, climatology, oceanology, glaciology and others that also deal with the water cycle in their own specific domains, namely the atmosphere, the oceans, the ice masses, etc., of the Earth; at the same time, however, hydrology integrates and links these other geosciences, in that through the global water balance it is also concerned with the exchanges of water between all these separate compartments.

With this definition it is now also possible to delineate the practical scope of hydrologic analysis in engineering and in other applied disciplines. It consists of the determination of the amount and/or flow rate of water that will be found at a given location and at a given time under natural conditions, without direct human control or intervention. The latter specification, that no human control be involved, is necessary to distinguish hydrology from the related discipline of hydraulics. Hydraulics is concerned with the study of controlled fluid motion in well-defined and often in human-made environments. For instance, problems involving pipe flow, irrigation water distribution or pumping of groundwater are not hydrologic in nature, but are more properly assigned to the realm of hydraulics.

1.2   THE HYDROLOGIC CYCLE

The water cycle, also called the hydrologic cycle, refers to the pathway of water in nature, as it moves in its different phases through the atmosphere, down over and through the land, to the ocean and back up to the atmosphere. When atmospheric water vapor condenses and precipitates over land, initially it moistens the surface and some amount of it is stored as *interception*, which later evaporates. As *precipitation* (and in a similar way *snowmelt*) continues, part of it may flow over the surface in the form of *overland flow* or *surface runoff*, and part of it may enter into the soil as *infiltration*. This surface runoff soon tends to collect locally, either in puddles or small ponds as *depression storage*, or in gullies or larger channels where it continues as *streamflow*, which ultimately ends up in a larger water body, such as a lake or the ocean. Streamflow is normally described by a *hydrograph*, that is the rate of flow at a gaging station as a function of time. The infiltrated water may flow rapidly through the near-surface soil layers to exit into springs or adjacent streams, or it may percolate more slowly through the profile to join the *groundwater*, which sooner or later seeps out into the natural river system, lakes and other open water bodies; part of the infiltrated water is retained in the soil profile by capillarity and other factors, where it is available for uptake by the roots of vegetation.

Soil layers and other geologic formations, whose pores and interstices can transmit water, are called *aquifers*. When an aquifer is in direct contact with the land surface, it is referred to as *unconfined*. The locus of points in an unconfined aquifer, where the water pressure is atmospheric, is called the *water table*. Although the water table is not a true free surface separating a saturated zone from a dry zone, it is sometimes assumed to be the upper boundary of the groundwater in an unconfined aquifer. The partly saturated zone in an unconfined aquifer, between the water table and the ground surface, is sometimes referred to as the *vadose zone*. In an unconfined aquifer, the term *groundwater* refers usually to the water found below the water table; *soil water* or *soil moisture* refers to the water above the water table. A water bearing geologic formation, that is separated from the surface by an impermeable layer, is referred to as a *confined aquifer*. Streamflow is fed both by surface runoff and by subsurface flow from riparian (i.e. located along the banks) aquifers. The streamflow, resulting from groundwater outflow is often called *base*

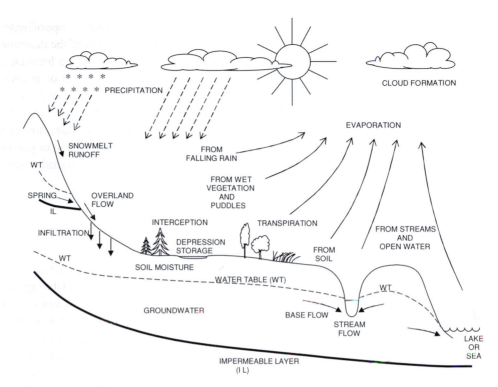

Fig. 1.1   Sketch of some of the main processes in the land phase of the water cycle.

flow; in the absence of *storm flow* or *storm runoff* caused by precipitation, base flow is also referred to as *drought flow* or *fair weather flow*.

Finally, the hydrologic cycle is closed by *evaporation*, which returns the water, while in transit in the different flow paths and stages of storage along the way, back into the atmosphere. When evaporation takes place through the stomates of vegetation, it can be referred to as *transpiration*. Direct evaporation from open water or soil surfaces and transpiration of biological water from plants are not easy to separate; therefore the combined process is sometimes called *evapotranspiration*. Evaporation of ice is commonly referred to as *sublimation*. While these distinctions are useful at times, the term evaporation is usually adequate to describe all processes of vaporization. Some of the main processes are drawn schematically in Figure 1.1.

1.3   SOME ESTIMATES OF THE GLOBAL WATER BALANCE

Numerous studies have been carried out to estimate the magnitude of the most important components of the water budget equations on a global scale. Because the available data base required for this purpose is still far from adequate, several of the methods used in these estimates may be open to criticism. Nevertheless, there is a fair agreement among

Table 1.1 Estimates of world water balance (m y^{-1})

| Reference | Land (1.49 × 10^8 km^2) | | | Oceans (3.61 × 10^8 km^2) | | Global |
	P	R	E	P	E	P = E
Budyko (1970, 1974)	0.73	0.31	0.42	1.14	1.26	1.02
Lvovitch (1970)	0.73	0.26	0.47	1.14	1.24	1.02
Lvovitch (1973)	0.83	0.29	0.54	–	–	–
Baumgartner and Reichel (1975)	0.75	0.27	0.48	1.07	1.18	0.97
Korzun et al. (1978)	0.80	0.315	0.485	1.27	1.40	1.13

some of the calculated values and, within certain limits, they provide a useful idea of the long-term average balance in different climatic regions of the world.

As shown in Table 1.1, the average annual precipitation and also evaporation are of the order of 1 m for the entire Earth. Over the landsurfaces the average precipitation intensity, P, is about 0.80 m y^{-1}, whereas the corresponding average evaporation, E, is around 0.50 m y^{-1} or about 60% to 65% of the precipitation. Under steady conditions, that is for long time periods, the remainder can be considered to be runoff from the landsurfaces into the oceans, R (expressed as height of water column per unit of time), or

$$R = P - E \tag{1.1}$$

Averaged over all continents and over long time periods, the annual runoff R is therefore around 35% to 40% of the precipitation. Except for South America and Antarctica (see Table 1.2), the values for the individual continents are not very different from the global values. Precipitation and streamflow runoff measurements have been and are being made routinely in many places on Earth. In contrast, evaporation has not received as much attention and no systematic measurements are available.

Estimates of the average distribution of water in different forms expressed as depth of water covering the globe, assumed to be a perfect sphere, are given in Table 1.3. These indicate that the 1 m of average annual precipitation is relatively large as compared to the active fresh water on Earth, that is the water which is not stored in permanent ice and deep groundwater. This means that the turnover of the active part of the hydrologic cycle is rather fast, and that the residence times in some of the major compartments of the water cycle are relatively short; the mean residence time can be taken as the ratio of the storage and the flux in or out of storage. For example, a continental runoff rate of 0.30 m y^{-1} (Table 1.1) and a storage in the rivers of (0.003/0.29) m of water on the 29% of the world occupied by land, gives a mean residence time of the order of 13 days for the rivers of the world. Similarly, a global evaporation rate of 1 m y^{-1}, with 0.025 m of storage in the atmosphere, leads to a mean residence time of the order of 9 days for the

Table 1.2 Some estimates of the mean precipitation (and river runoff) from available data for the continents (in m y^{-1}) (*)

	Europe	Asia	Africa	North America	South America	Australia and Oceania	Antarctica
Percent of land area	6.7	29.6	20.0	16.2	12.0	6.0	9.5
Reference							
Lvovitch (1973)	0.734	0.726	0.686	0.670	1.648	0.736	–
	(0.319)	(0.293)	(0.139)	(0.287)	(0.583)	(0.226)	–
Baumgartner and	0.657	0.696	0.696	0.645	1.564	0.803	0.169
Reichel (1975)	(0.282)	(0.276)	(0.114)	(0.242)	(0.618)	(0.269)	(0.141)
Korzoun et al.	0.790	0.740	0.740	0.756	1.600	0.791	0.165
(1977)	(0.283)	(0.324)	(0.153)	(0.339)	(0.685)	(0.280)	(0.165)

*The corresponding evaporation values can be determined with Equation (1.1).

Table 1.3 Estimates of different forms of global water storage (as depth in m over entire earth surface)

Source of data	Lvovitch (1970)	Baumgartner and Reichel (1975)	Korzun et al. (1978)
Oceans	2686	2643	2624
Ice caps and glaciers	47.1	54.7	47.2
Total ground water	117.6	15.73	45.9
		(excluding Antarctica)	
(Active ground water)	(7.84)	(6.98)	—
Soil water	0.161	0.120	0.0323
Lakes	0.451	0.248	0.346
Rivers	0.002 35	0.002 12	0.004 16
Atmosphere	0.0274	0.0255	0.0253

atmosphere. These are very short residence times. Moreover, as the oceans occupy about 71% of the Earth surface, the active fresh water in the hydrologic cycle is continually being distilled anew through ocean evaporation.

Maps depicting the approximate distribution of components of the water balance in different parts of the world have been presented by Lvovitch (1973), Budyko (1974), Baumgartner and Reichel (1975), Korzoun et al. (1977), and Choudhury et al. (1998). The relative and absolute magnitudes of the main components of the hydrologic cycle, namely P, R and E, can vary over a wide range from one location to another. Obviously, the long-term mean values of all three are negligible in desert locations. At the other extreme, maximal annual precipitation values of up to 26.5 m have been recorded in a mountainous monsoon environment (Cherrapunji, Meghalaya). Maximal mean evaporation values of

Table 1.4 Estimates of mean global heat budget at the earth surface in W m^{-2}

Reference	Land			Oceans			Global		
	R_n	$L_e E$	H	R_n	$L_e E$	H	R_n	$L_e E$	H
Budyko (1974)	65	33	32	109	98	11	96	80	16
Baumgartner and Reichel (1975)	66	37	29	108	92	16	96	76	20
Korzun et al. (1978)	65	36	29	121	109	12	105	89	16
Ohmura (2005)	62	36	26	125	110	15	104	85	19

up to 3.73 m y^{-1} have been inferred for the Gulf Stream in the western Atlantic (Bunker and Worthington, 1976) and up to 4 or even 5 m y^{-1} for the Gulf of Aqaba (Assaf and Kessler, 1976). Much research has been directed in recent years to study the evolution of today's climate in response to increasing greenhouse gases in the atmosphere. Although the issue is far from resolved, there are some indications of an accelerating hydrologic cycle in several regions (see, for example, Brutsaert and Parlange, 1998; Karl and Knight, 1998; Lins and Slack, 1999).

The strong linkage between the water cycle and climate is further illustrated by the estimates of the mean global surface energy budget in Table 1.4. Over large areas and over sufficiently long periods, when effects of unsteadiness, melt and thaw, photosynthesis and burning, and lateral advection can be neglected, this surface energy balance can be written as

$$R_n = L_e E + H \tag{1.2}$$

where R_n is the specific flux of net incoming radiation, L_e is the latent heat of vaporization, E is the rate of evaporation, and H is the specific flux of sensible heat into the atmosphere. The major portion of the incoming radiation is absorbed near the surface of the Earth, and it is transformed into internal energy. The subsequent partition of this internal energy into long-wave back radiation, upward thermal conduction and convection of sensible heat H, and latent heat $L_e E$, is one of the main processes driving the atmosphere. Table 1.4 indicates that the net energy is mainly disposed of as evaporation. Over the oceans, the latent heat flux $L_e E$ is on average larger than 90 percent of the net radiation. But even over the land surfaces of the Earth, $L_e E$ is on average still larger than half of R_n.

Because the global patterns of heating force the circulation of the planetary atmosphere, the implications of this large latent heat flux are clear. As a result of the relatively large latent heat of vaporization L_e, evaporation of water involves the transfer and redistributiuon of large amounts of energy under nearly isothermal conditions. Because, even at saturation, air can contain only relatively small amounts of water vapor, which can easily be condensed at higher levels, the air can readily be dried out; this release of energy through condensation and subsequent precipitation is the largest single heat source for the atmosphere. Thus processes in the water cycle play a central role in governing weather and climate.

1.4 METHODOLOGIES AND PROCEDURES

This book aims primarily to describe the occurrence and the transport of water in its continuous circulation over the landsurfaces of the Earth. Before starting this task, it is worthwhile to review briefly the different strategies that are available and that can be used for this purpose.

1.4.1 *Statistical analysis and data transformation*

As observed in Section 1.2, one of the main practical objectives of hydrologic analysis is the determination of the quantity of water, in storage or in transit, to be found at a given time and place, free of any direct human control. When a reliable record of observed hydrologic data is available, a great deal can be learned simply by a statistical analysis of this record. Although such an approach is proper for stationary systems in the prediction of long-term behavior for general planning purposes, it cannot be used for short-term and emergency forecasting, for example, during floods, or for day-to-day resource management decisions. Furthermore, reliable records are available for only a few locations over a limited period of time, and practically never where needed. Therefore in hydrology the problem is often such, that a method must be devised to transform some available data, which are of no direct interest, to the required hydrologic information. For instance, the problem may consist of determining the rate of flow in a river at a given location either from a known flow rate at some other point upstream or downstream, or from a known rainfall distribution over the upstream river basin. In other cases, the problem may consist of deducing the basin evapotranspiration from soil and vegetation on the basis of available meteorological data.

1.4.2 *The "physical" versus the "systems" approach*

The hydrologic literature is replete with attempts at classifying the methodologies and paradigms that have been used to transform hydrologic input into hydrologic output information. Until a few years ago it had become customary to consider two contrasting approaches, namely the "physical" approach and the "systems" approach. In the physical approach the input–output relationship is sought by the solution of the known conservation equations of fluid mechanics and thermodynamics with appropriate boundary conditions to describe the flow and transport of water throughout the hydrologic cycle. This approach has obvious limitations; the physiographic and geomorphic characteristics of most hydrologic systems are so complicated and variable, and the degree of uncertainty in the boundary conditions so large, that solutions are feasible only for certain highly simplified situations. In other words, the properties of natural catchments can never be measured accurately enough, and solutions, based on internal descriptions starting from first principles of fluid mechanics, can be obtained only for grossly idealized conditions, which are coarse approximations of any real situation.

The hydrologic "systems" (also "operational" or "empirical") approach is presumably based on a diametrically opposite philosophy. In this approach the physical structure of the various components of the hydrologic cycle and their inner mechanisms are not considered; instead, each component, however it may be defined, is thought of as a "black box," and the analysis focuses on discovering a mathematical relationship between the external input (e.g. rainfall, air temperature, etc.) and the output (e.g. river flow, soil moisture, evaporation, etc.). The structure of this mathematical relationship is mostly quite remote from the physical structure of the prototype phenomena in nature. This lack of correspondence between the inner physical mechanisms and the postulated functional formalisms makes this approach quite general operationally, because it permits the use of well-known algorithms and objective criteria in identification and prediction. However, this also underlies the main limitations of this approach. First, in assigning cause and effect the definition of input and output variables is mostly based on intuition guided by past experience, and the danger exists that some important phenomena are overlooked. Second, the best that can ever be expected with a black box approach is a satisfactory reproduction of a previously obtained input–output record; even when such data are available, it is difficult to accommodate fully the nonstationary effects in the system, and it is impossible to anticipate subsequent hydrologic changes, such as those resulting from urbanization, deforestation, reclamation, or climate change.

Because many hydrologic methods do not really fit in this physical-versus-empirical classification, a third possible approach was taken to be an intermediate one. In this view the performance of a hydrologic unit, say a catchment, is represented in terms of some idealized components or "grey boxes," which correspond to recognizable elements in the prototype, whose input–output response functions are structured after solutions of some tractable or suitably simplified situations of the physical processes perceived to be relevant. This third way was often called the "conceptual model" approach.

At first sight, a classification based on three distinct approaches, namely physical, empirical and conceptual, may appear reasonable. However, it is less than obvious how this classification can be applied to specific cases. Indeed, one might ask what the difference is between physical and empirical. After all, the essence of physical science is experimentation and conceptualization. Moreover, the physical approach of one discipline is usually the empiricism or the conceptual model of another. For example, Newton's "law" of viscous shear constitutes the physical basis of a wide area of fluid mechanics, whereas it represents a mere black box simplification in molecular physics. Darcy's law is the physical basis of much of groundwater hydrology, but in fluid mechanics it can be considered an operational approach, to avoid the complexity of flow analysis in an irregular and ill-defined pore network. The same dilemma is inherent in most other special concepts used in hydrology. This ambiguous difference between physical, empirical and conceptual shows that the classification of the methodologies should be based on other criteria.

1.4.3   Spatial scale and parameterization

General approach

All natural flow phenomena are governed by the principles of conservation of mass, momentum and energy, which can be expressed by a number of equations to provide a mathematical description of what goes on. However, because there are normally more dependent variables than available conservation equations, in order to close the system, additional relationships must be introduced. These closure relationships, also called *parameterizations*, relate some of the variables with each other to describe certain specific physical mechanisms; the mathematical form of these relationships, and the values of the material constants or *parameters* are usually based on experimentation.

A second point is that any physical phenomenon must be considered at a given scale; this scale is the available (depending on the data) or chosen (depending on the objectives of the study) resolution. It will become clear later on in this book that, while the fundamental conservation equations remain unaffected by the scale at which the phenomenon is being considered, most closure relationships in them are quite sensitive to scale. Indeed, a parameterization can be considered as a mathematical means of describing the subresolution (or microscale) processes of the phenomenon, in terms of resolvable scale (or macroscale) variables; these macroscale variables are the ones, which can be treated explicitly in the analysis or for which measured records are obtainable. Thus, the details of the microscale mechanisms are not considered explicitly, but their statistical effect is formulated mathematically by a parameterization in terms of macroscale variables.

All this suggests that a sound criterion to distinguish, in principle at least, one approach from another, may be the spatial scale at which the internal mechanisms are parameterized. For example, Newton's equation for viscous shear stress (see Equation (1.12) below) is a parameterization in terms of variables typically at the millimeter to centimeter scale; however, it reflects momentum exchanges at molecular scales, which are orders of magnitude smaller. The hydraulic conductivity is a parameter at the so-called Darcy scale (see Chapter 8), namely a scale somewhere intermediate between the Newtonian viscosity (or Navier–Stokes) scale for water and air inside the soil pores, on the one hand, and the field scales for infiltration and drainage, on the other. Several spatial scales are illustrated with the corresponding characteristic temporal scales in Figure 1.2 for some general types of water transport processes as they have been considered in hydrologic studies.

On the land surfaces of the Earth, the catchment or river basin sizes appear as scales of central importance. The terms *basin, catchment, watershed* and *drainage area* are roughly synonymous and are often used interchangeably. A basin can be defined as all of the upstream area, which contributes to the open channel flow at a given point along a river. The size of the basin depends on the selection of the point in the river system under consideration. Usually this point is taken, where the river flows into a large water body, such as a lake or the ocean, or where it changes its name as a tributary into a larger river. However, a basin or catchment can also defined by any point along the river, where the river flow is being measured. Basins are delineated naturally by the land surface topography, and topographic ridges are usually taken as their boundaries; they can be considered as the natural conveyance systems for mass and energy on the land surfaces of the Earth. In meteorology, the concern is more on atmospheric motions and weather systems, and this has led to a somewhat different scale classification; an example of a commonly used classification is shown in Table 1.5.

Table 1.5 A common scale classification in the atmosphere (after Orlanski, 1975)

Nomenclature	Scale range
Micro γ	<20 m
Micro β	20–200 m
Micro α	200 m–2 km
Meso γ	2 km–20 km
Meso β	20 km–200 km
Meso α	200 km–2000 km
Macro γ	>2000 km

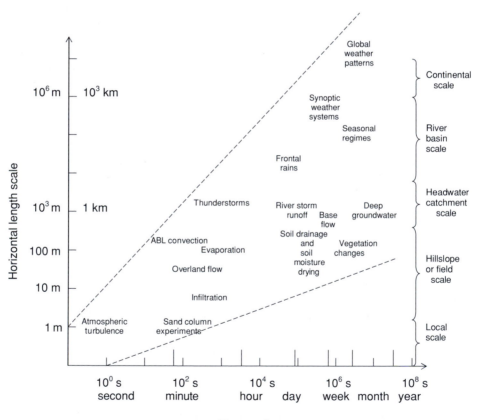

Fig. 1.2 Approximate ranges of spatial and temporal scales of some common physical processes that are relevant in hydrology.

To summarize, these observations indicate that, in deciding on a strategy to describe a hydrologic phenomenon, the relevant question is probably not so much whether a physical, a black box or a conceptual approach should be used. Rather, it is more useful to determine what scales are appropriate for the available and measurable data, and for the problem at hand. In other words, what is the appropriate level of parameterization?

Spatial variability and effective parameters

As mentioned above, a parameterization can be defined as a functional relationship between the variables describing the phenomenon in question. This relationship invariably contains one or more constant terms, reflecting material and fluid properties and vegetational, geomorphic, geologic and other physiographic features; these are called parameters and they are normally determined by experiment. Most hydrologic parameters tend to be highly variable in space. It stands to reason, therefore, that the experimental determination of any such space-dependent parameter must be carried out at the scale at which it is to be applied to describe the flow.

A second important issue is that any given parameterization is usually valid only over a certain finite range of spatial scales, and that the computational scale, that is the integration domain or the discretization of the equations, must lie within that range. Because the necessary data may be available only at a coarser resolution, in practical application a parameterization may have to be applied at scales, for which it was not intended originally and which are larger than permissible. This means in such a case that the spatial variability of the parameters at the finer scales, which is normally present in the natural environment, cannot be accounted for with the available data. This difficulty is often resolved by assuming that the parameterization is still valid at the larger scale, and that it can be implemented with averaged or *effective* values of the parameters. This approach is not always satisfactory, and it is still the subject of intense research. Some aspects of this issue related to land surface–atmosphere interactions are discussed elsewhere (Brutsaert, 1998).

Requirements

To be useful, a parameterization must satisfy several requirements. First, a parameterization must be *valid*, that is, it must be able to give a faithful description of the phenomenon in question. The word comes from the Latin "validus," which means healthy or strong, and thus reliable. Validation is the term, which refers to the testing of a parameterization, and it consists usually of the application of some goodness of fit measure to results of calculations with the parameterization relative to observations. Beside being valid, a useful parameterization must satisfy the dual requirements of parsimony and robustness.[1] A

---

[1] The law of parsimony, also known as Ockham's razor, comes to mind here. Actually, the principle was already promulgated by Aristotle, and the razor is essentially Ockham's (1989; pp. 17, 20, 128) paraphrase of it. More than 2300 years ago Aristotle (1929) wrote, for instance, in *Physics* (I, 6, 189a,15) "inasmuch as it can be done from the limited, the limited is better," and in (VIII, 6, 259a,10) "for when the outcome is the same, the limited is always to be preferred, and indeed in matters of nature, the limited, being the better, occurs more when possible."

given parameterization is said to be more *parsimonious* than another one, when it needs fewer variables and parameters to describe the phenomenon.

A parameterization can be called *robust* when the outcome is relatively insensitive to its structure and to errors and uncertainties in the input variables and parameters. In hydrology, a *model* usually refers to a combination of several parameterizations to simulate more complicated phenomena and their interactions.

1.5   Conservation laws: the equations of motion

1.5.1   *Rate of change of fluid properties*

Consider a fluid in motion with a velocity field $\mathbf{v} = u\mathbf{i} + v\mathbf{j} + w\mathbf{k}$, in which (u, v, w) are the velocity components and $(\mathbf{i}, \mathbf{j}, \mathbf{k})$ are the unit vectors in the (x, y, z) directions, respectively, and let $C(x, y, z, t)$ denote some property of this fluid. The rate at which this property changes for a given particle of the fluid located at (x, y, z) at time t, can be determined by tracking the particle to its new position $(x + u\delta t, y + v\delta t, z + w\delta t)$, a small distance away at time $t + \delta t$. The fluid property has then become

$$C(x + u\delta t, y + v\delta t, z + w\delta t) = C + \frac{\partial C}{\partial x}u\delta t + \frac{\partial C}{\partial y}v\delta t + \frac{\partial C}{\partial z}w\delta t + \frac{\partial C}{\partial t}\delta t$$

Thus after the small displacement, the property of the fluid assumes the new value $C + (DC/Dt)\delta t$. This shows that the rate of change of the property C of the moving fluid particle is given by

$$\frac{DC}{Dt} = \frac{\partial C}{\partial t} + u\frac{\partial C}{\partial x} + v\frac{\partial C}{\partial y} + w\frac{\partial C}{\partial z} \tag{1.3}$$

DC/Dt is commonly referred to as the *substantial* time derivative, and is also variously called the *fluid mechanical* time derivative, the time derivative *following the motion*, or the *material* or *particle* derivative. Physically, Equation (1.3) is the total rate of change in the property, as seen by an observer moving with the fluid. The first term on the right describes the changes taking place locally at (x, y, z). The last three terms describe the changes observed while moving between locations with different values of C; the rate of change depends on the speed of the motion, (u, v, w).

1.5.2   *Conservation of mass: the continuity equation*

Because hydrology is concerned with amounts of water observed at different times and locations, conservation of mass is the main governing principle. There are several ways of deriving a formulation that embodies this principle.

At a point

One way, after Euler's 1755 derivation (Lamb, 1932), is to consider an element of fluid mass, which occupies a small volume $\delta\forall = (\delta x \ \delta y \ \delta z)$ at time t, shown as ABCD in two dimensions in Figure 1.3, and whose center moves at a velocity $\mathbf{v} = u\mathbf{i} + v\mathbf{j} + w\mathbf{k}$. If the mass of fluid per unit volume, that is its density, is ρ, the mass of the element is given by $(\rho\delta\forall)$. In the absence of chemical reactions or sources and sinks, the mass of this element

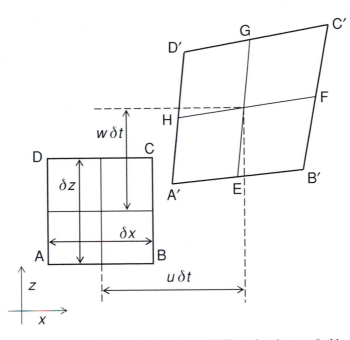

Fig. 1.3 At time t the mass $\rho\delta\forall$ occupies the volume ABCD, and at time $t + \delta t$ this same fluid mass has moved to A′B′C′D′. The center of the volume has moved from (x_0, y_0, z_0) to $(x_0 + u\delta t, y_0 + v\delta t, z_0 + w\delta t)$. The figure is shown in two dimensions for clarity; the third coordinate y can be imagined as pointing into the plane of the drawing.

does not change and must remain the same. Therefore, if the fluid property is taken as the mass of the fluid, or $C = (\rho\delta\forall)$, one has with Equation (1.3) that

$$\frac{D(\rho\delta\forall)}{Dt} = 0$$

or   (1.4)

$$\rho\frac{D(\delta\forall)}{Dt} + \delta\forall\frac{D\rho}{Dt} = 0$$

The rate of change of the fluid element volume $D(\delta\forall)/Dt$ can be derived by tracking the fluid element, shown in Figure 1.3, as it moves from ABCD to A′B′C′D′ during the small time interval δt. The point H is then located at

$$x = x_0 - \frac{\delta x}{2} + \left(u - \frac{\partial u}{\partial x}\frac{\delta x}{2} + \frac{\partial^2 u}{\partial x^2}\left(\frac{\delta x}{2}\right)^2\frac{1}{2} - \cdots\right)\delta t$$

and at

$$z = z_0 + \left(w - \frac{\partial w}{\partial x}\frac{\delta x}{2} + \frac{\partial^2 w}{\partial x^2}\left(\frac{\delta x}{2}\right)^2\frac{1}{2} - \cdots\right)\delta t$$

The point F is at

$$x = x_0 + \frac{\delta x}{2} + \left(u + \frac{\partial u}{\partial x}\frac{\delta x}{2} + \frac{\partial^2 u}{\partial x^2}\left(\frac{\delta x}{2}\right)^2\frac{1}{2} + \cdots\right)\delta t$$

and

$$z = z_0 + \left(w + \frac{\partial w}{\partial x} \frac{\delta x}{2} + \frac{\partial^2 w}{\partial x^2} \left(\frac{\delta x}{2} \right)^2 \frac{1}{2} + \cdots \right) \delta t$$

Therefore, the length of the segment HF, as projected on the x-axis, is given by $[\delta x + (\partial u/\partial x)\delta x \delta t]$; in a similar way, one obtains for the projection of the length of the segment EG on the z-axis the value $[\delta z + (\partial w/\partial z)\delta z \delta t]$, and for the length of the segment in the y-direction (not shown in Figure 1.3) $[\delta y + (\partial v/\partial y)\delta y \delta t]$. If δx, δy, δz and δt are all small enough, higher-order terms can be neglected, and the volume occupied by the mass at time $t + \delta t$ can be taken as the product of these three segments. Thus the change in volume during δt becomes

$$\frac{D(\delta x \delta y \delta z)}{Dt} \delta t = \left(1 + \frac{\partial u}{\partial x} \delta t \right) \delta x \left(1 + \frac{\partial v}{\partial y} \delta t \right) \delta y \left(1 + \frac{\partial w}{\partial z} \delta t \right) \delta z - \delta x \delta y \delta z$$

so that

$$\frac{D(\delta \forall)}{Dt} = \left(\frac{\partial u}{\partial x} + \frac{\partial v}{\partial y} + \frac{\partial w}{\partial z} \right) \delta \forall \tag{1.5}$$

In more concise vector notation this can also be written as

$$\frac{D(\delta \forall)}{Dt} = \nabla \cdot \mathbf{v} \, \delta \forall \tag{1.6}$$

where ∇ is the operator $\nabla = (\partial/\partial x)\,\mathbf{i} + (\partial/\partial y)\,\mathbf{j} + (\partial/\partial z)\,\mathbf{k}$. Equations (1.5) and (1.6) show that the divergence $\nabla \cdot \mathbf{v}$ is indeed, as its name suggests, the fractional rate of change of the fluid element volume. With this result, Equation (1.4) can be written as

$$\frac{D\rho}{Dt} + \rho \left(\frac{\partial u}{\partial x} + \frac{\partial v}{\partial y} + \frac{\partial w}{\partial z} \right) = 0 \tag{1.7}$$

or again, in vector notation, as

$$\frac{\partial \rho}{\partial t} + \nabla \cdot (\rho \mathbf{v}) = 0 \tag{1.8}$$

Equations (1.7) and (1.8) are forms of the classical continuity equation. The form of (1.8) is applicable to describe the conservation of mass of any substance at a given point (x, y, z), provided $(\rho \mathbf{v})$ is made to represent the specific mass flux $\mathbf{F}$, that is the transport of mass of that substance per unit cross sectional area and per unit time. Whenever the density of the substance in question can be considered constant, the continuity equation assumes its well-known form

$$\nabla \cdot \mathbf{v} = 0 \tag{1.9}$$

Note that the continuity equation is not usually derived this way; the present derivation is used to maintain unity and consistency of treatment with the conservation of momentum in Section 1.5.3; moreover, it is relevant to the study of deforming porous media later on in Chapter 8. A more common way to derive the continuity equation consists of setting up a mass balance for a certain volume fixed in space, also called a *control volume*. The mass balance states that the sum of all the inflow rates into the control volume minus the sum of all the outflow rates is equal to the time rate of change of mass stored in the control volume. For an infinitesimally small control volume this procedure also yields Equation (1.8).

Regardless of the derivation, however, it should be remembered that Equations (1.8) and (1.9) describe the flow at a point; therefore, in principle, the integration of (1.8) or (1.9) should allow the determination of the distribution of the amount and transport of water in space and in time.

Finite control volume

In a second but equally valid approach, the control volume is assumed to occupy the entire flow domain by integrating out the spatial dependence of the flux terms. Thus all flux terms are located on the boundaries of the flow domain and they can be grouped into bulk inflow rates Q_i and outflow rates Q_e. As a result, the continuity equation takes on the *lumped* form of the *storage equation*, as follows

$$Q_i - Q_e = \frac{dS}{dt} \tag{1.10}$$

where S is the amount of water stored in the control volume, and the ordinary derivative indicates that the time t is the only remaining independent variable. When Equation (1.10) describes the flow of liquid water with an assumed constant density, these variables can have the dimensions of $[Q] = [L^3/T]$ and $[S] = [L^3]$, where L and T represent the basic dimensions of length and time, respectively; if the Q-terms include precipitation and evaporation, it is often convenient to take these dimensions as $[Q] = [L/T]$ and $[S] = [L]$. In the lumped formulation of Equation (1.10), all interior variables and parameters represent spatial averages over the entire control volume.

1.5.3 Conservation of momentum: Euler and Navier–Stokes equations

The flow of a fluid is also subject to the principle of conservation of momentum. Again, there are several ways of obtaining a mathematical formulation of this principle.

At a point

The simplest method is probably to consider, as before, a small element of an ideal fluid with a mass $(\rho \delta \forall)$, as illustrated in Figure 1.4, and to apply Newton's second law to it. This states that the rate of change of momentum is equal to the sum of the impressed forces. The pressure and the velocity at the center (x, y, z) of this element are $p(x, y, z, t)$ and $\mathbf{v}(x, y, z, t)$, respectively. Accordingly, the property of the fluid element in this case is its momentum, or $C = (\rho \delta \forall \, \mathbf{v})$, and the rate of change is $D(\rho \delta \forall \mathbf{v})/Dt$, as given by Equation (1.3). Because the fluid is assumed to be ideal, the only relevant forces are those owing to pressure and to the acceleration of gravity. The latter is a vector, $\mathbf{g} = \mathbf{i}g_x + \mathbf{j}g_y + \mathbf{k}g_z$, whose direction defines the local vertical, and whose absolute value is commonly denoted by g; the coordinates shown in Figure 1.4 are oriented in such a way that $g_x = -g \sin \theta$, $g_y = 0$ and $g_z = -g \cos \theta$. As illustrated in Figure 1.4, the x-component of the net force acting on the fluid element is the sum of the forces acting on AD and BC plus the force due the Earth's gravity; this sum equals $-[(\partial p/\partial x) + \rho g \sin \theta]\delta x \delta y \delta z$. Similarly, the sum of the impressed forces in the z-direction is $-[(\partial p/\partial z) + \rho g \cos \theta]\delta x \delta y \delta z$. Adding to these an analogous y-component, one has in vector notation,

$$\frac{D}{Dt}(\rho \delta \forall \mathbf{v}) = -(\nabla p - \rho \mathbf{g})\delta \forall$$

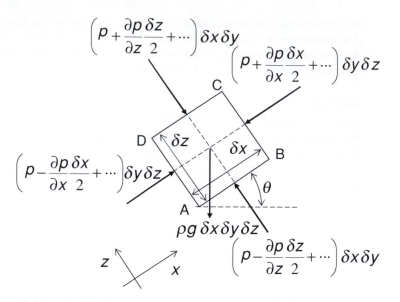

Fig. 1.4 Definition sketch for the conservation of momentum of a fluid element occupying the volume $\delta\forall = (\delta x\ \delta y\ \delta z)$, with its center at (x, y, z). The element is subject to pressure forces and the acceleration of gravity. The y-coordinate, which is not shown, points into the plane of the drawing.

Making use of Equation (1.4), one obtains immediately

$$\frac{D\mathbf{v}}{Dt} = -\frac{1}{\rho}\nabla p + \mathbf{g} \qquad (1.11)$$

which is a form of Euler's equation. Inclusion of the effect of viscosity into Euler's equation produces the Navier–Stokes equation; expanding the substantial derivative according to its definition (Equation (1.3)), one can write it as follows

$$\frac{\partial \mathbf{v}}{\partial t} + \mathbf{v}\cdot\nabla\mathbf{v} = -\frac{1}{\rho}\nabla p + \mathbf{g} + \mathbf{f} \qquad (1.12)$$

where $\mathbf{f}$ denotes the frictional force (per unit mass); for an incompressible Newtonian fluid it can be shown that this is given by $\mathbf{f} = \nu\nabla^2\mathbf{v}$, where ν is the kinematic viscosity. To repeat briefly, the first term on the left represents the change in momentum (per unit mass) of the fluid due to local acceleration, i.e. changes in velocity at the point (x, y, z) under consideration. The second term represents the momentum changes resulting from acceleration (or deceleration) experienced by the fluid as it moves between points with different velocities. The first term on the right represents the force resulting from the pressure gradient, and the second the force resulting from the gravity field of the Earth. If the z-axis represents the vertical and points upward (or $\theta = 0$ in Figure 1.4), Equation (1.12) can be written as

$$\frac{\partial \mathbf{v}}{\partial t} + \mathbf{v}\cdot\nabla\mathbf{v} = -\frac{1}{\rho}\nabla p - g\mathbf{k} + \mathbf{f} \qquad (1.13)$$

in which, it should be recalled, $\mathbf{k}$ is the unit vector in the z-direction.

The viscosity of a liquid at standard pressure depends on temperature. For most practical purposes, in the range $0 \leq T \leq 40 \,°C$ the kinematic viscosity of water (in units of $m^2\,s^{-1}$) as a function of temperature (in $°C$) can be calculated with sufficient accuracy by means of $v = 10^{-6}\,(1.785 - 0.057\,89T + 0.001\,128T^2 - 0.9671 \times 10^{-5}T^3)$; this yields roughly $1.00 \times 10^{-6}\,m^2\,s^{-1}$ at $20\,°C$. This expression is based on measurements at the National Bureau of Standards, made available by J. F. Swindells. Similarly, for most applications in hydrology the density of water (in units of $kg\,m^{-3}$) at 1 atmosphere can be calculated in the same temperature range by means of $\rho = (999.8505 + 0.060\,01T - 0.007\,917T^2 + 4.1256 \times 10^{-5}T^3)$.

Finite control volume

Like Equations (1.8) and (1.9), also here Equations (1.11), (1.12) and (1.13) describe the flow phenomenon at a point. Again, they can be extended to a larger control volume by integrating out the spatial dependence of the terms. This can be accomplished by multiplication of each term in Equation (1.13) by the differential volume $d\mathbf{s} \cdot d\mathbf{A}$ (in which $d\mathbf{s}$ and $d\mathbf{A}$ represent the differential flow path and cross-sectional area vectors, the latter pointing in the direction of flow) and by subsequent integration along all flow paths inside the control volume and across all areas of entry and exit of the control volume. For example, in the case of a conduit fixed in space occupied by a fluid volume S of constant density ρ, this yields for, say, the x-direction, approximately,

$$\rho \frac{d(S\,\overline{V_x})}{dt} + \rho(Q_e V_{xe} - Q_i V_{xi}) = \mathbb{F}_x \qquad (1.14)$$

where $\mathbb{F}$ is the sum of all forces acting on the fluid in the control volume, Q_i and Q_e are the inflow and outflow rates of the control volume, $\overline{V}$ is the average fluid velocity inside the control volume, V_i and V_e the average fluid velocity over the entry and exit section, respectively, of the control volume, and the subscript x denotes the component direction of the momentum and of the forces.

1.5.4 The kinematic approach

In principle, the description of fluid flow phenomena should involve conservation of mass, conservation of momentum, and conservation of energy. However, whenever, the relevant phenomena are isothermal, most of the energy is mechanical, and the energy conservation equation becomes redundant, so that it is often not included in the formulation. In this book, the energy conservation equation will be used only in relation to atmospheric phenomena, where it will be discussed further. In hydrologic applications, whenever both mass and momentum conservation principles are made use of, the mathematical description of the flow phenomena is called a *dynamic* formulation. However, in some situations, momentum changes, both temporal and spatial, are so small that they can be neglected. In such cases, the terms on the left-hand side of Equation (1.12) can be omitted and this greatly simplifies the formulation. In practice, the right-hand side of Equation (1.12) can then often be parameterized by an explicit functional relationship between the flow velocities in the system and some other variables such as pressure, water depth or water level height. Whenever only the continuity equation is required, and the momentum equation can be replaced by this type of relationship, the mathematical description is referred to as a *kinematic* formulation. The same idea can also be applied to larger control volumes. In this case, the combination of the storage equation (1.10) with a simple functional relationship between S and Q_e and/or Q_i is called a *lumped kinematic* formulation.

REFERENCES

Aristotle (1929). *The Physics*, with an English translation by P. H. Wicksteed and F. M. Cornford, 2 volumes. London: W. Heinemann, Ltd.; and Cambridge, MA: Harvard University Press.

Assaf, G. and Kessler, J. (1976). Climate and energy exchange in the Gulf of Aqaba (Eilat). *Mon. Weather Rev.*, **104**, 381–385.

Baumgartner, A. and Reichel, E. (1975). *The World Water Balance*. Amsterdam and NY: Elsevier Scientific Publishing Company.

Brutsaert, W. (1998). Land-surface water vapor and sensible heat flux: spatial variability, homogeneity, and measurement scales. *Water Resour. Res.*, **34**, 2433–2442.

Brutsaert, W. and Parlange, M. B. (1998). Hydrologic cycle explains the evaporation paradox. *Nature*, **396** (No. 6706, Nov. 5), 30.

Budyko, M. I. (1970). The water balance of the oceans. *Symposium on world water balance, Proc. Reading Sympos.*, Vol. I, Int. Assoc. Sci. Hydrology, Publication No. 92, pp. 24–33.
  (1947). *Climate and Life*. New York: Academic Press.

Bunker, A. F. and Worthington, L. V. (1976). Energy exchange charts of the North Atlantic Ocean. *Bull. Amer. Met. Soc.*, **57**, 670–678.

Choudhury, B. J., DiGirolamo, N. E., Susskind, J., Darnell, W. L., Gupta, S. K. and Asrar, G. (1998). A biophysical process-based estimate of global land surface evaporation using satellite and ancillary data II. Regional and global patterns of seasonal and annual variations. *J. Hydrol.*, **205**, 186–204.

Eagleson, P. S. (Chair) (1991). *Opportunities in the Hydrologic Sciences*, Committee on Opportunities in the Hydrologic Sciences, National Research Council. Washington, DC: National Academy Press.

Karl, T. R. and Knight, R. W. (1998). Secular trends of precipitation amount, frequency, and intensity in the USA. *Bull. Amer. Met. Soc.*, **79**, 231–241.

Korzoun, V. I. *et al.* (eds.) (1977). *Atlas of World Water Balance*, USSR National Committee for the International Hydrological Decade. Paris: UNESCO Press.

Korzun, V. I. *et al.* (eds.) (1978). *World Water Balance and Water Resources of the Earth*, USSR National Committee for the International Hydrological Decade. Paris: UNESCO Press.

Lamb, H. (1932). *Hydrodynamics*, sixth edition. Cambridge: Cambridge University Press (also 1945, NY: Dover Publication).

Lins, H. F. and Slack, J. R. (1999). Streamflow trends in the United States. *Geophys. Res. Letters*, **26**, 227–230.

Lvovitch, M. I. (1970). World water balance. *Symposium on world water balance*, *Proc. Reading Sympos.*, Vol. II, Int. Assoc. Sci. Hydrology, Publication No. 93, pp. 401–415.
  (1973). The global water balance. *Trans. Amer. Geophys. Un.*, **54**. (US-IHD Bull. No. 23), 28–42.

Ockham, William of (1989). *Brevis Summa Libri Physicorum* (or *Ockham on Aristotle's Physics*: a translation by J. Davies). Franciscan Institute, St. Bonaventure University.

Ohmura, A. (2005). Energy budget at the Earth's surface. In *Observed Global Climate*, ed. M. Hantel, Vol. XX, chapter 10, Landolt-Börnstein Handbook. Berlin: Springer Verlag.

Orlanski, I. (1975). A rational subdivision of scales for atmospheric processes. *Bull. Amer. Met. Soc.*, **56**, 527–530.

Price, W. E. and Heindl, L. A. (1968). What is hydrology? *Trans. Amer. Geophys. Un.*, **49**, 529–533.

PROBLEMS

1.1   Assuming that the average volume of water storage in the soil equals 0.05 m (if spread over the entire Earth surface; see Table 1.3), and that the average precipitation on the land surfaces of the Earth is equal to 0.8 m y^{-1} (Table 1.1), give an estimate (expressed in days) for the mean residence time of soil water. Consider conditions to be steady, so that precipitation is in balance with runoff and evaporation, and assume that all the precipitation infiltrates into the soil. What would this residence time be, if it is assumed that only one half of the precipitation infiltrates directly into the soil, and that the remaining half immediately evaporates as interception or runs off on the surface?

1.2   Recent estimates of the average surface energy fluxes at the global scale (Table 1.4) are as follows; the net radiation $R_n = 104$ W m^{-2}, the latent heat flux $L_e E = 85$ W m^{-2}, and the turbulent sensible heat flux $H = 19$ W m^{-2}. Express these fluxes as equivalent quantities of liquid water evaporated in units of mm y^{-1}. Assume that the latent heat of vaporization of water is roughly $L_e = 2.5 \times 10^6$ J kg^{-1} and the density $\rho_w = 10^3$ kg m^{-3}.

I WATER IN THE ATMOSPHERE

2 WATER ALOFT: FLUID MECHANICS OF THE LOWER ATMOSPHERE

On account of the short residence time and great mobility of water vapor in air, the lower atmosphere is one of the critical pathways in the global hydrologic cycle; it transports water and energy around the globe without regard to continental boundaries and thus links the continents, the upper atmosphere, and the oceans. The transport and distribution of water vapor in the lower atmosphere, where it is most abundantly present, are among the main factors controlling precipitation and evaporation from the surface; these processes, in turn, determine soil and groundwater storage and the different runoff phenomena.

2.1 WATER VAPOR IN AIR

2.1.1 Global features

The global amount of water vapor contained in the air is roughly (see Table 1.3) equivalent with a layer of liquid water covering the earth, with a thickness of around 25 mm on average. The thickness of this layer, which is the total liquid equivalent of water vapor in the atmospheric column at a given location, is also called the *precipitable water*, W_p. However, this quantity of water vapor is not distributed uniformly and it can greatly vary over a wide range of scales in space and in time. For instance, the water vapor content of the atmosphere, just like the temperature, generally tends to decrease with increasing latitude. Available data (Randel *et al.*, 1996) show that the precipitable water is more likely to be well below 5 mm near the Poles, and close to 50 mm near the Equator. But this is not always the case; even at similar latitudes there can be huge regional variations, the most extreme example being the warm dry deserts of the world. Most of the atmospheric water vapor is found relatively close to the ground, and at any given location water vapor decreases sharply with height; typically, about half the total water vapor in the atmospheric column can be found below a height of 1 or 2 km.

Because the global annual evaporation is around $E = 1$ m, the average atmospheric residence time of water vapor W_p/E is only about 9 days. This time scale governs the hydrologic interactions and water transfers between the atmosphere and the other two compartments of the global system, the oceans and the continents. This time scale is especially fundamental to the transport of atmospheric water vapor from its source regions – mainly evaporation from the oceans – to sinks in precipitating weather systems. Indeed, the excess precipitation on the continents, which does not evaporate, ultimately runs off to the seas and oceans of the world. A balance is maintained in the global system by the fact that over the oceans the situation is reversed and that evaporation is generally larger than precipitation, allowing the excess oceanic water vapor to be transported back to the continents. This transport of water from the oceans to the land areas, also called advection, takes place mostly in the form of water vapor and not as clouds; actually, in the atmosphere the total amount of water in the liquid and ice phases is less than 0.5% of the water in the vapor phase.

But beside its central role in the hydrologic cycle, water vapor strongly affects other aspects of the Earth's weather and climate as well. It is one of the main agents in the overall energy budget of the atmosphere in a number of ways. Globally, as seen in Table 1.4, the phase changes from liquid and solid to vapor are the main energy transfer mechanisms from the Earth's surface to the atmosphere; the subsequent condensation of this vapor in the air furnishes a large portion of the energy needed for the circulation of the atmosphere. Thus, the large-scale transport of water vapor as latent heat is one of the main redistribution mechanisms for the uneven radiative input from the Sun. In addition, the concentration and spatial distribution of the atmospheric water vapor are major factors controlling the amount and type of cloud, which in turn determine the solar radiation reaching the Earth's surface. Finally, as the most abundant greenhouse gas, water vapor absorbs and thus "traps" terrestrial infrared radiative energy, and then re-emits it at a lower temperature.

2.1.2 Some physical properties

For many practical purposes, the air of the lower atmosphere can be considered as a mixture of perfect gases; in the present context these may conveniently be assumed to be dry air of constant composition and water vapor. The water vapor content of the air can be expressed in terms of the mixing ratio, defined as the mass of water vapor per unit mass of dry air,

$$m = \rho_v / \rho_d \tag{2.1}$$

where ρ_v is the density of the water vapor and ρ_d the density of the air without the water vapor. The specific humidity is defined as the mass of water vapor per unit mass of moist air,

$$q = \rho_v / \rho \tag{2.2}$$

where $\rho = \rho_v + \rho_d$. The relative humidity is the ratio of the actual mixing ratio and the mixing ratio in water vapor saturated air at the same temperature and pressure,

$$r = m / m^* \tag{2.3}$$

This is nearly equal to (e/e^*), the ratio of the actual vapor pressure and the saturation vapor pressure; the latter is the pressure of the vapor, when it is in equilibrium with a plane surface of water or ice at the same temperature and pressure.

According to Dalton's law, the total pressure in a mixture of perfect gases equals the sum of the partial pressures, and each of the component gases obeys its own equation of state. Thus, the density of the dry air component is

$$\rho_d = \frac{p - e}{R_d T} \tag{2.4}$$

where p is the total pressure in the air, e is the partial pressure of the water vapor, T is the ("absolute") temperature, and R_d, which is given in Table 2.1, is the specific gas constant

Table 2.1 **Some physical constants**

Dry air
Molecular weight: 28.966 g mol^{-1}
Gas constant: $R_d = 287.04$ J kg^{-1}K^{-1}
Specific heat: $c_{pd} = 1005$ J kg^{-1}K^{-1}
$\qquad\qquad c_{vd} = 716$ J kg^{-1}K^{-1}
Density: $\rho = 1.2923$ kg m^{-3}
($p = 1013.25$ hPa, $T = 273.16$ K)

Water vapor
Molecular weight: 18.016 g mol^{-1}
Gas constant: $R_w = 461.5$ J kg^{-1}K^{-1}
Specific heat: $c_{pw} = 1846$ J kg^{-1} K^{-1}
$\qquad\qquad c_{vw} = 1386$ J kg^{-1}K^{-1}

Note. The values listed in Tables 2.1, 2.4 and 2.5, are adapted from the Smithsonian Meteorological Tables (List, 1971), where the original references are cited.

for dry air. Similarly, the density of water vapor is

$$\rho_v = \frac{0.622e}{R_d T} \tag{2.5}$$

where $0.622 = (18.016/28.966)$ is the ratio of the molecular weights of water and dry air.

The density of moist air from Equations (2.4) and (2.5) is

$$\rho = \frac{p}{R_d T}\left(1 - \frac{0.378e}{p}\right) \tag{2.6}$$

showing that it is smaller than that of dry air at pressure p. This means that water vapor stratification plays a role in determining the stability of the atmosphere. The equation of state of moist air can be obtained by eliminating e from Equations (2.4) and (2.5)

$$p = \rho T R_d (1 + 0.61\, q) \tag{2.7}$$

This indicates that the air mixture behaves as a perfect gas provided it has a specific gas constant

$$R_m = R_d(1 + 0.61\, q) \tag{2.8}$$

that is a function of the water vapor content. Therefore, Equation (2.7) is often expressed as

$$p = R_d \rho\, T_V \tag{2.9}$$

where T_V is the virtual temperature defined by

$$T_V = (1 + 0.61\, q)\, T. \tag{2.10}$$

The virtual temperature is the temperature that dry air should have in order to have the same density as moist air with given q, T and p.

Table 2.2 Some useful units

	SI (mks)	cgs
Length	meter	centimeter
	m	cm
Mass	kilogram	gram
	kg	g
Time	second	second
	s	s
Force	newton	dyne
	$N = kg\ m\ s^{-2}$	$dyn = g\ cm\ s^{-2}$
Pressure	pascal	microbar
	$Pa = N\ m^{-2}$	$\mu bar = dyn\ cm^{-2}$
Energy	joule	erg
	$J = N\ m$	$erg = dyn\ cm$
Power	watt	
	$W = J\ s^{-1}$	$erg\ s^{-1}$

Table 2.3 Conversion factors

Pressure	millibar	$1\ mb = 10^2\ Pa = 1\ hPa = 10^3\ \mu bar = 10^3 dyn\ cm^{-2}$
	millimeter mercury	$1\ mm\ Hg = 1.333\ 224\ hPa$
	atmosphere	$1\ atm = 1.013\ 25\ 10^5 Pa$
Energy	calorie (IT)	$1\ cal = 4.1868\ J = 4.1868\ 10^7\ erg$
(Energy/area)	(langley)	$(1\ ly = 1\ cal\ cm^{-2})$

The precipitable water is the total mass of water vapor in a vertical atmospheric column; if it is assumed that the pressure is negligible at the top of the atmosphere, it can be written as

$$W_p = \int_0^{p_0} q\ dp/g \qquad (2.11)$$

where p_0 is the surface pressure. Recall that the basic dimensions of these variables are $[q] = [M_w/M_a]$, $[p] = [M_a L^{-1} T^{-2}]$, and $[g] = [LT^{-2}]$, in which it is convenient to distinguish between the mass of air M_a and the mass of water substance M_w. Therefore the basic dimensions of precipitable water are $[W_p] = [M_w L^{-2}]$, i.e. water mass per unit area. In SI units this can be expressed in kg m^{-2}, which is roughly equivalent with mm of vertical liquid water column, because the density of liquid water is around 1000 kg m^{-3}.

For convenient reference, some common units and conversion factors are listed in Tables 2.2 and 2.3.

Table 2.4 Some properties of water

Temperature (°C)	c_w (J kg^{-1} K^{-1})	L_e (10^6 J kg^{-1})	e^* (hPa)	de^*/dT (hPa K^{-1})
−20	4354	2.549	1.2540	0.1081
−10	4271	2.525	2.8627	0.2262
0	4218	2.501	6.1078	0.4438
5	4202	2.489	8.7192	0.6082
10	4192	2.477	12.272	0.8222
15	4186	2.466	17.044	1.098
20	4182	2.453	23.373	1.448
25	4180	2.442	31.671	1.888
30	4178	2.430	42.430	2.435
35	4178	2.418	56.236	3.110
40	4178	2.406	73.777	3.933

c_w: specific heat; L_e: latent heat of vaporization; e^*: saturation vapour pressure.

Table 2.5 Some properties of ice

Temperature (°C)	c_i (J kg^{-1} K^{-1})	L_{fu} (10^6 J kg^{-1})	L_s (10^6 J kg^{-1})	e_i^* (hPa)	de_i^*/dT (hPa K^{-1})
−20	1959	0.2889	2.838	1.032	0.09905
−15	—	—	—	1.652	0.1524
−10	2031	0.3119	2.837	2.597	0.2306
−5	—	—	—	4.015	0.3432
0	2106	0.3337	2.834	6.107	0.5029

c_i: specific heat; L_{fu}: latent heat of fusion; L_s: latent heat of sublimation; e_i^*: saturation vapor pressure over ice.

2.1.3 *Saturation vapor pressure*

The saturation vapor pressure depends only on the temperature, or $e^* = e^*(T)$. Some values are presented in Tables 2.4 and 2.5. These values were obtained from the Goff–Gratch formulation (see List, 1971), which has been used as the international standard for some time. For water this formulation is

$$\log e^* = -7.902\,98(T_{st}/T - 1) + 5.028\,08\ \log(T_{st}/T)$$
$$- 1.3816 \times 10^{-7}\left(10^{11.344(1-T/T_{st})} - 1\right)$$
$$+ 8.1328 \times 10^{-3}\left(10^{-3.191\,49(T_{st}/T-1)} - 1\right) + \log e_{st}^* \tag{2.12}$$

where log( ) denotes the decimal logarithm, T is the temperature in K, T_{st} is the steam-point temperature 373.16 K, and e_{st}^* is the saturation vapor pressure at the steam-point temperature, i.e. 1013.25 hPa; the relationship is also sketched in Figure 2.1. For ice the saturation water

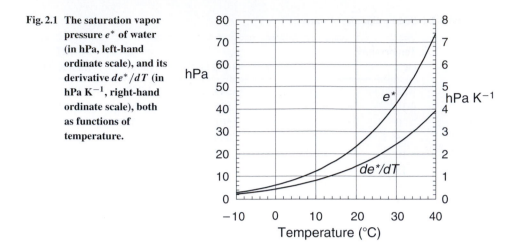

Fig. 2.1 The saturation vapor
pressure e^* of water
(in hPa, left-hand
ordinate scale), and its
derivative de^*/dT (in
hPa K^{-1}, right-hand
ordinate scale), both
as functions of
temperature.

vapor is

$$\log e_i^* = -9.09718(T_0/T - 1) - 3.56654 \log(T_0/T) \tag{2.13}$$
$$+ 0.876793 \, (1 - T/T_0) + \log e_{io}^*$$

where T_0 is the ice-point temperature 273.16 K, and e_{io}^* the saturation vapor pressure at
the ice-point temperature, i.e. 6.1071 hPa. Lowe (1977), who has also compared other
currently used expressions for the saturation vapor pressure, has presented polynomials for
e^*, de^*/dT, e_i^*, and de_i^*/dT, which are quite accurate and suitable for rapid computation.
For computational speed these polynomials should be used in nested form; for e^* the
representation takes the form

$$e^* = a_0 + T(a_1 + T(a_2 + T(a_3 + T(a_4 + T(a_5 + Ta_6))))) \tag{2.14}$$

where the polynomial coefficients are as follows when T is in K,

$a_0 = 6984.505\ 294$, $a_1 = -188.903\ 931\ 0$, $a_2 = 2.133\ 357\ 675$,
$a_3 = -1.288\ 580\ 973 \times 10^{-2}$, $a_4 = 4.393\ 587\ 233 \times 10^{-5}$,
$a_5 = -8.023\ 923\ 082 \times 10^{-8}$, and $a_6 = 6.136\ 820\ 929 \times 10^{-11}$.

2.2   HYDROSTATICS AND ATMOSPHERIC STABILITY

The first law of thermodynamics states that the heat added to a system equals the sum
of the change in internal energy and the work done by the system on its surroundings. If
these quantities are taken per unit mass and in differential form, this is for partly saturated
air

$$dh = du + pd\alpha \tag{2.15}$$

where $\alpha = \rho^{-1}$ is the specific volume, ρ is the density of the air, and (in this Section 2.2
only) u represents the internal energy. The equation of state (2.7), which on account

of (2.8) can also be written as

$$p = R_m T / \alpha \tag{2.16}$$

relates the three variables, α, the temperature T and the pressure p; thus only two of the three are needed to define the state. If α and T are chosen as these independent variables, Equation (2.15) becomes

$$dh = \left(\frac{\partial u}{\partial T}\right)_\alpha dT + \left[\left(\frac{\partial u}{\partial \alpha}\right)_T + p\right] d\alpha \tag{2.17}$$

Since by definition the specific heat capacity for constant volume is $c_v = (\partial u / \partial T)_\alpha$ and since it can be shown that $(\partial u / \partial \alpha)_T = 0$, combination of the differential form of Equations (2.16) with (2.17) produces

$$dh = (c_v + R_m) dT - \alpha dp \tag{2.18}$$

or

$$dh = c_p dT - \alpha dp \tag{2.19}$$

where by definition $c_p = (\partial h / \partial T)_p$ is the specific heat for constant pressure. With the hydrostatic law, giving the pressure change with height in a fluid at rest, i.e.

$$dp = -\rho g dz \tag{2.20}$$

one finally obtains from Equation (2.19)

$$dh = c_p dT + g dz \tag{2.21}$$

Equation (2.21) is derived here by a combination of the principle of conservation of energy with the equation of state and the hydrostatic equation. This result was obtained for air containing water vapor; however, the moisture content dependency of the specific heat at constant pressure, namely $c_p = q c_{pw} + (1 - q) c_{pd}$, is very weak and therefore Equation (2.21) is usually applied with the specific heat for dry air, i.e. c_{pd}.

The criterion for the stability of an atmosphere at rest can be obtained by the following thought experiment. Consider a small parcel of air with a temperature T_1 that undergoes a small vertical displacement without mixing with the surrounding body of air; this displacement is sufficiently small and fast, so that the pressure of the particle adjusts to its new environment in an adiabatic fashion, that is as a reversible process without heat exchange with its surroundings. Two cases are of interest, depending on the degree of saturation of the air.

2.2.1 Stability of a partly saturated atmosphere

Dry adiabatic lapse rate
If the atmosphere can be assumed to remain partly saturated during this process, there is no vaporization or condensation, and the change in heat of the parcel is $dh = 0$. With

Fig. 2.2   **Unstable atmosphere in which the temperature decrease with elevation (solid line) is larger than the dry adiabatic lapse rate (dashed line), so that $\Gamma > \Gamma_d$.**

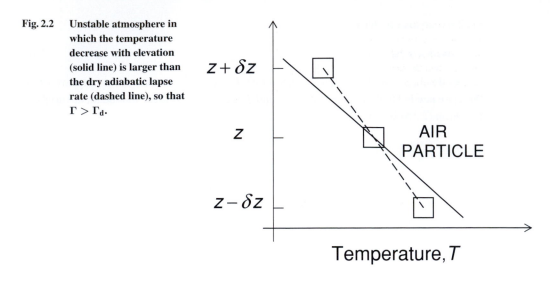

Equation (2.21) this yields for the temperature change of the parcel, as it moves up or down,

$$dT_1/dz = -g/c_p \tag{2.22}$$

which is of the order of 9.8 K km^{-1}. The vertical rate of decrease in temperature of the atmosphere, $-dT/dz$, is the lapse rate of the air, denoted here by Γ. A lapse rate of the atmosphere, that is equal to g/c_p, is called a dry adiabatic lapse rate, Γ_d.

Whenever the actual lapse rate in the atmosphere Γ is larger than Γ_d, a particle undergoing a small upward displacement δz and changing its temperature according to Equation (2.22), will be warmer and therefore lighter than its surroundings; this means that it will have a tendency to continue its upward motion (see Figure 2.2). By the same token, a particle undergoing a small downward displacement δz under the same lapse rate conditions, will be colder and thus heavier than the surrounding air; again, it will have a tendency to continue its downward movement. In both situations, once displaced however slightly, air parcels tend to continue their motion and amplify their displacements both upward and downward: under such lapse rate conditions the atmosphere is unstable. Conversely, in an atmosphere with $\Gamma < \Gamma_d$, the upward-moving parcel, whose temperature change is given by Equation (2.22), becomes surrounded by air that is relatively warmer (see Figure 2.3); thus it is heavier and it will tend to return to its original position, where it was in equilibrium with its surroundings. In this case, the parcel resists being moved away from its original position and vertical displacements are suppressed: the atmosphere is said to be stable. Under conditions when the actual lapse rate in the air is dry adiabatic, the atmospheric stability is neutral. In summary, one has for unsaturated air the following criteria:

$$\Gamma > \Gamma_d \qquad \text{unstable}$$
$$\Gamma = \Gamma_d \qquad \text{neutral}$$
$$\Gamma < \Gamma_d \qquad \text{stable}$$

Fig. 2.3   Stable atmosphere in which the temperature decrease with elevation (solid line) is smaller than the dry adiabatic lapse rate (dashed line), so that $\Gamma < \Gamma_d$.

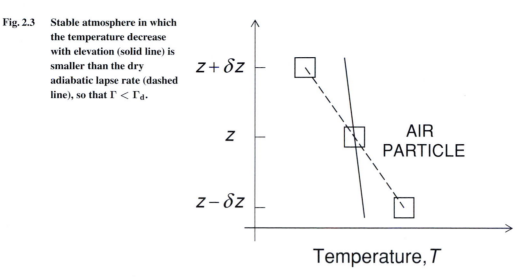

Unstable conditions typically occur whenever the atmosphere is being heated by the surface below, for example, as a result of solar radiation on days with clear sky, or as relatively cold air moves over a relatively warmer surface, such as a lake or the ocean in fall and early winter. An unstable atmosphere is subject to more intense mixing and turbulence than a neutral one; this also results in larger turbulent transport. Under certain conditions an unstable atmosphere will even give rise to various types of organized motions, with scales ranging from mere local updrafts and dust devils to large tropical storms. The atmosphere is often stable when the air is being cooled from below. This typically occurs at night under clear skies, when the surface is cooled by outgoing long-wave radiation or when relatively warm air flows over a relatively colder lake or ocean surface in spring. In addition, stable conditions, also called inversions, may result from larger-scale weather patterns, when relatively warmer air masses move over colder layers. Not surprisingly, stable conditions have the opposite effect of unstable conditions. Thus mixing and turbulence are suppressed, and atmospheric transport is normally smaller. Under extreme conditions, the turbulence may be eliminated altogether and the flow of the air may be laminar. Such conditions are sometimes visible in the evening of a calm sunny day, around sunset, when the air near the ground becomes chilly and the smoke from a chimney can be seen moving slowly through the tree tops of a forest. On a somewhat larger scale, as a result of the reduced turbulence and dispersion, inversion conditions also tend to aggravate pollution problems in populated areas.

Potential temperature

To repeat briefly, during small displacements, parcels of air undergo adiabatic temperature changes in accordance with Equation (2.22). In a perfectly neutral atmosphere the lapse rate of the atmosphere is also $-g/c_p$; therefore under such conditions a displaced parcel will, on average at least, always be surrounded by air at the same temperature and, as a result, there is no net exchange of heat. This means that even though there is a vertical temperature difference under neutral conditions, the heat flux is zero; consequently, the temperature used in heat transfer formulations should be corrected to account for that fact. This can be done by using the potential temperature θ instead of the actual temperature T. The potential temperature is the temperature that would result if air were brought adiabatically to a standard pressure level $p_0 = 1000$ hPa; for such a process $dh = 0$, and Equation (2.19), after substitution of α with (2.16), can be integrated to yield Poisson's equation

$$\theta = T(p_0/p)^{R_d/c_p} \tag{2.23}$$

which can serve to define the potential temperature θ and also to calculate it for a given pressure p and temperature T; note that in Equation (2.23) R_m is replaced by R_d for convenience. During an adiabatic process the potential temperature is conserved and therefore under perfectly neutral conditions in a dry atmosphere, or when the specific humidity is constant with height, the potential temperature should be a constant. A dry atmosphere is unstable when θ decreases, and stable when it increases with height. Nevertheless, the difference between T and θ is usually rather small, especially in the lower layers of the air near the ground surface, where most measurements are made. Therefore in many situations, when the height difference of the temperature measurements is only a few meters, the use of the actual temperature T is allowed; otherwise θ must be used in heat transfer formulations.

Density stratification due to water vapor

In the above considerations of atmospheric stability, the density stratification due to vertical humidity gradient $\partial q/\partial z$ was not taken into account. Under some conditions this can be an important factor, but it can be readily shown (see, for example, Brutsaert, 1982) that this may be incorporated in the analysis by means of the virtual potential temperature, defined as $\theta_v = (1 + 0.61q)\theta$; the virtual potential temperature is related to the potential temperature, in the same way the virtual temperature is related to the actual temperature, as indicated in Equation (2.10). Thus, strictly speaking, in the presence of humidity stratification, the atmosphere can be considered statically neutral, not when θ is constant, but only when θ_v is constant; it is unstable when θ_v decreases, and stable when it increases with height. Put differently, under such conditions the stability criterion for an atmosphere is not the lapse rate of the temperature, but the lapse rate of the virtual temperature; in practice, however, this difference is often ignored.

2.2.2 Stability of a saturated atmosphere

When the air is saturated, any increase in heat content dh of a parcel of air during an adiabatic process can only be the result of condensation, that is a decrease in the water vapor content of the air; this can be written as $dh = -L_e dq$, in which L_e is the latent heat of vaporization and q, defined in Equation (2.2), is the specific humidity. With Equation (2.21) one obtains now

$$-\frac{dT_1}{dz} = \Gamma_d + \frac{L_e}{c_p}\frac{dq}{dz} \tag{2.24}$$

Fig. 2.4   **Conditional instability in the atmosphere. A partly saturated air particle, which is raised, initially undergoes a rate of temperature decrease Γ_d (dashed line) which is larger than that of the environment, Γ. As it becomes saturated at z_C, the rate of temperature decrease is reduced to the saturated adiabatic lapse rate Γ_s (dashed line). Above z_F, the free convection level, conditions become unstable.**

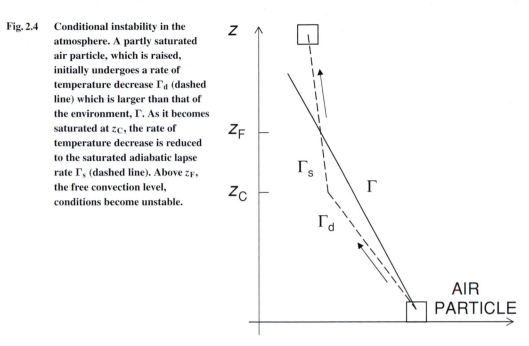

The quantity on the right-hand side of (2.24) is called the saturated adiabatic lapse rate, Γ_s. Since normally $dq/dz < 0$, the saturated adiabatic lapse rate must be smaller than the dry adiabatic lapse rate. Moreover, $(L_e/c_p)(dq/dz)$ depends on the temperature. At high temperatures, near the Equator, it is of the order of $\Gamma_s \approx 0.35\,\Gamma_d$, whereas at lower temperatures, for example around $-30\,^{\circ}\mathrm{C}$, it has approximately the same value as Γ_d, i.e. $9.8\,^{\circ}\mathrm{C\,km^{-1}}$. In the lower layers of the atmosphere it is on average of the order of $5.5\,^{\circ}\mathrm{C\,km^{-1}}$. If in a rising air mass the condensed moisture is being removed (e.g. through precipitation), the rate of temperature decrease is called the pseudo-adiabatic lapse rate. However, under most conditions the heat loss by the removal of this condensed water is fairly negligible, and the saturated adiabatic lapse rate is a satisfactory approximation. Thus, for saturated air, one has the following stability criteria:

$\Gamma > \Gamma_s$ unstable
$\Gamma = \Gamma_s$ neutral
$\Gamma < \Gamma_s$ stable

2.2.3   *Conditional instability*

It often happens that the actual lapse rate in the atmosphere is intermediate between the dry and the saturated adiabatic lapse rate, that is $\Gamma_s < \Gamma < \Gamma_d$; this case is referred to as conditional instability. When a partly saturated parcel of air is raised in such an atmosphere, it will initially change its temperature at the dry adiabatic rate in accordance with Equation (2.8), and thus remain colder than the surroundings (see Figure 2.4).

This situation is still stable. However, if the particle is made to rise further, and continues to cool down, it may reach the level z_C, where condensation takes place; above the condensation level z_C it will change its temperature at the saturated adiabatic lapse rate. If the rise continues, eventually above the level z_F the temperature of the particle will exceed that of the surroundings; the rising air is now lighter than its surroundings, and an unstable situation has been established. The air, which was originally forced to rise, will now take off by free convection and continue to rise without any outside agent. The height z_F is the free convection level. Thus whether or not a vertical displacement results in instability depends largely on the moisture content of the atmosphere. In a moist atmosphere the condensation level is low, and relatively small vertical displacements readily produce unstable conditions. In a dry atmosphere, the level z_C is higher, and the atmosphere is more likely to remain stable, even with relatively large vertical displacements.

2.3   TURBULENT TRANSPORT OF WATER VAPOR

The flow of the atmosphere is almost invariably turbulent. In a turbulent flow molecular diffusion can usually be neglected, and water vapor is moved from one place to another by advective transport, that is, by being linked to the motions of the air that contains it. One exception, when molecular diffusion may be of some consequence, occurs near a wall where the no slip condition reduces the velocity of the moving air to zero and the turbulence is largely suppressed. Thus in turbulent air flow, the specific mass flux of water vapor is given by

$$\mathbf{F}_v = \rho_v \mathbf{v} = \rho q \mathbf{v} \tag{2.25}$$

where $\mathbf{v}$ is the velocity of the air, ρ_v is the water vapor density, and q is the specific humidity. The variables $\mathbf{F}_v = \mathbf{i}F_{vx} + \mathbf{j}F_{vy} + \mathbf{k}F_{vz}$ and $\mathbf{v} = \mathbf{i}u + \mathbf{j}v + \mathbf{k}w$ are both vectors, with x denoting the direction of the mean wind near the ground and z the vertical.

Note that the transport described by Equation (2.25) can also be referred to as *convection* in fluid mechanics. However, this usage may lead to some confusion because, especially in the atmospheric sciences, convection is commonly reserved to describe transport involving gravity effects, resulting from unstable density stratification. To avoid such confusion in this regard, in this book any transport that is linked to the motion of the fluid is called *advection*.

Turbulent flux of water vapor
In turbulent flow the detailed description of the velocity field and also the temperature, the content of water vapor, or other admixtures of the air, at any given point in time and space, is practically impossible and it can only be accomplished in a statistical sense. The simplest and probably most important statistic is the mean. Accordingly, ever since Reynolds introduced the idea, it has been common practice in the analysis of turbulent flow phenomena to decompose the relevant variables into a mean and a turbulent fluctuation, namely $F_{vx} = \overline{F_{vx}} + F'_{vx}, \ldots, u = \bar{u} + u', \ldots, q = \bar{q} + q'$, etc.

After applying the customary time averaging over a suitable time period, one obtains from Equation (2.25) for the mean flux components of water vapor

$$\overline{F_{vx}} = \rho(\overline{u}\,\overline{q} + \overline{u'q'})$$
$$\overline{F_{vy}} = \rho(\overline{v}\,\overline{q} + \overline{v'q'}) \qquad (2.26)$$
$$\overline{F_{vz}} = \rho(\overline{w}\,\overline{q} + \overline{w'q'})$$

The first terms on the right of these three equations represent the advective transport of water vapor by the mean motion of the air. The second terms are the components of the advective vapor transport by the turbulence; they are also often called the Reynolds fluxes, and statistically speaking, they are covariances. The estimation and parameterization of these flux components is one of the core problems of hydrology.

Conservation equation of water vapor

The standard procedure for a more thorough analysis of the water vapor transport consists of combining the expressions for the fluxes, Equations (2.26), with the principle of mass conservation (1.8) applied to water vapor. This is accomplished by substituting ρ_v for ρ and $\mathbf{F}_v$ for $(\rho\mathbf{v})$ in Equation (1.8); since in this derivation, the bulk air itself is of less concern, it can be assumed to have a constant density, which allows use of Equation (1.9) for the mean velocity $\overline{\mathbf{v}}$. Thus, one obtains the conservation equation for the mean specific humidity, $\overline{q}$ (see Brutsaert, 1982), as follows

$$\frac{\partial \overline{q}}{\partial t} + \overline{u}\frac{\partial \overline{q}}{\partial x} + \overline{v}\frac{\partial \overline{q}}{\partial y} + \overline{w}\frac{\partial \overline{q}}{\partial z} = -\left[\frac{\partial}{\partial x}\left(\overline{u'q'}\right) + \frac{\partial}{\partial y}\left(\overline{v'q'}\right) + \frac{\partial}{\partial z}\left(\overline{w'q'}\right)\right] \qquad (2.27)$$

in which, it should be noted again, the molecular diffusion term is neglected. In principle, it should be possible to solve Equation (2.27) with appropriate boundary conditions to describe water vapor transport in the atmosphere. However, this equation presents several difficulties, which make its solution extremely difficult. First, since the fluxes in (2.26) are intrinsically dependent on the velocity of the air and the turbulence, it is necessary to consider the dynamics of the flow and to include the conservation equations for momentum and temperature in the solution process as well. A second and more fundamental difficulty is that this conservation equation for the mean specific humidity contains not only $\overline{q}$ as a dependent variable, which is the first moment, but also the covariances of q' with the velocity fluctuations u', v' and w', which are second moments. This means that Equation (2.27) has more than one unknown; this fact is an instance of the notorious closure problem of turbulence and it indicates that, without additional relationships, this equation cannot be solved mathematically.

Fortunately, it is possible to simplify the general problem, as formulated with the above fluxes, considerably and still obtain meaningful results. This is accomplished, first, by assuming that the atmosphere nearest the surface can be considered as a steady boundary layer above a quasi-homogeneous surface (Section 2.4), and, second, by the application of similarity assumptions to alleviate the turbulence closure problem by appropriate parameterization (Section 2.5).

2.4   THE ATMOSPHERIC BOUNDARY LAYER

2.4.1   *Quasi-homogeneous conditions*

In the atmosphere the largest changes in wind velocity, temperature and humidity are usually found in the vertical direction and in a distinct region near the surface. In contrast, horizontal changes are relatively mild, and tend to occur over distances of the order of tens of kilometers. For this reason the air near the surface may be regarded as a boundary layer, a concept set forth by Prandtl (1904) for momentum transport in the neighborhood of a solid wall. The atmospheric boundary layer (or ABL) can be defined as the lower part of the atmosphere, where the nature and properties of the surface affect the turbulence directly. Accordingly, the horizontal scales of most atmospheric flow phenomena of interest in hydrology are much larger than the vertical, so that the horizontal gradients are usually small compared to the vertical gradients, and the vertical velocities are small relative to the horizontal velocities. Thus, many problems can be solved by simply assuming that

$$\left(\frac{\partial}{\partial x}, \frac{\partial}{\partial y} \right) = 0 \quad \text{and} \quad \overline{w} = 0 \tag{2.28}$$

In addition, since x is the direction of the mean wind velocity near the ground, the mean velocity in the lateral y-direction can also be discarded, or $\overline{v} = 0$. Strictly speaking (2.28) is valid only when the surface is perfectly homogeneous or uniform. Such conditions are rare, and the properties of most natural landsurfaces are spatially quite variable; fortunately, in many situations of interest they can be considered to be at least statistically homogeneous (see, for example, Brutsaert, 1998), and the assumptions of (2.28) can still be used to describe the flow.

More generally (2.28) is tantamount to assuming that, as the air moves parallel to a homogeneous surface, on average (in the turbulence sense) the concentration of any property or admixture advected by the air changes only in the vertical and remains constant in the horizontal direction. The fact that the mean concentrations change only vertically is evidence that there is a source or a sink of the admixture at the surface, and thus the only turbulent fluxes of consequence are the vertical components. In the case of humidity with mean concentration (per unit mass of bulk air) $\overline{q}$, Equations (2.26) are thus effectively reduced to

$$F_{vz} = \rho \, \overline{w'q'} \tag{2.29}$$

in which henceforth the overbar on F_{vz} is omitted for convenience of notation.

While mathematically Equations (2.26) and (2.29) are unambiguous, a more intuitive sense of their physical significance can be obtained by considering the mechanism sketched in Figure 2.5. A particle of air, which undergoes a vertical velocity fluctuation w', travels a distance $w'\delta t$ during a time interval δt. After that air parcel has risen a small distance $w'\delta t$ from a level, where the mean specific humidity is $\overline{q}$, it has a specific humidity which is q' larger than the mean of its new environment; thus the rate (distance per unit time) at which this particular parcel transports absolute humidity upward is $(\rho \, q'w')$ times its volume. There are many such parcels – or eddies – in turbulent flow

Fig. 2.5   Sketch illustrating a small fluid parcel, which rises a small distance $w'\delta t$, to a level, where the mean humidity $\bar{q}$ is lower by an amount q' than at its original position.

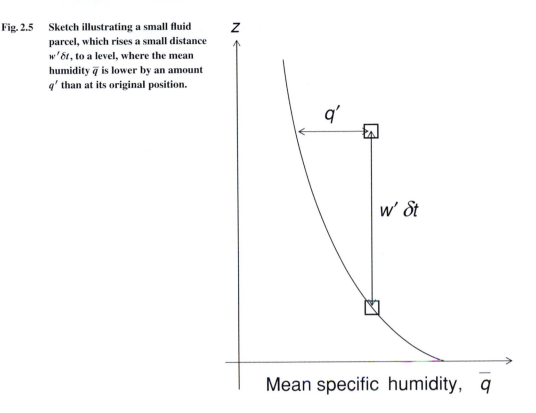

moving in all directions and the transport rate by all of them, i.e. the vertical transport of water vapor mass per unit horizontal area and per unit time is on average as indicated in Equation (2.29).

Similar expressions can be written for the fluxes of other properties or admixtures of the flow. The vertical flux component of horizontal momentum, with mean concentration $\bar{u}$, is

$$F_{\mathrm{mz}} = \rho \overline{w'u'} \tag{2.30}$$

and that of sensible heat, with mean concentration $c_{\mathrm{p}}\bar{\theta}$, can be written as

$$F_{\mathrm{hz}} = \rho c_{\mathrm{p}} \overline{w'\theta'} \tag{2.31}$$

Under steady conditions in the lowest few meters of the air above a uniform surface, on account of continuity the inflow rate equals the outflow rate, which means that these vertical fluxes must be constant with elevation. Hence the water vapor flux in Equation (2.29) is in fact equal to the rate of evaporation E from the surface, or $F_{\mathrm{vz}} = \rho \overline{w'q'}_0 \equiv E$, in which the 0 subscript denotes the value near the surface. In the case of momentum, there is a sink at the surface in the form of a shear stress, so that close to the surface it can also be assumed that $F_{\mathrm{mz}} \equiv -\tau = -\tau_0$, in which τ_0 is the shear stress at the surface. Similarly the flux in Equation (2.31) equals the sensible heat flux H at the surface, or $F_{\mathrm{hz}} = \rho c_{\mathrm{p}} \overline{w'\theta'}_0 \equiv H$. For convenience of notation, the surface shear stress, which is

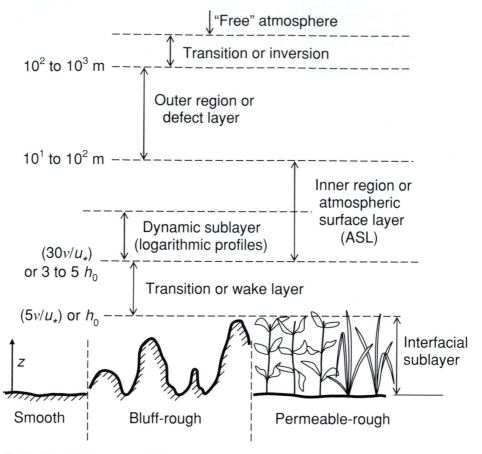

Fig. 2.6 Sketch of the typical structure of the atmospheric boundary layer (ABL) above three different types of uniform surfaces. The atmospheric surface layer (ASL) is the region where Monin–Obukhov similarity (MOS) is usually valid; h_0 is a typical height of the roughness obstacles. Under unstable conditions the outer region is called the mixed layer, and it is capped by an inversion layer. (The vertical axis scale is distorted.)

introduced here, is often expressed also as the friction velocity defined as

$$u_* \equiv (\tau_0/\rho)^{1/2} \tag{2.32}$$

This shows that in light of Equation (2.30), under steady or nearly steady conditions, one has to a good approximation near the ground that $u_*^2 = -\overline{w'u'}$.

2.4.2   *General structure of the ABL*

In the analysis, it is convenient to assume that the atmospheric boundary layer consists of a number of sublayers, in which different sets of variables are important to different degrees in governing turbulent transport. The main subdivision is into an inner and an outer region. In the *outer region* or *defect layer* the flow is strongly dependent on the

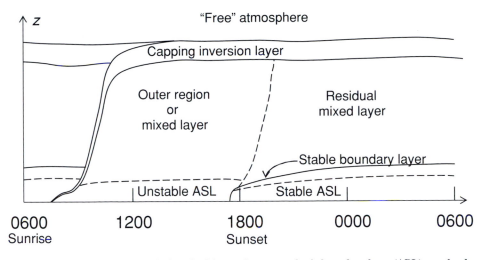

Fig. 2.7 Sketch of the typical diurnal evolution of a fair weather atmospheric boundary layer (ABL) over land under clear sky around the time of the equinox. The inner region or atmospheric surface layer (ASL) is unstable during the daytime as a result of solar heating at the surface; the ASL is stable at night as a result of radiative cooling. During the day the outer region is characterized by convective turbulence, fed by heating through the surface layer; after sunset this outer layer becomes virtually uncoupled from the surface, by the development of the stable nocturnal boundary layer.

free stream velocity outside the boundary layer, whereas in the *inner region*, also called variously *atmospheric surface layer* (or ASL), *Prandtl layer* or *wall layer*, the flow is more strongly affected by the nature of the surface (see Figure 2.6).

In the atmosphere, under conditions not very different from neutral, the outer region is affected both by the pressure gradients, reflecting larger scale weather patterns, and by the Coriolis forces, reflecting the effect of the rotation of the Earth. Under unstable conditions, the effects of the pressure and Coriolis forces are relatively small, and the outer region is more characterized by thermal convective turbulence; the outer region may then be referred to as the *mixed layer* or the *convection layer*. The upper limit of the unstable boundary layer is typically indicated by a sharp inversion, that is, a layer of stable air. Over land, the thickness of the boundary layer tends to vary in the course of the day. Consider for instance a typical evolution at mid-latitudes, in the absence of rapidly changing weather with passing fronts or precipitation activity. As stable conditions develop during the night, the boundary layer may range from a few tens of meters in the evening to about 500 m by early morning; then after sunrise, a new unstable boundary layer develops which may eventually reach a thickness of 1–2 km at full maturity by the middle of the day. This evolution is sketched in Figure 2.7. Figure 2.8 shows an example of the evolution of the temperature profile in the course of a sunny day. As a rule of thumb, the thickness of a typical boundary layer can be assumed to be of the order of 1 km; it is usually larger under unstable than under neutral conditions.

The thickness of the atmospheric surface layer (or ASL) is usually taken as the lower one tenth of the boundary layer. While there are several ways of defining this thickness

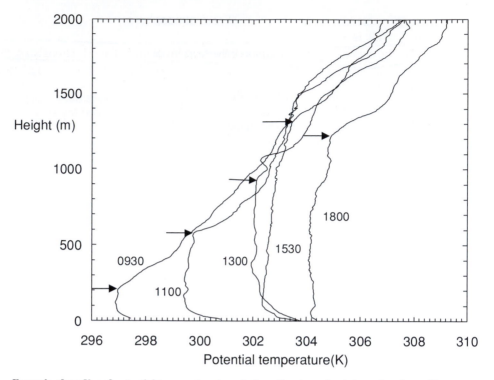

Fig. 2.8   **Example of profiles of potential temperature in and above the atmospheric boundary layer. The approximate heights of the inversion overlying the boundary layer are indicated by arrows. The measurements were made over gently rolling terrain ($z_0 = 0.45$ m, $d_0 = 8.9$ m; see Asanuma *et al.*, 2000) by means of radiosoundings at the indicated times (Central Daylight Savings Time) on June 13, 1992, in the Washita River Basin in Oklahoma.**

(see below), it coincides approximately with the region where the direction of the wind remains constant with height; this absence of "turning" confirms that it is indeed the region where the effect of the rotation of the Earth is of little consequence. The surface sublayer is also sometimes assumed to be the layer where the vertical turbulent fluxes do not change appreciably from their value at the surface, say less than 10%. Although the ASL occupies the lower part of the turbulent boundary layer, it does not extend all the way down to the surface. As illustrated in Figure 2.6, the height of the lower limit of the ASL can be assumed to be of the order of $30v/u_*$ in the case of smooth flow, and of the order of 3 to $5h_0$ in the case of rough flow; h_0 is the characteristic height of the roughness obstacles.

In general, under non-neutral conditions the air flow and the momentum transport are greatly affected by the transport of sensible heat and, to a lesser extent, water vapor, and vice versa. However, in the lower part of the atmospheric surface layer it is found that sensible heat and water vapor may be considered as merely passive admixtures, and that the effects of the density stratification resulting from temperature and humidity gradients are negligible. This lower region of the atmospheric surface layer is referred to as the *dynamic sublayer*. Under neutral conditions, the whole surface layer behaves as a dynamic layer.

Finally, in the immediate vicinity of the surface, the turbulence is strongly affected by the structure of the roughness elements, or it is greatly damped by the viscous effects; in most cases it is subjected to both effects. The region nearest to the surface where these effects are most important, is sometimes referred to as the *interfacial (transfer) sublayer*. In the case of smooth flow, as may occur for example over snow, water or salt flats, it is referred to as the *viscous sublayer*. Experiments have shown that its thickness is of the order of $5\nu/u_*$, in which ν is the viscosity of the air; the flow may be considered smooth when $(u_* h_0/\nu) < 1$, approximately, in which again h_0 is the average height of the surface roughness elements. Experiments have also shown that a surface can be considered rough, when $(u_* h_0/\nu) > 15$, approximately; in this case the interfacial sublayer may be referred to as a *roughness sublayer*, and its thickness is of the order of the mean height of the roughness obstacles. When the roughness obstacles consist of vegetation, which is more or less porous or permeable for the air stream, the interfacial sublayer is commonly referred to as the *canopy sublayer*.

2.5   TURBULENCE SIMILARITY

Over the past century or so, various turbulence closure schemes have been proposed, essentially by invoking similarity on the basis of dimensional analysis. In this type of approach, after the relevant physical quantities are identified, either from the governing equations or simply by inspection, they are organized into a reduced number of dimensionless quantities. Dimensional analysis only establishes the possible existence of a functional relationship between these dimensionless quantities, and it is incapable of providing the specific form of the functional relationship; the form of that function must usually be determined by experiment or on the basis of some conceptual transport model or other theoretical considerations. This section does not present an exhaustive review but only a few ideas that will be useful in the determination of evaporation in Chapter 4.

2.5.1   *Parameterization of the turbulent transport*

Most similarity formulations of turbulent flux have the common feature, that the mean of the product of temporal fluctuations in expressions such as (2.29), (2.30) and (2.31), i.e. the second moment, is replaced simply by the product of the spatial changes of the corresponding mean quantities, i.e. of the first moments. In the case of the specific humidity flux this is in general

$$\overline{w'q'} = -\mathrm{Ce}(\overline{u}_2 - \overline{u}_1)(\overline{q}_4 - \overline{q}_3) \tag{2.33}$$

where the subscripts 1 through 4 refer to the measurement heights above the surface and Ce is a dimensionless parameter, also called the water vapor transfer coefficient, or the Dalton number; Ce depends on the heights of the reference levels 1 through 4, beside a number of other (dimensionless) factors, as will be shown below; the minus sign indicates that the flux points in the direction of negative increments of $\overline{q}$. Note that the four heights in Equation (2.33) need not all be different; thus levels 4 and 3 could be the same as 2 and 1, respectively. In the case of the vertical momentum flux, one obtains in the same way

$$\overline{w'u'} = -Cd(\overline{u}_2 - \overline{u}_1)^2 \tag{2.34}$$

where Cd is the transfer coefficient for momentum, also called the drag coefficient; in the case of the vertical sensible heat flux, one has similarly

$$\overline{w'\theta'} = -Ch(\overline{u}_2 - \overline{u}_1)(\overline{\theta}_4 - \overline{\theta}_3) \tag{2.35}$$

where Ch is the heat transfer coefficient, also called the Stanton number.

In many applications the lowest reference level of the wind speed is taken at the surface where $\overline{u} = 0$. When in addition the vertical water vapor flux refers to that at the ground surface, namely E, Equation (2.33) assumes the common form

$$E = -Ce\,\rho\overline{u}\Delta\overline{q} \tag{2.36}$$

where $\overline{u}$ is the wind speed at a certain reference height above the ground and $\Delta\overline{q}$ is the difference between the mean specific humidity at two other reference heights (one of which may also be at the water or ground surface level), whose values will, again, affect the magnitude of Ce. In the same way, for the surface shear stress, Equation (2.34) becomes

$$\tau_0 = Cd\,\rho\overline{u}^2 \tag{2.37}$$

and, for the surface sensible heat flux, Equation (2.35) becomes

$$H = -Ch\,\rho c_p\overline{u}\Delta\overline{\theta} \tag{2.38}$$

Recall that the difference between T and θ is often small in the lower layers of the surface layer, where most measurements are made. Therefore in many situations, when the height difference of the temperature measurements is only a few meters, in expressions like (2.35) and (2.38) the use of $\overline{T}$ is allowed instead of $\overline{\theta}$.

2.5.2   *Some specific implementations: flux-profile functions*

The dimensionless transfer coefficients Ce, Cd and Ch, and their dependence on other dimensionless variables, have been the subject of much research. Major progress was made in the thirties by means of mixing length theory, as a result of contributions by Prandtl, von Karman, and Taylor in the framework of the turbulent diffusion approach; this led initially to the formulation of the logarithmic profile equations for the mean wind speed, the potential temperature, the specific humidity and other admixtures of the flow (see Monin and Yaglom, 1971; Brutsaert, 1982; 1993) and subsequently to further developments by Monin and Obukhov and others. In this section a few similarity approaches are reviewed that have been useful in the practical estimation of surface fluxes.

Neutral atmospheric surface layer

It is now generally agreed, and almost accepted by definition, that in the dynamic sublayer, and under neutral conditions in the whole atmospheric surface layer, the concentration of any admixture of the flow is a logarithmic function of height above the ground. Many different derivations of this relationship have appeared in the literature but the simplest is no doubt that given by Landau and Lifshitz (1959) in the 1944 edition of their book (see also Monin and Yaglom, 1971). The derivation is based strictly on dimensional analysis and on the observation that in plan-parallel flow an increase in velocity in the z-direction, $(d\bar{u}/dz)$, is evidence of a downward momentum flux and a sink at the surface. Thus, the mean velocity gradient in a fluid of density, ρ, is determined by the shear stress at the wall, τ_0, and the distance from the wall, $(z - d_0)$; in the last variable the (zero-plane) displacement height d_0 is introduced to account for the uncertainty of the position of the wall in the case of irregular and uneven surfaces. These variables can be combined into a single dimensionless quantity as follows,

$$\frac{u_*}{(z - d_0)(d\bar{u}/dz)} = k \tag{2.39}$$

where u_* is defined as in Equation (2.32). Experimentally, this combination k has been found to be nearly invariant and close to 0.4 under many different conditions; it is referred to commonly as von Karman's constant. The logarithmic profile follows upon integration of Equation (2.39).

In general, this logarithmic profile can be written as

$$\bar{u}_2 - \bar{u}_1 = \frac{u_*}{k}\ln\left(\frac{z_2 - d_0}{z_1 - d_0}\right) \tag{2.40}$$

where the subscripts 1 and 2 refer to two levels within the neutral surface layer. This result produces immediately the drag coefficient, as it appears in Equation (2.34), namely $Cd = \{k/\ln[(z_2 - d_0)/(z_1 - d_0)]\}^2$. Equation (2.39) can also be integrated as follows

$$\bar{u} = \frac{u_*}{k}\ln\left(\frac{z - d_0}{z_0}\right) \tag{2.41}$$

where z_0 is an integration constant, whose dimensions are length; it is usually referred to as the momentum roughness parameter or the roughness length. Its value depends on the conditions at the lower boundary of the region of validity of Equation (2.39). Graphically, it may be visualized as the zero velocity intercept of the straight line resulting from a semi-logarithmic plot of mean velocity data versus height in a neutral surface layer (see Figure 2.9). Equation (2.41) leads to the drag coefficient, as it appears in Equation (2.37), namely $Cd = \{k/\ln[(z - d_0)/z_0]\}^2$.

Dimensional arguments, similar to those leading to the profiles of the mean wind speed, produce for the mean specific humidity gradient

$$\frac{E/\rho}{u_*(z - d_0)(d\bar{q}/dz)} = -k \tag{2.42}$$

Fig. 2.9 Schematic illustration
of the mean wind
profile $\bar{u} = \bar{u}(z)$ in the
dynamic sublayer and
in the atmospheric
surface layer (ASL,
also called the surface
sublayer).

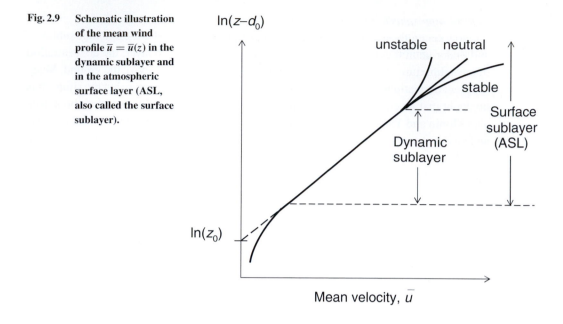

Once again, integration yields a logarithmic profile as follows,

$$\bar{q}_1 - \bar{q}_2 = \frac{E}{ku_*\rho} \ln\left(\frac{z_2 - d_0}{z_1 - d_0}\right) \tag{2.43}$$

This result, combined with Equations (2.33) and (2.37), produces a mass transfer coefficient for water vapor; in the case where wind speed and specific humidity are measured at the same two levels z_1 and z_2 one obtains Ce $= \{k/\ln[(z_2 - d_0)/(z_1 - d_0)]\}^2$; it is remarkable that this transfer coefficient has the same form as that for momentum, i.e. Ce $=$ Cd, as derived above. The fact, that under certain conditions transfer coefficients of different admixtures in turbulent flow are the same, is also referred to as the *Reynolds analogy*. The alternative form of Equation (2.43), when one of the specific humidity values is taken at the surface, $z = 0$, is

$$q_s - \bar{q} = \frac{E}{ku_*\rho} \ln\left(\frac{z - d_0}{z_{0v}}\right) \tag{2.44}$$

where q_s is the value of $\bar{q}$ at the surface and z_{0v} is the (scalar) roughness for water vapor (see Figure 2.10). In this case the transfer coefficient can be written as Ce $= k^2/\{\ln[(z_2 - d_0)/z_0]\ln[(z_1 - d_0)/z_{0v}]\}$, in which the subscript 2 refers to the height of the wind measurement and the subscript 1 refers to that of the specific humidity. In this formulation Ce would be equal to Cd only if the two roughness parameters z_0 and z_{0v} have the same value, which is rarely the case above land.

It would be possible to define a similar logarithmic relationship between the temperature and the surface sensible heat flux H; however, since under neutral conditions the temperature differences and the sensible heat flux are relatively small, this is not very meaningful. In what follows under non-neutral conditions the scalar roughness for sensible heat in the temperature profile will be denoted by z_{0h}.

Table 2.6 Typical roughness values for various surfaces

Surface description	z_0 (m)
Large water surfaces ("average")	
Snow, mud flats	0.0001–0.0005
Smooth runways	
Short grass	0.008–0.02
Long grass, prairie	0.02–0.06
Short agricultural crops	0.05–0.10
Tall agricultural crops	0.10–0.20
Prairie or short crops with scattered bushes and tree clumps	0.20–0.40
Continuous bushland	
Bushland in rugged and hilly (50–100 m) terrain	1.0–2.0
Mature pine forest	0.80–1.5
Tropical forest	1.5–2.5
Fore-Alpine terrain (200–300 m) with scattered tree stands	3.0–4.0

Fig. 2.10 Schematic illustration of the mean specific humidity profile $\bar{q} = \bar{q}(z)$ in the dynamic sublayer and in the atmospheric surface layer; $\bar{q}_s$ is its value at the surface.

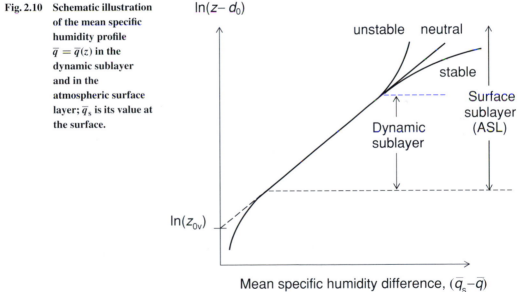

In practical applications the roughness parameters z_0, z_{0h}, z_{0v} and d_0 are best deter-mined experimentally for each specific surface. However, in the absence of measure-ments, it may be necessary to estimate them from simple geometric characteristics of the surface; numerous such relationships have appeared in the literature (e.g. Brutsaert, 1982). Wieringa (1993) has presented a review of available experimental determinations of z_0 over homogeneous terrain. A few typical values of z_0 taken from the literature for various surfaces are given in Table 2.6. As a useful first approximation for surfaces with densely placed obstacles such as natural vegetation with average height h_0, the momentum roughness z_0 can be assumed to be of the order of $h_0/10$, d_0 of the order of $h_0/2$ to $2h_0/3$, and z_{0h} and z_{0v} of the order of $h_0/100$ or smaller. The scalar roughness parameters z_{0h} and z_{0v} continue to be the subject of research (see, for example, Brutsaert and Sugita, 1996; Qualls and Brutsaert, 1996; Sugita and Brutsaert, 1996; Cahill *et al.*, 1997).

Monin–Obukhov similarity in the surface layer

Neutral conditions occur only seldom in the atmospheric boundary layer. Therefore, it is practically always necessary to include the effect of the stability, i.e. the density stratification, of the atmosphere in the formulation of the profile equations and of the corresponding transfer coefficients. One of the more common ways of doing this is based on the Monin–Obukhov (1954) approach, which assumes that the effect of the density stratification of the flow can be represented by the production rate of turbulent kinetic energy, resulting from the work of the buoyancy forces; it can be shown (see Monin and Yaglom, 1971; Brutsaert, 1982) that near the ground this rate is given by $(g/T_a)[(H/c_p\rho) + 0.61\ T_a E/\rho]$. The dimensionless variables in Equations (2.39) and (2.42) have the variables $(z - d_0)$ and u_* in common. Accordingly one can hypothesize that in a stratified turbulent flow any dimensionless characteristic of the turbulence depend only on the following: the height above the virtual surface level, $(z - d_0)$; the shear stress at the surface, τ_0; the density, ρ and the turbulent energy production rate by the buoyancy. These four quantities, which can be expressed in terms of three basic dimensions, viz. time, length and air mass, can be combined into one dimensionless variable. This variable, which was proposed by Monin and Obukhov (1954) (originally for $d_0 = 0$), is

$$\zeta = \frac{z - d_0}{L} \tag{2.45}$$

where L is known as the Obukhov stability length, defined by

$$L = \frac{-u_*^3}{k(g/T_a)(\overline{w'\theta'}_0 + 0.61\ T_a\overline{w'q'}_0)} \tag{2.46}$$

in which T_a is a mean reference temperature (in K) of the air near the ground and the subscript 0 refers to near-surface values of the fluxes, so that by definition these fluxes represent $(H/c_p\rho)$ and (E/ρ), respectively. In the original formulation of L the turbulent water vapor flux term did not appear; although in many cases the effect of the water vapor on the density stratification can be neglected, it is still is advisable to include it whenever possible.

With this hypothesis the dimensionless gradients of the mean wind, of the temperature and of the humidity, can be written as

$$\frac{k(z - d_0)}{u_*}\frac{d\overline{u}}{dz} = \phi_m(\zeta) \tag{2.47}$$

$$-\frac{ku_*(z - d_0)}{\overline{w'\theta'}_0}\frac{d\overline{\theta}}{dz} = \phi_h(\zeta) \tag{2.48}$$

$$-\frac{ku_*(z - d_0)}{\overline{w'q'}_0}\frac{d\overline{q}}{dz} = \phi_v(\zeta) \tag{2.49}$$

in which the subscripts m, h and v refer to momentum, sensible heat and water vapor, respectively. To be consistent with Equations (2.39) and (2.42), in the dynamic sublayer or under neutral conditions, when $\zeta \ll 1$ (but $z - d_0 \gg z_0$) these ϕ-functions become equal to unity. It is usually assumed that $\phi_v = \phi_h$, and thus that Reynolds's analogy is valid for scalar admixtures of the flow.

Equations (2.47)–(2.49) are formulated in terms of the gradients; these are not easy to determine from field measurements, which more often than not tend to be noisy. To avoid this problem, Equations (2.47)–(2.49) can be expressed in integral form as follows

$$\bar{u}_2 - \bar{u}_1 = \frac{u_*}{k} \left[\ln(\zeta_2/\zeta_1) - \Psi_m(\zeta_2) + \Psi_m(\zeta_1) \right] \tag{2.50}$$

$$\bar{\theta}_1 - \bar{\theta}_2 = \frac{\overline{w'\theta'}_0}{ku_*} \left[\ln(\zeta_2/\zeta_1) - \Psi_h(\zeta_2) + \Psi_h(\zeta_1) \right] \tag{2.51}$$

$$\bar{q}_1 - \bar{q}_2 = \frac{\overline{w'q'}_0}{ku_*} \left[\ln(\zeta_2/\zeta_1) - \Psi_v(\zeta_2) + \Psi_v(\zeta_1) \right] \tag{2.52}$$

in which each of the Ψ-functions, with its respective subscript, is defined by

$$\Psi(\zeta) = \int_0^\zeta [1 - \phi(x)] dx/x \tag{2.53}$$

and x is the dummy integration variable. Under neutral conditions, when $|L| \to \infty$ and $\zeta \to 0$, the Ψ-functions approach zero and Equations (2.50) and (2.52) reduce to the logarithmic profiles (2.40) and (2.43). It is also clear that, whenever $\bar{u}_1, \bar{\theta}_1$ and $\bar{q}_1$ refer to the surface values 0, θ_s and q_s, the dimensionless height ζ_1 must be taken as $z_0/L, z_{0h}/L$ and z_{0v}/L, respectively (as can be seen for the analogous neutral case in (2.41) and (2.44)). In the present case, Equations (2.50), (2.51) and (2.52) assume the form

$$\bar{u} = \frac{u_*}{k} \left[\ln \left(\frac{z - d_0}{z_0} \right) - \Psi_m \left(\frac{z - d_0}{L} \right) + \Psi_m \left(\frac{z_0}{L} \right) \right] \tag{2.54}$$

$$\theta_s - \bar{\theta} = \frac{H}{ku_*\rho c_p} \left[\ln \left(\frac{z - d_0}{z_{0h}} \right) - \Psi_h \left(\frac{z - d_0}{L} \right) + \Psi_h \left(\frac{z_{0h}}{L} \right) \right] \tag{2.55}$$

$$q_s - \bar{q} = \frac{E}{ku_*\rho} \left[\ln \left(\frac{z - d_0}{z_{0v}} \right) - \Psi_v \left(\frac{z - d_0}{L} \right) + \Psi_v \left(\frac{z_{0v}}{L} \right) \right] \tag{2.56}$$

The profiles described by Equations (2.54) and (2.56) are illustrated as the non-neutral, i.e. both stable and unstable, curves in Figure 2.9 and 2.10, respectively.

The nature of the "universal" functions, especially ϕ_m and ϕ_h, but less so ϕ_v, has been the subject of much theoretical and experimental research. One of the earliest forms of these ϕ-functions, intended for near-neutral conditions, i.e. small $|\zeta|$, was proposed by Monin and Obukhov (1954) simply by a series expansion and retention of the first term only, or $\phi = (1 + \beta_s\zeta)$, in which β_s is a constant. Subsequent experimental investigations have revealed, however, that this form of ϕ is applicable only under stable conditions, but not under unstable conditions. It was also observed later on (see Webb, 1970; Kondo et al., 1978) that this form can describe experimental data only over the range $0 \leq \zeta \leq 1$ with a β_s value of the order of 5, but that ϕ remains approximately constant for $\zeta > 1$. Accordingly, on

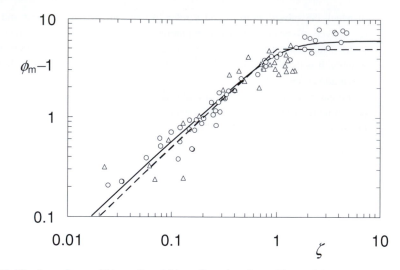

Fig. 2.11 The dependence of $(\phi_\mathrm{m} - 1)$ and $(\phi_\mathrm{h} - 1)$ on ζ under stable conditions, as determined in Cheng and Brutsaert (2005) from experimental wind profile data (circles) and temperature profile data (triangles) over a flat grassy surface ($z_0 = 0.0219$ m, $d_0 = 0.110$ m) in Kansas in October, 1999. The solid curve represents Equation (2.60) and the dashed straight line segments represent Equation (2.57).

the basis of the data then available (see Brutsaert, 1982) for stable conditions, the following was assumed

$$\phi_\mathrm{m}(\zeta) = \phi_\mathrm{h}(\zeta) = \phi_\mathrm{v}(\zeta) \begin{cases} = 1 + 5\zeta & \text{for } 0 \le \zeta \le 1 \\ = 6 & \text{for } \zeta > 1 \end{cases} \tag{2.57}$$

Equation (2.57) can be integrated with (2.53) to yield the stability correction functions Ψ needed for (2.50)–(2.52). These integral functions are

$$\Psi_\mathrm{m}(\zeta) = \Psi_\mathrm{h}(\zeta) = \Psi_\mathrm{v}(\zeta) \begin{cases} = -5\zeta & \text{for } 0 \le \zeta \le 1 \\ = -5 - 5\ln \zeta & \text{for } \zeta > 1 \end{cases} \tag{2.58}$$

Equations (2.57) and (2.58) can be compared with some more recent experimental data in Figures 2.11 and 2.12. With these same data a single formulation was proposed by Cheng and Brutsaert (2005) to cover the entire stable range $\zeta \ge 0$, namely

$$\Psi_\mathrm{m}(\zeta) = -a \ln\left[\zeta + (1 + \zeta^b)^{1/b}\right] \tag{2.59}$$

in which a and b are constants, whose values were found to be $a = 6.1$ and $b = 2.5$. Equation (2.59) is also illustrated in Figure 2.12. It can be seen that Equation (2.59) exhibits nearly the same behavior as the first of Equation (2.58) for small ζ, and nearly the same as the second for large values of ζ. The corresponding ϕ-function for the wind profile can be obtained by differentiation, as indicated by (2.53), to yield

$$\phi_\mathrm{m}(\zeta) = 1 + a\frac{\zeta + \zeta^b(1 + \zeta^b)^{-1+1/b}}{\zeta + (1 + \zeta^b)^{1/b}} \tag{2.60}$$

As illustrated in Figure 2.11, this equation behaves like $(1 + a\zeta)$ for small values of ζ and it approaches a constant $(1 + a)$ for large ζ, in accordance with (2.57). Figure 2.11 also

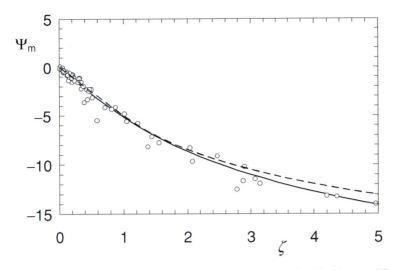

Fig. 2.12 The dependence of Ψ_m on ζ under stable conditions, as determined in Cheng and Brutsaert (2005) from experimental wind profile data over grass ($z_0 = 0.0219$ m, $d_0 = 0.110$ m) in Kansas in October, 1999. The solid curve represents Equation (2.59) and the dashed curve represents Equation (2.58).

indicates that, although the $\phi_h(\zeta)$ data points for temperature exhibit more scatter, Equation (2.60) can represent these points practically as well as the $\phi_m(\zeta)$ data points for wind speed; this suggests that it is safe to assume that under stable conditions the ASL similarity functions for sensible heat and for momentum are the same. Moreover, experimental and theoretical evidence by Dias and Brutsaert (1996) supports the turbulence similarity of scalars under stable conditions. Thus the Reynolds analogy appears to be valid and one can put $\phi_m(\zeta) = \phi_h(\zeta) = \phi_v(\zeta)$ and $\Psi_m(\zeta) = \Psi_h(\zeta) = \Psi_v(\zeta)$ for a stably stratified ASL.

For unstable conditions, Kader and Yaglom (1990) used a more fundamental approach; they reasoned, and were able to support with experimental evidence, that the surface layer can be subdivided into three sublayers, namely a dynamic, a dynamic–convective and a convective sublayer, for each of which they derived simple power laws to describe the turbulence. However, the resulting ϕ-functions cover only certain ranges, corresponding to these sublayers. Again, to cover the entire ζ-range, an interpolation formulation should be developed; accordingly, Brutsaert (1992; 1999) combined the functional behavior of ϕ_m and ϕ_h in each sublayer, and proposed the following expressions

$$
\begin{aligned}
\phi_m(\zeta) &= (a + by^{4/3})/(a + y) \quad && \text{for } y \le b^{-3} \\
\phi_m(\zeta) &= 1.0 \quad && \text{for } y > b^{-3}
\end{aligned}
\tag{2.61}
$$

and

$$
\phi_h(\zeta) = (c + dy^n)/(c + y^n)
\tag{2.62}
$$

in which $y = -\zeta = -(z - d_0)/L$, and a, b, c, d and n are constants. After considering available data collections, the constants were assigned the following values $a = 0.33$, $b = 0.41$, $c = 0.33$, $d = 0.057$ and $n = 0.78$. Figure 2.13 shows these ϕ-functions. The corresponding stability correction functions can be obtained in integral form by means of

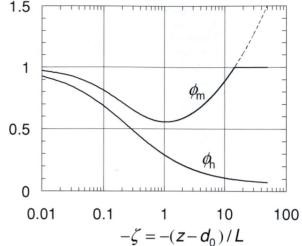

Fig. 2.13 **Flux-profile functions for momentum $\phi_m(\zeta)$ and for sensible heat $\phi_h(\zeta)$ under unstable conditions, corresponding with Equations (2.61) and (2.62).**

Equation (2.53) as follows,

$$\Psi_m(-y) = \ln(a + y) - 3by^{1/3} + \frac{ba^{1/3}}{2}\ln\left[\frac{(1 + x)^2}{(1 - x + x^2)}\right]$$

$$+ 3^{1/2}ba^{1/3}\tan^{-1}[(2x - 1)/3^{1/2}] + \Psi_0 \qquad \text{for } y \leq b^{-3}$$

$$\Psi_m(-y) = \Psi_m(b^{-3}) \qquad\qquad\qquad\qquad\qquad \text{for } y > b^{-3} \qquad (2.63)$$

and

$$\Psi_h(-y) = [(1 - d)/n]\ln[(c + y^n)/c] \qquad\qquad\qquad\qquad\qquad (2.64)$$

in which $x = (y/a)^{1/3}$ and, as before, $y = -\zeta = -(z - d_0)/L$ and a, b, c, d and n are constants. The symbol Ψ_0 is a constant of integration, given by $\Psi_0 = (-\ln a + 3^{1/2}ba^{1/3}\pi/6)$; in applications it is usually unimportant, because it cancels out in Equations (2.50) and (2.54). Figure 2.14 shows Equations (2.63) and (2.64) with the values of the constants given above behind (2.62). Also for unstable conditions, it is usually assumed that $\phi_h(\zeta) = \phi_v(\zeta)$ and $\Psi_h(\zeta) = \Psi_v(\zeta)$.

There is still no universal agreement on the vertical extent of the surface layer. However, numerous experimental observations (see, for example, Brutsaert, 1998, 1999) mostly under neutral and unstable conditions suggest that the lower limit z_{sb} can be estimated from $(z_{sb} - d_0) = \alpha_b z_0$, in which α_b is of the order of 50, ranging roughly between 40 and 60. Its upper limit z_{st} can be estimated by the rule of thumb, specifying that it is either at $(z_{st} - d_0) = \alpha_t h_i$, in which $\alpha_t = 0.12$, or at $(z_{st} - d_0) = \beta_t z_0$, in which $\beta_t = 120$, whichever is larger; the variable h_i is the height of the bottom of the inversion capping the atmospheric boundary layer. Note that the former value of z_{st} is for a moderately rough surface, whereas the latter is for very rough terrain; with a typical value of $h_i = 1000$ m, the cut-off value between very rough and moderately rough terrain is around $z_0 = (\alpha_t/\beta_t)h_i = 1$ m.

Bulk ABL similarity formulation

As mentioned, the atmospheric surface layer typically occupies only the lowest 10% or so of the boundary layer. Numerous attempts have also been made to formulate similarity

Fig. 2.14 Integral form of the flux-profile functions for momentum $\Psi_m(\zeta)$ and for sensible heat $\Psi_h(\zeta)$ under unstable conditions, as given by Equations (2.63) and (2.64).

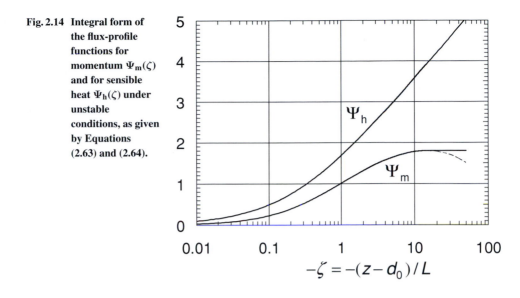

hypotheses for the entire boundary layer. In this approach, the surface fluxes are commonly related to "bulk" variables, namely values of the variables at the top and bottom of the ABL, or their averages over all or part of the ABL. The basic form of the equations is essentially similar to that of Equations (2.50)–(2.52), or (2.54)–(2.56), but extended for larger heights aloft above the surface layer. Ideas on the application of similarity to the entire ABL, including the outer region were put forth early on by Rossby and Montgomery (1935) and Lettau (1959), and subsequent developments can be traced through the work of Kazanski and Monin (1961), Clarke and Hess (1974), Zilitinkevich and Deardorff (1974), Yamada (1976), Garratt *et al.* (1982), Brutsaert (1982), Sugita and Brutsaert (1992), and Jacobs *et al.* (2000), among others. The various versions of this approach can be written in a general form as follows,

$$u_b = \frac{u_*}{k}\left[\ln\left((h_b - d_0)/z_0\right) - B\right]$$

$$v_b = -\frac{u_*}{k}A \tag{2.65}$$

$$\theta_s - \theta_b = \frac{\overline{w'\theta'}_0}{ku_*}\left[\ln\left((h_b - d_0)/z_{0h}\right) - C\right] \tag{2.66}$$

where A, B and C are functions of a number of dimensionless variables that affect transport in the outer region and where the subscript b indicates bulk or characteristic scale variables of the ABL. Thus h_b denotes a characteristic thickness or height scale of the ABL; the variables u_b and v_b are characteristic horizontal wind velocity components in the x- and y-directions, respectively (x is the direction of the near-surface wind; because it may involve the Earth's rotation, usually y points to the left of x in the Northern Hemisphere, and to the right in the Southern Hemisphere), such that $u_b^2 + v_b^2 = V_b^2$, in which V_b is a characteristic wind speed aloft. These bulk variables have been given different definitions in the past, depending on the specific implementation of the approach. In the early applications u_b, v_b and θ_b were taken as the values of these variables near the top of the ABL, in general, or just below the capping inversion, under unstable conditions.

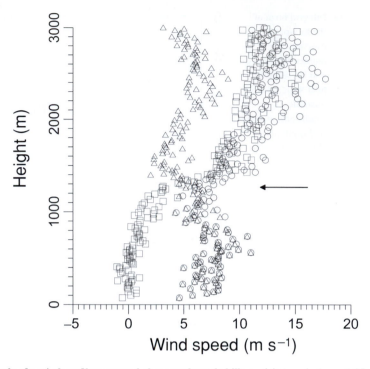

Fig. 2.15 Example of a wind profile measured above moderately hilly prairie terrain ($z_0 = 1.05$ m, $d_0 = 26.9$ m) in Kansas at 1500 CDT on August 14, 1987, by means of a radiosonde; the circles represent the wind speed, and the triangles and squares represent the x- and y-components of the wind velocity, respectively. The arrow indicates the height of the inversion. (From Brutsaert and Sugita, 1991).

The more recent implementations (see, for example, Brutsaert, 1999) have been mostly for unstable conditions with mean values of the variables in the mixed layer, and with the wind speed as a scalar. The rationale for this choice of bulk variables is that, indeed, as illustrated in Figure 2.15, owing to convection with vertical mixing the y-component of the velocity is nearly negligible, so that the x-component is practically equal to the wind speed; moreover, wind speed measurements aloft can be noisy, so that a height-averaged value is likely to be more robust. Figure 2.16 shows the corresponding temperature profile. Thus, with this choice of variables, the formulation for momentum and sensible heat can be written as

$$V_\mathrm{m} = \frac{u_*}{k} \left[\ln \left((h_\mathrm{i} - d_0)/z_0 \right) - B_\mathrm{w} \right] \tag{2.67}$$

$$\theta_\mathrm{s} - \theta_\mathrm{m} = \frac{\overline{w'\theta'}_0}{ku_*} \left[\ln \left((h_\mathrm{i} - d_0)/z_{0\mathrm{h}} \right) - C \right] \tag{2.68}$$

in which V_m and θ_m are the mean wind speed and potential temperature, respectively, in the mixed layer of the unstable ABL; h_i is the height of the top of the mixed layer, that is the bottom of the inversion above the ground, and B_w has been given a subscript w to indicate that the wind speed V is used, instead of the wind velocity components u and v.

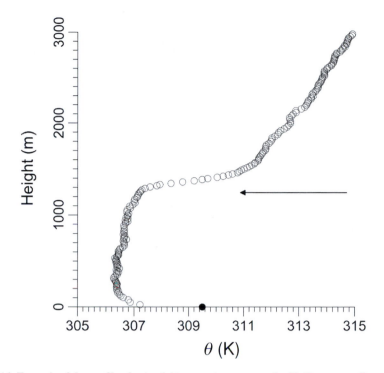

Fig. 2.16 **Example of the profile of potential temperature measured with the same radiosonde as that of the wind shown in Figure 2.15. The solid circle shows the median value of the surface temperature. The arrow indicates the height of the inversion. (From Brutsaert and Sugita, 1991.)**

Until now, no general definitive form has been derived for these functions B_w and C. An example of a formulation for unstable conditions, that has produced good results (Brutsaert, 1999), is summarized in what follows. It is based on the assumption of an ABL consisting of two layers, namely a surface layer, with profiles given by Equations (2.63) and (2.64), and above it a mixed layer as a slab with uniform profiles; it is further also based on the assumption explained behind (2.64) regarding the position of the top of the surface layer, where it meets the mixed layer. For moderately rough terrain, i.e. when $z_0 \leq (\alpha_t/\beta_t)h_i$, the resulting functions are

$$
\begin{aligned}
B_w &= -\ln(\alpha_t) + \Psi_m(\alpha_t(h_i - d_0)/L) - \Psi_m(z_0/L) \\
C &= -\ln(\alpha_t) + \Psi_h(\alpha_t(h_i - d_0)/L) - \Psi_h(z_{0h}/L)
\end{aligned} \tag{2.69}
$$

For very rough terrain, when $z_0 > (\alpha_t/\beta_t)h_i$, the functions are

$$
\begin{aligned}
B_w &= \ln((h_i - d_0)/(\beta_t z_0)) + \Psi_m(\beta_t z_0/L) - \Psi_m(z_0/L) \\
C &= \ln((h_i - d_0)/(\beta_t z_0)) + \Psi_h(\beta_t z_0/L) - \Psi_h(z_{0h}/L)
\end{aligned} \tag{2.70}
$$

The similarity functions B_w and C given by Equations (2.69) and (2.70) are illustrated in Figures 2.17 and 2.18. In the derivation of Equations (2.69) and (2.70) it was assumed that the outer region is a perfectly mixed slab layer; this assumption has its limitations. In fact,

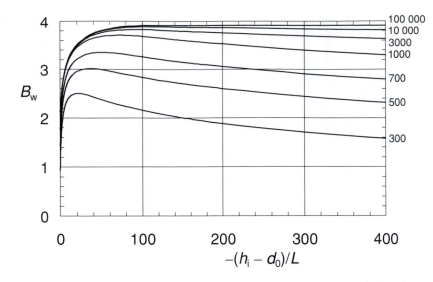

Fig. 2.17 Dependence of the bulk similarity function B_w on $[-(h_i - d_0)/L]$ and on $[(h_i - d_0)/z_0]$; the latter values are shown as numbers at the corresponding curves. The curves for $[(h_i - d_0)/z_0] \geq 10^3$ are obtained with Equation (2.69), and those for $[(h_i - d_0)/z_0] \leq 10^3$ are obtained with Equation (2.70); it is assumed that $\alpha_t = 0.12$ and $\beta_t = 120$.

under unstable conditions the potential temperature often tends to increase slightly with elevation, roughly from about the middle of the mixed layer (e.g. Figures 2.8 and 2.16); this is mostly the result of entrainment of warmer air into the ABL from above. Similarly the wind speed $\bar{u}$ is often affected by this entrainment. Therefore, in the practical application of these equations, it may be advisable to obtain the mean wind and temperature difference from measurements over the lower half of the mixed layer, that is below $(h_i/2)$, in order to minimize any possible entrainment effects.

The temperature difference term in Equation (2.68) has the lower value at the surface. As in (2.51), the lower value can also be taken at some level ζ in the surface layer; the proper formulation for this case can be obtained by simply subtracting (2.55) from (2.66) or (2.68), so that θ_s is eliminated. As a further alternative, an example of the application of the bulk ABL similarity approach with the lower value not at the surface, but at shelter level, has been presented by Qualls *et al.* (1993).

The bulk ABL similarity approach (also called BAS), as formulated here for unstable conditions, has several features which make it attractive to obtain surface fluxes u_* and H from soundings in the upper reaches of the boundary layer. First, the mixed layer variables V_m and θ_m, which are averages over the mixed layer, are more robust than the profiles $\bar{u}(z)$ and $\bar{\theta}(z)$; such profiles often tend to be erratic and noisy. Second, since these mixed layer variables are averages over a layer extending roughly between heights of the order of 100 m and 1 km above the ground, they reflect mean surface conditions over upwind distances of the order of 1–10 km; this provides the main justification and appeal of this approach to describe surface fluxes at the mesogamma scale (see Table 1.5), which is often the relevant spatial scale for hydrologic catchments.

The bulk similarity approach can also be applied to water vapor. However, because water vapor is not as well mixed in the outer region as potential temperature or wind speed, it is

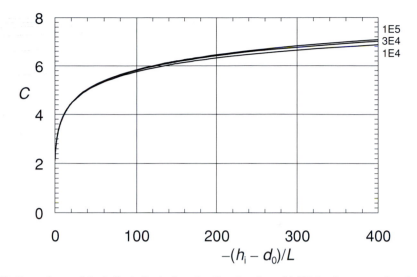

Fig. 2.18 Dependence of the bulk similarity function C on $[-(h_i - d_0)/L]$ for three example values of $[(h_i - d_0)/z_{0h}]$, as obtained with Equation (2.69) for moderately rough terrain. The values of $[(h_i - d_0)/z_{0h}]$, namely 10^4, 3×10^4 and 10^5 are indicated at the curves; clearly, C is not very sensitive to this variable and also the values obtainable for very rough terrain with Equation (2.70) fall mostly inside the outermost curves shown here. It is assumed that $\alpha_t = 0.12$ and $\beta_t = 120$.

less meaningful to use the average specific humidity $\bar{q}_m$. The approach has only been used with $q_b = \bar{q}_i$, the value of $\bar{q}$ at $z = h_i$, as follows

$$q_s - \bar{q}_i = \frac{\overline{w'q'}_0}{ku_*} [\ln((h_i - d_0)/z_{0v}) - D] \qquad (2.71)$$

where, as before in the case of B_w and C, D is a function of a number of variables; the only one that has been considered so far is $(h_i - d_0)/L$, but beside this effect, Equation (2.71) has been studied very little (see Brutsaert, 1982).

2.6   SURFACE BOUNDARY CONDITION: THE ENERGY BUDGET CONSTRAINT

The turbulent fluxes of water vapor and sensible heat near the Earth–atmosphere interface are linked not only by similarity relationships in the turbulent air, but also by the energy budget. Indeed, both evaporation E, as a latent heat flux, and the related sensible heat flux H require the supply of some other form of energy. Therefore their magnitudes are constrained by this available energy. The question can be treated quantitatively by considering the energy budget for a layer of surface material. Depending on the nature of the surface, this layer may consist of water, or of some other substrate like soil, plant canopy or snow; although this layer can be taken to be infinitesimally thin, it may

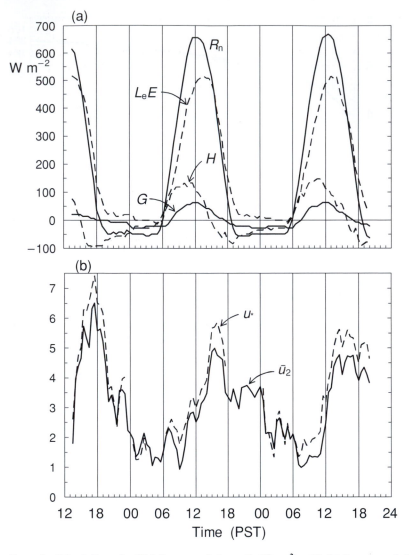

Fig. 2.19 Example of the daily cycle of (a) the energy balance (in W m^{-2}), with (b) the mean wind speed at 2 m, $\bar{u}_2$ (in m s^{-1}, solid line), and the friction velocity u_* (in dm s^{-1}, dashed line), for an irrigated grass covered surface at Davis, California, on June, 2–4, 1965. The balance equation was assumed to be $R_n = L_e E + H + G$. The evaporation was measured with a weighing lysimeter (Pruitt and Angus, 1960) and the surface shear stress with a floating drag plate lysimeter (Goddard, 1970). The data are drawn from Brooks and Pruitt (1966); the roughness of the grassy surface was estimated to be $z_0 = 0.97 \pm 0.14$ cm (Morgan *et al.* 1971).

sometimes even comprise a lake or a vegetational canopy over its entire depth. For many practical purposes, the energy budget equation can be written as

$$R_n - L_e E - H + L_p F_p - G + A_h = \frac{\partial W}{\partial t} \qquad (2.72)$$

In words, Equation (2.72) states that the difference between incoming and outgoing energy fluxes is equal to the rate of increase of the energy stored in the layer under consideration; the sign convention is such that the energy fluxes toward the layer are taken as positive and those away from it as negative. In (2.72) the quantity R_n is the net radiative flux density at the upper surface of the layer, L_e is the latent heat of vaporization, L_p is the thermal conversion factor for fixation of carbon dioxide, F_p is the specific flux of CO_2, G is the specific energy flux leaving the layer at the lower boundary, A_h is the energy advection into the layer expressed as specific flux, and $\partial W/\partial t$ is the rate of energy storage per unit horizontal area in the layer; in the case of an ice or snow layer this last term may include the energy consumed by fusion, and L_e may have to be replaced by L_s, the heat of sublimation. At present in the SI system all these surface energy fluxes are commonly expressed in units of W m^{-2}.

Example 2.1. Some features of the surface energy budget

The order of magnitude and the diurnal variation of the main terms in the energy budget for different surfaces are illustrated in Figures 2.19–2.22. Figure 2.19 shows the terms in an irrigated environment under clear sky in the summer. Figure 2.20a illustrates the response of the turbulent heat fluxes in response to varying cloudiness in the course of a spring day, whereas Figure 2.20b shows a typical clear sky situation, which is generally similar to Figure 2.19. In contrast to what happens over land, Figure 2.21 shows how over deep water the turbulent heat fluxes $L_e E$ and H do not follow the diurnal cycle of the solar radiative energy supply; as a result of the large heat capacity of the water body, the surface temperature tends to remain more constant, and less affected by the radiative energy input. Figure 2.22 illustrates the gradual evolution of the three main terms of the energy budget in natural prairie during a period of prolonged drying in the fall season during the First ISLSCP Field Experiment. As the soil moisture content is decreasing the evaporation rate exhibits a steady decrease. On the other hand, the sensible heat flux is not increasing in the same steady way, as one might expect if the available energy were constant; it is more erratic and is more responsive to the vagaries of the weather while the radiation is steadily declining as winter approaches.

2.6.1   Net radiation

This quantity can be broken down into several components, viz.

$$R_n = R_s(1 - \alpha_s) + \varepsilon_s R_{ld} - R_{lu} \tag{2.73}$$

where R_s is the (global) short-wave radiation, α_s is the albedo of the surface, R_{ld} is the downward long-wave or atmospheric radiation, ε_s is the emissivity of the surface and R_{lu} is the upward long-wave radiation. The downward long-wave radiation is multiplied by the emissivity ε_s, because this is equal to the absorptivity, which is the fraction of the incoming long-wave radiation absorbed by the surface. The net radiation can be measured directly, and at present fairly reliable instruments are available for this purpose. In the absence of direct measurements, or when great accuracy is required, R_n can be obtained from measurements of its components on the right-hand side of Equation (2.73). When

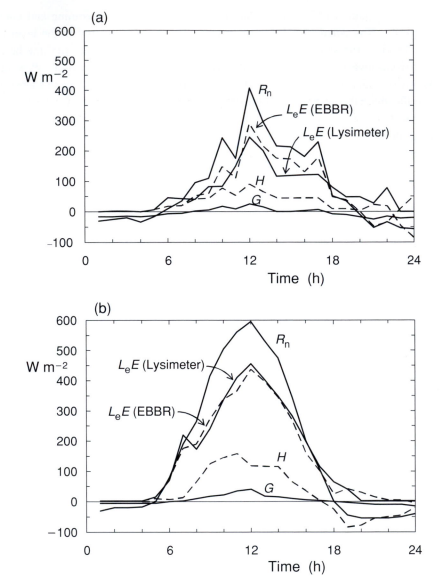

Fig. 2.20   Example of the daily cycle of the energy balance for a young (a) and a mature (b) maize canopy near Versailles, France. EBBR indicates the latent heat flux obtained by the Bowen ratio method (after Perrier *et al.*, 1976).

these measurements are not available, the components can be obtained by theoretical methods or simpler empirical formulae.

Short-wave radiation

R_s is the radiant flux resulting directly from the solar radiation. This incoming solar radiation has most of its energy contained in the wavelength range from 0.1 to 4 μm. At the outside of the atmosphere this flux, i.e. the solar constant, has been measured on

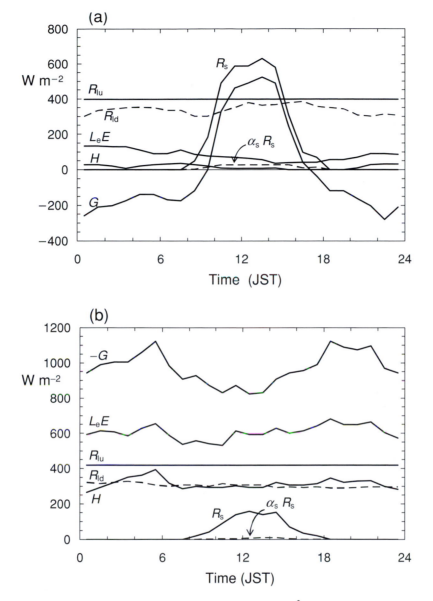

Fig. 2.21 Examples of the daily course of the energy balance in (W m^{-2}) at the surface of a deep water body. The data were obtained over the East China Sea during the Air Mass Transformation Experiment (AMTEX) on February 15(a) and 25(b), 1974; the time is Japanese Standard Time. (After Yasuda, 1975.)

satellites (e.g. Liou, 2002) to be of the order of $R_{so} = 1366$ W m^{-2} (or around 1.958 cal min^{-1}cm^{-2}). As it passes through the atmosphere, the solar radiation is modified by scattering, absorption, and reflection by different types of molecules and colloidal particles; thus at the Earth's surface the global short-wave radiation consists of direct solar radiation and diffuse sky radiation. The short-wave radiation can be measured and

Fig. 2.22 Example of the daily course of three terms in the surface energy budget, namely the latent heat flux L_eE, the sensible heat flux H, and the net radiation R_n, over a period of drying from September 19 (DOY 262) through October 12 (DOY 285), 1987, in a hilly prairie region in northeastern Kansas, as used in the study of Brutsaert and Chen (1996). The turbulent fluxes were measured by means of the eddy correlation method at Station 26 of the FIFE experiment. The different curves represent the flux on different days of the year, DOY, namely 262 (diamonds), 266 (squares), 270 (triangles), 273 (circles), 276 (times), 280 (minus), and 285 (plus). The time is Central Daylight Savings Time.

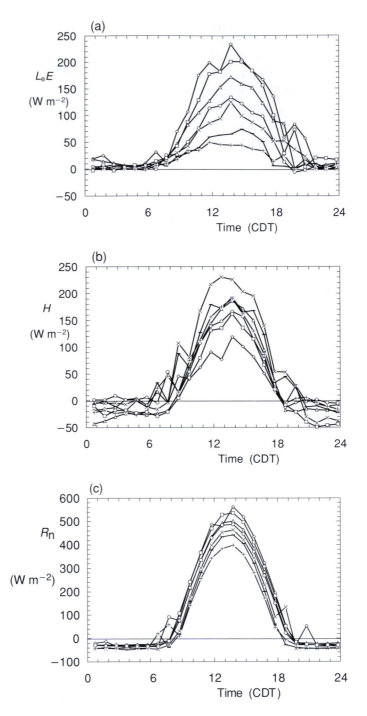

data are available from national weather services and agricultural agencies. In the event that suitable data are not available, it may be necessary to make an estimate by means of one of several theoretical models or simpler empirical formulae that relate short-wave radiation with other physical factors, such as extraterrestrial radiation, optical air mass, turbidity, water vapor content of the air, amount and type of cloud cover. However, these should be used with caution.

A simple equation which can be used for daily averages, was proposed by Prescott (1940) in terms of daily total extraterrestrial radiation Q_{se} as follows

$$Q_s = Q_{se}[a + b\,(n/N)] \tag{2.74}$$

where a and b are constants which depend on the location, the season and the state of the atmosphere; their values have been determined for many locations and on average they appear to be around $a = 0.25$ and $b = 0.50$. In Equation (2.74) n is the actual number of hours of bright sunshine and N the number of daylight hours; as a first approximation for steady weather conditions n/N can be related to the mean fractional cloud cover m_c by

$$a(n/N) + bm_c = 1 \tag{2.75}$$

in which a and b are different constants, for which values averaging around 1.1 and 0.85, respectively, have been observed in the Netherlands and in Japan (e.g. De Vries, 1955; Kondo, 1967).

Many other regression equations like (2.74), also for instantaneous values R_s, have been proposed in the literature, but such simple equations can be only poor substitutes for direct measurements. Nevertheless, it is possible to obtain fairly accurate radiation estimates by better empirical and partly theoretical methods, which are, however, more difficult to apply. Examples of such methods, which can give useful results, are those presented by among others, Kondo (1967; 1976), Paltridge and Platt (1976, p. 137) and Meyers and Dale (1983). Because such approaches often rely on the extraterrestrial radiation, it is appropriate to take a quick look at it.

Extraterrestrial radiation

The extraterrestrial radiation R_{se} can readily be calculated for a given latitude, time of day and day of the year from the solar constant. For a horizontal surface, instantaneous values can be calculated from

$$R_{se} = R_{so}(d_{so}/d_s)^2 \cos \beta \tag{2.76}$$

in which β is the zenith angle, that is the angular distance between the sun and the vertical, and in which d_s and d_{so} are the instantaneous distance and the annual mean distance of the Earth from the Sun, respectively; however, d_s and d_{so} differ by at most 3.5%, so that this effect is often neglected in hydrologic applications. It can be readily shown that the zenith angle can be calculated as follows

$$\cos \beta = \cos \phi \cos h \cos \delta + \sin \phi \sin \delta \tag{2.77}$$

where ϕ is the latitude and h is the hour angle, such that its origin $h = 0$ is local noon or 1200, and 24 h $= 2\pi$. The angle δ is the solar declination, that is the angular distance of the Sun north (or south when negative) of the Equator. Daily values of Equation (2.76)

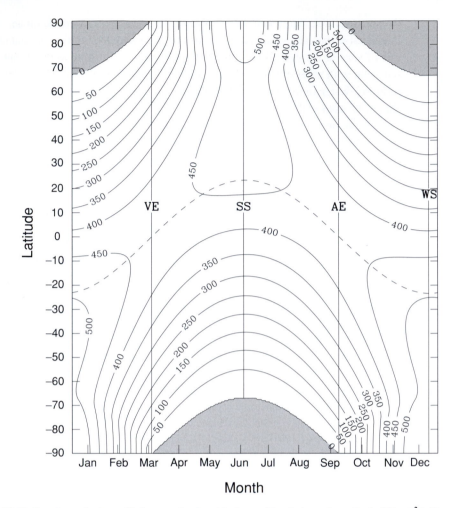

Fig. 2.23 Daily values of solar radiation on a horizontal plane without atmosphere Q_{se} in (W m^{-2}). The solar constant was taken as $R_{so} = 1366$ W m^{-2}. The vernal equinox (VE), the summer solstice (SS) and the autumnal equinox (AE) are indicated by solid vertical lines; the dashed line shows the solar declination. (From Liou, 2002.)

can be obtained by integration of (2.77) over $dt = dh/\omega$ between sunrise $h = -h_s$ and sunset $h = +h_s$, in which $\omega = 2\pi$ rad d$^{-1} = (\pi/12)$ rad h^{-1}. This yields for a horizontal surface

$$Q_{se} = (2R_{so}/\omega)(\cos\phi \sin h_s \cos\delta + h_s \sin\phi \sin\delta) \tag{2.78}$$

in which the variation in distance from the sun has been neglected. The sunrise and sunset angle h_s can be calculated by putting $\beta = \pi/2$ or $\cos\beta = 0$; this produces $\cos h_s = -\tan\phi \tan\delta$. The declination δ moves between its extreme values of roughly plus and minus $23°17'$ between the solstices on approximately June 21 and December 21. It can be calculated from $\sin\delta = \sin\varepsilon \sin\lambda$, in which ε is the oblique angle ($23°17'$) and λ is the true longitude angle of the earth with respect to the sun, which varies between 0 at the spring equinox and π at the fall equinox; the matter is somewhat complicated by the fact that the Earth moves on an ellipse around the Sun, but as noted above the eccentricity of this ellipse is small. The declination is usually determined as a function of day of the year (see Paltridge and Platt, 1976; Liou, 2002). Figure 2.23 gives an idea of the variability of the daily totals of solar radiation Q_{se}; the Earth is closest to the Sun in the month of January, so that the curves are somewhat asymmetric between North and South, with the maximal radiation occurring in the South.

Surface albedo

This is the ratio of the global short-wave reflected radiative flux and the flux of the corresponding incident radiation; in contrast to the term reflectivity, the albedo also includes the diffuse portion of the radiation. In energy budget studies the albedo usually refers to an integral value over all wave lengths; however, sometimes, to distinguish it from the spectral albedo, it is called the integral albedo. In the case of an ideal rough surface, the albedo should be independent of the direction of the primary beam. For most natural surfaces the fraction of directly and diffusely reflected radiation depends on the direction of the incoming beam. Therefore, on days with sunshine, the albedo of most surfaces depends on the altitude of the Sun, but this dependence decreases with increasing cloudiness. For example, for water surfaces it appears that the albedo can be represented well by a power function of the solar altitude (see Anderson, 1954; Payne, 1972). The albedos of other surfaces obey similar relationships. However, for daily totals it is common practice to use a mean value of the albedo. Table 2.7 presents a brief summary of mean albedo values for various surfaces obtained from summaries of available data (see Van Wijk and Scholte-Ubing, 1963; Kondratyev, 1969; List, 1971; Budyko, 1974)

Long-wave or terrestrial radiation

Also sometimes called nocturnal radiation, this is the radiant flux resulting from the emission of the atmospheric gases and the land and water surfaces of the Earth. All materials on Earth and around it have a much lower temperature than the Sun, so that the radiation they emit has much longer wavelengths than the global radiation. There is practically no overlap, since most of the radiation emitted by the Earth is contained in the range from 4 to 100 μm. Long-wave radiation can be measured, but the needed measurements for a particular area of interest are rarely available, so that it must often be calculated from other measurements. It is convenient to consider two components of the terrestrial radiation at the Earth's surface separately, namely a component of upward radiation from the surface R_{lu}, and that of downward radiation from the atmosphere R_{ld}.

The upward component is usually obtained by assuming that the ground, the canopy or the water surface under consideration is equivalent with an infinitely deep grey body of uniform temperature and emissivity ε_s which is close to unity. This allows the following formulation

$$R_{lu} = \varepsilon_s \sigma T_s^4 \tag{2.79}$$

in terms of the (absolute) surface temperature T_s; $\sigma (= 5.6697 \times 10^{-8}$ W m^{-2} K^{-4} $= 1.354 \times 10^{-12}$ cal cm^{-2} s^{-1} K^{-4}) is the Stefan–Boltzmann constant. Table 2.8

Table 2.7 Approximate mean albedo values for various natural surfaces

Nature of surface	Albedo
Deep water	0.04–0.08
Moist dark soils; ploughed fields	0.05–0.15
Gray soils, bare fields	0.15–0.25
Dry soils, desert	0.20–0.35
White sand; lime	0.30–0.40
Green grass and other short vegetation (e.g. alfalfa, potatoes, beets)	0.15–0.25
Dry grass; stubble	0.15–0.20
Dry prairie and savannah	0.20–0.30
Coniferous forest	0.10–0.15
Deciduous forest	0.15–0.25
Forest with melting snow	0.20–0.30
Old and dirty snow cover	0.35–0.65
Clean, stable snow cover	0.60–0.75
Fresh dry snow	0.80–0.90

Table 2.8 Values of the emissivities ε_s of some natural surfaces

Nature of surface	Emissivity
Bare soil (mineral)	0.95–0.97
Bare soil (organic)	0.97–0.98
Grassy vegetation	0.97–0.98
Tree vegetation	0.96–0.97
Snow (old)	0.97
Snow (fresh)	0.99

summarizes a few values of ε_s for different surfaces compiled from the literature (see, for example, Van Wijk and Scholte-Ubing, 1963; Kondratyev, 1969). In many practical applications it is simply assumed that $\varepsilon_s = 1$. Moreover, since T_s is rarely known, for daily or longer averages over land, Equation (2.79) is often applied by using the air temperature T_a instead of the surface temperature T_s.

The downward long-wave radiation R_{ld} can be calculated accurately on the basis of vertical profile data of humidity and temperature. Such data are not always available where the long-wave radiation is needed; as a result simpler methods have been developed that rely on readily available measurements, such as air temperature and humidity near the ground. For clear sky conditions, they are mostly based on an equation of the type

$$R_{ldc} = \varepsilon_{ac} \sigma T_a^4 \tag{2.80}$$

Fig. 2.24 **Comparison between measured longwave radiation R_{ldcs} under clear sky and radiation R_{ldc} estimated by means of Equations (2.80) and (2.81) with the original constants $a = 1.24$ and $b = 1/7$. (From Sugita and Brutsaert, 1993.)**

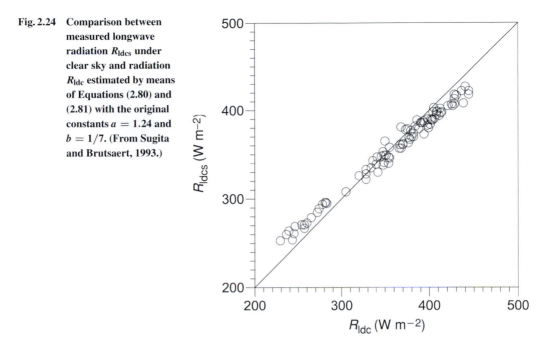

where T_{a} is the air temperature near the ground, usually taken at shelter level, and $\varepsilon_{\mathrm{ac}}$ is the atmospheric emissivity under clear skies.

Several expressions have been proposed for this emissivity. Most of these are strictly empirical, but it is also possible to derive $\varepsilon_{\mathrm{ac}}$ from physical considerations. In one such derivation (Brutsaert, 1975; 1982), the equation for radiative transfer in a plane stratified atmosphere is solved by assuming first, a simple power function slab emissivity, and second, a near-Standard Atmosphere to obtain the temperature and humidity profiles. With typical average values of the parameters, the resulting atmospheric emissivity can be written as

$$\varepsilon_{\mathrm{ac}} = a(e_{\mathrm{a}}/T_{\mathrm{a}})^{b} \qquad\qquad (2.81)$$

where a and b are constants; these were derived to be $a = 1.24$ and $b = 1/7$, when the vapor pressure of the air e_{a} is in hPa $(= \mathrm{mb})$ and T is in K. Equation (2.81) has been found (see Mermier and Seguin, 1976; Aase and Idso, 1978; Daughtry *et al.*, 1990) to yield satisfactory results under conditions which, on average, are fairly well represented by a Standard Atmosphere. With instantaneous measurements in the Great Plains of the USA, Equation (2.81) with the original constants was also found to perform well (see Figure 2.24), but somewhat better with values of the constants $a = 0.980$ and $b = 0.0687$ (Sugita and Brutsaert, 1993). Culf and Gash (1993) using the same derivation of (2.81) as in Brutsaert (1975), but with actually recorded profiles (instead of a Standard Atmosphere), obtained $a = 1.31$ and $b = 1/7$ for the dry season in Niger. Crawford and Duchon (1999) were able to improve the performance of (2.81) with $b = 1/7$ in

Fig. 2.25 Comparison between measured long-wave radiation R_{lds} under various sky conditions and radiation R_{ld} estimated by means of Equations (2.80) and (2.82) with $a = 0.0496$ and $b = 2.45$ and with cloudiness values from visual sky inspection. (From Sugita and Brutsaert, 1993.)

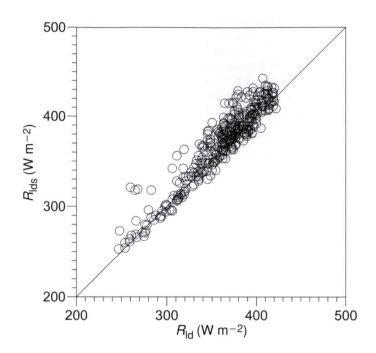

the Great Plains by expressing a as an empirical sinusoidal function of month of the year.

The downward long-wave radiation is affected by cloudiness. Several empirical methods of incorporating this effect (see Bolz, 1949; Budyko, 1974) can be expressed in the form

$$R_{ld} = R_{ldc} \left(1 + a m_c^b\right) \tag{2.82}$$

where m_c is the fractional cloudiness and a and b are (different) constants. On the basis of measurements in Germany, Bolz (1949) obtained $b = 2$ and different values of a depending on cloud type, with an average of $a = 0.22$. More recently, with visual cloudiness observations in the Great Plains, Sugita and Brutsaert (1993) derived values $a = 0.0496$ and $b = 2.45$, without consideration of cloud type, and different values of a and b for different cloud types. Their analysis also showed that the standard error of prediction with Equation (2.80) was of the order of 10–15 W m^{-2} for clear sky conditions and of the order of 20–25 W m^{-2} for various sky (including cloudy) conditions without cloudiness correction; incorporation of a cloudiness correction with (2.82) and these constants improved the R_{ld} estimate with (2.80), i.e. reduced the standard error, by roughly 5 W m^{-2} on average (see Figure 2.25), and by an additional amount of roughly the same magnitude when also information was included on the type of cloud. Deardorff (1978) proposed a simple weighting parameterization, namely an atmospheric emissivity for cloudy sky given by $\varepsilon_a = [m_c + (1 - m_c)\varepsilon_{ac}]$; this is equivalent to (2.82) with $a = [(1/\varepsilon_{ac}) - 1]$ and $b = 1$.

2.6.2   The energy flux at the lower boundary of the layer

The nature of G and the optimal method of its determination depend on the type of substrate layer to which the energy budget equation is applied. For a thin layer of soil, for a vegetational canopy or for a whole lake or stream, the term G in Equation (2.72) represents the heat flux into the ground. For a water surface, G is the heat flux into the underlying water body. Over land covered with vegetation the daily mean value of G, that is the ground heat flux, is often one or more orders of magnitude smaller than the major terms in the energy budget, R_n, H and $L_e E$. The main reason for this is that positive daytime values of G (warming) often tend to be compensated by negative nighttime values (cooling). Therefore, in design calculations, the daily values of G are often neglected.

Measurement of the soil heat flux

Several methods are available to determine G for a landsurface (see Brutsaert, 1982), but a detailed review is beyond the scope of this book. One of the more reliable methods to measure G considers changes in heat storage in the upper layer of the soil, as described by the equation

$$Q_{H1} - Q_{H2} = \int_{z_1}^{z_2} C_s(z) \frac{\partial T}{\partial t} dz \qquad (2.83)$$

where Q_{H1} and Q_{H2} are the heat flux densities at levels z_1 and z_2, respectively, C_s is the volumetric heat capacity of the soil and T is the temperature in the soil. On the basis of a compilation of thermal properties of soil components by De Vries (1963), this heat capacity (in J m^{-3} K^{-1}) can be calculated as follows

$$C_s = (1.94\theta_m + 2.50\theta_c + 4.19\theta) \times 10^6 \qquad (2.84)$$

where θ_m, θ_c and θ are the volume fractions of mineral soil, organic matter and water, respectively. Thus if z_1 refers to the soil surface and z_2 to some lower level where Q_{H2} is known, the surface heat flux $G = Q_{H1}$, during a certain time interval, may be calculated by numerical integration of (2.83) for measured soil temperature and moisture content profiles at the beginning and at the end of the interval. If the depth z_2 is large enough, Q_{H2} can be assumed to be negligible; if it is not sufficiently large to allow this assumption, the heat flux Q_{H2} must be determined. In the so-called combination method, suggested by C. B. Tanner of Wisconsin, Q_{H2} is measured by means of a heat flux plate placed at a depth of 5–10 cm below the surface. The integral in Equation (2.83) is then determined from successive temperature profile measurements above the level of the heat flux plate (see also Hanks and Tanner, 1972).

Empirically based methods to estimate the soil heat flux

When necessary measurements are not available, the surface soil heat flux may be estimated on the basis of empirical relationships. The simplest assumption is that it is proportional to some other term in the energy budget equation. An obvious choice is the sensible heat flux into the air; thus

$$G = c_H H \qquad (2.85)$$

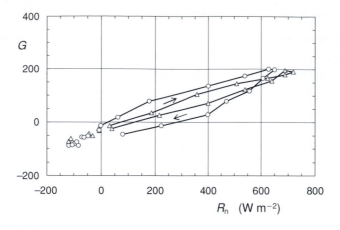

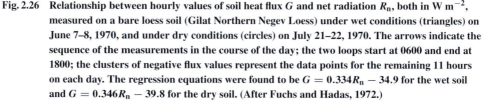

Fig. 2.26 Relationship between hourly values of soil heat flux G and net radiation R_n, both in W m^{-2}, measured on a bare loess soil (Gilat Northern Negev Loess) under wet conditions (triangles) on June 7–8, 1970, and under dry conditions (circles) on July 21–22, 1970. The arrows indicate the sequence of the measurements in the course of the day; the two loops start at 0600 and end at 1800; the clusters of negative flux values represent the data points for the remaining 11 hours on each day. The regression equations were found to be $G = 0.334R_n - 34.9$ for the wet soil and $G = 0.346R_n - 39.8$ for the dry soil. (After Fuchs and Hadas, 1972.)

where c_H is a constant; for bare soil Kasahara and Washington (1971) have taken $c_H = 1/3$. The soil heat flux can also be assumed to be proportional to net radiation, or

$$G = c_R R_n \tag{2.86}$$

where, again, c_R is an empirical constant. From available experimental observations, it appears that on average for bare soil c_R can be given a value around 0.3 (see, for example, Fuchs and Hadas, 1972; Nickerson and Smiley, 1975; Idso *et al.*, 1975) (see also Figure 2.26); however, for any given soil it can be expected to vary with moisture content. For surfaces covered with vegetation c_R will normally be smaller and it will depend not only on the soil moisture state of the soil but also on the type of vegetation; for example, a value of 0.2 has been measured for maize (Perrier, 1975), and most measurements for grass have yielded values around $c_R = 0.1$. To reduce the dependency of c_R on the type of vegetation, Choudhury *et al.* (1987) considered the attenuation of the radiation by the plant canopy and found that the following empirical adjustment

$$c_R = c_{R0} \exp(-a\mathrm{La}) \tag{2.87}$$

yields improved results. The parameter c_{R0} is the value of c_R in (2.86) for bare soil, La is the leaf area index, which is the area (one side) of foliage per unit area of ground surface, and a is a parameter; with data over the midday hours, they found that $c_{R0} = 0.4$ and $a = 0.5$ for wheat, with an inferred variation between 0.45 and 0.65 for different types of vegetation. With measurements around midday on bare soil, soybeans, alfalfa and cotton, Kustas *et al.* (1993) obtained $c_{R0} = 0.34$ and $a = 0.46$ for La < 4, and $c_R = 0.07$ on average for larger values of La. A few typical values of the leaf area index La are listed in Table 2.9 for a number of plant communities from the data collection of Scurlock *et al.*, 2001. Because the leaf area index is not always easily estimated, several studies have also investigated the use

Table 2.9 Leaf area index by biome

Biome	Number of observations	Mean	Standard deviation	Minimum	Maximum
ALL	878	4.51	2.52	0.002	12.1
FOREST Boreal deciduous broadleaf	53	2.58	0.73	0.6	4.0
FOREST Boreal evergreen needleleaf	86	2.65	1.31	0.48	6.21
CROPS Temperate and tropical	83	3.62	2.06	0.2	8.7
DESERTS	6	1.31	0.85	0.59	2.84
GRASSLANDS Temperate and tropical	25	1.71	1.19	0.29	5.0

PLANTATIONS
(managed forests)
Temperate deciduous
   broadleaf, Temperate
   evergreen needleleaf,
   and Tropical deciduous
   broadleaf
	77	8.72	4.32	1.55	18.0

SHRUBLAND
Heath or
   Mediterranean-type
   vegetation
	5	2.08	1.58	0.4	4.5

FOREST Boreal temperate deciduous needleleaf	17	4.63	2.37	0.5	8.5
FOREST Temperate deciduous broadleaf	184	5.06	1.60	1.1	8.8
FOREST Temperate evergreen broadleaf	57	5.70	2.43	0.8	11.6
FOREST Temperate evergreen needleleaf	199	5.47	3.37	0.002	15.0
FOREST Tropical deciduous broadleaf	18	3.92	2.53	0.6	8.9
FOREST Tropical evergreen broadleaf	60	4.78	1.70	1.48	8.0
TUNDRA Circumpolar and alpine	11	1.88	1.47	0.18	5.3
WETLANDS Temperate and tropical	6	6.34	2.29	2.5	8.4

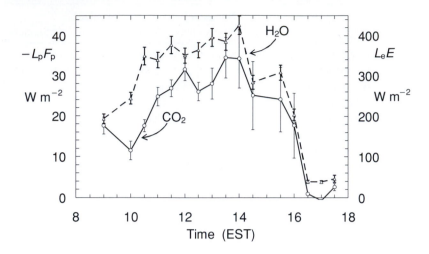

Fig. 2.27 CO_2 **flux density,** $-L_p F_p$ **(circles and left-hand ordinate scale), into a stand of maize in W m^{-2} estimated by means of the energy budget method. The experiment took place near Ithaca, NY, on August 13, 1970. The** $-L_p F_p$ **values shown were around 8% of the estimated latent heat flux** $L_e E$ **(triangles and right-hand ordinate scale), which can be seen to follow a similar diurnal variation. The time is Eastern Standard Time. The vertical bars are error estimates of the flux values. (After Sinclair, 1971.)**

of remotely sensed surrogates for La, such as the normalized difference vegetation index (N_{DVI}), and other measures of surface greenness; for instance, with measurements over the same types of vegetation, Kustas *et al.* (1993) derived $c_R = 0.40 - 0.33 N_{DVI}$.

The daytime variations of the major energy fluxes at land surfaces are often quite similar, exhibiting some kind of self-preservation, which keeps them proportional to each other through the day (see Section 4.3.4). Nevertheless, both (2.85) and (2.86) are oversimplifications, since G is related not to one but to all terms in Equation (2.72); therefore, such simple relationships should be calibrated anew for each given problem, and the values of the constants can be considered accurate only for certain specific conditions. One point in their favor is that, much more so than the other major fluxes in the energy budget R_n, H and $L_e E$, the soil heat flux G tends to be highly variable in space (see Kustas *et al.*, 2000), so that a dense network of measurements would be needed to obtain a meaningful areal average. Therefore, expressions like (2.85) and (2.86) can be useful to obtain averages over larger areas, especially, when used with remotely sensed observations. In the past, attempts have also been made to determine G on the basis of analytical solutions of the linearized heat flow equation with effective parameters for the thermal conductivity and the specific heat of the soil profile (see Brutsaert, 1982, p. 151). However, also this approach can produce only rough estimates.

2.6.3 *Minor terms in the energy budget*

Although they may be quite important under certain conditions, the energy absorption by photosynthesis, the energy advection, and the rate of change of energy storage are usually relatively small in most applications in hydrology.

The flux of CO_2 is usually neglected, although under favorable conditions, say on a sunny summer day, it can be of the order of 5% of the global radiation, and up to 8% to

10% of the latent heat flux. An example of the fluxes of carbon dioxide and water vapor, estimated by means of the energy budget method with the Bowen ratio (cf. Equation (4.11)) on a day of intense photosynthetic activity, is shown in Figure 2.27.

The advected energy term A_h comprises all the energy changes resulting from water flowing in or out of the system to which Equation (2.72) is applied. Precipitation is a source of advection; rainfall may be important in the case of the energy budget of a snow cover, and snowfall may affect the energy balance of a warm lake. Advection by river flows may sometimes have to be considered in the energy budget of a lake, especially when it is shallow.

The term $(\partial W / \partial t)$ can be omitted from (2.72) when it is applied to a thin layer of water, soil or canopy. In the case of tall vegetation, however, it may have to be considered; for example, it has been observed (Stewart and Thom, 1973) that this term can be especially significant after sunrise and near sunset, when it may be of the same order of magnitude as the net radiation R_n. Still, on a daily basis it can be safely neglected. When the layer under consideration is a snow pack, this term is generally important (McKay and Thurtell, 1978) since, as formulated in Equation (2.72), it includes the energy used in fusion. In the case of an entire lake, $(\partial W / \partial t)$ can be determined from successive water temperature profile surveys.

2.6.4 Global climatology of the surface energy budget

To give a rough idea of their magnitude, Figure 2.28 shows the global long term averages of the main components of the energy budget calculated by Ohmura (2005) from the

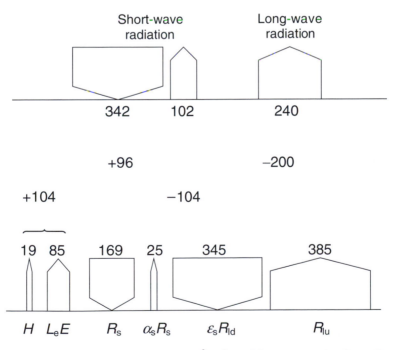

Fig. 2.28 Mean global energy budget fluxes in W m^{-2}, estimated from an extensive observational data base by Ohmura (2005). The incoming short-wave radiation at the top of the atmosphere is 342 W m^{-2}, which is one quarter of the solar constant.

available data. Also shown are the radiative flux values at the top of the atmosphere and their attenuation on their way down. It can be seen that globally the net radiation at the Earth's surface is $R_n = 104$ W m^{-2}, which is also roughly equal to the sum of the sensible and latent turbulent heat fluxes $H + L_e E$. The evaporative flux is shown to be $L_e E = 85$ W m^{-2}; because 1 W m^{-2} produces an evaporation rate of roughly 1.07 mm of water per month, this is equivalent with an annual evaporation of 1.09 m, which agrees with the values listed in Table 1.1. If it is assumed that there is no global warming (or cooling), the sum of the incoming and outgoing radiative fluxes at the top of the atmosphere must also be zero. The ratio of the outgoing and incoming short-wave radiation shows that the average albedo of the Earth–atmosphere system for extra-terrestrial radiation is of the order of 0.3. Globally, the atmosphere is being warmed by the short-wave radiation at a rate of $342 - 169 + 25 = 96$ W m^{-2}; however, the rate of net long-wave radiation input into the atmosphere is $385 - 345 - 240 = -200$ W m^{-2}, which results in a cooling. Thus the net cooling rate of the global atmosphere due to radiation is $-96 + 200 = 104$ W m^{-2}, and this is balanced by the energy input into the atmosphere by the surface turbulent heat fluxes $H + L_e E$.

The energy fluxes at the surface and at the top of the atmosphere, as shown in Figure 2.28, were derived with the constraint that they would exhibit a perfectly balanced steady state. In fact, the separately measured fluxes do not exhibit a perfect balance, but the discrepancy is only of the order of a few W m^{-2}; the issue continues to be the subject of intense investigations.

REFERENCES

Aase, J. K. and Idso, S. B. (1978). A comparison of two formula types for calculating long wave radiation from the atmosphere, *Water Resour. Res.*, **14**, 623–625.

Anderson, E. R. (1954). *Energy-budget studies*. Water loss investigations: Lake Hefner studies, Tech. Report, Prof. Paper 269, pp. 71–119. Geol. Survey, US Dept. Interior.

Asanuma, J., Dias, N. L., Kustas, W. P. and Brutsaert, W. (2000). Observations of neutral profiles of wind speed and specific humidity above a gently rolling landsurface. *J. Met. Soc. Japan*, **78**, 719–730.

Bolz, H. M. (1949). Die Abhängigkeit der infraroten Gegenstrahlung von der Bewölkung, *Z. Met.*, **3**, 201–203.

Brooks F. A. and Pruitt, W. O. (1966). *Investigation of energy, momentum and mass transfers near the ground*, Final Rept. 1965, (DA Task IVO-14501-B53A-08, Defense Doc. Ctr., Cameron Station, Alexandria, VA 22314). Dept. Water Sci. & Engin., University of California, Davis.

Brutsaert, W. (1975). On a derivable formula for long wave radiation from clear skies. *Water Resour. Res.*, **11**, 742–744.

(1982). *Evaporation Into the Atmosphere: Theory, History and Applications*. Boston, MA: D. Reidel Publ. Co.

(1992). Stability correction functions for the mean wind speed and temperature in the unstable surface layer *Geophys. Res. Letters*, **19**(5), 469–472.

(1993). Horton, pipe hydraulics and the atmospheric boundary layer. *Bull. Amer. Met. Soc.*, **74**, 1131–1139.

(1998). Land-surface water vapor and sensible heat flux: spatial variability, homogeneity, and measurement scales. *Water Resour. Res.*, **34**, 2433–2442.

(1999). Aspects of bulk atmospheric boundary layer similarity under free-convective conditions, *Revs. Geophys.*, **37**, 439–451.

Brutsaert, W. and Chen, D. (1996). Diurnal variation of surface fluxes during thorough drying (or severe drought) of natural prairie. *Water Resour. Res.*, **32**, 2013–2019.

Brutsaert, W. and Sugita, M. (1991). A bulk similarity approach in the atmospheric boundary layer using radiometric skin temperature to determine regional surface fluxes. *Bound.-Layer Met.*, **55**, 1–23.

(1996). Sensible heat transfer parameterization for surfaces with anisothermal dense vegetation. *J. Atmos. Sci.*, **53**, 209–216.

Budyko, M. I. (1974). *Climate and Life*. NY: Academic Press.

Cahill, A. T., Parlange, M. B. and Albertson J. D. (1997). On the Brutsaert temperature roughness length model for sensible heat flux estimation. *Water Resour. Res.*, **33**, 2315–2324.

Cheng, Y. and Brutsaert, W. (2005). Flux-profile relationships for wind speed and temperature in the stable atmospheric boundary layer. *Bound.-Layer Met.*, **114**, 519–538.

Choudhury, B. J., Idso, S. B. and Reginato, R. J. (1987). Analysis of an empirical model for soil heat flux under a growing wheat crop for estimating evaporation by an infrared-temperature based energy balance equation. *Agric. Forest Met.*, **39**, 283–297.

Clarke, R. H. and Hess, G. D. (1974). Geostrophic departure and the functions A and B of the Rossby-number similarity theory. *Bound.-Layer Met.*, **7**, 267–287.

Crawford, T. M. and Duchon, C. E. (1999). An improved parameterization for estimating effective atmospheric emissivity for use in calculating daytime downwelling longwave radiation. *J. Appl. Met.*, **38**, 474–480.

Culf, A. D. and Gash, J. H. C. (1993). Longwave radiation from clear skies in Niger: a comparison of observations with simple formulas. *J. Appl. Met.*, **32**, 539–547.

Daughtry, C. S. T., Kustas, W. P., Moran, M. S., Pinter, P. J., Jackson, R. D., Brown, P. W., Nichols, W. D. and Gay, L. W. (1990). Spectral estimates of net radiation and soil heat flux. *Remote Sens. Environ.*, **32**, 111–124.

Deardorff, J. W. (1978). Efficient prediction of ground surface temperature and moisture, with inclusion of a layer of vegetation. *J. Geophys. Res.*, **83**, 1889–1903.

De Vries, D. A. (1955). Solar radiation at Wageningen, *Meded. Landbouwhogeschool, Wageningen*, **55**, 277–304.

(1963). Thermal properties of soils. In *Physics of Plant Environment*, ed. W. R. Van Wijk. Amsterdam: North-Holland Pub. Co., pp. 210–235.

Dias, N. L. and Brutsaert W. (1996). Similarity of scalars under stable conditions. *Bound.-Layer Met.*, **80**, 355–373.

Fuchs, M. and Hadas, A. (1972). The heat flux density in a non-homogeneous bare loessial soil, *Bound.-Layer Met.*, **3**, 191–200.

Garratt, J. R., Wyngaard, J. C. and Francey, R. J. (1982). Winds in the atmospheric boundary layer – Prediction and observation. *J. Atmos. Sci.*, **39**, 1307–1316.

Goddard, W. B. (1970). A floating drag-plate lysimeter for atmospheric boundary layer research. *J. Appl. Met.*, **9**, 373–378.

Hanks, R. J. and Tanner, C. B. (1972). Calorimetric and flux meter measurements of soil heat flow. *Soil Sci. Soc. Amer. Proc.*, **36**, 537–538 .

Idso, S. B., Aase, J. K. and Jackson, R. D. (1975). Net radiation – soil heat flux relations as influenced by soil water content variations. *Bound.-Layer Met.*, **9**, 113–122.

Jacobs, J. M., Coulter, R. L. and Brutsaert, W. (2000). Surface heat flux estimation with wind-profiler/RASS and radiosonde observations. *Adv. Water Resour.*, **23**, 339–348.

Kader, B. A. and Yaglom A. M. (1990). Mean fields and fluctuation moments in unstably stratified turbulent boundary layers. *J. Fluid Mech.*, **212**, 637–662.

Kasahara, A. and Washington, W. M. (1971). General circulation experiments with a six-layer NCAR model, including orography, cloudiness and surface temperature calculation. *J. Atmos. Sci.*, **28**, 657–701.

Kazanski, A. B. and Monin, A. S. (1961). On the dynamic interaction between the atmosphere and the earth's surface. *Bull Acad. Sci. USSR, Geophys. Ser., Engl. Transl.*, **5**, 514–515.

Kondo, J. (1967). Analysis of solar radiation and downward long-wave radiation data in Japan. *Sci. Rep. Tohoku Univ.,* Sendai, Japan. *Ser. 5, Geophys.*, **18**, 91–124.

    (1976). Heat balance of the East China Sea during the Air Mass Transformation Experiment. *J. Met. Soc. Japan*, **54**, 382–398.

Kondo, J., Kanechika, O. and Yasuda, N. (1978). Heat and momentum transfers under strong stability in the atmospheric surface layer. *J. Atmos. Sci.*, **35**, 1012–1021.

Kondratyev, K.Ya. (1969). *Radiation in the Atmosphere*. New York: Academic Press.

Kustas, W. P., Daughtry, C. S. T. and Van Oevelen, P. J. (1993). Analytical treatment of the relationships between soil heat flux/net radiation ratio and vegetation indices. *Remote Sens. Environ.*, **46**, 319–330.

Kustas, W. P., Prueger, J. H., Hatfield, J. L., Ramalingam, K. and Hipps, L. E. (2000). Variability in soil heat flux from a mesquite dune site. *Agric. Forest Met.*, **103**, 249–264.

Landau, L. D. and Lifshitz, E. M. (1959). *Fluid Mechanics*. London: Pergamon Press.

Lettau, H. (1959). Wind profile, surface stress and geostrophic drag coefficients in the atmospheric surface layer. *Adv. Geophys.*, **6**, 241–257.

Liou, K. N. (2002). *An Introduction to Atmospheric Radiation*, second edition. New York: Academic Press.

List, R. J. (1971). *Smithsonian Meteorological Tables*, sixth edition, fifth reprint. City of Washington: Smithsonian Institution Press.

Lowe, P. R. (1977). An approximating polynomial for the computation of saturation vapor pressure. *J. Appl. Met.*, **16**, 100–103.

McKay, D. C. and Thurtell, G. W. (1978). Measurements of the energy fluxes involved in the energy budget of a snow cover. *J. Appl. Met.*, **17**, 339–349.

Mermier, M. and Seguin, B. (1976). Comment on 'On a derivable formula for long-wave radiation from clear skies' by W. Brutsaert. *Water Resour. Res.*, **12**, 1327–1328.

Meyers, T. P. and Dale, R. F. (1983). Predicting daily insolation with hourly cloud height and coverage. *J. Appl. Met.*, **22**, 537–545.

Monin, A. S. and Obukhov, A. M. (1954). Basic laws of turbulent mixing in the ground layer of the atmosphere. *Tr. Geofiz. Instit. Akad. Nauk, S.S.S.R.*, No. 24 (151), 163–187. (German translation: 1958, *Sammelband zur Statistischen Theorie der Turbulenz*, H. Goering, (ed.). Berlin: Akademie Verlag.)

Monin, A. S. and Yaglom, A. M. (1971). *Statistical Fluid Mechanics*: *Mechanics of Turbulence*, Vol. 1. Cambridge, MA: The MIT Press.

Morgan, D. L., Pruitt, W. O. and Lourence, F. J. (1971). *Analyses of energy, momentum and mass transfer above vegetative surfaces*. Research and Development Tech. Rept. E COM 68-G10-F. Davis, CA: Department of Water Science and Engineering, University of California.

Nickerson, E. C. and Smiley, V. E. (1975). Surface layer energy parameterizations for mesoscale models. *J. Appl. Met.*, **14**, 297–300.

Ohmura, A. (2005). Energy budget at the Earth's surface. In *Observed Global Climate*, ed. M. Hantel, Vol. XX, chapter 10, Landolt-Börnstein Handbook. Berlin: Springer Verlag.

Paltridge, G. W. and Platt, C. M. R. (1976). *Radiative Processes in Meteorology and Climatology*. Amsterdam: Elsevier.

Payne, R. E. (1972). Albedo of the sea surface. *J. Atmos. Sci.*, **29**, 959–970.

Perrier, A. (1975). Assimilation nette, utilisation de l'eau et microclimat d'un champ de maïs. *Ann. Agron.*, **26**, 139–157.

Perrier, A., Itier, B., Bertolini, J. M. and Katerji, N. (1976). A new device for continuous recording of the energy balance of natural surfaces. *Agric. Met.*, **16**, 71–84 .

Prandtl, L. (1905). Ueber Flüssigkeitsbewegung bei sehr kleiner Reibung, *Verhandl. III. Internat. Math. Kong., Heidelberg, 1904*. Leipzig: Teubner. (Also in (1961) *Gesammelte Abhandlungen*, Vol. 2. Berlin: Springer-Verlag, pp. 575–584; in English in *NACA Tech. Mem. No.* 452.)

Prescott, J. A. (1940). Evaporation from a water surface in relation to solar radiation. *Trans. Roy. Soc. South. Aust.*, **64**, 114–125.

Pruitt, W. O. and Angus, D. E. (1960). Large weighing lysimeter for measuring evapotranspiration. *Trans. Amer. Soc. Agric. Eng.*, **3**, 13–15.

Qualls, R. J. and Brutsaert, W. (1996). Effect of vegetation density on the parameterization of scalar roughness to estimate spatially distributed sensible heat fluxes. *Water Resour. Res.*, **32**, 645–652.

Qualls, R. J., Brutsaert, W. and Kustas, W. P. (1993). Near-surface air temperature as substitute for skin temperature in regional surface flux estimation. *J. Hydrol.*, **143**, 381–393.

Randel, D. L., Vonder Haar, T. H., Ringerud, M. A., Stephens, G. L., Greenwald, T. J. and Combs, C. L. (1996). A new global water vapor dataset. *Bull. Amer. Met. Soc.*, **77**, 1233–1246.

Rossby, C. G. and Montgomery, R. B. (1935). The layers of frictional influence in wind and ocean currents, *Pap. Phys. Oceanogr. Met.*, **3**(3), 101 pp. Cambridge, MA: Massachussetts Institute of Technology.

Scurlock, J. M. O., Asner, G. P. and Gower, S. T. (2001). *Worldwide Historical Estimates and Bibliography of Leaf Area Index, 1932–2000*, ORNL Technical Memorandum TM-2001/268, Oak Ridge National Laboratory, Oak Ridge, Tennessee, USA.

Sinclair, T. R. (1971). An evaluation of leaf angle effect on maize photosynthesis and productivity. PhD Thesis, Cornell University, Ithaca NY. (See also, Sinclair, T. R., Allen, L. H. and Lemon, E. R. (1975). An analysis of errors in the calculation of energy flux densities above vegetation by a Bowen-ratio method. *Bound.-Layer Met.*, **8**, 129–139.)

Stewart, J. B. and Thom, A. S. (1973). Energy budgets in pine forest. *Quart. J. Roy. Met. Soc.*, **99**, 154–170.

Sugita, M. and Brutsaert, W. (1992). The stability functions in the bulk similarity formulation for the unstable boundary layer. *Bound.-Layer Met.*, **61**, 65–80.

  (1993). Cloud effect in the estimation of instantaneous downward longwave radiation. *Water Resour. Res.*, **29**, 599–605.

  (1996). Optimal measurement strategy for surface temperature to determine sensible heat flux from anisothermal vegetation. *Water Resour. Res.*, **32**, 2129–2134.

Van Wijk, W. R. and Scholte-Ubing, D. W. (1963). Radiation. In *Physics of Plant Environment*, ed. W. R. Van Wijk. Amsterdam: North Holland Pub. Co., pp. 62–101.

Wieringa, J. (1993). Representative roughness parameters for homogeneous terrain. *Bound.-Layer Met.*, **63**, 323–363.

Webb, E. K. (1970). Profile relationships: the log-linear range, and extension to strong stability. *Quart. J. Roy. Met. Soc.*, **96**, 67–90.

Yamada, T. (1976). On the similarity functions A, B and C of the planetary boundary layer. *J. Atmos. Sci.*, **33**, 781–793.

Yasuda, N. (1975). The heat balance of the sea surface observed in the East China Sea. *Sci. Rept. Tohoku Univ.*, Sendai, Japan. *Ser. 5, Geophys.*, **22**, 87–105.

Zilitinkevich, S. S. and Deardorff, J. W. (1974). Similarity theory for the planetary boundary layer of time-dependent height. *J. Atmos. Sci.*, **31**, 1449–1452.

PROBLEMS

2.1   Multiple choice. Indicate which of the following statements are correct. An unstable atmosphere:
(a)   causes turbulence to be damped;
(b)   usually results in a vertical profile of the horizontal wind velocity that is more uniform than that of a stable atmosphere;
(c)   causes the mean horizontal wind velocity to be larger (on a regional scale) than that of a neutral atmosphere;
(d)   in the surface layer is favorable to disperse the pollutants;
(e)   is likely to be found over a deep lake in the spring, when warm air blows over the water;

2.2   Multiple choice. Indicate which of the following statements are correct. Stable conditions in the atmosphere near the Earth's surface:
(a)   result in increased turbulent mixing (as compared to unstable conditions);
(b)   are necessarily the result of smoothness of the surface;
(c)   are often observed under nearly windless conditions with surface cooling by long-wave radiation;
(d)   would be expected over an extensive and deep-water body in early spring under a cloudy sky, when warm air moves over the water;
(e)   are more likely to be accompanied by dew (negative evaporation) than unstable conditions (over land);
(f)   indicate that there is a high likelihood for thunderstorm activity.

2.3   For the ocean and for large lakes, typical values of the drag coefficient and the water vapor transfer coefficient (both with wind and specific humidity measurements at 10 m above the surface) are of the order of $Cd_{10} = 1.4 \times 10^{-3}$ and $Ce_{10} = 1.2 \times 10^{-3}$, respectively. Calculate the roughness parameters z_0 and z_{0v} from these transfer coefficients for neutral conditions.

2.4   From observations at an ocean station, it has been determined that the estimation of the drag coefficient, Cd_{10} (with wind speed measurements, in m s^{-1}, at 10 m above the surface), can be improved by assuming that it is a function of the wind speed, namely, $Cd_{10} = (0.80 + 0.05 \, \bar{u}_{10}) \times 10^{-3}$. In contrast, the heat transfer coefficient, Ch_{10} (with wind speed, in m s^{-1}, and temperature both measured at 10 m), is a constant, namely $Ch_{10} = 1.2 \times 10^{-3}$. Determine the range of the roughness parameters, z_0 and z_{0h}, in the wind speed range, $4 \leq \bar{u} \leq 21$ m s^{-1}. Assume neutral atmospheric conditions.

2.5   Solve the previous problem but with a drag coefficient assumed to be given by $Cd_{10} = 0.50 \, (\bar{u}_{10})^{0.45} \times 10^{-3}$, again, with the wind speed in m s^{-1}.

2.6   For wind speed measurements at z_1 and z_2, and specific humidity measurements at z_3 and z_4, derive an expression for the water vapor transfer coefficient, Ce, in terms of z_1, z_2, z_3 and z_4, valid under neutral conditions. Make use of Equations (2.40) and (2.43).

2.7   Evaporation measurements on Lake Ontario have revealed that the mass transfer coefficient in Equation (2.36) is Ce $= 1.1 \times 10^{-3}$, on average, under neutral conditions, with measurements at 10 m above the surface. If the momentum roughness of a water surface can be assumed to be $z_0 = 0.02$ cm, estimate the scalar roughness for water vapor, z_{0v}.

2.8   In practical applications, the wind profile in the lower atmosphere is sometimes approximated by a power-type equation, as follows: $\bar{u} = C_p u_*(z/z_0)^m$. In this equation, C_p and m are constants, whose values can be assumed to be around 6 and $(1/7)$, respectively, under neutral conditions. (a) Derive an expression for the drag coefficient, Cd, with this equation. (b) Calculate its magnitude if the surface roughness is $z_0 = 0.02$ cm and the wind speed, $\bar{u}$, is measured at 10 m above the surface. (c) Next, consider the mass transfer coefficient, Ce, as given by Equation (2.36) in which $-\Delta q = \bar{q}_s - \bar{q}$, q_s is the value of the specific humidity at the surface and $\bar{q}$ is the value at height z. Under conditions of light winds above open water, it can be assumed that Ce $=$ Cd. With this assumption, derive the power-type equation for the specific humidity profile $(\bar{q}_s - \bar{q})$, which is the analog of that for wind, given above.

2.9   The stability of the lower atmosphere is commonly characterized by means of the dimensionless variable, $\zeta = (z - d_0)/L$, defined in Equation (2.45). An alternative variable to do this is the Richardson number, defined as Ri $= (g/T_a)[(d\bar{\theta}/dz)/(d\bar{u}/dz)^2]$. Derive the relationship between ζ and Ri, in terms of ϕ_m and ϕ_h, defined in (2.47) and (2.48). Assume that the water vapor flux term, $\overline{w'q'}$, in L is negligible.

2.10 Derive an expression for the specific humidity profile, similar to Equation (2.43), but applicable to stable conditions. Note that (2.43) is valid only for neutral conditions. Assume that the flux profile relationship for stable conditions is given by (2.49) with (2.57). Check your result by comparing with (2.52) and (2.58).

2.11 During a field experiment above a grassy surface, the following mean values were measured over a 1 h period: the temperature at 1.5 m above the ground, $T_{1.5} = 31.29$ °C; at 3.0 m above the ground, $T_3 = 30.87$ °C; and the wind speed at 2.0 m, $u_2 = 3$ m s^{-1}. The surface roughness was estimated to be $z_0 = 0.01$ m, and the displacement height, d_0, was found to be negligible. (a) Calculate, by iteration, the friction velocity, u_*(m s^{-1}), and the sensible heat flux, H (W m^{-2}). The evaporation term $\overline{w'q'}$ in L can be assumed to be negligible. (b) If the net radiation is $R_n = 392$ W m^{-2}, calculate the rate of evaporation from $L_e E = R_n - H$, first in W m^{-2} and then in mm month^{-1}.

2.12 In a neutral atmosphere, a northerly wind (i.e. blowing to the south) has a velocity, $\bar{u} = 8$ m s^{-1}, at 2 m above prairie terrain; the surface parameters are $z_0 = 0.09$ m and $d_0 = 0.50$ m. (a) Calculate the x- and y-components and the direction of the "free stream wind," (i.e. u_b and v_b at $z = h_b$) by using $A = 4.5$ and $B = 1.5$ in Equation (2.65). Assume that the ABL has a thickness, $h_b = 800$ m. (b) If the rate of evaporation is 0.6 mm h^{-1}, the air temperature is 15 °C, and the relative humidity is 70% at 2 m, what is the specific humidity at $z = h_b = h_i = 800$ m above the ground? Assume $D = 0$ in (2.71). Hint: combine (2.71) with (2.44).

2.13 Derive an expression for $(\bar{\theta}_1 - \bar{\theta}_m)$ over very rough terrain as the analog of (2.68) in which θ_s is replaced by $\bar{\theta}_1$; the latter is the potential temperature at a height, z_1, within the atmospheric surface layer. Hint: subtract (2.55) from (2.68) and substitute (2.70) for C.

2.14 The magnitude of the solar declination angle depends on the day of the year; it varies between roughly $\pm 23.28°$ on the solstices and $0°$ on the equinoxes. (a) What is the zenith angle of the Sun, i.e. β, at noon when $h = 0$? (b) When and where on Earth is the day length 12 hours, i.e., $h = \pi/2$ when the Sun rises and sets? Show with (2.77) that there are two possible situations for this to occur.

2.15   Prove Equation (2.78) by integration of (2.77) over one day.

2.16   Calculate the total daily solar radiation (W m^{-2}) on a horizontal plane in the absence of an atmosphere for a latitude of 45° on June 21. Compare with the value shown in Figure 2.23.

2.17   The following data are averages for a typical summer day in a temperate climate: air temperature, $T_a = 17.94$ °C; relative humidity, 66%; and incoming, short-wave radiation, $R_s = 468$ cal cm^{-2} d^{-1}. Calculate the net radiation, R_n (in W m^{-2}), for a surface covered with vibrant, but short vegetation. (To a first approximation, assume that the daily average of the surface temperature, T_s, is the same as the air temperature, and that cloudiness does not affect the long-wave radiation, when it is derived from T_a.)

2.18   Same as Problem 2.17 but with the following data: air temperature, $T_a = 20.45$ °C; relative humidity, 64%; and incoming, short-wave radiation, $R_s = 477$ cal cm^{-2} d^{-1}.

2.19   The following data are available for a deep lake in a temperate climate (latitude 42.5° N) for typical days respectively in the months of December (not frozen) and July: mean air temperature, $T_a = -2.78$ and 20.56 °C; mean water surface temperature, $T_s = 6.12$ °C and 19.20 °C; relative humidity for the region, 76% and 64%; and fraction of sunshine hours for the region, $n/N = 0.33$ and 0.63. Estimate the daily incoming short-wave radiation, Q_s(in W m^{-2}), by using climatological methods with (2.74) and Figure 2.23. With this value as an estimate of R_s, calculate the mean daily net radiation, R_n(in W m^{-2}). Assume a surface emissivity of unity, and that the temperatures remain roughly constant through the day.

2.20   During the night, a cloud layer, whose temperature is 4 °C, moves over an area covered with snow, whose temperature is -3 °C. Calculate the maximal rate of evaporation of the snow cover, which is due to the radiation from the clouds, if the absorptivity of the cloud is 0.93 and that of snow, 0.99. Assume that the atmosphere is transparent, that the air temperature is also -3 °C and that steady conditions prevail. The latent heat of sublimation is 2.8×10^6 J kg^{-1}. Give the result in W m^{-2} and mm day^{-1}.

2.21   Give an estimate of typical values of $c_R = G/R_n$ in Equation (2.86), that can be expected for (a) cropland and (b) grassland, on the basis of (2.87).

2.22   Show why the global incoming short-wave radiation is one fourth of the solar constant, as indicated in Figure 2.28.

3 PRECIPITATION

Because the entire hydrologic cycle is basically driven by it, precipitation has to be considered the main component. Indeed, it is a truism that wherever there is no precipitation, there is also not much of a hydrologic cycle. The detailed study of precipitation and of all its aspects is properly the domain of meteorology. In hydrology, precipitation is primarily of interest after it reaches the ground surface, and this is reflected in the organization of this chapter. However, to gain a better understanding of the occurrence and distribution of precipitation and its temporal and spatial scales, it is also useful to have a knowledge of at least some elementary aspects of its generation mechanisms and of its major types.

3.1 FORMATION OF PRECIPITATION

3.1.1 *Mechanisms*

Several processes take place jointly in the formation of precipitation. In brief, these are the production of supersaturation of the air, condensation of water vapor into ice crystals and droplets, the subsequent growth of these condensation products, and the supply of moist air to where the first three processes occur. These processes involve a number of different mechanisms, which are briefly reviewed in what follows.

Cooling of the air
As indicated in the previous chapter, the water-holding capacity of air decreases, as its temperature decreases. Thus, air can become supersaturated by being cooled down. Such cooling can occur by advection, for instance, as warmer air moves over a colder surface, by radiative cooling or also as a result of the mixing of two different air masses; but these are generally not very effective mechanisms and only capable at most of fog formation or light drizzle. A more effective mechanism of cooling of the air consists of its being lifted to higher elevations; in the generation of precipitation this is by far the main mechanism. Air can be forced to rise by being heated from below, by moving over mountainous terrain, or by frontal activity, that is by having to move over relatively heavier, that is colder air. When the vertical air motions are relatively weak and gentle, for example in the case of stable air in a warm front, the lifting may result in what is referred to as *stratiform* precipitation. On the other hand, when the air is already unstable, and the vertical motions are relatively strong, the resulting precipitation tends to be of the so-called *convective* type. Under certain conditions the movement of air over mountainous terrain can involve a combination of both stratiform and convective precipitation.

Condensation and growth of condensation products

When the air reaches or exceeds saturation, water vapor may start to condense in the form of liquid drops or ice crystals on the small nuclei, such as dust, smoke and various types of salt particles, that are invariably present in the atmosphere. Initially, these condensation products are small enough to be kept afloat in the atmosphere as cloud. Condensation generates latent heat. Therefore, any further growth of these droplets and ice crystals depends on the rate of diffusion of water vapor to their surface from the surrounding air and on the rate of conduction of latent heat away from their surface into the air; in the case of liquid droplets, there are also hygroscopic effects due to some nuclei, surface tension and for the larger sizes, accretion. It can be shown (see, for example, Fleagle and Businger, 1963) that the growth of droplets is primarily controlled by condensation up to radius sizes of the order of 15 μm. Full-grown raindrop development to radii larger than 20 μm requires accretion by collisions and coalescence with neighboring drops of different sizes and fall velocities; this process, in turn, is affected by the turbulence of the air, the temperature distribution in the clouds and electrical effects. In contrast, the growth of ice particles depends only on vapor diffusion and on heat removal, and not on accretion; however, on account of the lower vapor pressure characteristics of ice, the water vapor diffusion process is much more effective than in the case of liquid water. If the particle sizes continue to increase, eventually they become too heavy and precipitation takes place. The onset of precipitation on the ground depends on a number of conditions; most obvious among them is that the condensation products must be large enough to fall down against the entrainment of updrafts, and to overcome evaporation while falling down to the ground. A rough rule of thumb for the demarcation between precipitation and cloud particle diameters is around 0.1 mm. Substantial rates of precipitation usually can take place only when the cloud thickness is 1200 m or more.

Blanchard (1972) has written an enlightening history of the discovery of the main mechanisms of raindrop and ice crystal formation.

Moisture supply

Calculation of the precipitable water, as defined in Equation (2.11), shows that even under the most favorable conditions an atmospheric column at rest can hold only a very limited quantity of water vapor. For instance (List, 1971), for a near-surface temperature of 20 °C and a surface pressure of about $p_0 = 1000$ hPa, a saturated atmosphere with a pseudoadiabatic lapse rate can at most hold an amount, which is equivalent with about 5 cm of liquid water; for 10 °C this precipitable water is only about half as much. Heavy precipitation amounts regularly exceed such values. But even so, it is well known that the humidity of the air tends to remain relatively constant during precipitation events. This means that it is not so much the local precipitable water, but the horizontal influx of moist air into an area, that controls the local intensity and the total amount of precipitation. The specific nature of this moisture influx depends on the weather system.

Water recycling

In the study of regional water budgets over seasonal or longer time periods it is often of interest to determine the origin of the water vapor producing the precipitation. Part of this

Table 3.1 **Annual precipitation recycling ratio for different regions**

Region	Square root of area (km)	Recycling ratio	Reference
Amazon	2300	0.25	Brubaker *et al.* (1993)
Amazon	2500	0.25–0.35	Eltahir and Bras (1994)
Mississippi	1800	0.10	Benton *et al.* (1950)
Mississippi	1400	0.24	Brubaker *et al.* (1993)
Eurasia	2200	0.11	Budyko (1974)
Eurasia	1300	0.13	Brubaker *et al.* (1993)
Sahel	1500	0.35	Brubaker *et al.* (1993)

water vapor originates from evaporation outside the region, while the remainder results from evaporation within the region in question. The precipitated water produced by the evaporation inside the region can be referred to as recycled water. Water recycling has been the subject of intensive investigations (see, for example, Eltahir and Bras, 1996; Gong and Eltahir, 1996). Table 3.1 shows some estimates for a few regions of the world. Recycling of precipitation, or lack thereof, resulting from the soil moisture conditions can be a strong feedback mechanism leading to persistence of weather and climate patterns.

3.1.2   Types of precipitation

Precipitation can reach the ground surface in different forms.

Drizzle is a very light, usually uniform, precipitation consisting of numerous minute droplets with diameters in excess of 0.1 mm but smaller than 0.5 mm.

Rain is precipitation consisting of water drops larger than 0.5 mm. It can be classified as light rain when the intensity is smaller than 2.5 mm h^{-1}, moderate when it is between 2.5 and 7.5 mm h^{-1}, and heavy when it exceeds 7.5 mm h^{-1}.

Snow is precipitation in the form mainly of branched hexagonal or star-like ice crystals, resulting from direct reverse sublimation of the atmospheric water vapor; snow particles can reach the ground as single crystals, but more often than not they do so after agglomerating as snowflakes. These flakes tend to be larger at temperatures close to freezing. The specific gravity of snow can vary over a wide range (Judson and Doesken, 2000), but as a rule of thumb for fresh snow it is often taken around 0.1.

Sleet (North American usage) is precipitation consisting of fairly transparent pellets or grains of ice, formed as a result of the passage of raindrops through a layer of colder air near the ground. In British usage the word sleet refers to precipitation consisting of melting snow or a mixture of snow and rain.

Glaze or freezing rain is ice deposited by drizzle or rain on cold surfaces.

Snow pellets (also called *granular snow* or *graupel*) are a form of precipitation consisting of white, opaque, small grains with diameters between roughly 0.5 and 5 mm.

Small hail is precipitation consisting of white, semitransparent or translucent grains with diameters ranging from about 2 to 5 mm. These grains are mostly round, and sometimes conical in shape, and they have a glazed appearance. Small hail falls usually accompanied by rain, when the temperature is above freezing.

Soft hail consists of round, opaque grains in the same size range as small hail, but they are softer in appearance and tend to disintegrate more easily.

Hail consists of balls or irregular chunks of ice with diameters between 5 and 50 mm, or even larger. These lumps of ice can be transparent or they can consist of concentric layers of clear and opaque ice; such layered structure is the result of the alternating rising and falling movements during the hail formation. Hail usually falls during violent and prolonged convective storms under above-freezing temperature conditions near the ground; it can cause severe damage.

Dew consists of moisture in the form of liquid drops on the ground surface and on the vegetation and other surface elements, as a result of direct condensation of atmospheric water vapor. It typically occurs at night on surfaces that have been cooled by outgoing long-wave radiation.

Hoar frost forms in the same way as dew, but the water vapor condenses directly into ice. These ice crystals can assume a wide variety of shapes.

3.2   MAJOR PRECIPITATION WEATHER SYSTEMS

3.2.1   *Extratropical cyclones and fronts*

These types of systems normally result from the interaction of two contrasting air masses. An air mass can be defined as a body of air with approximately uniform physical characteristics such as (potential) temperature and humidity. The interface between two different air masses is called a *cold front* when relatively colder air displaces and moves beneath relatively warmer air, and a *warm front* in the opposite case. Cold fronts tend to be relatively steep, with average slopes on the order of 0.015; in the Northern Hemisphere they are often oriented from the southwest to the northeast and move toward the east and southeast. An approaching cold front is usually announced by increasing wind speeds and the appearance of altocumulus clouds (Figure 3.1). All the while the pressure decreases, and lower clouds, mainly of the cumulonimbus type, move in with the onset of precipitation. As the front comes closer the precipitation intensity increases. After the passage of the front the pressure rapidly rises and the temperature falls sharply; the wind direction changes, typically from a southerly or southwesterly direction to a more westerly or northerly direction. Cold fronts are often followed by drier and cooler weather. The stability of the warm air mass determines the type of precipitation generated by a cold front. If the warm air is stable the clouds are of the stratiform type. The clouds are

Fig. 3.1   Cross section of a typical
cold front. (Vertical scale
is exaggerated.)

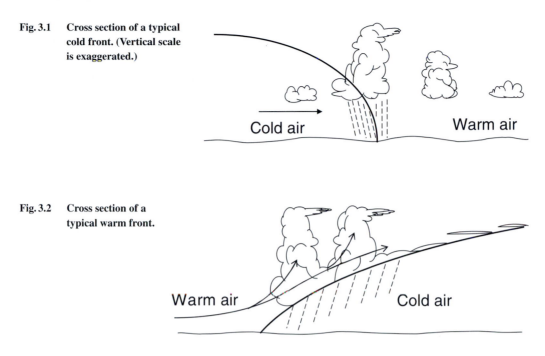

Fig. 3.2   Cross section of a
typical warm front.

of the cumuliform type and the precipitation convective (see Section 3.2.2), if the warm air is conditionally unstable (see Figure 3.1). In this case, scattered thunderstorms and showers may develop, and in extreme cases the front may evolve into a continuous line of thunderstorms, called a *squall line*.

Warm fronts are usually not as steep, with slopes that are on average of the order of 0.01; they also move more slowly than cold fronts, and are not as well defined. As the warm air moves over the cold air a broad band of clouds develops (Figure 3.2), extending up to several hundreds of kilometers ahead of the front's position on the ground. Also in the case of a warm front, it is the stability of the warm air that determines the type of precipitation produced by the front. When the warm air in the approaching frontal air mass is moist and stable, the sequence of cloud types is cirrus, cirrostratus, altostratus, and nimbostratus, and the precipitation increases gradually. When the air is moist and conditionally unstable, the same sequence may occur, but altocumulus and cumulonimbus, often with thunderstorms, will also be observed.

The interface between contrasting air masses tends to be unstable and it often evolves further, through the rotation of the Earth, into a spiraling stream called a *cyclone*. A cyclone is a large low-pressure zone, and it is usually accompanied by cloud systems and precipitation. An anti-cyclone is the opposite case, namely a high-pressure zone, which usually brings fair weather; it is also normally characterized by *subsidence*, a slow downward air motion resulting from the horizontal divergence of the air away from the pressure high.

Although there is an infinite variability in the occurrence of cyclones, they have certain features in common; a typical life cycle of a cyclone is sketched in Figures 3.3, 3.4

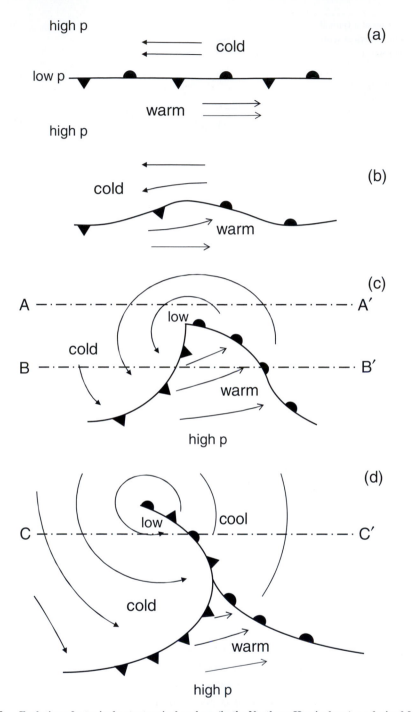

Fig. 3.3 Evolution of a typical extratropical cyclone (in the Northern Hemisphere), as derived from surface weather maps, with the geostrophic wind velocity vectors shown parallel to the isobars; (a) illustrates an assumed initial state as a stationary front with wind shear; this interface is unstable and it gradually develops into a frontal wave with growing amplitude as shown in (b) and (c). In (d) the warm front has been overtaken by the cold front and the cyclone has become occluded. From then on the cyclone loses its strength and the fronts gradually dissolve.

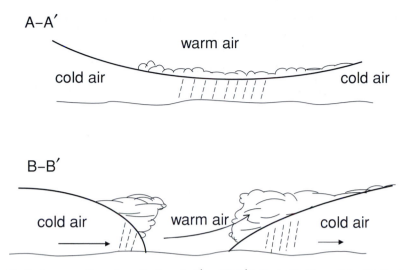

Fig. 3.4 Vertical cross sections along the lines A–A' and B–B' indicated in Figure 3.3c. (Vertical scale is exaggerated.)

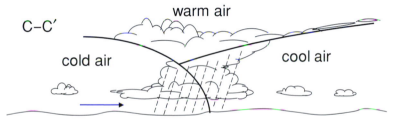

Fig. 3.5 Vertical cross section along the line C–C' indicated in Figure 3.3d, showing a cold-front occlusion.

and 3.5. In the initial stage the winds on both sides of a stationary front are blowing in parallel but opposite directions. As a result of the shear and small disturbances, surface roughness or heating irregularities, the front may gradually assume a wave-like shape, which may persist and increase in amplitude, and eventually evolve into a counterclockwise (in the Northern Hemisphere) flow pattern, called a *frontal wave*. By now there is a well-defined cold front and a warm front, and the cyclonic circulation continues to intensify. The cold front section usually moves faster and eventually overtakes the warm front. At this point, which is the time of maximal intensity of the cyclone, the combined front is referred to as an *occlusion* or an *occluded front*. In the later stages of the occlusion, the intensity of the cyclone and the frontal movement gradually decrease; finally the occlusion vanishes while a new stationary front may be formed. Observe that, in contrast to the cold and warm fronts sketched in Figures 3.1 and 3.2, those shown in Figure 3.4 are assumed to involve a stable warm air mass, so that the clouds shown are of the stratiform type. Many different factors control the evolution of occluded fronts. It has been shown (Stoelinga *et al.*, 2002) that, more than the temperature contrasts, it is the stability contrasts across the fronts which govern the dynamics of occlusions.

Frontal cyclones dominate the weather in the mid- and high latitudes, mainly in the colder season, when the contrasts between the equatorial and the polar regions are the most pronounced. In the warmer seasons, such cyclonic systems are generally weaker. They typically involve length scales of the order of 10^3 km, also referred to as the mesoalpha (see Table 1.5), macro, or synoptic scale and their highest incidence is around 55° latitude.

3.2.2   *Extratropical convective weather*

Unstable atmospheric conditions have the potential of generating organized systems of convective quasi-vortical movement of air, over a wide range of scales. Recall that an ideal vortex is a flow in which the streamlines are concentric circles. In the atmosphere, however, convective systems are considerably more complex than an inverted bathtub vortex. Under the right moisture conditions of the atmosphere, these systems can develop into thunderstorms, and may consist of a single storm cell, or of several cells as part of a mesoscale convective system. The spatial extent of these systems tends to cover mainly the mesobeta into the mesoalpha scales, typically ranging from about 50 to 500 km, but individual cells can be as small as only a few kilometers. Individual cells are characterized by strong local updrafts and downdrafts. Simply put, the updrafts are a manifestation of the unstable conditions of the air (see Figures 2.2 and 2.4) and lead to condensation in the cooling air resulting in precipitation. The downdrafts, on the other hand, result not only from the entrainment by falling precipitation and some evaporative cooling, but also from return flows required by continuity to compensate for the upward motions (Vonnegut, 1997); some of these are produced after updrafts reach their highest level and then fall back as downward currents. Most systems of this type are accompanied by a specific surface pressure pattern, first described by Fujita (1955) from time-to-space conversion of barograph data. In brief, this pattern consists of a high-pressure zone or *mesohigh*, trailed by a low-pressure zone, also called a *mesodepression* or *wake low*. Some of the mechanisms involved have subsequently been further elucidated (see Johnson, 2001) and are sketched in Figures 3.6 and 3.7. It is generally believed that the high pressure is a result of evaporative cooling in precipitation downdrafts below cloud base; additional effects may be caused by the impinging of the downdrafts on the ground, causing a pressure *nose*, and by hydrometeor loading. Williams (1963) showed that the observed pressure deficit could be the result of a descending, or subsiding, dry current to the rear of the convective air, but the causes for this remain unclear. The nose, also known as a cold air outflow leading edge or gust front, often assumes the form of a surge; some 20 cases have been studied with a 461 m tower by Goff (1976) (see Figures 3.8 and 3.9), which indicate that they have some features in common with the open channel surges discussed in Chapter 7 (see also Simpson, 1997). At present, the details of storm development, and the possible roles of gravity currents and gravity waves in the observed pressure patterns, are still not completely understood.

Mesoscale convective systems of thunderstorms can be organized as squall lines or as mesoscale convective complexes. Squall lines or instability lines are relatively narrow bands of convective elements, like that illustrated in Figure 3.7; they are often

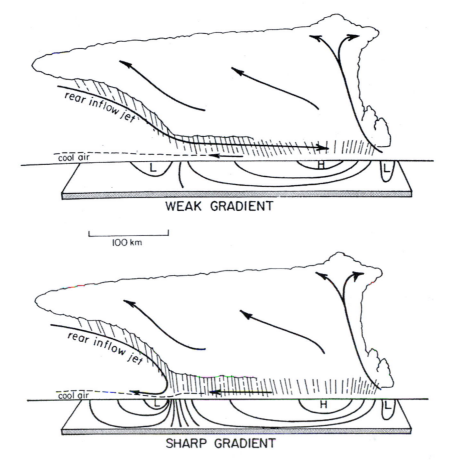

Fig. 3.6   **Typical surface pressure pattern and vertical structure of a convective system. The top shows the case of weak and the bottom the case of sharp pressure gradients, in association with rear-inflow jets that proceed forward toward the leading convective line or are blocked, respectively. (From Johnson, 2001.)**

accompanied by brief and sudden wind storms or "squalls," and tend to occur along sharp cold fronts. Mesoscale convective complexes are a major mechanism for the production of heavy precipitation at the midlatitudes during the warmer seasons of the year (Maddox, 1980; Fritsch *et al.*, 1986; Houze *et al.*, 1989).

3.2.3   *Seasonal tropical systems*

These systems occur in the zone of convergence of the northerly and southerly trade winds and as wave-like structures in the zone between the subtropical high-pressure belts and the Equator; they are mostly responsible for the well-known tropical rainfall and the abundant natural vegetation in those regions. They typically produce deep cloud clusters and, intermittent with fair weather, heavy precipitation of the convective type. These systems tend to have a seasonal character, as they follow the sun between the two tropics.

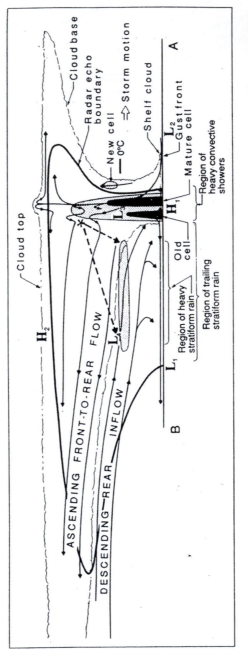

Fig. 3.7 Sketch of a vertical cross section through a mesoscale convective system (squall line) parallel to its motion. (From Houze *et al.*, 1989.)

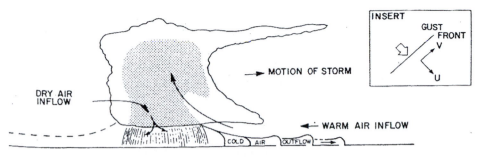

Fig. 3.8 Sketch of thunderstorm circulations associated with cold air outflow at the leading edge. Several surges are shown. The darkened area consists of falling or suspended precipitation. The dashed line is the upper boundary of the outflow at the rear in the wake of the storm. The insert shows the horizontal wind coordinates relative to the front. (From Goff, 1976.)

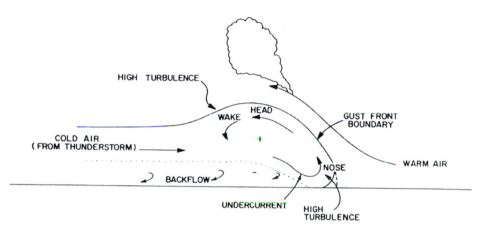

Fig. 3.9 Sketch of the leading edge of the cold air outflow. The presence of cloud depends on the height of the condensation level. The indicated flows are relative to the gust front. Plus and minus signs indicate direct and indirect circulations, and the dotted line indicates separation due to local shear. (From Goff, 1976.)

Where they are combined with monsoon winds and orographic effects, as in Meghalaya in the eastern parts of India, they have produced some of the largest long-term rainfall amounts on record. Monsoons are large-scale quasi-steady wind regimes, often resulting from specific geographic and topographic features of the regions where they occur, and characterized by a seasonal reversal of wind direction. In response to the differential heating by the surface, they blow from land to sea in winter, and from sea to land in summer, producing a wet–dry season cycle.

3.2.4 *Large-scale tropical convective systems*

These are well-developed low-pressure systems, of tropical ocean origin, which can travel long distances accompanied by strong winds and heavy rainfall. As they move away from their origins, they can cause severe weather in the coastal regions. As long as the wind speeds remain below 40 km h^{-1}, they are referred to as *tropical depressions*; they are called *tropical storms* for wind speeds between 40 and 120 km h^{-1}, and *hurricanes* for winds above that range. In the western Pacific Ocean they are known as *typhoons*. Such systems have also generated some of the largest rainfalls ever recorded. Amounts of 15–25 cm in a 24 h period are not uncommon over level land.

3.2.5   Orographic effects

The precipitation resulting from each of the general weather types discussed here can be markedly affected by topographic features, such as elevation, slope and aspect of the land surface. The result tends to be increased precipitation on windward slopes, and smaller precipitation on leeward slopes, also called rain shadows. In some regions with identifiable prevailing wind directions, such as the coastal ranges of western North America or the foothills of Meghalaya in eastern India, the windward slopes can be readily identified. On the other hand, as noted by Gilman (1964), in the Appalachian mountains, the windward and leeward sides can be quite variable, depending on the wind direction. Smith (1979) specified that there are three independent mechanisms of orographic precipitation, as follows. (i) Large-scale upslope precipitation, which is generated by forced vertical motion of the stratiform type or by triggered convection as the air moves over rising terrain. (ii) Small-scale redistribution of precipitation from pre-existing clouds by small hills; over the hill tops the precipitation is increased, because their higher surface can intercept the falling drops before they evaporate, and apparently also because the drops undergo increased accretion by washout of low-level clouds. (iii) Generation of upslope winds in a conditionally unstable air mass as a result of slope heating by the Sun; these develop into rising thermals, which in turn can grow into cumulonimbus clouds above the lifting condensation level.

In general, because there are several other factors beside elevation, the effect of orography by itself in causing increased precipitation is not always obvious; physically, its main effect is as a trigger mechanism for convective activity. Accordingly, as observed by Suzuki *et al.* (2002), the relationship between precipitation and elevation is usually more pronounced for convective than for stratiform rainfall. The relationship is also stronger and more apparent for larger accumulated rainfall amounts, for longer accumulation time scales, and for larger rainfall intensities. For instance, in the analysis of hourly rainfall data, the random effects of other factors may dominate the measured precipitation, so that the effect of topography may go undetected. With daily rainfall data topographic effects gradually emerge, albeit with large variations from one day to the next. With monthly data the effects of the other factors tend to become averaged out and the effect of elevation is more apparent. Numerous studies reviewed by Daly *et al.* (1994) have reported linear relationships between precipitation and elevation, but other relationships, such as loglinear functions have also been documented. In the mid-latitudes the climatological precipitation maxima tend to occur at or near the crest of mountainous barriers. However, in warmer regions (e.g. Hawaii), or in large-scale precipitation events (e.g. the Sierras in California) the maximal precipitation may occur somewhat lower, ahead

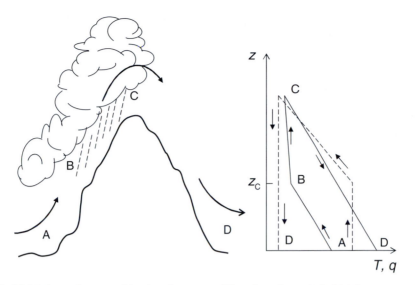

Fig. 3.10 **Moist air passing over rising terrain may result in a downslope wind which is warmer and drier. The graph on the right shows the changes in temperature T (solid line) and specific humidity q (dashed line) of a parcel of air with height z; the lapse rate of the solid line segments AB and CD is very close to dry adiabatic, and that of the segment BC is close to saturated adiabatic. The height z_C is the condensation level, where precipitation starts and the rate of temperature decrease changes from dry adiabatic to saturated adiabatic.**

of the topographic barrier; this may be due to upwind "rainout." On the other hand, in some situations of steep and narrow mountain ridges, the precipitation maxima may be delayed, and may occur downwind from the crest of the barrier.

Example 3.1. Passage over mountain barriers

In the previous chapter the concepts of adiabatic lapse rate and the resulting atmospheric stability were in introduced. Figure 2.4 illustrates how air, which is initially stable, can be made unstable by being forced to rise. This is called conditional instability. As another illustration, consider this time an air mass that is forced to rise by moving over rising terrain, as sketched in Figure 3.10. Again, initially the temperature of an air parcel will decrease at a rate roughly equal to the dry adiabatic lapse rate. Above the condensation level this rate will become smaller, and roughly equal to the saturated adiabatic lapse rate. As the saturated air continues to rise, it cools further down and its moisture gradually precipitates out. After it passes the peak, and goes back down, through its descent the air warms up at the dry adiabatic lapse rate and gradually becomes less and less saturated. Finally, after its passage downwind from the mountain the air is both warmer and drier than it was originally upwind. This type of phenomenon is variously referred to as *Föhn* in the Alpine regions of Europe, *Santa Ana* in southern California and *Chinook* east of the Rocky Mountains of North America. Again, as was the case with frontal systems, when the entire incoming air mass is initially stable and moves

Fig. 3.11 Sketch illustrating the
          application of the
          Thiessen polygon method
          to estimate the subareas
          A_i assigned to the
          precipitation gages on the
          map of a catchment. The
          subareas are bounded by
          the boundaries of the
          catchment and by the
          lines drawn midway
          between the stations. The
          locations of the stations
          are indicated by the
          numbered circles.

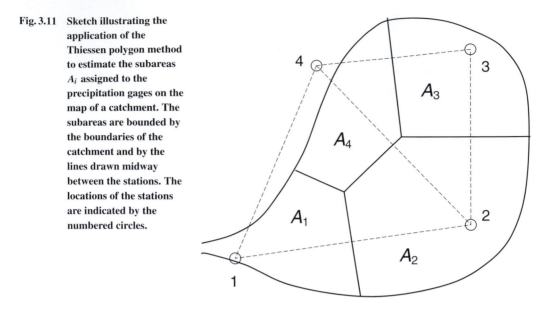

uniformly over the rising terrain, the resulting precipitation can be expected to be of the stratiform type. However, in situations when the incoming air is already unstable, much higher convective type precipitation intensities develop; this may be especially the case with irregular topography and uneven surface heating, when some parts of the air mass have higher temperatures than others at the same height, so that locally the convection mechanisms, as illustrated in Figure 2.4, come into play.

3.3   PRECIPITATION DISTRIBUTION ON THE GROUND

3.3.1   *Spatial distribution*

Areal average from precipitation gages
In hydrologic analyses at the basin- or catchment-scale, the input is of necessity taken as the average precipitation over the entire area. Different weighting methods have been used in the past to estimate this average from the available precipitation gage network. When no other information is available, the only possible method is to take the regular average value, i.e. the *arithmetic mean*, with equal weights assigned to all gage stations. When the locations of the stations are known on a map, the *Thiessen polygon* method (Thiessen, 1911) has been commonly used. Here each gage represents a subarea, A_i, which is determined as the area bounded by the perpendicular bisectrices between the station and those surrounding it (see Figure 3.11); the spatial average is calculated by weighting the individual stations with their representative area, namely

$$\langle P \rangle = \frac{1}{A} \sum_{i=1}^{n} A_i P_i \qquad\qquad (3.1)$$

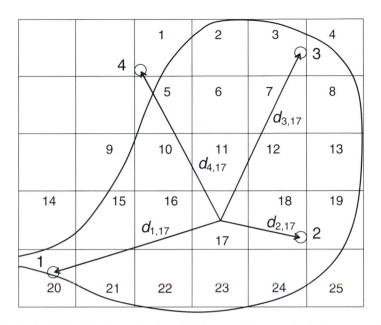

Fig. 3.12 Sketch illustrating the application of the inverse distance method. For example, the precipitation over subarea 17 is determined as $\sum_{i=1}^{4} d_{i,17}^{-2} P_i / \sum_{i=1}^{4} d_{i,17}^{-2}$. The average precipitation over the entire catchment is then obtained as the weighted average of all subarea values, as shown in Equation (3.2). The locations of the stations are indicated by the numbered circles.

where n is the number of rain gage stations in the area, and A is the surface area of the catchment, that is the sum of the subareas, or $A = \sum_{i=1}^{n} A_i$.

The *inverse distance* method is equally simple in principle, but it is easier to implement. It is based on the assumption that the precipitation at any given point is influenced by all stations in the area, each weighted by the inverse of a power of its distance from the point. Note, as an aside, that the principle can also be used to calculate missing data. To obtain the areal average, the method is applied by subdividing the area into m rectangular subareas, each with an assumed uniform precipitation as calculated for the point at its center; the resulting mean precipitation is then

$$\langle P \rangle = \frac{1}{A} \sum_{j=1}^{m} A_j \left(\sum_{i=1}^{n} d_{ij}^{-b} \right)^{-1} \sum_{i=1}^{n} d_{ij}^{-b} P_i \tag{3.2}$$

in which A_j is the surface area of the jth subarea, A is the total surface area of the catchment, and n the total number of precipitation stations; d_{ij} is the distance of the center of the jth subarea from the ith rain gage in the catchment and b is a constant, which in most applications has been taken as 2. It can be seen that, for $b = 0$, Equation (3.2) produces the arithmetic mean. Dean and Snyder (1977) found that $b = 2$ yielded the best results in the Piedmont region of the southeastern United States, whereas Simanton and Osborn (1980) concluded from measurements in Arizona that b can range between 1 and 3 without significantly affecting the results. Figure 3.12 illustrates the application of this method.

Fig. 3.13   **Sketch illustrating the application of the isohyetal method to estimate the average precipitation over a catchment. The subareas A_i are bounded by the isohyets and by the boundaries of the catchment. The locations of the stations are indicated by the numbered circles.**

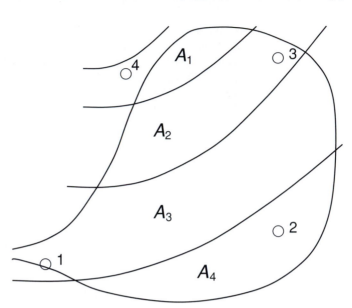

Still another graphical procedure is known as the *isohyetal* method (see Reed and Kincer, 1917), which consists of drawing isohyets or contour lines of equal precipitation (Figure 3.13), by interpolation between the measured values at the rain gage stations. The method can be applied with Equation (3.1), in which the A_i values are the areas bounded by the isohyets and by the boundaries of the catchment, and the P_i values are the average precipitation of the two isohyets bounding the corresponding A_i. Some of the difficulties in the derivation of areal averages in mountainous terrain have been addressed by Peck and Brown (1962).

Many more methods have been proposed in the literature. Although they can be quite different in principle, a comparison of 13 of them has shown (Singh and Chowdhury, 1986) that over longer, say monthly or annual, periods they all produce comparable results; the shorter the time period, the more they can be expected to produce different results.

Objective analysis

The simple averaging methods just described are fairly arbitrary in their design and not based on well-defined criteria. However advances, in what is variously called *objective analysis* (see Gandin, 1963; Kagan, 1997) and *geostatistics* (Journel and Huijbregts, 1978; Delhomme, 1978; Kitanidis, 1997), have led to a more objective interpolation technique, namely the *best linear unbiased estimator*, also known as *kriging*. This is a weighting procedure, in which the weights are determined on the basis of the spatial structure of the rainfall fluctuations, and on the basis of the dual criteria that the estimation error, i.e. difference between the estimated value and the true but unknown value, at any point be zero on average, and that the corresponding mean square error be minimal. For details on the application of this method the reader is referred to the specialized literature.

Complex and mountainous terrain poses some additional challenges in the estimation of the precipitation distribution and its mapping. Attempts have been made to incorporate orographic effects in objective analysis in the derivation of areal averages (Chua and Bras, 1982; Phillips *et al.*, 1992). In a different approach, Daly *et al.* (1994) relied on digital elevation data, including height and aspect, to derive local regression equations with which to distribute point measurements of monthly and annual precipitation to regularly spaced grid cells.

Some distribution functions

The areal distribution of precipitation over different time scales has been the subject of intensive research and numerous relationships have been proposed in the literature (see, for example, Court, 1961; Burns, 1964; Huff, 1966; Fogel and Duckstein, 1969) to reduce the estimated point precipitation, when it is to represent a larger area. In some of these, correlation studies were carried out relating the decay with increasing distance from a gage at a central location (Huff and Shipp, 1969; Hutchinson, 1969). For design purposes for use with point rainfall frequency data (and not with individual storms), in the United States the reduction factors shown graphically in Figure 3.14, developed by the US Weather Bureau (1957–1960; Miller, 1963; Myers and Zehr, 1980), have been widely used. Various aspects of areal rainfall reduction procedures have subsequently been studied by Rodriguez-Iturbe and Mejia (1974), Eagleson *et al.* (1987), Smith and Karr (1990), Omolayo (1993), Sivapalan and Blöschl (1998), Asquith and Famiglietti (2000), DeMichele *et al.* (2001), and Allen and DeGaetano (2005) among others. Several of these investigations have shown that the reduction factor dependency on area is also a function of the severity, i.e. the return period of the event. However, it was also generally found that for several reasons the curves shown in Figure 3.14 are on the safe side, and therefore are likely to result in a more conservative design. Thus actual precipitation tends to decrease more rapidly with area covered, as the return period increases, than indicated in the figure; also, convective type storms exhibit a more rapid decrease with area than those of the stratiform type. Finally, although the reduction factors shown in the figure appear to level off at around 1000 km^2, it has been found that in fact the reduction factor continues to decrease in an exponential manner as the area increases beyond that, even up to $20\,000 \text{ km}^2$.

3.3.2   Temporal distribution

Precipitation is normally recorded on an hourly or daily basis, and in the data records it can be reported over different averaging periods. The description of the evolution of precipitation over time depends largely on the adopted temporal resolution. In applied hydrology, a record of precipitation intensity with time for individual storm events is commonly referred to as a *hyetograph*; a hyetograph is usually presented as a bar graph with an hourly time step. The accumulated precipitation over time, is called a *mass curve*.

A *double mass curve* is a graph of seasonal or annual accumulated precipitation at a given station plotted against the mean accumulated precipitation for a number of neighboring or surrounding stations. Double mass analysis was introduced by Merriam (1937) to check the consistency of the record at a station that has undergone changes

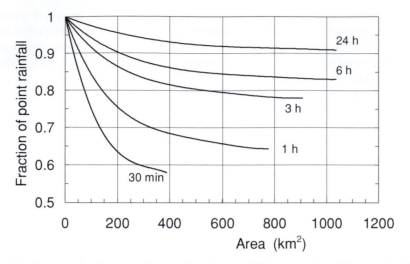

Fig. 3.14 Depth–area reduction curves for use with point rainfall frequency data. (After standard NOAA charts; Hershfield, 1961.)

Fig. 3.15 Example of a double mass analysis, showing the accumulated precipitation at a single station (Spencer, West Virginia) against the mean accumulated values of 13 neighboring stations in the same climatic division area. Exposure conditions at Spencer changed around 1964 (after Chang and Lee, 1974).

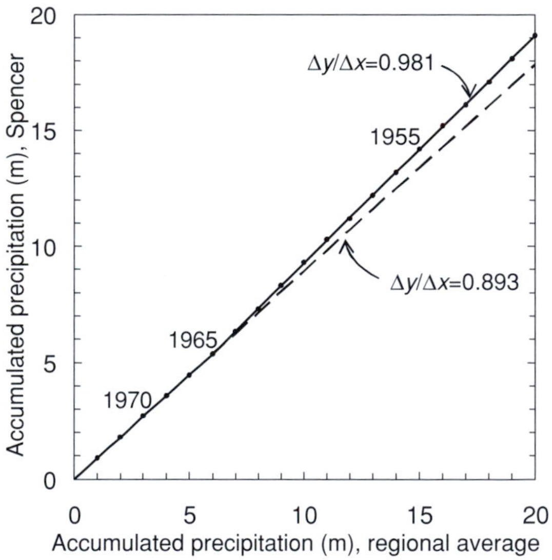

in exposure and location of the gage, in instrumentation or in measurement procedures (see also Chang and Lee, 1974).

Example 3.2. Double mass curve

Figure 3.15 shows accumulated annual precipitation at Spencer, West Virginia, against the mean accumulated precipitation values for 13 stations within the same climatic area in southwestern West Virginia. The figure illustrates that a change in measuring conditions occurred around 1964; the data prior to that date can be adjusted to these new conditions by multiplying them by a factor in accordance with the change in slope $\Delta y/\Delta x$, as shown.

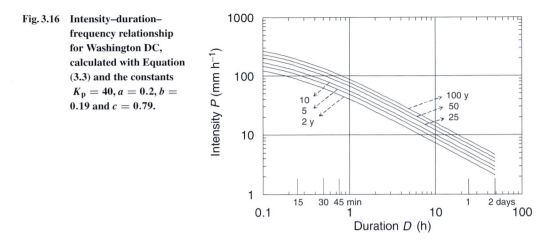

Fig. 3.16 Intensity–duration–frequency relationship for Washington DC, calculated with Equation (3.3) and the constants $K_p = 40$, $a = 0.2$, $b = 0.19$ and $c = 0.79$.

Double mass curve analysis can also be used to interpolate missing data (Paulhus and Kohler, 1952), and it has been applied to other types of hydrologic data, such as streamflow, sediment records and precipitation–runoff relations (Searcy and Hardison, 1960).

3.3.3   Runoff design rainfall data

For engineering design purposes, mainly related to the rational method or the unit hydrograph (see Chapter 12), point rainfall data are often organized according to the intensity, the duration and the frequency of the storm events. Such analyses have been published for many different locations. Hershfield (1961), Miller (1963) and Bruce (1968) have published maps covering North America. Similar maps have been produced for other regions of the world as well. Because rainfall-generating mechanisms can be quite similar in different hydrologic regions, many attempts have been made to generalize this type of information in the form of empirical functions. Some pertinent results can be found, for example, in the studies by Bell (1969), Chen (1983), Ferreri and Ferro (1990), Kothyari and Garde (1992), Ferro (1993), Pagliara and Viti (1993) and Alila (2000). This topic continues to be the subject of research (see, for example, Madsen *et al.*, 2002).

A widely used equation, whose evolution can be traced through the work of Meyer (1917, p. 149), Sherman (1931) and Bernard (1932), can be written as follows

$$P = K_p \frac{T_r^a}{(D + b)^c} \tag{3.3}$$

in which P is the intensity [L/T] of a rainfall episode of duration D [T] and with a return period T_r (see Chapter 13), and K_p, a, b, and c are constants for a given location.

Example 3.3. Intensity–duration–frequency relationship

Figure 3.16 shows an example of Equation (3.3) with the constants $K_p = 40$, $a = 0.2$, $b = 0.19$ h and $c = 0.79$ (with P in mm h^{-1}, D in h, and T_r in years) for

Table 3.2 Greatest known observed point rainfall

Duration	(mm)	Location	Date
1 min	38	Barot, Guadeloupe	Nov. 26, 1970
5 min	63	Haynes Camp, California	Feb. 2, 1976
8 min	126	Fussen, Bavaria	May 25, 1920
15 min	198	Plumb Point, Jamaica	May 12, 1916
20 min	206	Curtea-de-Arges, Roumania	Jul. 7, 1889
30 min	280	Sikeshugou, Hebei	Jul. 3, 1974
42 min	305	Holt, Missouri	Jun. 22, 1947
60 min	401	Shangdi, Inner Mongolia	Jul. 3, 1975
1 h 12 min	440	Gaoj, Gansu	Aug. 12, 1985
2 h 30 min	550	Bainaobao, Hebei	Jun. 25, 1972
2 h 45 min	559	D'Hanis, Texas (17 miles NNW)	May 31, 1935
3 h	600	Duan Jiazhuang, Hebei	Jun. 28, 1973
4 h 30 min	782	Smethport, Pennsylvania	Jul. 18, 1942
6 h	840	Muduocaidang, Inner Mongolia	Aug. 1, 1977
10 h	1400	Muduocaidang, Inner Mongolia	Aug. 1, 1977
18 h	1589	Foc Foc, Reunion	Jan. 7–8, 1966
24 h	1825	Foc Foc, Reunion	Jan. 7–8, 1966
2 d	2467	Aurere, Reunion	Apr. 8–10, 1958
3 d	3240	Grand Ilet, Reunion	Jan. 24–27, 1980
4 d	3721	Cherrapunji, Meghalaya	Sep. 12–15, 1974
5 d	3951	Commerson, Reunion	Jan. 23–27, 1980
7 d	4653	Commerson, Reunion	Jan. 21–27, 1980
10 d	5678	Commerson, Reunion	Jan. 18–27, 1980
15 d	6083	Commerson, Reunion	Jan. 14–28, 1980
31 d	9300	Cherrapunji, Meghalaya	Jul. 1–31, 1861
2 mon	12767	Cherrapunji, Meghalaya	Jun.–Jul. 1861
4 mon	18738	Cherrapunji, Meghalaya	Apr.–Jul. 1861
6 mon	22454	Cherrapunji, Meghalaya	Apr.–Sep. 1861
1 y	26461	Cherrapunji, Meghalaya	Aug. 1860–Jul. 1861
2 y	40768	Cherrapunji, Meghalaya	1860–1861

Source: World Meteorological Organization (1986).

Washington, DC; these values were obtained from the data presented in US Weather Bureau (1955). Interestingly, some 20 years earlier Bernard (1932) reported for Washington the values $K_p = 34.4$, $a = 0.2$, $b = 0$ and $c = 0.78$ for rain events longer than 1 h; this illustrates how P values, with a certain probability of occurrence, may increase as the available period of record becomes longer.

While the constants in (3.3) can be expected to change from place to place, the values reported in the literature for a, b and c vary within relatively narrow ranges, namely $0.15 < a < 0.3$, $5 < b < 10$ min and $0.6 < c < 0.8$. Thus when no other

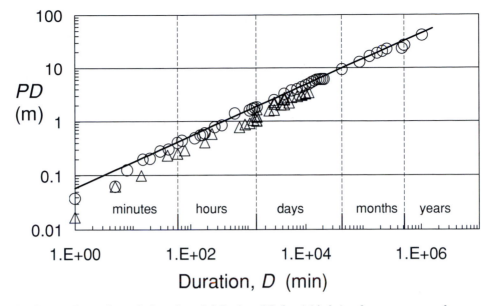

Fig. 3.17   The largest observed cumulative point rainfall values PD (in m) (circles) and some near-record values (triangles). Most of the largest values are also listed in Table 3.2. The upper envelope is given by $PD = 0.0584\,D^{0.48}$ (in m), with D in min (adapted from World Meteorological Organization, 1986).

information is available, for durations $D \leq 24$ h the typical values $a = 0.2$, $c = 0.7$ can be adopted; b is of the order of minutes only, so that it is usually omitted from (3.3) for durations D in excess of 1 or 2 h. Attempts have been made to relate K_p to climate indices. For instance, Kothyari and Garde (1992) showed that it can be related with the rainfall depth of $D = 24$ h and $T_r = 2$ y, namely $(24\,P_{24}^2)$, as follows

$$K_p = C_p \left(24\,P_{24}^2\right)^{0.33} \tag{3.4}$$

where C_p is a local constant, which covers the relatively narrow range $6 < C_p < 9$, if P is in mm h^{-1}, D in h, and T_r in years. When they applied Equation (3.3) with $a = 0.2$ and $c = 0.7$ and with $K_p = 40.1$ to the data from 78 stations in India, the resulting multiple correlation coefficient was 0.90; however, application of (3.3) with (3.4), in which $C_p = 8.31$, produced a much improved multiple correlation coefficient of 0.96.

As a reference, which may serve as guide for the maximum possible precipitation, Table 3.2 (from World Meteorological Organization, 1986) and Figure 3.17 show some of the largest point rainfall values ever observed for different durations. The upper envelope shown in Figure 3.17 is given by

$$P = 416.6\,D^{-0.52} \tag{3.5}$$

if the rainfall rate P in mm h^{-1} and the duration D is in hours (h).

3.4 Interception

3.4.1 Definition and observed magnitudes

Interception is the part of precipitation that moistens the different surface elements, mainly vegetation, and is temporarily stored on them. When the surface elements are fully saturated, so that they have reached their full *interception storage capacity*, any excess intercepted water on them flows or drips down to the ground. In practice, the interception storage capacity is usually defined more specifically as the amount of water left on the canopy at the end of a storm, under conditions of zero evaporation and after all drip has ceased; thus during a storm the stored depth of water can exceed the storage capacity. The precipitation that reaches the ground is often called *net precipitation*. In the case of vegetation, most of the net precipitation filters through the canopy as *throughfall*; a small part flows down along major branches and stems as *stemflow*, and tends to concentrate over the roots. The interception of precipitation by a vegetation canopy can greatly affect the hydrologic budget at the ground surface. The water held by the foliage elements that evaporates before it can reach the ground is thus no longer available for infiltration and runoff. Therefore, the amount of intercepted precipitation, that returns to the atmosphere by evaporation, is often called the *interception loss*.

Most interception studies have focused on forested surfaces, where the largest values occur. Both the type of vegetation and the type of precipitation appear to play a role. Indeed, tall or dense vegetation tends to incur larger interception losses than short or sparse vegetation. Also, interception losses as a fraction of precipitation are usually larger when the precipitation events are of moderate intensity and longer duration, than when they occur in the form of short intense bursts and downpours. For example, in tall dense forest vegetation at temperate latitudes interception losses have been observed that are as large as 30% to 40% of the gross precipitation (Gash *et al.*, 1980). In tropical forests with high intensity rainfall, however, the observed losses have tended to be more of the order of 10% to 15% (see Lloyd and Marques, 1988; Lloyd *et al.*, 1988; Ubarana, 1996), even though the evaporation rates during rainfall were not very different. Similarly, sparse forests also tend to have lower values of interception, namely around 10% to 20% of the precipitation (see Gash *et al.*, 1995; Valente *et al.*, 1997). Interception losses in heather and shrub covered terrain are smaller than one third of the values in dense forest (Calder, 1990) under the same climatic conditions.

3.4.2 Interception loss mechanisms in vegetation

For a single precipitation event the total interception loss is the sum of the evaporation from the wet vegetation during the event and the evaporation of the water remaining on the vegetation after the precipitation has ceased. Horton (1919) was probably the first to formulate this, for a storm duration D, which is long enough to saturate the vegetation, as follows

$$L_\mathrm{i} = \int_0^D E_\mathrm{i} dt + S_\mathrm{ic} \tag{3.6}$$

where E_i is the rate of evaporation of intercepted water [L/T], and S_ic the interception storage capacity of the vegetation [L]. When the precipitation ceases before the vegetation

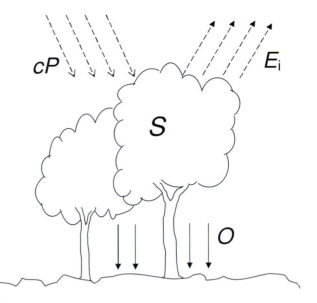

Fig. 3.18 Sketch illustrating the water balance of a vegetation canopy.

is fully saturated, the precipitation amount lost by interception is

$$
L_i = \int_0^D E_i \, dt + S \tag{3.7}
$$

in which S is the amount of water stored on the partly saturated vegetation. Before this equation can be used to estimate the loss, the variables S and E_i must be known.

A common way to determine S is to treat the vegetation as one or more storage elements, representing the canopy and the trunks, to which the lumped storage equation (1.10) can be applied. In the simplest approach, when a single element is assumed to represent the evolving canopy storage S, one can take the precipitation as the inflow, and the evaporation and liquid drainage as the two outflow rates (Figure 3.18); thus,

$$
\frac{dS}{dt} = cP - E_i - O \qquad \text{for } 0 \le S \le S_{ic} \tag{3.8}
$$

where c is the horizontal density or fractional cover of the intercepting vegetation, P the precipitation intensity, and O the liquid drainage outflow rate from the vegetation. Note that Equation (3.8) should be applied only to stands of vegetation that are sufficiently uniform at the scales under consideration; thus it would not be applicable, for example, in the case of chessboard-like surfaces consisting of forest stands and clearings with different vegetation, or to sparse stands of trees in a grassy or bare soil environment. In such situations the analysis may have to be applied separately to each type of land cover and the results can then be weighted according to the fraction of the area each one occupies.

Integration of Equation (3.8) yields for the storage

$$
S = c \int_0^D P \, dt - \int_0^D E_i \, dt - \int_0^D O \, dt \qquad \text{for } S \le S_{ic} \tag{3.9}
$$

Hence for short precipitation events that are not long enough to saturate the vegetation, Equation (3.7) produces a loss

$$L_i = c \int_0^D P \, dt - \int_0^D O \, dt \qquad \text{for } S \le S_{ic} \tag{3.10}$$

Equation (3.10) is valid as long as the cumulative precipitation is smaller than the amount needed to saturate the vegetation, and (3.6) is valid after that, when $S = S_{ic}$.

To allow the practical implementation of Equations (3.6) and (3.10), various assumptions have been proposed by different authors, regarding c, E_i, and O. The main difficulties in assessing these assumptions are the complexity of the vegetation cover precluding more thorough analysis and the absence of experimental support for most of the processes involved in interception. Some of these assumptions are briefly discussed in the following.

Some common assumptions

The fractional vegetation cover c is often assumed to be simply related with the free through-fall coefficient p, as $c = (1 - p)$; this coefficient is the portion of the precipitation that reaches the ground without hitting the canopy (Gash and Morton, 1978). Both c and p can be measured (see Section 3.4.3). The drainage rate O has been estimated in various ways. The simplest way to describe it is with the assumption that as long as the canopy is partly saturated there is no drip, and that once it is saturated at the end of the storm the amount of water on the canopy rapidly falls to its storage capacity S_{ic} (Gash, 1979; Noilhan and Planton, 1989). These can be written as

$$O = 0 \qquad \text{for } S < S_{ic} \tag{3.11}$$

and, from Equation (3.8)

$$cP - E_i - O = 0 \qquad \text{for } S = S_{ic} \tag{3.12}$$

The rate of evaporation from the intercepting vegetation E_i is the most critical but also the most difficult variable to determine. For operational purposes, it is now commonly (Noilhan and Planton, 1989; Gash *et al.*, 1995) assumed that, when the vegetation is saturated, it can be estimated by means of a suitably chosen potential evaporation E_{po} (see Chapter 4) from the fraction c of the surface occupied by intercepting vegetation and that the evaporation from the remaining fraction $(1 - c)$ can be ignored. For partly saturated surfaces during the wetting up phase of the interception process, it has mostly (see Rutter *et al.*, 1971) been assumed that the evaporation is proportional to the relative saturation (S/S_{ic}). (This assumption is an application of Equation (4.33) with (4.34).) Both assumptions can be combined as

$$E_i = c(S/S_{ic})E_{po} \tag{3.13}$$

The main problem with these underlying assumptions is that it is still not very clear exactly how this potential evaporation should be defined or estimated; this issue will require further study. In any event, these assumptions lead now to the following expressions for the interception loss. If t_0 denotes the time to saturation, the loss for short precipitation events follows directly from (3.10) with (3.11), namely

$$L_i = c \int_0^D P \, dt \qquad \text{for } S \leq S_{ic} \text{ and } D \leq t_0 \tag{3.14}$$

For long events Equation (3.6) can be rewritten as

$$L_i = \int_0^{t_0} E_i \, dt + S_{ic} + \int_{t_0}^D E_i \, dt \qquad \text{for } S = S_{ic} \text{ and } D > t_0 \tag{3.15}$$

or, upon substitution of the first two terms on the right by (3.9) with (3.11), and of the third by (3.13),

$$L_i = c \left(\int_0^{t_0} P \, dt + \int_{t_0}^D E_{po} \, dt \right) \qquad \text{for } S = S_{ic} \text{ and } D > t_0 \tag{3.16}$$

Equations (3.14) and (3.16) can be readily solved numerically by also keeping track of S by means of (3.9).

Lumped kinematic solution

The assumption that the vegetation, as a hydrologic flow system, can be represented by a storage element governed by Equation (3.8) (with $O = 0$), and with a storage–outflow relationship given by Equation (3.13), is a perfect example of the lumped kinematic approach. Gash (1979; Gash *et al.*, 1995) made use of this simple structure to derive a closed form solution for the evolution of S with time; by assuming constant (or averaged) values of P and E_{po} during the precipitation event of duration D, he obtained

$$D = \frac{S_{ic}}{c E_{po}} \ln \left(1 - \frac{E_{po} S}{P S_{ic}} \right) \tag{3.17}$$

The time to saturation is therefore

$$t_0 = (S_{ic}/c E_{po}) \ln \left[1 - (E_{po}/P) \right] \tag{3.18}$$

which can be used immediately with (3.14) and (3.16) to estimate the interception loss. For constant (or averaged) values of P and E_{po} during the precipitation event, (3.14) and (3.16) can be written simply as

$$L_i = c P D \qquad \text{for } S \leq S_{ic} \text{ and } D \leq t_0 \tag{3.19}$$

and

$$L_i = c \left[P t_0 + (D - t_0) E_{po} \right] \qquad \text{for } S = S_{ic} \text{ and } D > t_0 \tag{3.20}$$

where (PD) is the cumulative precipitation at the end of the precipitation event, and $(P t_0)$ is the cumulative precipitation needed to saturate the vegetation.

Stemflow and interception loss from the trunks

In several past analyses of interception the water balance of the trunks and stems has been treated separately from that of the leaves (Rutter *et al.*, 1975; Gash *et al.*, 1995). Since the evaporation from the trunks is usually very small compared with the evaporation from the canopy, the resulting losses consist mainly of the evaporation of the water remaining on the trunks after the end of stemflow. Thus, for precipitation events long enough to saturate the trunk storage, the total loss is equal to the maximal trunk storage S_{tic}. When the precipitation does not quite saturate the trunk storage, by analogy with (3.14) or (3.19), the loss may be taken as ($p_t \, PD$), where p_t is the proportion of the precipitation that is diverted to stemflow.

In most situations, however, these losses are considerably smaller than those from the leaves of the canopy. For example, for pine forests in Great Britain (Gash, 1979; Gash *et al.*, 1980), the trunk losses were found to be about 2% to 9% of the total interception loss; for the Amazonian rain forest a value of about 9% was observed (Lloyd *et al.*, 1988).

3.4.3 *Experimental determination of the vegetation structure parameters*

The main surface parameters controlling the interception loss are S_{ic} and c; the surface roughness z_0 probably plays only a minor role through its effect on the evaporation rate. The storage capacity S_{ic} is usually estimated with a procedure proposed by Leyton *et al.* (1967; see also Gash and Morton, 1978). The method is based on the observation that, as indicated in Equation (3.6), the loss is equal to the canopy storage when evaporation is equal to zero. As before, let P represent the average rainfall rate during an event of duration D. Then, in a plot of the gross, i.e. total precipitation (PD), versus the net precipitation, i.e. throughfall [$(1 - c)PD$], for a number of observed precipitation events, S_{ic} can be taken as the intercept of the lower envelope with a slope of unity; the lower envelope represents the events with minimal E_i, so that $PD = (1 - c)PD + S_{ic}$. The data points must be taken from events of sufficiently long duration, to ensure that the canopy is fully saturated. This is illustrated in Figure 3.19. Observe, however, that the vertical axis should represent $(1 - p_t)PD$, instead of PD, to account for stemflow, but the difference is usually small and can be neglected. The free throughfall coefficient p can be determined from throughfall measurements for small storms insufficient to saturate the canopy (Gash and Morton, 1978) and the canopy cover can then be obtained by assuming $c = (1 - p)$.

Typical values of the specific canopy storage capacity (S_{ic}/c) (i.e. the storage capacity per unit area of cover) and of c are respectively, 0.8–1.2 mm and 0.68–1.00 for dense pine forest (Gash and Morton, 1978; Gash *et al.*, 1980), around 0.56 mm and 0.45 for sparse pine forest (Gash *et al.*, 1995), 0.8 mm and 0.92 for Amazonian rain forest (Lloyd *et al.*, 1988), and 0.64 mm and 0.64 for sparse pine forest and 0.35 mm and 0.60 for eucalyptus forest (Valente *et al.*, 1997). For grasses ranging in height between 0.1 m and 0.5 m, (S_{ic}/c) values ranging between 0.43 and 2.8 mm have been reported (Merriam, 1961).

The stemflow parameters S_{tic} and p_t can be determined as, respectively, the mean slope and the intercept of the regression of stemflow versus precipitation for each tree on which measurements are made (see, for example, Gash and Morton, 1978). Typical values for S_{tic} and p_t are, respectively, 0.014–0.74 mm and 0.016–0.29 for dense pine forest, 0.17 mm and 0.0275 for sparse pine forest, 0.15 mm and 0.036 for Amazonian forest, 0.019 mm and 0.0038 for Mediterranean sparse pine forest and 0.027 mm and 0.017 for eucalyptus forest.

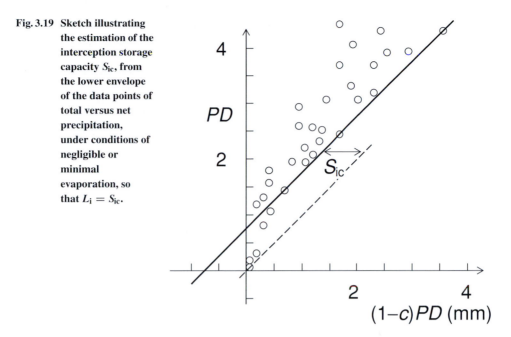

Fig. 3.19 Sketch illustrating the estimation of the interception storage capacity S_{ic}, from the lower envelope of the data points of total versus net precipitation, under conditions of negligible or minimal evaporation, so that $L_i = S_{ic}$.

3.4.4 Empirical equations

In the past many attempts have been made to relate interception empirically to the cumulative precipitation (PD) by linear regression equations (Helvey and Patric, 1965; Jackson, 1975), which can produce useful results in certain applications (Gash, 1979). Comparison of Equation (3.19) (or (3.14)) with (3.20) (or (3.16)) suggests that this may be a reliable approach for the early stages of a storm, but not later on, after the canopy is saturated. Once the canopy is saturated, Equation (3.20) indicates that the duration of the storm may be a better predictor than the amount of precipitation. The issue has been discussed by Horton (1919), who felt that expressing interception in terms of shower duration would be more logical than in terms of amount of precipitation; nevertheless he proposed a linear regression equation in terms of cumulative precipitation, after he found that this is close to linearly related with shower duration.

As a rough estimate for interception over longer time periods, Equation (3.6) suggests that the following may be useful with standard hourly rainfall data

$$L_i = n(S_{ic} + c\overline{E_{po}D}) \tag{3.21}$$

where $\overline{D}$ is the average duration of the n precipitation events during the period, and $\overline{E_{po}}$ the average rate of evaporation from a wet surface during the same events. Detailed calculations with more complex formulations have shown that in the growing season on average E_{po} is relatively invariant over a wide range of climatic conditions (Gash et al., 1980; Lloyd et al., 1988; Valente et al., 1997) and that good results can be obtained with values mostly around 0.2 mm h^{-1} and ranging only between 0.15 and 0.30 mm h^{-1}. For c and S_{ic} the typical values can be used that are mentioned above. The vegetation cover fraction c is usually

Table 3.3 Effective film thickness f_t for different plants (kg m^{-2} or mm)

Species	Thickness
Big bluegrass	0.203
Slough grass	0.102
Monterey pine	0.0762
Baccharis pilularis (Coyote Brush, evergreen ground-cover shrub)	0.1778
Chaparral (dense thicket, California)	0.152
Annual ryegrass	0.127

between 0.6 and 1.0; a rough estimate of S_{ic} can be obtained from

$$S_{ic} = c \, f_t \text{La} \tag{3.22}$$

where La is the leaf area index, which is the area (one side) of foliage per unit area of ground surface, and f_t is the maximum storage of water per unit area of foliage. Table 3.3 shows a few values of f_t for different plant species, collected by Merriam (1961); these values suggest that 0.2 kg m^{-2} can be taken as an upper value. Table 2.9 shows values of the leaf area index La for a number of plant communities.

   Thus, with these assumptions, Equation (3.21) can also be written in terms of the leaf area index as

$$L_i = 0.2 \, nc \, (\text{La} + \overline{D}) \tag{3.23}$$

3.5 RELIABILITY OF OPERATIONAL PRECIPITATION MEASUREMENTS

Precipitation was probably the first hydrologic variable to be measured regularly on a routine basis, and in many places in the world such measurements started more than a century ago. Thus for a variety of purposes in hydrology, the availability of this historic data base presents a useful opportunity. In principle the measurement of precipitation should be a simple matter. It is important to be aware, however, that most of the available records of precipitation from the past suffer from substantial systematic error and that caution is required in their use. This has, of course, been known for a long time (see, for example, Larson and Peck, 1974; McGuiness and Vaughan, 1969; Neff, 1977; Golubev et al., 1992; Duchon and Essenberg, 2001), but it is only in relatively recent years that steps have been taken to remedy the situation. Although much remains to be done to solve the archival precipitation data correction problem (Groisman and Legates, 1994), a better understanding is gradually emerging.

   Standard precipitation gages are usually placed with their orifice at some height above the ground (from 0.5 m on up, depending on the type), primarily for convenience and to avoid raindrop splash and snow drift. Thus one of the main factors is the distortion of the wind field by the presence of the precipitation gage as an obstacle, which results in a wind

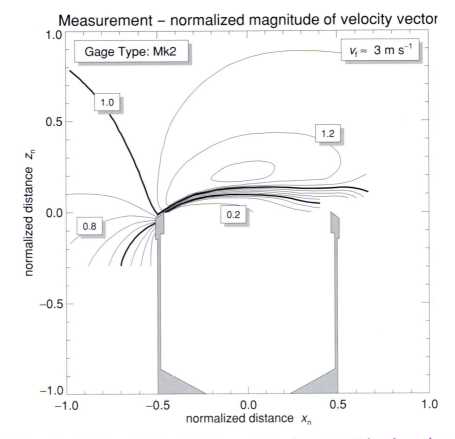

Fig. 3.20 Normalized velocity contour lines, derived from wind-tunnel measurements in and around a standard British Mk2 precipitation gage, in its vertical plane of symmetry parallel to the flow. The reference air velocity in the tunnel was 3 m s^{-1} from left to right; the dimensions are normalized with the outer diameter of the gage, which is 136.6 mm. The wind velocities above the gage orifice can be seen to be about 35% higher than the free-stream velocity. (From Nespor and Sevruk, 1999.)

speed increase above its orifice and the development of wake eddies around it. This, in turn, tends to carry the finer precipitation particles over the orifice, thus decreasing their number entering the gage. This effect increases with height of the orifice. Therefore, it can be expected that the discrepancy between actual and measured precipitation will increase with increasing wind speed, with decreasing precipitation intensity and with increasing height of the gage orifice above ground level. Figure 3.20 illustrates the distortion of the wind field above a rain gage, as derived from wind-tunnel measurements by Nespor (1993; see also Nespor *et al.*, 1994; Nespor and Sevruk, 1999); it can be seen how in this case the velocity of the air above the gage is about 20% to 30% higher than in the approaching undisturbed wind field. The losses in precipitation due to wind range on average between 2% and 10% for rain, and between 20% and 50% for snow; however, they may be much larger during individual precipitation events. Typical catch deficiencies in relation to wind as the only factor are shown in Figure 3.21. A more detailed analysis

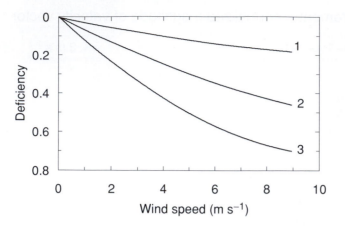

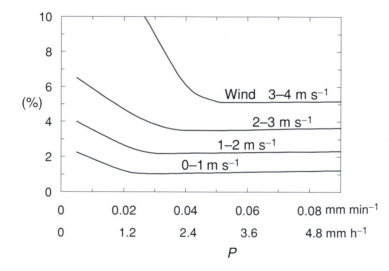

Fig. 3.21 Curves of gage catch deficiency against wind speed for liquid precipitation (curve 1); for solid precipitation with a single shield around the gage (curve 2); and for solid precipitation with an unshielded gage (curve 3). The curves represent a summary of data from different sites in the United States, Russia and England, collected by Larson and Peck (1974). For rain the catches of shielded and unshielded gages were nearly the same.

Fig. 3.22 Class averaged percentage difference (%) between precipitation measurements by elevated (at 1 and 1.5 m) Hellmann gages, and a ground-level gage, observed by Sevruk (1993a) as a function of mean precipitation intensity P for different wind speeds. Based on data obtained at Les Avants, Switzerland, April–September, 1938–1947.

of rainfall data by Sevruk (1993a) allowed him to derive the average wind induced error as a function of rainfall intensity, as shown in Figure 3.22; these observations indicate that the error increases rapidly with wind speed for small intensities, but more slowly for larger intensities.

Different measurement techniques have been tried to solve this problem (see Rodda, 1967; Robinson and Rodda, 1969; Sevruk, 1974). The best is probably to use pit gages

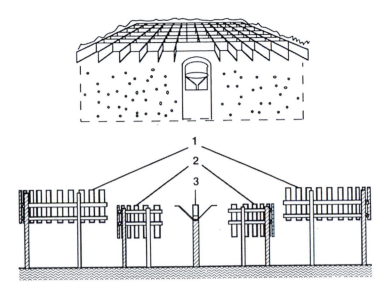

Fig. 3.23 Cut-away sketch of the two reference standard precipitation gages recommended by the World Meteorological Organization (WMO) for calibration purposes. The pit gage, which is surrounded by an anti-splash grid, is for rain; the shielded gage (3), which is surrounded by two fences of 4 and 12 m diameter (2 and 1), is for snow; the heights of the fences are 3 and 3.5 m. (From Sevruk, 1993b.)

(Duchon and Essenberg, 2001) with their orifice at ground level. Wind shields are also being used, but most of the designs can only alleviate the problem and never eliminate it; while they are useful in reducing the catch deficiency for snow, they appear to have much less of an effect for rain. For the purpose of calibration of precipitation gages of different types of design, the World Meteorological Organization (WMO) has recommended the use of two reference standard gages, which have been found to have negligible wind error (Sevruk, 1993b) and which are shown schematically in Figure 3.23. The reference gage for rain consists of a Mk2 gage (British Meteorological Office), installed in a pit with the orifice flush at ground level and surrounded by a grid to avoid splash. The reference gage for snow is a Tretyakov gage (Russian Meteorological Services) with a shield and surrounded by two octagonal lath-fences, respectively 3 m high and 4 m diameter, and 3.5 m high and 12 m diameter.

Additional losses may be caused by initial wetting of the gage (i.e. interception), evaporation, and by the mechanisms of recording gages. Methods have been developed to correct for these systematic errors of undercatch (see, for example, Legates and DeLiberty, 1993; Sevruk, 1993a; 1996; Sevruk and Nespor, 1998; Yang et al., 1998; Nespor and Sevruk, 1999). These studies have led to the consensus that beside mean wind speed, also the rainfall rate and the drop size distribution should be considered in applying corrections to the measured data. Some of these effects are illustrated in Figures 3.22 and 3.24. A study by Habib et al. (1999) showed that the averaging time should be considered in the error estimation; an hourly time scale or smaller was recommended.

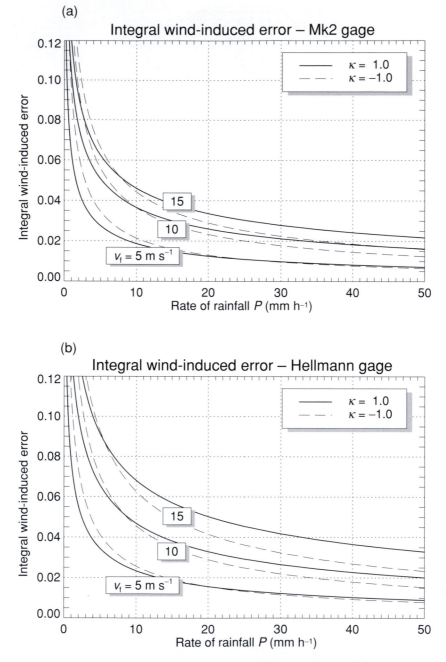

Fig. 3.24 Example of calculated wind-induced error for the (a) Mk2, (b) Hellmann, and (c) the tipping bucket ASta (Automatic Station) precipitation gages, as a function of the rate of rainfall P in mm h^{-1}. The error was calculated for three wind speeds, as indicated, and for two types of raindrop size distributions, namely orographic ($\kappa = -1$), and thunderstorm ($\kappa = 1$). (From Nespor and Sevruk, 1999.)

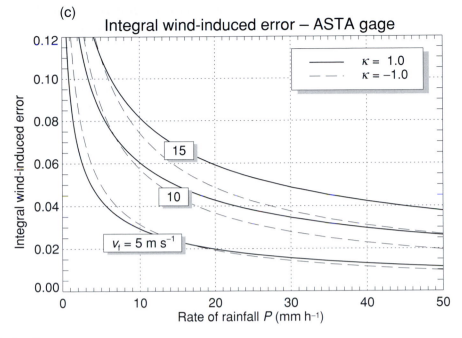

Fig. 3.24   *(cont.)*

REFERENCES

Alila, Y. (2000). Regional rainfall depth-duration-frequency equations for Canada. *Water Resour. Res.*, **36**, 1767–1778.

Allen, R. J. and DeGaetano, A. T. (2005). Areal reduction factors for two eastern U.S. regions with high rain gauge density. *J. Hydrol. Eng. (ASCE)*, **10**.

Asquith, W. H. and Famiglietti, J. S. (2000). Precipitation areal-reduction factor estimation using an annual-maxima centered approach. *J. Hydrol.*, **230**, 55–69.

Bell, F. C. (1969). Generalized rainfall-duration-frequency relationships. *J. Hydraul. Div., Proc. ASCE.*, **95** (HY1), 311–327.

Benton, G. S., Blackburn, R. T. and Snead, V. O. (1950). The role of the atmosphere in the hydrologic cycle. *Trans. Amer. Geophys. Un.*, **31**, 61–73.

Bernard, M. M. (1932). Formulas for rainfall intensities of long duration. *Trans. Amer. Soc. Civil Eng.*, **96**, 592–606, 617–624.

Blanchard, D. C. (1972). Bentley and Lenard: pioneers in cloud physics. *Amer. Scientist*, **60**, 746–749.

Brubaker, K. L., Entekhabi, D. and Eagleson, P. S. (1993). Estimation of continental precipitation recycling. *J. Clim.*, **6**, 1077–1089.

Bruce, J. P. (1968). *Atlas of rainfall intensity-duration frequency data for Canada*. Climatol. Studies No. 8. Toronto: Met. Branch, Dept. of Transport.

Budyko, M. I. (1974). *Climate and Life*. New York: Academic Press.

Burns, F. (1964). The relationship between point and areal rainfall in prolonged heavy rain. *Met. Mag., London*, **98**, 289–293.

Calder, I. R. (1990). *Evaporation in the Uplands*. Chichester: John Wiley.

Chang, M. and Lee, R. (1974). Objective double-mass analysis. *Water Resour. Res.*, **10**, 1123–1126.

Chen, C. L. (1983). Rainfall intensity-duration-frequency formulas. *J. Hydraul. Eng.*, **109**, 1603–1621.

Chua, S.-H. and Bras, R. L. (1982). Optimal estimators of mean areal precipitation in regions of orographic influence. *J. Hydrol.*, **57**, 23–48.

Court, A. (1961). Areal-depth rainfall formulae. *J. Geophys. Res.*, **66**, 1823–1831.

Daly, C., Neilson, R. P. and Phillips, D. L. (1994). A statistical-topographic model for mapping climatological precipitation over mountainous terrain. *J. App. Met.*, **33**, 140–158.

Dean, J. D. and Snyder, W. M. (1977). Temporally and areally distributed rainfall. *J. Irr. and Drain. Div., Proc. ASCE.*, **103** (IR2), 221–229.

Delhomme, J. P. (1978). Kriging in the hydrosciences. *Adv. Water Resour.*, **1**, 251–266.

DeMichele, C., Kottegoda, N. T. and Rosso, E. (2001). The derivation of areal reduction factor of storm rainfall from its scaling properties. *Water Resour. Res.*, **37**, 3247–3252.

Duchon, C. E. and Essenberg, G. R. (2001). Comparative rainfall observations from pit and aboveground rain gauges with and without wind shields. *Water Resour. Res.*, **37**, 3253–3263.

Eagleson, P. S., Fennesey, N. M., Qinliang, W. and Rodriguez-Iturbe, I. (1987). Application of spatial Poisson models to air mass thunderstorm rainfall. *J. Geophys. Res.*, **92** (D8), 9661–9678.

Eltahir, E. A. B. and Bras, R. L. (1994). Precipitation recycling in the Amazon basin. *Quart. J. Roy. Met. Soc.*, **120**, 861–880.

    (1996). Precipitation recycling. *Rev. Geophysics*, **34**, 367–378.

Ferreri, G. B. and Ferro, V. (1990). Short-duration rainfalls in Sicily. *J. Hydraul. Eng.*, **116**(3), 430–435.

Ferro, V. (1993). Discussion of "Rainfall intensity–duration–frequency formula for India". *J. Hydraul. Eng.*, **119**(8), 960–962.

Fleagle, R. G. and Businger, J. A. (1963). *An Introduction to Atmospheric Physics*. NY: Academic Press.

Fogel, M. M. and Duckstein, L. (1969). Point rainfall frequencies in convective storms. *Water Resour. Res.*, **5**, 1229–1237.

Fritsch, J. M., Kane, R. J. and Chelius, C. R. (1986). The contribution of mesoscale convective weather systems to the warm season precipitation in the United States. *J. Clim. Appl. Met.*, **25**, 1333–1345.

Fujita, T. (1955). Results of detailed synoptic studies of squall lines. *Tellus*, **7**, 405–436.

Gandin, L. (1963). *Objective Analysis of Meteorological Fields*. Leningrad: Gidrometeorologichoskoe Isdatel'stvo. English translation (1965), Jerusalem: Israel Program for Scientific Translation.

Gash, J. H. C. (1979). An analytical model of rainfall interception by forests. *Quart. J. R. Met. Soc.*, **105**, 43–55.

Gash, J. H. C. and Morton, A. J. (1978). An application of the Rutter model to the estimation of the interception loss from Thetford Forest. *J. Hydrol.*, **38**, 49–58.

Gash, J. H. C., Wright, I. R. and Lloyd, C. R. (1980). Comparative estimates of interception loss from three coniferous forests in Great Britain. *J. Hydrol.*, **48**, 89–105.

Gash, J. H. C., Lloyd, C. R. and Lachaud, G. (1995). Estimating sparse forest rainfall interception with an analytical model. *J. Hydrol.*, **170**, 79–86.

Gilman, C. S. (1964). Rainfall. Section 9. In *Handbook of Applied Hydrology*, ed. V. T. Chow, Section 9. New York: McGraw-Hill Book Co., pp. 9.1–9.68.

Goff, R. C. (1976). Vertical structure of thunderstorm outflows, *Mon. Weather Rev.*, **104**, 1429–1440.

Golubev, V. V., Groisman, P. Ya. and Quayle, R. G. (1992). An evaluation of the U.S. standard 8-inch nonrecording rain gage at the Valdai polygon, USSR. *J. Atmos. Oceanic Technol.*, **49**, 624–629.

Gong, C. and Eltahir, E. (1996). Sources of moisture in West Africa. *Water Resour. Res.*, **32**, 3115–3121.

Groisman, P. Ya. and Legates, D. R. (1994). The accuracy of United States precipitation data. *Bull. Amer. Met. Soc.*, **75**, 215–227.

Habib, E., Krajewski, W. F., Nespor, V. and Kruger, A. (1999). Numerical simulation studies of rain gage data correction due to wind effect. *J. Geophys. Res.*, **104** (D16), 19 723–19 733.

Helvey, J. D. and Patric, J. H. (1965). Canopy and litter interception of rainfall by hardwoods of the eastern United States. *Water Resour. Res.*, **1**, 193–206.

Hershfield, D. M. (1961). Rainfall frequency atlas of the United States, for durations from 30 minutes to 24 hours and return periods from 1 to 100 years, Tech. Paper No. 40. Washington, DC: Weather Bureau, US Dept Commerce.

Horton, R. E. (1919). Rainfall interception. *Mon. Wea. Rev.*, **47**, 603–623.

Houze, R. A., Rutledge, S. A., Biggerstaff, M. I. and Smull, B. F. (1989). Interpretation of Doppler weather radar displays of midlatitude mesoscale convective systems. *Bull. Amer. Met. Soc.*, **70**, 608–619.

Huff, F. A. (1966). Rainfall gradients in warm seasonal rainfall. *J. App. Met.*, **5**, 437–453.

Huff, F. A. and Shipp, W. L. (1969). Spatial correlations of storm, monthly and seasonal precipitation. *J. App. Met.*, **8**, 542–550.

Hutchinson, P. (1969). Estimation of rainfall in sparsely gauged areas. *Bull. Int. Ass. Sci. Hydrology*, **14**, 101–199.

Jackson, I. J. (1975). Relationships between rainfall parameters and interception by tropical forest. *J. Hydrol.*, **24**, 215–238.

Johnson, R. H. (2001). Surface mesohighs and mesolows. *Bull. Amer. Met. Soc.*, **82**, 13–31.

Journel, A. G. and Huijbregts, C. J. (1978). *Mining Geostatistics.* San Diego, CA: Academic Press.

Judson, A. and Doesken, N. (2000). Density of freshly fallen snow in the Central Rocky Mountains. *Bull. Amer. Met. Soc.*, **81**, 1577–1587.

Kagan, R. L. (1997). *Averaging of Meteorological Fields.* Dordrecht: Kluwer Academic Publishers.

Kitanidis, P. K. (1997). *Introduction to Geostatistics, Applications to Hydrogeology.* Cambridge: Cambridge University Press.

Kothyari, U. C. and Garde, R. J. (1992). Rainfall intensity–duration–frequency formula for India. *J. Hydraul. Eng.*, **118**(2), 323–336.

Larson, L. W. and Peck, E. L. (1974). Accuracy of precipitation measurements for hydrologic modeling. *Water Resour. Res.*, **10**, 857–863.

Legates, D. R. and DeLiberty, T. L. (1993). Precipitation measurement biases in the United States. *Water Resour. Bull.*, **29**, 855–861.

Leyton, L. E., Reynolds, R. C. and Thompson, F. B. (1967). Rainfall interception in forest and moorland. *Int. Symp. on Forest Hydrology*, ed. W. E. Sopper and H. W. Lull. Oxford: Pergamon Press, pp. 163–178.

List, R. J. (1971). *Smithsonian Meteorological Tables*, sixth edition, fifth reprint. City of Washington: Smithsonian Institution Press.

Lloyd, C. R. and Marques, A. de O. (1988). Spatial variability of throughfall and stemflow measurements in Amazonian rainforest. *Agric. Forest Met.*, **42**, 63–73.

Lloyd, C. R., Gash, J. H. C., Shuttleworth, W. J. and Marques, A. de O. (1988). The measurement and modelling of rainfall interception by Amazonian rain forest. *Agric. Forest Met.*, **43**, 277–294.

Maddox, R. A. (1980). Mesoscale convective complexes. *Bull. Amer. Met. Soc.*, **61**, 1374–1387.

Madsen, H., Mikkelsen, P. S., Rosbjerg, D., and Harremoes, P. (2002). Regional estimation of rainfall intensity-duration-frequency curves using generalized least squares regression of partial duration series statistics. *Water Resour. Res.*, **38**(11), 1239; doi:10.1029/2001WR001125.

McGuiness, J. L. and Vaughan, G. W. (1969). Seasonal variation in rain gauge catch. *Water Resour. Res.*, **5**, 1142–1146.

Merriam, C. F. (1937). A comprehensive study of the rainfall on the Susquehanna Valley. *Trans. Amer. Geophys. Un.*, **18**, 471–476.

Merriam, R. A. (1961). Surface water storage on annual ryegrass. *J. Geophys. Res.*, **66**, 1833–1838.

Meyer, A. F. (1917). *The Elements of Hydrology.* New York: John Wiley & Sons, Inc.

Miller, J. F. (1963). Probable maximum precipitation and rainfall-frequency data for Alaska, Tech. Paper No. 47. Washington, DC: Weather Bureau, US Dept. Commerce.

Myers, V. A. and Zehr, R. M. (1980). A methodology for point-to-area rainfall frequency ratios, NOAA Tech. Report NWS 24. Washington, DC: Nat. Weather Service, Nat. Oc. Atmos. Admin., US Dept. Commerce.

Neff, E. L. (1977). How much rain does a rain gage gage? *J. Hydrol.*, **35**, 213–220.

Nespor, V. (1993). Comparison of measurements and flow simulation: The Mk2 precipitation gauge. In *Aktuelle Aspekte in der Hydrologie/Current Issues in Hydrology; Festschrift zum 60. Geburtstag von H. Lang*, ed. D. Grebner, no 53. Zurich: Zuercher Geographische Schriften, Swiss Federal Institute of Technology, pp. 114–119.

Nespor, V. and Sevruk, B. (1999). Estimation of wind-induced error of rainfall gauge measurements using a numerical simulation. *J. Atmos. Oceanic Technol.*, **16**, 450–464.

Nespor, V., Sevruk, B., Spiess, R. and Hertig, J.-A. (1994). Modelling of wind-tunnel measurements of precipitation gauges. *Atmos. Environ.*, **28**, 1945–1949.

Noilhan, J. and Planton, S. (1989). A simple parameterization of land surface processes for meteorological models. *Mon. Wea. Rev.*, **117**, 536–549.

Omolayo, A. S. (1993). On the transposition of areal reduction factors for rainfall frequency estimation. *J. Hydrol.*, **145**, 191–205.

Pagliara, S. and Viti, C. (1993). Discussion of "Rainfall intensity-duration-frequency formula for India". *J. Hydraul. Eng.*, **119**(8), 962–966.

Paulhus, J. L. H. and Kohler, M. A. (1952). Interpolation of missing precipitation records. *Mon. Weath. Rev.*, **80**, 129–133.

Peck, E. L. and Brown, M. J. (1962). An approach to the development of isohyetal maps for mountainous areas. *J. Geophys. Res.*, **67**, 681–693.

Phillips, D. L., Dolph, J. and Marks, D. (1992). A comparison of geostatistical procedures for spatial analysis of precipitation in mountainous terrain. *Ag. For. Met.*, **58**, 119–141.

Reed, W. G. and Kincer, J. B. (1917). The preparation of precipitation charts. *Mon. Wea. Rev.*, **45**, 233–235.

Robinson, A. C. and Rodda, J. C. (1969). Wind, rain and the aerodynamic characteristics of rain gauges. *Met. Mag., London*, **98**, 113–120.

Rodda, J. C. (1967). *The rainfall measurement problem.* Proc. Gen. Assembly, Intern. Assoc. Sci. Hydrology, Berne, IASH Pub. 78, pp. 215–231.

Rodriguez-Iturbe, I. and Mejia, J. M. (1974). On the transformation of point rainfall to areal rainfall. *Water Resour. Res.*, **10**, 729–735.

Rutter, A. J., Kershaw, K. A., Robins, P. C. and Morton, A. J. (1971). A predictive model of rainfall interception in forests. I. Derivation of the model from observations in a plantation of Corsican Pine. *Ag. Met.*, **9**, 367–384.

Rutter, A. J., Morton, A. J. and Robins, P. C. (1975). A predictive model of rainfall interception in forests. II. Generalization of the model and comparison with observations in some coniferous and hardwood stands. *J. Appl. Ecol.*, **12**, 367–380.

Searcy, J. K. and Hardison, C. H. (1960). Double-mass curves. Geological Survey Water Supply Paper 1541-B. Washington, DC: US Department of the Interior, pp. 31–66.

Sevruk, B. (1974). The use of stereo, horizontal, and ground level orifice gages to determine a rainfall–elevation relationship. *Water Resour. Res.*, **10**, 1138–1141.

   (1993a). Wind-induced measurement error for high-intensity rains. In *Precipitation Measurement*, ed. B. Sevruk, Proc. Int. Workshop on Precipitation Measurement, St. Moritz, Switzerland, 3–7 Dec., 1989. Zurich: Institute of Geography, Swiss Federal Institute of Technology, pp. 199–204.

   (1993b). WMO precipitation measurement intercomparisons. In *Precipitation Measurement and Quality Control*, ed. B. Sevruk and M. Lapin, Proc. Symposium on Precipitation and Evaporation, Vol. 1, pp. 120–121. Bratislava, Slovakia: Slovak Hydrometeorological Institute, and Zurich, Switzerland: Swiss Federal Institute of Technology, Department of Geography.

   (1996). Adjustment of tipping-bucket precipitation gauge measurements. *Atmos. Res.*, **42**, 237–246.

Sevruk, B. and Nespor, V. (1998). Empirical and theoretical assessment of the wind induced error of rain measurement. *Water Sci. Technol.*, **37**, 171–178.

Sherman, C. W. (1931). Frequency and intensity of excessive rainfalls at Boston, Massachusetts. *Trans. Amer. Soc. Civil Eng.*, **95**, 951–960, 966–968.

Simanton, J. R. and Osborn, H. B. (1980). Reciprocal-distance estimate of point rainfall. *J. Hydraul. Div., Proc. ASCE*, **106**(HY7), 1242–1246.

Simpson, J. E. (1977). *Gravity Currents in the Environment and in the Laboratory*, second edition. Cambridge: Cambridge University Press.

Singh, V. P. and Chowdhury, P. K. (1986). Comparing some methods of estimating mean areal rainfall. *Water Resour. Bull.*, **22**, 275–282.

Sivapalan, M. and Blöschl, G. (1998). Transformation of point rainfall to areal rainfall: intensity-duration-frequency curves. *J. Hydrol.*, **204**, 150–167.

Smith, J. A. and Karr, A. F. (1990). A statistical model of extreme storm rainfall. *J. Geophys. Res.*, **95**(D3), 2083–2092.

Smith, R. B. (1979). The influence of mountains on the atmosphere. *Adv. Geophysics*, **21**, 87–230.

Stoelinga, M. T., Locatelli, J. D. and Hobbs, P. V. (2002). Warm occlusions, cold occlusions, and forward-tilting cold fronts. *Bull. Amer. Met. Soc.*, **83**, 709–721.

Suzuki, Y., Nakakita, E. and Ikebuchi, S. (2002). A study of dependence properties of rainfall distribution on topographic elevation. *J. Hydrosc. & Hydraul. Eng. (JSCE)*, **20**, 1–11.

Thiessen, A. H. (1911). Precipitation averages for large areas. *Mon. Wea. Rev.*, **39**, 1082–1084.

US Weather Bureau (1955). *Rainfall Intensity–Duration–Frequency Curves, For Selected Stations in the United States, Alaska, Hawaiian Islands, and Puerto Rico*, Technical paper No. 25. Washington, DC: US Dept. Commerce.

   (1957–1960). *Rainfall Intensity–Frequency Regime*, Technical paper No. 29, Parts 1–5. Washington, DC: US Dept. Commerce.

Valente, F., David, J. S. and Gash, J. H. C. (1997). Modelling interception loss for two sparse eucalypt and pine forests in central Portugal using reformulated Rutter and Gash analytical models. *J. Hydrol.*, **190**, 141–162.

Vonnegut, B. (1977). Quaint cumulus convection conviction. *Eos, Trans. AGU*, June 10.

Ubarana, V. N. (1966). Observation and modelling of rainfall interception loss in two experimental sites in Amazonian forest. In *Amazonian Deforestation and Climate*, ed. J. H. C. Gash, C. A. Nobre, J. M. Roberts and R. L. Victoria. Chichester: J. Wiley.

Williams, D. T. (1963). *The thunderstorm wake of May 4, 1961*. Nat. Severe Storms Project Rept.18. Washington, DC: US Dept. Commerce [NTIS PB-168223].

World Meteorological Organization (1986). *Manual for Estimation of Probable Maximum Precipitation*, second edition, Operational Hydrology Rept. No. 1, WMO-No.332. Geneva, Switzerland: Secretariat of the WMO. (Table updated by the National Weather Service, Office of Hydrology, Hydrometeorological Branch, 1992.)

Yang, D., Goodison, B. E. and Metcalf, J. R. (1998). Accuracy of the NWS 8" standard nonrecording precipitation gauge: results and application of WMO intercomparison. *J. Atmos. Oceanic Technol.*, **15**, 54–67.

PROBLEMS

3.1 Sketch the potential temperature versus height, z, for the air parcel undergoing the changes depicted in Figure 3.10. Indicate the different segments of the curve with the same letters A, B, C and D as it passes over the ridge.

3.2 Show how the inverse distance method expression (3.2) produces the arithmetic mean for $b = 0$.

3.3 Using the principle underlying the inverse distance method (see Equation (3.2)), derive an expression to calculate missing precipitation data at one of the n rain gages in the area under consideration. In other words, estimate the precipitation at the, say, pth station, with the missing data, from the measured precipitation at the other $(n-1)$ rain gages, which are separated from it by respective distances, $d_{1,p}, d_{2,p}, \ldots d_{n-1,p}$.

3.4 Estimate the return period, T_r (in years), of a rain storm in which 60 mm fell over a period of 90 min, at a location where K_p in Equation (3.3) is of the order of 36. Use typical values for the parameters a and c.

3.5 Suppose that rainfall intensity–duration–frequency data, for a region with a humid, temperate climate, can be described by Equation (3.3) with the constants, $K_p = 30$, $a = 0.2$, $b = 0.05$ h and $c = 0.70$ (with P in mm h^{-1}, D in h, and T_r in years). Estimate the 50 y rainfall with a duration of 70 min.

3.6 Estimate the maximal depth ever recorded on Earth for a duration of 90 min.

3.7 By comparing Figures 3.21 and 3.22, estimate roughly what the average rainfall intensity was in the data from which Figure 3.21 was derived; in other words, for what value(s) of P are these two figures in agreement? Assume that the various gages, whose measurements were used in these two figures, have similar hydrodynamic characteristics affecting the gage deficiencies.

4 EVAPORATION

In terms of the water quantities transported on a global basis, evaporation is the second most important component of the hydrologic cycle, after precipitation. The general climatology of the hydrologic cycle reviewed in Chapter 1, indicates that over the land-surfaces of the Earth evaporation amounts on average to approximately 60% to 65% of the average precipitation. But this estimate provides only an idea of the order of magnitude to be expected; the actual evaporation rate at any given time and place is likely to be quite different from the climatological mean, and more thorough analysis is often called for.

4.1   EVAPORATION MECHANISMS

As a physical phenomenon, evaporation is the transition of water from the liquid phase to the vapor phase. This transition requires first, an energy supply to provide water molecules the necessary kinetic energy to escape from the liquid surface; and second, some mechanism to remove the escaped molecules from the immediate vicinity of the liquid surface thus preventing that they would return to condense (see Figure 4.1). These two requirements have traditionally given rise to two classes of methods to describe evaporation, namely

(i)   mass transfer or aerodynamic formulations, which consist primarily of the description of the water vapor transport mechanisms in the near-surface air of the atmosphere, and

(ii)   energy budget formulations, in which the main focus is on the energy supply aspects of the phenomenon.

Actually, this classification scheme is somewhat unsatisfactory, because it is almost never possible to consider mass transfer and energy aspects of evaporation in isolation from each other; as will become clear below, energy budget methods usually cannot avoid mass transfer considerations in their application, and vice versa. Nevertheless, this classification will be used in what follows, mainly for historic reasons. In addition, a third class of methods is considered, namely

(iii) water budget formulations, in which evaporation is treated as the unknown rest term in the continuity equation (1.7) or (1.8) for various types of control volumes that include the landsurface–atmosphere interface as a boundary.

Among these three, the formulations in class (i) are based on the most direct description of the water vapor transport mechanisms, so that whenever possible they should

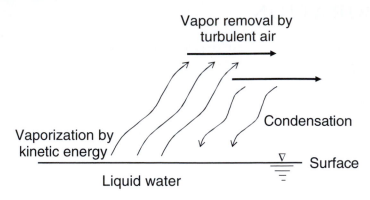

Fig. 4.1   **Liquid water molecules with sufficient kinetic energy escape from the liquid surface by vaporization.**
A removal mechanism is needed to prevent the establishment of an equilibrium state in which
vaporization would become balanced by condensation.

have priority. The formulations in class (ii) are indirect since they make use of quantities that are physically quite different from water vapor transport, as such; however, since none of these quantities is part of the hydrologic cycle, energy budget methods still allow independent estimates of the water vapor flux. This is not the case for the formulations in class (iii). Although conceptually the most obvious and appealing among the three, water budget formulations require a knowledge of all the other components in the water budget; thus they are unsuitable whenever the objective is closure of the hydrologic cycle and independent estimates of all its components are required.

4.2   MASS TRANSFER FORMULATIONS

4.2.1   *In terms of turbulent fluctuations*

Direct or eddy-correlation method
Equation (2.29) constitutes the basis for the direct measurement of the rate of evaporation $[ML^{-2}T^{-1}]$, or

$$E = \rho \, \overline{q'w'} \tag{4.1}$$

In practice E is determined by measuring the fluctuations w' and q' and then computing the cross-correlation over a suitable averaging period, which is usually taken on the order of 15–30 min, sometimes up to 1 h at most. While the theoretical basis of this method is straightforward, the requirements on the instrumentation are quite stringent. For instance, for measurements at a few meters above the ground, the upper frequency response limit should be at least of the order of 5–10 Hz. Therefore, it is only in recent years that sufficient progress has been made in the development of suitable instrumentation, that is commercially available. However, because the instrumentation is costly and requires special skill to operate, at present this method is a realistic option only in special experimental settings.

Variance methods

When the data needed for Equation (4.1), and temperature fluctuations θ' are available, they can also be used to calculate the variances of the fluctuations

$$\sigma_q^2 = \overline{(q')^2} \; ; \sigma_w^2 = \overline{(w')^2} \; ; \sigma_\theta^2 = \overline{(\theta')^2} \tag{4.2}$$

These may then be related to the covariance given in (4.1), i.e. the rate of evaporation, by means of simple similarity assumptions. These relationships form the basis of the variance method, which can be used as a complement or as an alternative to the eddy correlation method to determine turbulent surface fluxes E, u_* and H. The variance method was probably first proposed by Tillman (1972) and further elaborated upon by Wesely (1988) and others (see, for example, Asanuma and Brutsaert, 1999; Eng *et al.*, 2003). One disadvantage of the eddy correlation technique is that (4.1) is very sensitive to the vertical orientation of the velocity sensor to measure w'; variance-based techniques do not suffer this drawback. The dissipation method is another alternative method that makes use of the same kind of turbulence measurements to derive the surface fluxes (see Champagne *et al.*, 1977; Brutsaert, 1982).

4.2.2   *Methods in terms of mean variables*

Over a uniform surface with adequate fetch, formulations in terms of mean variables are based directly on the similarity principles for the atmospheric boundary layer discussed in Chapter 2. The word "mean" as used here, refers to the fact that the $\overline{q}$, $\overline{u}$ and $\overline{\theta}$ data are obtained by averaging over a certain time period, in the same way as was explained for the second moments in the previous section. These methods can be classified into two general types, namely bulk transfer methods and mean profile methods.

Bulk-transfer approach

In this approach the flux is determined by means of equations, whose general form is given by (2.33) for water vapor, and by (2.34) and (2.35) for its analogs momentum and temperature, respectively. One of the more common forms of (2.33) used in practical applications is as follows

$$E = \text{Ce} \, \rho \, \overline{u}_1 (\overline{q}_s - \overline{q}_2) \tag{4.3}$$

where the subscripts 1 and 2 refer to measurement levels z_1 and z_2 above the ground, the subscript s refers to the ground surface at $z = 0$ and Ce can be determined theoretically or empirically. The specific humidity at the surface $\overline{q}_s$ must be known in Equation (4.3); therefore it is used mostly over water where $\overline{q}_s$ can be taken simply as $q^*(T_s)$, the saturation value at the temperature of the water surface. The main practical advantage of this mass-transfer approach, usually with a constant-known coefficient Ce, lies in the fact that it can be applied on a routine basis with regular and easily obtainable data of mean wind speed, water surface temperature and humidity of the air.

As discussed already in Section 2.5.2, Equation (4.3) can be justified readily by the form of the flux-profile functions (2.41), (2.44), (2.54), (2.55), and (2.56). However, these functions also show that any empirical mass transfer coefficient Ce for data taken in the atmospheric boundary layer, can only be constant if the roughness parameters are constant, and either if the atmosphere is neutral, or if the effect of stability as reflected in ζ is negligible or constant.

Example 4.1. Mass-transfer coefficient in neutral atmosphere

Under neutral conditions by virtue of (2.41) and (2.44), the water vapor transfer coefficient, as it appears in Equation (4.3), is simply

$$Ce = \frac{k^2}{\ln\left[(z_2 - d_0)/z_{0v}\right]\,\ln\left[(z_1 - d_0)/z_0\right]} \tag{4.4}$$

in which z_1 and z_2 are the heights of the measurement of the wind speed and of the specific humidity, respectively, and in which d_0 can be taken to be zero over a water surface. Within a certain range of normal wind speeds, neutral conditions are apparently often satisfied over ocean and sea surfaces. Indeed, numerous experimental determinations have shown that on average the ocean transfer coefficients are of the order of $Ce_{10}(\cong Ch_{10}) \cong 1.2\ (\pm 0.30) \times 10^{-3}$, in which the subscript indicates that the measurements are taken at $z_1 = z_2 = 10\,$m above the surface. Generally, the corresponding drag coefficient is a little larger, and of the order of $Cd_{10} \cong 1.4\ (\pm 0.3) \times 10^{-3}$, on average; it also tends to be more sensitive to the sea state.

The scatter among many of the experimental estimates of the transfer coefficients Ce_{10}, Ch_{10} and Cd_{10} over water is considerable. This means that when accurate results are required the use of some average coefficient may not be adequate and it may be necessary to include the effects of atmospheric stability and of the roughness lengths, and therefore in the case of water surfaces, also of sea state. Numerous expressions have been proposed relating Cd_{10} to wind speed or surface shear stress for large water surfaces (see Brutsaert, 1982). Over water surfaces of limited size, such as small lakes, Ce can be expected to depend on fetch, that is the distance from the upwind shore. However, in the case of medium size lakes, with fetches of the order of 1–10 km, Ce is quite insensitive to fetch, provided the specific humidity of the air and the wind speed are determined over the center of the lake surface. Thus Equation (4.3) with $Ce_{10} = 1.2 \times 10^{-3}$ can also be used for such conditions as a first approximation. For more accurate results, however, it may be advisable to calibrate Ce in (4.3) for each individual lake.

The form of (4.3) is, in a sense, also suggestive of many other types of mass transfer equations, mostly empirical, which have been proposed in the past. One such evaporation equation, originally proposed by Stelling in 1822 (see Brutsaert, 1982) and still in use today, can be written as

$$E = (a + b\,\bar{u}_1)(\bar{e}_s - \bar{e}_2) \tag{4.5}$$

where $\bar{e}$ is the mean vapor pressure and the subscripts refer to the heights of the measurements. From the definition of the specific humidity $q = \rho_v/\rho$, with the equation of state for water vapor (2.5) and for bulk air (2.6), it follows that to a good approximation

$$q = 0.622\,e/p \tag{4.6}$$

This shows that the vapor pressure e is closely proportional to the specific humidity q. The introduction of the additional constant a in (4.5) may be viewed as a means of improving the curve-fit between the mean wind speed and the rate of evaporation. Although their theoretical justification is marginal, equations like Stelling's (4.5) have been found useful to describe evaporation from water or wet surfaces. Some examples for various problems and surfaces can be found in papers by Penman (1948, 1956), Brutsaert and Yu (1968), Shulyakovskiy (1969) and Neuwirth (1974) among many others. Mass-transfer equations, in terms of the vapor pressure difference, are sometimes written in a more general form as follows

$$E = f_e(\overline{u}_1)(e_s - \overline{e}_2) \tag{4.7}$$

where, as before, the subscripts refer to the levels above the surface z_1 and z_2 at which the measurements are made, and $f_e(\overline{u})$ is called the wind function, which can be obtained experimentally or from similarity; obviously, in the case of (4.5), one has $f_e(\overline{u}) = a + b\overline{u}$.

Mean profile methods

The available flux-profile functions for the boundary layer given in Section 2.5.2 allow the calculation of the surface fluxes from measurements of mean concentration at two or more levels. The specific form of the profile functions depends on the level above the surface, i.e. the specific sublayer, where the measurements are made (see Figure 2.6).

Profile methods are most useful in the atmospheric surface layer, where they can be based on the Monin–Obukhov similarity. Recall that the surface sublayer is the fully turbulent layer, located between a height z_{sb}, which is well above the surface roughness elements – say at least four to five times their height h_0 – and a height z_{st}, which is roughly of the order of one tenth of the thickness of the boundary layer; a more precise estimate of the extent of the surface layer is presented in Section 2.5.2. The profiles in this layer are given by Equations (2.50)–(2.52) (or (2.54)–(2.56)). The subscripts 1 and 2 in these equations refer to a lower and upper level at which the respective measurements of $\overline{q}$, $\overline{u}$ and $\overline{\theta}$ are made; clearly, these elevations need not be the same in all three equations. The Ψ-functions appearing in (2.50)–(2.52) are given in (2.58), (2.59), (2.63) and (2.64).

In this approach, the flux of any admixture, be it E, u_* or H, cannot be calculated simply from measurements of its corresponding concentration, $\overline{q}$, $\overline{u}$ or $\overline{\theta}$, only; indeed, except under neutral conditions, each of Equations (2.50)–(2.56) contains also the momentum flux u_*, and the Obukhov length L, defined in (2.46), which, in turn, contains the three fluxes. In practice there are two alternative methods of closing a flux determination problem.

The first method consists of the simultaneous solution of Equations (2.50)–(2.52) (or (2.54)–(2.56)) for the three unknown surface fluxes u_*, H and E, with known measurements at least at two levels of mean specific humidity, mean wind speed and mean temperature. This numerical problem may be solved in different ways. One simple way is by iteration, as follows; it is assumed initially that the profiles are logarithmic, i.e. that $L = \infty$ so that the Ψ-functions are zero. This permits the calculation of a first estimate of the fluxes with (2.50)–(2.52) (or (2.54)–(2.56)), from which a first estimate can be made of the Obukhov length L in (2.46). This first estimate of L allows next the calculation of a second estimate of the fluxes by means of (2.50)–(2.52) (or (2.54)–(2.56)), which in turn allow the calculation of a second estimate of L, and so on. The iteration can be stopped when successive estimates cease to change appreciably. When measurements of $\bar{u}$, $\bar{q}$, and $\bar{\theta}$ are available at more than two elevations, at each iteration u_*, E and H can be obtained as the slopes from (2.50)–(2.52) (or (2.54)–(2.56)) by least squares regression through the origin.

Example 4.2. Evaporation by profile method in a neutral atmosphere

In a neutrally stratified atmosphere, the turbulent heat flux is relatively small; therefore the Obukhov length L, defined in (2.46), is large, and thus ζ small, so that the Ψ-functions become negligible. As a result, Equations (2.50)–(2.52) reduce to the logarithmic profile equations (2.40) and (2.43). Combination of these two equations allows the direct calculation of the evaporation rate by means of the following expression

$$E = \frac{k^2 \rho (\bar{u}_2 - \bar{u}_1)(\bar{q}_3 - \bar{q}_4)}{\ln\left(\dfrac{z_2 - d_0}{z_1 - d_0}\right)\ln\left(\dfrac{z_4 - d_0}{z_3 - d_0}\right)} \qquad (4.8)$$

in terms of measurements of the wind speed at levels z_1 and z_2, and measurements of specific humidity at levels z_3 and z_4 above the ground. An equation similar to this result was first presented by Thornthwaite and Holzman (1939). While this derivation provides a good didactic illustration of the profile method, it should be noted that Equation (4.8) is of limited practical applicability, because over land the atmosphere is only rarely neutral. Thus, in most cases the profile method requires solution of the full set of equations (2.50)–(2.52) (or (2.54)–(2.56)).

The second method consists of using the known mean profile and the surface flux of another but similar scalar, in addition to the mean profile of the scalar under consideration. The requirement of similarity refers in this context to the equality of the transfer coefficients Ce and Ch in Equations (2.33) and (2.35) or in (2.36) and (2.38) for the scalars; in this sense, it also refers to the equality of the functions Ψ_h and Ψ_v in the profile equations (2.51) and (2.52) (or (2.55) and (2.56)). Probably the oldest application of this principle is the Bowen ratio (Bowen, 1926)

$$\text{Bo} = H/L_e E \qquad (4.9)$$

in which L_e is the latent heat of vaporization of water. Hence, if similarity is valid, this ratio, which is used mostly in the energy budget method (see Section 4.3) can also be written in terms of profile measurements as follows

$$\text{Bo} = \frac{c_p(\bar{\theta}_1 - \bar{\theta}_2)}{L_e(\bar{q}_1 - \bar{q}_2)} \qquad (4.10)$$

Over water the surface values $\bar{\theta}_s$ and $\bar{q}_s$ are commonly used instead of values in the air $\bar{\theta}_1$ and $\bar{q}_1$, respectively. The Bowen ratio concept thus leads to a simple expression for evaporation in terms of the sensible heat flux and of mean specific humidity and mean temperature measurements in the surface layer, as follows

$$E = \frac{H(\bar{q}_1 - \bar{q}_2)}{c_p(\bar{\theta}_1 - \bar{\theta}_2)} \tag{4.11}$$

As an aside, in a similar way the surface flux F of any other passive admixture of the air (e.g. CO_2) can be estimated by means of (4.11), by replacing the measurements of $\bar{q}$ in this expression by measurements of the mean concentration $\bar{c}$ of the admixture under consideration. Alternatively, the surface flux F of any admixture can also be expressed in terms of, say, measurements of mean specific humidity and concentration $\bar{c}$, and a known rate of evaporation, as follows

$$F = \frac{E(\bar{c}_1 - \bar{c}_2)}{(\bar{q}_1 - \bar{q}_2)} \tag{4.12}$$

So far in this section, the mean profile method has been explained for application with profile data in the surface layer. In principle, the method can also be applied with upper air data measured in the outer region of the boundary layer by means of bulk ABL similarity equations of the type given by Equations (2.65), (2.66) (or (2.67) and (2.68)) and (2.71). For the same reason, the simultaneous solution of these bulk similarity equations may require an iteration method, like the one described above for the surface layer (Mawdsley and Brutsaert, 1977). A more recent formulation of the functions suitable for this purpose under unstable conditions is given in Equations (2.69) and (2.70). One unresolved difficulty with this approach, however, is that in the outer region temperature and humidity do not exhibit similarity, so that, as shown by Brutsaert and Chan (1978), C is not equal to D.

4.3   ENERGY BUDGET AND RELATED FORMULATIONS

4.3.1   *Standard application*

When the main objective is the determination of evaporation E (or the sensible heat flux into the air H), it is convenient to rewrite the energy budget equation (2.72) as

$$L_e E + H = Q_n \tag{4.13}$$

where Q_n is defined as the available energy flux density

$$Q_n = R_n - G + L_p F_p + A_h - \partial W/\partial t \tag{4.14}$$

whose terms are discussed in Section 2.6. As mentioned, in many applications the last three terms in (4.14) are of little consequence, so that it is often sufficiently accurate to put $Q_n = R_n - G$.

In hydrology it is common practice to express the specific energy fluxes as equivalent rates of evaporation; Equation (4.13) can then be written as

$$E + H_e = Q_{ne} \tag{4.15}$$

where $H_e = H/L_e$ and $Q_{ne} = Q_n/L_e$. Observe, however, that with a typical value of $L_e = 2.466 \times 10^6$ J kg^{-1} (at 15 °C in Table 2.4), 1 W m^{-2} is roughly equivalent with an evaporation of 1.07 kg m^{-2} per month. Thus as a rule of thumb, to have a rough idea of the magnitudes of the fluxes involved, energy flux units of W m^{-2} can be interchanged with hydrologic units of millimeters per month (mm mo^{-1}) of liquid water evaporation.

When Q_n and either H or E can be determined independently, Equation (4.13) provides directly the remaining unknown flux. Usually, however, both H and E are unknown, and an indirect method must be used. From the methodological point of view, these indirect energy budget methods are analogous to the mean profile methods of Section 4.2.2. In both, essentially three equations are used which contain three unknowns E, u_* and H implicitly. In the profile methods these are the equations for $\overline{q}$, $\overline{u}$ and $\overline{\theta}$. In the energy budget methods, (4.13) is used either with equations for $\overline{q}$ and $\overline{\theta}$, or with equations for $\overline{u}$ and $\overline{\theta}$ or $\overline{q}$, as will be shown next.

With Bowen ratio (EBBR)

When Q_n is known, the combination of the energy budget equation (4.13) with the Bowen ratio defined in Equation (4.9) produces

$$E = \frac{Q_{ne}}{1 + \text{Bo}} \qquad (4.16)$$

Similarly, for the sensible heat flux one has

$$H_e = \frac{\text{Bo}\ Q_{ne}}{1 + \text{Bo}} \qquad (4.17)$$

Bo can be determined as shown in (4.10), from profile data of temperature and specific humidity in the atmospheric surface layer. As discussed in Section 4.2.1, these data should be taken as averages over 15–30 min, approximately. Equation (4.16) shows that the energy budget with Bowen ratio (EBBR) method is most accurate when Bo is small. Both (4.16) and (4.17) produce a singularity when Bo $= -1$; but, as pointed out by Tanner (1960), over an active vegetation this is not a problem, as this situation usually occurs when H is low, around sunrise, sunset and occasionally at night. The situation does occur more often over cold water, and it may be necessary to use an alternative method when $-1 < \text{Bo} < -0.5$ to avoid the problem of a very small denominator in Equations (4.16) and (4.17). Tanner (1960) suggested the use of a bulk-transfer method for these special conditions. Another way consists of using mean values of Bo corrected by means of wind measurements, as outlined by Webb (1964); this method is especially useful when some terms in the available energy Q_n are only known for daily periods or longer.

The EBBR method has the advantage that no similarity functions for the atmospheric turbulence appear explicitly in the formulation. With Equation (4.10) no measurements of turbulence or of the mean wind speed are required, and the formulation, as written in (4.16) with (4.10), is independent of atmospheric stability. In addition, when Bo is small, the EBBR method may be less susceptible, albeit not immune, to imperfect fetch conditions, than mean profile methods, in which such effects are more directly apparent. The validity of the EBBR method depends critically on the similarity of the temperature and humidity profile; for the surface layer this requires the equality of the terms in the square brackets of Equations (2.51) and (2.52) (or (2.55) and (2.56)). The latest evidence supports the view that $\Psi_v = \Psi_h$, even under stable conditions (see Dias and Brutsaert, 1996).

With profiles of wind and of a scalar (EBWSP)

If the profile data, either of the mean temperature or of the mean specific humidity, are lacking to apply the EBBR, the energy budget method can be applied instead with the mean wind speed profile. In fact, this procedure is potentially more powerful than the Bowen ratio method, because it yields not only E and H but also u_*.

As an illustration of this method, suppose that the specific humidity measurements are not available. It is then possible to use Equation (4.13) together with profile equations (2.50) and (2.51) (or (2.54) and (2.55)) in the surface layer, as a system of three equations with three implicit unknowns E, u_* and H. This system can be solved with measurements of Q_n, $\overline{\theta}_1 - \overline{\theta}_2$ (or $\theta_s - \overline{\theta}$) and $\overline{u}_2 - \overline{u}_1$ (or $\overline{u}$ and z_o). The method can equally be applied with measurements higher up aloft in the mixed layer of the boundary layer. Thus in this case the system of three equations is (4.13) with (2.67) and (2.68). Similarly, if only humidity but no temperature measurements are available, to apply the method with surface layer data, the system of equations can consist of (4.13) with (2.50) and (2.52) (or (2.54) and (2.56)), or in the case of mixed layer data even with (2.67) and (2.71).

This EBWSP method and its simpler derivatives (see next section) are sometimes referred to as combination methods on the grounds, that both energy budget and hydrodynamic aspects of evaporation are considered. But this is misleading, since the Bowen ratio method is no less dependent on the validity of the hydrodynamics underlying (say) Equations (2.50)–(2.52) (or (2.54)–(2.56)), than the formulation of the mean wind speed profile.

4.3.2   *Evaporation from wet surfaces: simplified expressions*

The EBWSP method with measurements at one level

When the surface is wet, the surface specific humidity may be assumed to be the saturation value at the surface temperature, i.e. $q_s = q^* (T_s)$. This allows an approximation, first introduced by Penman (1948) and given in Equation (4.20) below; the main advantage of this approximation is that it eliminates the need for measurements of $\overline{q}$, $\overline{u}$ and $\overline{\theta}$ at two levels, as in the profile methods (Section 4.2.2) and standard energy budget methods (Section 4.3.1) and that measurements at one level suffice.

The equation derived by Penman (1948) was intended for an open water surface. Here a somewhat more general derivation is presented, which is applicable to any wet surface, but which retains the essential features. By virtue of Equation (4.6), the Bowen ratio (4.10) can also be written in terms of the vapor pressure; with lower measurements at the surface, where $e_s = e^* (T_s)$, the Bowen ratio is

$$\mathrm{Bo} = \gamma \frac{(\overline{T}_s - \overline{T}_a)}{(\overline{e}_s - \overline{e}_a)} \tag{4.18}$$

where e_a and T_a are the vapor pressure and temperature in the air, respectively, at some reference level, and where

$$\gamma = \frac{c_p\, p}{0.622\, L_e} \tag{4.19}$$

Table 4.1 Values of (γ/Δ) at 1000 mb (γ is defined by Equation (4.19) and Δ can be obtained from Table 2.4)

Air temperature T_a (°C)	(γ/Δ)
−20	5.864
−10	2.829
0	1.456
5	1.067
10	0.7934
15	0.5967
20	0.4549
25	0.3505
30	0.2731
35	0.2149
40	0.1707

is commonly referred to as the psychrometric constant; at 20 °C and $p = 1013.25$ hPa it is $\gamma = 0.67$ hPa K^{-1}. Note that the θ difference is replaced by that of T, since they are often practically the same in the surface layer. The crucial step in Penman's analysis is the assumption

$$\frac{e_s^* - e_a^*}{T_s - T_a} = \Delta \tag{4.20}$$

where $\Delta = (de^*/dT)$ is the slope of the saturation water vapor pressure curve $e^* = e^*(T)$ at the air temperature T_a (see Figure 2.1); $e_a^* = e^*(T_a)$ is the corresponding saturation vapor pressure and $e_s^* = e^*(T_s)$ is the vapor pressure at the temperature of the surface, as indicated by the subscript. Since e_s for a wet surface is the value at saturation, the Bowen ratio (4.18) is thus, approximately

$$\mathrm{Bo} = \frac{\gamma}{\Delta}\left[1 - \frac{(e_a^* - \bar{e}_a)}{(\bar{e}_s - \bar{e}_a)}\right] \tag{4.21}$$

In this expression Δ depends only on temperature and γ depends on both temperature and pressure. Values of (γ/Δ) for different temperatures at $p = 1000$ hPa are presented in Table 4.1 and Figure 4.2; they were obtained by means of (4.19) and values of Δ and L_e listed in Table 2.4. Substitution of (4.21) into (4.16) produces

$$Q_{ne} = \left(1 + \frac{\gamma}{\Delta}\right)E - \frac{\gamma}{\Delta}\left(\frac{e_a^* - \bar{e}_a}{\bar{e}_s - \bar{e}_a}\right)E \tag{4.22}$$

In the second term on the right of Equation (4.22), a bulk-transfer equation can be used, such as (4.7), to replace the unknown $E/(\bar{e}_s - \bar{e}_a)$ by a wind function $f_e(\bar{u}_r)$. Thus (4.22) yields the desired result, the Penman (1948) equation in its usual form

$$E = \frac{\Delta}{\Delta + \gamma}\,Q_{ne} + \frac{\gamma}{\Delta + \gamma}E_A \tag{4.23}$$

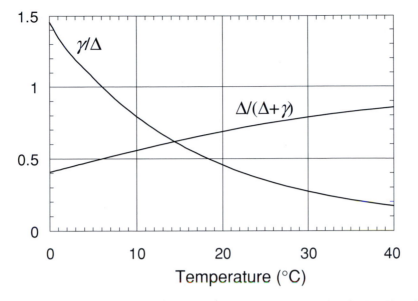

Fig. 4.2 Temperature dependence of (γ/Δ) and $\Delta/(\Delta + \gamma)$ at 1000 hPa; γ is defined by Equation (4.18) and $\Delta = de^*/dT$ is shown in Figure 2.1, and can be obtained from Equation (2.12) or from Table 2.4.

where E_A, a drying power of the air, is defined by

$$E_A = f_e(\overline{u}_r)(e_a^* - \overline{e}_a) \tag{4.24}$$

The ratio $\Delta/(\Delta + \gamma)$ is illustrated in Figure 4.2 for a pressure of 1000 hPa. Note that in Penman's (1948) original derivation it was assumed that $Q_{ne} = R_n/L_e$ and that all the other terms in Equation (4.14) are negligible. As mentioned, from the practical point of view, the main feature of this result is that it requires measurements of mean specific humidity, wind speed and temperature at one level only. This is a direct consequence of the approximation introduced in (4.20). For this reason, Penman's equation is useful when measurements at more than one level, needed for profile methods or standard energy budget methods, are unavailable or impractical.

Equation (4.23) has been widely used, but there is still no generally accepted way to formulate $f_e(\overline{u}_r)$, the wind function in E_A. Its definition in (4.24) suggests that any suitable mass transfer coefficient can be used for this purpose (see Section 4.2.2). Penman (1948) originally proposed an equation of the Stelling-type (4.5) as follows

$$f_e(\overline{u}_2) = 0.26\,(1 + 0.54\,\overline{u}_2) \tag{4.25}$$

where $\overline{u}_2$ is the mean wind speed at 2 m above the surface in m s^{-1}, and the constants require that E_A in Equation (4.24) is in mm d^{-1} and the vapor pressure is in hPa. There are indications that Equation (4.25) yields reasonable results for natural terrain with small to moderate roughness (see Thom and Oliver, 1977); on the basis of experimental observations, it has also been suggested (Doorenbos and Pruitt, 1975) that for irrigated crops, the constant 0.54 should be replaced by 0.86. In calculations of long-term mean values of E_A with equations like (4.25), to a first approximation the wind speed at 2 m can be estimated by assuming a power dependency on height, or

$$\bar{u}_2 = \bar{u}_r \, (2/z_r)^{1/7} \tag{4.26}$$

where z_r is the height (in m) at which the available wind data are measured.

A more fundamental approach to determine the wind function is based on turbulence similarity. Thus in terms of the bulk water vapor transfer coefficient as defined, for example, in (4.3), in which z_1 is the height of the measurement of $\bar{u}_1$ and z_2 that of $\bar{e}_a$, one obtains by virtue of (4.6), the wind function

$$f_e(\bar{u}_1) = 0.622 \, \rho p^{-1} \mathrm{Ce} \, \bar{u}_1 \tag{4.27}$$

Ce can be determined by means of the similarity profile functions of Chapter 2. Under neutral conditions, on account of Equations (4.4), (4.6) and (4.7) this is (to a good approximation)

$$f_e(\bar{u}_1) = \frac{0.622 \, k^2 \, \bar{u}_1}{R_d T_a \ln\left[(z_2 - d_0)/z_{0v}\right] \, \ln\left[(z_1 - d_0)/z_0\right]} \tag{4.28}$$

where, again, z_1 is the level of the wind speed measurement and z_2 that of the water vapor pressure.

When Penman's equation is applied to calculate mean values of E over periods of a day or longer, the use of wind functions like (4.25), (4.27) or (4.28) may be adequate. However, when hourly values are required, the effect of atmospheric stability, which varies through the day, may be important. It is possible to include the effect of the atmospheric stability in the wind function, by writing the drying power of the air (4.24) in a form similar to (2.56) (see also Brutsaert, 1982) as follows

$$E_A = k u_* \rho (q_a^* - \bar{q}_a) \left[\ln \left(\frac{z_a - d_0}{z_{0v}} \right) - \Psi_v \left(\frac{z_a - d_0}{L} \right) + \Psi_v \left(\frac{z_{0v}}{L} \right) \right]^{-1} \tag{4.29}$$

where $\bar{q}_a$ and q_a^* are the specific humidity of the air and the saturation specific humidity at air temperature, respectively. The problem can be solved by the following iteration procedure. An initial value of E is calculated in the usual way by means of Equation (4.23) using a neutral E_A, say (4.24) with (4.28); it is also possible to use (4.29) with $\Psi_v = 0$, and u_* is calculated by means of (2.54) with $\Psi_m = 0$. The initial value of E is used to obtain H by means of (4.13). These initial values of E, u_* and H provide a first estimate of the Obukhov length L by means of (2.46). This value of L allows now the calculation of a second estimate of u_* by means of (2.54) and a second estimate of E_A by means of (4.29), which produces a second estimate of E by means of (4.23), and so on. An example of the application of this method has been presented by Katul and Parlange (1992).

Evaporation from wet surfaces in the absence of advection

The two-term structure of Equation (4.23) suggests an interpretation which may serve as an aid in understanding the effect of regional or large-scale advection. When the air has been in contact with a wet surface over a very long fetch, it could be argued that it may tend to become vapor saturated, so that E_A, shown in (4.24), should tend to zero.

Accordingly, Slatyer and McIlroy (1961) reasoned that the first term on the right of Equation (4.23) may be considered a lower limit for evaporation from moist surfaces. Thus

$$E_e = \frac{\Delta}{\Delta + \gamma} Q_{ne} \tag{4.30}$$

was referred to as equilibrium evaporation, and the second term of (4.23) may be interpreted a departure from that equilibrium. In the absence of cloud condensation or radiative divergence, this departure would stem from large-scale or regional advection effects, involving horizontal variation of surface or atmospheric conditions.

Subsequent investigations have shown, however, that over wet surfaces, true equilibrium conditions are encountered only rarely, if ever. The main reason for this is that the atmospheric boundary layer is never a perfectly homogeneous boundary layer, as would be the case in channel flow; rather, it is continually responding to unsteady large-scale weather patterns, involving condensation aloft and dry air entrainment, which tend to maintain a humidity deficit even over the ocean. Nevertheless, the idea underlying Equation (4.30) has led Priestley and Taylor (1972) to use equilibrium evaporation as the basis for an empirical relationship to describe evaporation from a wet surface under conditions of minimal advection, E_{pe}. With data obtained over ocean and moist land surfaces they concluded that it is roughly proportional to E_e, that is

$$E_{pe} = \alpha_e \frac{\Delta}{\Delta + \gamma} Q_{ne} \tag{4.31}$$

where α_e is a constant, which they found to be about 1.26. This value was later confirmed in many other studies (see Brutsaert, 1982) and α_e is now generally accepted to be of the order of 1.20–1.30, on average, for advection-free water surfaces and moist landsurfaces with short vegetation. Equation (4.31) is equivalent with a Bowen ratio

$$Bo_{pe} = \alpha_e^{-1} \left[(\gamma/\Delta) + 1 \right] - 1 \tag{4.32}$$

which is illustrated in Figure 4.3 for different α_e values, together with some experimental data points.

These values of α_e indicate that over the ocean or other moist surfaces the second term of (4.23), that is the large-scale advection, accounts on average for about 20% to 23% of the evaporation rate. But this is only an average and large variations have been observed in different experimental settings. Still, it is remarkable that so many landsurfaces covered with fairly short vegetation, such as grass, which is not actually wet but with ample water available to the roots, yield about the same average values, ranging between 1.20 and 1.30, as open water surfaces. This may be the result of a fortuitous compensation of the specific humidity of non-wet leaf surfaces, which is lower than saturation, by a larger effective roughness, and thus transfer coefficient, of the vegetative surface. Still, in some studies drastically different values of α_e have been reported. This has been especially the case for very rough surfaces; for instance McNaughton and Black (1973) obtained $\alpha_e = 1.05$ for a young, 8 m high fir forest.

Fig. 4.3   Variation of Bowen
ratio Bo_{pe} for moist
surfaces as given by
Equation (4.32);
the solid curve
represents an α_e
value of 1.26 and the
two dashed curves
represent values of
1.20 and 1.30. The
data points (for daily
values) were
collected by Davies
and Allen (1973)
from different
sources.

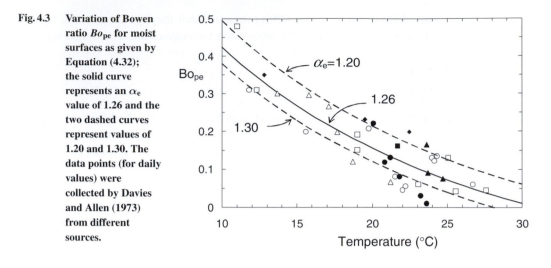

Related empirical equations, but with R_s instead of Q_n as in (4.31), have been pro-posed by Makkink (1957), Jensen and Haise (1963), and Stephens and Stewart (1963). The short-wave radiation is often well correlated with the net radiation, which is the main component of Q_n over daily periods or longer. Such equations, which provide a good alternative to (4.31) when only short-wave radiation and temperature are available, have been used to determine irrigation requirements and as climatological indices of potential evaporation. However, the physical significance of such indices is not always clear, as is shown next.

Potential evaporation

Because several of the simple energy budget-type methods for wet surfaces are often used as measures of potential evapotranspiration, a few comments are in order on this concept. The term potential evapotranspiration appears to have been introduced by Thornthwaite (1948) in the context of the classification of climate. It is now generally understood to refer to the maximal rate of evaporation from a large area covered completely and uniformly by an actively growing vegetation with adequate moisture at all times. The area is specified as large to avoid the possible effects of advection. Although the concept is widely used, it has also caused confusion, because it does not encompass all possible conditions and it involves several ambiguities. The concept requires closer specification if it is to serve as an unequivocal parameter.

Transpiration, even at the potential rate, involves such biological effects as stomatal impedance to the diffusion of water vapor, and the stage in the growth cycle of the vegetation. For this reason, the term potential evaporation is probably preferable. It can be defined to refer to the evaporation from any large uniform surface that is sufficiently moist or wet, so that the air in contact with it is fully saturated. Note that a wet or moist surface is not the same as one that has an adequate moisture supply for the roots of an actively growing vegetation; over short non-wet vegetation with adequate moisture the evapotranspiration is often fairly similar to the evaporation from open water under the same conditions. As mentioned above, a possible explanation for this is that the stomatal impedance to water vapor diffusion may be compensated by the larger roughness values, resulting in larger transfer coefficients, of the vegetational surface.

Another point of ambiguity is that potential evaporation is often estimated by means of meteorological data observed under nonpotential conditions. Because the air interacts with the underlying surface, this is not the same rate as that which would be calculated or observed, if the surface had been moist or adequately supplied with water. There- fore, potential evaporation estimated on the basis of measurements carried out under nonpotential conditions should be called "apparent" to reflect this fact. Examples of apparent potential evaporation are the estimates made by means of an evaporation pan or by means of the Penman equation (4.23), on the basis of measurements in the actual, i.e. non-potential or arid, environment. Another example of apparent potential evaporation would be that obtained by means of Equation (4.3), in which q_s at the dry surface is assumed to be given by $q^*(T_s)$, i.e. the saturation specific humidity at the temperature of that surface. In what follows the "true" potential evaporation will be denoted by E_{po}, and the apparent potential evaporation by E_{pa}.

4.3.3   Operational methods for landsurfaces

Many operational procedures used in applied hydrology to predict evaporation involve some type of potential evaporation, used in conjunction with a procedure to derive the actual evaporation from it for the prevailing non-potential conditions.

Proportional fluxes with surface moisture "bucket"
Probably the oldest method, which follows work by Budyko (1955; 1974) and Thornth- waite and Mather (1955), is based on the following proportionality

$$E = \beta_e E_p \qquad\qquad (4.33)$$

where E_p is a potential evaporation rate, and β_e a reduction factor reflecting the mois- ture availability. As mentioned above, potential evaporation is a somewhat ambiguous concept; not surprisingly, therefore, in practice Equation (4.33) has been applied with two different classes of E_p, such as E_{pa}, the apparent potential evaporation as defined in the previous section, and E_{pe}, the Priestley–Taylor equation given in (4.31).

The reduction factor β_e is often assumed to be a function of soil water content. In the application of (4.33) with such expressions for apparent potential evaporation E_{pa} as (4.3) and (4.23), a common assumption has been

$$\begin{aligned}
\beta_e &= 1 && \text{for } w > w_0 \\
\beta_e &= (w - w_c)/(w_0 - w_c) && \text{for } w \le w_0
\end{aligned} \qquad (4.34)$$

where w_0 is a critical soil water content above which E equals E_p, and w_c is a lower cut-off value below which E is zero. This is illustrated in Figure 4.4. The value of w can be determined on the basis of a soil water budget (see Thornthwaite and Mather,

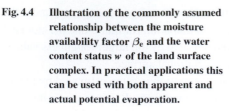

Fig. 4.4   Illustration of the commonly assumed relationship between the moisture availability factor β_e and the water content status w of the land surface complex. In practical applications this can be used with both apparent and actual potential evaporation.

1955; Budyko, 1974, p. 335; Manabe, 1969; Carson, 1982). The values of w_0 and w_c must be determined by calibration; for a surface soil layer with an assumed thickness of about 1 m, w_0 is generally taken to be of the order of 10–20 cm of water. The reduction factor β_e can also be related to some other surface moisture indices beside w, such as the accumulated actual evaporation minus precipitation (Priestley and Taylor, 1972), the local near-surface soil moisture content (see Davies and Allen, 1973; Crago and Brutsaert, 1992; Chen and Brutsaert, 1995), the soil moisture deficit (Grindley, 1970) and the antecedent precipitation index (Choudhury and Blanchard, 1983; Mawdsley and Ali, 1985; Owe *et al.*, 1989), again through calibration of the model with available data.

In some implementations of the same idea, the actual evaporation E is expressed in terms of the equilibrium evaporation E_e, by combining Equation (4.33) with (4.31), as follows

$$E = (\beta_e \alpha_e) E_e \tag{4.35}$$

in which E_e can be determined by means of (4.30). For instance, Figure 4.5 shows the results of Davies and Allen (1973) expressed as the product $(\beta_e \alpha_e)$ versus volumetric water content of the upper 5 cm of the soil. Although a nonlinear function is fitted to the data, it is similar to Equation (4.34). With data measured over prairie terrain it was found by Chen and Brutsaert (1995) that, with θ_{10} as the volumetric moisture content in the upper 10 cm of the soil profile, the relationship between E and E_e can be described by the following linear function

$$(\beta_e \alpha_e) = 1.26(\theta_{10} - 0.05)/0.22 \tag{4.36}$$

in the range $0.05 \leq \theta_{10} \leq 0.27$, and $(\beta_e \alpha_e) = 1.26$ for higher moisture contents; however, it was also observed that the relationship could be improved markedly by making $(\beta_e \alpha_e)$ not only dependent on soil moisture content but also on the density of the grassy vegetation cover, as expressed by the leaf area index La and the green vegetation fraction.

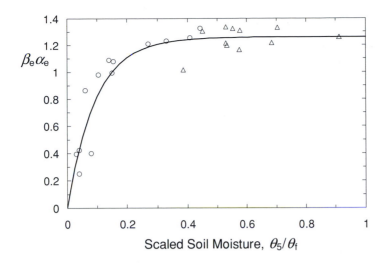

Fig. 4.5 Variation of $(\beta_e \alpha_e)$ with the water content in the upper 5 cm of the soil θ_5, expressed relative to its value at field capacity θ_f for a sandy loam covered with perennial ryegrass in Ontario. The curve represents the function $\beta_e \alpha_e = 1.26[1 - \exp(-10.563\theta_5/\theta_f)]$. (After Davies and Allen, 1973.)

One difficulty in applying the formulation (4.33) with an apparent potential evaporation E_{pa} is that, as the surface dries out, the two quantities on the right-hand side of (4.33) move in opposite directions. Indeed, whenever β_e approaches zero, E_{pa} tends to become large; this may lead to an unstable product of a large with a small quantity, each with considerable noise. On the other hand, E_{pe} depends mainly on radiation and temperature, and not on the dryness of the air; hence application of (4.33) with E_{pe} is likely to be more robust and therefore preferable.

Surface resistance concept

A second procedure of reducing E_p to E is based on the realization that the release of water vapor from a vegetation is controlled by the stomata of the leaves. This is illustrated schematically in Figure 4.6. The underlying idea is that the air is assumed to be saturated with water vapor inside the stomatal cavities but not at the outside surface of the leaves, and the stomata provide an obstruction or a resistance to the diffusion of the water vapor from the inside to the outside of the leaves. This is often referred to a stomatal resistance r_{st}. Because evaporation also takes place from the soil surface, beside the leaves, the basic idea is usually extended to include this transport as well; thus the soil air at some depth below the surface can be assumed to be saturated and the soil pores can then be visualized as providing a resistance to the diffusion of water vapor to the soil surface. Hence more generally, the resistance approach is based on the concept of one or more resistance parameters in parallel and/or in series, which may account for the moisture stress in the vegetation and/or soil, and which relate saturation specific humidity q_s^*, at the temperature T_s of the evaporating surface, to the actual (non-saturated) specific humidity q_s at the evaporating surface.

Several such resistance parameters have been used for this purpose (see, for example, Monteith, 1973). The one given by Thom (1972) is instructive as an illustration and can be

Fig. 4.6 **Sketch of a typical cross section through the underside of a leaf. Photosynthesis occurs in the mesophyll cells in the interior. The epidermis consists usually of a single layer of cells, covered by the protective cuticle; some of these cells are guard cells, which surround and control the size of the stomata by shrinking and swelling. The stomata are slit-like openings through which gas exchange and water vapor loss take place.**

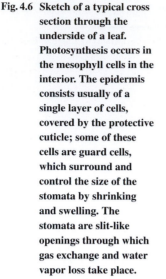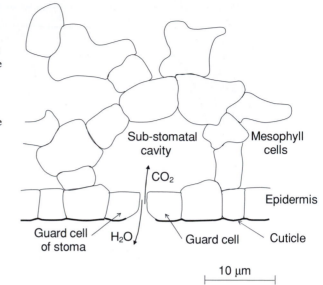

Fig. 4.7 **Schematic diagram showing the resistance parameters, which can be used to describe transfer to and from a vegetational surface. The subscripts 1 and 2 refer to the levels of the wind speed and specific humidity measurements, respectively, in the atmospheric surface layer. In this context, Equation (4.3) is sometimes written in resistance notation as $E = \rho(\bar{q}_s - \bar{q}_2)/r_{av}$, which defines the aerodynamic resistance parameter for water vapor r_{av} in terms of the mass transfer coefficient Ce, as shown.**

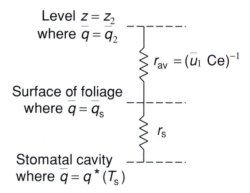

defined by

$$r_s = \rho\,(q_s^* - q_s)/E \tag{4.37}$$

in which r_s is the surface resistance, and q_s is the actual (not saturated) mean specific humidity at the evaporating surface; the basic concept is illustrated in Figure 4.7. For practical use there have generally been two types of evaporation equation based on the resistance concept. In the derivation of the first type, q_s, which is unknown, is eliminated between Equation (4.37) and the standard mass transfer equation (4.3) to yield the expression

$$E = \frac{\text{Ce}\,\bar{u}_1}{(1 + r_s\text{Ce}\,\bar{u}_1)}\rho\,(q_s^* - \bar{q}_2) \tag{4.38}$$

In the derivation of the second type, Equation (4.38) is used (instead of (4.3)) to obtain an expression in a way analogous to (4.23), namely

$$E = \frac{\Delta Q_{ne} + \gamma \text{Ce}\,\bar{u}_1\rho\,(q_2^* - \bar{q}_2)}{[\Delta + \gamma(1 + r_s\text{Ce}\,\bar{u}_1)]} \tag{4.39}$$

Equation (4.39) is in the form of the Penman–Monteith equation (see, however, Monteith, 1973; 1981; Thom, 1975)

Numerous experiments have been conducted to determine resistance values for different types of vegetation. This has been mostly done in the context of expressions related to Equation (4.39). A few examples are beans (Black *et al.*, 1970), sugar beets (Brown and Rosenberg, 1977), tropical rainforest (Dolman *et al.*, 1991), eucalyptus forest (Dunin and Greenwood, 1986), pine forest (Gash and Stewart, 1975; Lindroth, 1985), maize (Mascart *et al.*, 1991), barley (Monteith *et al.*, 1965), sorghum (Szeicz *et al.*, 1973), and fir forest (Tan and Black, 1976). In addition, many attempts have been made to relate resistance parameters with such factors as Bowen ratio, soil moisture suction in the root zone, soil moisture deficit, humidity deficit in the air, solar radiation, temperature, leaf area index and others (see VanBavel, 1967; Szeicz and Long, 1969; Federer, 1977; Garratt, 1978; Lindroth, 1985; Stewart, 1988; Gash *et al.* 1989). The relationships developed so far are mainly statistical, so that they are vegetation and site dependent. Therefore, the resistance formulation is probably not yet sufficiently general to be practical for predictive purposes, but it has been useful as a diagnostic index in certain simulation studies (for example, to calculate missing data).

As a note of caution, in previous studies the resistance formulation has not always been used with consistent definitions for Ce (or r_{av}) and r_s (Thom, 1972; Brutsaert, 1982, p. 111). For instance, the drag coefficient Cd (or the related so-called aerodynamic conductance) is often used instead of Ce, as required in the rigorous derivation of Equation (4.23) with (4.24) and (4.27). This drag coefficient is defined in (2.37). Because it is not likely that above vegetation $z_0 = z_{0v}$, nor that $\Psi_m = \Psi_v$(or Ψ_h), Cd is rarely equal to Ce. As a result of this inappropriate use of Cd (instead of Ce), it is not clear how the turbulence aspects of the transport, normally embodied in Ce, can be partitioned or separated from the strictly vegetational and/or soil moisture aspects of the transport supposedly embodied in r_s. This has undoubtedly contributed to the difficulty in deriving general relationships for both Ce and r_s on the basis of (4.39).

Although the resistance formulation with r_s may appear conceptually quite different from Equation (4.33) with the reduction factor β_e, both approaches are, in fact, practically the same. Indeed, (4.38) is equivalent with (4.33) (in which (4.3) is used to represent E_p for a wet surface) and a reduction factor

$$\beta_e = (1 + r_s Ce \, \overline{u}_1)^{-1} \tag{4.40}$$

Similarly, (4.39) is the same as (4.33) with (4.23) and a reduction factor

$$\beta_e = [1 + r_s Ce \, \overline{u}_1 \, \gamma/(\Delta + \gamma)]^{-1} \tag{4.41}$$

and as (4.33) with (4.31) and a reduction factor

$$\beta_e = \alpha_e^{-1}[1 + \gamma Ce \, \overline{u}_1 \rho \, (q_2^* - \overline{q}_2)/\Delta Q_{ne}] \, [1 + r_s Ce \, \overline{u}_1 \gamma/(\Delta + \gamma)]^{-1} \tag{4.42}$$

In practical applications of Equations (4.38) and (4.39) a knowledge of the parameters Ce and r_s is essential. The physical nature of Ce is well understood and based on sound turbulence theory. But the conceptual significance of the resistance concept remains problematic, in spite of the many studies devoted to it.

Complementary fluxes with advection–aridity

This concept was first proposed by Bouchet (1963), who postulated, almost in diametrical opposition to Equation (4.33), a certain complementary relation between the actual evaporation E, and what we now recognize as the apparent potential evaporation E_{pa}. The underlying argument may be developed as follows. If for one or other reason, independent of the available energy, the actual evaporation E decreases below its true potential value E_{po}, a certain amount of energy not used up in evaporation becomes available. This manifests itself as an increase in the sensible heat flux ΔH, or

$$E_{po} - E = \Delta H \tag{4.43}$$

At the regional scale this decrease of E, relative to E_{po}, affects primarily the temperature, humidity and turbulence of the air near the ground, but it probably has a smaller effect on the net radiation. This increased sensible heat flux ΔH, causes an increase in the apparent potential evaporation E_{pa} inferred for these drier and warmer conditions. In general, to a first approximation this increase can be assumed to be proportional to ΔH, so that one has

$$E_{pa} = E_{po} + \varepsilon_a \Delta H \tag{4.44}$$

in which ε_a is an effectiveness parameter, which may depend on the adopted definition of E_{pa}. Combination of Equations (4.43) and (4.44) yields then $E_{pa} + \varepsilon_a E = (1 + \varepsilon_a)E_{po}$. In the original derivation, Bouchet (1963) assumed that in (4.44) E_{pa} is increased by exactly ΔH; in this case, combination of (4.43) and (4.44) yields immediately the simple complementary relationship

$$E + E_{pa} = 2\, E_{po} \tag{4.45}$$

This result can be rearranged to yield the actual evaporation in dimensionless form

$$\frac{E}{E_{po}} = \frac{2E/E_{pa}}{1 + E/E_{pa}} \tag{4.46}$$

and similarly the apparent potential evaporation

$$\frac{E_{pa}}{E_{po}} = \frac{2}{1 + E/E_{pa}} \tag{4.47}$$

In Equations (4.46) and (4.47) the ratio (E/E_{pa}) may be considered as a moisture or humidity index, which depends on such factors as soil moisture and vegetation density; both relationships are illustrated in Figure 4.8. It can be seen that the dependence of (E/E_{po}), as given by Equation (4.46) and shown in Figure 4.8, has a similar trend as those shown in Figures 4.4 and 4.5.

Applications of Equation (4.45) have been made over different time scales, namely monthly (Morton, 1976; 1983), daily (Brutsaert and Stricker, 1979) and hourly (Parlange and Katul, 1992). In the application of Equation (4.45) by Brutsaert and Stricker (1979), E_{pa} can be estimated by means of (4.23), and E_{po} can be taken as E_{pe} and estimated by (4.31). Thus it was assumed that the effect of the aridity on the performance of (4.23) under non-potential conditions would mainly show up in the second term, and

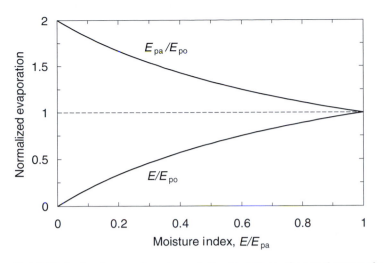

Fig. 4.8   Sketch illustrating the complementary relationship between the actual evaporation E and the apparent potential evaporation E_{pa}, for varying conditions of moisture availability, as expressed by their ratio E/E_{pa}. Both E and E_{pa} are normalized with the true potential evaporation E_{po}.

not at all in the first term on the right-hand side. Substitution of (4.23) and (4.31) into (4.45) yields the practical result

$$E = (2\alpha_e - 1)\frac{\Delta}{\Delta + \gamma}Q_{ne} - \frac{\gamma}{\Delta + \gamma}E_A \qquad (4.48)$$

As before, there are several ways of estimating E_A, the drying power of the air. It can simply be estimated in terms of the vapor pressure deficit by means of Equation (4.24) with Penman's wind function (4.25). It can also be estimated in terms of the specific humidity deficit of the air by means of the mass transfer coefficient Ce defined in Equation (4.3). In this case the actual evaporation is given by

$$E = (2\alpha_e - 1)\frac{\Delta}{\Delta + \gamma}Q_{ne} - \frac{\gamma}{\Delta + \gamma}Ce\,\bar{u}_1\rho(q_2^* - \bar{q}_2) \qquad (4.49)$$

The main advantage of (4.48), (4.49) and other equations like them based on the complementary approach, is that they do not require any information related to soil moisture, canopy resistance, or other measures of aridity, because they rely on meteorological parameters only. The main limitation is that, while the idea underlying (4.45) is simple and plausible, it was not arrived at in a rigorous theoretical or experimental way. Equations like (4.48) and (4.49) have been applied in a number of studies in widely different climates. The approach appears to perform best under conditions with relatively mild advection. However, under strongly advective conditions with large saturation humidity deficits, it seems that the validity of the assumptions on which it is based become questionable. Sugita *et al.* (2001) concluded that (4.45) is only approximately valid in most cases; Hobbins *et al.* (2001) showed how the basic approach can be improved by calibration of the parameters α_e and E_A (or Ce) in Equation (4.49). Although the complementary approach shows specific promise as a practical tool, it still awaits a definitive physical analysis to make it fully effective.

A unified parametric formulation

As reviewed here, several of the procedures in current practice, can be cast in a single form as follows

$$E = \beta_e \left[a \frac{\Delta}{\Delta + \gamma'} Q_{ne} + b \frac{\gamma}{\Delta + \gamma'} Ce\, \bar{u}_1 \rho (q_2^* - \bar{q}_2) \right]$$
(4.50)

in which a and b are weighting constants for the first and second terms, respectively, which together with the remaining parameters β_e and γ' depend on the chosen model. As before, β_e is used if actual evaporation is obtained by reduction of potential evaporation E_{pe} or E_{pa}; in the potential evaporation given by Penman's equation (4.23), the remaining parameters are $a = b = 1$, and $\gamma' = \gamma$, whereas in that given by Priestley and Taylor's equation (4.31) they are $a = \alpha_e$, $b = 0$, and $\gamma' = \gamma$. In the Penman–Monteith equation (4.39), for actual evaporation, $\beta_e = a = b = 1$ and $\gamma' = \gamma(1 + r_s\, Ce\, \bar{u}_1)$. In the advection–aridity version of Brutsaert and Stricker (4.49) they are $\beta_e = 1$, $a = 2\alpha_e - 1$, $b = -1$ and $\gamma' = \gamma$. Equation (4.50) indicates that the different formulations are related, but the parameters can vary considerably.

4.3.4 *Diurnal cycle over land: the self-preservation approximation*

It is well known that under certain favorable conditions, when horizontal advection is not too strong, the daytime variations of the major energy fluxes at land surfaces are quite similar. This similarity in the diurnal cycle of the different energy flux components over land is illustrated in Figures 2.19 and 2.20, and in Figure 4.9. This means that during any given day the ratios of these fluxes remain approximately constant, which may be considered a manifestation of some kind of "self-preservation." Because evaporation is usually relatively small during the night, this self-preservation can sometimes be useful to relate daily averages with instantaneous or hourly values. This idea was made use of by Jackson *et al.* (1983) by means of $(L_e E / R_s)$, in order to estimate the total daily latent heat flux on the basis of a one-time-of-day value. The idea was also used for the same purpose by Shuttleworth *et al.* (1989), Sugita and Brutsaert (1991), and Nichols and Cuenca (1993) by means of the evaporative fraction $EF = L_e E / (R_n - G)$ or $EF = L_e E / (L_e E + H)$. Crago (1996) applied the idea with still another dimensionless evaporation rate, namely $\alpha_e = E / E_e$, where E_e is the equilibrium evaporation defined in Equation (4.30).

In more general terms (Brutsaert and Sugita, 1992), the assumption of self-preservation requires that the evaporative flux ratio,

$$ER = L_e E / F$$
(4.51)

be taken as a constant during the daytime hours. In (4.51) F is some other flux term (beside $L_e E$) in the surface energy budget, which exhibits a similar diurnal cycle as $L_e E$, so that it can serve as a reference. The assumption of similarity can be assessed for different flux terms in Figures 4.10 (mainly the curves with the open symbols) and 4.11, with data measured during FIFE, the First ISLSCP Field Experiment conducted in hilly tallgrass terrain in eastern Kansas; it appears to work well when F is taken as the available energy

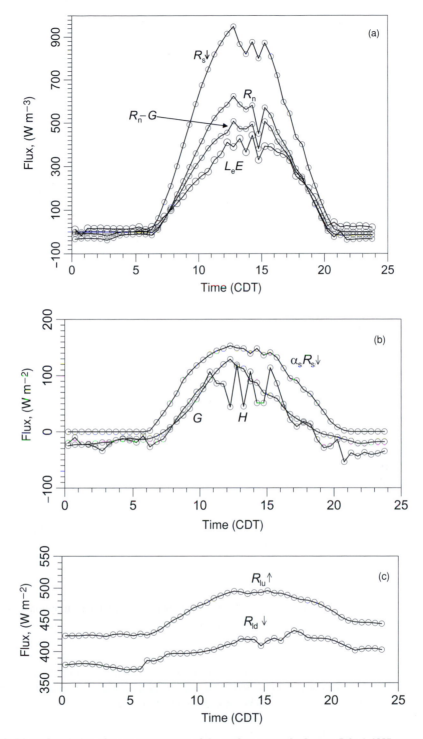

Fig. 4.9 Diurnal variation of mean components of the surface energy budget on July 6, 1987, measured at six stations over a 15 km × 15 km experimental area in a hilly prairie region in northeastern Kansas. (a) L_eE, R_n, $R_n - G$ and R_s; (b) G, H and $\alpha_s R_s$; (c) R_{lu} and R_{ld}. The time is Central Daylight Savings Time. (From Brutsaert and Sugita, 1992.)

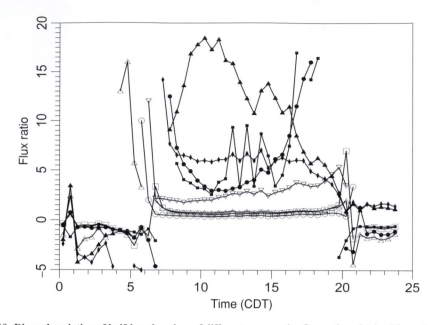

Fig. 4.10 Diurnal variation of half-hourly values of different evaporative flux ratios, obtained from the flux data shown in Figure 4.9; $L_e E / \alpha_s R_s$ (inverted open triangles); $L_e E / (R_n - G)$ (open triangles); $L_e E / R_n$ (open squares); $L_e E / R_s$ (open circles); $L_e E / H (\equiv Bo^{-1})$, solid squares); $L_e E / G$ (solid circles); $L_e E / (R_{ld} - R_{ld(night)})$ (solid triangles); and $L_e E / (R_{lu} - R_{lu(night)})$ (solid diamonds). Note that the curves of the inverted open triangles, but especially those of the open triangles, the open squares and the open circles are nearly coincident during the daytime hours, so that they may not be easy to distinguish. (From Brutsaert and Sugita, 1992.)

flux $(R_n - G)$ or $(L_e E + H)$, net radiation R_n, incoming short-wave radiation R_s and also, but less so, the reflected short-wave radiation $\alpha_s R_s$. The fact that equilibrium evaporation E_e is strongly related with the available energy flux $(R_n - G)$ (see Equation (4.30)), explains that Equation (4.51) also works well when F is taken as that quantity. Note that when F is taken as the sensible heat flux H, so that ER^{-1} is the Bowen ratio Bo, self-preservation appears to be considerably less robust than in the case of $F = (R_n - G)$. It was shown in Crago and Brutsaert (1996) that this is caused by the different error propagation properties of EF and Bo. Figure 4.11 also shows that self-similarity is not applicable at night. As a further illustration, Figure 4.12 shows the evolution of the daytime evaporative fraction EF during a long drying period over the same tallgrass prairie terrain at station 26; on any day EF remained fairly invariant during the daytime hours, but it decreased from day to day as the soil moisture was gradually declining (see also Figure 2.22).

In practical applications, this concept of self-similarity is implemented as follows. If ER is sufficiently constant during the day, the instantaneous evaporation rate at any moment during the daytime can be estimated with (4.51), that is

$$L_e E_i = ER_d F_i \qquad\qquad (4.52)$$

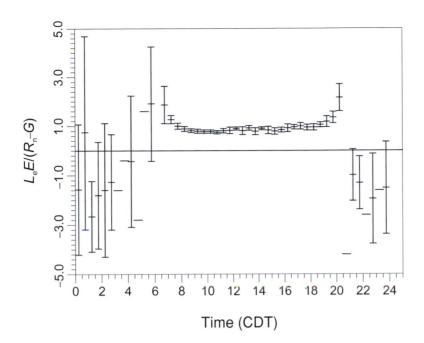

Fig. 4.11 Same as Figure 4.10, but only the evaporative fraction, i.e. the evaporative flux ratio with $F = (R_n - G)$, is shown. The error bars indicate the mean and standard deviation for the six measuring stations. (From Sugita and Brutsaert, 1991.)

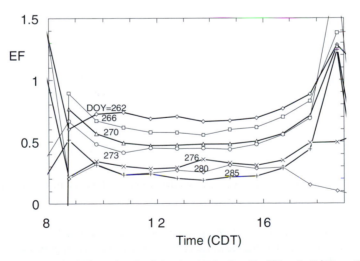

Fig. 4.12 Evolution of the diurnal cycle of the evaporative fraction $EF = L_e E/(R_n - G)$ over a period of drying from Sept. 19 (DOY 262) through Oct. 12 (DOY 285), 1987, in a hilly prairie region in northeastern Kansas. The time is Central Daylight Savings Time and the line numbers indicate the day of the year, DOY. (From Brutsaert and Chen, 1996.)

Conversely the total daytime evaporation rate can be estimated from

$$L_e E_d = \text{ER}_i F_d \tag{4.53}$$

in which subscript i refers to the instantaneous values and d refers to the daytime totals, such that $E_d = \Sigma E_i$ and $F_d = \Sigma F_i$ represent the sums of their respective instantaneous (in practice, hourly or half-hourly) values during the day.

4.4   Water budget methods

Water budget methods are based on the principle of conservation of mass applied to some part of the hydrologic cycle. Conservation of mass can be formulated as a budget equation, such as Equation (1.10), for a suitable control volume; evaporation can then be determined as the only unknown rest term among the outflow rates Q_o in the budget equation, if all the other terms in Q_i, Q_o and S can be determined independently. Although water budget methods are by far the simplest in principle, their application is often difficult and impractical. Therefore, they are less commonly used than mass transfer or energy budget methods. Still, their conceptual simplicity is an appealing feature and they can be very useful for certain purposes, such as climatological calculations or the validation of other methods over longer periods.

4.4.1   Terrestrial water budget

Over a landsurface of area A, the mean evaporation rate can be expressed in terms of the water balance equation (1.10), which for the present purpose can be rewritten as follows

$$E = P + [(Q_{ri} + Q_{gi}) - (Q_{ro} + Q_{go})] - \frac{dS}{dt} \tag{4.54}$$

where P is the areal mean rate of precipitation, Q_{ri} and Q_{ro} are the total surface inflow and outflow rates (in the river system), Q_{gi} and Q_{go} are the total groundwater inflow and outflow rates, respectively, all per unit area, and S is the water volume stored per unit area. If the area is a natural river basin or some other hydrologic catchment, bounded by natural divides, the groundwater terms are normally negligible and the surface inflow Q_{ri} is zero or, in case of artificial interbasin water exchange, it is usually known exactly. Hence, if $R = (Q_{ro} - Q_{ri})$ is the mean net surface runoff rate per unit area from the basin, Equation (4.54) can be simplified to

$$E = P - R - \frac{dS}{dt} \tag{4.55}$$

Even with reliable data on precipitation and runoff, Equation (4.55) still contains the two unknowns E and S; thus to close (4.55), either it must be applied over sufficiently long periods, so that dS/dt becomes less important, or an additional equation is required to determine S. The water storage in the basin is not easily determined. It can often be assumed that the storage returns to the same value at the end or the beginning of the same season in the previous year; therefore, an annual period is usually considered long enough to make dS/dt negligible. Several methods have been proposed to apply Equation (4.55) to periods shorter than a year by using indirect methods to estimate the change in storage.

Closure by relating storage to evaporation

In this class of methods the additional equation is obtained by relating E to S. For instance, Budyko (1955; 1974, p. 97) took as the additional equation (4.33) with (4.34), in which w was assumed to represent the water storage S in the basin and the value of w_0 was taken as a layer of 10–20 cm water, to be obtained by calibration with seasonal and regional variations. The method can be applied with average monthly values of E, P, R, $S = (S_1 + S_2)/2$ and $(dS/dt) = (S_2 - S_1)$, in which the subscripts 1 and 2 refer to the beginning and the end of the month. The calculations can be carried out by successive approximations as follows. An initial value of S_1 is chosen at random for the first month. Substitution of (4.33) into (4.55) yields an equation for S_2 and without E; with an initial value of S_1 chosen at random for the first month this produces a first estimate of S_2, which when substituted in (4.33) produces E for the first month. The same procedure is carried out for the second month, with S_2 of the first serving as S_1 of the second, and so on. The sum of all these monthly E values can be compared with the total annual value of $(P - R)$. The ratio of the two should allow a proportional adjustment of the assumed value of S_1 of the first month, and the process can be started over again and continued until the calculated annual E equals the recorded $(P - R)$. The main weakness of any method based on a relationship such as (4.33) is, beside the question of the validity of this proportionality, first the unknown value of the maximal water content parameter w_0, and second the rather ambiguous meaning of the potential evaporation concept. Budyko's method has been applied extensively over various regions of the former USSR. A very similar method has also been proposed by Thornthwaite and Mather (1955; see also Steenhuis and VanderMolen, 1986).

Closure by relating storage to stream flow

In a second class of methods the additional equation is obtained by relating S with the runoff R from the basin, usually during recessions of the river flow, i.e. during drought flows in the absence of precipitation, so that P does not have to be considered in Equation (4.55). In past studies (see Tschinkel, 1963; Daniel, 1976; Brutsaert, 1982) this has mostly been done by means of kinematic functions, which can be written in the form

$$S = K_n R^m \tag{4.56}$$

where K_n and m are constants (cf. Equation (12.48)). This combination of (4.55) and (4.56) is another example of the lumped kinematic approach; after elimination of S, in principle it should thus be possible to determine E from streamflow data. The parameters K_n and m can be determined by calibration under conditions of negligible E. The main drawback of the application of Equation (4.56) in this context, is that the storage S in (4.56) refers mainly to groundwater storage and not to near-surface soil moisture which feeds most of the evaporative processes in the basin. This means that recession flows are sensitive only to evaporation from areas, where the roots of the vegetation are in direct contact with the water table. Hence the evaporation determined this way originates mostly from the riparian zone, and not from areas further away from the stream channels, where the vegetation and the groundwater are essentially uncoupled. The S variable in Equation (4.56), which drives the streamflow R, can be assumed to represent total basin storage, only after the soil moisture has been totally depleted, that is after a long recession.

This difficulty can be avoided, as shown by Dias and Kan (1999), by integrating Equation (4.55) over sufficiently long budget periods Δt, at the end of which most of the water

Fig. 4.13 **Mean monthly climatology of precipitation, streamflow and seasonal water budget evaporation (SWB) estimates for the Jangada river basin (1055 km^2) in Parana (26°30′ S, 50°20′ W), which is mainly covered by forest vegetation. Winter and summer are well defined, but precipitation is fairly uniform throughout the year. (From Dias and Kan, 1999.)**

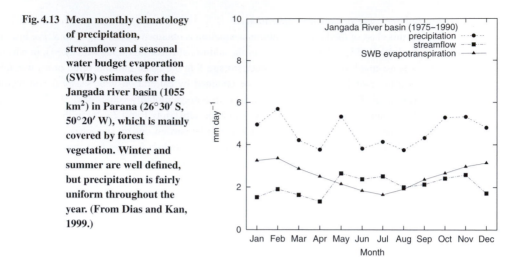

storage consists of groundwater only, so that (4.56) can be used to determine S. The beginning of the budget period can be taken as the end of the previous recession, when again (4.56) can be used to determine S. The integral of (4.55) is simply the equation rewritten with values of E, P and R averaged over the budget period Δt and with dS/dt replaced by $(S_f - S_i)/\Delta t$. Thus S_i and S_f, which are the initial and final values of the storage, can be determined from streamflow data by means of (4.56), and E remains as the only unknown in (4.55). In this approach the budget periods Δt are of variable duration, and are taken to be at least one month long. The method was implemented by assuming that the system is linear, that is $m = 1$ in (4.56), and K_n was determined from the lower envelope of a plot of daily flows versus the flow on the next day, i.e. Q_i versus Q_{i+1} during long recessions, while $P = 0$ (cf. Figure 10.30); the lower envelope was assumed to represent conditions of small or negligible evaporation from the basin. Some results obtained with this approach are illustrated in Figure 4.13; in this particular case, it can be seen that even though the rainfall does not display a seasonal dependency, the seasonal signal is well reflected in the calculated evaporation. A similar approach to derive evaporation was used by Wittenberg and Sivapalan (1999).

Finally, in a third class of methods the additional equation needed to close (4.55) is derived from the atmospheric water budget. In this approach the atmospheric water convergence allows the estimation of $(P - E)$, which then yields dS/dt for basins where R is known. This is discussed next.

4.4.2 *Atmospheric water budget*

In this method evaporation is determined as the, preferably only, unknown term in the water budget equation for a suitably chosen finite-size control volume in the atmosphere. Just like the terrestrial water budget, it is based on an integral form of Equation (1.10), which states that the total inflow minus outflow of water mass equals the time rate of change of stored water in the control volume. Thus for a control volume consisting of an atmospheric column of base area A with periphery C, this can readily (Brutsaert, 1982) be shown to be

$$E - P = \frac{\partial W}{\partial t} + \frac{1}{Ag} \int_{p_t}^{p_s} \int_C (q\, V_n)\, dC\, dp \qquad (4.57)$$

where E and P are evaporation and precipitation intensity averaged over A, W is the total water vapor content per unit surface area of the control volume, q is the specific humidity along the vertical boundary, V_n is the horizontal wind component normal to the same boundary pointing outward, p is the pressure, with subscripts s and t referring to the surface and the top of the column, and the water content in the solid and liquid phases is neglected. Equation (4.57) states that the difference between the average rate of evaporation and precipitation over a given area of the Earth's surface equals the rate of increase of water vapor over the area plus the total flux directed away from the region.

In the past this method has been used primarily over water surfaces (for a review of the early work, see Brutsaert, 1982). Among the first applications over land was the analysis of Benton and Estoque (1954) for the North American continent, followed by those of Rasmusson (1971) and Magyar et al. (1978) for the USA. In the early studies, the data needed were usually obtained from the operational radiosounding network, with twice-daily observations, so that $\partial W / \partial t$ was taken as the difference over 12 h. This world-wide grid was originally not designed for the purpose of budget calculations, but rather to observe synoptic-scale features with time scales of a few days and length scales of the order of 10^3 km. Figure 4.14 shows a comparison between results obtainable with this method over water using radiosonde observations every 6 h at four stations with an enclosed area $A = 17 \times 10^4$ km^2, and values obtained using a mean-profile method by Kondo (1976). Rasmusson (1977) has made a detailed analysis of the errors in flux divergence computations resulting from the usual limits of spatial and temporal resolution and of instrumental accuracy in typical radiosonde observations in networks of different scales. He concluded that with the operational network and current observational schedules, the applicability of the method to basins with an area smaller than 25×10^4 km^2 is limited, and that the results are likely to be unreliable; with such data the method can yield good results for areas of the order of 25×10^4 to 10^6 km^2, but it is best suited for areas larger than 10^6 km^2. In recent years, however, the situation has been changing. With the advent of improved data assimilation schemes, in which observational data can be combined with model calculation output, it is expected that the accuracy of the method may be improved considerably. As a result the method has been receiving renewed interest in the past decade (see, for example, Brubaker et al., 1994; Oki et al., 1995; Rasmusson and Mo, 1996; Berbery et al., 1996; Yarosh et al., 1996). The appeal of the method stems mainly from the simplicity of the budget concept. It is the only approach that can estimate evaporation over larger areas, and it can be useful to compare or extend the results of more local methods.

4.4.3 Soil profile water budget

With soil moisture measurements

It is also possible to consider the soil profile as a control volume, to determine its water budget. In this case, the integral of the continuity equation (1.8) (see also Equation (8.54)) for a soil column of thickness h_{so} and unit horizontal area, without lateral inflows or outflows

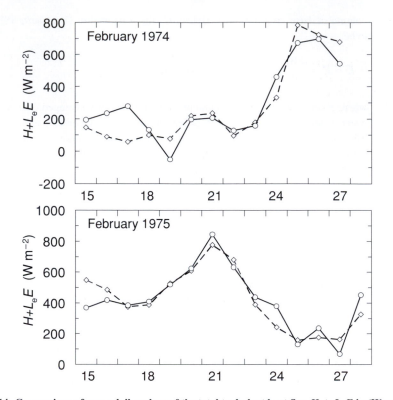

Fig. 4.14 **Comparison of mean daily values of the total turbulent heat flux** $H + L_e E$ **in** (W m^{-2}) **at the sea surface obtained by means of the atmospheric water budget (solid lines) by Nitta (1976) and by Murty (1976), with the averaged values obtained by a mean-profile method over water (dashed lines) at several stations in the East China Sea. Because the data were not taken at the same locations, the areas to which both methods were applied coincided only approximately. The experiment took place in February of 1974 and 1975, and the area enclosed for the atmospheric water budget was of the order of** $17 \times 10^4 \text{ km}^2$**; its shape was roughly rectangular with center at Okinawa. (After Kondo, 1976.)**

and with all flux terms taken as mean values over a sampling period, gives an evaporation rate

$$E = -\frac{1}{h_{so}} \int_0^{h_{so}} \frac{\partial \theta}{\partial t} \, dz + P - q_{zd} \qquad (4.58)$$

where θ is the soil water content as volume fraction, z is the vertical coordinate pointing down from $z = 0$ at the surface, P is the rate of precipitation (or irrigation), and q_{zd} the rate of downward seepage or drainage through the lower boundary of the soil layer at $z = h_{so}$. Mean values of the finite difference form of $(\partial \theta / \partial t)$ over a sampling period, as a function of z, can be determined by various methods. In the earlier field experiments related to irrigation of agricultural crops (see Israelsen, 1918; Edlefsen and Bodman, 1941), the method consisted of soil sampling and gravimetric analysis before and after drying of the samples in an oven. More recently, the neutron scattering method, TDR and other techniques (see, for example, Schmugge *et al.*, 1980) have become available that allow *in situ* soil moisture measurements.

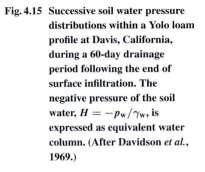
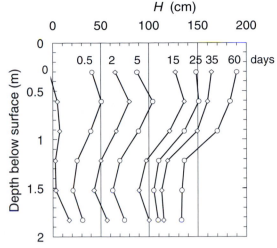

Fig. 4.15 Successive soil water pressure distributions within a Yolo loam profile at Davis, California, during a 60-day drainage period following the end of surface infiltration. The negative pressure of the soil water, $H = -p_w/\gamma_w$, is expressed as equivalent water column. (After Davidson *et al.*, 1969.)

The method is probably most useful in situations where q_{zd} is negligible, so that evaporation is the only depletion mechanism of the moisture content of the soil profile. Still, with some additional information it may be possible to obtain reliable estimates of q_{zd}. If data are available on the vertical water pressure gradient $(\partial p_w/\partial z)$ (in finite difference form) and on the hydraulic conductivity $k = k(z)$, the downward drainage rate can in principle be calculated with Darcy's law (Equation (8.19)). However, in several field studies (see Davidson *et al.*, 1969; Nielsen *et al.*, 1973) it has been observed that during the vertical redistribution of soil water at depths of 1 m or more, where there is no direct influence of surface evaporation, the hydraulic gradient is rarely very different from unity. An example of this phenomenon is shown in Figure 4.15. This allows the approximation of Equation (8.9) by

$$q_{zd} = k \tag{4.59}$$

Thus in such a case, q_{zd} may be estimated with only a measurement of the soil water content at $z = h_{so}$, provided, of course, $k = k(\theta)$ is known. In many situations, however, especially during the second stage of drying (see Chapter 9), the downward drainage rate at some depth may simply be neglected (see Jackson *et al.*, 1973); but this needs to be checked in each particular case.

Measurements of soil water content and water pressure at several levels in the profile are not easy and they require many precautions. The soil water depletion method is probably only useful for special experimental situations under favorable conditions, and it is clearly not generally applicable on a routine basis. It may be hard, if not impossible, to apply when the following conditions are present: a water table close to the surface, frequent and large precipitation, non-negligible or net lateral inflows, a large drainage rate, and considerable variability in the soil properties. Thus the accuracy obtainable with this method depends largely on the local conditions. Some other practical aspects of the soil sampling have been discussed by Jensen (1967).

A water budget-based instrument: the lysimeter

A lysimeter is a container placed in the field and filled with soil, on which vegetation can be maintained for the purpose of studying various soil–water–plant relationships under natural conditions. The rate of evapotranspiration from this instrument is obtained by solution of Equation (4.58). In order to produce the same rates of evapotranspiration as the surrounding area, a lysimeter should be representative of the conditions of the natural soil profile and of the vegetation around it. In other words, when designing and installing a lysimeter, care must be taken to insure the same water fluxes at the soil surface and the same developments of the plant roots in the soil profile. This means that its surface should be flush with the surrounding ground surface, and that it should be at least as deep as the rooting depth of the vegetation; moreover, the profiles of the soil structure, soil texture, soil water content and soil temperature in the lysimeter must be made as similar as possible to those on the outside. These conditions are not always easily met (Brutsaert, 1982). Various methods can be used to determine the different terms in Equation (4.58). One of the more complete designs (Pruitt and Angus, 1960) has a circular surface area of 6.1 m diameter and a depth of 0.91 m and it is equipped with temperature control; the integral term in (4.58) is determined by continuous monitoring of weight changes and q_{zd} is measured as the outflow by maintaining soil water suction control at the bottom of the container.

Other parameterizations

Water budget considerations can also be used to express soil surface evaporation as a capillary rise phenomenon in terms of soil properties and other variables beside soil water content. However, such derivations require some understanding of the physics of flow in partly saturated soils and also the solution of the Richards equation for various boundary conditions. This topic is therefore delayed until Chapter 9, after the principles of flow in porous media have been presented in Chapter 8.

4.5   EVAPORATION CLIMATOLOGY

In Chapter 1 it was pointed out that on a global basis the annual evaporation is of the order of 1 m, which balances the annual precipitation. Table 1.1 indicates that the evaporation from the land surfaces of the Earth is around 0.5 m, which is roughly two-thirds of the mean annual precipitation.

Interestingly, practical experience and folk wisdom suggest values similar to those given in Tables 1.1 and 1.2. For example, irrigation engineers, when lacking better information, sometimes use the rule of thumb that the duty of water for a well-irrigated crop is around $1\,l\,s^{-1}\,ha^{-1}$. Similarly, farmers in the northeastern United States are said to require a weekly rainfall of about one inch, that is 2.5 cm, to maintain field crops in good condition during their active growing period. With typical irrigation efficiencies of 25% to 40%, and a growing season of 4–5 months, both practical estimates of the evaporative requirements in agriculture are consistent with global climatological values of around $0.50\,m\,y^{-1}$.

However, because precipitation and the radiative energy supply are highly variable over the surface of the Earth, the actual evaporation is usually quite different from these long-term climatological mean values. Over periods shorter than a year, deviations of evaporation from the mean can be characterized by a cyclic or periodic behavior, namely with a daily and with a seasonal time scale. In the extreme case of an arid, warm climate, with a pronounced dry and wet season, the seasonal evaporation cycle is similar to the

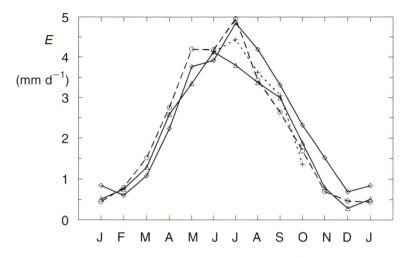

Fig. 4.16 Measured monthly evapotranspiration rates in mm/d for meadow covers on lysimeters at four locations in the eastern United States. The data at Seabrook, NJ (diamonds), Waynesville, NC (triangles) and Raleigh, NC (plus signs) represent maximal or near-maximal values, but the data at Coshocton, OH (circles) are actual values, possibly affected by irregular rainfall and moisture deficits in the soil profile. The annual mean values are 2.44 mm d^{-1} at Seabrook, 2.16 mm d^{-1} at Waynesville, and 2.29 mm d^{-1} at Coshocton. (After VanBavel, 1961.)

rainfall cycle. In a humid climate, or over water, the seasonal march of the evaporation rate follows closely the cycle of energy available for evaporation. In most climates over land the seasonal evaporation cycle is affected both by the available water and by the available energy. As an example, in Figure 4.16 the monthly mean evaporation rates are shown for several locations in the eastern United States. Thus, the cyclic behavior here is similar to the solar radiation input and to that of the air temperature. The same holds true for shallow water bodies. But over deep water bodies the evaporation cycle does not coincide with the solar winter–summer cycle. In contrast to a landsurface, a water body can store and release large amounts of heat and thus its temperature responds only slowly to energy inputs, not unlike the way a fly wheel responds to torque; as a result the cycle of available energy for evaporation may lag several months behind the solar input cycle. For example, the rate of evaporation from Lake Ontario is maximal in fall and early winter, and minimal in late spring and early summer (Phillips, 1978), as is also shown in Figure 4.17; the corresponding net radiation and heat storage are shown in Figure 4.18.

The daily evaporation cycle is usually more pronounced over land than over water. Over land, where much less heat is conducted below the surface, the daily cycle generally follows the daily march of the solar radiation. Illustrations of the daily cycles of evapotranspiration from different surfaces are shown in Figures 2.19–2.22 and in Figure 4.9, together with other components of the surface energy budget. In Figure 4.19 an example is shown of the daily cycle of evaporation from bare soil. This figure also illustrates the general behavior of evaporation after a rainfall or after irrigation, when the available water stored in the soil profile is gradually being depleted. Because the experiment took place during a drying period, the daily cycle is superimposed on a trend of decreasing

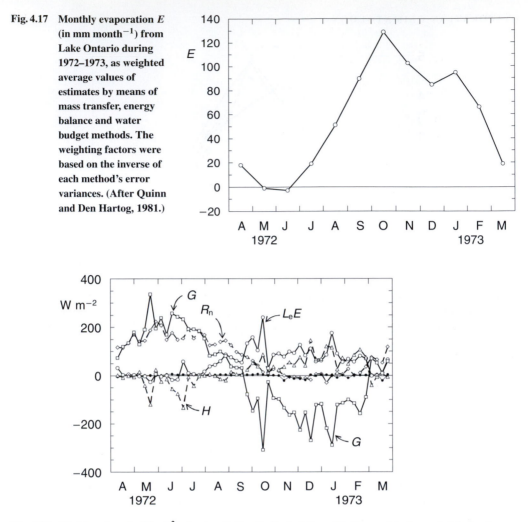

Fig. 4.17 Monthly evaporation E (in mm month^{-1}) from Lake Ontario during 1972–1973, as weighted average values of estimates by means of mass transfer, energy balance and water budget methods. The weighting factors were based on the inverse of each method's error variances. (After Quinn and Den Hartog, 1981.)

Fig. 4.18 Weekly values (in W m^{-2}) of net radiation R_n (dashed line with diamonds), change in lake heat content G (solid line with squares), advective heat inflow rate by rivers and precipitation (thin line with solid circles), latent heat flux $L_e E$ (solid line with circles) and sensible heat flux H (dashed line with triangles) for Lake Ontario during 1972–1973 from data in Pinsak and Rodgers (1981).

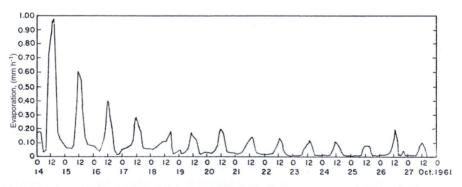

Fig. 4.19 Rate of evaporation from a bare soil surface during a drying cycle, measured with a weighing lysimeter in Arizona. (From Van Bavel and Reginato, 1962.)

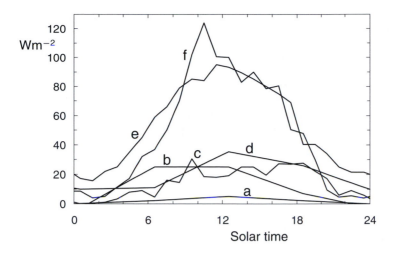

Fig. 4.20 Typical diurnal changes in hourly evaporation (in W m^{-2}) from arctic tundra for representative seasons. The dry snow period is represented by line a in late April, by line b in May, and by line c in early June. The melt period is represented by line d in mid to late June, and the postmelt period by line e in July. Line f shows the measurements in August 1969. The measurements were made as part of an energy balance program on Axel Heiberg Island (79°25′ N, 90°45′ W, 200 m ASL) in 1969 and 1970. The total annual evaporation was estimated at 140 mm, which is roughly 80% of the total precipitation. (After Ohmura, 1982)

daily mean evaporation; this general feature is similar to the drying shown in Figures 2.22 and 4.12, and it provides the justification for the assumption of self-preservation discussed above in Section 4.3.4. Finally, Figure 4.20 shows an example of the cyclic behavior of evaporation as observed under the extreme conditions of an arctic tundra environment. Remarkably, in spite of the harsh environment, the annual evaporation of 140 mm still amounts to more than one quarter of the global mean value over land.

REFERENCES

Asanuma, J. and Brutsaert, W. (1999). Turbulence variance characteristics of temperature and humidity in the unstable atmospheric surface layer above a variable pine forest. *Water Resour. Res.*, **35**, 515–521.

Benton, G. S. and Estoque, M. A. (1954). Water vapor transfer over the North American continent. *J. Meteor.*, **11**, 462–477.

Berbery, E. H., Rasmusson, E. M. and Mitchell, K. E. (1996). Studies of North American continental hydrology using ETA model forecast products. *J. Geophys. Res.*, **101**(D3), 7305–7319.

Black, T. A., Tanner, C. B. and Gardner, W. R. (1970). Evapotranspiration from a snap bean crop. *Agronomy J.*, **62**, 66–69.

Bouchet, R. J. (1963). Evapotranspiration réelle, évapotranspiration potentielle, et production agricole. *Ann. Agron.*, **14**, 743–824.

Bowen, I. S. (1926). The ratio of heat losses by conduction and by evaporation from any water surface. *Phys. Rev.*, **27**, 779–787.

Brown, K. W. and Rosenberg, N. J. (1977). Resistance model to predict evapotranspiration and its application to a sugarbeet field. *Agronomy J.*, **65**, 341–347.

Brubaker, K. L., Entekhabi, D. and Eagleson, P. S. (1994). Atmospheric water vapor transport and continental hydrology over the Americas. *J. Hydrology*, **155**, 409–430.

Brutsaert, W. (1982). *Evaporation into the Atmosphere: Theory, History and Applications*. Dordrecht, Holland/Boston, USA: D. Reidel Pub. Co.

Brutsaert, W. and Chan, F. K.-F. (1978). Similarity functions D for water vapor in the unstable atmospheric boundary layer. *Bound.-Layer Met.*, **14**, 441–456.

Brutsaert, W. and Chen, D. (1996). Diurnal variation of surface fluxes during thorough drying (or severe drought) of natural prairie. *Water Resour. Res.*, **32**, 2013–2019.

Brutsaert, W. and Stricker, H. (1979). An advection-aridity approach to estimate actual regional evapotranspiration. *Water Resour. Res.*, **15**, 443–450.

Brutsaert, W. and Sugita, M. (1992). Application of self-preservation in the diurnal evolution of the surface energy budget to determine daily evaporation. *J. Geophys. Res.*, **97**(D17), 18 377–18 382.

Brutsaert, W. and Yu, S.-L. (1968). Mass transfer aspects of pan evaporation. *J. Appl. Met.*, **7**, 563–566.

Budyko, M. I. (1955). On the determination of evaporation from the land surface. *Meteorol. Gidrol.*, **1**, 52–58 (in Russian).

    (1974). *Climate and Life*. New York: Academic Press.

Carson, D. J. (1982). Current parametrizations of land-surface processes in atmospheric general circulation models. In *Land Surface Processes in Atmospheric General Circulation Models*, ed. P. S. Eagleson. New York: Cambridge University Press, pp. 67–108.

Champagne, F. H., Friehe, C. A., LaRue, J. C. and Wyngaard, J. C. (1977). Flux measurements, flux estimation techniques and fine-scale turbulence measurements in the unstable surface layer over land. *J. Atmos. Sci.*, **34**, 515–530.

Chen, D. and Brutsaert, W. (1995). Diagnostics of land surface spatial variability and water vapor flux. *J. Geophys. Res.*, **100**(D12), 25, 595–525, 606.

Choudhury, B. J. and Blanchard, B. J. (1983). Simulating soil water recession coefficients for agricultural watersheds. *Water Resour. Bull.*, **19**, 241–247.

Crago, R. D. (1996). A comparison of the evaporative fraction and the Priestley–Taylor parameter α for parameterizing daytime evaporation. *Water Resour. Res.*, **32**, 1403–1409.

Crago, R. D. and Brutsaert, W. (1992). A comparison of several evaporation equations. *Water Resour. Res.*, **28**, 951–954.

    (1996). Daytime evaporation and the self-preservation of the evaporative fraction and the Bowen ratio. *J. Hydrol.*, **178**, 241–255.

Daniel, J. F. (1976). Estimating groundwater evapotranspiration from streamflow records. *Water Resour. Res.*, **12**, 360–364.

Davidson, J. M., Stone, L. R., Nielsen, D. R. and La Rue, M. E. (1969). Field measurement and use of soil-water properties. *Water Resour. Res.*, **5**, 1312–1321.

Davies, J. A. and Allen, C. D. (1973). Equilibrium, potential and actual evaporation from cropped surfaces in southern Ontario. *J. Appl. Met.*, **12**, 649–657.

Dias, N. L. and Brutsaert, W. (1996). Similarity of scalars under stable conditions. *Bound.-Layer Met.*, **80**, 355–373.

Dias, N. L. and Kan, A. (1999). A hydrometeorological model for basin-wide seasonal evapotranspiration. *Water Resour. Res.*, **35**, 3409–3418.

Dolman, A. J., Gash, J. H. C., Roberts, J. M. and Shuttleworth, W. J. (1991). Stomatal and surface conductance of tropical rainforest. *Agr. Forest Meteor.*, **54**, 303–318.

Doorenbos, J. and Pruitt, W. O. (1975). *Crop water requirements*. Irrigation and Drainage Paper No. 24. Rome: FAO (United Nations).

Dunin, F. X. and Greenwood, A. N. (1986). Evaluation of the ventilated chamber for measuring evaporation from a forest. *Hydrol. Process.*, **1**, 47–61.

Edlefsen, N. E. and Bodman, G. B. (1941). Field measurements of water movement through a silt loam soil. *J. Amer. Soc. Agron.*, **33**, 713–731.

Eng, K., Coulter, R. L. and Brutsaert, W. (2003). Vertical velocity variance in the mixed layer from radar wind profilers. *J. Hydrol. Engineering (ASCE)*, **8**, 301–307.

Federer, C. A. (1977). Leaf resistance and xylem potential differ among broadleaved species. *Forest Sci.*, **23**, 411–419.

Garratt, J. R. (1978). Transfer characteristics for a heterogeneous surface of large aerodynamic roughness. *Quart. J. Roy. Met. Soc.*, **104**, 491–502.

Gash, J. H. C. and Stewart, J. B. (1975). The average resistance of a pine forest derived from Bowen ratio measurements. *Bound.-Layer Met.*, **8**, 453–464.

Gash, J. H. C., Shuttleworth, W. J., Lloyd, C. R., Andre, J.-C., Goutorbe, J.-P. and Gelpe, J. (1989). Micrometeorological measurements in Les Landes forest during HAPEX-Mobilhy. *Agr. Forest Met.*, **46**, 131–147.

Grindley, J. (1970). Estimation and mapping of evaporation. Symposium on Water Balance, Vol. I, Intern. Assoc. Hydrol. Sci., Publ. **92**, 200–213.

Hobbins, M. T., Ramirez, J. A. and Brown, T. C. (2001). The complementary relationship in estimation of regional evapotranspiration: an enhanced advection-aridity model. *Water Resour. Res.*, **37**, 1389–1403.

Israelsen, O. W. (1918). Studies on capacities of soils for irrigation water, and on a new method of determining volume weight. *J. Agric. Res.*, **13**, 1–37.

Jackson, R. D., Kimball, B. A., Reginato, R. J. and Nakayama, F. S. (1973). Diurnal soil-water evaporation: time–depth–flux patterns. *Soil Sci. Soc. Amer. Proc.*, **37**, 505–509.

Jackson, R. D., Hatfield, J. L., Reginato, R. J., Idso, S. B. and Pinter, P. J. (1983). Estimation of daily evapotranspiration from one time of day measurements. *Agric. Water Manage.*, **7**, 351–362.

Jensen, M. E. (1967). Evaluating irrigation efficiency. *J. Irrig. Drain. Div., Proc. ASCE*, **93**(IR1), 83–98.

Jensen, M. E. and Haise, H. R. (1963). Estimating evapotranspiration from solar radiation. *J. Irrig. Drain. Div., Proc. ASCE*, **89**(IR4), 15–41.

Katul, G. G. and Parlange, M. B. (1992). A Penman–Brutsaert model for wet surface evaporation. *Water Resour. Res.*, **28**, 121–126.

Kondo, J. (1976). Heat balance of the East China Sea during the Air Mass Transformation Experiment. *J. Met. Soc. Japan*, **54**, 382–398.

Lindroth, A. (1985). Canopy conductance of coniferous forests related to climate. *Water Resour. Res.*, **21**, 297–304.

Magyar, P., Shahane, A. N., Thomas, D. L. and Bock, P. (1978). Simulation of the hydrologic cycle using atmospheric water vapor transport data. *J. Hydrol.*, **37**, 111–128.

Manabe, S. (1969). Climate and ocean circulation, 1. The atmospheric circulation and the hydrology of the earth's surface. *Mon. Weath. Rev.*, **97**, 739–774.

Makkink, G. F. (1957). Ekzameno de la formulo de Penman. *Netherl. J. Agric. Sci.*, **5**, 290–305.

Mascart, P., Taconet, O., Pinty, J.-P. and BenMehrez, M. (1991). Canopy resistance formulation and its effect in mesoscale models – a HAPEX perspective. *Agr. Forest Met.*, **54**, 319–351.

Mawdsley, J. A. and Ali, M. F. (1985). Estimating nonpotential evapotranspiration by means of the equilibrium evaporation concept. *Water Resour. Res.*, **21**, 383–391.

Mawdsley, J. A. and Brutsaert, W. (1977). Determination of regional evapotranspiration from upper air meteorological data. *Water Resour. Res.*, **13**, 539–548.

McNaughton, K. G. and Black, T. A. (1973). A study of evapotranspiration from a Douglas fir forest using the energy balance approach. *Water Resour. Res.*, **9**, 1579–1590.

Monteith, J. L. (1973). *Principles of Environmental Physics*. New York: American Elsevier Publ. Co.
(1981). Evaporation and surface temperature. *Quart. J. Roy. Met. Soc.*, **107**, 1–27.

Monteith, J. L., Szeicz, G. and Waggoner, P. E. (1965). The measurement and control of stomatal resistance in the field. *J. Appl. Ecol.*, **2**, 345–355.

Morton, F. (1976). Climatological estimates of evapotranspiration. *J. Hydraul. Div., Proc. ASCE*, **102**, 275–291.
(1983). Operational estimates of areal evapotranspiration. *J. Hydrology*, **66**, 1–76.

Murty, L. K. (1976). Heat and moisture budgets over AMTEX area during AMTEX '75. *J. Met. Soc. Japan*, **54**, 370–381.

Neuwirth, F. (1974). Über die Brauchbarkeit empirischer Verdunstungsformeln dargestellt am Beispiel des Neusiedler Sees nach Beobachtungen in Seemitte und in Ufernähe. *Arch. Met. Geophys. Bioklim Ser. B*, **22**, 233–246.

Nichols, W. E. and Cuenca, R. H. (1993). Evaluation of the evaporative fraction for parameterization of the surface energy balance. *Water Resour. Res.*, **29**, 3681–3690.

Nielsen, D. R., Biggar, J. W. and Erh, K. T. (1973). Spatial variability of field-measured soil-water properties. *Hilgardia*, **42**, 215–259.

Nitta, T. (1976). Large-scale heat and moisture budgets during the Air Mass Transformation Experiment. *J. Met. Soc. Japan*, **54**, 3–14.

Ohmura, A. (1982). Evaporation from the surface of the arctic tundra on Axel Heiberg Island. *Water Resour. Res.*, **18**, 291–300.

Oki, T., Musiake, K., Matsuyama, H. and Masuda, K. (1995). Global atmospheric water balance and runoff from large river systems. *Hydrol. Processes*, **9**, 655–678.

Owe, M., Choudhury, B. J. and Ormsby, J. P. (1989). Large area variability in climate-based soil moisture estimates and implications for remote sensing. *GeoJ.*, **19**(2), 177–183.

Parlange, M. B. and Katul, G. G. (1992). An advection-aridity evaporation model. *Water Resour. Res.*, **28**, 127–132.

Penman, H. L. (1948). Natural evaporation from open water, bare soil and grass. *Proc. R. Soc. London*, A **193**, 120–146.
(1956). Evaporation: an introductory survey. *Netherl. J. Agric. Sci.*, **4**, 9–29.

Phillips, D. W. (1978). Evaluation of evaporation from Lake Ontario during IFYGL by a modified mass transfer equation. *Water Resour. Res.*, **14**, 197–205.

Pinsak, A. P. and Rodgers, G. K. (1981). Energy balance. In *IFYGL – The International Field Year for the Great Lakes*, ed. E. J. Aubert and T. L. Richards. Ann Arbor, MI: NOAA, Great Lakes Envir. Res. Lab., US Dept. Commerce, pp. 169–197.

Priestley, C. H. B. and Taylor, R. J. (1972). On the assessment of surface heat flux and evaporation using large-scale parameters. *Mon. Weath. Rev.*, **100**, 81–92.

Pruitt, W. O. and Angus, D. E. (1960). Large weighing lysimeter for measuring evapotranspiration. *Trans. Amer. Soc. Agric. Eng.*, **3**, 13–15.

Quinn, F. H. and Den Hartog, G. (1981). Evaporation synthesis. In *IFYGL – The International Field Year for the Great Lakes*, ed. E. J. Aubert and T. L. Richards. Ann Arbor, MI: NOAA, Great Lakes Envir. Res. Lab., US Dept. Commerce, pp. 221–245.

Rasmusson, E. M. (1971). A study of the hydrology of eastern North America using atmospheric vapor flux data. *Mon. Weath. Rev.*, **99**, 119–135.

(1977). *Hydrological application of atmospheric vapor-flux analyses*. Operational Hydrol. Rept No. 11, WMO-No. 476, World Meteor. Org.

Rasmusson, E. M. and Mo, K. C. (1996). Large-scale atmospheric moisture cycling as evaluated from global NMC analysis and forecast products. *J. Climate*, **9**, 3276–3297.

Schmugge, T. J., Jackson, T. J. and McKim, H. L. (1980). Survey of methods for soil moisture determination. *Water Resour. Res.*, **16**, 961–979.

Shulyakovskiy, L. G. (1969). Formula for computing evaporation with allowance for temperature of free water surface. *Soviet Hydrol. Selec. Papers*, No. **6**, 566–573.

Shuttleworth, W. J., Gurney, R. J., Hsu, A. Y. and Ormsby, J. P. (1989). FIFE: The variation in energy partition at surface flux stations. Int. Assoc. Sci. Hydrol., Publ. **186**, 67–74.

Slatyer, R. O. and McIlroy, I. C. (1961). *Practical Microclimatology*. Melbourne, Australia: CSIRO.

Steenhuis, T. S. and VanderMolen, W. H. (1986). The Thornthwaite–Mather procedure as a simple engineering method to predict recharge. *J. Hydrology*, **84**, 221–229.

Stephens, J. C. and Stewart, E. H. (1963). A comparison of procedures for computing evaporation and evapotranspiration. General Assembly Berkeley. Int. Assoc. Sci. Hydrol., Publ. **62**, 123–133.

Stewart, J. B. (1988). Modeling surface conductance of pine forest. *Agr. Forest Meteor.*, **43**, 19–37.

Sugita, M. and Brutsaert, W. (1991). Daily evaporation over a region from lower boundary layer profiles measured with radiosondes. *Water Resour. Res.*, **27**, 747–752.

Sugita, M., Usui, J., Tamagawa, I. and Kaihotsu, I. (2001). Complementary relationship with a convective boundary layer model to estimate regional evaporation. *Water Resour. Res.*, **37**, 353–365.

Szeicz, G. and Long, I. F. (1969). Surface resistance of crop canopies. *Water Resour. Res.*, **5**, 622–633.

Szeicz, G., VanBavel, C. H. B. and Takami, S. (1973). Stomatal factor in the water use and dry matter production of sorghum. *Agr. Met.*, **12**, 361–389.

Tan, C. S. and Black, T. A. (1976). Factors affecting the canopy resistance of a Douglas-fir forest. *Bound.-Layer Met.*, **10**, 475–489.

Tanner, C. B. (1960). Energy balance approach to evapotranspiration from crops. *Soil Sci. Soc. Amer. Proc.*, **24**, 1–9.

Thom, A. S. (1972). Momentum absorption by vegetation. *Quart. J. Roy. Met. Soc.*, **97**, 414–428.

(1975). Momentum, mass and heat exchange of plant communities. In *Vegetation and the Atmosphere, Vol. I, Principles*, ed. J. L. Monteith. London: Academic Press, pp. 57–109.

Thom, A. S. and Oliver, H. R. (1977). On Penman's equation for estimating regional evaporation. *Quart. J. Roy. Met. Soc.*, **103**, 345–357.

Thornthwaite, C. W. (1948). An approach toward a rational classification of climate. *Geograph. Rev.*, **38**, 55–94.

Thornthwaite, C. W. and Holzman, B. (1939). The determination of evaporation from land and water surfaces. *Mon. Weath. Rev.*, **67**, 4–11.

Thornthwaite, C. W. and Mather, J. R. (1955). *The Water Balance*. Publications in Climatology, 8, No. 1. Centerton, NJ: Lab. of Climatology.

Tillman, J. (1972). The indirect determination of stability, heat and momentum fluxes in the atmospheric boundary layer from simple scalar variables during dry unstable conditions. *J. Appl. Met.*, **11**, 783–792.

Tschinkel, H. M. (1963). Short-term fluctuation in streamflow as related to evaporation and transpiration. *J. Geophys. Res.*, **68**, 6459–6469.

VanBavel, C. H. M. (1961). Lysimetric measurements of evapotranspiration rates in the eastern United States. *Soil Sci. Soc. Amer. Proc.*, **25**, 138–141.

    (1967). Changes in canopy resistance to water loss from alfalfa induced by soil water depletion. *Agr. Met.*, **4**, 165–176.

VanBavel, C. H. M. and Reginato, R. J. (1962). Precision lysimetry for direct measurement of evaporative flux. *Intern. Sympos. Methodol. of Plant Eco-Physiol.*, Montpellier, France, pp. 129–135.

Webb, E. K. (1964). Further note on evaporation with fluctuating Bowen ratio. *J. Geophys. Res.*, **69**, 2649–2650.

Wesely, M. (1988). Use of variance techniques to measure dry air–surface exchange rates. *Bound.-Layer Met.*, **44**, 13–31.

Wittenberg, H. and Sivapalan, M. (1999). Watershed groundwater balance estimation using streamflow recession analysis and baseflow separation. *J. Hydrol.*, **219**, 20–33.

Yarosh, E. S., Ropelewsky, C. F. and Mitchell, K. E. (1996). Comparisons of humidity observations and ETA model analyses and forecasts for water balance studies. *J. Geophys. Res.*, **101**(D18), 23 289–23 298.

PROBLEMS

4.1   Derive Equation (4.4) from (2.41) and (2.44).

4.2   The local vertical vapor flux over a large, uniform, grass surface is measured to be 4 mm d^{-1}. At 2.0 m above the surface, the air temperature, the relative humidity and the wind speed are 20 °C, 60% and 5 m s^{-1}. Calculate the value of the wind speed and the specific humidity at 10 m, assuming a neutral practically isothermal atmosphere and a surface roughness of 1 cm. Ignore the displacement height.

4.3   Suppose you are given the local evaporation rate, E (in mm d^{-1}), from a wet surface at ground level. The surface temperature and the surface specific humidity are T_s and q_s, respectively. The corresponding values at 2 m elevation are T_2 for temperature and q_2 for specific humidity. Derive an expression for the local turbulent sensible heat flux, H (in W m^{-2}), near the ground in terms of these variables. (Do not use Bowen's ratio or the psychrometric constant as a variable in your final expression.)

4.4   Prove Equation (4.6).

4.5   Show a derivation of Equation (4.28) from (4.4), (4.6) and (4.7).

4.6   Multiple choice. Indicate which of the following statements are correct. The empirical mass-transfer approach consists of determining mean evaporation from a water surface by means of an equation of the type of (4.7), in which $f_e(\bar{u}_1)$ is some known function of the mean wind speed at a given elevation.
        (a)   For long-term (say 1 d or longer) averages, the wind function, which appears in Equation (4.7), is also needed to calculate the Bowen ratio.
        (b)   Equation (4.7) is less useful over crop-covered land, because it requires the determination of q_s at a surface, which is often irregular and ill defined.

(c)   Equation (4.7) can be adapted to determine the turbulent sensible heat flux at the surface, if the temperatures at the surface and in the air are known.

(d)   In principle (not in practice), the wind function should incorporate the effect of the surface roughness, z_0.

(e)   But, on an hourly basis, $f_e(\bar{u}_1)$ cannot be improved by including the effect of the stability of the atmosphere.

4.7   Multiple choice. Indicate which of the following statements are correct. In the case of a uniform, horizontal surface, evaporation can be expressed by the following equations. The subscripts 1 and 2 refer to the reference levels, z_1 and z_2, respectively, within an atmospheric-surface layer under neutral conditions. Assume $d_0 = 0$.

(a) $E = \dfrac{u_*^2 \rho(\bar{q}_1 - \bar{q}_2)}{(\bar{u}_2 - \bar{u}_1)}$

(b) $E = -(ku_* z)\rho \dfrac{d\bar{q}}{dz}$

(c) $E = -u_*^2 \rho \dfrac{(d\bar{q}/dz)}{(d\bar{u}/dz)}$

(d) $E = \dfrac{\rho k^2 (\bar{u}_2 - \bar{u}_1)(\bar{q}_1 - \bar{q}_2)}{[\ln(z_2/z_1)]^2}$

(e) $E = \dfrac{\rho k^2 \bar{u}_2 (q_s - \bar{q}_2)}{[\ln(z_2/z_0)][\ln(z_2/z_{0v})]}$

4.8   Consider the empirical mass transfer equation (4.24) as developed by Penman (1948). Calculate the value of the scalar roughness, z_{0v}, which is implied by that equation for a neutral atmosphere above a surface with a momentum roughness of $z_0 = 0.05$ m, for an air temperature of 20 °C, and for a typical wind speed of $\bar{u}_2 = 5$ m s^{-1}.

4.9   How many millimetres per month (mm month^{-1}) of liquid water can be evaporated at 25 °C by an energy supply of 1 W m^{-2}? Prove your answer.

4.10   The following measurements were recorded at a micrometeorological site: net radiation, $R_n = 200$ W m^{-2}; heat flux into the ground, $G = 40$ W m^{-2}; and evaporation rate, $E = 5 \times 10^{-8}$ m s^{-1}. (a) Calculate the turbulent sensible heat flux, H, in W m^{-2}. (b) How large was Bowen's ratio? (c) Was the atmosphere stable or unstable? Why? (d) Was the soil warming up or cooling? Why?

4.11   The following measurements were recorded at a micrometeorological site: net radiation, $R_n = 250$ W m^{-2}; heat flux into the ground, $G = 30$ W m^{-2}; and sensible, heat flux, $H = \rho c_p \overline{w'T'} = 55$ W m^{-2}. (a) Calculate the evaporation rate, E, in kg m^{-2}s^{-1} and in mm d^{-1}. (b) How large was Bowen's ratio? (c) Was the atmosphere stable or unstable? Why? (d) Was the soil warming up or cooling? Why?

4.12   Calculate, in mm d^{-1}, for a typical summer day, the different measures of evaporation listed below. The data for that day and, thus also, R_n, are given and calculated in Problem 2.17, and the wind speed at 10 m above the ground was measured as 10.4 km h^{-1}. Assume that the average daily ground heat flux, G, is negligible. (a) Potential evaporation by means of Penman's method; use Penman's wind function. (b) Potential evaporation under conditions of minimal advection according to

Priestley and Taylor (1972); use $\alpha_e = 1.27$. (c) Actual evaporation, assumed to be given by the equilibrium evaporation; use $\alpha_e = 1.0$. (d) Actual evaporation by the advection–aridity approach of Equation (4.48).

4.13   Same as Problem 4.12 but with the data of Problem 2.18; assume that the wind speed at 10 m above the ground was 8.96 km h^{-1}.

4.14   Give an expression for the Bowen ratio, implied by the equilibrium evaporation concept, as formulated in Equation (4.30)? What is the numerical value of this Bowen ratio at 25 °C?

4.15   Multiple choice. Indicate which of the following statements are correct. The Penman equation (4.23):
   (a)   has the advantage, in practical applications, that only measurements at one level (instead of vertical gradients or differences) above the ground are needed;
   (b)   is well suited to calculate actual watershed evapotranspiration under drought conditions, because it takes account of the moisture saturation deficit of the air;
   (c)   should, in principle, be adjusted for any given surface, as a function of the surface roughness, z_0, and of the atmospheric stability to yield an accurate result;
   (d)   is applicable even in the tropics;
   (e)   for calculations over land covered with vegetation for periods of a day or longer, the ground heat flux, G, can often be neglected.

4.16   Consider the same lake as in Problem 2.19, and the following additional data for December and July: mean heat flux into the lake water body, $G = -430$ and 390 cal cm^{-2} d^{-1}; and mean wind speed at 10 m above the water surface, $\bar{u}_{10} = 15.3$ and 10.1 km h^{-1}. Calculate the mean evaporation, in mm d^{-1}, from the lake for these typical days in December and July, by means of: (a) the energy budget method with the Bowen ratio; (b) Penman's method; (c) Priestley and Taylor's method.

4.17   (a) Calculate the mean evaporation from the lake considered in Problems 2.19 and 4.16 for the same two days, by using the mass-transfer method (in mm d^{-1}). To a first approximation, assume that conditions are neutral, so you can use transfer coefficients, Ce$_{10} = 1.2 \times 10^{-3}$. (b) Is this a reasonable assumption for these two cases? Why not?

4.18   Derive the expression for β_e, as given by Equation (4.40), by equating (4.38) and (4.33) (with (4.3) to represent E_p for a wet surface).

4.19   Estimate the time of local (i.e. solar) noon in Central Daylight Savings Time (CDT) for the location of the measurements shown in Figures 4.9–4.12. The longitude of this location is approximately 96°31′ W and CDT is 5 h behind Universal Time, or (UTC−0500).

4.20   Suppose the following measurements are available in the atmospheric surface layer under unstable conditions: net radiation, R_n; ground heat flux, G; mean wind speed at two levels, $\bar{u}_1, \bar{u}_2$; and potential temperature at two levels, $\bar{\theta}_1$ and $\bar{\theta}_2$. Write down three equations that will allow you to solve for the three unknowns, viz., the rate of evaporation, E; the sensible heat flux, H; and the surface shear stress, u_*. Make sure that the answer contains, beside some constants, only the variables listed here (or functions thereof).

II   WATER ON THE SURFACE

5 WATER ON THE LAND SURFACE: FLUID MECHANICS OF FREE SURFACE FLOW

Owing to the irregular topography of the Earth's continents, surface runoff, that is the flow of water over land, takes place in many different ways. When for some reason, such as rainfall, snowmelt, the overtopping of small depressions, or the emergence of groundwater at a source, surface flow is initiated, it may at first proceed as a thin sheet flow; however, as a result of local irregularities, the flow soon gathers in small gullies and rills, which in turn join to form rivulets in the fashion of a tree-like network. Eventually these merge with others to become larger rivers, which finally end up in some lake or in the ocean. Thus the flow system consists of an intricate combination of many different types of flow regimes, in channels of different geometries and sizes. For purposes of analysis, to describe the basic hydraulic elements of landsurface runoff, it is convenient and useful to distinguish between two major types of free surface flow; these are first, sheet flow or overland flow, which is most likely to occur under conditions of heavy precipitation in source areas where runoff is being generated which feeds into streams; and second, the flow that occurs in larger permanent open channels. Both types of flow are usually unsteady and spatially varied. In this chapter a general description is given of free surface flow. The general principles are then applied to overland flow and to channel flow and streamflow routing in the next two chapters.

5.1 FREE SURFACE FLOW

The flow of water on a solid surface is governed by the usual conservation equations of fluid mechanics, namely the continuity equation for mass and the Navier–Stokes equations for momentum. One important condition on the boundaries can be formulated by observing that once a fluid particle is on an impermeable surface, it stays on it (see Lamb, 1945, p. 7). In other words, it moves with the surface, and its velocity relative to the surface is either purely tangential or zero (in the case of no slip), for otherwise a finite flow of fluid would take place across the surface. Thus, if the surface is described by a function $F = F(x, y, z, t) = 0$, then any displacement occurring with the fluid particles should leave that function unchanged, i.e.

$$\frac{DF}{Dt} = 0 \tag{5.1}$$

The operator D/Dt, already defined in Equation (1.3), is the time derivative *following the motion*, also called the *fluid mechanical* time derivative, the *substantial* time derivative,

Fig. 5.1 **Definition sketch for two-dimensional free surface flow. FS indicates the free surface of the flowing water.**

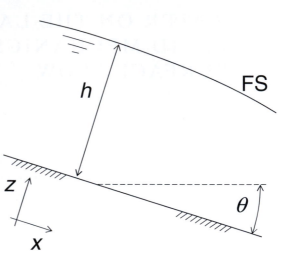

or the *material* or *particle* derivative; for the mean motion it can be defined as

$$\frac{D}{Dt} = \frac{\partial}{\partial t} + \overline{u}\frac{\partial}{\partial x} + \overline{v}\frac{\partial}{\partial y} + \overline{w}\frac{\partial}{\partial z} \qquad (5.2)$$

where, as in Chapter 2, $\overline{u}, \overline{v}$ and $\overline{w}$ are the mean (in the turbulence sense) velocity components of the fluid in the x, y and z directions, respectively, of the velocity vector $\mathbf{v} = (\overline{\mathbf{v}} + \mathbf{v}')$.

To simplify the argument, consider a two-dimensional motion of water with a free surface, which is located at a distance, taken normally to the bottom surface, $z = z_s(x, t)$ from an arbitrary reference; the water is flowing over a bottom, which is located at a normal distance $z = z_b(x, t)$ from that same reference (see Figure 5.1). Observe that, contrary to its usage in Chapter 2, here the z-axis is not vertical, but has an angle θ with it. For the situation shown in the figure the function defining the position of the water surface is $F(x, z, t) = [z_s(x, t) - z] = 0$; therefore, condition (5.1) becomes for the water surface

$$\overline{u}\frac{\partial z_s}{\partial x} - \overline{w} + \frac{\partial z_s}{\partial t} = 0 \qquad \text{at } z = z_s \qquad (5.3)$$

Similarly, the bottom surface can be described by $F(x, z, t) = [z_b(x, t) - z] = 0$, in which the time dependency allows, in principle at least, for bottom sediment accretion or erosion; thus Equation (5.1) leads to an analogous condition for the bottom interface of the fluid, which looks the same as Equation (5.3), but with the subscript s replaced by a subscript b. Usually, however, the bottom can be treated as a solid wall without slip, so that this bottom condition reduces to $\overline{u} = \overline{w} = 0$. With the latter bottom condition, the condition for the free surface (5.3) can also be written in terms of the water depth as follows

$$\overline{u}\frac{\partial h}{\partial x} - \overline{w} + \frac{\partial h}{\partial t} = 0 \qquad \text{at } z = h \qquad (5.4)$$

in which the water depth is defined as $h = z_s - z_b$ and the reference level $z = 0$ is placed at the bottom.

5.2 HYDRAULIC THEORY: SHALLOW WATER EQUATIONS

In most situations of free surface flow encountered in hydrology, it is possible to make certain simplifications. The main assumption is that the path- or streamlines are only slightly curved so that the accelerations normal to the direction of mean flow are negligible. This means that the pressure distribution may be taken as hydrostatic along the z-direction, i.e. normally to the bottom, or

$$\frac{\partial p}{\partial z} + \gamma \cos \theta = 0 \tag{5.5}$$

where θ is the slope angle of the bottom and $\gamma (\equiv \rho g)$ is the specific weight of the water. With $z = 0$ at the bottom surface, the integral of (5.5) is

$$p = \gamma \cos \theta (h - z) \tag{5.6}$$

where as before $h = h(x, t) = (z_s - z_b)$ is the water depth measured normally to the bottom. If the bed slope angle θ is constant in the main direction of flow, Equation (5.6) yields immediately upon differentiation

$$\frac{\partial p}{\partial x} = \gamma \cos \theta \frac{\partial h}{\partial x} \tag{5.7}$$

As this pressure gradient is not a function of z, the corresponding acceleration of the water particles is independent of z as well. Therefore the velocity parallel to the bottom $\bar{u}$ preserves its dependence on z, independently of x and t. Accordingly, it is permissible to replace $\bar{u} = \bar{u}(x, z, t)$ by its average over z, namely $V = V(x, t)$ defined by

$$V = \frac{1}{h} \int_0^h \bar{u} \, dz \tag{5.8}$$

These two simplifications, namely the hydrostatic pressure distribution and the assumption of an average velocity V, constitute the basis of the so-called hydraulic theory of free surface flow; as will become clear below, it reduces the two-dimensional problem to a one-dimensional problem. The theory is usually referred to as shallow water theory or the theory of long waves. It consists of reducing the continuity and the momentum or Reynolds equations to the shallow water equations. This will be shown in what follows.

5.2.1 Equation of continuity

The equation of continuity of an incompressible fluid is given by Equation (1.9). For turbulent flow this can be equally applied to the mean and to the turbulent velocity components; if there is also a source inflow ϕ_1 at the point under consideration, the equation of continuity for the mean velocity components becomes in the case of two-dimensional motion

$$\frac{\partial \bar{u}}{\partial x} + \frac{\partial \bar{w}}{\partial z} - \phi_1 = 0 \tag{5.9}$$

Integration over z produces

$$\int_0^h \frac{\partial \overline{u}}{\partial x} dz + [\overline{w}]_0^h - \int_0^h \phi_1 dz = 0 \tag{5.10}$$

and insertion of condition (5.4) for a solid bottom surface

$$\int_0^h \frac{\partial \overline{u}}{\partial x} dz + \overline{u}|_{z=h} \frac{\partial h}{\partial x} + \frac{\partial h}{\partial t} - \int_0^h \phi_1 \, dz = 0 \tag{5.11}$$

By virtue of Leibniz's rule (see Appendix) for the differentiation of an integral, the first term can be rewritten as

$$\frac{\partial}{\partial x} \int_0^h \overline{u} dz - \overline{u}|_{z=h} \frac{\partial h}{\partial x} \tag{5.12}$$

The equation of continuity becomes finally

$$\frac{\partial h}{\partial t} + \frac{\partial}{\partial x} (Vh) - i = 0 \tag{5.13}$$

in which i is the net lateral inflow per unit width of flow, which results from the integration of ϕ_1 in (5.11). Equation (5.13) was probably first derived by Dupuit (1863; p. 149) for $i = 0$.

5.2.2 Conservation of momentum

The conservation of momentum at a point in a moving Newtonian fluid is given by the Navier–Stokes equation. When the flow is turbulent, this is conveniently transformed into the Reynolds equation for the mean quantities. The Reynolds equation can be readily obtained from Equation (1.12), by replacing each of the dependent variables by the sum of its mean and fluctuation, both in the turbulence sense, and by subsequently applying the time-averaging operation over a suitable time period. For the two-dimensional case of incompressible flow under consideration, and with a source inflow ϕ_1, the component of Equation (1.12) parallel to the bottom surface can be written as follows

$$\frac{\partial \overline{u}}{\partial t} + \overline{u} \left(\frac{\partial \overline{u}}{\partial x} + \phi_1 \right) + \overline{w} \frac{\partial \overline{u}}{\partial z} = -g \sin \theta - \frac{1}{\rho} \frac{\partial p}{\partial x} + \nu \nabla^2 \overline{u} - (\nabla \cdot \overline{\mathbf{v}'}) u' \tag{5.14}$$

in which $\mathbf{v}' = (u'\mathbf{i} + v'\mathbf{j} + w'\mathbf{k})$ is the turbulent fluctuation in the velocity vector $\mathbf{v} = (\mathbf{v} + \mathbf{v}')$. Observe that (5.14), without the last two terms on the right-hand side, is in the form of Euler's equation (1.11); these two terms represent respectively the stresses due to viscosity and the Reynolds stresses due to the turbulence. To obtain a momentum equation in terms of the average velocity V defined in (5.8), it is necessary to integrate (5.14) over z, as follows. For convenience, first, the zero quantity, consisting of (5.9) multiplied by $\overline{u}$, is added to (5.14) to obtain

$$\frac{\partial \overline{u}}{\partial t} + \frac{\partial}{\partial x} (\overline{u}^2) + \frac{\partial}{\partial z} (\overline{w}\,\overline{u}) = -g \sin \theta - \frac{1}{\rho} \frac{\partial p}{\partial x} + \nu \nabla^2 \overline{u} - (\nabla \cdot \overline{\mathbf{v}'}) u' \tag{5.15}$$

By using Leibniz's rule (see Appendix), one can write the integral of the first term on the left of Equation (5.15) as

$$\int_0^h \frac{\partial \bar{u}}{\partial t} dz = \frac{\partial}{\partial t}(Vh) - \bar{u}|_{z=h} \frac{\partial h}{\partial t} \tag{5.16}$$

In the same way, by also using surface condition (5.4) and the assumption that

$$\int_0^h \bar{u}^2 dz = V^2 h \tag{5.17}$$

one obtains the integral of the second term on the left of (5.15) as

$$\int_0^h \frac{\partial(\bar{u}^2)}{\partial x} dz = \frac{\partial}{\partial x}(V^2 h) - \overline{w}\,\bar{u}|_{z=h} + \bar{u}|_{z=h} \frac{\partial h}{\partial t} \tag{5.18}$$

On account of the definition of V in (5.8), the assumption of (5.17) can be valid only if $\bar{u}$ is uniform, that is, constant along z. With a no-slip condition at $z = 0$, this is never the case; nevertheless, in turbulent open channel flow, which is well mixed in the vertical, it is usually an acceptable approximation. However, for laminar and transitional flows, a correction coefficient (often associated with the name of Boussinesq; see Bakhmeteff, 1941), namely,

$$\beta_c = \int_0^h (\bar{u}/V)^2 dz/h \tag{5.19}$$

may have to be applied to the first term on the right-hand side of (5.18), i.e. the advective acceleration term.

The integration of the other terms in Equation (5.15) is straightforward. The third term on the left yields upon integration simply the product of $\overline{w}$ and $\bar{u}$ at $z = h$. If it can be assumed that θ is small, the first term on the right can be approximated by replacing $-\sin\theta$ by $-\tan\theta = S_0$, which is the slope of the bottom surface, so that its integral becomes $(g S_0 h)$. Similarly, the pressure gradient in the second term on the right can be replaced by the water depth gradient on account of (5.7), provided the slope angle θ is small enough.

In the hydraulic approach to free surface flow, the integral of the last two terms of Equation (5.15) is usually expressed in terms of the friction slope S_f, as a closure parameterization to account for the effects of the viscosity and the turbulence. For the present case of two-dimensional flow, this is

$$\int_0^h \nu\left(\frac{\partial^2 \bar{u}}{\partial x^2} + \frac{\partial^2 \bar{u}}{\partial z^2}\right) dz - \int_0^h \left(\frac{\partial(\overline{u'u'})}{\partial x} + \frac{\partial(\overline{w'u'})}{\partial z}\right) dz = -gh S_f \tag{5.20}$$

With (5.16), (5.18) and (5.20), the integral of (5.15) becomes finally

$$\frac{\partial}{\partial t}(Vh) + \frac{\partial}{\partial x}(V^2 h) + hg\left(\frac{\partial h}{\partial x} + S_{\mathrm{f}} - S_0\right) = 0 \tag{5.21}$$

which is the momentum equation of the hydraulic theory of free surface flow. This result is often written in an alternative form, which is obtained by subtraction of continuity (5.13) multiplied by V, and by subsequent division by h; thus the momentum equation is often written as follows

$$\frac{\partial V}{\partial t} + V\frac{\partial V}{\partial x} + g\left(\frac{\partial h}{\partial x} + S_{\mathrm{f}} - S_0\right) + \frac{iV}{h} = 0 \tag{5.22}$$

Equations (5.13 and 5.22) are known as the shallow water equations; a simpler version was first presented by Saint Venant in the nineteenth century, so that they are often named after him. To recapitulate briefly, the shallow water equations are based on the following assumptions. (i) The pressure distribution in the water is hydrostatic leading to Equation (5.5); (ii) the bed slope S_0 is constant and small, which leads from Equation (5.6) to (5.7), and allows replacement of $\sin\theta$ by $\tan\theta = -S_0$; (iii) the effects of viscous and turbulent stresses can be parameterized and combined in a friction slope S_{f}, defined in Equation (5.20); (iv) the velocity is not very dependent on z, so that β_{c} in Equation (5.19) can be taken as unity.

Example 5.1. Steady flow

The meaning of the different terms in Equation (5.22) can be illustrated by considering steady flow conditions in the absence of lateral inflow. Thus, after putting both $\partial V/\partial t$ and i equal to zero, (5.22) can be readily integrated over a flow distance δx to yield

$$\frac{V_1^2}{2} + gh_1 + gS_0\delta x = \frac{V_2^2}{2} + gh_2 + gS_{\mathrm{f}}\delta x \tag{5.23}$$

in which the subscripts 1 and 2 refer to the entrance and exit of the flow section δx. Figure 5.2 shows the balance of the left- and right-hand sides of this equation. Recall that the bed slope is sufficiently small, so that x, the coordinate along the bed, can be represented as horizontal in the figure. Because the integration of a force (or rate of momentum change) over distance yields work, the terms of Equation (5.23) may be considered as different forms of energy. In open channel hydraulics the energy per unit weight with respect to the channel bottom is called the *specific energy*; in the present notation this is $[h + V^2/(2g)]$. As shown in Figure 5.2, its elevation defines the *energy grade line* (EGL); the friction slope S_{f} is the slope of the energy grade line. The quantity $[z + p/(\rho g)]$ in any cross section defines the *hydraulic grade line* (HGL); because this is equal to the water depth h, the HGL coincides with the water surface.

Equations (5.13) and (5.22) were derived for two-dimensional flow, i.e. an infinitely wide channel. It can, however, readily be shown that for a channel with finite cross section of arbitrary shape, but wide enough so that the flow is approximately still

Fig. 5.2 Illustration of the different terms in Equation (5.23), i.e. the integrated shallow water momentum equation under steady conditions.

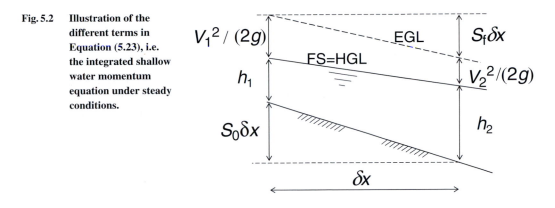

two-dimensional, they assume the form (see, for example, Stoker, 1957)

$$\frac{\partial A_c}{\partial t} + \frac{\partial}{\partial x}(V A_c) - Q_1 = 0 \tag{5.24}$$

and

$$\frac{\partial V}{\partial t} + V\frac{\partial V}{\partial x} + g\left(\frac{\partial h}{\partial x} + S_f - S_0\right) + \frac{Q_1 V}{A_c} = 0 \tag{5.25}$$

where A_c is the wetted cross-sectional area and Q_1 the lateral inflow per unit length of channel.

In hydrology the average velocity is often of less importance than the rate of flow, $Q = V A_c$. For a wide channel of surface width $B_s = (\partial A_c/\partial h)$, (5.13) and (5.22) assume the form

$$\frac{\partial A_c}{\partial t} + \frac{\partial Q}{\partial x} - Q_1 = 0 \tag{5.26}$$

and

$$A_c^2\frac{\partial Q}{\partial t} + 2A_c Q\frac{\partial Q}{\partial x} + \left(g A_c^3 - Q^2 B_s\right)\frac{\partial h}{\partial x} + g A_c^3(S_f - S_0) = 0$$

5.3   FRICTION SLOPE

In the determination of the friction slope in the shallow water equations, it is commonly assumed that the resistance to flow, resulting from the last two terms of Equation (5.15), acts in the same way in unsteady nonuniform flow as it does in steady uniform flow. Thus, from inspection of (5.21) or (5.22) for such conditions one has

$$S_f = S_0 \tag{5.27}$$

The definition of S_f for two-dimensional flow, as given in Equation (5.20), can be written for uniform flow, when $\partial()/\partial x = 0$, as

$$S_f = \frac{-1}{\rho g h}\int_0^h \frac{\partial \tau_{zx}}{\partial z}dz \tag{5.28}$$

in which here

$$\tau_{zx} = \rho \left(\nu \frac{\partial \bar{u}}{\partial z} - \overline{(w'u')} \right) \tag{5.29}$$

is the shear stress and in which the subscript z denotes the direction normal to the plane on which the stress acts and the subscript x denotes the direction of this stress itself. Equation (5.28) is readily integrated with the boundary condition that the shear stress at the water surface is zero. If τ_0 is the shear stress at the bottom wall where $z = 0$, this integration yields

$$S_f = \frac{\tau_0}{\gamma h} \tag{5.30}$$

or, in terms of the friction velocity $S_f = u_*^2/gh$. Unfortunately, an expression for S_f in terms of the shear stress at the wall is not of much help at this point. Instead, to be able to solve the shallow water equations an expression is required in terms of the main dependent variables, namely h and V. Hence, to repeat briefly, first a relationship is obtained relating the slope S_0 with the flow variables h and V (or the analogous variables, such as A_c and Q) under uniform steady conditions. In accordance with Equation (5.27), this relationship obtained for S_0 is then used in the shallow water momentum equation to parameterize S_f in terms of the same flow variables. In the following two sections, relationships are presented for laminar and turbulent flow.

5.3.1   Laminar flow

The case of two-dimensional steady uniform flow, that is plan-parallel flow down a plane surface, can be solved exactly for laminar conditions, when both u' and w' are zero. For such conditions all terms in Equation (5.14) (or (5.15)) are zero, except the first and third on the right-hand side. Integrating these remaining two terms twice (with the conditions that $\partial u/\partial z = 0$ at $z = h$ and that $u = 0$ at $z = 0$) and making use of Equation (5.27) or $-\sin\theta = S_f$, one obtains the velocity profile

$$u = \frac{g S_f}{\nu}(hz - z^2/2) \tag{5.31}$$

After normalization with the maximal velocity u_h at $z = h$, this velocity profile can be written as $u/u_h = 2(z/h) - (z/h)^2$, which is illustrated in Figure 5.3. Integrating (5.31) over z, according to (5.8), one obtains the average velocity,

$$V = \frac{g S_f h^2}{3\nu} \tag{5.32}$$

In the absence of lateral inflow by precipitation the applicability of Equation (5.32) depends mainly on the Reynolds number $\mathrm{Re} \equiv (Vh/\nu)$; as Re increases the flow will become turbulent, but the transition may also depend on the smoothness of the surface, the uniformity and stationarity of the flow, and possibly other factors. Experimentally, Equation (5.32) has been found to fail for Re values as low as 300, and for flows over smooth surfaces without lateral inflow it has also been observed to be valid up to Re $=$ 1000 (see, for example, Chow, 1959; Woo and Brater, 1961). An upper limit of

Fig. 5.3   **Velocity profile for steady laminar free surface flow on a sloping plane. The velocity u_h is the maximal value at the water surface where $z = h$.**

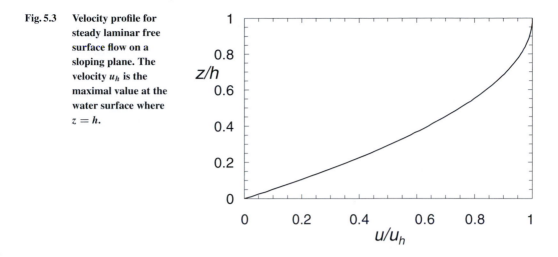

Re $\equiv (Vh/\nu) = 500$ can be taken as a typical value. The case of rectangular channels with finite width has been studied by Woerner *et al.* (1968).

The effect of rainfall impact at the free surface on the friction slope was considered by Yoon and Wenzel (1971). Their experimental results over smooth surfaces can be summarized by a simple formula proposed in Brutsaert (1972),

$$V = \frac{g S_f h^2}{\nu\left(3 + c S_f^d P^e\right)} \qquad \text{for Re} < 800, \ S_0 \le 0.03 \tag{5.33}$$

where P is the rainfall intensity in cm h^{-1} and $c, d, \ e$ are constants, whose values were estimated to be $c = 5.36, d = 0.16$ and $e = 0.36$. With additional data Shen and Li (1973) subsequently concluded that $c = 2.32, d = 0$ and $e = 0.40$.

5.3.2   Turbulent flow

For Reynolds numbers larger than 1000, free surface flow in nature may generally be considered fully turbulent and rough, with little or no effect from the impact of lateral inflow. Unlike laminar flow, however, even when it is steady and uniform, there is no exact solution available for turbulent flow. As was the case for the analogous turbulence closure problem in the atmosphere discussed in Chapter 2, the main difficulty here is the presence of the second moments of the velocity fluctuations, that is the Reynolds stress, in Equations (5.14) and (5.29). At present, the only practical way to eliminate (or determine) this unknown shear stress τ_{zx} is either to invoke similarity theory or to rely entirely on empirical results.

Similarity for two-dimensional turbulent flow

Uniform two-dimensional turbulent flow can be considered as a fully developed boundary layer. For rough flow conditions, which are usually the case for flow over a natural land-surface, the velocity profile in the inner region of the neutral boundary layer is given by Equations (2.40) or (2.41). Strictly speaking, one can apply this logarithmic profile only in the lower 10% to 20% of the boundary layer thickness; in the outer reaches of the boundary layer, some type of velocity defect law is more appropriate. However, in two-dimensional free surface flow this defect law (see, for example, Keulegan, 1938) is often assumed to be

$$\overline{u}_h - \overline{u} = -\frac{u_*}{k} \ln \left(\frac{z - d_0}{h - d_0} \right) \tag{5.34}$$

which implies that Equation (2.40) can be used over the whole depth of the flow. This is apparently only an approximation, but in laboratory type turbulent boundary layers the difference between Equations (2.40) or (5.34) and observed velocity profiles in the outer region is quite small (see, for example, Hinze, 1959, p. 473; Monin and Yaglom, 1971, pp. 300–301, 315–317; Kisisel *et al.*, 1973). Thus, in light of (2.41) and (5.30), the mean velocity can be expressed roughly (for $h \gg z_0$) as

$$V = \frac{(g S_f)^{1/2}(h - d_0)}{k h^{1/2}} \left[\ln \left(\frac{h - d_0}{z_0} \right) - 1 \right] \tag{5.35}$$

or (when also $h \gg d_0$),

$$V = \frac{(g S_f h)^{1/2}}{k} \left[\ln \left(h/z_0 \right) - 1 \right] \tag{5.36}$$

As an aside, it should be mentioned that one possible problem with the application of (2.40) to open channel flow is that z_0 may be a function of Froude number $\text{Fr} = V/(gh)^{1/2}$ (Iwagaki, 1954; Chow, 1959); this means that the structure of the turbulent boundary layer of water flowing down a slope may also be affected by gravity, g, beside the variables that were considered in the derivation of Equation (2.40). A second potential difficulty is the effect of the raindrop impact. For instance, Kisisel *et al.* (1973) showed that for shallow flows with h values around 15 mm and a high rainfall intensity of $P = 125 \text{ mm h}^{-1}$, the measured velocity profiles $(\overline{u}_h - \overline{u})$ were logarithmic, but the resulting V values would be only about half the magnitude predicted by (5.36). For less shallow flows, however, the effect of rainfall impact is likely to be small.

For certain applications it has on occasion been found convenient to express the turbulent velocity profile by a simple power function of height, instead of by (2.40). Among the more recent forms, for $z \gg d_0$, the following has been used

$$\overline{u} = C_p u_* \left(\frac{z}{z_0} \right)^m \tag{5.37}$$

in which C_p and m are constants. The use of power functions to describe wind speed profiles in the lower atmosphere goes back at least to the work of Stevenson in the 1870s (see Brutsaert, 1982, for a review). An equation similar to (5.37) was implicit in the work of Prandtl and Tollmien (1924; see also Brutsaert, 1993), and it has subsequently been applied by many in the solution of various turbulent transport problems. The parameters C_p and m may be determined by fitting Equation (5.37) to the more accurate (2.41) over the range of heights z that are of interest; values of m tend to lie in the range from 1/6 to 1/8, with a typical value of 1/7, and C_p is of the order of m^{-1}. Integration of (5.37) according to (5.8) and substitution of the friction velocity u_* (see (2.32)) by (5.30) yields the average velocity

$$V = \left[\frac{C_p g^{1/2}}{(m + 1) z_0^m} \right] S_f^{1/2} h^{m+1/2} \tag{5.38}$$

A derivation of Equation (5.38), somewhat different from the present one, was first published by Keulegan (1938). The main point of interest is that (5.38) can serve as a theoretical basis for some of the empirical equations for S_f that will be reviewed next.

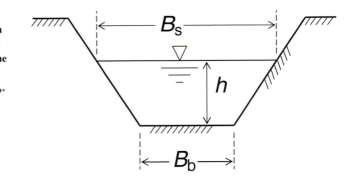

Fig. 5.4 An open channel with a trapezoidal cross section. In this case the cross-sectional area is $h(B_s + B_b)/2$ and the wetted perimeter is $P_w = [(B_s - B_b)^2 + 4h^2]^{1/2} + B_b$.

Empirical equations

Most empirical equations derived from uniform steady flow measurements are of the form

$$V = C_r R_h^a S_f^b \tag{5.39}$$

where C_r is a resistance factor, which ideally depends only on the nature of the channel, and a and b are constants; R_h is the hydraulic radius defined as

$$R_h = \frac{A_c}{P_w} \tag{5.40}$$

in which A_c is the cross-sectional area of the channel, and P_w the wetted perimeter of that area. In two-dimensional flow situations and wide channels, the hydraulic radius equals the depth of the flow, or $R_h = h$.

Example 5.2

This can be seen in the case of a channel with a trapezoidal cross section with a water depth h, a bottom width B_b and a water surface width B_s (see Figure 5.4); according to (5.40), this has a hydraulic radius

$$R_h = \frac{h(B_s + B_b)/2}{[(B_s - B_b)^2 + 4h^2]^{1/2} + B_b}$$

which approaches h as both B_s and B_b become large (compared to h) and practically equal to each other.

Probably the oldest form of open channel equation (5.39) is the one developed in France by Chézy around 1770 in which $a = b = \frac{1}{2}$ (Mouret, 1921). Numerous expressions have been proposed for C_r (see Chow, 1959, p. 94, for a review). Some insight can be gained in the nature of C_r, by comparing Chézy's equation with the more theoretically based expressions for turbulent flow. It can be seen from the similarity expression for wide channels (5.36), that in the Chézy equation the resistance factor C_r is given by $C_r = (g/k)^{1/2}[\ln(h/z_0) - 1]$; this indicates that C_r depends not only on the roughness but also on the hydraulic radius. It can also be seen from the power function expression (5.38), that C_r is independent of h, and therefore of R_h, only if $m = 0$, that is when

Table 5.1 Some typical values of the roughness coefficient *n* for natural channels

Channel type	n
Earth, straight, with short vegetation	0.02–0.03
Gravel bed, straight	0.03–0.04
Earth and gravel, winding with some weeds on banks	0.03–0.05
Sand and gravel bed with boulders or with brush and overhanging trees on banks	0.035–0.06
Boulders and banks of exposed rock	0.05–0.08
Earth overgrown with weeds	0.07–0.09

the velocity is uniform along the vertical. This suggests that for a given channel C_r can only be truly constant for highly turbulent flow, or for very high Reynolds numbers $Re \equiv Vh/\nu$.

Another very popular formula is the Gauckler–Manning (GM) equation, named after the two engineers who contributed most to its development (see Powell, 1962; 1968; Williams, 1970; 1971). It is usually written as

$$V = \frac{1}{n} S_f^{1/2} R_h^{2/3} \tag{5.41}$$

The constant n is referred to as the channel roughness coefficient, when the variables are expressed in SI units. Numerous experiments have been carried out to determine it for all kinds of channels and surfaces. Some values are shown in Table 5.1, but more detailed results for a wider range of conditions can be found in Chow (1959) and in Barnes (1967). Comparison with Equation (5.38) shows that the GM formula (5.41) can be derived theoretically by assuming a power law such as (5.37) with $m = 1/6$. This indicates that the GM formula can be expected to be valid over a range of lower Reynolds numbers than Chézy's, which requires the extreme value $m = 0$ for a perfectly uniform velocity profile. Equation (5.38) also shows that n is directly proportional to $z_0^{1/6}$; it should be recalled from Section 2.5.2 that, as a first approximation, z_0 may be assumed to be of the order of one tenth of the size of the roughness elements of the wall. In any event, the power law assumption, on which the GM equation is implicitly based, should be adequate for most practical applications. This is illustrated in Figure 5.5, which gives a comparison between the dependence of the average velocity V on the water depth (h/z_0), as calculated with the logarithmic profile (5.36) and that calculated with the power profile (5.38). The two curves display satisfactory agreement for the value of the constant $C_p = 5.4$. Comparison of the GM equation (5.41) with (5.38) produces then the following relationship between the channel roughness coefficient and the boundary layer roughness height,

$$n = 0.0690 z_0^{1/6} \tag{5.42}$$

in which z_0 is expressed in metres.

Fig. 5.5 **Comparison between the average stream velocity V, as described by the logarithmic equation (5.36) (heavy line) and by the power equation (5.38) (thin line). The constants in the power equation are $m = (1/6)$ and $C_p = 5.4$.**

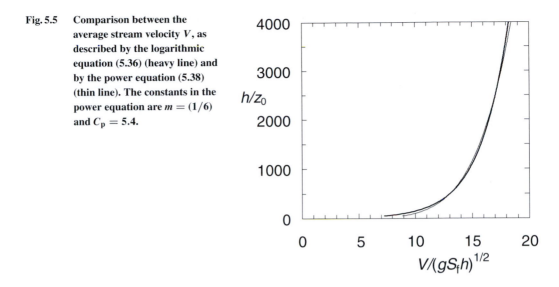

The equations of Chézy and Gauckler–Manning suggest that a in (5.39) normally lies in a range between 0.5 and 0.7 under fully turbulent flow conditions. On the other hand, Equation (5.32) shows that $a = 2$ for laminar flow. In some studies of high Reynolds number sheet flow over surfaces covered with short vegetation, such as grass, it has been concluded that a, as the power of h, may have an intermediate value close to unity. Horton (1938) adopted $a = 1$ to derive the rising hydrograph; he conjectured that this might represent a flow that is 75% turbulent and 25% laminar. Horner and Jens (1942) derived this value of $a = 1$ from experimental data by different investigators. An analysis by Henderson and Wooding (1964) of data published by Hicks (1944) confirmed that a may indeed be close to unity over a very rough or grass-covered surface. Wooding (1965) interpreted this phenomenon by noting that, owing to fluctuations in depth and roughness over an irregular surface, the flow regime can vary spatially and temporally between laminar and turbulent; in addition, even when the flow near the water surface is turbulent, the flow within the lower layers between the grass stems and leaves may be more like laminar seepage through a porous medium.

5.3.3 *Summary of friction slope parameterizations*

Equation (5.39) can be used as a general expression for the friction slope S_f. For two-dimensional flow or for wide channels, it assumes the form

$$V = C_r h^a S_f^b \qquad (5.43)$$

which can also be formulated conveniently in terms of the rate of flow per unit width, $q = Vh$. The values of the parameters for laminar and turbulent flow are summarized in Table 5.2

Table 5.2 Values of the parameter constants in the friction slope S_f as given by Equations (5.39) or (5.43)

Parameter	Laminar		Turbulent	
	Without rain	With rain	(Gauckler–Manning)	(Chézy)
C_r	$\dfrac{g}{3\nu}$	$\dfrac{g}{\nu(3 + cS_f^d P^e)}$	n^{-1}	C_r
a	2	2	2/3	1/2
b	1	1	1/2	1/2

Note: P is the rainfall intensity and values of the constants c, d and e are given following Equation (5.33).

5.4 GENERAL CONSIDERATIONS AND SOME FEATURES OF FREE SURFACE FLOW

The solution of the shallow water equations (5.13) and (5.22) (or (5.24) and (5.25)) is not easy, and most flow problems encountered in natural situations have to be analyzed by numerical methods. The availability of digital computation technology in recent decades has greatly facilitated this, and rapid advances have been made in this field (see Liggett and Cunge, 1975; Cunge *et al.*, 1980; Tan, 1992; Montes, 1998). Nevertheless for a better understanding of their structure and the physical implications, it is useful to consider simpler forms of these equations; these are valid in certain special situations, for which solutions may be more easily obtainable, or for which important features of the flow can be deduced by inspection.

5.4.1 *Complete system of the shallow water equations: small disturbances*

As can be seen in the second term and in the term containing S_f, Equation (5.22), describing the conservation of momentum, is a nonlinear partial differential equation. However, if the flow is a small departure from an initially uniform steady state, it is possible to linearize the shallow water equations, which greatly facilitates the solution. More importantly, however, not only is the solution easier, but it also brings out clearly some of the general physical features, regarding the coexistence of different wave types, which are inherent in the nonlinear system as well.

Consider for this purpose a small departure from uniform steady flow, by the substitution of $V = V_0 + V_p$ and $h = h_0 + h_p$, in which the subscript 0 refers to uniform steady conditions and the subscript p refers to a small perturbation or disturbance. This is illustrated in Figure 5.6 for the water depth h. Thus Equations (5.13) and (5.22) become, after retention of the first-order terms,

$$V_0 \frac{\partial h_p}{\partial x} + h_0 \frac{\partial V_p}{\partial x} + \frac{\partial h_p}{\partial t} - i = 0$$

and   (5.44)

$$\frac{\partial V_p}{\partial t} + V_0 \frac{\partial V_p}{\partial x} + g \frac{\partial h_p}{\partial x} + g(S_f - S_0) = 0$$

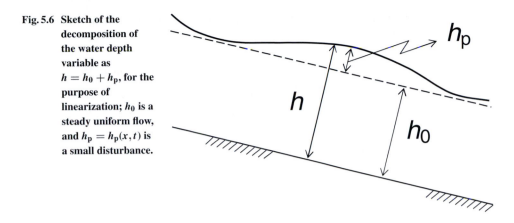

Fig. 5.6 Sketch of the decomposition of the water depth variable as $h = h_0 + h_p$, for the purpose of linearization; h_0 is a steady uniform flow, and $h_p = h_p(x, t)$ is a small disturbance.

Solutions of the dynamic part

If the departures V_p and h_p are assumed to be small, one can, as a further approximation, neglect the effect of the additional turbulence resulting from the perturbation, and put $S_f = S_0$. This leaves then only the purely dynamic part of the momentum equation, i.e. the first three terms of the second of Equations (5.44). Differentiating the second of (5.44) with respect to x, and eliminating $(\partial V_p/\partial x)$ by means of the first, one can combine these two equations into one, or in the absence of lateral inflow,

$$\frac{\partial^2 h_p}{\partial t^2} + 2V_0 \frac{\partial^2 h_p}{\partial x \partial t} + \left(V_0^2 - gh_0\right) \frac{\partial^2 h_p}{\partial x^2} = 0 \tag{5.45}$$

Because the undisturbed flow is uniform and steady, it is convenient to describe the motion of the small disturbance relative to a reference moving with the velocity V_0 of this undisturbed flow. This can be done by substituting $x_m = (x - V_0 t)$ and $t_m = t$, where the subscript m refers to the moving reference; thus the partial derivatives are $\partial/\partial x = \partial/\partial x_m$ and $(\partial/\partial t) = (\partial/\partial t_m) - V_0(\partial/\partial x_m)$, and Equation (5.45) becomes

$$\frac{\partial^2 h_p}{\partial t_m^2} - gh_0 \frac{\partial^2 h_p}{\partial x_m^2} = 0 \tag{5.46}$$

Similarly, by differentiating the first of (5.44) with respect to x, substituting $(\partial h_p/\partial x)$ from the second of (5.44) and by using the same coordinate transformation, one obtains

$$\frac{\partial^2 V_p}{\partial t_m^2} - gh_0 \frac{\partial^2 V_p}{\partial x_m^2} = 0 \tag{5.47}$$

The same type of equation can also be derived for the rate of flow $q = Vh$. Putting $q = q_0 + q_p$, one has $q_0 = V_0 h_0$ and $q_p = V_0 h_p + V_p h_0$, because $V_p h_p$ is negligible. Thus one obtains, by adding Equations (5.46) and (5.47), after multiplying each by V_0 and h_0, respectively,

$$\frac{\partial^2 q_p}{\partial t_m^2} - gh_0 \frac{\partial^2 q_p}{\partial x_m^2} = 0 \tag{5.48}$$

Equations (5.46), (5.47) and (5.48) are in the form of the classical linear wave equation

$$\frac{\partial^2 y}{\partial t_m^2} - c_0^2 \frac{\partial^2 y}{\partial x_m^2} = 0 \tag{5.49}$$

in which the dependent variable y can represent h_p, V_p or q_p, and in which

$$c_0 \equiv (gh_0)^{1/2} \tag{5.50}$$

is a constant introduced at this point for conciseness of notation. The general solution of this wave equation can be written as

$$F_1 [x_m - c_0 t_m] + F_2 [x_m + c_0 t_m] \tag{5.51}$$

where F_1 and F_2 are arbitrary functions, which must be determined from the initial and boundary conditions. It can be readily verified that Equation (5.51) is indeed a solution by substituting it back into (5.49) and carrying out the differentiations. It can also be shown, that the solution must have the form of (5.51), as follows. Consider for this purpose the coordinate transformations $\xi = x_m - c_0 t_m$ and $\eta = x_m + c_0 t_m$, which lead from y as a function of x_m and t_m to (say) Y as a function of ξ and η, or

$$y(x_m, t_m) = Y(\xi, \eta)$$

By applying the chain rule of differentiation to (5.49) with this equality, one obtains the differential equation for Y, or

$$\frac{\partial^2 Y}{\partial \xi \partial \eta} = 0$$

This shows that $(\partial Y/\partial \xi)$ is independent of η, and conversely that $(\partial Y/\partial \eta)$ is independent of ξ. Hence if Y is to depend on both ξ and η it must have the form $Y = F_1(\xi) + F_2(\eta)$, which is the same as (5.51).

The form of Equation (5.51) describes actually two waves, each with a speed of propagation c_0 (relative to a reference moving with the fluid velocity V_0 of the undisturbed flow), but traveling in opposite directions. To distinguish the speed of propagation of a disturbance or of a wave, from the velocity of the fluid itself, it is common to refer to it as *celerity*. As an illustration, consider now the case where y represents the water surface elevation h_p, which is the easiest to visualize. In this case, initially at $t_m = 0$, the function $F_1(x_m - c_0 t_m)$ defines a water surface configuration $h_p = F_1(x_m)$; at a later time $t_m = t_{m1}$ it describes the configuration $h_p = F_1(x_m - c_0 t_{m1})$. However the water surface shape is still the same, except that during the t_{m1} units of time it has moved to the right without distortion over a distance $c_0 t_{m1}$. The same can be said about the function $F_2(x_m + c_0 t_m)$, which defines a water surface configuration moving in the opposite direction, with the same celerity c_0. The actual displacement of the water surface is the sum of these two waves. Equation (5.50) is commonly referred to as Lagrange's celerity equation.

Example 5.3. Long channel with arbitrary initial conditions

To determine the arbitrary functions F_1 and F_2, initial and boundary conditions are needed. Consider a wide uniform channel extending to $x_m = \pm\infty$ from the (moving) origin at $x_m = 0$. Assume as initial conditions that the variable $y = f(x_m)$ and also its time derivative $(\partial y/\partial t_m) = g(x_m)$ are known for any value of x_m. Thus with (5.51) one has

$$\begin{aligned} y(x_m, 0) &= f(x_m) = F_1(x_m) + F_2(x_m) \\ \frac{\partial y}{\partial t_m}\bigg|_{x_m, 0} &= g(x_m) = c_0 \left[-F_1'(x_m) + F_2'(x_m) \right] \end{aligned} \tag{5.52}$$

Division of the second of Equations (5.52) by c_0 followed by integration and subsequent combination with the first of (5.52), yields the following form for the two functions

$$F_1(x_m) = \frac{1}{2} f(x_m) - \frac{1}{2c_0} \int_{x_{m0}}^{x_m} g(s)ds$$

$$F_2(x_m) = \frac{1}{2} f(x_m) + \frac{1}{2c_0} \int_{x_{m0}}^{x_m} g(s)ds$$

(5.53)

in which s is the dummy integration variable. Thus the solution (5.51) can be written for this case as

$$y = \frac{1}{2} [f(x_m - c_0 t_m) + f(x_m + c_0 t_m)] + \frac{1}{2c_0} \int_{x_m - c_0 t_m}^{x_m + c_0 t_m} g(s)ds$$

(5.54)

which is commonly attributed to d'Alembert. This result (5.54) is equally applicable to the water surface depth h_p, the velocity V_p and the rate of flow q_p. However, in each case the functions $f(x_m)$ and $g(x_m)$ should represent the initial conditions of the intended variable.

Example 5.4. Infinitely long channel with zero initial time derivative

If the value of the time derivative of the variable in question is initially equal to zero, or $g(x_m) = 0$, d'Alembert's solution (5.54) becomes simply

$$y = \frac{1}{2} f(x_m - c_0 t_m) + \frac{1}{2} f(x_m + c_0 t_m)$$

(5.55)

As an illustration, consider the following function to describe the initial disturbance,

$$f(x_m) = \left[\alpha\left(1 + 10x_m^2\right)\right]^{-1}$$

(5.56)

In this expression α is a constant, which should be large to ensure that the perturbation is small, compared to the undisturbed part of the variable, i.e. h_0, V_0 or q_0, describing the steady uniform flow. With this initial condition, the solution is, in accordance with Equation (5.55),

$$y(x_m, t_m) = \frac{1}{2\alpha} \left[\left(1 + 10(x_m - c_0 t_m)^2\right)^{-1} + \left(1 + 10(x_m + c_0 t_m)^2\right)^{-1}\right]$$

(5.57)

This solution is shown in Figure 5.7 for the values $c_0 t_m = 0, 0.2, 0.4, 0.8$ and 1.2, and illustrates the propagation of the disturbance with time, relative to the reference point $x_m = 0$, which is moving with the velocity V_0.

If the initial disturbance is a unit impulse $y = \delta(x_m)$ (see Appendix) (and $\partial y/\partial t = 0$), Equation (5.55) yields the unit response function for an infinitely long channel as

$$u = \frac{1}{2} [\delta(x_m - c_0 t_m) + \delta(x_m + c_0 t_m)]$$

or, in the original coordinate system

$$u = \frac{1}{2} [\delta(x - (V_0 + c_0)t) + \delta(x - (V_0 - c_0)t)]$$

(5.58)

This describes a translatory motion of two delta functions, one with, and one against the flow V_0. Equation (5.56) is not a delta function, but Figure 5.7 gives an idea of how the two unit impulses in (5.58) proceed.

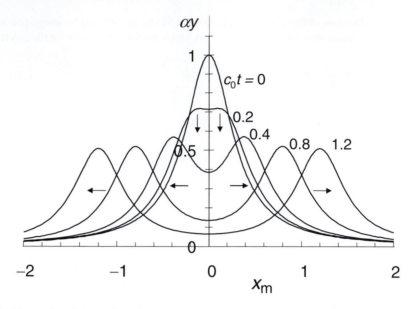

Fig. 5.7 **Illustration of the motion of a disturbance, whose initial distribution is given by Equation (5.55). This disturbance is made up of two components, which are one half of the original disturbance and which are moving away in opposite directions from the origin $x_m = 0$. The origin itself is moving downstream with a velocity V_0. The curves are obtained with Equation (5.56) for $c_0 t_m = 0, 0.2, 0.4, 0.8$, and 1.2.**

Example 5.5. Semi-infinite channel with known upstream inflow

A situation which has been the subject of many studies is that of a channel reach, with a given inflow at its upstream end. If the upstream end is taken as the origin, where $x = 0$, the boundary and initial conditions can be written as

$$y(0, t) = y_u(t); x = 0, t > 0$$

$$y(\infty, t) \text{ finite}; x \to \infty, t > 0$$

$$y(x, 0) = 0; x > 0, t = 0 \tag{5.59}$$

$$\frac{\partial y(x, 0)}{\partial t} = 0; x > 0; t = 0$$

These conditions can now be used with Equation (5.49) to solve the problem. However, rather than to attempt a frontal attack, it is convenient here to make use of the d'Alembert solution (5.58) for the unit impulse derived in the previous example, and by inspection of Figure 5.7. This shows immediately that the component of that solution moving to the left away from $x = 0$ can be considered an image, and that the component moving into the domain $x > 0$ is the desired result. Thus one has the unit response function, that is the solution for an input $\delta(t)$ at $x = 0$,

$$u = u(x, t) = \delta(x - (V_0 + c_0)t) \tag{5.60}$$

If the input at $x = 0$ is actually $y = y_u(t)$ instead of a delta function, the solution can be readily obtained by convolution. For instance, in the case where $y(t)$ represents the flow rate disturbance q_p, the solution resulting from an inflow $q_u(t)$ at $x = 0$ is

$$q = q(x, t) = q_0 + \int_0^t q_u(\tau)\delta(x - (V_0 + c_0)(t - \tau))d\tau \tag{5.61}$$

or, upon integration,

$$q = q(x, t) = q_0 + q_u(t - x/(V_0 + c_0)) \tag{5.62}$$

To repeat briefly the results of this section, general solutions of the dynamic part of the shallow water equations result in two waves; their speed of propagation $c_0 = (gh_0)^{1/2}$ relative to the mean motion, is also known as Lagrange's celerity equation (cf. Equation (5.50)). One of these "dynamic" waves is moving in the direction of the current and the other against the current. Thus two observers, one moving downstream with a velocity $c_{01} \equiv (V_0 + c_0)$ and the other with a velocity against the current $c_{02} \equiv (V_0 - c_0)$, would see the small disturbance as a stationary, i.e. non-moving, displacement of the surface from equilibrium. Recall in this context, that the ratio V_0/c_0 defines the Froude number for steady uniform flow, that is

$$\mathrm{Fr}_0 = \frac{V_0}{(gh_0)^{1/2}} \tag{5.63}$$

Therefore, when $c_{02} < 0$, or $\mathrm{Fr}_0 < 1$, the flow is subcritical and this disturbance (or the observer) actually moves upstream; when $c_{02} > 0$, the flow is supercritical, and while this disturbance still moves against the current, it is smaller than V_0 and is thus swept downstream. The paths of these two observers traced on the x–t plane are called characteristics. From the above analysis it follows that these characteristics can be defined by the ordinary differential equations

$$\frac{dx}{dt} = V_0 + (gh_0)^{1/2} = c_{01}$$

and   (5.64)

$$\frac{dx}{dt} = V_0 - (gh_0)^{1/2} = c_{02}$$

The concept of characteristics, which is being introduced here in an offhand way, arises formally in the theory of partial differential equations. However, this is beyond the scope of the present discussion. For an introduction to the mathematics, the reader is referred to such books as Sommerfeld (1949), and on the application of characteristics to free surface flow to Stoker (1957) or Abbott (1975).

Solutions of complete system: two types of wave

If the last term in the second of Equations (5.44) is not neglected, it must also be expressed in terms of the initial steady uniform flow variables and their perturbations; making use of (5.39) for flow in a wide channel, with $R_h = h$ (or (5.43)), one obtains

$$S_f = \frac{(V_0 + V_p)^2}{C_r^2(h_0 + h_p)^{2a}} = \frac{V_0^2}{C_r^2 h_0^2}[(1 + 2V_p/V_0 + \cdots)(1 - 2ah_p/h_0 + \cdots)]$$

or, neglecting higher-order terms,

$$S_f - S_0 = b^{-1} S_0 [(V_p / V_0) - (a h_p / h_0)] \tag{5.65}$$

for which values of a and b can be taken from Table 5.2. In many applications in hydrology, the discharge rate q is of greater interest than the water depth h or the velocity V. Hence, replacing V by q/h and making use of $q = q_0 + q_p$, one can rewrite (5.44) for turbulent flow with $b = 1/2$ as follows

$$\frac{\partial h_p}{\partial t} + \frac{\partial q_p}{\partial x} - i = 0$$

and $\qquad\qquad\qquad\qquad\qquad\qquad\qquad\qquad\qquad\qquad\qquad$ (5.66)

$$\left(g h_0^3 - q_0^2\right) \frac{\partial h_p}{\partial x} + 2 q_0 h_0 \frac{\partial q_p}{\partial x} + h_0^2 \frac{\partial q_p}{\partial t} + 2 g h_0^3 S_0 \left(\frac{q_p}{q_0} - \frac{(1+a) h_p}{h_0} \right) = 0$$

These may be combined into one equation by operating with $(\partial / \partial t)$ on the second of (5.66) and then substituting $(\partial h_p / \partial t)$ from the first, or

$$h_0^2 \frac{\partial^2 q_p}{\partial t^2} + 2 q_0 h_0 \frac{\partial^2 q_p}{\partial t \partial x} + \left(q_0^2 - g h_0^3\right) \frac{\partial^2 q_p}{\partial x^2} + \left(2 g h_0^3 S_0 / q_0\right) \frac{\partial q_p}{\partial t}$$

$$+ 2 (1 + a) g S_0 h_0^2 \frac{\partial q_p}{\partial x} = \left(q_0^2 - g h_0^3\right) \frac{\partial i}{\partial x} + 2 (1 + a) g h_0^2 S_0 i \tag{5.67}$$

This equation reduces to the one first derived by Deymie (1938) for the special case without lateral inflow, i.e. for $i = 0$, and with Chézy's formula, i.e. for $a = 1/2$. The solution of Deymie's equation for the propagation of a disturbance resulting from a known $q_p = q_u(t)$ at $x = 0$, has been obtained by different methods (see Deymie, 1939; Massé, 1939; Lighthill and Whitham, 1955; Dooge and Harley, 1967).

The more general solution, for an inflow $q_p = q_u(t)$ at $x = 0$ and an arbitrary non-zero lateral inflow $i = i(x, t)$, has been presented by Brutsaert (1973) and the reader is referred to the journal article for the mathematical details. The conditions, that must be satisfied by (5.67) to describe this situation, remain the same as in (5.59), and can be applied to q_p, or

$$q_p(0, t) = q_u(t); x = 0, t > 0$$

$$q_p(\infty, t) \text{ finite}; x \to \infty, t > 0$$

$$q_p(x, 0) = 0; x > 0, t = 0 \tag{5.68}$$

$$\frac{\partial q_p(x, 0)}{\partial t} = 0; x > 0; t = 0$$

The solution of this problem is

$$q_p(x, t) = \int_0^t \int_0^\infty G(\xi, \tau; x, t) i(\xi, \tau) d\xi d\tau$$

$$- \left(g h_0^3 - q_0^2\right) \int_0^t q_u(\tau) \left[\frac{\partial G(\xi, \tau; x, t)}{\partial \xi} \right]_{\xi=0} d\tau \tag{5.69}$$

The symbol $G()$ denotes Green's function, which can be shown (Brutsaert, 1973) to be in this case

$$G(\xi, \tau; x, t) = -\left(4gh_0^5\right)^{-1/2} \exp[d_1(x - \xi) - d_2(t - \tau)]$$

$$\times \left\{ \begin{array}{l} I_0\left[d_3\left(t - \tau - \dfrac{(x - \xi)}{c_{01}}\right)^{1/2}\left(t - \tau - \dfrac{(x - \xi)}{c_{02}}\right)^{1/2}\right] \\[2ex] \times H\left[t - \tau - \dfrac{V_0(x - \xi)}{c_{01}c_{02}} + (gh_0)^{1/2}\dfrac{|\xi - x|}{c_{01}c_{02}}\right] \\[2ex] - I_0\left[d_3\left(t - \tau - \dfrac{x}{c_{01}} + \dfrac{\xi}{c_{02}}\right)^{1/2}\left(t - \tau - \dfrac{x}{c_{02}} + \dfrac{\xi}{c_{01}}\right)^{1/2}\right] \\[2ex] \times H\left[t - \tau - \dfrac{x}{c_{01}} + \dfrac{\xi}{c_{02}}\right] \end{array} \right\} \qquad (5.70)$$

The constants in (5.70) are $d_1 = (aS_0/h_0)$; $d_2 = (S_0V_0/h_0)(a\mathrm{Fr}_0^2 + 1)/\mathrm{Fr}_0^2$; and $d_3 = (S_0V_0/h_0)[(1 - \mathrm{Fr}_0^2)(1 - a^2\mathrm{Fr}_0^2)]^{1/2}/\mathrm{Fr}_0^2$; the steady uniform Froude number Fr_0 is defined in Equation (5.63) and the dynamic wave celerities c_{01} and c_{02} in (5.64). The symbol $H()$ is the Heaviside step function (see Appendix) and $I_0()$ is the modified Bessel function of the first kind of order zero.

Example 5.6. Semi-infinite channel with known upstream inflow

In many situations of practical interest, the lateral inflow does not have a large effect on the solution; therefore, to bring out the most important features of the solution (5.69), in what follows its simplest form is considered, that is the case $i = 0$. When the lateral flow i is absent, only the second term on the right of (5.69) remains. After carrying out the operations the result can be given as a simple convolution integral (see Appendix)

$$q(x, t) = q_0 + \int_0^t q_u(\tau)u(x, t - \tau)d\tau \qquad (5.71)$$

As before, $u(x, t)$ denotes the unit response of this channel, that is the flow rate $q_p(x, t)$ at any time t and at any point x, resulting from an upstream inflow at $x = 0$ given by a unit impulse (or Dirac delta function) $q_u(t) = \delta(t)$. This can be written as consisting of two parts

$$u = u_1 + u_2 \qquad (5.72)$$

The first part of u is given by

$$u_1 = \exp(-d_4x)\delta\left(t - \dfrac{x}{c_{01}}\right) \qquad (5.73)$$

where

$$d_4 = \dfrac{S_0}{h_0}\dfrac{(1 - a\,\mathrm{Fr}_0)}{(\mathrm{Fr}_0 + \mathrm{Fr}_0^2)}$$

and the Froude number for steady uniform flow is defined in Equation (5.63). The second part of u in (5.72) is given by

$$u_2 = \dfrac{d_3}{2t_0}\left(\dfrac{x}{c_{01}} - \dfrac{x}{c_{02}}\right)\exp(d_1x - d_2t)\,I_1(d_3t_0)\,H\left(t - \dfrac{x}{c_{01}}\right) \qquad (5.74)$$

where $t_0 = [(t - x/c_{01})(t - x/c_{02})]^{1/2}$, and where d_1, d_2 and d_3 are constants already defined below Equation (5.70). The symbol $I_1(\ )$ denotes the modified Bessel function of the first kind of order one, and $H(\ )$ is the Heaviside step function (see Appendix).

The solution (5.72) consists of two wave like motions that modify the steady uniform flow q_0. The first part, given by Equation (5.73), is identical to the solution of the analogous dynamic case given in (5.60), except that it also contains an exponential decay term in x. This part retains the form of a delta function and the argument of this delta function shows that it has the celerity of a dynamic wave, namely the same as given by Equation (5.60) or the first of (5.64). The amplitude of u_1 decreases exponentially, provided $d_4 > 0$. However, when $d_4 < 0$, that is when $Fr_0 > a^{-1}$, the wave grows exponentially; this is the well-known criterion for bore formation, namely $Fr_0 > 2$ (in the case of Chézy) or $Fr_0 > 3/2$ (in the case of GM). For small Froude numbers the amplitude of u_1 decays as $\exp[-S_0 x/(Fr_0 h_0)] = \exp[-g^{1/2} S_0 x/(h_0^{1/2} V_0)]$. This means that the dynamic part of the disturbance decays rapidly and becomes unimportant over short distances x, whenever the bed slope is large and the flow velocity small. Interestingly, a very similar result was obtained by an analysis of the nonlinear equations (5.13) and (5.22) by Lighthill and Whitham (1955; see also Stoker, 1957, p. 505), who showed that the dynamic front of any surface disturbance decays as $\exp(-g S_0 t/V_0)$, for small Froude numbers.

The second part, u_2 given by Equation (5.74), constitutes the main body of the wave. Mathematically, the Heaviside step function eliminates singularities from this part of the solution. Physically, the unit step function guarantees that this part will never forge ahead faster or further than the position of the dynamic front $x = (c_{01}t)$, given by Equation (5.73), and it ensures that the first part given by (5.73) represents indeed the leading edge of the disturbance caused by the unit impulse at $x = 0$. The celerity of the main body of the wave can be determined from the mean travel time of the wave. This mean time of occurrence of a wave is the first moment about the origin, denoted as m_1' or μ; this is also called its centroid or center of area. Therefore, the travel time is the difference between the mean time of occurrence of the wave at the point of observation and that at the origin $x = 0$, that is the time for its center of gravity to reach x. The upstream inflow at $x = 0$ is given by a delta function, whose first moment about the origin is zero. Therefore, mathematically the mean travel time is the first moment of the outflow rate $q_p(x, t)$ about $t = 0$ for a given x, or

$$m_1' = \int_{-\infty}^{+\infty} tq(x, t)dt \tag{5.75}$$

The Laplace transform of (5.72) can be used as a moment generating function (Dooge, 1973). Accordingly, the first moment is the first derivative of the Laplace transform at the origin in the transform domain, and it can be readily shown that it is given by

$$m_1' = \frac{x}{(1+a)\, V_0} \tag{5.76}$$

This yields a celerity x/m_1' for the main body of the disturbance

$$c_{k0} = (1 + a)\, V_0 \tag{5.77}$$

For reasons which will become clear in Section 5.4.3, a wave with a celerity $[(1 + a)V]$ is referred to as a kinematic wave; Equation (5.77) represents its linearized form, as indicated by the 0 subscript.

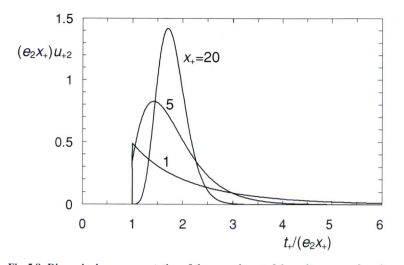

Fig. 5.8 **Dimensionless representation of the second part of the unit response function**
$u_{+2} = h_0 u_2 / (S_0 V_0)$ **for the upstream inflow problem, as obtained by solution of the linearized**
complete shallow water equations. Each curve is shown as a function of time t_+ at different
distances x_+ downstream from the release point of the unit impulse. The values of the ordinate
and the abscissa are multiplied and divided, respectively, by $(e_2 x_+)$ to allow convenient
comparison between the three curves. The constant is $e_2 = Fr_0 / (Fr_0 + 1)$ and the Froude
number is taken as $Fr_0 = 0.5$.

Several features of the solution can be brought out in a more general way by considering
it in a similarity framework. Inspection of the solution suggests that for this purpose one
can define a dimensionless distance along the channel, and a dimensionless time, as follows

$$x_+ = \frac{S_0 x}{h_0} \quad \text{and} \quad t_+ = \frac{S_0 V_0 t}{h_0} \tag{5.78}$$

Recalling the definitions given in Equations (5.50), (5.63) and (5.64), one can readily express
the two parts of the solution in terms of these dimensionless variables. Thus one obtains
from (5.73),

$$u_1 = \exp(-e_1 x_+)\delta(t_+ - e_2 x_+) \tag{5.79}$$

in which the constants depend only on the Froude number, namely $e_1 = (1 - a Fr_0)/(Fr_0^2 + Fr_0)$ and $e_2 = Fr_0/(Fr_0 + 1)$. In a similar way, (5.74) yields the second part of the unit
response in terms of the dimensionless variables

$$u_2 = \frac{S_0 V_0}{h_0} \frac{e_4 x_+}{\tau_+} \exp(a x_+ - e_5 t_+) \, I_1(e_6 \tau_+) \, H(t_+ - e_2 x_+) \tag{5.80}$$

where $\tau_+ = [(t_+ - e_2 x_+)(t_+ - e_3 x_+)]^{1/2}$, a is defined in Equations (5.39) and (5.43),
and the remaining constants are $e_3 = Fr_0/(Fr_0 - 1)$, $e_4 = [(1 - a^2 Fr_0^2)/(1 - Fr_0^2)]^{1/2}/Fr_0$,
$e_5 = (a Fr_0^2 + 1)/Fr_0^2$, and $e_6 = [(1 - a^2 Fr_0^2)(1 - Fr_0^2)]^{1/2}/Fr_0^2$. This second part is shown in
Figure 5.8 as a function of t_+, as it would be observed at different distances x_+ from the
point of release of the unit impulse. To allow an easier display of the evolution of the shape
of this wave and a comparison between the different curves, in the figure t_+ is scaled in
the abscissa with x_+, and u_2 is scaled in the ordinate with x_+^{-1}; therefore the horizontal

axis is made to shrink and the vertical axis is made to stretch, as x_+ increases. As mentioned, the first part u_1 becomes negligible very quickly. For the case shown in Figure 5.8, Equation (5.79) indicates that at $x_+ = (0, 1, 5, 20)$, the relative volume of u_1 decreases as $\exp(-e_1 x_+) = (1.0, 0.411, 0.0117, 1.9 \times 10^{-8})$.

The main point of this simple example of a linear analysis, leading to the solution (5.72) with (5.73) and (5.74) or with (5.79) and (5.80), is to show how in the propagation of free surface disturbances there are two main types of wave, namely dynamic and kinematic waves. As shown in Equations (5.60) and (5.64), the former are the result of the first three terms in (5.22) (or in the second of (5.44)); as shown in (5.77), the latter arise when also the two slope terms are included in the analysis. Except in unusual cases, such as supercritical flow (when $Fr_0 > 1$) or dynamic shock (see Chapter 7), the former are normally faster than the latter, but they tend to decay relatively quickly; comparing (5.64) with (5.77), one sees that the kinematic wave is faster than the dynamic wave, only when $(aV_0) > (gh_0)^{1/2}$, or $Fr_0 < (1/a)$; as noted, two paragraphs earlier, this is also the criterion for bore formation or dynamic shock. Moreover, when $1 < Fr_0 < (1/a)$, both e_4 and e_6 become imaginary; this changes the modified Bessel function $I_1(\ )$ to a regular Bessel function $J_1(\ )$, which exhibits oscillatory behavior. In the linear analysis the two types of waves appear separately, and the total disturbance is the result of their simple superposition, as shown in (5.72). Because the momentum shallow water equation (5.22) is quite nonlinear, in real world situations one can expect these two special types of propagation to interact with each other. Nevertheless, the linear analysis has clearly illustrated some of their most important features.

As an aside, it is of interest to point out that the ratio of the celerity of the dynamic waves relative to the mean velocity V, and the relative celerity of the kinematic wave (aV) is also referred to as the Vedernikov number, namely $Ve = (gh)^{1/2}/(aV)$ (Vedernikov, 1946; Chow, 1959). As shown above for the linear case, $Ve > 1$ is the criterion for bore formation.

5.4.2   The diffusion analogy: a first approximation

In many situations encountered in nature, the flow velocities change relatively slowly, so that the acceleration (inertia or dynamic) terms $(\partial V/\partial t)$ and $V(\partial V/\partial x)$ often are rather small compared to the other terms in the governing equations. For example, it was noted by Iwasaki (1967), that in the upper Kitakami, a river some 195 km long draining an area of 7860 km^2 in northern Honshu, these inertia terms were observed to be at most 1.5%, and usually smaller than 1% of the stage gradient $g[(\partial h/\partial x) - S_0]$. Similarly, the following values were presented in the *Flood Studies Report* (Natural Environment Research Council, 1975) as being typical for British rivers.

$$\frac{S_f}{S_0} \sim 0.9 \qquad \frac{\partial h/\partial x}{S_0} \sim 2 \times 10^{-2}$$

$$\frac{\partial V/\partial t}{g S_0} \sim \frac{V \partial V/\partial x}{g S_0} \sim 1.7 \times 10^{-3} \tag{5.81}$$

Recall that the governing equations of free surface flow describe the conservation of mass and the conservation of momentum. In this section the consequences are considered of neglecting these inertia terms in the momentum equations (5.22) or (5.25). However, the continuity equations (5.13) or (5.24) are left intact.

The diffusion equation of free surface flow

Consider again the case of a very wide channel. Omission of the acceleration terms reduces the momentum equation (5.22), in the absence of lateral inflow, to

$$\frac{\partial h}{\partial x} + S_f - S_0 = 0 \tag{5.82}$$

If Equation (5.39) can be assumed to be valid, the friction slope can be written concisely as

$$S_f = \alpha_r q^{1/b} \tag{5.83}$$

where $\alpha_r = (C_r h^{a+1})^{-1/b}$, and where as before $q = (Vh)$ is the flow rate per unit width. Thus (5.82) can be written as

$$\alpha_r q^{1/b} - S_0 + \frac{\partial h}{\partial x} = 0 \tag{5.84}$$

Proceeding in the same way as for Equation (5.67), that is applying $\partial/\partial t$ to the momentum equation (5.84) and $\partial/\partial x$ to the continuity equation (5.13), and subtracting one from the other, one obtains

$$b^{-1}\alpha_r q^{-1+1/b}\frac{\partial q}{\partial t} - \frac{\partial^2 q}{\partial x^2} + q^{1/b}\frac{\partial \alpha_r}{\partial t} = 0 \tag{5.85}$$

Because α_r depends only on the geometry of the cross section, which in turn is related to the water depth h, the derivative in the third term becomes

$$\frac{\partial \alpha_r}{\partial t} = \frac{d\alpha_r}{dh}\frac{\partial h}{\partial t}$$

Making use of the continuity equation (5.13) to replace this partial time derivative of h, i.e. $(\partial h/\partial t) = -(\partial q/\partial x)$, and making use of (5.83) to eliminate $q^{1/b}$, one obtains from (5.85)

$$\frac{\partial q}{\partial t} - \left(\frac{bq}{\alpha_r}\frac{d\alpha_r}{dh}\right)\frac{\partial q}{\partial x} = \left(\frac{bq}{S_f}\right)\frac{\partial^2 q}{\partial x^2} \tag{5.86}$$

The same derivation carried out for a channel with wide rectangular cross section $A_c = (B_c h)$ can be shown to yield a similar result, namely

$$\frac{\partial Q}{\partial t} - \left(\frac{bQ}{\alpha_r}\frac{d\alpha_r}{dA_c}\right)\frac{\partial Q}{\partial x} = \left(\frac{bQ}{B_c S_f}\right)\frac{\partial^2 Q}{\partial x^2} \tag{5.87}$$

Equations (5.86) and (5.87) are in the form of a nonlinear advective diffusion equation. Accordingly, the term $D = bq/S_f$ (or $bQ/B_c S_f$) can be referred to as diffusivity; for convenient reference, the term $c_d = -[(bq/\alpha_r)(d\alpha_r/dh)] = S_f^b(d\alpha_r^{-b}/dh)$ (or $c_d = -[(bQ/\alpha_r)(d\alpha_r/dA_c)] = B_c S_f^b(d\alpha_r^{-b}/dA_c)$) will henceforth be called the *advectivity*. As usual, the dimensions of the diffusivity are $[L^2\,T^{-1}]$, and those of the advectivity $[L\,T^{-1}]$. In general descriptive terms, the magnitude of the advectivity reflects the speed of propagation of a flow disturbance (in q or h), whereas the diffusivity is related to the speed with which this disturbance will spread out streamwise or, which is equivalent, dissipate its magnitude.

Solution of the linearized equation

In most practical applications of this approach, a linearized form of the equation has been used. This is readily obtained from (5.87) by proceeding as before in Section 5.4.1; thus, decompose the variables into a uniform steady part and a perturbation, or $q = q_0 + q_p$ and $h = h_0 + h_p$, and retain only the first-order terms, to obtain

$$\frac{\partial q_p}{\partial t} + \left(\frac{dq_0}{dh_0}\right)\frac{\partial q_p}{\partial x} = \left(\frac{bq_0}{S_0}\right)\frac{\partial^2 q_p}{\partial x^2} \tag{5.88}$$

This is a linear diffusion equation, in which the constant diffusivity is

$$D_0 = \left(\frac{bq_0}{S_0}\right) \tag{5.89}$$

with $b = 1/2$ for turbulent flow, and the constant advectivity is

$$c_{d0} = \frac{dq_0}{dh_0} \tag{5.90}$$

In a linear diffusion channel, this advectivity is clearly the same as the celerity $(a + 1)V_0$ of the main body of the wave of the complete linear solution, given by Equation (5.77). This will become clear in Section 5.4.3, but it can already be verified by determining (dq_0/dh_0) from (5.43) for uniform steady flow. This means that, as will be shown below, the advectivity of the diffusion equation is also the kinematic wave celerity c_{k0}, or

$$c_{d0} = c_{k0} \tag{5.91}$$

The linear diffusion formulation in (5.88) was derived for a very wide channel; for a channel cross section $A_c = A_{c0} + A_{cp}$ with a flow rate $Q = Q_0 + Q_p$, the basic equation can be written as

$$\frac{\partial Q_p}{\partial t} + c_{k0}\frac{\partial Q_p}{\partial x} = D_0\frac{\partial^2 Q_p}{\partial x^2} \tag{5.92}$$

which is the linearized form of (5.87). The diffusivity is now

$$D_0 = \left(\frac{bQ_0}{B_c S_0}\right) \tag{5.93}$$

and the advectivity is, in light of (5.91),

$$c_{k0} = \frac{dQ_0}{dA_{c0}} \tag{5.94}$$

This can again be approximated by $c_{k0} = (a + 1)V_0$, if the channel is wide enough.

The diffusion approximation of free surface flow has been the subject of numerous investigations (see Schönfeld, 1948; Hayami, 1951; Appleby, 1954; Daubert, 1964; Van de Nes and Hendriks, 1971; Dooge, 1973). The general case, with an inflow $q_u(t)$ at the upstream boundary of the channel at $x = 0$ and a nonzero lateral inflow $i = i(x, t)$, has been presented by Brutsaert (1973) as a special case of (5.69) with (5.70). To allow a comparison with the solution of the complete shallow water equations, consider again the same example as before with zero lateral inflow and with a known value of the upstream inflow $q_u(t)$.

Example 5.7. Semi-infinite channel with known upstream inflow

In the diffusion approach the governing equation is now (5.88), but the boundary conditions remain essentially the same as given in Equation (5.68) (or (5.59)). As was the case for the complete formulation in Section 5.4.1, the solution of the linear diffusion analogy is most conveniently given by its unit response. This is the outflow rate $q_p(x, t) = u(x, t)$ at a distance x downstream from the point $x = 0$ where the flow rate perturbation is a delta function or $q_p(0, t) = q_u(t) = \delta(t)$. The unit response can then be used with (5.71) to calculate the result for any arbitrary function $q_u(t)$. The diffusion equation has been thoroughly studied (see, for example, Carslaw and Jaeger, 1986) for various boundary conditions. It can be readily shown that this unit response function can be written as

$$u = \frac{x}{(4\pi D_0 t^3)^{1/2}} \exp\left(\frac{-(x - c_{k0}t)^2}{4D_0 t}\right) \tag{5.95}$$

This solution has a form which is closely related to the Gaussian or normal distribution in which the position of the mean is $x = (c_{k0}t)$ (and the spatial variance $\sigma^2 = (2D_0t)$. Thus this unit response confirms that the main body of this wave, that is its centroid, moves downstream with a celerity, given by the advectivity c_{k0} of Equation (5.91). It also shows how the diffusivity causes dispersion or spreading of the wave as it moves along.

To generalize this result it is useful to express it in terms of the dimensionless variables introduced in (5.78). Substitution of c_{k0} and D_0 (with $b = 1/2$) from (5.77) and (5.89), respectively, yields

$$u = \frac{S_0 V_0}{h_0} \frac{x_+}{\left(2\pi t_+^3\right)^{1/2}} \exp\left\{\frac{-[x_+ - (a + 1)t_+]^2}{2t_+}\right\} \tag{5.96}$$

This solution of the diffusion approach is compared in Figure 5.9 with the main unit response u_2 in the analogous solution of the complete shallow water equations, as given in (5.80). It can be seen that there is little difference between the two formulations for larger values of the dimensionless distance x_+. In fact, Figure 5.9 suggests that the diffusion approximation should be adequate in practical calculations for $x_+ > 5$. However, comparison of Equations (5.79) and (5.80) with (5.96) also indicates that the agreement between the two sets of curves in Figure 5.9 is perforce affected by the magnitude of the Froude number Fr_0. As mentioned earlier, the solution of the complete shallow water equations exhibits singular behavior as $Fr_0 \geq 1$; therefore it can be expected that the diffusion approach becomes less accurate as the Froude number approaches unity, i.e. as the flow velocity V_0 becomes critical. Finally, Figure 5.9 illustrates how for small values of x_+ the volume under the u_2 curve is smaller than unity; for instance, the u_2 curve at $x_+ = 1.0$ is lower than the corresponding diffusion result. This merely illustrates the fact that for small values of x_+ the dynamic wave part u_1 is still not negligible; for instance, as mentioned above, at $x_+ = 1.0$, in Equation (5.79) the term $\exp(-e_1 x_+)$ is still 0.411; but it rapidly decays with increasing x_+.

The main point of this analysis of the diffusion analogy of free surface flow is that it has further illustrated how the inertia terms, i.e. the first two terms in Equation (5.22) which represent acceleration, are the ones that generate the dynamic waves, shown in (5.64) and (5.73). These waves are absent from the solution given by (5.95) and (5.96). The motion of the main body of the wave is essentially controlled by the last three terms of Equation (5.22); both the solution of the complete system and that of the diffusion analogy indicate that the main body moves with the celerity of a kinematic wave; this will be further discussed in Section 5.4.3.

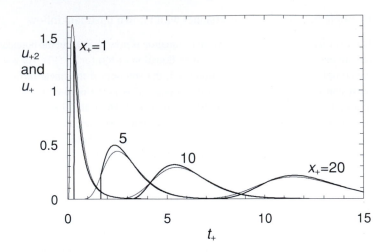

Fig. 5.9 **Comparison between the unit response function $u_{+2} = h_0 u_2 / (S_0 V_0)$ for the upstream inflow problem, obtained with the complete shallow water equations (solid line) and $u_+ = h_0 u / (S_0 V_0)$ obtained with the diffusion approximation (thin line). Each unit response curve is shown as a function of time t_+ at different distances x_+ downstream from the release point of the unit impulse. The constant is $a = 2/3$ and the Froude number is taken as $\mathrm{Fr}_0 = 0.5$. The shallow water results for $x_+ = 1,\ 5$ and 20 are the same as those shown in Figure 5.8.**

An improved linear diffusivity

Equation (5.88) was derived here simply by neglecting the first two terms in the momentum equation (5.22) and then linearizing. This way it can be seen that it is mainly the third term in (5.22), namely $g(\partial h/\partial x)$, which is responsible for the diffusion character of the resulting equation. Indeed, when this third term is also omitted, the resulting formulation loses its diffusion character, as will be shown in Section 5.4.3. However, it should be noted that this is not the only way to obtain a linear diffusion equation for free surface flow. It is also possible to start from the complete linear system (5.67) and to modify (rather than eliminate) the second derivatives involving time by means of the quasi-steady-uniform flow or kinematic wave assumption (see Section 5.4.3; also Brutsaert, 1973). As will become clear below (see Equation (5.118)) this assumption allows the substitution $(\partial q/\partial t) = (a + 1)V_0[-(\partial q/\partial x) + i]$ in the second derivatives involving time in (5.67). This leads then to a diffusion equation with a diffusivity, different from (5.89), namely

$$D_0 = \left(\frac{bq_0}{S_0}\right)\left(1 - a^2 \mathrm{Fr}_0^2\right) \tag{5.97}$$

or, for a channel of width B_c,

$$D_0 = \left(\frac{bQ_0}{B_c S_0}\right)\left(1 - a^2 \mathrm{Fr}_0^2\right) \tag{5.98}$$

In the case of the upstream inflow treated in Example 5.7, the unit response (5.96) becomes with this improved diffusivity

$$u = \frac{S_0 V_0}{h_0} \frac{x_+}{\left[2\pi\left(1 - a^2 \mathrm{Fr}_0^2\right)t_+^3\right]^{1/2}} \exp\left\{\frac{-[x_+ - (a + 1)t_+]^2}{2\left(1 - a^2 \mathrm{Fr}_0^2\right)t_+}\right\} \tag{5.99}$$

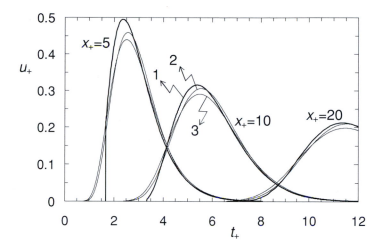

Fig. 5.10 Comparison between the unit response function $u_+ = h_0 u / (S_0 V_0)$ for the upstream inflow problem, obtained with the complete shallow water equations (1) and obtained with the diffusion approximation (2 and 3). The curves indicated by 2 are obtained with the diffusivity $D_0 = (bq_0/S_0)(1 - a^2 \mathrm{Fr}_0^2)$ and those indicated by 3 with $D_0 = (bq_0/S_0)$. Each unit response curve is shown as a function of time t_+ at different distances x_+ downstream from the release point of the unit impulse. The constant is $a = 2/3$ and the Froude number is taken as $\mathrm{Fr}_0 = 0.5$. Some of the curves indicated by 1 and 3 are also shown in Figures 5.8 and 5.9.

As can be seen in Figure 5.10, for $\mathrm{Fr}_0 = 0.5$, the diffusivity (5.97) leads to an improved agreement with the solution of the full shallow water equations. However, for smaller values of Fr_0, or for large values of x_+, the effect of this difference in diffusivity between Equations (5.89) and (5.97) can be expected to be small.

5.4.3   The quasi-steady-uniform flow approach: a second approximation

It was indicated earlier (see (5.81), for example) that typically the first three terms in the momentum equation (5.22) (or (5.25)), tend to be some two to three orders of magnitude smaller than those representing the effects of gravity and of friction, namely S_0 and S_f. In the previous section the dynamic terms were omitted, but the term $\partial h/\partial x$ was kept in the formulation, and this was shown to lead to the diffusion analogy. Often, however, under conditions that turn out to be quite common in nature, it is possible to neglect that term as well, and to keep only S_0 and S_f. In other words, it is assumed that water flows downhill but it is prevented from accelerating or decelerating very much, because the frictional resistance of the bed is overwhelming all other factors. Thus under such conditions the momentum equation (5.22) (or (5.25)) can be simplified to the following

$$S_f = S_0 \qquad\qquad\qquad (5.100)$$

Again, as in the previous section, the continuity equation (5.13) (or (5.24)) is left intact. This is the basis of the so-called quasi-steady-uniform flow or kinematic wave approximation.

The kinematic wave approach for free surface flow

In view of the form of the various expressions for the friction slope discussed in Section 5.3, Equation (5.100) is tantamount to assuming that, at a given location x along the flow channel, the average velocity V is a function of the hydraulic radius only, that is

$$V = V(R_h) \tag{5.101}$$

For a given cross section, the hydraulic radius is uniquely related to the mean depth of flow and to the cross-sectional area; thus V can also be expressed as a function of one of those variables, and one can use

$$V = V(h) \quad \text{or} \quad V = V(A_c) \tag{5.102}$$

as well.

As before, the most important features of the flow can be deduced by considering a wide channel. Equation (5.102), which now represents the momentum equation, indicates that the flow rate per unit width $q = (Vh)$ is a function of h only; thus, the continuity equation (5.13) can be written as

$$\frac{\partial h}{\partial t} + \frac{dq}{dh}\frac{\partial h}{\partial x} = i \tag{5.103}$$

Conversely, on account of (5.102) one has $h = h(q)$ as well, and therefore substituting this in Equation (5.13) one obtains in a similar manner

$$\frac{\partial q}{\partial t} + \frac{dq}{dh}\frac{\partial q}{\partial x} = \frac{dq}{dh}i \tag{5.104}$$

Equations (5.103) and (5.104) both have a structure, which is similar to that of the total time derivative of h and of q, namely

$$\frac{\partial h}{\partial t} + \frac{dx}{dt}\frac{\partial h}{\partial x} = \frac{dh}{dt} \tag{5.105}$$

and

$$\frac{\partial q}{\partial t} + \frac{dx}{dt}\frac{\partial q}{\partial x} = \frac{dq}{dt} \tag{5.106}$$

Hence it follows that

$$\frac{dq}{dh} = \frac{dx}{dt} \tag{5.107}$$

defines a wave speed; for the sake of conciseness this can be denoted by $dx/dt = c_k$, so that

$$c_k = \frac{dq}{dh} \tag{5.108}$$

The speed of this wave represents the rate of displacement of any point along x, where the depth (or the cross-sectional area) and the rate of flow increase respectively as

$$\frac{dh}{dt} = i \quad \text{and} \quad \frac{dq}{dt} = c_k i \tag{5.109}$$

Put differently, comparison of Equations (5.103) and (5.104) with (5.105) and (5.106), respectively, shows that to an imaginary observer, moving along x with a velocity defined by (5.108) and (5.107), it would appear that the cross-sectional area and the flow rate change as indicated by (5.109). In the absence of lateral inflow, when $i = 0$, the velocity defined by (5.108) and (5.107) is the speed of propagation of points where $dh/dt = 0$ and where $dq/dt = 0$, that is the speed of propagation of points of a given value of h and q. These observations are somewhat similar to the comments made in connection with Equations (5.51), (5.55) and (5.64). Thus (5.107), describing the path of the observer, defines the characteristics of the problem. Because (5.103) and (5.104) are of the first order, there is only one set of characteristics, namely in the forward direction.

One practical result obtainable with this approach is the celerity of a small monoclinal rising wave in an open channel. In a channel with an arbitrary cross section, in light of (5.24), the differential equations (5.103) and (5.104) assume the following form

$$\frac{\partial A_c}{\partial t} + c_k \frac{\partial A_c}{\partial x} = Q_l \tag{5.110}$$

and

$$\frac{\partial Q}{\partial t} + c_k \frac{\partial Q}{\partial x} = c_k Q_l \tag{5.111}$$

where the wave celerity (5.108) is now given by

$$c_k = \frac{dQ}{dA_c} \tag{5.112}$$

For a wide channel $R_h = h$ and $A_c = (B_c h)$ to a good approximation; thus, with V in q or Q given by (5.39), both (5.108) and (5.112) yield immediately

$$c_k = (a + 1)\, V \tag{5.113}$$

in which $(a + 1)$ is of the order of 1.5 to 1.7, depending on whether Chézy or GM is adopted to describe the flow. However, when the cross section does not have a wide rectangular shape, the wave celerity (5.112) yields, with (5.39),

$$c_k = (a + 1)\, V - \frac{aQ}{B_s P_w} \frac{dP_w}{dh} \tag{5.114}$$

where B_s is the width of the channel at the water surface; dP_w/dh is the rate of increase in the wetted perimeter P_w with depth, which is zero for a wide channel.

Apparently, the quasi-steady-uniform flow approximation has been used as early as 1857 by Kleitz (1877, p. 172) and his fellow engineers on the Rhone River, and by Breton in 1867 (Forchheimer, 1930). Equation (5.108) was also applied successfully with gage heights on the Mississippi and Missouri Rivers by Seddon (1900); it is now sometimes referred to as the Kleitz–Seddon law. The full implications of the approximation were investigated by Lighthill and Whitham (1955). They called the wave motion *kinematic*, because it arises from the elimination of the dynamic aspects of the momentum equation, namely the first three terms of Equation (5.22), leading to the assumption of (5.101) and (5.102).

Significance of the approximation

Within the validity range of hydraulic theory, Equations (5.13) and (5.22) (or (5.24) and (5.25)) describe the phenomenon of free surface flow; therefore, as the solutions of (5.44) and (5.67) indicate, dynamic waves always occur, but as shown in (5.73), gravity modifies their amplitude. Thus in general, small forerunners of a disturbance move with velocities given approximately by (5.64). However, as a result of gravity and friction, the main part of the disturbance usually moves with a much smaller velocity, namely as given by (5.107), (5.108) and (5.113) (or (5.77) and (5.91) for the linearized case). When $(g S_0 t / V)$ is large, the dynamic waves are damped sufficiently that the kinematic waves, which usually move at a slower speed, assume the dominant role. It is under such conditions that Equations (5.105) and (5.106) describe the flow. As will be shown in Chapters 6 and 7, the kinematic wave approach is useful in the solution of several problems of practical interest.

Solution of the linearized equation

As before, when the disturbances around a steady uniform reference flow are not excessive, one can decompose the variables into an undisturbed part and a perturbation, and the continuity equation is the first of (5.66), rewritten here for convenience

$$\frac{\partial h_p}{\partial t} + \frac{\partial q_p}{\partial x} = i \tag{5.115}$$

According to (5.43) with (5.100), the rate of flow can be written in terms of the decomposed variables as

$$q_0 + q_p = C_r S_0^b (h_0 + h_p)^{a+1} = q_0 \left(1 + \frac{h_p}{h_0} \right)^{a+1}$$

Hence, because $q_0 = V_0 h_0$, and presumably $h_p \ll h_0$, one can write

$$q_p = (a + 1) V_0 h_p \tag{5.116}$$

This shows that, since q_p is a function of h_p only (V_0 is constant), just like (5.103) and (5.104), (5.115) can be written as a total time derivative of h_p, that is

$$\frac{\partial h_p}{\partial t} + \frac{dq_p}{dh_p} \frac{\partial h_p}{\partial x} = i \tag{5.117}$$

or, alternatively as a total time derivative of q_p, as

$$\frac{\partial q_p}{\partial t} + \frac{dq_p}{dh_p} \frac{\partial q_p}{\partial x} = \frac{dq_p}{dh_p} i \tag{5.118}$$

From (5.117) and (5.118) one can define a wave celerity (dq_p/dh_p); however, by virtue of (5.116) and (5.43), this is equal to (dq_0/dh_0). Hence one has

$$\frac{dq_p}{dh_p} = \frac{dq_0}{dh_0} = c_{k0} \tag{5.119}$$

or

$$c_{k0} = (a + 1) V_0 \tag{5.120}$$

As anticipated, this result is the same as the celerities given by (5.77) and (5.91).

This result was obtained for a wide channel of rectangular cross section; for other cross-sectional geometries, one can formulate the rate of flow in terms of decomposed variables in a way similar to Equation (5.116), namely

$$Q_0 + Q_p = C_r S_0^b (A_{c0} + A_{cp})^{a+1} (P_{w0} + P_{wp})^{-a}$$

$$= Q_0 \left(1 + \frac{A_{cp}}{A_{c0}}\right)^{a+1} \left(1 + \frac{P_{wp}}{P_{w0}}\right)^{-a}$$

Again, because $A_{cp} \ll A_{c0}$ and $P_{wp} \ll P_{w0}$, one obtains the analog of (5.116)

$$Q_p = (a+1)\, V_0 A_{cp} - a Q_0 \frac{P_{wp}}{P_{w0}} \tag{5.121}$$

With the equation of continuity one can, as before, define a celerity, (dQ_p/dA_{cp}) which now assumes the form

$$c_{k0} = (a+1)\, V_0 - \frac{a Q_0}{P_{w0}} \frac{dP_w}{dA_c} \tag{5.122}$$

Note that this result shows that also here $(dQ_p/dA_{cp}) = (dQ_0/dA_{c0})$.

Example 5.8. Semi-infinite channel with known upstream inflow

The general solution of Equation (5.118) (and of (5.117)) is especially simple in the absence of lateral inflow, when $i = 0$, namely

$$q_p = q_p(x - c_{k0} t) \tag{5.123}$$

Equation (5.123) shows that, in a linear kinematic channel, an upstream disturbance is merely translated downstream. Unlike a disturbance in a linear dynamic channel and in a linear diffusion channel, it does not undergo any deformation as it propagates downstream. Thus, in contrast to Equations (5.72) and (5.95), the unit response that is the outflow, at a time t and at a distance x downstream from a point $x = 0$, where the flow disturbance at $t = 0$ is a unit impulse $\delta(x, t)$, is now simply

$$u(t) = \delta(x - c_{k0} t) \tag{5.124}$$

This describes a translation of the input without distortion, as it moves down along the channel. But it should be remembered that this represents a disturbance over and above the steady uniform flow q_0.

5.4.4 *The lumped kinematic approach for free surface flow: a third approximation*

Besides the approximations that led to the analysis of the kinematic wave in Section 5.4.3, the lumped formulation has the additional feature that the spatial dependency of the continuity equation (5.13) (or (5.24)) is eliminated. This is accomplished by integrating out the x-variable, so that q or Q becomes located on the boundaries of the flow domain, in the form of inflows, and outflows, and $\partial h/\partial t$ becomes the rate of change of the water depth averaged over the entire flow domain, that is the stored water. As already explained in Chapter 1, this produces the (lumped) storage equation

$$Q_i - Q_e = \frac{dS}{dt} \tag{5.125}$$

in which all inflow and outflow rates are grouped in Q_i and Q_e, respectively, and S represents the stored water volume in the flow domain under consideration. By analogy with (5.101) and (5.102), a kinematic relationship between S and the flow rate Q_i and /or Q_e is then invoked to allow solution of Equation (5.125) (i.e. (1.10)). Applications of this concept will be presented in Chapters 6 and 7.

REFERENCES

Abbott, M. B. (1975). Method of characteristics. In *Unsteady Flow in Open Channels*, ed. K. Mahmood and V. Yevjevich, Vol. I, chapter 3. Fort Collins, CO: Water Resources Pubs., pp. 63–88.

Appleby, F. W. (1954). Runoff dynamics – a heat conduction analogue of storage flow in channel networks, Assemblée Générale de Rome, 1954, Internat. Assoc. Scient. Hydrology, Publication No. 38, pp. 338–348.

Bakhmeteff, B. A. (1941). Coriolis and the energy principle in hydraulics. In *Theodore Von Karman Anniversary Volume: Contributions to Applied Mechanics*. Pasadena: Calif. Inst. Technology, pp. 59–65.

Barnes, H. H. (1967). *Roughness characteristics of natural channels*. Geol. Surv., Water-Supply Paper 1849, 213 pp.

Brutsaert, W. (1972). Discussion of "Mechanics of sheet flow under simulated rainfall". *J. Hydraul. Div., Proc. ASCE*, **98**(HY2), 406–407.

(1973). Review of Green's functions for linear open channel. *J. Engin. Mechanics Div., Proc. ASCE*, **99**(EM6), 1247–1257.

(1982). *Evaporation into the Atmosphere: Theory, History and Applications*. Boston, MA: D Reidel Publishing Company.

(1993). Horton, pipe hydraulics and the atmospheric boundary layer. *Bull. Amer. Met. Soc.*, **74**, 1131–1139.

Carslaw, H. S. and Jaeger, J. C. (1986). *Conduction of Heat in Solids*, second edition. Oxford: Clarendon Press.

Chow, V. T. (1959). *Open-Channel Hydraulics*. New York: McGraw-Hill.

Cunge, J. A., Holley, F. M. and Verwey, A. (1980). *Practical Aspects of Computational River Hydraulics*. Boston, MA: Pitman Advanced Pub. Program.

Daubert, A. (1964). Quelques aspects de la propagation des crues. *La Houille Blanche*, **19**(3), 341–346.

Deymie, P. (1939). Propagation d'une intumescence allongée (Problème aval), in *Proc. Fifth Internat. Congress for Applied Mechanics*, Cambridge, MA. New York: John Wiley and Sons, Inc., pp. 537–544.

Dooge, J. C. I. (1973). *Linear Theory of Hydrologic Systems*, Tech. Bull. 1468, Agr. Res. Serv., US Dept. Agric.

Dooge, J. C. I. and Harley, B. M. (1967). Linear routing in uniform open channels. In *Proc. Internat. Hydrology Symposium*, Vol. 1. Fort Collins: Colorado State University, pp. 57–63.

Dupuit, J. (1863). *Études Théoriques et Pratiques sur le Mouvement des Eaux Dans les Canaux Découverts et à Travers les Terrains Perméables*, 2me édition. Paris: Dunod.

Forchheimer, Ph. (1930). *Hydraulik*, 3. Aufl. Leipzig and Berlin: B. G. Teubner.

Hayami, S. (1951). On the propagation of flood waves. Bulletin No. 1. Disaster Prevention Research Institute, Kyoto University, Kyoto, Japan.

Henderson, F. M. and Wooding, R. A. (1964). Overland flow and groundwater flow from a steady rainfall of finite duration. *J. Geophys. Res.*, **69**, 1531–1540.

Hicks, W. I. (1944). Discussion of "Preliminary report on analysis of runoff resulting from simulated rainfall on a paved plot" by C. F. Izzard and M. T. Augustine. *Trans. Amer. Geophys. Un.*, **25**, 1039–1041.

Hinze, J. D. (1959). *Turbulence*. New York: McGraw-Hill.

Horner, W. W. and Jens, S. W. (1942). Surface runoff determination from rainfall without using coefficients. *Trans. Amer. Soc. Civ. Engrs.*, **107**, 1039–1075.

Horton, R. E. (1938). The interpretation and application of runoff plat experiments with reference to soil erosion problems. *Soil Sci. Soc. Amer. Proc.*, **3**, 340–349.

Iwagaki, Y. (1954). On the law of resistance to turbulent flow in open rough channels, *Proc. 4th Japan Nat. Congress Appl. Mechanics*, pp. 229–233.

Iwasaki, T. (1967). Flood forecasting in the River Kitakami. In *Proc. Internat. Hydrology Symposium*, Sept. 6–8, 1967, Vol. 1. Fort Collins: Colorado State University, pp. 103–112.

Keulegan, G. H. (1938). Laws of turbulent flow in open channels. *J. Res. Nat. Bur. Standards*, **21**, 707–741.

Kisisel, I. T., Rao, R. A. and Delleur, J. W. (1973). Turbulence in shallow water under rainfall. *J. Eng. Mech. Div., Proc. ASCE*, **99**(EM1), 31–53.

Kleitz, M. (1877). Note sur la théorie du mouvement non permanent des liquides. *Annales des Ponts et Chaussées*, 5^{e}serie, **14**, 2me Semestre, 133–196.

Lamb, H. (1932). *Hydrodynamics*, sixth edition. Cambridge: Cambridge University Press (also 1945, New York: Dover Publications).

Liggett, J. A. and Cunge, J. A. (1975). Numerical methods of solution of the unsteady flow equations. In *Unsteady Flow in Open Channels*, ed. K. Mahmood and V. Yevjevich, Vol. I, chapter 4. Fort Collins, CO: Water Resources Pubs., pp. 89–182.

Lighthill, M. J. and Whitham, G. B. (1955). On kinematic waves, I. Flood movement in long rivers. *Proc. R. Soc. London*, A**229**, 281–316.

Massé, P. (1939). Recherches sur la théorie des eaux courantes. In *Proc. Fifth Internat. Congress for Applied Mechanics*, Cambridge, MA, 1938. New York: John Wiley and Sons, Inc., pp. 545–549.

Monin, A. S. and Yaglom, A. M. (1971). *Statistical Fluid Mechanics: Mechanics of Turbulence*, Vol. 1. Cambridge, MA: The MIT Press.

Montes, S. (1998). *Hydraulics of Open Channel Flow*. Reston, VA: ASCE Press.

Mouret, M. G. (1921). *Antoine Chézy – histoire d'une formule d'hydraulique* (*Extrait des Annales des Ponts et Chaussées*, II, 165–268). Paris: A. Dumas, Ed.

Natural Environment Research Council (1975). *Flood Studies Report*, Vol. III, Flood routing studies (by R. K. Price). London.

Powell, R. W. (1962). Another note on Manning's formula. *J. Geophys. Res.*, **67**, 3634–3635.
    (1968). The origin of Manning's formula. *J. Hydraul. Div., Proc. ASCE*, **94**(HY4), 1179–1181.

Prandtl, L. and Tollmien, W. (1924). Die Windverteilung über dem Erdboden, errechnet aus den Gesetzen der Rohrströmung. *Zeits. Geophysik*, **1**, 47–55.

Schönfeld, J. C. (1948). Voortplanting en verzwakking van hoogwatergolven op een rivier. *De Ingenieur*, **60**(B1), 1–7.

Seddon, J. A. (1900). River hydraulics. *Trans. Amer. Soc. Civ. Engrs.*, **43**, 179–243.

Shen, H. W. and Li, R.-M. (1973). Rainfal effect on sheet flow over smooth surface. *J. Hydraul. Eng., Proc. ASCE*, **99**(HY5), 771–792.

Sommerfeld, A. (1949). *Partial Differential Equations in Physics*, translated by E. G. Strauss. New York: Academic Press Inc.

Stoker, J. J. (1957). *Water Waves*. New York: Interscience Publishers, Inc.

Tan, W. (1992). *Shallow Water Hydrodynamics*. Beijing: Water & Power Press.

Van de Nes, T. J. and Hendriks, M. H. (1971). Analysis of a Linear Distributed Model of Surface Runoff, Report No. 1. Laboratory of Hydraulics and Catchment Hydrology, Agricultural University, Wageningen, Netherlands.

Vedernikov, V. V. (1946). Characteristic features of a liquid flow in an open channel. *Dokl. Akad. Nauk SSSR*, **52**(3), 207–210.

Williams, G. P. (1970). Manning formula – a misnomer? *J. Hydraul. Div., Proc. ASCE*, **96**(HY1), 193–200.

(1971). Manning formula – a misnomer? Closure. *J. Hydraul. Div., Proc. ASCE*, **97**(HY5), 733–735.

Woerner, J. L., Jones, B. A. and Fenzl, R. N. (1968). Laminar flow in finitely wide rectangular channels. *J. Hydraul. Div., Proc. ASCE*, **94**(HY3), 691–704.

Woo, D.-C. and Brater, E. F. (1961). Laminar flow in rough rectangular channels. *J. Geophys. Res.*, **66**, 4207–4217.

Wooding, R. A. (1965). A hydraulic model for the catchment-stream problem. *J. Hydrol.*, **3**, 257–267.

Yoon, Y. N. and Wenzel, Jr., H. G. (1971). Mechanics of sheet flow under simulated rainfall. *J. Hydraul. Div., Proc. ASCE*, **97**(HY9), 1367–1386.

PROBLEMS

5.1   Derive the x-component of the Reynolds equations, namely (5.14), from the corresponding component of the Navier–Stokes equations (1.12).

5.2   Derive Equation (5.35) from (5.34); show the intermediate steps.

5.3   Calculate the Boussinesq correction factor, β_c, for the logarithmic velocity profile in a wide open channel. Thus, integrate (5.19), in which the velocities are given by (5.34) and (5.36). What is its value for $(h/z_0) = 100$?

5.4   Calculate the Boussinesq correction factor, β_c, for the power-type velocity profile in a wide, open channel. Thus, integrate (5.19), in which the velocities are given by (5.37) and (5.38). What is its value for $m = 1/6$?

5.5   Write down the expression for the hydraulic radius, R_h using Equation (5.40) for a channel (a) with a triangular cross section ($B_b = 0$),, and (b) with a rectangular cross section ($B_s = B_b$).

5.6   What would be the values of the powers a and b in Equation (5.43), if the velocity profile in a very wide, open channel were given by (5.37) with the classical value, $m = 1/7$.

5.7   Consider flow in a channel with a trapezoidal cross section and side banks having a slope, 1 vertical to 2 horizontal, and a bottom width, $B_b = 5$ m; the channel roughness coefficient is $n = 0.015$, and the bed slope is $S_0 = 0.001$. Given the water depth, at the center as $h = 2$ m, under uniform, steady-state conditions, calculate the velocity, V; the rate of flow, $Q = V A_c$; and the Reynolds number, $\mathrm{Re} = V R_h/\nu$.

5.8   For a channel with the same characteristics as in Problem 5.7, with a flow rate of $Q = 60$ m^3 s^{-1}, calculate the depth of flow, h, at the center of the channel. Use trial and error.

5.9   Consider a very wide open channel in which the velocity profile is given by the logarithmic equation (5.34); assume that the displacement height is negligible or $d_0 = 0$. At what fraction of h does the mean velocity V occur? (Note: in hydrologic practice, this depth is often taken as $z = 0.4\,h$.)

5.10   Consider a very wide channel in which the velocity distribution is given by the power-type equation (5.37); assume $m = 1/6$. At what fraction of the depth h does the mean velocity V occur? (Note: in hydrologic practice, this depth is often taken as $z = 0.4\,h$.)

5.11   In hydrologic practice, it is common to determine the mean velocity, V, as the average of measurements at $0.2\,h$ and at $0.8\,h$; thus, $V = (\bar{u}_{0.2} + \bar{u}_{0.8})/2$. Calculate what the error would be if the velocity profile were logarithmic as given by Equation (5.34); assume that $d_0 = 0$.

5.12   Demonstrate that Equation (5.51) is a solution of (5.49).

5.13   Consider the diffusion approximation describing flow in a wide open channel, as given by (5.87) with $\alpha_r = (C_r\,h^{a+1})^{-1/b}$, in which the parameters are defined in (5.43). Show for a linearized channel that the coefficient of the second term on the left, namely $(bQ/\alpha_r)\,(d\alpha_r/dA_c)$, is equivalent with the Kleitz–Seddon equation for the celerity, i.e. $c_{k0} = (dQ_0/dA_{c0})$. Hint: linearization is accomplished by putting $A_c = A_{c0} + A_{cp}$ and $Q = Q_0 + Q_p$, and by assuming A_{cp} and Q_p to be relatively small perturbations.

5.14   The diffusion equation for open channel flow can be written in the form of (5.92) with (5.93) and (5.94). Consider a flood routing problem, in which the equation becomes (in m and s):

$$\frac{\partial Q}{\partial t} + 2.17\,\frac{\partial Q}{\partial x} = 1365\,\frac{\partial^2 Q}{\partial x^2}$$

(a) Give a rough estimate of the mean velocity of the flow, V (not the flood wave), from this equation. (b) Show how you derive the equation, which you use to estimate V. (c) What is the significance of the magnitude of the coefficients (namely 2.17 and 1365 in this case) in regard to the shape of the flood wave? In other words, what would be the effect of each coefficient, if it were larger or smaller, on the evolution of the shape of the flood wave? Discuss the effect of each coefficient separately (one sentence each).

5.15   (a) Derive Equation (5.114) from the Kleitz–Seddon equation (5.112). (b) Implement (5.114) to calculate the celerity of a kinematic wave in a channel with a triangular cross section. Make use of the GM equation, (5.41), in which $S_f = S_0$.

5.16   Multiple choice. Indicate which of the following statements are correct. The derivation of the shallow-water equations, (5.13) and (5.22), requires the following.
(a)   The lateral inflow does not depend on x; in other words, it must be uniform along the direction of flow.
(b)   The pressure distribution is hydrostatic along z, the direction normal to the bottom.
(c)   The velocity profile obeys a power law, $\bar{u} = a\,z^m$, where a and m are constants.
(d)   The slope of the channel bottom must be small, in the direction of flow, to allow substitution of $-\sin\theta$ by S_0.
(e)   The roughness of the channel must be constant in the direction of flow.

6   OVERLAND FLOW

This type of flow, also variously called sheet flow or shallow flow, is likely to occur in the initial stages of surface runoff. It is usually observed on surfaces with low permeability and in areas with a saturated soil profile and with the water table close to the surface. Overland flow has been one of the central problems in urban hydrology and it has been the subject of much research. Interest in this phenomenon has been largely the result of its relevance in the design of small engineering structures for roads, highways, airports and other urban and industrial settings and also in the design of some surface irrigation systems.

6.1   THE STANDARD FORMULATION

The main objective in the analysis is usually the determination of the flow at the downstream end of a sloping plane for a known lateral inflow or outflow, for example, owing to rainfall, irrigation and/or infiltration. This type of situation is sketched in Figure 6.1. In general, flow of water with a free surface over a plane bed, with a constant slope S_0 and with a length L, receiving a uniform, but possibly unsteady lateral inflow $i = i(t)$, whose velocity component in the direction of flow is negligible, can be described by the shallow water equations (5.13) and (5.22); for the present purpose these can be rewritten here as

$$\frac{\partial h}{\partial t} + \frac{\partial}{\partial x}(Vh) - i = 0 \tag{6.1}$$

for the conservation of mass, and

$$\frac{\partial V}{\partial t} + V\frac{\partial V}{\partial x} + g\left(\frac{\partial h}{\partial x} + S_f - S_0\right) + \frac{iV}{h} = 0 \tag{6.2}$$

for the conservation of momentum. The dependent variables are the vertically averaged velocity V and the height h of the water surface above the bed; g is the acceleration of gravity and S_f is the friction slope. As the lateral inflow rate may consist of rainfall P (or irrigation), and infiltration f, one has $i = P - f$.

When the plane is initially dry, the essential features of the problem can be captured by the following boundary conditions

$$0 \le x \le L, \qquad t = 0, \qquad V = 0, \qquad h = 0$$

and $\hspace{11cm}$ (6.3)

$$x = 0 \qquad t > 0 \qquad V = 0, \qquad h = 0$$

Although analytical solutions can be obtained for certain special conditions (see Brutsaert, 1968), the complete solution can be obtained only by numerical methods.

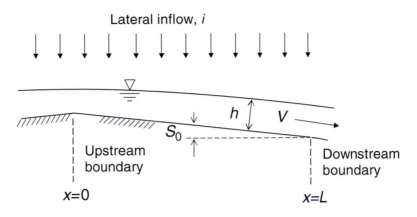

Fig. 6.1   Definition sketch of a plane with lateral inflow.

Using this approach Woolhiser and Liggett (1967) and Liggett and Woolhiser (1967) have presented a thorough analysis of this problem for a steady uniform lateral inflow i.

For the general problem defined by the boundary conditions (6.3), it is convenient to scale the variables with the length of the plane L and with the normal depth h_{0L} and the corresponding velocity V_{0L} at the downstream end $x = L$, where $(h_{0L} V_{0L}) = (iL)$. (Recall that the *normal depth* is the depth produced by uniform flow at a given discharge rate q, as given by Equation (5.43) with $S_f = S_0$) This scaling leads to the following dimensionless variables, $x_+ = (x/L)$, $t_+ = (V_{0L}t/L)$, $h_+ = (h/h_{0L})$ and $V_+ = (V/V_{0L})$. Equations (6.1) and (6.2) assume then the dimensionless form

$$\frac{\partial h_+}{\partial t_+} + \frac{\partial (V_+ h_+)}{\partial x_+} - 1 = 0$$

and   (6.4)

$$\frac{\partial V_+}{\partial t_+} + V_+ \frac{\partial V_+}{\partial x_+} + \frac{1}{Fr_{0L}^2} \frac{\partial h_+}{\partial x_+} + Ki_0 \left[\left(\frac{V_+}{(h_+^a)} \right)^{1/b} - 1 \right] + \frac{V_+}{h_+} = 0$$

In (6.4), the symbol Ki_0 represents the kinematic flow number at $x = L$, defined as

$$Ki_0 = \frac{S_0 L}{Fr_{0L}^2 h_{0L}}$$ (6.5)

and Fr_{0L} is the corresponding Froude number (cf. Equation (5.63))

$$Fr_{0L} = \frac{V_{0L}}{(gh_{0L})^{1/2}}$$ (6.6)

Except for the presence of the Froude number, Ki_0 has nearly the same form as $(d_4 x)$ in (5.73) or $(e_1 x_+)$ in (5.79), where it plays essentially the same role, in that it is a measure of the rate of attenuation of the dynamic waves. In dimensionless equations all terms normally tend to be of order one, except those that involve a dimensionless number like Fr or Ki. Thus the second of Equations (6.4) shows that the motion becomes kinematic when $Ki_0 \gg 1$.

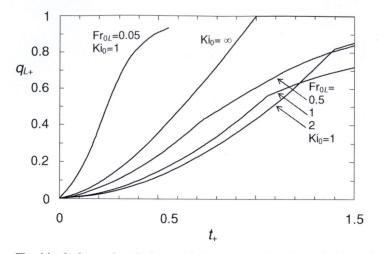

Fig. 6.2   The rising hydrograph at the lower end of a plane $x = L$, calculated with the full shallow water equations for $Ki_0 = 1$ and for different values of the Froude number as indicated at each curve. The hydrograph is given as the scaled rate of flow per unit width $q_{L+} = (V_{L+}h_{L+})$ against scaled time t_+. The kinematic wave result is indicated by $Ki_0 = \infty$. (After Woolhiser and Liggett, 1967.)

Fig. 6.3   The same as Figure 6.2, for $Ki_0 = 10$. (After Woolhiser and Liggett, 1967.)

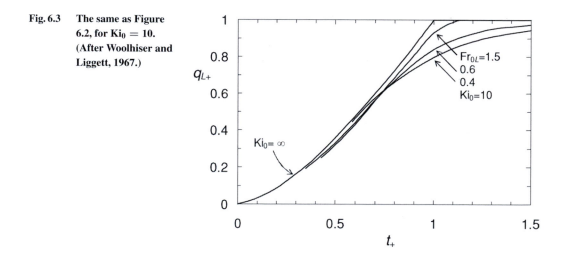

On the basis of numerical simulations, Woolhiser and Liggett (1967) reported that for large values of Ki_0 the kinematic approach can produce practically the same accuracy in the solution as the full system of the shallow water equations (6.4); when $Ki_0 = 10$, the maximal error in the outflow hydrograph was found to be of the order of 10%, but also to decrease rapidly as Ki_0 increases. They concluded from this that the kinematic wave result is a good approximation for most overland flow situations, because Ki_0 rarely falls below 10. Some of their results are illustrated in Figures 6.2 and 6.3. Because the kinematic approach is considerably simpler and allows closed-form solutions, it is investigated next.

6.2   KINEMATIC WAVE APPROACH

With the kinematic approximation, only the two slope terms in Equation (6.2) (or the second of (6.4)) are of any importance (cf. Equation (5.100)); thus the momentum equation becomes

$$S_f = S_0 \tag{6.7}$$

Physically, Equation (6.7) states that the water surface is assumed to be parallel to the bed. Substitution of (6.7) into (5.43) yields immediately the kinematic relationship

$$q = K_r h^{a+1} \tag{6.8}$$

where $q = (Vh)$ is the rate of flow per unit width of plane $[L^2T^{-1}]$, and K_r can be defined with (5.43) as $K_r = C_r S_0^b$, in which the values C_r, a and b are listed in Table 5.2 for different flow conditions.

Equation (6.8) implies a unique relationship between q (or V) and h. Hence, as already indicated in Section 5.4.3, the continuity equation (6.1) is suggestive of the mathematical form of a total derivative

$$\frac{\partial h}{\partial t} + \frac{dx}{dt}\frac{\partial h}{\partial x} = \frac{dh}{dt} \tag{6.9}$$

with the following equalities

$$\frac{dx}{dt} = \frac{dq}{dh} \quad \text{and} \quad \frac{dh}{dt} = i \tag{6.10}$$

The quantity dx/dt defines a kinematic wave celerity, which by virtue of (6.8) is

$$\frac{dx}{dt} = c_k = (a+1)\, K_r h^a = (a+1)\, K_r^{1/(a+1)} q^{a/(a+1)} \tag{6.11}$$

To an observer moving forward at this rate dx/dt, both equalities in (6.10) will appear to hold. Recall from Chapter 5, that the path of such an imaginary observer traced on the x–t plane is called a *characteristic* of the wave motion.

6.2.1   *Unsteady lateral inflow*

Consider first the case of an arbitrary unsteady, but uniform, lateral inflow $i = i(t)$, and consider an imaginary observer moving with a velocity dx/dt given by Equation (6.11). To this observer, that is along the characteristics, on account of (6.10), it will appear

(i)   that the water depth h changes at a rate $i = i(t)$, so that at any time this depth is the integral of i, namely

$$h = \int_{t_0}^{t} i\, dt \tag{6.12}$$

or, with (6.8),

$$q = K_r \left(\int_{t_0}^{t} i \, dt \right)^{a+1} \tag{6.13}$$

in which t_0 is the starting point of the characteristic; and

(ii)   that $dq/dt = (dq/dh)(dh/dt) = i(dx/dt)$, which yields the following integral for the rate of flow

$$q = \int_{x_0}^{x} i \, dx \tag{6.14}$$

where again the lower limit x_0 is the starting point of the characteristic.

(iii)   The equation of the characteristics $x = x(t)$ is obtained by integration of (6.11), or with (6.13),

$$x = (a + 1) K_r \int_{t_0}^{t} d\tau \left(\int_{t_0}^{\tau} i \, d\sigma \right)^{a} + x_0 \tag{6.15}$$

where τ and σ are dummy variables of integration.

The integrals presented in (6.13), (6.14) and (6.15) were first derived by Ishihara and Takasao (1959) in a critical analysis of the unit hydrograph concept. Smith and Woolhiser (1971) studied overland flow on an infiltrating surface; they obtained numerical solutions for the kinematic wave formulation with a lateral inflow $i(t)$ as the difference between rainfall rate and infiltration rate obtained from numerical solution of Richards's equation. Parlange *et al.* (1981) and also Giraldez and Woolhiser (1996) considered different cases of unsteady lateral inflow and infiltration, i.e. $i = i(t)$, and derived analytical solutions. The runoff resulting from a steady inflow rate, which was first studied by Henderson and Wooding (1964), is the key to understanding more general situations. This case is treated next.

6.2.2   Steady lateral inflow

When the lateral inflow remains constant with time, there are two phases of hydrologic interest. The first is the buildup of the flow on a plane that is initially dry in accordance with Equation (6.3); the second is the subsidence of the flow after the lateral inflow has ceased and $i = 0$.

Buildup phase: the rising hydrograph
Since $h = 0$ for $t = 0$ according to Equation (6.3), the integral of (6.12) is simply

$$h = it \tag{6.16}$$

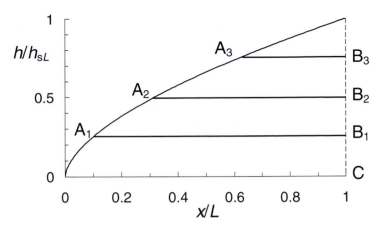

Fig. 6.4 Water depth profiles $0A_1B_1C$, $0A_2B_2C$, etc., during the buildup phase, obtained with the kinematic wave approach (for fully turbulent flow with $a = 2/3$); the profiles are shown as functions of downstream distance at different times after the start of the lateral inflow i. The water depth is normalized with the equilibrium depth at $x = L$, which is given by Equation (6.19), or $h_{sL} = (iL/K_r)^{1/(a+1)}$.

On the other hand, the integral of (6.14) is

$$q = i(x - x_0) \tag{6.17}$$

in which x_0 is the starting point of the characteristic (i.e. the initial position of the "observer" invoked above) at $t = t_0 = 0$. Because x_0 can assume any value over the length of the plane $0 \leq x \leq L$, there is an infinity of characteristics on which (6.17) is valid, each depending on x_0. The boundary characteristic starting at $x = x_0 = 0$, is however of special interest. On that particular characteristic (6.17) assumes the form

$$q = ix \tag{6.18}$$

By virtue of Equation (6.8), (6.18) gives the position of a given depth, $x = x(h)$, as

$$x = (K_r/i)\, h^{a+1} \tag{6.19}$$

Thus on this particular characteristic starting at $t = 0$ and $x = 0$ in the $x-t$ plane, i.e. at $h = 0$ and $x = 0$ on the physical $h-x$ plane, both (6.16) and (6.19) hold. This trajectory on the $h-x$ plane is shown in Figure 6.4 as going from 0 to A_1, A_2, etc., for different values of t. For all the other characteristics, at x values larger than given by (6.19), (6.17) is not very useful, because x_0 is left unspecified, but (6.16) still indicates the water depth h as a function of time, independently of x_0. Therefore, downstream from the point x, given by (6.19), h is independent of x (see Figure 6.4).

Actually, Equations (6.18) and (6.19) also represent the continuity condition that must be satisfied under equilibrium conditions, that is when the flow rate at any point x equals the total lateral inflow upstream from that point. This means that equilibrium conditions are established upstream from any point x where the boundary characteristic has passed and that the entire plane is at equilibrium as soon as that characteristic has reached $x = L$. From then on, (6.18) and (6.19) are valid over the entire flow domain, namely $0 \leq x \leq L$. The time required to reach this steady state equilibrium is obtained by combining (6.16) with (6.19) at $x = L$, or

$$t_s = (L/K_r i^a)^{1/(a+1)} \tag{6.20}$$

When the duration of lateral inflow exceeds the time to equilibrium, the outflow hydrograph at $x = L$ can be readily obtained. Prior to the time to equilibrium, it is obtained from (6.8) combined with (6.16); once equilibrium is established it can be obtained from (6.18). Thus the rising hydrograph at $x = L$ is given by

$$q_L = \begin{cases} K_r i^{a+1} t^{a+1} & \text{for } t \leq t_s \\ iL & \text{for } t \geq t_s \end{cases} \tag{6.21}$$

Equation (6.21) is the main result of this section. The performance of (6.20) has been compared with experimental data on turbulent sheet flow from the literature by McCuen and Spiess (1995); they concluded that its use should be restricted by the criterion $(nL/\sqrt{S_0}) < 30$ m.

To generalize the result shown in Equation (6.21), it is useful to express it in terms of dimensionless variables. The simplest way to proceed here is to take the equilibrium discharge rate (iL) from the plane and the time to equilibrium t_s as scaling variables. This reduces Equation (6.21) to

$$q_{L+} = \begin{cases} (t_+)^{a+1} & \text{for } t_+ \leq 1 \\ 1 & \text{for } t_+ \geq 1 \end{cases} \tag{6.22}$$

where now $q_{L+} = (q_L/q_{sL})$ and $t_+ = (t/t_s)$, in which $q_{sL} = (iL)$ is the equilibrium outflow rate at $x = L$; this rising hydrograph is illustrated in Figure 6.5. Figures 6.6 and 6.7 show a comparison between the kinematic wave rising hydrograph and experimental data of Izzard (1944, 1946) scaled in the same manner. It can be seen that some of the hydrographs in Figure 6.6 initially start out as laminar flow, and change to turbulent flow later on around $t_+ = 0.4$; also, around $t_+ = 0.9$ some dynamic effects, which are neglected in the kinematic formulation, appear to enter into play.

Decay phase: recession hydrograph after rain stops
As soon as $i = 0$, according to Equation (6.10) one has $(dh/dt) = 0$. Hence, to the observer moving at a celerity given by (6.11) it now appears that h remains constant. In other words, (6.11) is the velocity of a point of the water surface with the given value of h. Thus on the $h-x$ plane the characteristics describe straight lines parallel to the surface of the plane where $h = 0$. One such characteristic is shown in Figure 6.8, as going from A_1 to A_2, A_3, etc., for successive values of time t after the lateral inflow has ceased. Because h remains constant, Equation (6.11) can be integrated immediately to yield

$$x = (a + 1) K_r h^a t + x_0 \tag{6.23}$$

where x_0 is the starting value of x, i.e. its initial value at the time $t = 0$, when the rain stops and the recession starts.

In case the duration of the rain is longer than the time to equilibrium t_s, i.e. $D > t_s$, initially the water surface has an equilibrium profile as given by Equation (6.19), so that

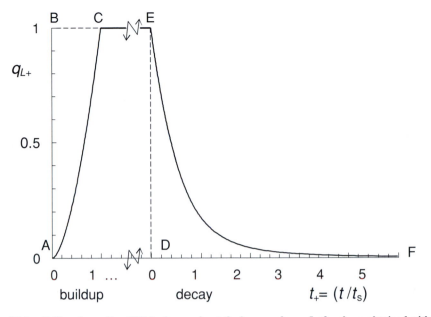

Fig. 6.5 Rising (AC) and receding (EF) hydrographs at the lower end $x = L$ of a plane, obtained with the kinematic wave approach (for turbulent flow with $a = 2/3$). The rate of flow is scaled with the equilibrium rate of flow $q_{sL} = iL$ and the time is scaled with the time to equilibrium given by Equation (6.20), so that $q_{L+} = (q_L/q_{sL})$ and $t_+ = (t/t_s)$. The area ABC represents the volume stored on the plane under equilibrium flow conditions, and it is equal to the area DEF.

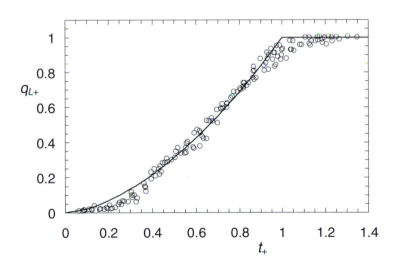

Fig. 6.6 Comparison between scaled rising hydrograph obtained with the kinematic wave approach (for turbulent flow with $a = 2/3$) and scaled experimental data obtained by Izzard (1944) on a plane covered with turf. The solid line represents $q_{L+} = t_+^{5/3}$. The data points are derived from several different experimental combinations, namely rainfall intensities $P = i = 91.4$ and 45.7 mm h^{-1}, slopes $S_0 = 0.01, 0.02$ and 0.04, and plane lengths $L = 22, 15, 7.3$ and 3.7 m. (After Morgali, 1970.)

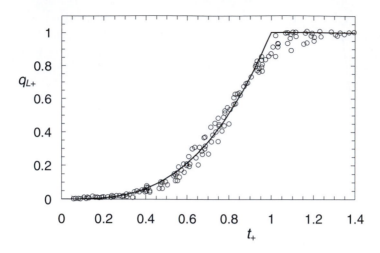

Fig. 6.7 Comparison between scaled rising hydrograph obtained with the kinematic wave approach (for laminar flow with $a = 2$) and scaled experimental data obtained by Izzard (1944) on a plane covered with asphalt. The solid line represents $q_{L+} = t_+^3$. The data points come from several different experimental combinations, namely rainfall intensities $P = i = 91.4$ and 45.7 mm h^{-1}, slopes $S_0 = 0.001, 0.005, 0.01$ and 0.02, and plane lengths $L = 22, 15, 7.3$ and 3.7 m. (After Morgali, 1970.)

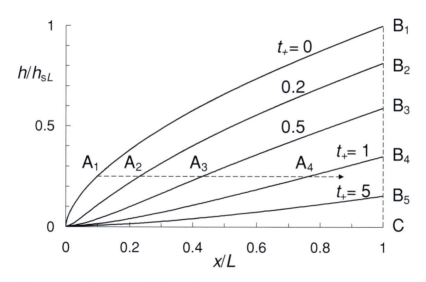

Fig. 6.8 Water depth profiles $0A_1 B_1 C$, $0A_2 B_2 C$, etc., during the decay phase, obtained with the kinematic wave approach (with $a = 2/3$) ; the profiles are shown as functions of downstream distance at different times after the cessation of the lateral inflow i. The water depth is normalized with the equilibrium depth at $x = L$, which is given by Equation (6.19) or $h_{sL} = (iL/K_r)^{1/(a+1)}$. The initial profile is the equilibrium, i.e. steady state, profile shown in Figure 6.4. The characteristic starting at A_1 successively passes A_2, A_3, etc., and maintains a constant h, until it is swept off the plane at $x = L$.

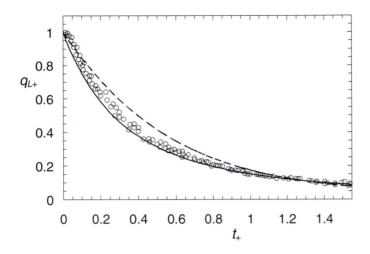

Fig. 6.9   **Comparison between scaled receding hydrograph obtained with the kinematic wave approach (6.28) (solid line for laminar flow with $a = 2$ and dashed line for turbulent flow with $a = 2/3$) and experimental data obtained by Izzard (1944) on a plane covered with turf. The solid line represents $t_+ = (1/3)q_{L+}^{-2/3}(1 - q_{L+})$ and the dashed line $t_+ = (3/5)q_{L+}^{-2/5}(1 - q_{L+})$. The experimental data points were obtained for the same experimental combinations as those of Figure 6.6. (After Morgali, 1970.)**

(6.23) becomes

$$x = (a + 1) K_r h^a t + (K_r/i)h^{a+1} \tag{6.24}$$

As before, it is convenient to recast this in dimensionless form, by scaling the water depth with the equilibrium depth (i.e. the initial depth prior to the recession phase) at $x = L$. Thus Equation (6.24) assumes the form

$$\frac{x}{L} = (a + 1)h_+^a t_+ + h_+^{a+1} \tag{6.25}$$

where t_+ is defined behind (6.22); the dimensionless water depth is $h_+ = (h/h_{sL})$ in which the equilibrium depth at the outlet is $h_{sL} = (iL/K_r)^{1/(a+1)}$, in accordance with (6.19). Equation (6.25) is illustrated in Figure 6.8, and shows successive water surface profiles for increasing values of the time t after the cessation of the lateral inflow i.

Upon substitution of h by means of Equation (6.8), at the outflow point $x = L$, (6.24) becomes

$$L = (a + 1) K_r^{1/(a+1)} q_L^{a/(a+1)} t + q_L/i \tag{6.26}$$

This allows the calculation of the recession hydrograph $q_L = q_L(t)$, or rather in this case implicitly as $t = t(q_L)$,

$$t = \left[(a + 1)K_r^{1/(a+1)} i q_L^{a/(a+1)}\right]^{-1}(iL - q_L) \tag{6.27}$$

which is the main result of this analysis. Henderson and Wooding (1964) found that (6.21) and (6.27) gave a good description of the experimental data for a grass-covered surface published by Hicks (1944) and that the best fit for his three cases was obtained with $a = 0.8, 0.8$ and 1.0, respectively.

Again, with the same scaled variables as in (6.22), (6.27) can be expressed in a more universal way, as follows

$$t_+ = \left[(a+1)(q_{L+})^{a/(a+1)}\right]^{-1}(1 - q_{L+}) \qquad (6.28)$$

The recession hydrograph described by Equation (6.28) is illustrated graphically in Figure 6.5 for the case of fully turbulent flow with $a = 2/3$. Figure 6.9 shows a comparison between (6.28), both for laminar flow with $a = 2$ and for turbulent flow with $a = 2/3$, and experimental data obtained by Izzard (1946) on a turf surface for the same experimental set-up combinations shown in Figure 6.6; it can be seen that while the rising hydrographs exhibited turbulent flow, the recessions were somewhat closer to the laminar curve, except initially.

Runoff sequence for a short rainfall burst

In the case that the rainfall duration D is shorter than the time to equilibrium, i.e. for $D < t_s$, the water surface profile at the end of the rain, (i.e. the initial profile at the beginning of the decay phase) is typically represented by one of the profiles 0ABC shown in Figure 6.4. Let in what follows the reference $t = 0$ indicate the beginning of the rainfall. If $h = h_0 (= iD)$ (cf. Equation (6.16)) denotes the largest depth achieved during the buildup phase, once the rainfall stops, the point A moves downstream at a constant velocity $[(a+1)K_r h_0^a]$ and it will reach $x = L$ at a time (see Equation (6.24))

$$D + t_p = D + \left(L - (K_r/i)h_0^{a+1}\right) \big/ \left((a+1)\,K_r h_0^a\right) \qquad (6.29)$$

Thus as long as $D \leq t < D + t_p$, the water depth and the outflow rate at $x = L$ remain constant at, respectively, $h = h_0$ and

$$q_L = K_r h_0^{a+1} \qquad (6.30)$$

After that, for $t \geq t_p + D$, the outflow rate is given by (6.27), but with the addition of a time shift D to account for the duration of the lateral inflow.

To summarize, the hydrograph sequence for the case, when the rain stops before full equilibrium is reached, is as follows in terms of scaled variables. As the lateral inflow starts, the outflow rate at $x = L$ is given by the first of Equations (6.22). At the moment $t = D$, that is $t_+ (= t/t_s) = D_+$, when the lateral inflow ceases, the outflow rate is

$$q_{L+} = (D_+)^{a+1} \qquad (6.31)$$

where, as before, $q_{L+} = (q_L/q_{sL})$, in which $q_{sL} = iL$, and $D_+ = (D/t_s)$. The rate of flow at $x = L$ will remain constant at the value given by (6.31) for a duration (cf. (6.29))

$$t_{p+} = \left(1 - h_{0+}^{a+1}\right)\left[(a+1)h_{0+}^a\right]^{-1} \qquad (6.32)$$

where $t_{p+} = (t_p/t_s)$, and $h_{0+} = (h_0/h_{sL})$. Because $h_0 = (iD)$, this duration of constant flow can also be expressed more conveniently in terms of the relative rainfall duration D_+, as follows

$$t_{p+} = \left(1 - D_+^{a+1}\right)\left[(a+1)D_+^a\right]^{-1} \qquad (6.33)$$

From then on, i.e. for $t_+ \geq (D_+ + t_{p+})$ after the onset of the rain, the rate of flow is given by Equation (6.28). Since the time reference $t = 0$ is taken at the onset of the rain, here (6.28)

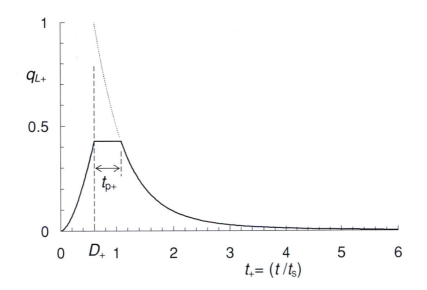

Fig. 6.10 Hydrograph at the lower end $x = L$ of a plane (heavy line), resulting from a uniform rainfall of duration D_+, obtained with the kinematic wave approach (with $a = 2/3$). The rate of flow is scaled with the equilibrium rate of flow $q_{sL} = (iL)$ and the time is scaled with the time to equilibrium given by Equation (6.20), so that $qL_+ = (qL/q_{sL}), t_+ = (t/t_s)$ and $D_+ = D/t_s$. In this example $D_+ = 0.6$ and $t_{p+} = 0.483$.

should be recast as

$$t_+ = D_+ + \left[(a+1)(q_{L+})^{a/(a+1)}\right]^{-1}(1 - q_{L+}) \tag{6.34}$$

As an example, this outflow sequence is shown in Figure 6.10 for a rainfall duration $D = 0.6t_s$; in this case (6.33) yields $t_{p+} = 0.483$.

Effect of raindrop impact

In the analysis so far it was assumed that K_r in (6.8) is unaffected by the impact of the raindrops on the water surface. Under conditions of turbulent flow this assumption may be a good approximation but, as seen in Table 5.2, under conditions of laminar flow the additional resistance may be considerable. This effect can be incorporated in the recession analysis as follows. Let K_{rr} denote the parameter in Equation (6.8) under conditions of rainfall impact, and K_{rn} the same parameter in the absence of rain. Both parameters can be determined, for example by means of Equations (5.33) and (5.32), respectively. Equation (6.24) must now be adjusted to

$$x = (a+1)K_{rn}h^a t + (K_{rr}/i)h^{a+1} \tag{6.35}$$

which, at $x = L$ becomes, instead of (6.26),

$$L = (a+1)K_{rn}^{1/(a+1)}q_L^{a/(a+1)}t + (K_{rr}/K_{rn})q_L/i \tag{6.36}$$

As before, this result yields immediately the outflow hydrograph, i.e. $t = t(q_L)$, as follows

$$t = \left[(a+1)K_{rn}^{1/(a+1)}iq_L^{a/(a+1)}\right]^{-1}\left[iL - (K_{rr}/K_{rn})q_L\right] \tag{6.37}$$

Equation (6.37) indicates how at $t = 0$, when the rainfall ceases abruptly, the rate of flow q_L immediately becomes larger than iL by an amount (K_{rn}/K_{rr}); this increase is caused by the sudden decrease of the flow resistance in the absence of the impact of the rainfall drops. With the expressions given in Table 5.2, this increased flow rate is roughly $(1 + cS_0^d P^e)$ times the equilibrium flow rate iL; for instance, with a slope $S_0 = 0.001$ and a rainfall intensity $P = 0.3$ cm h^{-1}, this indicates a sudden increase of 38%. But in actual flow situations this sudden increase is unlikely to be that large, and the value predicted by Equation (6.37) can only be considered as an upper limit. Indeed, a sudden change in shear stress resulting from the cessation of the rainfall, must also involve accelerations, which are neglected in the kinematic approach leading to (6.37), and which will tend to offset this effect. Moreover, even if it were to occur, the spike is rapidly dissipated. Finally, natural rainfall events never cease suddenly, but they tend to decrease rather gradually. Brief increases in runoff, upon the cessation of rainfall have been observed experimentally and reported by Izzard (1946), but they were much smaller than those predicted here by Equation (6.37).

6.3 Lumped kinematic approach

Although this approach is now dated, as it was developed prior to the more fundamental analyses described above, the lumped kinematic approach is still of some interest because it has often been used as the framework to analyze valuable experimental data. It was developed by Horton (1938) in his pioneering analysis of overland flow; it was subsequently applied by Izzard (1944) in processing the data from his extensive experimental investigations on rain runoff from paved and grassy surfaces. In this approach the continuity equation is replaced by the storage equation (1.10) or (5.125). In the notation of overland flow this storage equation can be written as

$$iL - q_L = L\frac{d\langle h\rangle}{dt} \tag{6.38}$$

where

$$\langle h\rangle = \frac{1}{L}\int_0^L h\,dx \tag{6.39}$$

denotes the spatial average of the water depth over the plane. To close (6.38), q_L must be related with $\langle h\rangle$; this can be done for steady equilibrium flow conditions by combining (6.8) and (6.18) to obtain

$$q_L = K_1\langle h\rangle^{a+1} \tag{6.40}$$

where $K_1 = \{[(a + 2)/(a + 1)]^{(a+1)}K_r\}$. If it is now assumed that (6.40) is also valid under non-steady conditions during buildup or subsidence as well, its substitution in (6.38) yields

$$iL - q_L = LK_1^{-1/(a+1)}\frac{dq_L^{1/(a+1)}}{dt} \tag{6.41}$$

To determine the outflow rate q_L at the downstream end of the surface, (6.41) must be integrated for the imposed input $i = i(t)$. The essential features of the problem can again be obtained readily by considering the buildup phase and the decay phase for a lateral inflow rate i, which is constant in time.

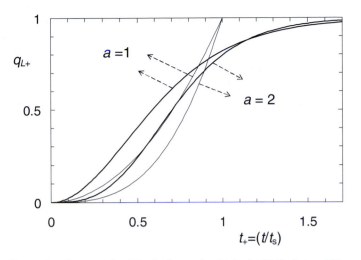

Fig. 6.11 **Comparison between the rising hydrographs obtained with the lumped kinematic approach (heavy lines) and the kinematic approach (thin lines). The same scaling is used in both cases. The heavy curve for $a = 1$ represents the solution proposed by Horton (1938), and the heavy curve for $a = 2$ is essentially the same as the solution used by Izzard (1946) to develop his dimensionless hydrograph.**

The rising hydrograph

For the case when i is applied starting at $t = 0$ on an initially dry plane, Equation (6.41) can be rewritten in terms of the scaled variables defined behind (6.22); in fact, mathematically this scaling appears to be the most obvious option. The resulting differential equation is

$$1 - q_{L+} = \frac{(a+1)d}{(a+2)} \frac{\left(q_{L+}^{1/(a+1)}\right)}{dt_+} \tag{6.42}$$

with the condition that $q_{L+} = 0$ for $t_+ = 0$. Equation (6.42) can be integrated in closed form for values of $(a + 1)$ equal to 1, 2, 3, 4, 3/2 and 4/3, but only 2 and 3 appear to be relevant for surface runoff. As indicated in Chapter 5, the value $(a + 1) = 2$ has been derived from several data sets of overland flow on grass covered surfaces (see Wooding, 1965), and $(a + 1) = 3$ is the theoretical value for laminar flow.

For $a = 1$ the solution of (6.42) is

$$q_L^+ = \tanh^2 (1.5t_+) \tag{6.43}$$

Similarly, for $a = 2$ the solution can be shown to be

$$t_+ = 0.125 \ln[(1 + y + y^2)(1 - y)^{-2}] + (\sqrt{3}/4)\tan^{-1}[(2y + 1)/\sqrt{3}]$$
$$-(\sqrt{3}/4)\tan^{-1}[1/\sqrt{3}] \tag{6.44}$$

in which $y = q_{L+}^{1/3}$. Both (6.43) and (6.44) are shown in Figure 6.11, where they can be compared with the results obtainable with Equation (6.22). Prior to the development of the kinematic wave approach, these two solutions have been widely used in practical design.

Horton (1938) proposed Equation (6.43), with the justification that $(a + 1) = 2$ represents a flow, which is 75% turbulent. The equation was subsequently used as the basis for the design of urban and airport storm drainage facilities (see Horner and Jens, 1942; Hathaway, 1945; Jens, 1948). The study by Izzard (1946), which built on Horton's approach, is also well known and its results have been widely applied (Linsley *et al.*, 1975). On the basis of extensive experiments, Izzard (1946) concluded that for $iL < 3.8 \ \text{m}^2 \text{h}^{-1}$ the flow can be described by the lumped kinematic formulation with Equation (6.40) in which $(a + 1) = 3$ for laminar flow. The value of the other parameter, namely K_1, was derived from his experimental data as a function of surface roughness and of rainfall intensity. For the rising phase of the outflow he obtained the solution presumably by numerical means, and presented it graphically as a dimensionless hydrograph. The time variable t was scaled with the time to equilibrium, which Izzard took as the time required to produce an outflow rate which is 97% of the equilibrium outflow rate, or $q_L = 0.97(iL)$. By assuming (see the curve for $a = 2$ in Figure 6.11) that the volume of water detained in surface storage on the plane is roughly half of the volume of inflow during the time required to reach equilibrium t_e, he was able to propose the following expression

$$t_e = \frac{2 \langle h_s \rangle}{i} \tag{6.45}$$

in which t_e is the time to equilibrium after the start of the rain and $\langle h_s \rangle$ is the average water depth after equilibrium has been reached.

Figure 6.11 illustrates that the lumped kinematic approach does not really produce a very good mathematical description of overland flow, as compared with the kinematic wave solution, which is known to provide a close approximation to the exact solution. Thus the question can be raised how the experimental results could be described so well by the lumped approach in Izzard's (1946) study. The explanation of this discrepancy probably lies in the method used to scale the experimental rising hydrographs. As illustrated in Figure 6.11, $q_{L+}(\equiv q_L/iL)$ approaches unity asymptotically in the lumped kinematic solution, so that with noisy data the identification of $q_{L+} = 0.97$, to determine the time to equilibrium, is not easy. However, Izzard (1946, p. 148) noted that with the above definition of t_e in (6.45), for $a = 2$ the lumped kinematic solution indicates that at the time $t = 0.5t_e$ the outflow rate is roughly $q_{L+} = 0.55$; therefore, he decided instead to non-dimensionalize the experimental rising hydrographs with the criterion $t_e = 2t_{0.55}$, in which $t_{0.55}$ is the time at which the outflow is 0.55 the equilibrium value. As shown by Woolhiser and Liggett (1967; Fig. 8), with this time scaling the agreement is improved considerably. This should not be surprising because this way the curves are forced to coincide at t/t_e equal to 0 and to 0.55.

The recession hydrograph

After the rain stops $i = 0$, and Equation (6.41) can immediately be integrated for any value of a. Again, in dimensionless terms, this can be written as

$$q_{L+} = \left(\frac{a(a+2)}{(a+1)} t_+ + (q_{Li+})^{-a/(a+1)} \right)^{-(a+1)/a} \tag{6.46}$$

in which the subscript i indicates the initial value of the dimensionless outflow rate, that is at $t = 0$, when the rain stops; for the case $D \gg t_s$, when the rainfall duration is much larger than t_s (see Equation (6.20)) this initial outflow rate can be taken equal to one. Izzard (1946) used a recession function, which is essentially the same as (6.46) with $a = 2$, but with different scaling, to analyze his experimental data.

REFERENCES

Brutsaert, W. (1968). The initial phase of the rising hydrograph of turbulent free surface flow with unsteady lateral inflow. *Water Resour. Res.*, **4**, 1189–1192.

Giraldez, J. V. and Woolhiser, D. A. (1996). Analytical integration of the kinematic equation for runoff on a plane under constant rainfall rate and Smith and Parlange infiltration. *Water Resour. Res.*, **32**, 3385–3389.

Hathaway, G. A. (1945). Design of drainage facilities (Military airfields: a symposium). *Trans. Amer. Soc. Civ. Engrs.*, **110**, 697–733.

Henderson, F. M. and Wooding, R. A. (1964). Overland flow and groundwater flow from a steady rainfall of finite duration. *J. Geophys. Res.*, **69**, 1531–1540.

Hicks, W. I. (1944). Discussion of "Preliminary report on analysis of runoff resulting from simulated rainfall on a paved plot" by C. F. Izzard and M. T. Augustine. *Trans. Amer. Geophys. Un.*, **25**, 1039–1041.

Horner, W. W. and Jens, S. W. (1942). Surface runoff determination from rainfall without using coefficients. *Trans. Amer. Soc. Civ. Engrs.*, **107**, 1039–1075.

Horton, R. E. (1938). The interpretation and application of runoff plat experiments with reference to soil erosion problems. *Soil Sci. Soc. Amer. Proc.*, **3**, 340–349.

Ishihara, T. and Takasao, T. (1959). Fundamental researches on the unit hydrograph method and its application. *Trans. Japan Soc. Civil Eng.*, No. 60 (Extra Paper 3–3), 1–34. (In Japanese; English translation by K. Hoshi, Ed., Kyoto University, 1996.)

Izzard, C. F. (1944). The surface profile of overland flow. *Trans. Amer. Geophys. Un.*, **25**, 959–968.
  (1946). Hydraulics of runoff from developed surfaces. *Proc. 26th Ann. Meeting, Highway Res. Board, Nat. Res. Council*, **26**, 129–150.

Jens, S. W. (1948). Drainage of airport surfaces – some basic design considerations. *Trans. Amer. Soc. Civ. Engrs.*, **113**, 785–809.

Liggett, J. A. and Woolhiser, D. A. (1967). Difference solutions of the shallow-water equation. *J. Eng. Mechanics Div., Proc. ASCE*, **93** (EM2), 39–71.

Linsley, R. K., Kohler, M. A. and Paulhus, J. L. H. (1975). *Hydrology for Engineers*, second edition. New York: McGraw-Hill Book Co.

McCuen, R. H. and Spiess, J. M. (1995). Assessment of kinematic wave time of concentration. *J. Hydraul. Eng.*, **121**, 256–266.

Morgali, J. R. (1970). Laminar and turbulent overland flow hydrographs. *J. Hydraul. Div., Proc. ASCE*, **96**, 441–460.

Parlange, J.-Y., Rose, C. W. and Sander, G. (1981). Kinematic flow approximation of runoff on a plane: an exact analytical solution. *J. Hydrol.*, **52**, 171–176.

Smith, R. E. and Woolhiser, D. A. (1971). Overland flow on an infiltrating surface. *Water Resour. Res.*, **7**, 899–913.

Wooding, R. A. (1965). A hydraulic model for the catchment-stream problem. *J. Hydrol.*, **3**, 254–267.

Woolhiser, D. A. and Liggett, J. A. (1967). Unsteady, one-dimensional flow over a plane- The rising hydrograph. *Water Resour. Res.*, **3**, 753–771.

PROBLEMS

6.1   Derive Equations (6.4) from (6.1) and (6.2) by making use of the appropriate scaling variables.

6.2   Consider a smooth plane with slope, S_0, on which it has been raining for a long time, so that steady flow has been established. Derive a relationship between the rainfall rate, P, and the downstream distance, x, where the flow will change from laminar to turbulent. Assume that this occurs at a critical Reynolds number, $V h / \nu = 500$.

6.3   (a) Derive the water surface profile, $h = h(x)$, once steady state has been established, on a plane, $L = 30$ m long, with a slope, $S_0 = 0.0015$, resulting from a rainfall rate, $P = 37$ mm h^{-1}. Assume that the kinematic wave assumption is valid. (b) What is the depth, h, of the flow, at $x = L = 30$ m, in this analysis?

6.4   In the previous problem, use is made of the kinematic wave assumption. If this simplification were not valid, which terms would still be negligible in the shallow water equations (6.1) and (6.2)? Thus, write these equations in their simplest form, which would allow the solution of the same steady-state problem, but without the kinematic wave assumption. In doing so, express S_f in terms of the dependent variables, V and h, for a wide channel.

6.5   Assume that free-surface flow over a plane uniform surface resulting from a steady rain intensity, P, can be described by the kinematic-wave method, so that (6.1) is the governing equation, with $i = P$. Assume that the flow is fully turbulent, and that its dynamics can be described by the GM equation, (5.41). With what celerity does a point on the free surface, with a given depth, $h = D_g$, move downstream, after the rain ceases? Assume that the roughness coefficient n is not affected by the impact of the raindrops. Show how you obtain this answer.

6.6   Show that the area, ABC, in Figure 6.5 represents, indeed, the volume stored on the plane under equilibrium flow conditions, and that it is equal to the area DEF. (Hint: perform the integration $\int q \, dt = \int t \, dq$ between the appropriate limits.)

6.7   A concrete pavement is $L = 40$ m long, and it has a slope of $S_0 = 0.0015$. A uniform rainfall starts at $t = 0$, and continues for a long time at a steady rate of $P = 25$mm h^{-1}. (a) First, determine whether the maximum flow, at $x = L$, is laminar or turbulent. (b) Compute the rising hydrograph (in mm h^{-1} to make it comparable to the rainfall rate), at the lower end of the pavement, after this rainfall starts. Use the kinematic wave method, with and without the effect of the rain. (c) Plot the two results on one figure as $q = q(t)$ in mm h^{-1}, and t in hours. Note: under turbulent conditions, the effect of the rainfall on the flow is usually neglected. Under laminar conditions, (5.33) can be used to incorporate this effect.

6.8   Same as the previous problem for a plane soil surface with a short grassy vegetation and with $L = 45$ m, $S_0 = 0.02$, and $P = 85$ mm h^{-1}.

6.9   A concrete pavement is $L = 35$ m long, and it has a slope of $S_0 = 0.005$. A steady rainfall lasts for a long time at a rate of $P = 30$ mm h^{-1}. (a) First, determine whether the maximum flow, at $x = L$, is laminar or turbulent. (b) Compute the recession hydrograph (in cm h^{-1}) at the lower end of the pavement after this rainfall stops. Use the kinematic wave method, and neglect the effect of the raindrop impact and the temporary increase at $t = 0$ (when the rain stops). Plot the hydrograph in mm h^{-1} to match the units of P. (c) On the basis of the result obtained in (b), what is the flow rate, in mm h^{-1}, at $x = L$, 15 min after the rain stops?

6.10   Consider the same situation as described in the previous problem. Initially, after the rain stops, the runoff, at $x = L$, is likely to increase a little on account of the decreased resistance. Calculate the maximal value of this brief increase, as a fraction of the equilibrium flow, using the empirical formula (5.33) in (6.36) or (6.37).

6.11   For a steady rainfall on a uniform, plane surface of length L, the average depth of the water on the plane, $\langle h \rangle$, is related to the depth at the outlet $(x = L)$, as: $\langle h \rangle = [(a + 1)/(a + 2)]\, h_L$, where a is the power defined in (5.43). Show the validity of this relationship. Note: this relationship is the basis for Equation (6.40) to develop the lumped, kinematic approach.

6.12   Multiple choice. Indicate which of the following statements are correct. The kinematic wave method differs from the approach based on the complete (i.e. shallow water or St. Venant) equations, because:
(a)   the equation of continuity is always replaced by the lumped storage equation;
(b)   the flow is assumed to be laminar;
(c)   it is assumed that there exists a single-valued (i.e. unique) relationship between rate of flow and water depth;
(d)   gravity effects are assumed to be balanced completely by frictional effects;
(e)   it is only applicable if dynamic (i.e. acceleration of the fluid) effects are negligible.

7   STREAMFLOW ROUTING

Also called flood routing and channel routing, this is one of the classical problems in applied hydrology. The word *routing* refers in general to the mathematical procedure of tracking or following water movement from one place to another; as such, the word also includes the description of the conversion of precipitation into various subsurface and surface runoff phenomena. However, *streamflow routing* refers specifically to the description of the behavior of a flood wave as it moves along in a well-defined open channel. In practical terms, the problem consists of the determination of the discharge hydrograph $Q = Q(t)$ at a given point along a stream, from a known hydrograph further up- or downstream and from a knowledge of the physical characteristics of the channel. The wave may be the result of inflows into the channel following various events such as heavy rainfall, snowmelt, failure or overtopping of natural or artificial dams due to landslides or earthquakes, tidal interactions and releases from artificial reservoirs.

Over the years different methods have evolved. The more fundamental approaches are based on hydraulic theory of open channel flow and consist of ad-hoc solutions of some form of the complete shallow water equations by numerical techniques on digital computer. A detailed treatment of such techniques is beyond the scope of this book and good reviews of available methods to solve the complete Saint Venant equations have been presented by, among others, Liggett and Cunge (1975) and others in Mahmood and Yevjevich (1975), Cunge *et al.* (1980), and Montes (1998). When accuracy is of primary concern, a good numerical technique with the complete shallow water equations should be the method of choice. However, to use any of the available numerical codes, it is essential to have a thorough understanding of the underlying fluid mechanical principles.

Some crucial aspects of the routing problem can be clarified by analytical solution and inspection of simplified formulations that are applicable in certain special situations. Simplified approaches can also be adequate, and sometimes even preferable for planning and preliminary design purposes. In this chapter major attention is given to one such scheme, the Muskingum method; it was developed, in principle, by means of the lumped kinematic approach and has been found to yield good results under a wide range of conditions of practical interest in natural rivers. It also has some theoretical ramifications, which are relevant for a deeper understanding of the complete shallow water equations. First, however, the behavior is considered of the shallow water equations for the two extremes of momentum transfer, namely under dynamic and kinematic conditions, in the case of large waves.

7.1   TWO EXTREME CASES OF LARGE FLOOD
WAVE PROPAGATION

In Chapter 5 it was shown how different types of wave can result from small disturbances of the water surface, depending on which terms are important and which terms can be omitted in the momentum equation. Special types of large wave can also be simulated, again depending on which are the main factors controlling the momentum budget of the flow.

7.1.1   *Surge or dynamic shock*

Under certain conditions the water surface exhibits an obvious and visible disturbance, also variously called an abrupt wave, a surge, a bore or a moving hydraulic jump. Similar phenomena occur in other situations in the environment as well, and they constitute the front of what are broadly referred to as gravity currents (Simpson, 1997). For instance, in Chapter 3 analogous surges were seen to occur as the gust front of thunderstorms (see Figures 3.8 and 3.9).

   Any such disturbance may be interpreted, if not as a discontinuity, at least as a point where the water depth h is not a smooth function; this interpretation means that the derivative $(\partial h/\partial x)$ is discontinuous, i.e. indeterminate. In general, it depends on the nature of the flow around such a wave, whether it will amplify or decay. As already noted in Chapter 5 on the basis of the solution of the linear case, the criterion for bore formation is Fr $> a^{-1}$ (where a is the power in Equations (5.39) and (5.43)), or Fr $> 1.5 \sim 2$. However, the instantaneous speed of propagation c_s of such a wave depends solely on the magnitude of the discontinuity itself, as will now be shown.

Types of abrupt wave
In general, there are four different types of abrupt wave. A simple way of visualizing them makes use of a thought experiment illustrated in Figure 7.1. Consider a sluice gate under well-established steady flow conditions, whose opening is suddenly changed. If the gate is raised, the downstream flow can be seen to develop situation A and the upstream flow develops situation C. If the sudden change is downward, situations B and D develop, respectively upstream and downstream from the gate. Waves in class A move downstream as positive surges; these types of bore are the ones of primary interest in hydrology, as they can transport large amounts of water and have caused some major flooding events in the past. Waves of type B are also positive surges, but in contrast to A, they advance upstream. They are typical for tidal bores, which are observed in some estuaries and rivers affected by tidal action. Such bores can be formed when the rising tide reverses the river flow and the tidal water enters into a gradually narrowing and shallowing channel, usually with a small bed slope; this narrowing environment slows down the leading edge of the tide, and allows it to be overtaken by the deeper and faster traveling water of the continually rising tide coming up from the rear. Although the survey was not totally exhaustive, at last count (Bartsch-Winkler and Lynch, 1988),

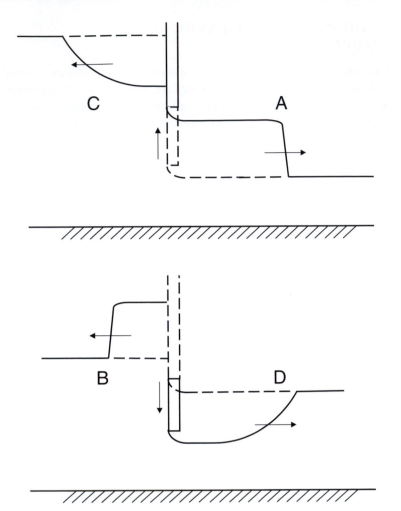

Fig. 7.1 Generation of the four different types of abrupt wave by the sudden movement of the sluice gate. Waves of type A and B are stable positive surges. Waves of type C and D are relatively short lived (i.e. unstable) negative surges. The dashed lines indicate the original positions prior to the movement of the gate.

some 67 localities were identified worldwide, where well-defined tidal bores occur, with reported heights ranging between 0.2 and 6.0 m. Waves of type C are upstream moving negative surges, which have on occasion been observed in hydropower supply canals, when the water demand is drastically stepped up. Waves of type D are downstream moving negative surges, which can result on occasion from a suddenly decreased water supply in an open channel.

As will be seen below, waves travel faster in deeper water than in shallow water. In the positive surges of type A and B, which result in a higher water surface, the deeper water tends to overtake the more shallow leading edge water; therefore, the surge maintains itself and can be considered stable. In the case of the negative (or receding) surges of

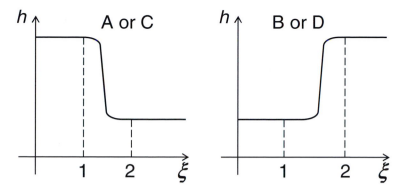

Fig. 7.2   Definition sketch for the analysis of the four types of abrupt surges.

type C and D, which result in a depressed water surface, the deeper leading edge flees away from the more shallow water which follows behind more slowly; therefore, the sharp front cannot maintain itself and soon spreads out: the wave is unstable. In spite of these differences, the initial stages of all four types of waves can be analyzed in the same way, as will be shown next.

Analysis of flood wave

The problem is most easily analyzed from a reference moving at the same velocity c_s as the wave, so that the phenomenon becomes one of steady state. One may thus assume that h and V become functions of a single variable $\xi = (x - c_s t)$, and the partial derivatives become ordinary derivatives as follows

$$\frac{\partial(\ )}{\partial t} = \frac{d(\ )}{d\xi}\frac{\partial \xi}{\partial t} = -c_s\frac{d(\ )}{d\xi} \quad \text{and} \quad \frac{\partial(\ )}{\partial x} = \frac{d(\ )}{d\xi}\frac{\partial \xi}{\partial x} = \frac{d(\ )}{d\xi} \tag{7.1}$$

Because the velocity V and depth h vary markedly and abruptly across the wave, the inertia and hydrostatic pressure gradient terms are predominant in the momentum equation and, as a first approximation, the friction term gS_f and the gravity term gS_0 can both be neglected. In the absence of lateral inflow, the momentum equation for a wide channel, (5.22), becomes then with (7.1)

$$-(c_s - V)\frac{dV}{d\xi} + g\frac{dh}{d\xi} = 0 \tag{7.2}$$

Upon multiplication by $(h\,d\xi)$ and integration between two points 1 and 2, respectively, some small but finite distance upstream and downstream from the abrupt wave (see Figure 7.2), Equation (7.2) assumes the form

$$-\int_1^2 (c_s - V)\,h\frac{dV}{d\xi}\,d\xi + \frac{g}{2}\int_1^2 \frac{dh^2}{d\xi}\,d\xi \tag{7.3}$$

The same operations can be applied to the continuity equation for a wide channel, (5.13), without lateral inflow. Thus, with the coordinate transformation (7.1), Equation (5.13) becomes

$$-c_s \frac{dh}{d\xi} + \frac{d(Vh)}{d\xi} = 0 \tag{7.4}$$

After integration across the wave, this becomes in turn

$$(c_s - V)\,h \;=\; \text{constant}$$

or $\hspace{17cm}$ (7.5)

$$(c_s - V_1)\,h_1 \;=\; (c_s - V_2)h_2$$

Equation (7.5) can now be substituted into (7.3) to integrate the latter. According to (7.5) part of the first term in (7.3) is independent of ξ. Therefore after taking that part outside the integral sign, (7.3) becomes

$$(c_s - V_1)\,h_1(V_1 - V_2) + \frac{g}{2}\left(h_2^2 - h_1^2\right) = 0 \tag{7.6}$$

Equation (7.5) can also be used to eliminate V_2 from (7.6); after substitution of $V_2 = [c_s - (c_s - V_1)h_1/h_2]$ into (7.6) and some algebra, one obtains finally

$$c_s = V_1 \pm \left(\frac{g\,h_2(h_2 + h_1)}{2\,h_1}\right)^{1/2} \tag{7.7}$$

The square root term in Equation (7.7) represents the celerity of the bore relative to the velocity at cross section 1. Because of symmetry, the subscripts 1 and 2 in Equation (7.7) can be interchanged, to yield the celerity relative to the velocity at section 2. Whenever sections 1 and 2 are defined respectively as the section upstream and the downstream from the abrupt wave, the plus sign in Equation (7.7) describes the downstream motion of surges of type A and D, and the minus sign upstream moving surges of type B and C. Thus flood waves, as encountered in hydrology require the plus sign in (7.7).

It can be verified that, if the analysis had been carried out for a channel with arbitrarily shaped cross sections at points 1 and 2, the result would have been

$$c_s = V_1 \pm \left(\frac{g\,(A_2h_2 - A_1h_1)}{2A_1(1 - A_1/A_2)}\right)^{1/2} \tag{7.8}$$

When h is large, or when the disturbance is small, $h_1 \approx h_2$, and (7.7) relative to V_1 ($\approx V_2$), reduces to Lagrange's celerity equation (5.50), as was to be expected. When $c_s = 0$, Equations (7.7) and (7.8) describe a stationary hydraulic jump.

The analysis presented here is simplified considerably, in that the effects of bed slope and resistance have been omitted. When the wave travels over large distances in a natural river these factors can play an important role; nevertheless Equation (7.8) can sometimes provide worthwhile first order information on the main features of such waves.

Disastrous floods
Abrupt waves of type A have been associated with some extreme flooding events in the past. For instance, in the United States, the Johnstown flood is still among the largest natural disasters on record (see McCullough, 1968; Degen and Degen, 1984). The flood

Fig. 7.3 **The broken South Fork dam, as seen from inside the empty reservoir. The opening was about 130 m wide near the top; the spillway can be seen on the right just below the bridge. It took the flood surge roughly 53 min to reach Johnstown. Drawing by Schell and Hogan. (From *Harper's Weekly*, 1889.)**

took place in the afternoon of May 31, 1889, after a night of heavy rain, and was caused by the failure of a badly maintained dam on South Fork Creek, some 23 km upstream along the Little Conemaugh River and some 135 m higher than the town itself. After the dam suddenly gave way completely (at 1510) (Figure 7.3), a wall of water, exceeding 15 m in some places, raged down the valley; as it reached Johnstown (at 1607), in a little less than an hour, it spread out somewhat over a wider area, but its center was still at least 10 m high. While the main event in Johnstown was over in 10 min, it left more than 2200 dead in its aftermath and near total destruction of the city.

Example 7.1. Some features of the Johnstown flood

Interestingly, the reported features of the Johnstown flood are not unreasonable in light of Equation (7.7), and some of them can be reconstructed with a few rough estimates of the effective parameters. Dam breach problems like this are highly unsteady in nature, and they have been the subject of intensive study (see, for example, Yevjevich, 1975). However, assume for the present example that the reservoir was large enough, resulting in a steady inflow into the river channel after the dam had failed. From the eyewitness accounts, the height of the surge appears to have been of the order of $h_1 = 10$ m. With a mean slope of $S_0 = (135/23\,000)$ and an assumed roughness of $n = 0.07$ (see Table 5.1), Equation (5.41) yields a velocity of the water behind the surge of $V_1 = 5.08$ m s^{-1}.

Assuming also (as a first approximation) that the valley cross-sectional area did not vary appreciably along the flow path, and (for want of better information) that in the valley ahead of the surge the effective water depth was of the order of $h_2 = 1.0$ m, according to Equation (7.7) one finds that the surge came down at $c_s = 7.4$ m s^{-1}; thus, it would have taken the wave about 52 min to cover the 23 km, which in fact it did. Admittedly, this result is obtained with unknown and therefore assumed values of n and h_2, and it would be possible to obtain the same value of c_s with other equally plausible combinations. For instance, roughly the same result would be obtained with assumed values of $n = 0.05$ and $h_2 = 0$ m, representing a smoother channel with negligible depth of flow prior to the disaster (cf. Problem 7.1). It would require a more detailed field survey and analysis to estimate which values of n, h_1 and h_2 would be appropriate for the dam-break event that took place in this valley.

A notorious but less disastrous flash flood occurred on Willow Creek in the town of Heppner in Oregon, in the late afternoon on Sunday June 14, 1903. According to Morrow County records (see also Taylor and Hatton, 1999), it was generated by a cloudburst, which was later estimated to amount to about 35 mm over an area of some 50 km^2, mainly around Balm Fork, in the hills 10–15 km south of town. The flood waters raced into town around 1700, causing the death of some 247 inhabitants, nearly 20% of the population, and destroying one third of all structures. The *Heppner Gazette* of June 18, 1903, reported that the flood struck, "without a second's warning, a leaping, foaming wall of water, 40 feet in height." This surge height may well have been an overestimate. Without detailed information on the temporal and spatial distribution of that rainfall event, it is hard to know exactly how the runoff was funneled into a surge in the Willow Creek valley. Nevertheless, the suddenness and power of the flood were no doubt also exacerbated by the presence of a laundry at the upstream end of town; combined with accumulating debris, this structure at first blocked passage of the water somewhat, only to give way after a short while, abruptly releasing the built-up water mass.

Other examples of this type of flood are the glacier lake outbursts caused by ice avalanches in Peru, where they are known as aluviónes (see Lliboutry *et al.*, 1977; Morales-Arnao, 1999). Typically, they are triggered with almost no warning and bring down ice blocks, boulders and mud, leaving death and destruction in their path. In the past three centuries more than 22 major outburst floods have destroyed a number of towns and villages in the region. In 1941 one such aluvión, caused by the failure of a moraine dam higher up in the Cordillera Blanca, destroyed about one third of the city of Huaraz and resulted in an estimated 5000–7000 deaths.

7.1.2   *Monoclinal rising wave or kinematic shock*

While the surge considered in the previous section is obtained by considering only the dynamic terms in the momentum equation (5.22), the monoclinal rising wave is obtained from a balance of the other two terms in that equation, i.e. $S_f = S_0$ (in the absence of lateral inflow). In other words, the accelerations are considered to be so gentle that the flow can be assumed to be quasi-steady-uniform and that the only forces of any consequence

Fig. 7.4   Monoclinal wave, as a
gentle transition
between two uniform
quasi-steady regimes.
Its rate of
propagation is given
by Equation (7.9).

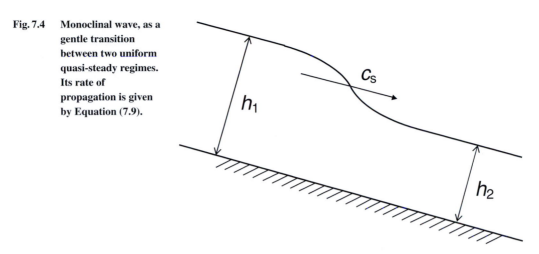

are frictional resistance and gravity. It is often considered the kinematic equivalent of the dynamic shock, because the wave front constitutes a moving transition between two regions of essentially uniform flow.

Again, the problem can be analyzed within a reference frame moving at the same velocity as the wave (Figure 7.4). This yields, as before, Equation (7.5) for the continuity equation, or

$$c_s = \frac{q_2 - q_1}{h_2 - h_1} \tag{7.9}$$

In a perfect kinematic system this result suggests a discontinuity in h or in its derivative, because in order to keep c_s a constant, dq/dh must also be constant, even though $q = q(h)$ is nonlinear. This is the reason why the monoclinal rising wave has also been referred to as a kinematic shock. In actual rivers, however, flood rises, which are subject mainly (but never only) to the effects of gravity and friction, usually do not display any such discontinuities and they are quite continuous and smooth; they usually extend over large distances and the transition or shock thickness may be considerable. In view of this ambiguity, the question arises whether the type of motion predicted by Equation (7.9) can even exist or maintain itself in the real world. The matter can be resolved by considering the full equation of motion (5.22) to determine under what conditions diffusion and wave steepening might be in balance, to allow the stability of the uniformly progressive flow assumed in the derivation of (7.9). By means of a moving coordinate system it can be shown that the monoclinal wave profile is actually stable, as assumed, provided $h_1 > h_2 > h_{cr}$, in which h_{cr} is the critical depth, and provided the wave front extends an infinite distance downstream. It can also be shown that in most practical cases on large rivers the wave will rise to $0.90(h_1 - h_2)$ within a distance of the order of h_2/S_0, which is usually of the order of some tens of kilometers; this means that the monoclinal rising wave is often well approximated on long rivers. The details of the analysis, although straightforward, are beyond the present scope and various aspects can be found elsewhere (Lighthill and Whitham, 1955; Henderson, 1966).

For a small rise, Equation (7.9) reduces to (5.108), that is, the Kleitz–Seddon principle, and the kinematic shock velocity c_s approaches the kinematic wave celerity c_k, as defined in (5.108) and (5.112). Thus one obtains (5.113), or

$$c_k = (a + 1) V \tag{7.10}$$

for a wide channel, or (5.114) otherwise; values of a are listed in Table 5.2.

7.2   A LUMPED KINEMATIC APPROACH: THE MUSKINGUM METHOD

This streamflow routing method is named after the Muskingum Watershed Conservancy District in Ohio, where it was evidently first applied. The Muskingum River is a tributary of the Ohio River. Ever since it was presented by McCarthy (1938), this method and several of its derivatives and variants have been widely used in hydrologic applications. There are several other streamflow routing methods that are based on the lumped kinematic approach, among which the lag-and-route method (Meyer, 1941) and the method of Kalinin and Milyukov (see, for example, Apollov *et al.*, 1964, p. 53) are probably the better known (see also Chow, 1959; Dooge, 1973). The Muskingum method has been the subject of much research and its performance characteristics and limitations are well understood so that it can serve as the prototype of this approach.

7.2.1   *Conceptual derivation*

The method is based on the storage equation (1.10) or (5.125), which can be rewritten here for convenience as

$$Q_i - Q_e = \frac{dS}{dt} \tag{7.11}$$

where Q_i and Q_e are the rate of flow, respectively, at the inlet and at the outlet section of the channel reach under consideration. When the hydrographs of Q_i and Q_e are plotted on the same graph, S, the water stored in the reach during the passage of the flood wave, is the cumulative area between the two curves, as shown in Figure 7.5. In the Muskingum method, closure of (7.11) is achieved by the assumption that S is a weighted function of both inflow rate and outflow rate of the channel reach control volume. This function can be derived as follows.

Muskingum storage function

At the point of inflow the flow rate can often be approximated by a power function of the cross sectional area, and the same holds true for the point of outflow, or

$$Q_i = \alpha_i A_{ci}^{\beta} \quad \text{and} \quad Q_e = \alpha_e A_{ce}^{\beta} \tag{7.12}$$

where α and β are constants and the subscripts denote the inflow and outflow section of the channel reach under consideration. For example, according to Equation (5.39) for a wide channel with an average width B_c, these constants are $\alpha = (C_r S_0^b B_c^{-a})$ and $\beta = (a + 1)$. The water volume stored in the reach S is equal to its length Δx times the

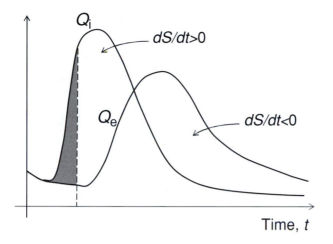

Fig. 7.5 Illustration of the inflow hydrograph $Q_i = Q_i(t)$ and the outflow hydrograph $Q_e = Q_e(t)$ in a channel reach, together with the rate of change of water stored in the reach during the passage of a flood wave. As long as the inflow rate is larger than the outflow rate, i.e. $Q_i > Q_e$ the storage is increasing, and as soon as the opposite occurs, i.e. when $Q_i < Q_e$, the storage decreases. The shaded area indicates how the storage itself increases with time, as $S = \int (Q_i - Q_e)\, dt$.

average cross sectional area of the reach, or

$$S = [X\, A_{ci} + (1 - X)A_{ce}]\Delta x \tag{7.13}$$

where X is a constant reflecting the relative weight, which the cross sections at the inflow end and at the outflow end exert on the average cross section (see Figure 7.6). Substituting (7.12) in (7.13), one obtains

$$S = \left[X\, \alpha_i^{-1/\beta} Q_i^{1/\beta} + (1 - X)\, \alpha_e^{-1/\beta} Q_e^{1/\beta}\right]\Delta x \tag{7.14}$$

If it is further assumed that the cross sections at the inflow and outflow end exhibit similarity, so that all the constants can be combined into one constant, say K, and that the system is linear so that $\beta = 1$, one obtains finally the Muskingum storage function

$$S = K[X\, Q_i + (1 - X)\, Q_e] \tag{7.15}$$

The constant K is also referred to as the storage coefficient. Substitution of (7.15) into the storage equation (7.11) yields the governing differential equation

$$Q_e + K(1 - X)\frac{dQ_e}{dt} = Q_i - KX\frac{dQ_i}{dt} \tag{7.16}$$

Interpretation of the parameters
Some insight can be gained in the nature of the parameters K and X by considering the two main features of the flood wave in progress. These features are the time it takes for the flood wave to pass through the reach, and the change in shape the flood wave

Fig. 7.6   Water stored in the
           channel reach control
           volume of length Δx.
           The cross-sectional
           areas are A_{ci} and A_{ce}
           at the inflow and at
           the outflow end,
           respectively. The
           corresponding
           average water depths
           at these sections are h_i
           and h_e.

Fig. 7.7   The time of travel t_t of
           the flood wave through
           the reach is the time
           between the centroids
           of the inflow and
           outflow hydrographs.
           In the linear
           Muskingum storage
           function this is given
           by $t_t = K$.

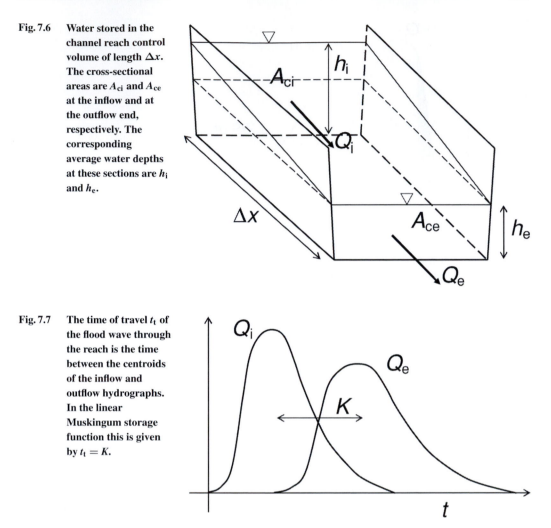

undergoes, as it travels through the reach. The moments, which are used in what follows, are defined in Chapter 13.

The mean time of occurrence of a flood wave is the first moment about the origin, denoted as m'_1 or μ; this quantity is also referred to as the centroid or center of area of the hydrograph. Therefore, the travel time is the difference between the mean time of occurrence of the wave at the exit section and that at the inflow section of the reach, that is $t_t = m'_{e1} - m'_{i1}$, as shown in Figure 7.7; as before, the subscripts e and i refer to the outflow and inflow sections of the reach. In other words, that travel time is the time between the centers of area of the inflow wave $Q_i(t)$ and of the outflow wave $Q_e(t)$, or

$$t_t = \frac{\int_0^\infty t\, Q_e\, dt}{\int_0^\infty Q_e\, dt} - \frac{\int_0^\infty t\, Q_i\, dt}{\int_0^\infty Q_i\, dt} \tag{7.17}$$

The integrals in the denominators of (7.17), which are the zeroth moments, are required to normalize Q; ideally, they should equal each other, if there are no lateral inflows or outflows in the reach. Substitution of (7.16) in (7.17) produces

$$t_t = -K \int_0^\infty t \frac{d}{dt}[X Q_i + (1 - X) Q_e] dt \Bigg/ \int_0^\infty Q_i \, dt \tag{7.18}$$

Integration by parts, and imposition of the condition that both Q_i and Q_e are zero for t at infinity, leads finally to the desired result

$$t_t = K \tag{7.19}$$

In words, Equation (7.19) states that the parameter K can be interpreted as a measure of the lag or the time of travel t_t of the flood wave through the reach. Accordingly, when the channel reach has a length Δx, the celerity of a Muskingum wave is

$$c_m = \Delta x / K \tag{7.20}$$

The width or average duration of a flood wave hydrograph, which is one of the more obvious measures of its shape, can be conveniently characterized by its standard deviation σ, that is, the square root of its second moment about the mean, $\sqrt{m_2}$. Thus, the change in shape of a flood wave hydrograph, after passing through a channel reach, can be described by the difference between the second moments of the outflow and inflow hydrographs, namely $(m_{e2} - m_{i2})$. Since the second moment about the mean is related to the moments about the origin as indicated in Equation (13.12), this difference can be written as

$$(m_{e2} - m_{i2}) = (m'_{e2} - m'_{i2}) - (m'_{e1})^2 + (m'_{i1})^2 \tag{7.21}$$

in which the difference between the two moments about the origin is

$$(m'_{e2} - m'_{i2}) = \frac{\int_0^\infty t^2 Q_e \, dt}{\int_0^\infty Q_e \, dt} - \frac{\int_0^\infty t^2 Q_i \, dt}{\int_0^\infty Q_i \, dt} \tag{7.22}$$

As before, in (7.22) the two terms in the denominators should be equal to each other, when there are no additional in- or outflows in the reach. Substitution of Equation (7.16) into (7.22) produces now

$$(m'_{e2} - m'_{i2}) = -K \int_0^\infty t^2 \frac{d}{dt}[X Q_i + (1 - X)Q_e] dt \Bigg/ \int_0^\infty Q_i \, dt \tag{7.23}$$

Integrating (7.23) by parts, and making use of the same operations that led from (7.18) to (7.19), one finds

$$(m'_{e2} - m'_{i2}) = 2K(m'_{e1} - KX) \tag{7.24}$$

Finally, substituting (7.24) into (7.21), and recalling that according to (7.19) $m'_{e1} - m'_{i1} = K$, one obtains

$$(m_{e2} - m_{i2}) = K^2(1 - 2X) \tag{7.25}$$

In the conceptual derivation of Equation (7.15), X was introduced simply to weight the relative effects of the inflow and outflow sections. Accordingly, in cases of pure reservoir action, that is, when the flood passes through a level pool whose stage (or level) is controlled by a spillway at the downstream end, the storage S should be independent of the inflow rate, and therefore $X = 0$. On the other hand, in a uniform rectangular channel with a plane water surface the two sections should be weighted equally and ideally $X = 0.5$. Equation (7.25) allows now a fuller interpretation of X. Equation (7.19) already showed that the parameter K is the mean residence time of the flood wave in the reach; both second moments and K^2 in (7.25) have the basic dimensions $[\text{T}^2]$. Hence $\sqrt{1 - 2X}$ reflects the rate of increase of the (streamwise) width of the wave as it travels through the reach; because mass is conserved, $\sqrt{1 - 2X}$ must also reflect the rate of decrease of its height, that is, the rate of subsidence of the peak discharge of the flood hydrograph. According to (7.25), the difference between the two second moments is maximal when $X = 0$, that is under conditions of pure reservoir action. On the other hand, (7.25) indicates that, when $X = 1/2$, the wave does not undergo deformation, but it retains its original shape as it travels. Because the peak of a flood wave normally decreases along its path, in principle X should be smaller than 0.5.

7.2.2 Analytical solution

The ordinary differential equation (7.16) can be readily solved. One common technique is to multiply both sides by $\exp[t/K (1 - X)]$. This allows it to be written as

$$\frac{d}{dt}\left(e^{t/K(1-X)}K(1 - X)Q_e\right) = -KX\frac{d}{dt}\left(e^{t/K(1-X)}Q_i\right) + \frac{K}{1 - X}e^{t/K(1-X)}Q_i \qquad (7.26)$$

Finally, the integral of (7.26) provides the outflow rate Q_e resulting from a given inflow rate into the reach $Q_i = Q_i(t)$, as follows

$$Q_e = \frac{e^{-t/K(1-X)}}{K(1 - X)^2}\int Q_i(\tau)e^{\tau/K(1-X)}d\tau - \frac{X}{(1 - X)}Q_i(t) + \text{constant} \qquad (7.27)$$

in which the value of the constant depends on the values of the flow rates at some reference time.

Unit response function

This is the outflow from the channel reach in response to a unit impulse inflow into the reach at the inflow section at $t = 0$ (see Appendix). Thus with a Dirac delta function inflow, $Q_i = \delta(t)$, Equation (7.27) immediately yields

$$u(t) = \frac{e^{-t/K(1-X)}}{K(1 - X)^2} - \frac{X\,\delta(0)}{(1 - X)} \qquad (7.28)$$

The first two moments of the unit response

An alternative way to describe a function is by means of its moments. These can be determined for the unit response function (7.28) as follows. The first moment of the unit response about the origin is (see Chapter 13)

$$m'_{u1} = \frac{\int\limits_0^\infty t\,u\,dt}{\int\limits_0^\infty u\,dt} \tag{7.29}$$

With the unit response (7.28), one can check that the integral in the denominator of (7.29), i.e. the zeroth moment, equals one, as it should; moreover, the first moment of the delta function $\delta(0)$ is zero. Therefore, after insertion of (7.28), Equation (7.29) can be rewritten as

$$m'_{u1} = \int\limits_0^\infty t\,\frac{e^{-t/K(1-X)}}{K(1-X)^2}\,dt \tag{7.30}$$

After integration by parts, one finds that the first moment of the unit response is simply equal to the Muskingum parameter K, or

$$m'_{u1} = K \tag{7.31}$$

The second moment of the unit response about the origin is

$$m'_{u2} = \frac{\int\limits_0^\infty t^2 u\,dt}{\int\limits_0^\infty u\,dt} \tag{7.32}$$

Proceeding in the same way as for the first moment, one finds that this second moment about the origin can be written in terms of the Muskingum parameters as

$$m'_{u2} = 2K^2(1-X) \tag{7.33}$$

Since the second moment about the mean is related to the first two moments (see Equation (13.12)), one obtains finally

$$m_{u2} = K^2(1-2X) \tag{7.34}$$

The higher moments can be derived in the same way.

As an aside, comparison of Equations (7.31) and (7.34) with (7.19) and (7.25), respectively, reveals that $m'_{u1} = (m'_{e1} - m'_{i1})$ and $m_{u2} = (m_{e2} - m_{i2})$. This is not unexpected. Indeed, the Muskingum channel reach is a linear system, to which the theorem of moments, as given by Equations (A22) and (A28) should be fully applicable.

7.2.3   Standard implementation

Numerical calculations

Although the analytical solution provides insight into the structure of the Muskingum formulation, it is difficult to use with observed streamflow data. In hydrologic practice, the Muskingum method is normally applied over finite time increments Δt; for this purpose Equation (7.11) can be approximated as

$$\frac{1}{2}(Q_{i1} + Q_{i2})\Delta t - \frac{1}{2}(Q_{e1} + Q_{e2})\Delta t = S_2 - S_1 \tag{7.35}$$

in which the subscripts 1 and 2 refer to the beginning and the end of the time period Δt. Upon substitution of (7.15) this becomes

$$\frac{1}{2}(Q_{i1} + Q_{i2})\Delta t - \frac{1}{2}(Q_{e1} + Q_{e2})\Delta t$$

$$= K\{[X\,Q_{i2} + (1 - X)Q_{e2}] - [X\,Q_{i1} + (1 - X)Q_{e1}]\} \tag{7.36}$$

After collecting the terms, (7.36) can be written as a coefficient equation

$$Q_{e2} = c_0 Q_{i2} + c_1\,Q_{i1} + c_2\,Q_{e1} \tag{7.37}$$

in which

$$c_0 = \frac{-2X + \Delta t/K}{2(1 - X) + \Delta t/K}, \qquad c_1 = \frac{2X + \Delta t/K}{2(1 - X) + \Delta t/K} \quad \text{and}$$

$$c_2 = \frac{2(1 - X) - \Delta t/K}{2(1 - X) + \Delta t/K} \tag{7.38}$$

with the obvious requirement that $(c_0 + c_1 + c_2) = 1$.

Constraints on the parameters

In practical applications, the parameters in the Muskingum method must satisfy a number of constraints, if it is to perform well. When the method was originally developed not much attention was paid to this issue, as the basic underlying assumptions were not fully understood; therefore, the method sometimes produced unreasonable results (such as negative flow rates). The values of the weighting parameter X, of the time step Δt, and of the length of the channel reach Δx, affect the outcome of the calculations, so that some attention should be given to their choice.

(i)   As already discussed earlier, Equation (7.25) indicates that X should not exceed 0.5; indeed values in excess of 0.5 would indicate that the flood peak magnitude increases as it moves downstream; this never occurs in situations where the lumped kinematic approach is applicable. Moreover, a negative value of X would indicate in (7.15) that a larger inflow rate into the reach results in a smaller storage. Thus, one can constrain X as follows

$$0 \le X \le 0.5 \tag{7.39}$$

(ii)   The Muskingum method involves several time scales, namely the finite time step of the numerical solution Δt, the time of travel through the channel or lag K, and the characteristic life time of the incoming flood wave, say its time to peak t_p. To allow sufficient resolution of the temporal behavior of the flood wave, it stands to reason that Δt should be small compared with the life time of the incoming flood. Therefore it is usually assumed (see Jones, 1981; Ponce and Theurer, 1982) that

$$\Delta t \le a t_p \tag{7.40}$$

in which a is a number of the order of 4 to (preferably) 5.

Fig. 7.8 **The heavy line triangular region indicates the validity range of the Muskingum formulation. The area inside the triangle is defined by the conditions that $0 \leq X \leq 0.5$ and that the coefficients c_0, c_1 and c_2 be positive, as indicated in (7.41); c_m denotes the celerity of the wave. (After Miller and Cunge, 1975.)**

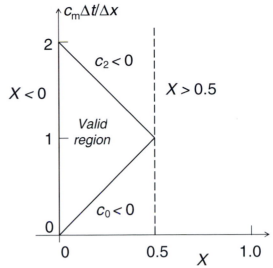

(iii)   Several studies have dealt with the optimal length of the length of the reach Δx, but there is no general consensus in the literature. As can be seen in Equation (7.20), Δx and K are related, so that the magnitude of Δx plays a role in the calculations with (7.37) and (7.38). In the original applications of the Muskingum method Δx was taken as the total length of channel through which the routing was applied. In more recent applications the total length of channel is usually subdivided in a number of subsections of length Δx, and the calculations are carried out for each of these subsections. One simple constraint, which would obviously avoid negative outflows and which was suggested by Miller and Cunge (1975), is that the coefficients in Equation (7.37) should all be positive. Hence, from (7.38) and (7.20) one can deduce that, strictly speaking, the following should be satisfied.

$$\frac{\Delta x}{\Delta t} \leq \frac{c_m}{2X}$$

and   (7.41)

$$\frac{\Delta x}{\Delta t} \geq \frac{c_m}{2(1 - X)}$$

With Equation (7.39) this allows construction of the diagram shown in Figure 7.8. In practical applications the flood hydrographs needed for calibration may not be available at the desired distances along the reach. However, a coarse spatial resolution in available hydrograph data can sometimes be remedied by interpolation, to generate a smaller step Δx. Laurenson (1959) has shown an example how such interpolation can be implemented to achieve improved results.

In some situations it may prove impossible to adhere strictly to these conditions (i), (ii) and (iii). Fortunately, however, the Muskingum method can be fairly robust and violation of criteria like (7.41) does not necessarily lead to useless results. For instance, Weinmann and

Laurenson (1979) reported that they did experience negative outflows in their calculations, but as long as the criterion

$$\Delta x \leq \frac{t_p c_m}{2X} \tag{7.42}$$

was satisfied, these negative outflows were found to be small and short enough to be ignored for practical purposes. Note that in light of Equation (7.40) this criterion is a relaxed version of the first of (7.41).

7.3   ESTIMATION OF THE MUSKINGUM PARAMETERS

7.3.1   *Calibration with previously recorded flood wave events*

Estimation with the Muskingum storage function
Especially in the early applications of the method, the parameters X and K were usually estimated from available data of inflow and outflow in a given reach of interest. In this approach, the flow data are used to determine the storage in the reach, from which the storage function (7.15) can be derived. In practice, the first step consists of obtaining S from Q_i and Q_e by means of Equation (7.35) as shown in Table 7.1. The initial value S_1 is unknown, but its value is immaterial; in the computations, it can usually be set at zero by using (7.15) with a zero intercept.

The next step consists of plotting the S values against the weighted values of the inflow and the outflow, that is $[X Q_i + (1 - X)Q_e]$, for different trial values of X. That value of X is then selected, which produces the relationship that most closely approximates a single-valued straight line without a loop. The value of K is the slope of that straight line, and ideally its intercept is $-S_1$. The main drawback of this method is that it does not rely on objective criteria to determine optimality; rather, it involves a trial-and-error adjustment of X by fitting a straight line to a loop, which is usually not straight but curved. Nevertheless, the procedure is simple and easy to carry out, as shown in the following example.

Table 7.1 Estimation of the storage S from inflow and outflow hydrographs

Inflow	Averaged inflow	Outflow	Averaged outflow	Difference in storage	Storage
Q_{i1}		Q_{e1}		from (7.35)	S_1
	$(Q_{i1} + Q_{i2}) / 2$		$(Q_{e1} + Q_{e2}) / 2$	$(S_2 - S_1)$	
Q_{i2}		Q_{e2}			S_2
	$(Q_{i2} + Q_{i3}) / 2$		$(Q_{e2} + Q_{e3}) / 2$	$(S_3 - S_2)$	
Q_{i3}		Q_{e3}			S_3
	$(Q_{i3} + Q_{i4}) / 2$		$(Q_{e3} + Q_{e4}) / 2$	$(S_4 - S_3)$	
Q_{i4}		Q_{e4}			S_4
etc.					

Table 7.2 Application of the Muskingum channel routing method in Example 7.2

time (h)	Q_i (m³ s⁻¹)	Q_e (m³ s⁻¹)	ΔS (m³h s⁻¹)	S (m³h s⁻¹)	$XQ_i + (1-X)Q_e$ $X = 0.1$	$X = 0.3$	$X = 0.5$
0	172	139		0	143	149	156
1	250	124	79	79	137	162	187
2	438	220	172	251	241	285	329
3	736	342	306	557	382	460	539
4	1077	542	464	1021	596	703	810
5	1622	805	675	1696	887	1050	1214
6	2090	1271	818	2514	1353	1517	1681
7	2294	1684	714	3229	1745	1867	1989
8	2247	1973	442	3670	2001	2055	2110
9	2090	2169	98	3768	2161	2145	2130
10	1622	2090	−273	3495	2044	1950	1856
11	1271	1895	−546	2948	1833	1708	1583
12	1015	1622	−615	2333	1561	1440	1318
13	844	1333	−548	1785	1285	1187	1089
14	711	1077	−428	1357	1040	967	894
15	627	891	−315	1042	864	812	759
16	549	759	−236	806	738	696	654
17	488	651	−186	620	634	602	569
18	433	558	−143	477	545	520	495
19	388	496	−116	360	485	464	442
20	343	434	−100	261	425	407	389
21	313	396	−87	174	388	371	355
22	283	350	−75	99	343	330	317
23	266	319	−60	39	314	303	293
24	249	296	−50	−12	291	282	272
25	236	265	−38	−50	263	257	251
26	224	235	−20	−70	234	232	230
27	213	220	−9	−79	219	218	216
28	201	204	−5	−84	204	203	203
29	192	197	−4	−88	196	195	194
30	182	189	−6	−94	188	187	186

Example 7.2. Standard application of the Muskingum method

Consider, as an illustration, the flow rates Q_i and Q_e for a channel reach; these are listed in Table 7.2. Also listed are the results of the calculations required to estimate the parameters as outlined already in Table 7.1. The values of incremental storage ΔS were calculated by means of Equation (7.35), and were summed in the next column to obtain the values of the storage S in the reach. These values can then be plotted against $X\,Q_i + (1 - X)Q_e$ for different trial values of the weighting parameter X. The value

Fig. 7.9   Storage in the channel reach S $(\mathrm{m^3 h\,s^{-1}})$, as a function of the weighted flow rate $XQ_i + (1-X)Q_e$ $(\mathrm{m^3\,s^{-1}})$ for values of $X = 0.1$ (triangles), 0.3 (circles), and 0.5 (squares) with the values of the flow rates of Example 7.2, listed in Table 7.2.

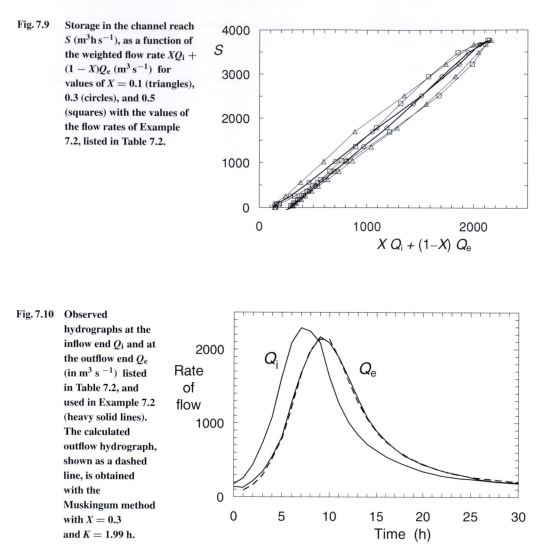

Fig. 7.10   Observed hydrographs at the inflow end Q_i and at the outflow end Q_e (in $\mathrm{m^3\,s^{-1}}$) listed in Table 7.2, and used in Example 7.2 (heavy solid lines). The calculated outflow hydrograph, shown as a dashed line, is obtained with the Muskingum method with $X = 0.3$ and $K = 1.99$ h.

$X = 0.3$ appeared to yield the best single valued relationship, required by (7.15). This is illustrated in Figure 7.9; also shown are the curves for the extreme values $X = 0.1$ and 0.5 to illustrate the evolution of the relationship as a function of X. The regression line in the form of (7.15) for the value of $X = 0.3$ is $S = 1.99\,(0.3\,Q_i + 0.7\,Q_e) - 489$. (This suggests that initially the storage in the reach could have been assumed to be $S = 489$ $\mathrm{m^3 h\,s^{-1}}$, instead of the value $S = 0$ adopted in Table 7.2, in order to force the regression in Figure 7.9 through the origin in accordance with Equation (7.15)). The time of travel through the reach is $K = 1.99$ h. With these values of X and K, Equation (7.37) can be written as $Q_{e2} = -0.051Q_{i2} + 0.579Q_{i1} + 0.472Q_{e1}$; the calculated outflow hydrograph can be compared in Figure 7.10 with the original data of Q_e (i.e. the values listed in Table 7.2) used in the calibration.

Estimation with optimization techniques: the method of least squares

For the purpose of parameter estimation, the Muskingum channel can also be treated as a black box, to which more objective systems techniques (cf. Section 12.2.1) can be applied. In what follows, the method of least squares is presented as an example.

The simple form of Equation (7.37) immediately suggests that it should be possible to estimate the constant coefficients by multiple linear regression. There is, however, one small complication, in that the coefficients must satisfy the constraint $c_0 + c_1 + c_2 = 1$. This constraint can be incorporated by defining the variables $y = Q_{e2} - Q_{e1}$, $x_1 = Q_{i2} - Q_{e1}$ and $x_2 = Q_{i1} - Q_{e1}$, and by recasting (7.37) as follows

$$y = c_0 x_1 + c_1 x_2 \tag{7.43}$$

which can be considered as a linear regression equation of y on x_1 and x_2, forced through the origin. By applying the method of least squares to Equation (7.43) it can readily be shown that optimal values of the coefficients can be calculated with the following

$$c_0 = \frac{\sum yx_1 \sum x_2^2 - \sum yx_2 \sum x_1 x_2}{\sum x_1^2 \sum x_2^2 - \left(\sum x_1 x_2\right)^2}$$

and   (7.44)

$$c_1 = \frac{\sum yx_2 \sum x_1^2 - \sum yx_1 \sum x_1 x_2}{\sum x_1^2 \sum x_2^2 - \left(\sum x_1 x_2\right)^2}$$

and, of course, $c_2 = 1 - c_0 - c_1$. The summations can be performed over the time range for which the hydrographs are available; this will yield values of the coefficients, which perform well "on average" (in the least squares sense). However, if certain features of the hydrograph require greater accuracy, such as for example the flows in the vicinity of the peak discharge, it may be desirable to perform the summation over a more narrow time range.

With these three coefficients c_0, c_1 and c_2 determined, Equation (7.37) can be applied to solve any routing problem. In case the Muskingum parameters are needed for one or other reason, they can be calculated from these coefficients by inversion of (7.38), that is by using

$$K = (c_1 + c_2)/(c_0 + c_1)$$

and   (7.45)

$$X = 0.5(c_1 - c_0)/(c_1 + c_2)$$

Example 7.3. Application of the multiple regression method

The Q_i and Q_e hydrographs of Example 7.2, as listed in Table 7.2 and shown in Figure 7.10, can be used to illustrate the method. The reader can verify that the sums needed in (7.44) have the following values: $\sum yx_1 = 2\,772\,315$, $\sum yx_2 = 1\,942\,872$, $\sum x_1^2 = 8\,383\,823$, $\sum x_2^2 = 3\,838\,571$, and $\sum x_1 x_2 = 5\,454\,548$ (in the units of Table 7.2 squared). With Equation (7.44) these yield the following values of the coefficients $c_0 = 0.018$,

$c_1 = 0.480$ and $c_2 = 0.502$; with Equation (7.45) the corresponding Muskingum parameters are $K = 1.97$ h and $X = 0.24$. All these values are close to the corresponding ones obtained in Example 7.2 using the standard trial-and-error procedure; in fact, a plot of the resulting hydrograph Q_e would be difficult to distinguish from the curve shown in Figure 7.10 for $K = 1.99$ h and $X = 0.3$.

7.3.2   *From physical characteristics of the channel*

Diffusive behavior of the Muskingum wave

Among the properties of the linear kinematic wave is the fact, explained in Section 5.4.3, that it propagates along the channel with a celerity c_{k0}, but without any change in shape. Therefore, it was puzzling to practitioners for some time in the past that this is apparently not the case for the Muskingum wave, even though it is based on the same approximation as outlined in Section 5.4.4. Indeed, as illustrated in Figure 7.10, a Muskingum flood wave undergoes not only translation in time but also a change in shape.

It is now realized that the change in shape of the calculated wave is not the result of the underlying physics, but rather the result of numerical diffusion caused by the approximation of derivatives by ratios of finite differences. Any lumped kinematic approach is based on the storage equation (7.11); that equation is a discretized form of the continuity equation, in which the spatial derivative is approximated by a difference over a distance Δx. As pointed out by Cunge (1969), it is this approximation that causes the spreading of the calculated wave.

The diffusion introduced by finite difference approximations can be determined by trying to recover the partial differential equation (5.111), without lateral inflow, from (7.37). Recall that the subscripts 1 and 2 refer to the beginning and end of the time interval Δt; similarly, the subscripts i and e refer to the inflow and exit end of the spatial reach Δx. For the present purpose, the four Q terms can be expressed in terms of the rate of flow $Q(x, t)$ by a Taylor expansion as follows

$$Q_{i1} = Q$$

$$Q_{i2} = Q + \frac{\partial Q}{\partial t}\Delta t + \frac{1}{2}\frac{\partial^2 Q}{\partial t^2}(\Delta t)^2 + \cdots$$

$$Q_{e1} = Q + \frac{\partial Q}{\partial x}\Delta x + \frac{1}{2}\frac{\partial^2 Q}{\partial x^2}(\Delta x)^2 + \cdots \tag{7.46}$$

$$Q_{e2} = Q + \frac{\partial Q}{\partial t}\Delta t + \frac{1}{2}\frac{\partial^2 Q}{\partial t^2}(\Delta t)^2 + \cdots$$

$$\cdots + \frac{\partial}{\partial x}\left(Q + \frac{\partial Q}{\partial t}\Delta t + \cdots\right)\Delta x + \frac{1}{2}\frac{\partial^2}{\partial x^2}(Q\cdots)(\Delta x)^2 + \cdots$$

where the terms of order higher than 2 have been neglected. Substitution of (7.46) into (7.37) and division by $[(1 - c_0)\Delta t]$ produces the following partial differential equation

$$\frac{\partial Q}{\partial t} + \frac{\Delta x(1 - c_2)}{\Delta t(1 - c_0)} \frac{\partial Q}{\partial x} + \frac{(\Delta x)^2(1 - c_2)}{2\Delta t(1 - c_0)} \frac{\partial^2 Q}{\partial x^2}$$

$$+ \frac{\Delta t}{2} \frac{\partial^2 Q}{\partial t^2} + \frac{\Delta x}{(1 - c_0)} \frac{\partial^2 Q}{\partial x \partial t} = 0 \qquad (7.47)$$

This result is obtained from Equation (7.37). In principle, (7.37) is merely a finite difference form of the kinematic wave equation (5.111) (in the absence of lateral inflow Q_1).
The kinematic wave equation contains only first-order derivatives. Hence, this suggests
that the first two terms of (7.47) must represent the corresponding derivative terms in
the kinematic wave equation (5.111); it also suggests that the additional three terms in
(7.47), which contain second-order derivatives, were somehow introduced spuriously by
the finite difference approximation. This can be verified as follows. Upon substitution of
(7.38) for the constants, the coefficient of the second term of (7.47) reduces to $(\Delta x / K)$;
since Δx is the length of the reach, and K the time of travel of this flood wave, their ratio
is the Muskingum celerity c_m, already defined in (7.20). The corresponding coefficient in
the kinematic wave equation (5.111) is c_k. This means that the Muskingum wave celerity
is in fact the kinematic wave celerity, or

$$c_m = c_k \qquad (7.48)$$

The remaining three terms in (7.47) can be combined into one term by means of the
following kinematic wave identities (obtained by taking derivatives of (5.111) (for $Q_1 = 0$)
and making use of (7.48)),

$$\frac{\partial^2 Q}{\partial x \partial t} = -(\Delta x / K) \frac{\partial^2 Q}{\partial x^2}$$

$$\frac{\partial^2 Q}{\partial t^2} = (\Delta x / K)^2 \frac{\partial^2 Q}{\partial x^2} \qquad (7.49)$$

Finally, with some algebra (7.47) becomes

$$\frac{\partial Q}{\partial t} + \frac{\Delta x}{K} \frac{\partial Q}{\partial x} - \frac{(\Delta x)^2(1 - 2X)}{2K} \frac{\partial^2 Q}{\partial x^2} = 0 \qquad (7.50)$$

This is the standard advective diffusion equation with an advectivity, given by the kinematic wave celerity,

$$c_{ko} = \Delta x / K \qquad (7.51)$$

and with a diffusivity

$$D_0 = c_{k0}(1 - 2X) \Delta x / 2 \qquad (7.52)$$

in which, as before, the subscript 0 indicates linearity. Equation (7.52) illustrates how the
discretization Δx is responsible for the diffusion effect inherent in the storage equation.
In the limit, when Δx is made to approach zero to obtain a derivative, the diffusivity

disappears; this is not the case for the celerity in (7.51), because the time of travel K then also goes to zero.

Physically based estimation of the parameters: the MCD method
The MCD acronym stands for the Muskingum–Cunge–Dooge method and it refers to the names of the two investigators who contributed independently to its development.

As proposed by Cunge (1969), the numerical diffusion in the Muskingum formulation can be put to good use in the estimation of the parameters X and K; this is done simply by requiring this numerical diffusion to be equal to the physical diffusion resulting from the hydraulic characteristics of the flow. Accordingly, equating (7.51) and (7.52) with (5.94) and (5.93), respectively, one obtains

$$K = \frac{\Delta x}{d Q_0 / d A_{c0}} \tag{7.53}$$

which for wide channels is equal to $\Delta x / [(a + 1) V_0]$, and

$$X = \frac{1}{2} - \frac{b Q_0}{c_{k0} B_c S_0 \Delta x} \tag{7.54}$$

The symbol Q_0 is a typical reference flow rate in the channel, b the parameter of (5.39) which is normally taken as $1/2$ for turbulent flow (see Table 5.2), c_{k0} is the kinematic wave celerity, B_c is the channel width, S_0 is the bed slope, and Δx is the length of the channel reach. With the more accurate expression (5.98) for the diffusivity, this is

$$X = \frac{1}{2} - \frac{b Q_0}{c_{k0} B_c S_0 \Delta x} \left(1 - a^2 \mathrm{Fr}_0^2 \right) \tag{7.55}$$

In cases when the channel is sufficiently wide, the Kleitz–Seddon principle (5.108) can be used to express (7.55) in even simpler terms, namely

$$X = \frac{1}{2} - \frac{b h_0}{(a + 1) S_0 \Delta x} \left(1 - a^2 \mathrm{Fr}_0^2 \right) \tag{7.56}$$

All these expressions for X indicate that it is normally smaller than 0.5.

In a different development, Dooge (1973) determined the parameters K and X of the Muskingum formulation by equating its first two moments (7.31) and (7.34) with the first two moments obtainable from the unit response (5.72); recall that this response function is obtained by the exact solution of the linearized complete shallow water equation (5.67). The details of this derivation are beyond the present scope, but it is easy to show that the resulting expressions are the same as (7.53) and (7.56). This indicates that the application of the Muskingum formulation with (7.53) and (7.56) will ensure that it produces a wave motion whose average speed of propagation and dispersion are the same as those obtainable with the exact solution. It also means that the expressions for the Muskingum parameters in Equations (7.53)–(7.56) are even better than would be suggested by a cursory review of Cunge's (1969) derivation. These expressions conform not only with the diffusion approximation, but with the exact solution of the linearized complete shallow water equation (5.67), as well.

Practical implementation: linear or nonlinear?

In the standard application of the Muskingum method the parameters K and X are usually considered as constants, and treated as characteristics of the channel reach in question. However, the physically based expressions (7.53) and (7.54) show how in reality K and X depend on the reference rate of flow Q_0, once Δx has been decided upon. As long as the actual flow rates Q are only small deviations from the reference flow Q_0, a linearized algorithm can be expected to perform well. Flood waves normally involve large deviations, and since the Muskingum method is essentially linear, the question arises what value should be assigned to Q_0 in Equations (7.53) and (7.54) to ensure optimal results. A few studies have focused on this issue.

Actually, the form of (7.53) and (7.54) opens up the possibility of applying the Muskingum method in a nonlinear way. Cunge (1969) already suggested that Q_0 be adjusted to allow extrapolation of the Muskingum parameters beyond the range of previously observed events. Miller and Cunge (1975, p. 226) subsequently treated the parameters as functions of time $K = K(Q(t))$ and $X = X(Q(t))$, by assigning to Q_0 and (dQ_0/dA_{c0}) the values corresponding to the flow rate Q at that time and adjusting them at every time step of the computation; they applied this technique to a channel with a compound cross section. In the same vein, Koussis (1978) proposed an adjustment of K at each computational time step, by means of the uniform rating curve to estimate (dQ_0/dA_{c0}) in (7.53); but while X can also be easily adjusted as a function of Q, he found from his analysis of wave propagation on the Rhine, that the results tend to be relatively insensitive to the exact value of X and that therefore a constant value should be adequate. On the other hand, Ponce and Yevjevich (1978) concluded that the overall difference between a linear and nonlinear application of Equations (7.53) and (7.54) in (7.37) is usually quite small. In addition, they obtained better results using an average value of Q to represent Q_0, but opined that the use of the peak value of the hydrograph as Q_0 might be easier to implement in practice.

Example 7.4. Application of the MCD method

Consider again the inflow and outflow hydrographs of Example 7.2 and illustrated in Figure 7.11. Assume that these flow rates took place in a river channel with an effective width of $B_c = 170$ m, an effective slope $S_0 = 0.0004$ and an effective roughness $n = 0.035$. The length of the reach, that is the distance between the inflow and outflow section was taken as $\Delta x = 23$ km. In Example 7.2 the Muskingum parameters were found to be $K = 1.99$ h and $X = 0.30$. It is easy to check that Equations (7.53) and (7.54) produce the same values of these parameters with an assumed reference value of the discharge rate of roughly $Q_0 = 2000$ m^3 s^{-1}; by means of the Gauckler–Manning equation (5.41) this can be shown to correspond with a water depth $h_0 = 6.14$ m and a reference velocity $V_0 = 1.92$ m s^{-1}. The hydrograph calculated with these values of K and X has already been compared in Figure 7.10 with the observed outflow hydrograph. The sensitivity of the MCD method to the value of the assumed reference discharge rate Q_0 can be tested by carrying out the calculations with two different values, say $Q_0 = 1500$ and 2500 m^3 s^{-1}. In the case of $Q_0 = 1500$ m^3 s^{-1}, in the same river channel (5.41) produces a velocity $V_0 = 1.71$ m s^{-1} and a water depth $h_0 = 5.17$ m. With these values one obtains with (7.53) and (7.54) the parameter values $K = 2.24$ h and $X = 0.34$. In the case of

Fig. 7.11   Observed hydrographs
at the inflow end Q_i and
at the outflow end Q_e (in
$m^3\,s^{-1}$) listed in Table
7.2, and used in
Example 7.2 (heavy
lines). The calculated
outflow hydrographs
(thin lines) are obtained
with the MCD method
with reference flow rate
$Q_0 = 2\,500\,m^3\,s^{-1}$
(triangles) and $Q_0 =$
$1500\,m^3\,s^{-1}$ (circles).

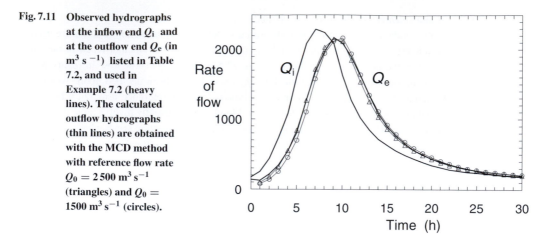

$Q_0 = 2500\,m^3\,s^{-1}$, the corresponding values are $V_0 = 2.09\,m\,s^{-1}$, $h_0 = 7.02\,m$, $K = 1.83\,h$ and $X = 0.27$. As illustrated in Figure 7.11, the calculated outflows are not strongly affected by the choice of the reference flow rate Q_0. Not surprisingly, a lower value of Q_0 results in a generally slower wave motion through the reach, and a somewhat delayed arrival of the peak discharge.

7.3.3 Adjustment of calibration parameters on physical grounds

The expressions for the Muskingum parameters derived in the previous section require a knowledge of the effective channel hydraulic parameters of the channel reach. In this case, "effective" means that the hydraulic parameters should have values, which produce the optimal results with the Muskingum procedure. Therefore, effective parameters may not be easy to ascertain from the available information on the channel characteristics or even from field surveys. Rather, for a given flood event, they are probably best obtained by calibration, by means of the procedures explained in Section 7.3.1. However, the parameters in the Muskingum formulation are not constants, so that for some other flood that must be routed through a reach for design or forecast purposes, they are not likely to be the same as those obtained by calibration with past events. Nevertheless, the physically based expressions (7.53)–(7.56) obtained in the previous section can still be used to adjust or scale the parameters obtained by calibration, to render them applicable to any other flood event with different flow rates. This can be done conveniently by taking the peak flow rate, or some other characteristic flow rate as reference.

The adjustment can be applied to channels of any cross-sectional shape, but it is especially simple for wide channels, so that $R_h = h$. Let the parameters obtained by calibration with a flood event on record be denoted by K_r and X_r, and those for the design or forecast flood, that is to be routed, by K_d and X_d; use the same subscripts also for the corresponding flow characteristics in these two events. For example, if the peak velocity V_{ipd} at the inflow end of the reach is known, one obtains immediately by means of (7.53) the time lag in the reach for the design or forecast event

$$K_d = K_r \frac{V_{ipd}}{V_{ipr}} \tag{7.57}$$

In (7.57), V_{ipr} is the flow velocity at the peak of the flood event on record at the inflow end of the reach. In a similar way, one obtains from Equation (7.54) the adjusted weighting parameter

$$X_{\text{d}} = \frac{1}{2} + (X_{\text{r}} - 0.5)\, h_{\text{ipd}}/ h_{\text{ipr}} \tag{7.58}$$

where h_{ipr} is the peak water depth at the inflow section of the reach, measured or calculated for the reference flood event used in the calibration, and h_{ipd} is the corresponding peak water depth at the entrance section for the flood event for which the parameter X_{d} is sought; values of a are listed in Table 5.2. With the more accurate expression (7.56) and (5.63) one obtains in a similar way

$$X_{\text{d}} = \frac{1}{2} + (X_{\text{r}} - 0.5) \left(\frac{g h_{\text{ipd}} - a^2 V_{\text{ipd}}^2}{g h_{\text{ipr}} - a^2 V_{\text{ipr}}^2} \right) \tag{7.59}$$

REFERENCES

Apollov, B. A., Kalinin, G. P. and Komarov, V. D. (1964). *Hydrological Forecasting (Gidrologicheskie prognozy)*, translated from Russian. Jerusalem: Israel Progr. Scient. Translation.

Bartsch-Winkler, S. and Lynch, D. K. (1988). *Catalog of Worldwide Tidal Bore Occurrences and Characteristics*. Denver, CO: US Geological Survey Circular 1022.

Carter, R. W. and Godfrey, R. G. (1960). Storage and flood routing. *Manual of Hydrology: Part 3. Flood Flow Techniques*. Geol. Survey Water-Supply Paper 1543-B, Washington, DC: US Dept. Interior, pp. 81–104.

Chow, V. T. (1951). A practical procedure of flood routing. *Civil Eng. Publ. Works Rev.*, **46** (No. 542), 586–588.

  (1959). *Open-Channel Hydraulics*. New York: McGraw-Hill.

Cunge, J. A. (1969). On the subject of the flood propagation computation method (Muskingum method). *J. Hydraul. Res.*, **7**, 205–230.

Cunge, J. A., Holly, F. M. and Verwey, A. (1980). *Practical Aspects of Computational River Hydraulics*. Boston: Pitman Adv. Publ. Program.

Degen, P. and Degen, C. (1984). *The Johnstown Flood of 1899: The Tragedy of the Conemaugh*. Philadelphia: Eastern Acorn Press.

Dooge, J. C. I. (1973). *Linear Theory of Hydrologic Systems*, Tech. Bull. 1468, Agr. Res. Serv., US Dept. Agric.

Harper's Weekly (June 15, 1889), **23**, No. 1695, p. 481.

Henderson, F. M. (1966). *Open Channel Flow*. New York: Macmillan Publ. Co.

Jones, S. B. (1981). Choice of space and time steps in the Muskingum–Cunge flood routing method. *Proc. Inst. Civ. Eng., part 2*, no. 71, 759–772.

Koussis, A. D. (1978). Theoretical estimations of flood routing parameters. *J. Hydraul. Div., Proc. ASCE*, **104** (HY1), 109–115.

Laurenson, E. M. (1959). Storage analysis and flood routing in long river reaches. *J. Geophys. Res.*, **64**, 2423–2431.

Liggett, J. A. and Cunge, J. A. (1975). Numerical methods of solution of the unsteady flow equations. In *Unsteady Flow in Open Channels*, ed. K. Mahmood and V. Yevjevich, Vol. I, chapter 4. Fort Collins, CO: Water Resource Pubs., pp. 89–182.

Lighthill, M. J. and Whitham, G. B. (1955). On kinematic waves, I. Flood movement in long rivers. *Proc. R. Soc. London*, A **229**, 281–316.

Lliboutry, L. A., Morales-Arnao, B., Pautre, A. and Schneider, B. (1977). Glaciological problems set by the control of dangerous lakes in Cordillera Blanca, Perú. *J. Glaciol.*, **18** (79), 239–290.

Mahmood, K. and Yevjevich, V. (eds.) (1975). *Unsteady Flow in Open Channels*, Vols I – III. Fort Collins, CO: Water Resource Publications.

McCarthy, G. T. (1938). The unit hydrograph and flood routing. Unpublished paper, Conference of the North Atlantic Division, US Corps of Engineers, New London, CN.

McCullough, D. G. (1968). *The Johnstown Flood*. New York: Simon and Schuster.

Meyer, O. H. (1941). Simplified flood routing. *Civil Eng.*, **11**, 306–307.

Miller, W. A. and Cunge, J. A. (1975). Simplified equations of unsteady flow. In *Unsteady Flow in Open Channels*, ed. K. Mahmood and V. Yevjevich, Vol. I, chapter 5. Fort Collins, CO: Water Resource. Pubs., pp. 183–257.

Montes, S. (1998). *Hydraulics of Open Channel Flow*. Reston, VA: ASCE Press.

Morales-Arnao, B., Glaciers of Peru. In *Glaciers of South America*; ed. R. S. Williams, Jr., and J. G. Ferrigno. Geological Survey Prof. Paper 1386-I. Denver, CO: US Dept. Interior.

Ponce, V. M. and Theurer, F. D. (1982). Accuracy criteria in diffusion routing. *J. Hydraul. Div., Proc. ASCE*, **108** (HY6), 747–757.

Ponce, V. M. and Yevjevich, V. (1978). Muskingum–Cunge method with variable parameters. *J. Hydraul. Div., Proc. ASCE.*, **104** (HY12), 1663–1667.

Simpson, J. E. (1997). *Gravity Currents in the Environment and in the Laboratory*, second edition. Cambridge: Cambridge University Press.

Taylor, G. H. and Hatton, R. R. (1999). *The Oregon Weather Book*. Corvallis, OR: Oregon State University Press.

Weinmann, P. E. and Laurenson, E. M. (1979). Approximate flood routing methods: a review. *J. Hydraul. Div., Proc. ASCE*, **105** (HY12), 1521–1536.

Yevjevich, V. (1975). Sudden water release. In *Unsteady Flow in Open Channels*, ed. K. Mahmood and V. Yevjevich, Vol. II, chapter 15. Fort Collins, CO: Water Resource Pubs., pp. 587–668.

PROBLEMS

7.1   In Example 7.1, it is assumed that the effective depth, h_2, in the river just ahead of the surge, was 1.0 m, and that the GM roughness in the river was $n = 0.07$. (a) What depth, h_2, would be required (i.e. prior to the sudden failure of the dam) if the GM roughness were in reality $n = 0.08$ (instead of 0.07) to obtain the same time of travel of the surge, namely 57 min, as observed in the case of the Johnstown flood? Keep the same values of the slope and of the height of the surge, as adopted in Example 7.1. (b) What would be the required depth, h_2, if the roughness were actually $n = 0.09$?

7.2   A steady flow, with depth $h = 2$ m, is maintained by means of a sluice gate in a concrete-lined canal with a uniform rectangular cross section; its width is $B_c = 5$ m, the bed slope is $S_0 = 0.0008$ and the GM roughness is $n = 0.015$. Calculate the downstream celerity of the surge caused by a sudden rise of the water level to $h = 4$ m by opening the gate.

7.3   Take the outflow, Q_e, listed in Table 7.2, as the inflow into the next reach downstream from the one, considered in Example 7.2. Using the values, $X = 0.3$ and $K = 2$ h, calculate the outflow from this next reach by means of the Muskingum method.

7.4   The following hydrographs, Q_i and Q_e, respectively, were measured at the entrance and exit sections of a reach on the Conecuh River in Alabama between Andalusia and Brooklyn in March–April of 1944 (Carter and Godfrey, 1960).

Date (at noon)	Q_i (m^3 s^{-1})	Q_e (m^3 s^{-1})	Date (at noon)	Q_i (m^3 s^{-1})	Q_e (m^3 s^{-1})
1944 March 16	120.64	118.38	29	982.70	1013.86
17	216.65	158.59	30	1280.06	1022.35
18	314.35	205.89	31	1390.51	994.03
19	472.94	272.44	1944 April 1	1169.62	1090.32
20	611.71	481.44	2	957.22	1263.07
21	594.72	518.26	3	580.56	1135.63
22	753.31	549.41	4	416.30	849.60
23	1302.72	719.33	5	322.85	583.39
24	1699.20	965.71	6	263.09	387.98
25	1634.06	1263.07	7	221.75	291.70
26	1356.53	1449.98	8	176.43	241.57
27	977.04	1469.81	9	172.19	218.35
28	614.54	1180.94			

(a) Determine the routing coefficients, c_0, c_1, and c_2, of Equation (7.37) by multiple regression. (b) Estimate the values of the Muskingum parameters, K and X, from these coefficients. (c) Using the coefficients obtained in (a), route the inflow through this reach, and compare the routed outflow with the measured outflow. Present this comparison graphically.

7.5   Consider the 1944 event listed in Problem 7.4. (a) Determine the values of the Muskingum parameters, K and X, by the technique illustrated in Example 7.2. (b) Route the measured inflow through this reach, and compare the routed outflow with the measured outflow. Present the results graphically.

7.6   Consider the hydrographs listed for Problem 7.4. For a design peak inflow rate, $Q_i = 3000$ m^3 s^{-1}, predict the design peak outflow rate by the following two methods. (a) As a first approximation, assume strict linearity and similarity between the design flood wave and the 1944 event. (b) Assume that the river has a constant and large width, that its bed slope is $S_0 = 0.002$, that its GM roughness is $n = 0.04$, that the length of the reach is 35 km, and that the Muskingum parameters for the 1944 event are $K_r = 2$ days and $X_r = 0.2$; find the design values of the parameters by using Equations (7.57) and (7.58). (c) With these design values, calculate the design peak outflow, Q_e resulting from an inflow hydrograph Q_i whose values are 1.766 times those listed in Exercise 7.4. Compare this result with that obtained in (a).

7.7   Making use of the data given in Problems 7.4 and 7.6, give an estimate of the cross-sectional area of the flow, A_c, at the time of the outflow peak at Brooklyn in 1944.

7.8   The Muskingum flood routing method is a finite difference implementation of the diffusion equation. Determine the numerical value of K, in the Muskingum storage equation, for the flow situation described in Problem 5.14, if the length of the channel section is $L = 2.5$ km.

7.9   A hydrograph at the downstream end of a river section (after a period of heavy rain over upland catchments) was recorded as follows.

Time (h)	0	2	4	6	8	10	12	14	16	18	20
Rate of flow $(m^3\ s^{-1})$	1.5	1.4	16.9	54.1	72.8	62.4	46.3	31.7	20.7	13.9	9.6

Assume that there were no intermediate inflows (nor losses) in this river channel. From previous events, the travel time in this reach is known to be 4 h, and the storage-weighting factor is $X = 0.25$. (a) Determine the flood hydrograph at the upstream end of the reach for this event by means of the Muskingum method. Hint: start computations at $t = 20$ h and go backwards. (b) If the width of the river is 15 m and the water depth (at the downstream end) at the time of peak is 3.5 m, what is the average velocity of the water (at the downstream end) at the time of peak? (Assume a rectangular cross section.) (c) From the available information, give an estimate of the length of the river reach.

7.10   Using the following hydrograph of a historical flood event at the entrance and exit sections of a river reach:

Day (at noon)	Q_i inflow $(m^3\ s^{-1})$	Q_e outflow $(m^3\ s^{-1})$	Day (at noon)	Q_i inflow $(m^3\ s^{-1})$	Q_e outflow $(m^3\ s^{-1})$
1	668	611	9	2109	2895
2	1685	785	10	1668	2256
3	4647	1951	11	1325	1776
4	7906	4641	12	1098	1402
5	7864	7102	13	962	1125
6	5547	7392	14	869	966
7	3792	5657	15	781	854
8	2721	4098	16	708	811

(a) Determine the routing coefficients, c_0, c_1 and c_2, of Equation (7.37) by multiple regression. (b) Estimate the Muskingum parameters, K and X, from these coefficients. (c) Use these coefficients to route the inflow through the reach, and compare the result with the measured outflow.

7.11   Assume that the following flood hydrograph was observed at the upstream end of a 40 km long river reach.

Time (h)	0	2	4	6	8	10	12	14	16	18	20
Discharge rate $(m^3\ s^{-1})$	0	16.7	53.7	72.2	61.9	45.9	31.4	20.5	13.7	9.49	0

(a) If you know that the channel is uniform with a rectangular cross section and that the wetted cross-sectional area at the peak discharge (i.e. after 6 h) is 42.0 m², show that the celerity of the flood wave is approximately 10 km h^{-1} (i.e. 2.78 m s^{-1}). (b) Assume that the celerity remains constant through the reach and that the weighting factor in the storage function is $X = 0.25$. First, estimate the value of the Muskingum parameter K from (a); then calculate the peak flow at the downstream end of the reach.

7.12   Derive the diffusion equation (7.50) from (7.46), via (7.47).

7.13   Derive the expression for the design weighting factor, X_d, given by (7.59) from (7.56).

7.14   Multiple choice. Indicate which of the following statements are correct.
        (a)   In general, the diffusion approach to describe flow in open channels can be considered intermediate (in regard to the underlying assumptions) between the case of the moving hydraulic jump (or bore) and that of the monoclinal rising wave.
               Some of the underlying assumptions are:
        (b)   that gravity effects are balanced completely by shear stress effects;
        (c)   that the lateral inflow rate must be negligible;
        (d)   that the channel is infinitely wide;
        (e)   that the slope of the water surface relative to the channel bottom is negligible.

7.15   Multiple choice. Indicate which of the following statements are correct. The Muskingum flood routing method makes use of storage function given by (7.15). This method:
        (a)   works often well to predict the propagation of flood waves, involving slow and gradual momentum changes;
        (b)   has the advantage of not requiring any previous flood data, since the necessary parameters can also be derived from information on channel roughness and channel geometry;
        (c)   in its general form requires that the peak of the outflow, Q_e, from a channel reach occur at the time when the inflow wave has the same rate of flow;
        (d)   is rather restricted in its applicability, because it is based on the assumption that storage is a function of the inflow and outflow with hysteresis;
        (e)   can be applied to an ungated linear reservoir, provided the storage is made a function of outflow only.

7.16   Multiple choice. Indicate which of the following statements are correct. During the passage of a flood wave through a reservoir (whose outflow rate is controlled by a fixed weir), the peak of the outflow hydrograph occurs (if both inflow, Q_i, and outflow hydrograph, Q_e, are plotted on the same graph):
        (a)   at the time of the inflection point of the inflow hydrograph;
        (b)   at the time $t = K$, where K is a constant in $S = K [XQ_i + (1 - x)Q_e]$;
        (c)   at the time where the slope of the outflow hydrograph goes from positive to negative;
        (d)   at the time of the intersection with the inflow hydrograph;
        (e)   at a time K units later than the peak of the inflow hydrograph.

7.17   Multiple choice. Indicate which of the following statements are correct. The motion of a flood wave in a medium-size river:
        (a)   is usually more like that of a bore (or moving hydraulic jump) than like that of a monoclinal rising wave;
        (b)   is always such, that the rate of flow is a unique, single-valued function of the water depth.

(c)   often takes place with a celerity that is approximately proportional to the square root of the bed slope;

(d)   usually has a sufficiently high Reynolds number so that the effect of bed roughness becomes negligible;

(e)   can be predicted fairly well by solving the continuity equation together with the complete shallow water momentum equation;

(f)   is usually not greatly affected by the impact of the raindrops on the flowing water in the river;

(g)   can be described by the kinematic wave approach only when the channel bed is relatively smooth, so that the GM roughness, n, is small;

(h)   can be described by the kinematic wave approach: the approximation is better for a channel with a small slope than for one with a larger slope (assume everything else is the same);

(i)   can be described by the shallow water equations even when the river is not very wide.

7.18   Multiple choice. Indicate which of the following statements are correct. The classical, Muskingum flood routing method:

(a)   is not suitable to describe diffusion (i.e. "broadening" and "flattening") of a flood wave as it travels downstream, because it is based on the kinematic wave assumption;

(b)   is not suitable to predict the propagation of bores or moving hydraulic jumps, because it neglects change-in-momentum effects (in other words, it is based on the assumption $(DV/Dt) = 0$;

(c)   can be used to describe a stationary hydraulic jump by adjusting the velocity of the moving coordinate system;

(d)   is based on the assumption that the storage of water in the channel is a linear function both of inflow rate and of outflow rate;

(e)   in its general form requires that the peak of the outflow flood occur at the time when the inflow flood has the same rate of flow.

III WATER BELOW THE SURFACE

8 WATER BENEATH THE GROUND: FLUID MECHANICS OF FLOW IN POROUS MATERIALS

8.1 POROUS MATERIALS

The great majority of all near-surface geologic formations, in which water is stored and transported, are unconsolidated porous rocks made up of particles of different sizes. This type of formation is usually referred to as a *soil* close to the surface, and as an *aquifer* at greater depths. However, the terms soil and aquifer material are often used interchangeably. Many of these formations consist of alluvial and colluvial deposits, which as riparian aquifers are major contributors to streamflow. Although in some regions underlain by limestone or karst formations, large quantities of water can be transported through solution channels and caves, globally they are of much less importance. Formations consisting of volcanic rock, shale and clay layers, which are porous but which transmit water relatively slowly, are often considered impermeable for hydrologic purposes; as such they are referred to as *aquicludes*.

The voids or open spaces between the particles of soils and other granular materials are referred to as pores. An important property of such water-bearing formations is their porosity. This can be defined as follows

$$n_0 = \lim_{\Delta\forall\to0} \left(\frac{\text{volume of voids in } \Delta\forall}{\Delta\forall} \right) \tag{8.1}$$

where $\Delta\forall$ is a small volume of porous material. This definition is subject to the continuum paradox: on the one hand, the limit is necessary to allow the description of phenomena at a point by means of infinitesimal calculus; on the other hand, the volume $\Delta\forall$ must be kept large enough so that n_0 represents a meaningful ensemble average over pores of many different sizes. The porosity of a soil depends primarily on its particle size distribution and on its structure. Some of these features are illustrated in Figures 8.1 and 8.2. A soil with a wide distribution of particle sizes tends to have a smaller porosity than a soil consisting of particles or grains of a more uniform size. The structure of a granular porous material refers to the arrangement of the particles among one another and to their aggregation into larger structures. Thus the porosity can be increased by agricultural operations, such as ploughing or raking, or by frost; these processes "open up" the soil simply by rearranging the relative positions of the particles. Similarly, the porosity of the soil can be decreased by compaction. In principle, in the case of soils consisting of inert material, their texture, that is the size of the particles, should not affect the porosity, as long as their structure, particle size distributions and chemical composition are similar. However, actual soils are not inert, but the surfaces of their particles carry electrical

Fig. 8.1   Illustration of the effect of texture and of particle size distribution on the porosity. In boxes (a), (c) and (d) the spherical particles of uniform size arranged in a cubic packing (i.e. a similar structure) result in aggregates with exactly the same porosity, regardless of the sizes of the particles. The particles in box (b), which are of different sizes, result in an aggregate with smaller porosity.

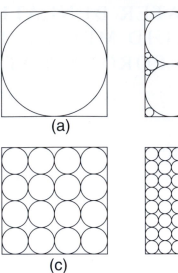
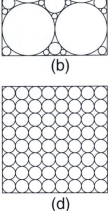

(a)   (b)

(c)   (d)

Fig. 8.2   Illustration of the effect of structure of spherical particles of uniform size on porosity. Among the regular arrangements, the cubic packing in (a) is the most open and the rhombohedral packing in (b) is the tightest; (c) shows how secondary structures (crumbs) can be formed, resulting in larger porosities.

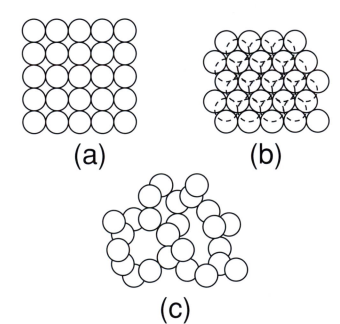

charges; these charges affect the structure of the soil, and are increasingly effective with decreasing particle size. Moreover, their chemical composition can vary widely. Some of the soil constituents, such as colloidal clay, organic matter, lime, and colloidal oxides of iron and aluminum, can act as cementing agents, which further the aggregation of particles into larger structures. As a result, clayey soils tend to have higher porosities than sandy soils.

The volumetric water content of a soil can be defined in a similar way as Equation (8.1), namely

$$\theta = \lim_{\Delta\forall \to 0} \left(\frac{\text{volume of water in } \Delta\forall}{\Delta\forall} \right) \tag{8.2}$$

Clearly, when a soil is fully saturated, its water content, denoted by θ_0, is by definition equal to the porosity, or $\theta_0 = n_0$.

8.2 HYDROSTATICS OF PORE-FILLING WATER IN THE PRESENCE OF AIR

Water in the pores of near-surface soils and other geologic formations is usually in intimate contact with atmospheric air. Although for most hydrologic purposes water and air can be treated as immiscible fluids, it is still necessary to consider their interaction in the description and formulation of the different water transport mechanisms. In this section hydrostatic conditions are explored, i.e. when both fluids, water and air, are at rest.

8.2.1 Pressure in relation to water content

As the water content of a soil is reduced, the pressure of the remaining soil water generally becomes smaller than that in the atmospheric air, which displaces the water in the pores. This process can be illustrated by the following thought experiment.

A thought experiment
Consider, as shown in Figure 8.3, a sample of soil SS to be tested; it is placed on a porous plate PP in a container C, to which a flexible tube FT is connected. (In a real experiment a Büchner filter with fritted disk with sufficiently small pores can be used for this purpose.) The pores of the porous plate PP are much smaller than those of the sample SS (see Figure 8.4). Set up the experiment in such a way that initially the entire system, i.e. soil sample, container and tube, are filled with water, and that the vertical distance d between the center of the sample and the outlet of the tube is zero. Assume for this simple experiment that the soil sample is incompressible so that it maintains its original volume throughout, and that the water has a constant density ρ_w. Now increase d stepwise by small increments and wait after each step until equilibrium is established, that is, until water stops flowing out at the lower end of the tube. Record after each step the value of d and the total volume $\forall_d$ of water drained during all previous steps. If $\Delta\forall$ is the volume of the sample SS, this total volume $\forall_d$ of drained water can be converted to the volumetric water content of the soil sample simply by $\theta = (n_0 - \forall_d/\Delta\forall)$. Note that, relative to atmospheric pressure, the pressure of the water in the soil pores is given by $p_w = -\gamma_w d$, where $\gamma_w = \rho_w g$ is the specific weight of the water. In practice, when the density of the water is constant, it is often convenient to express the pressure as equivalent height of water column ψ_w; in these units the pore water pressure in this experiment can simply be written as $\psi_w = p_w/\gamma_w$. Thus the pressure is negative relative

Fig. 8.3 Sketch of a thought experiment on a soil sample SS, supported on a porous plate PP in a receptacle C. The average pressure in the water in the sample is equal to $-(\rho_w g d)$. The blow-up section B-U is shown in Figure 8.4.

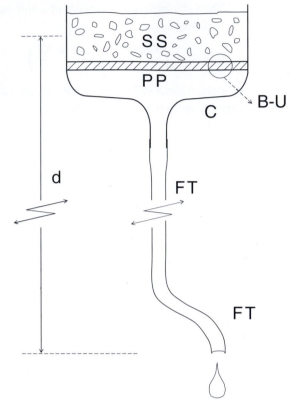

Fig. 8.4 Blow-up view of the porous plate (or membrane) supporting the porous material, as shown in Figure 8.3. The volume occupied by water is shown as dark area.

to the atmospheric pressure; this negative pressure is often also called soil water suction or soil water tension. In the remainder of this chapter the soil water suction, whenever expressed as a positive height of water column, will also be denoted by $H(= -\psi_w)$.

Typically, the relationship between the soil water pressure and the water content is of the form shown in Figure 8.5. This relationship is also variously referred to as the *soil water characteristic*, the *soil water suction relationship* and the *soil water retention relationship*. Some other examples of soil water characteristic curves are shown in Figure 8.6.

Equilibrium moisture content profile in a homogeneous soil
If the variable $z = d(= -p_w/\gamma_w = H)$ refers to the height above the water table, the soil water characteristic also provides a description of the water content distribution in a homogeneous soil profile, which occurs under hydrostatic conditions, that is when there is no flow. This is illustrated in Figure 8.7. As defined in Chapter 1, the water table is

Fig. 8.5 Soil water characteristic curve for a fine sand (Oso Flaco) measured during a stepwise drainage process, showing the degree of saturation, $S = \theta / n_0$, against negative pressure in the water $H = -p_w / \gamma_w$, expressed in cm of equivalent water column; in this particular experiment the porosity was $n_0 = 0.405$.

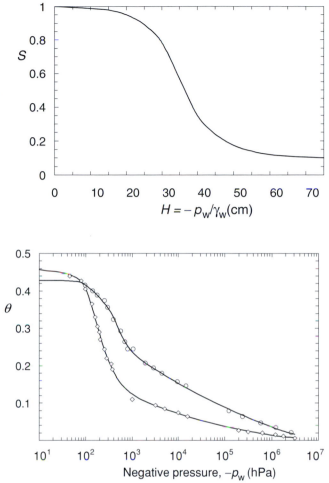

Fig. 8.6 Soil water characteristic curves for two loam soils (Adelanto, circles, and Pachappa, diamonds) during drainage (1 hPa is roughly equivalent with 1.02 cm of water column). (After Jackson *et al.*, 1965.)

the locus of points in the soil profile, where the water pressure is atmospheric, and thus $p_w = 0$, whenever atmospheric pressure can be taken as the reference. Below the water table the water pressure p_w is positive, and above the water table it is negative. The pressure of the water in the soil profile above a water table can be measured by means of a tensiometer; this instrument appears to have been introduced by Richards (1928). An example of a tensiometer is illustrated in Figure 8.8; in this case it is simply a tube filled with water, closed at one end by a sensing element consisting of a porous material, and connected at the other end to a manometer. The material of the sensing element must have pores that are sufficiently fine to ensure continuous contact (without air leakage) between the water in the soil and that in the tube; recall that this is the same requirement as that of the porous plate PP in Figure 8.3. The porous tip of the tensiometer is installed at the point where the water pressure is to be measured and that pressure can then be recorded with the manometer at the surface. Although many different types of tensiometer have been developed, with various types of pressure gauges, they all function in essentially the same way. A smaller soil water content is indicated by a larger suction, that is a larger negative pressure measurement by the tensiometer.

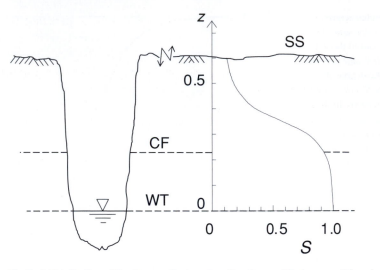

Fig. 8.7   Vertical distribution of the degree of saturation S under equilibrium conditions in a uniform soil profile. SS indicates the soil surface, WT is the position of the water table, and CF the approximate height of the capillary fringe. The height z above the water table is in meters; in this example the soil is a fine sand and the curve is the same as shown in Figure 8.5. "Equilibrium" indicates that there is no flow and that the soil water pressure distribution is hydrostatic.

Fig. 8.8   Sketch of a tensiometer installed in the field; a manometer fluid has a column height d_1 above that of the reservoir surface d_2; the porous cup at the end of the tube, which is placed at depth d_3, is filled with water in contact with the soil water; at A, the main tube can be opened to fill it with water or to bleed it of air bubbles.

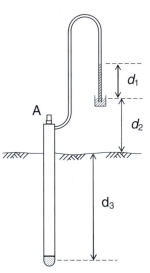

8.2.2   Mechanisms of water retention

Water or any other fluid can be held in natural soils by several types of mechanism, involving different forces at the molecular level. Some of the more important mechanisms can be ranked, according to increasing energy required to remove the water, as follows. (i) As water is removed from a non-shrinking saturated soil, it is replaced by air, and water–air interfaces develop in the pore space; the energy required to form these interfaces, and thus to withdraw the water, is directly related to surface tension.

(ii) Natural soils, especially those with finer texture, consist of particles that are not inert, but active as a result of electrical charges at their surfaces. In the presence of water such surface-active particles interact, repelling each other and attracting in the intervening spaces ions, and thus also more water to relieve the osmotic pressure; this osmotic pressure difference, between the water in the immediate vicinity of the particles and that in the larger pores of the soil, in turn gives rise to hydrostatic pressure differences. This may also be accompanied by swelling during wetting, and shrinking during subsequent drying. (iii) Water molecules also behave as dipoles, and as such they may undergo attraction by the electrical surface charges of the soil particles. (iv) When the soil contains clay particles (especially montmorillonite), water can be held between the "sandwich" layers of these clayey minerals in a quasi-crystalline fashion; actually, the water is held so strongly by this mechanism that it could be debated whether the water should be considered "free" or "chemically bound."

Surface tension

Among these mechanisms, those involving surface tension are probably the most important in the context of hydrology. Surface tension or capillarity acts at the interface between two fluids. In simple terms, one can visualize the molecules in a liquid as being subject to attraction by their neighbors. Far from boundaries the field is symmetrical and these molecular effects are balanced. However, near the interface with another fluid whose molecules are less attracting, the balance is broken and the molecules are pulled more toward their own bulk. In order to increase the interfacial area of the liquid in question, work is necessary. The surface tension is a measure of this work and, thus, of the energy required to maintain that interface. In the case of the interface between water and air, in the range of $0 \leq T \leq 30\,°C$, the surface tension can be estimated approximately by $\sigma = (75.6 - 0.14\,T) \times 10^{-3}\,\mathrm{J\,m^{-2}}$ (or $\mathrm{N\,m^{-1}}$), with the temperature T in $°C$ (degrees celsius). The word capillarity is derived from the Latin *capilla*, or hair, because surface tension is manifested by the phenomenon of the rise of water in hair-thin glass tubes. In the soil profile the zone above the water table, where the water content is near saturation even though the pressure is negative, is often referred to as the *capillary fringe* (see Figure 8.7). Similarly, whenever the soil water pressure p_w is negative, it is also called *capillary pressure*.

Equation of Laplace

As a result of surface tension, any pressure difference across the interface of two immiscible fluids that are in contact with each other is accompanied by a curvature of their interface. It can readily be shown that this phenomenon is described by

$$\Delta p = \sigma \left(\frac{1}{r_2} \pm \frac{1}{r_1} \right) \tag{8.3}$$

in which $\Delta p (= p_a - p_w)$ is the pressure difference across the interface, and r_1 and r_2 are the two principal radii of curvature of the interface; the plus sign between the two terms inside the brackets applies to *synclastic* interfaces, and the minus sign to anticlastic surfaces. A surface is said to be synclastic when the centers of the two radii of curvature

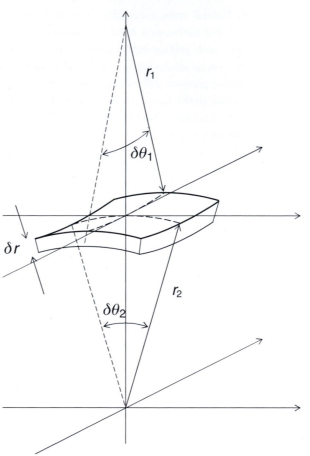

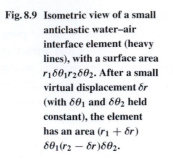

Fig. 8.9 **Isometric view of a small anticlastic water–air interface element (heavy lines), with a surface area $r_1\delta\theta_1 r_2\delta\theta_2$. After a small virtual displacement δr (with $\delta\theta_1$ and $\delta\theta_2$ held constant), the element has an area $(r_1 + \delta r)\delta\theta_1(r_2 - \delta r)\delta\theta_2$.**

are on the same side of the interface; it is called *anticlastic* when they are on opposite sides. Equation 8.3 is commonly attributed to Laplace and can be derived as follows.

Derivation of the Laplace equation

Because the surface tension σ represents the energy required to maintain the interface, it is convenient to make use of the principle of virtual work. Thus consider a small anticlastic surface element $(\delta x \delta y) = (r_1\delta\theta_1 r_2\delta\theta_2)$ as shown in Figure 8.9. Let this element now undergo a virtual displacement δr, such that the angles $\delta\theta_1$ and $\delta\theta_2$ are constrained to remain the same infinitesimal fraction of π, but the radii of curvature become $(r_1 + \delta r)$ and $(r_2 - \delta r)$. As a result of this displacement the area of the element becomes $[(r_1 + \delta r)\delta\theta_1(r_2 - \delta r)\delta\theta_2]$. With these preliminaries, the work required to perform this displacement can be equated with the energy required to maintain the change in surface area of the element, as follows

$$-\Delta p\,(r_1\delta\theta_1 r_2\delta\theta_2)\,\delta r = \sigma\delta\theta_1\delta\theta_2[(r_1 + \delta r)(r_2 - \delta r) - r_1 r_2] \qquad (8.4)$$

After canceling the superfluous terms, and after neglecting the remaining term in δr, Equation (8.3) results immediately. A similar derivation can be constructed for a synclastic interface.

Fig. 8.10 **Blow-up sketch of a water–air interface in a cylindrical capillary tube of inside radius R; the radius of curvature of this meniscus is $r = R/\cos\alpha$, in which α is the wetting angle. The pressure in the water at the meniscus in the center of the tube is $p_\mathrm{w} = -\gamma_\mathrm{w} H_\mathrm{c}$.**

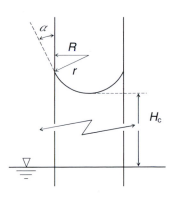

Example 8.1. Special case of a capillary tube

In the case of the capillary tube, in which water is in contact with air at atmospheric pressure, as illustrated in Figure 8.10, the pressure increase $\Delta p = (p_\mathrm{a} - p_\mathrm{w})$ in the transition from water to air equals the suction (i.e. negative pressure) in the water $(-p_\mathrm{w})$. If the radius of the tube R is small enough so that the interface can be assumed to have the curvature of a sphere, the radii of curvature are both equal to $R/\cos\alpha$, where α is the wetting angle of water with the glass. The wetting angle between water and quartz-like materials and other soil minerals is usually (except in the presence of impurities) small; the water pressure according to Equation (8.3) becomes then simply

$$p_\mathrm{w} = -2\sigma/R \qquad\qquad (8.5)$$

and the capillary rise is $H_\mathrm{c} = -p_\mathrm{w}/\gamma_\mathrm{w}(= H)$ or roughly $H_\mathrm{c} = (0.149/R)$ at $18\,^\circ\mathrm{C}$ when both H_c and R are in cm and the specific weight is constant. Equation (8.5) is valid for the ideal case of a tube of circular cross section with a radius R. Figure 8.4 illustrates how the water can be held in an analogous way in the pores of an irregular array of particles; for such pores of irregular cross section, Equation (8.5) can be used to define an effective radius: this is the radius R of a capillary tube of circular cross section, with the same value of $\Delta p = -p_\mathrm{w}$ across its water–air interface.

Pore size distribution

In numerous studies the effective radius of curvature R of the air–water interface in a pore, as used in Equation (8.5), has also been taken as a measure of the size of that pore; by analogy with pipes with a circular cross section, R is usually taken to be equal to twice the hydraulic radius R_h, as defined in Equation (5.40), that is the ratio of the cross-sectional area of the pore to its wetted perimeter. Equation (8.5) indicates that the pressure drop across the air–water interface in any pore is inversely proportional to the size of the pore. This means that with increasing negative pressures or suctions $-p_\mathrm{w}(= \gamma_\mathrm{w} H)$ increasingly smaller pores are being emptied. The soil water characteristic relates the suction H with the water content θ, that is the water still left in the soil. Hence, if it is assumed that at a given suction H all pores above size R are empty, the soil water characteristic with (8.5) is equivalent with a cumulative pore size distribution. In other

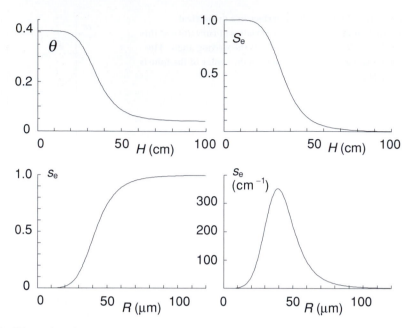

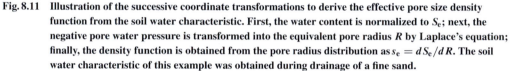

Fig. 8.11 Illustration of the successive coordinate transformations to derive the effective pore size density function from the soil water characteristic. First, the water content is normalized to S_e; next, the negative pore water pressure is transformed into the equivalent pore radius R by Laplace's equation; finally, the density function is obtained from the pore radius distribution as $s_e = dS_e/dR$. The soil water characteristic of this example was obtained during drainage of a fine sand.

words, the degree of saturation $S = (\theta/\theta_0)$ may be considered an index of the fraction of the total pore volume, that is occupied by pores smaller than R.

To avoid the limitations of the capillary model in the range of lower water contents, where other water retention forces are predominant, in this context it is convenient to use a linear transformation, defining an effective saturation,

$$S_e = \frac{\theta - \theta_r}{\theta_0 - \theta_r} = \frac{S - S_r}{1 - S_r} \tag{8.6}$$

where θ_0 is the water content at atmospheric pressure, i.e. at $H = 0$; the subscript r refers to the residual water content or residual degree of saturation, which is mainly a normalizing parameter but which may also be visualized as the moisture present in dead end pores or otherwise so strongly held that it is unavailable for flow. The effective pore size density $s_e = dS_e(R)/dR$ can be obtained by determining the slope of $S_e(R)$ as a function of R. Figure 8.11 illustrates how such effective pore size distribution and density functions can be obtained from the soil water characteristic by means of Equations (8.5) and (8.6).

This approach to obtain a pore-size distribution probably dates back to the work of Donat (1937), who used his results to characterize the structure and stability of soils. Similar early studies were also undertaken by Schofield (1938), Bradfield and Jamison (1938), Leamer and Lutz (1940), Childs (1942), Russell (1941) and Feng and Browning (1946). As discussed further in Section 8.3.4, this type of pore size distribution has also

Fig. 8.12 Illustration of hysteresis in capillary tubes with expansions. The water levels fall to positions 1 if the tubes are initially already filled with water when inserted into the water bath; the water levels rise to positions 2 if the tubes are initially empty when inserted into the water.

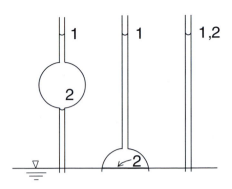

Fig. 8.13 Illustration of hysteresis in an aggregate of soil particles. The menisci indicated by 1 are obtained during drying, and those indicated by 2 during wetting. All menisci in the sketch have the same curvature, and therefore the fluid pressure at each of these interfaces is the same, even though the water content, when the fluid is at position 1, differs from that at position 2.

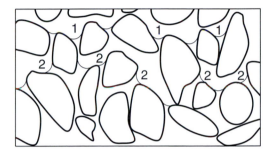

been related to permeability by, for example, Childs and Collis-George (1950), Marshall (1958), Mualem (1978) and many others (see Brutsaert, 1967, 1968a,b). The basic idea of a pore size distribution of a soil obtained in this way clearly has its shortcomings. As shown in Figure 8.4, the water–air interfaces, or menisci, tend to occur at the narrower openings or "necks" of the pores. It is thus incorrect to assume that at a given suction all pores with effective sizes larger than that given by Equation (8.5) are filled with air. Nevertheless, in the past the concept has been useful for comparative purposes and in order-of-magnitude estimates.

8.2.3 Hysteresis

The relationship between water content and capillary pressure exhibits hysteresis. This means that this relationship depends on the sequence of events of wetting and drying, by which the current water content of the soil is attained. It also means that single-valued soil water characteristic curves, such as those shown in Figures 8.5 and 8.6, are applicable only either for sustained drying events, or (but with a different shape) for sustained wetting or infiltration events, but not for situations involving alternate wetting and drying. The word hysteresis is derived from the ancient Greek word ὕστερος, meaning slow, lagging behind or delayed. Figure 8.12 shows that the water–air interface can be found at different levels in capillary tubes with expansions, depending on the manner, in which these tubes have been filled. Similarly, Figure 8.13 illustrates schematically how for the same water pressure, with menisci occurring at pore necks of the same size, it is possible to have a different water content, depending on whether the pressure is achieved by filling the pores upward or by emptying them downward. In Figure 8.14 some examples are

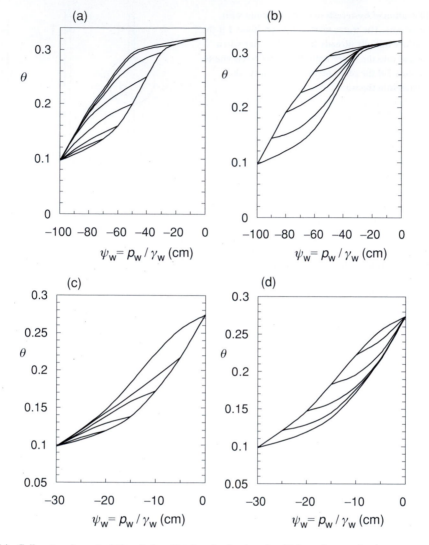

Fig. 8.14 Soil water characteristic relationship showing hysteresis with boundary and primary scanning curves for (a) Adelaide dune sand, draining; (b) Adelaide dune sand, wetting; (c) Molonglo sand, draining; and (d) Molonglo sand, wetting. The lines are best-fit through the data; ψ_w denotes pressure expressed as equivalent water column. (After Talsma, 1970.)

presented of hysteresis for different sandy soils. The bounding curves of the hysteresis regions shown in these figures are called *wetting and drying boundary curves*; any point inside the hysteresis region can be reached by scanning curves; the scanning curves starting from the drying and wetting boundary curves can be called *primary wetting and drying scanning curves*, respectively. It is obvious by now that there is an infinity of possible scanning curves in the hysteresis region. To describe these quantitatively, some type of interpolation scheme must be devised.

Fig. 8.15 Typical drying and wetting boundary curves of the soil water characteristic. The value of the water content θ_{01} is saturation, and θ_{02} is satiation; the difference between the two results from entrapped air in dead-end pores. The figure also illustrates how a fractional volume of water $\delta\theta$ enters the soil during wetting, as the soil water suction decreases from $H_w + \delta H_w$ to H_w.

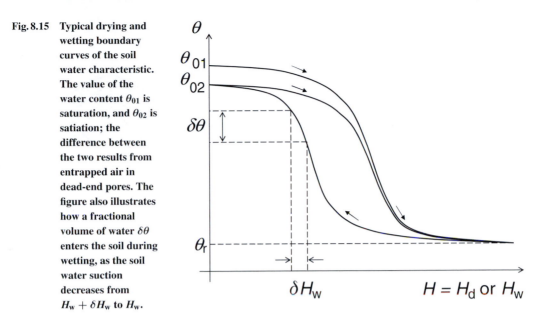

A general discussion of hysteresis in physical terms was already presented by Haines (1930). However, one of the first attempts to treat capillary hysteresis in soils quantitatively was published by Poulovassilis (1962), who made use of the concept of the independent domain, proposed by Néel (1942, 1943) and Everett (1954; 1955).

The effect of entrapped air

One obvious manifestation of hysteresis is often observed in experiments involving successive and repeated drying and wetting cycles. As an illustration, consider the following sequence illustrated in Figure 8.15. Assume that initially the soil is fully saturated and that its water content is $\theta = \theta_{01}$; thus all its pores are filled with water so that $\theta_{01} = n_0$ which is the true porosity as defined in Equation (8.1). Consider that, next, the soil is drained down to $\theta = \theta_r$ by imposing a negative water pressure, and that it is then subsequently rewetted by bringing the pressure back to zero. At this point the water content will invariably not be θ_{01}, but somewhat smaller, say θ_{02}. This difference can be attributed largely to the presence of entrapped air in dead-end pores, from which it cannot escape during the rewetting cycle. Usually, all subsequent drying and wetting cycles will continue to take place between θ_{02} and θ_r, and normally it will be impossible to recover the original water content at θ_{01}. In carefully controlled experimental set-ups in the laboratory the full saturation can only be achieved by taking special precautions, such as the use of deaerated water or the passage of CO_2 through the soil prior to the application of the water, or prolonged soaking or immersion. The water content θ_{02} is often referred to as *satiation*, to distinguish it from the full *saturation*, θ_{01}. The two terms are often used interchangeably. However, in field situations involving normal wetting and drying processes, the water content at $H = 0$ is usually satiation.

Fig. 8.16 Example of the independent domain function $F = F(H_d, H_w)$.

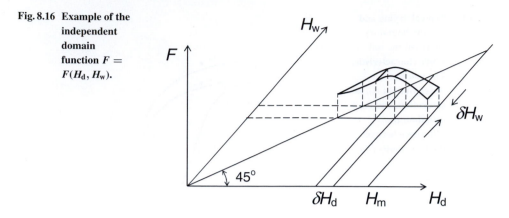

The independent domain approach

In brief, this concept involves the assumption that each element (or pore) of the total pore space is specified completely by the (negative) pressure range over which it is emptied, and that over which it is filled. Implicit in this is that any such element is either full or empty, with a transition sometimes referred to as a Haines jump. With this assumption, one can define a function $F = F(H_d, H_w)$, which represents the fraction of the pore space, which drains at a negative pressure or suction H_d and wets or is filled at a suction H_w. This function can be represented graphically by the isometric projection shown in Figure 8.16. All pores require either the same or a larger suction to be emptied than that required to fill them, so that $H_d \geq H_w$. The symbol H_m denotes the maximal tension to be experienced in the soil water. The function F can now be related to the soil water characteristic as follows.

As illustrated in Figure 8.15, the fluid volume that enters the soil during wetting between $H_w + \delta H_w$ and H_w amounts to

$$\delta\theta = \frac{\partial\theta}{\partial H_w}\delta H_w$$

On the other hand, as shown in Figure 8.16, in terms of F this volume equals

$$\delta\theta = \left[\int_{x=H_w}^{H_m} F(x, H_w)dx\right]\delta H_w \tag{8.7}$$

in which x is the dummy variable of integration. Thus, one obtains for the slope of the wetting boundary curve of the soil water characteristic

$$\frac{\partial\theta}{\partial H_w} = \int_{x=H_w}^{H_m} F(x, H_w)\,dx \tag{8.8}$$

Similar equalities can be obtained by considering the amount of water drained between H_d and $H_d + \delta H_d$, which yield the slope of the drying boundary curve of the soil water

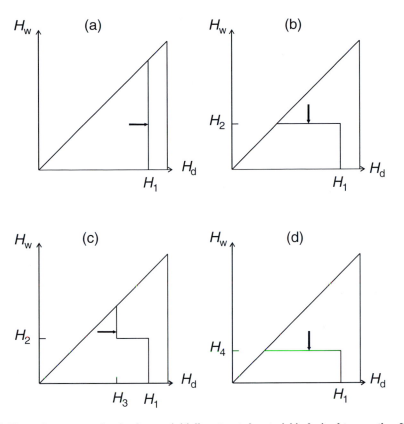

Fig. 8.17 Example sequence showing how an initially saturated material is drained to a suction H_1, after which it is refilled to H_2 and subsequently drained again to H_3 and rewetted to H_4. The area to the left of the line dividing the inside of the triangle represents empty pore space and the area to the right is pore space filled with water.

characteristic as

$$\frac{\partial \theta}{\partial H_{\mathrm{d}}} = \int\limits_{x=H_{\mathrm{d}}}^{0} F(H_{\mathrm{d}}, x)\, dx \tag{8.9}$$

Hence one has finally

$$F = \frac{\partial^2 \theta}{\partial H_{\mathrm{d}} \partial H_{\mathrm{w}}} \tag{8.10}$$

This can be integrated to yield different primary and secondary scanning curves during wetting and drying.

Example 8.2. Sequence of alternate wetting and drying

To illustrate the procedure, assume as shown in Figure 8.17, that an initially satiated material is drained to a suction H_1, after which it is refilled to H_2 and subsequently drained again to H_3 and rewetted to H_4. The successive integrations can be performed as follows to obtain

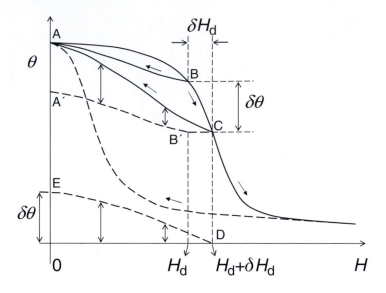

Fig. 8.18 Illustration of how the fractional volume of water $\delta\theta$, which is extracted during drying (BC), re-enters the soil during wetting, as the soil water suction decreases from $H + \delta H$ to 0. The vertical difference between the primary wetting scanning curve CA and the curve CB'A' is shown as the curve DE.

the successive water contents

$$\theta_1 = \theta_0 - \int_0^{H_1} \int_0^x F(x, y)\,dy\,dx$$

$$\theta_2 = \theta_1 + \int_{H_2}^{H_1} \int_{x=y}^{H_1} F(x, y)\,dx\,dy$$

$$\theta_3 = \theta_2 - \int_{H_2}^{H_3} \int_{H_2}^x F(x, y)\,dy\,dx$$

(8.11)

$$\theta_4 = \theta_3 + \int_{H_2}^{H_3} \int_{x=y}^{H_3} F(x, y)\,dx\,dy + \int_{H_4}^{H_2} \int_{x=y}^{H_1} F(x, y)\,dx\,dy$$

The hysteresis function F can be determined from the primary wetting scanning curves on the basis of the following considerations. In Figure 8.18, $\delta\theta$ is the water content drained between H_d and $H_d + \delta H_d$ and $(\delta\theta/\delta H_d)$ is the rate of drainage, that is the amount of water content drained per unit drainage suction increase. The primary wetting scanning curve BA shows how the water content which drained between 0 and H_d is redistributed during rewetting; similarly, the primary wetting scanning curve CA shows how the water content that drained between 0 and $H_d + \delta H_d$ is redistributed during rewetting. Hence, subtraction of the amount of water entering the pore space as described by curve BA from the amount entering, as described by curve CA, shows how the amount $\delta\theta$ redistributes itself during rewetting. Graphically, this difference is the vertical (in Figure 8.18) distance between curves CA and CB'A', which is also shown as the curve DE. It follows that the rate of change of this rate of drainage, namely $\delta(\delta\theta/\delta H_d)/\delta H_w$ is the increase in refilled water content per unit wetting suction decrease. In other words, for a given H_d, F is the value of the increase in refilled water content for each increment δH_w. Therefore, in Figure 8.18

Table 8.1 Example of the F distribution of the independent domain approach for the porous material studied by Poulovassilis (1962). F is expressed as drainable porosity in percent per $(4\ \text{cm})^2$

H_d (cm) →	0	4	8	12	16	20	24	28	H_w (cm)
							0.95		28
						2.38	0.01		24
					1.90	2.38	0.95		20
				1.43	4.29	6.19	1.90		16
			0.01	4.76	17.14	9.05	0.95		12
		0.95	4.29	8.57	8.57	3.81	0.95		8
	3.81	3.33	5.24	2.86	1.90	0.48	0.95		4
									0

(see also Equation (8.10)) for a given H_d, F is the slope of the curve DE. Table 8.1 shows an example of the F function obtained experimentally by Poulovassilis (1962); another example is presented in Problem 8.3.

The independent domain model was concluded to compare favorably with experimental data by Poulovassilis (1962) and Talsma (1970), but not so favorably by Topp (1971). An early example of the application of this model to the problem of intermittent infiltration with redistribution of soil water can be found in the numerical study of Ibrahim and Brutsaert (1968).

It is not easy to obtain the experimental data needed to estimate the F function. For this reason several attempts have been made to simplify the independent domain model with various similarity assumptions by, among others, Parlange (1976), Mualem and Miller (1979), and Braddock *et al.* (2001). In what follows a brief description is presented of Parlange's (1976) proposal, which has been useful in many practical applications.

The main assumption is that F is independent of H_w, so that $F(H_d, H_w)$ can be replaced by $F(H_d)$. This means that, for example in Table 8.1, the F values in each column can be replaced by their averages. In Table 8.1, the values in the column between H_d values 20 and 24 cm, all become equal to 4.05%. It also means that, for example in Figure 8.18, DE would be represented by a straight line. The main advantage of this simplification is that it becomes possible to calculate all the scanning curves from the boundary drying curve, which is also the easiest to determine experimentally. For instance, the function $F(H_d)$ can be determined immediately by integration of Equation (8.9) as

$$F = \frac{1}{H_d}\frac{\partial\theta}{\partial H_d} \tag{8.12}$$

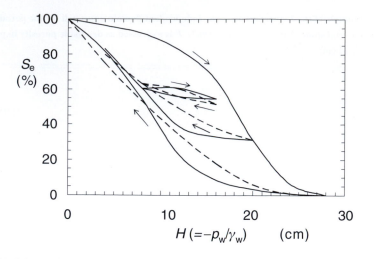

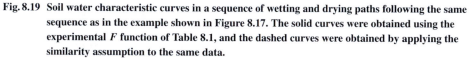

Fig. 8.19 Soil water characteristic curves in a sequence of wetting and drying paths following the same sequence as in the example shown in Figure 8.17. The solid curves were obtained using the experimental F function of Table 8.1, and the dashed curves were obtained by applying the similarity assumption to the same data.

The wetting boundary curve can be calculated from (8.8) as

$$\delta\theta = \left[\int_{x=H_w}^{H_m} F(x)\,dx \right] \delta H_w$$

or

$$\theta = \int_{y=H_w}^{H_m} \int_{x=y}^{H_m} F(x)\,dx\,dy \tag{8.13}$$

Other scanning curves can be calculated as before in the manner shown in the example with Equations (8.11).

Example 8.3. Numerical calculation

The numerical procedure can be illustrated by using the same sequence as shown in Figure 8.17. Assume for the present example $H_1 = 20$ cm, $H_2 = 8$ cm, $H_3 = 16$ cm, and $H_4 = 4$ cm. The calculations can be readily carried out by summing and subtracting the values in Table 8.1, as indicated in Figure 8.17. The results for the regular independent domain procedure using the values of Table 8.1 and those for Parlange's (1976) simplified procedure using averaged columns in Table 8.1 can be compared in Figure 8.19.

While the determination of hysteresis curves is difficult enough with data obtained in the laboratory, it is even more so with field data (see Royer and Vachaud, 1975; Watson *et al.*, 1975). Therefore, although the independent domain approach and its simpler versions may perhaps be considered crude approximations, they should be quite useful in practical simulations of soil water flow problems. Certainly, the error resulting from this approximation will be much smaller than the unavoidably large errors resulting from uncertainties

Fig. 8.20 **Soil water characteristic curve for a fine sand (Oso Flaco) during drainage, showing the effective saturation, $S_e = (\theta - \theta_r)/(n_0 - \theta_r)$, against negative pressure in the water $H = -p_w/\gamma_w$, expressed in cm of equivalent water column. The curve represents Equation (8.15) with $n_0 = 0.405$, $\theta_r = 0.0381$, $a = 0.0280$ cm^{-1}, and $b = 6.7$; the circles represent the experimental curve, already shown in Figure 8.5.**

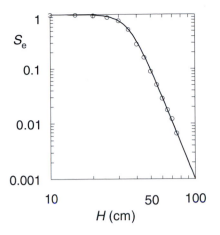

in the values of soil water flow parameters and from ignoring hysteresis altogether, as is currently still almost universal practice.

8.2.4 Some expressions for the soil water characteristic function

Several empirical functions have been proposed in the literature to describe the soil water characteristic. Among the better known is the power function introduced by Brooks and Corey (1964),

$$
\begin{aligned}
S_e &= 1 && \text{for } 0 < H < H_b \\
S_e &= (H/H_b)^{-b} && \text{for } H \geq H_b
\end{aligned}
\tag{8.14}
$$

where as usual $H[= (-p_w/\gamma_w) = -\psi_w]$ denotes the suction expressed as height of water column, where S_e is defined in (8.6), and where b and H_b are constants which are characteristic for a given soil; the latter is also referred to as the *bubbling* or *air entry* suction. Finer textured soils tend to have smaller values of b and larger values of H_b than coarser textured soils. A disadvantage of Equation (8.14) is its two-part structure and the singular behavior of its derivative at $H_w = H_b$.

In order to obtain a smooth transition from $S_e = 1$ (or $\theta = \theta_0$) down to $S_e = 0$ (or $\theta = \theta_r$), Brutsaert (1966) proposed instead

$$
S_e = (1 + (aH)^b)^{-1}
\tag{8.15}
$$

where again a and b are constants for a given soil. In Equation (8.15) the constant a has the dimensions of inverse length $[L^{-1}]$, which represents the inverse of negative capillary pressure expressed as height of water column; observe that the value of a^{-1} happens to be the capillary suction where the effective degree of saturation S_e is at 50%. Figure 8.20 illustrates the shape of Equation (8.15) for Oso Flaco sand, when plotted with logarithmic scales; the parameters used for this were found to be $n_0 = 0.405$, $S_r = 0.094$, $a = 0.0280$ cm^{-1} and $b = 6.7$. Equation (8.15) was extended by Van Genuchten (1980) by introducing an additional constant parameter c as follows

$$S_e = (1 + (aH)^b)^{-c} \tag{8.16}$$

The choice of the particular function will mostly have to depend on the desired balance between parsimony and flexibility of the parameterization. The price of greater flexibility is usually a larger number of parameters: Equations (8.14) and (8.15) require three parameters whereas Equation (8.16) needs the estimation of four parameters for its application.

8.3   WATER TRANSPORT IN A POROUS MATERIAL

8.3.1   *Dynamics of pore-filling fluids: the law of Darcy*

The original experiments
In a report on the public fountains and water supply for the City of Dijon in Burgundy, Darcy (1856) presented the results of his experiments on the seepage of water through a pipe filled with sand, with a 0.35 m inside diameter and a 3.00 m effective length (see Figure 8.21). In brief, he found that the rate of flow Q through the sand layer was directly proportional to the cross-sectional area A of the sand column and to the difference of hydraulic head h across the layer, and inversely proportional to the length ΔL of the sand column. In this notation his result can be formulated as

$$Q = kA(h_1 - h_2)/\Delta L \tag{8.17}$$

in which the subscripts 1 and 2 refer to the entrance and the exit section of the column, respectively. The symbol k represents a constant of proportionality, which is now commonly referred to as the *hydraulic conductivity*, and which has the dimensions [L T^{-1}]. In the experiments of Darcy the water had essentially a constant specific weight and the hydraulic head can be defined as usual, namely

$$h = z + \frac{p_w}{\gamma_w} \tag{8.18}$$

where z is the vertical coordinate. When the negative pressure is expressed as equivalent water column, Equation (8.18) can also be written concisely as $h = z - H$.

Any instrument used to measure the hydraulic conductivity k, that is similar to the set-up originally used by Darcy, is often referred to as a *permeameter*. Over the years many different designs have been developed, but they are all nearly the same in principle, in that they provide the measurements of Q and $(h_1 - h_2)$ needed to invert Equation (8.17) in order to estimate k. Some types of permeameters are also available commercially.

Formulation at a point
Under the assumption that the porous material can be treated as a continuum, both A and ΔL can be allowed to become infinitesimally small, so that Equation (8.17) describes the flow at a point and can be written concisely in common vector notation as

$$\mathbf{q} = -k\nabla h \tag{8.19}$$

where $\mathbf{q} = q_x \mathbf{i} + q_y \mathbf{j} + q_z \mathbf{k}$ is the specific volumetric flux, that is the volumetric rate of flow per unit area of porous material, $\nabla = (\partial/\partial x)\mathbf{i} + (\partial/\partial y)\mathbf{j} + (\partial/\partial z)\mathbf{k}$ the gradient

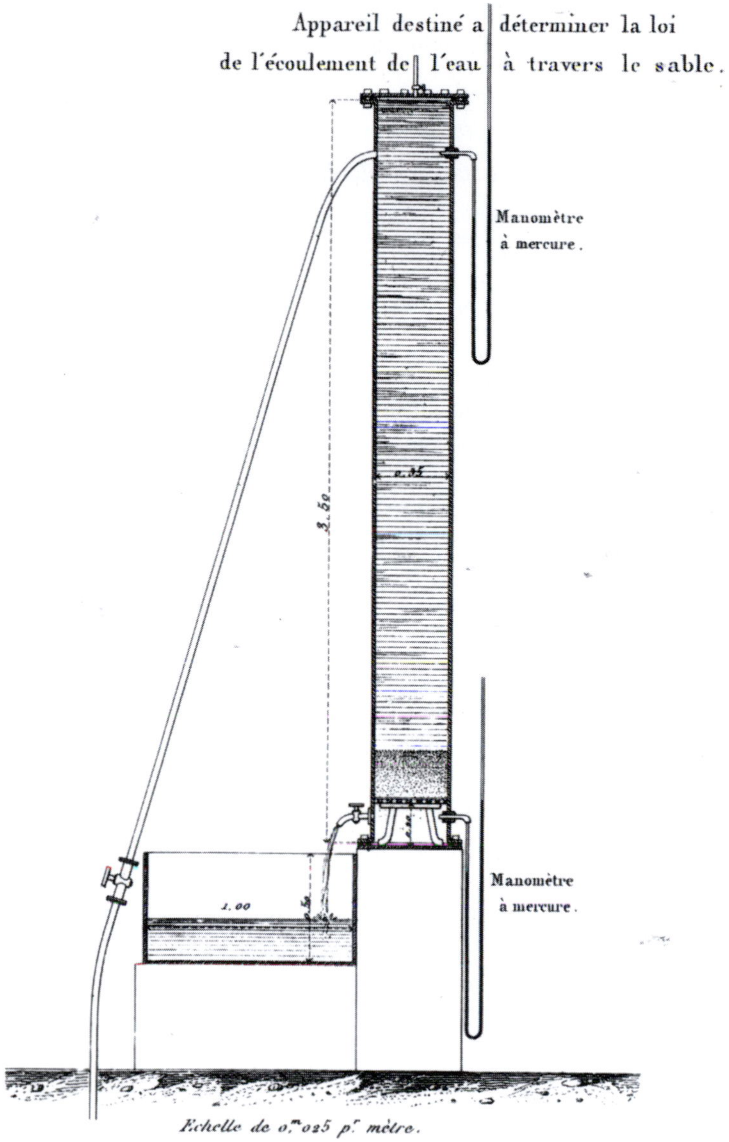

Fig. 8.21 Drawing of
the original
experimental
set-up as
presented by
Darcy (1856).

Appareil destiné a déterminer la loi
de l'écoulement de l'eau à travers le sable.

Manomètre
à mercure.

Manomètre
à mercure.

Échelle de 0.^m025 p.^r mètre.

operator. With subscripts representing the vector components, this is also often written
as

$$q_i = -k\frac{\partial h}{\partial x_i} \tag{8.20}$$

where x_i represents x, y and z for $i = 1$, 2 and 3, respectively. For a fluid of density ρ
at a pressure p in an infinitesimal control volume, the forces that drive the flow are (per
unit volume) the pressure gradient ∇p and gravity ($\rho g\mathbf{k}$), if z is the vertical coordinate
so that $\mathbf{k}$ is the unit vector in the vertical direction. Therefore, whenever the fluid density

varies over the flow region, the hydraulic head cannot be given by Equation (8.18), but it must be defined in gradient form as follows,

$$\nabla h = \mathbf{k} + \frac{1}{\rho g}\nabla p \tag{8.21}$$

In the same notation as Equation (8.20) this gradient can be written as

$$\frac{\partial h}{\partial x_i} = \frac{\partial x_3}{\partial x_i} + \frac{1}{\rho g}\frac{\partial p}{\partial x_i} \tag{8.22}$$

in which x_3 is specified as the vertical coordinate.

The (intrinsic) permeability

It stands to reason that the ease or difficulty with which a fluid flows through a porous medium depends on the sizes and the arrangement of the pores and also on the properties of the fluid. Therefore the hydraulic conductivity must be affected by both factors. Whenever it is desirable to separate the effects of the fluid from those of the porous matrix, use can be made of the permeability, also called intrinsic permeability, k' with dimensions $[L^2]$. This is defined as

$$k = \frac{k'\gamma}{\mu} = \frac{k'g}{\nu} \tag{8.23}$$

where $\gamma = \rho g$ is the specific weight, ρ the density, μ the dynamic viscosity and $\nu = \mu/\rho$ the kinematic viscosity of the fluid in question, respectively. One way to arrive at the form of Equation (8.23) is by simple dimensional analysis; thus one observes that the ease with which the fluid moves through the porous material, that is the hydraulic conductivity k, $[L\,T^{-1}]$, is in fact the ratio $-\mathbf{q}/\nabla h$; it is then reasonable to assume that this ratio should be affected by the following three variables: the area available for the flow in the pores, as characterized by some effective or average pore radius R_e, $[L]$, and the two fluid properties that govern the dynamics of low Reynolds number (or *creeping*) flows, namely the viscosity μ, $[M\,L^{-1}T^{-1}]$, and in light of (8.21) the specific weight γ, $[ML^{-2}T^{-2}]$. The only way to combine these three variables to obtain the same dimensions as the hydraulic conductivity is as $k = (\mathrm{Ge}\, R_e^2\gamma/\mu)$, in which Ge is a dimensionless constant introduced to represent the geometrical shape of the pores; hence, on dimensional grounds Darcy's equation assumes the form

$$q_i = -\frac{\mathrm{Ge}\, R_e^2\gamma}{\mu}\frac{\partial h}{\partial x_i} \tag{8.24}$$

The main point of (8.24) is that it shows first, how the fluid property effects on the flow can be separated from those of the porous matrix; and second, how the porous matrix itself affects the flow, not only by the sizes of the pores in terms of R_e, but also by their geometrical shape in terms of Ge. The sizes of the pores depend mainly on the sizes and the distribution of the sizes of the solid particles of the porous material. The geometrical shapes of the pores, in turn, depend largely on the arrangement of these same particles. This explains how the hydraulic conductivity of soils can be greatly reduced by compaction, as carried out for road or dam construction purposes, or also greatly

increased by ploughing, harrowing and other agricultural operations. But the concept
has its limitations, and should be used with caution, especially in the case of clayey
soils. Depending on the type of clay, the pore structure of some soils may be sensitive to
the type of electrolytes or salts that are in solution in the water; for instance, sodium is
notorious in this regard. This means that it is not always possible to separate the effects
of the fluid on the hydraulic conductivity completely from those of the porous matrix.

It is of interest to observe that in the form of Equation (8.24), Darcy's law can be
considered as a generalization of several well-known creeping flow equations, obtainable
for certain flow regions with a regular geometry, to the case of a totally irregular pore
geometry resulting from the random assemblage of the particles of soils and other granular materials. For instance, in the case of flow through a straight pipe of circular cross
section, creeping flow is described by the Hagen–Poiseuille equation; this is equivalent
with (8.24) if q_i is taken to represent the average velocity in the pipe, R_e is the radius of
the pipe and $Ge = (1/8)$. (Tubes with other cross-sectional shapes have been analyzed by,
among others, Boussinesq (1868; 1914) and Graetz (1880).) It is also easy to show (see,
for example, Lamb, 1932, p. 582) that Equation (8.24) can describe the flow between
parallel plates (as used in the Hele–Shaw model) by putting $Ge = (1/3)$ and R_e as half
the spacing of the plates, again if q_i is made to represent the average velocity. Similarly,
Equation (8.24) can be used to describe the average velocity of flow down a plane as in
(5.32), provided $Ge = 1/3$, R_e is taken as the depth of flow h and the hydraulic gradient
is the slope of the plane S_0. All three expressions just mentioned are exact solutions of
the Navier–Stokes equations (1.12) for creeping flow, and can be found in elementary
textbooks in fluid mechanics. (Recall that creeping flow is flow with a very low Reynolds
number so that $(D\mathbf{v}/Dt)$ becomes negligible.) In the literature there have been numerous
attempts to derive Darcy's law from the Navier–Stokes equations, mostly by analogy
with these exact solutions. On account of the irregular geometry of the pores resulting
from random packings of particles, to arrive at the desired result any such derivation must
involve some kind of ensemble averaging and other stochastic assumptions, which may
not always be valid. But regardless of such considerations, pragmatically it is probably
preferable to adopt Darcy's law simply as it is, that is as an experimentally obtained and
verified relationship, in which k or k' is best obtained from measurements.

True velocity

As defined in Darcy's equation, $\mathbf{q}$ or q_i is the volumetric rate of flow per unit bulk area
of porous material. Thus even though it has the basic dimensions of $[L\,T^{-1}]$ it does not
represent the average velocity of the fluid particles. The "true" average velocity inside
the pores is usually assumed to be given by (q_i/n_0) under fully saturated conditions, and
by (q_i/θ) under partly saturated conditions.

Anistropy

As formulated in Equation (8.19) (or (8.24)), q_i and $-\partial h/\partial x_i$ are vectors pointing in the
same direction and k (or k') is a scalar quantity, that is independent of direction. Porous
materials in which this holds true are referred to as *isotropic*. A material is said to be
anisotropic when its properties, such as the hydraulic conductivity or the permeability,

depend on direction. Most soils and other water-bearing formations tend to be anisotropic to some degree. In some cases larger permeabilities in the horizontal may be the result of layering of sediments during their deposition in the soil formation process; in other situations cracking during drying of clayey soils, and subsequent filling of the vertical cracks by coarser wind-blown loess, may have resulted in relatively larger permeabilities in the vertical.

In an anisotropic material the two vectors q_i and $-\partial h/\partial x_i$ do not necessarily point in the same direction. It can be shown that the only way to formulate a linear relationship between such vectors is by means of a second-order tensor or a dyad. Thus Darcy's law for anisotropic material must be of the form

$$q_x = -k_{xx}\frac{\partial h}{\partial x} - k_{xy}\frac{\partial h}{\partial y} - k_{xz}\frac{\partial h}{\partial z}$$
$$q_y = -k_{yx}\frac{\partial h}{\partial x} - k_{yy}\frac{\partial h}{\partial y} - k_{yz}\frac{\partial h}{\partial z} \qquad (8.25)$$
$$q_z = -k_{zx}\frac{\partial h}{\partial x} - k_{zy}\frac{\partial h}{\partial y} - k_{zz}\frac{\partial h}{\partial z}$$

Alternatively, this can be written more concisely in the subscript notation as

$$q_i = -\sum_{j=1}^{3} k_{ij}\frac{\partial h}{\partial x_j} \qquad (8.26)$$

In general, the second-order tensor k_{ij} has nine components. Symmetry of such a tensor quantity, i.e. $k_{ij} = k_{ji}$, is known to be a sufficient condition to allow it to be diagonalized, such that it has only three components, along three principal axes. The hydraulic conductivity tensor is usually assumed to be symmetrical.

Example 8.4. Directions of the gradient and of the flux vector

To bring out some of the implications of anisotropy, consider the flux vector $\mathbf{q}$ and the negative hydraulic gradient vector $-\nabla h$, which drives it, in a principal axes system, as shown in Figure 8.22. The gradient consists of two components, and can therefore be broken down as follows

$$\nabla h = \frac{\partial h}{\partial x}\mathbf{i} + \frac{\partial h}{\partial z}\mathbf{k} = |\nabla h|\,(\cos\alpha\mathbf{i} + \sin\alpha\mathbf{k}) \qquad (8.27)$$

Similarly, the specific flux can be written as

$$\mathbf{q} = q_x\mathbf{i} + q_z\mathbf{k} = |\mathbf{q}|\,(\cos\beta\mathbf{i} + \sin\beta\mathbf{k}) \qquad (8.28)$$

These two vectors are related by Darcy's law (8.25), so that by virtue of (8.27) the flux can also be written as

$$\mathbf{q} = -|\nabla h|(k_{xx}\cos\alpha\mathbf{i} + k_{zz}\sin\alpha\mathbf{k}) \qquad (8.29)$$

Fig. 8.22 Definition sketch of the directions of the specific flux and gradient vectors in an anisotropic material, in which $k_{xx} > k_{zz}$, and x and z are principal axes.

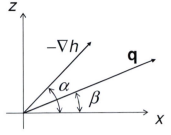

Thus, since the x and z components must be the same in Equations (8.28) and (8.29), respectively, one obtains

$$\tan \beta = \frac{k_{zz}}{k_{xx}} \tan \alpha \qquad (8.30)$$

This shows that the directions of the specific flux and of the (negative) hydraulic gradient can be the same only when the porous medium is isotropic. In an anisotropic material these two vectors point in different directions.

Scale dependence of the hydraulic conductivity

It has been noted in numerous applications of Darcy's law to field conditions, that the values of k, required to reproduce the observed flow rates, tend to depend on the size of the domain over which Equation (8.19) is integrated or averaged. In other words, the magnitude of k is scale dependent. Thus values of the hydraulic conductivity for a given soil are often larger when obtained from auger hole measurements in the field, than from small column measurements in the manner of Darcy (see Figure 8.21) or in other types of permeameters. Values of k obtained by inversion methods with data from small river catchments (see Brutsaert and Lopez, 1998) tend to be still larger. Permeameters involve a flow domain with typical length scales of the order of 1.0 m at most, while auger hole measurements and pumping tests tend to have zones of influence with length scales of the order of 10–10^2 m, respectively; small catchments usually involve scales of the order of 10^3 m or more.

There are several possible reasons for this scale dependence. One is that most permeameter measurements are carried out on disturbed samples; thus the soil is scooped up and placed in the permeameter in a way that usually does not replicate the original soil structure in the field. Moreover, under natural conditions in the field, most soils have *macropores* and other additional conduits resulting from decaying plant roots, from worms and from burrowing animals; even if undisturbed samples are used, it is nearly impossible to include such macropores and other larger channels within the relatively small confines of a permeameter. Finally, even within a supposedly homogeneous soil type, all soil properties display pronounced spatial variability. At field scales of the order of 10^2 m, the hydraulic conductivity is usually close to lognormally distributed (see Rogowski, 1972; Nielsen *et al.*, 1973; Hoeksema and Kitanidis, 1985); this means that the larger k values in the domain have a relatively larger effect on the overall flow than the smaller values. But regardless of the distribution, in a two- and three-dimensional flow situation, regions of smaller conductivity can be bypassed by the flow, resulting

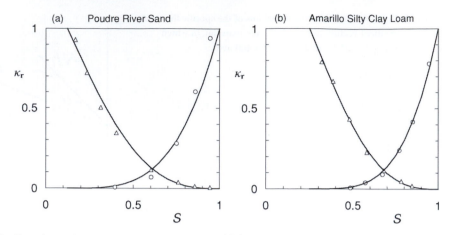

Fig. 8.23 **Experimental values of relative permeability k'/k_0' (or relative conductivity) of water (wetting fluid – circles) and of air (non-wetting fluid – triangles), as a function of the degree of saturation $S = \theta/\theta_0$, in a sandy soil and in a silty clay loam. The curves for water represent Equations (8.36) with (8.45) and those for air represent the analogous expression for a non-wetting fluid. (After Brooks and Corey, 1966.)**

in a larger effective conductivity for the entire flow domain, than a simple average value would suggest (see El-Kadi and Brutsaert, 1985). In flow simulation calculations, the variability of the soil characteristics is usually avoided by the assumption that the entire flow domain is homogeneous and that effective parameters can be used (see Section 1.4.3), to represent the flow domain. The scale dependency of the hydraulic conductivity is a direct result of this assumption.

8.3.2 Partly saturated flow

Extension of Darcy's law

It was probably Buckingham (1907) who first postulated that Darcy's law (8.19) is also valid for a soil that is only partly saturated with water, and that in this case the hydraulic conductivity is a function of water content, or $k = k(\theta)$. As the water content of the soil is reduced, k decreases. Reasons for this are that fewer pores are available for flow, and that the flow paths become more tortuous and thus longer, as the water can no longer move through emptied pores and must move around them. Because the larger pores are emptied first, the initial decrease in conductivity for a certain decrease in water content is larger than that later on at lower water contents.

Actually, the water of the empty pores has been replaced by air. Under such conditions Darcy's law is also valid to describe the flow of air; but it must be applied with an "air head" gradient as given by Equation (8.21) with the density of air. This is illustrated in Figure 8.23, which shows the relative permeabilities $\kappa_r = (k'/k_0')$ for water and for air in two soils, as examples. In what follows k_0 and k_0' will usually denote the hydraulic conductivity and permeability, respectively, at saturation or satiation. Under partly saturated conditions the hydraulic conductivity k is also often called the *capillary conductivity*. In hydrology mainly the water is of interest, because it can usually be assumed that the air movement takes place under negligibly small pressure gradients.

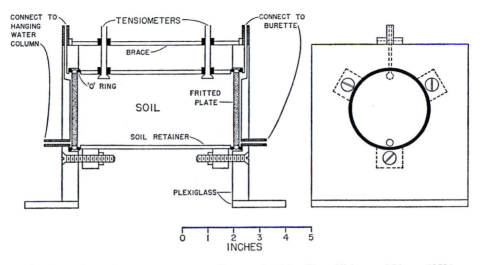

Fig. 8.24   Details of experimental set-up to measure capillary conductivity. (From Nielsen and Biggar, 1959.)

Fig. 8.25   Capillary conductivity of Columbia Silt Loam, as a function of pressure and water content, as measured with the apparatus shown in the previous figure. The difference between the first and second drying cycle was mainly due to the initial soil consolidation resulting from the negative pressure in the water (1 hPa is roughly equivalent to 1.02 cm of water column). (After Nielsen and Biggar, 1961.)

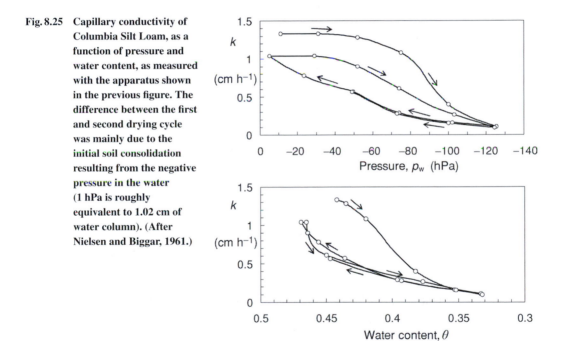

Determination of the capillary conductivity

In principle, the capillary conductivity can be measured by column experiments, similar to those conducted by Darcy, provided the water content θ or the mean water pressure p_w in the column can be measured or otherwise maintained at the desired value. An example of such an experimental set-up is shown in Figure 8.24 (Nielsen and Biggar, 1959), and the corresponding measured values of $k = k(-p_w)$ and $k = k(\theta)$ are shown in Figure 8.25. It can be seen that the relationship between capillary conductivity k

Fig. 8.26 Capillary conductivity as a
function of negative pressure
$k = k(-p_w)$ for the drainage
(desorption) cycle for different
soils: (1) Pachappa sandy loam,
(2) Indio sandy loam, (3) Fort
Collins loam, (4) Aiken clay
loam, (5) Chino clay (10 kPa is
roughly equivalent to 1.02 m of
water column). (After Gardner
and Miklich, 1962.)

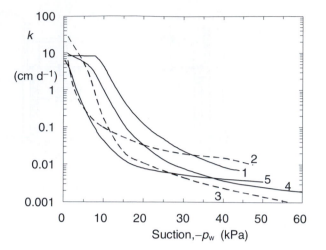

and water suction $H(= -p_w/\gamma)$ displays hysteresis; however, the relationship between conductivity and water content θ is fairly free of hysteresis. Many studies in the literature have confirmed that $k = k(\theta)$ exhibits very little, if any, hysteresis (see Jackson et al., 1965; Talsma, 1970; Topp, 1971). Some additional examples of $k = k(H)$ are presented in Figure 8.26.

For the purpose of simulating flow problems in nature, however, it is clearly preferable to determine $k(\theta)$ for the undisturbed soil profile. To date most experimental determinations have been restricted to vertical flow. Various studies, consisting mostly of the inverse application of finite difference forms of the governing differential equation (see Section 8.4.1), in the absence of precipitation and by preventing evaporation at the surface, have been carried out, for example, by Ogata and Richards (1957), Nielsen et al. (1973), Davidson et al. (1969), Baker et al. (1974), Libardi et al. (1980) and Katul et al., (1993). However, measurements of soil water content and water pressure, at several levels in the profile and over extended periods of time, are not easy and require many precautions. Thus, field methods are usually hard, if not impossible, to apply when one of the following conditions is present: a water table close to the surface, frequent and large precipitation, non-negligible or unknown net lateral inflows, a large vertical drainage rate at the lower end of the profile, and large variability in the soil properties. Because field methods are only feasible under exceptionally favorable conditions, many attempts have also been made to develop conceptual prediction methods. Some of these methods will be touched upon in Section 8.3.4.

Soil water diffusion formulation

In the solution of certain flow problems in partly saturated soils, it has been found convenient to reformulate Darcy's law as a diffusion equation. Thus the pressure gradient in (8.19) is replaced by a water content (i.e. concentration) gradient, and Darcy's law can be written as

$$q_i = -D_w \frac{\partial \theta}{\partial x_i} - k \frac{\partial x_3}{\partial x_i} \tag{8.31}$$

Fig. 8.27 **Soil water diffusivity as a function of water content, $D_w = D_w(\theta)$ during desorption for the same soils as in Figure 8.26. (After Gardner and Miklich, 1962.)**

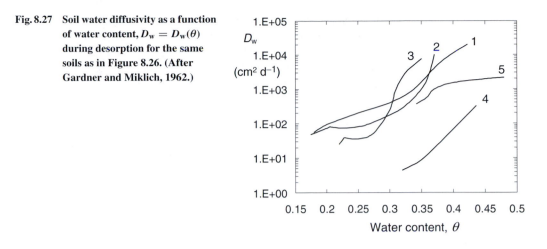

by defining (Klute, 1952) the soil water diffusivity $D_w(\theta)$ by

$$D_w = -k\frac{dH}{d\theta} \tag{8.32}$$

or also $D_w = k(d\psi_w/d\theta)$. From the physical point of view, Equation (8.31) does not contain any more information than (8.19), and in many practical simulations there may be no significant advantage in this diffusion formulation. In fact, when part of the flow domain is saturated, (8.31) with (8.32) may present some difficulties on account of the singular nature of D_w when θ is constant; these are avoidable with (8.19). Nevertheless, as will be seen in the next chapter, the diffusion formulation continues to be of interest because it has greatly simplified the analytical treatment of a number of important soil water problems. Some examples of the dependence of soil water diffusivity on water content are shown in Figure 8.27.

8.3.3 Limitations of Darcy's law

Upper limit

In light of the analogy between Darcy's law and other creeping flow equations of fluid mechanics, it should not be surprising that experiments have shown that, as the Reynolds number increases beyond a certain limit, the specific flux **q** gradually deviates from its linear proportionality with the hydraulic gradient ∇h. Indeed, by definition, creeping flow is flow for which the appropriate Reynolds number is sufficiently small, so that the acceleration terms, both temporal and advective, in the Navier–Stokes equations are negligible. Any Reynolds number definition requires the adoption of a characteristic velocity and of a characteristic length of the flow geometry. The definition of the permeability k' in Equation (8.23) indicates that it has the basic dimensions of $[L^2]$ and that it can be considered proportional to a characteristic or typical cross-sectional area of flow; thus, since the specific flux has the basic dimensions $[L\,T^{-1}]$, it is convenient to define the Reynolds number for flow in a porous medium as follows

$$Re_p = \frac{|q|\sqrt{k'}}{\nu} \tag{8.33}$$

Fig. 8.28 Converging section of a duct illustrating inertia effect in a pore with variable cross section and with the resulting convective acceleration $\partial V / \partial x$; V_i and V_o are the mean inflow and outflow velocity, respectively. At higher velocities this effect becomes more pronounced; the flow ceases to be creeping and Darcy's equation is no longer valid.

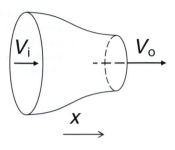

The Reynolds number is a measure of the relative magnitude of the inertial forces over the viscous forces; this means that Darcy's law cannot be expected to be valid for values of Re_p that are much in excess of order one. It should be emphasized that this initial deviation from Darcy's law is not caused by the onset of turbulence, but solely by the effects of fluid accelerations. These accelerations result from the irregular and tortuous flow paths in the pores, as illustrated in Figure 8.28.

Forchheimer (1930, p. 54) was among the first to analyze experimental data and in 1901 he proposed the following to describe them

$$|\nabla h| = \alpha \, |q| + \beta q^2 \tag{8.34}$$

where α and β are constants for any given soil. Some insight can be gained in the nature of these constants by dimensional inspection of the Navier–Stokes equation (1.12). The term on the left and the first term on the right in Equation (8.34) correspond to the three terms on the right of Equation (1.12); thus α represents the effect of viscosity and in terms of the permeability it is $\alpha = \nu/(gk')$. The second term on the right, βq^2, corresponds to the terms on the left-hand side of (1.12) and therefore, it represents the inertia effects on the flow. For steady conditions the left-hand side of (1.12) is $\mathbf{v} \cdot \nabla \mathbf{v}$. Dimensionally, the two velocity terms are proportional to q^2. On the other hand, the dimensions of ∇ are $[L^{-1}]$; because $\sqrt{k'}$ is the characteristic length scale of the pores, this suggests that β is inversely proportional to it, or

$$\beta = \frac{C}{g\sqrt{k'}} \tag{8.35}$$

where C is a constant, which can be expected to depend on the geometry and shape of the pore spaces. Equation (8.35) has been confirmed in numerous experimental investigations. For instance, Arbhabhirama and Dinoy (1973) have reported values of C ranging between approximately 0.6 (sand) and 0.2 (angular gravel).

Lower limit

Several experiments with clayey soils have shown that Darcy's law also fails to describe the measurements when the flow rates become very small (see Miller and Low, 1963; Swartzendruber, 1968). These experiments indicate that under conditions of low flow rates (or small h-gradients), the measured specific fluxes q are smaller than $-k\nabla h$ would require. The issue has been somewhat controversial (Olsen, 1966) and is still far from resolved (Neuzil, 1986). Nevertheless, this phenomenon may possibly result from the fact that in the neighborhood of clay particles, which invariably have electrical charges, the water molecules may become oriented in a quasi-regular fashion, on account of their dipole characteristics;

as a result the viscosity of the water may no longer be Newtonian, a necessary condition for the validity of the Navier–Stokes equations and Darcy's law. Also, in many situations additional driving mechanisms may be at play, beside the hydraulic gradient ∇h; these may cause flow even when $\nabla h = 0$.

Additional driving forces

For most purposes in hydrology it can be assumed that the driving forces are primarily mechanical in nature, which means that the flow is driven by gravity and by a pressure gradient in accordance with Darcy's law. However, in general, water transport in a porous medium can be influenced by several other factors, involving thermal, osmotic and sometimes even electrical effects. For instance, changes in temperature at some point in a partly saturated soil may result in changes in surface tension which affect the pressure p_w for a given water content, and thus the liquid transport. An input of heat can be accompanied by local vaporization, which in turn sets up a specific humidity gradient, and thus a water vapor transport in the air filled pore space; this vapor may then condense further down and affect the liquid water flux.

On account of the complexity of the various phenomena and their interaction within the soil, at present there is apparently no theory available, which is generally accepted. Nevertheless, in recent years many problems, involving simultaneous heat and water transport, have been studied within a framework developed by Philip and DeVries (1957; DeVries, 1958; DeVries and Philip, 1986), but with mixed results (Jackson et al., 1974; Kimball et al., 1976). Raats (1975) and Nakano and Miyazaki (1979) have explored the theoretical and practical compatibility of the formulation of Philip and DeVries with concepts of irreversible thermodynamics; more recently, Cahill and Parlange (1998) have clarified the role of the vapor transport. By numerical simulations Milly (1984) has investigated the relative importance of the temperature gradient on the water transport; he concluded that, for many practical purposes, it is sufficiently accurate to assume that the water transport is essentially isothermal and driven only by the hydraulic gradient ∇h.

8.3.4 Expressions for the conductivity and the soil water diffusivity functions

Conductivity of saturated materials

In the past, numerous equations have been proposed to predict the hydraulic conductivity of porous materials, mostly on the basis of measurements of the particle sizes or their distribution or also of the pore size distribution, as obtained from Equation (8.5). However, such equations were usually obtained from permeameter measurements of k and are therefore valid only at the local scale. As noted earlier, k tends to be scale dependent, so that these methods cannot be used when the hydraulic conductivity is to be used at the field or catchment scale, as is often the case in hydrology. For applications over larger areas it is therefore advisable to obtain k by means of inverse methods with measurements at the appropriate scale. One such inverse method, based on drought flow recession analysis, will be considered in Chapter 10.

Conductivity of partly saturated materials

As seen earlier, under partly saturated conditions the determination of $k(\theta)$ is not an easy task. However, while in many flow calculations the accuracy of k at high water contents is fairly critical, at lower water contents some inaccuracies can be tolerated. Therefore, it has been found useful to represent $k(\theta)$ by relatively simple parametric equations. As illustrated in Figure 8.23, it is not surprising that the following has been used widely,

$$k = k_0 S_e^n \tag{8.36}$$

where k_0 is the hydraulic conductivity at saturation and n is a constant; the effective saturation S_e is defined in (8.6). Equation (8.36) requires the determination of four parameters, namely k_0, θ_0, θ_r and n. Inspection of past determinations of n shows that it may be as low as 1 and as high as 20, but that typical values lie around 3–5; n appears to be small for materials with a narrow pore size distribution and larger for wider pore size distributions. Interestingly, this power form equation has been derived on the basis of some widely different theoretical models. It is tempting, therefore, to conclude that the power form is independent of its method of derivation. Some of these models are reviewed in Section 8.3.5. For instance, Averyanov (Polubarinova-Kochina, 1952) proposed Equation (8.36) with $n = 3.5$, and Irmay (1954) proposed it with $n = 3$; more recently, it was found (Brutsaert, 2000) that $n = (2 + 2.5/b)$, where b is the same as in Equation (8.14), provides the best description of the available experimental data.

Equation (8.36) is one of the oldest and still among the most widely used expressions today. Recently it has also received renewed theoretical interest because it arises naturally in the fractal characterization of soils. Other parameterizations have been proposed for $k(\theta)$, but they are all fairly similar to (8.36).

Since the water content θ is a function of the negative water pressure $H(= -p_w/\gamma_w)$, it is also possible to express k as a function of H. Gardner (1958) has proposed an empirical function, which can be fitted to data for many different soils, viz.

$$k = \frac{a}{b + H^c} \tag{8.37}$$

where a, b and c are constants; note that (a/b) is equal to k_0, the hydraulic conductivity at satiation and b is the value of H^c for $k = k_0/2$. The range of c was found to lie between about 2 for clayey soils and 4 or more for sandy soils. It can be seen that Equation (8.37) is of the right general shape to fit to experimental data such as shown in Figure 8.29. As already mentioned and illustrated in Figure 8.25, however, $k(H)$ normally exhibits marked hysteresis, so that the constants have to be adjusted to reflect this.

In some applications it is convenient to describe the hydraulic conductivity by an exponential function as follows

$$
\begin{aligned}
k &= k_0 \qquad \text{for } H \leq H_b \\
k &= k_0 \exp[-a(H - H_b)] \qquad \text{for } H > H_b
\end{aligned}
\tag{8.38}
$$

where a and H_b are constants for a given soil; Equation (8.38) was introduced by Gardner (1958) without H_b; this constant was added later by P. E. Rijtema to allow incorporation of the capillary fringe. The spatial variability and physical significance of the parameter a in (8.38) have been investigated (White and Sully, 1992); at the field scale, a appears to be lognormally distributed, like k_0.

Soil water diffusivity

A diffusivity function, which has been useful in the solution of a number of problems, is of the following exponential form

$$D_w = D_{wi} \exp[\beta(\theta - \theta_i)/(\theta_0 - \theta_i)] \tag{8.39}$$

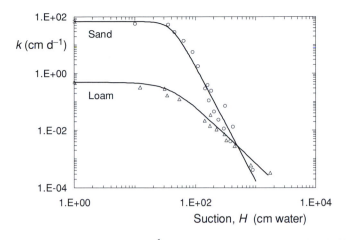

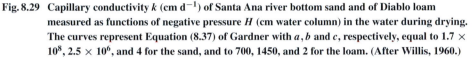

Fig. 8.29 Capillary conductivity k (cm d^{-1}) of Santa Ana river bottom sand and of Diablo loam measured as functions of negative pressure H (cm water column) in the water during drying. The curves represent Equation (8.37) of Gardner with a, b and c, respectively, equal to 1.7×10^8, 2.5×10^6, and 4 for the sand, and to 700, 1450, and 2 for the loam. (After Willis, 1960.)

where β is a constant, D_{wi} is the diffusivity at some initial or other reference moisture content θ_{i}, and θ_0 is the moisture content at satiation. Gardner and Mayhugh (1958) used Equation (8.39) in the numerical solution of the problem of sorption, or horizontal infiltration (see Chapter 9). Reichardt *et al.* (1972) made use of Equation (8.39) to scale experimental data on horizontal infiltration obtained from eight different air-dry soils, so that they could represent the results by a single regression equation in terms of dimensionless variables (see Figure 8.30). Miller and Bresler (1977), who reconsidered the analysis of Reichardt *et al.* (1972), showed that for many soil types β in Equation (8.39) may be fairly constant and not very different from 8. They also found by linear regression that, if θ_{i} is taken as the air-dry water content of the soil, D_{wi} is in fact related to the rate of advance of the wetting front during horizontal infiltration or sorption. It was subsequently shown by Brutsaert (1979), how theoretically this relationship is a direct consequence of the physical nature of sorption; in addition, it was shown that D_{wi} must also be related to the infiltrated volume of water during sorption and that the constants involved in these two relationships are unique functions of β. It should be mentioned that the value of $\beta = 8$ was obtained with repacked laboratory soil columns. Field measurements by Clothier and White (1981; 1982) yielded a much lower value; actually, while on average the data could be represented by Equation (8.39) with $\beta = 3$, in the moisture range $0.20 \leq \theta \leq 0.36$, D_{w} was found to be nearly constant. In any event, Equation (8.39) should be considered a two-parameter expression, as indicated. This issue will be reexamined in Chapter 9.

A second diffusivity equation has a simple power form and follows directly from Equation (8.32), implemented with (8.14) and (8.36). The result can be written as

$$D_{\mathrm{w}} = k_0 \alpha S_{\mathrm{e}}^{\beta} \tag{8.40}$$

where $\alpha = H_{\mathrm{b}}[(\theta_0 - \theta_{\mathrm{r}})b]^{-1}$, and $\beta = (n - b^{-1} - 1)$. A somewhat less simple form, which has been used to parameterize soil properties for hydrologic purposes (Brutsaert, 1968b), results from the similar combination of Equation (8.36) with (8.15) or (8.16), by means of

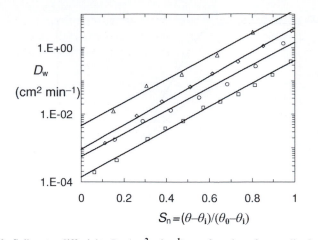

Fig. 8.30 Soil water diffusivity D_w (cm^2 min^{-1}) as a function of normalized water content $S_n = (\theta - \theta_i)/(\theta_0 - \theta_i)$ during horizontal infiltration (see Equation (9.25) and Figure 9.2) into uniform air-dry soil columns. The points are the experimental values and they represent Fresno fine sand (triangles), Hanford sandy loam (diamonds), Yolo clay loam (circles), and Sacramento clay (squares); the straight lines are the exponential approximations by means of Equation (8.39) with the parameters θ_0, θ_i, β and D_{wi}, respectively, equal to 0.31, 0.007, 7.97, and 0.004 58 for Fresno, 0.35, 0.012, 8.36, and 0.000 90 for Hanford, 0.42, 0.04, 7.85, and 0.000 56 for Yolo, and 0.55, 0.07, 8.02, and 0.000 14 for Sacramento. (Data from Reichardt *et al.*, 1972.)

(8.32), to wit

$$D_w = k_0 \alpha S_e^\beta \left(1 - S_e^\delta\right)^\gamma \tag{8.41}$$

where $\alpha = [(\theta_0 - \theta_r)ab]^{-1}$, $\beta = (n - b^{-1} - 1)$, $\gamma = (b^{-1} - 1)$ and $\delta = 1$ in the case of (8.15), or $\alpha = [(\theta_0 - \theta_r)abc]^{-1}$, $\beta = [n - (bc)^{-1} - 1]$, $\gamma = (b^{-1} - 1)$ and $\delta = c^{-1}$ in the case of Equation (8.16).

8.3.5 Some models for permeability

The experimental determination of the hydraulic conductivity is never easy for saturated soils, but it is especially difficult for unsaturated soils as a function of soil water content. For this reason, many attempts have been made to develop simple conceptualizations of the flow process and to represent $k = k(\theta)$ by simple parametric equations, in terms of other properties of the soil that are easier to determine.

Uniform pore size models

In one common approach the porous medium is assumed to be analogous to a bundle of uniform and parallel capillary tubes with circular cross section. Around 1950, Averyanov (Polubarinova-Kochina, 1952) analyzed the flow of an annulus of wetting fluid in a single tube, the central part of which was occupied by stagnant air. The solution of this flow problem yielded an equation that could be approximated closely by (8.36) with a power $n = 3.5$. By making the slightly different assumption that the non-wetting fluid at the center of the tube moves under the same pressure gradient as the wetting fluid along the wall, Yuster (1951) obtained Equation (8.36) with $n = 2$. In a different approach, use is made of the hydraulic radius concept; it was proposed originally by Kozeny (1927) for saturated materials and can be defined as the ratio of pore volume to particle area. Irmay (1954) extended this to unsaturated porous media and derived (8.36) with $n = 3$.

Parallel models

In this approach the pore system is assumed to be equivalent with a bundle of uniform capillary tubes of many different sizes. The distribution of the pore sizes is derived from the soil water characteristic, i.e. $S_e = S_e(H)$, through Equation (8.5), i.e. $H = H(R)$, as explained after Equation (8.6). The true mean velocity in each pore can be described by a Hagen–Poiseuille type equation for creeping flow, namely (8.24) with a value of Ge around $(1/8)$.

Because $s_e(R) = dS_e(R)/dR$ represents the pore size density, $\delta\theta(R) = (d\theta/dR)\delta R = \theta_0(1 - S_r)(dS_e/dR)\delta R = \theta_0(1 - S_r)s_e(R)\delta R$ is the portion of the pore volume occupied by "active" pores, with radius between $(R - \delta R/2)$ and $(R + \delta R/2)$, where δR denotes a very small increment of R. It follows that $[\theta_0(1 - S_r)s_e(R)\delta R]$ is also the area, per unit bulk cross-sectional area of porous material, occupied by openings whose sizes are between $(R - \delta R/2)$ and $(R + \delta R/2)$. The flow rate through this elemental area is, by virtue of (8.24),

$$-\frac{\text{Ge } g}{v}\frac{\partial h}{\partial x_\text{n}}\left[\theta_0\left(1 - S_r\right)\right]s_e(R)R^2\delta R \tag{8.42}$$

in which the subscript n refers to the direction normal to the area under consideration. With $s_e(R)dR = dS_e$ and with Laplace's equation (8.5), i.e. $R = 2\sigma/(\gamma H)$, one obtains the intrinsic permeabilty, by integration over all pores filled with water,

$$k' = (2\sigma/\gamma)^2\text{Ge}[\theta_0(1 - S_r)]\int_0^{S_e}[H(x)]^{-2}\,dx \tag{8.43}$$

where x is the dummy variable representing S_e.

Purcell (1949) and Gates and Tempelaar-Lietz (1950) were among the first to apply this approach, and came up with expressions similar to Equation (8.43). However, because (8.43) tended to yield values considerably larger than available experimental data, several subsequent authors included a tortuosity factor in the formulation to account for the limitations inherent in this model of straight parallel tubes. The tortuosity concept had originally been introduced by Carman (1937; 1956) as an improvement on the uniform hydraulic radius model of Kozeny (1927), and it can be expressed as $T = (L_e/L)^2$, in which L_e is the actual or microscopic path length of the fluid particles in the pores, and L is their apparent or macroscopic path length along the Darcy streamlines. In several studies this tortuosity was assumed to depend on the water content, i.e. S_e; in this case the relative permeability $\kappa_r = k'/k_0'(= k/k_0)$ can be written as

$$\kappa_r = \left[\beta\int_0^{S_e}[H(x)]^{-2}dx\right]\bigg/\left[\beta_0\int_0^1[H(x)]^{-2}dx\right] \tag{8.44}$$

where the variable $\beta = \beta(S_e)$ is related to the tortuosity and β_0 is its value at $S_e = 1.0$. Burdine (1953) proposed on the basis of his experimental data that $(\beta/\beta_0) = S_e^2$.

Example 8.5. Calculation with the power function

A well-known application of the parallel model is the result of Brooks and Corey (1966), who adopted Burdine's assumption in (8.44) and integrated it with (8.14) for the soil water characteristic, to derive Equation (8.36) with the value of the exponent

$$n = 3 + \frac{2}{b} \tag{8.45}$$

Observe that without Burdine's assumption for tortuosity the parallel model would have yielded $n = 1 + 2/b$ instead of (8.45). Equation (8.36) applied with (8.45) can be compared with experimental measurements for two soils in Figure 8.23. For the Poudre sand the wetting fluid curve was calculated with the parameters $S_r = 0125$ and $b = 3.4$; for the Amarillo silty clay loam the values used were $S_r = 0.250$ and $b = 2.3$.

Series-parallel models

The theoretical construction of these types of model also starts with a bundle of parallel pores, each with a different but uniform size. However, these pores are then cut normally to the direction of flow with two resulting faces, and finally after some random rearrangement of the tubes the faces are joined again. This way account is taken of the random variations of the pore sizes, not only in the plane normal to the direction of flow, but also along the direction of flow. In the original version of this model the discharge rate in each single pore, which consists of two sections in series, is assumed to be governed by the section with the smaller diameter. Again, the distribution of the pore sizes is derived from $S_e = S_e(H)$, by means of Laplace's equation for capillary rise (8.5), i.e. $H = 2\sigma/(\gamma R)$; also, the true velocity in each pore is obtained by means of a Hagen–Poiseuille like equation, as shown in (8.24). This approach was pioneered by Childs and Collis-George (1950) in a finite-difference scheme to calculate the permeability from experimental $H = H(\theta)$ data. The model was subsequently reformulated in integral form by Brutsaert (1968a), to allow the derivation of more concise analytical expressions for k; this formulation is presented next.

One of the implicit but basic assumptions of the original approach is that the sizes of the pores in flow sequence are completely independent of each other. As noted, this is visualized in the construction of the model, by considering that the porous medium is equivalent with an array of parallel tubes or flow channels of different sizes, which is first cut into two parts by a plane section normal to the direction of flow; this section produces two surfaces, which are subsequently joined together again after some random rearrangement of the tubes. It was shown earlier that $[\theta_0(1 - S_r)s_e(R)\delta R]$ is the area, per unit bulk cross-sectional area of porous material, which is occupied by openings whose sizes are between $(R - \delta R/2)$ and $(R + \delta R/2)$; this is also equal to the probability of a point in any cross section through the medium being found in a pore with that size. Therefore, the fraction of the area of this section occupied by the sequence of flow pores with size between $(y - \delta y/2)$ and $(y + \delta y/2)$ of the first surface with pores with size between $(z - \delta z/2)$ and $(z + \delta z/2)$ of the second surface is equal to

$$[\theta_0 (1 - S_r)]^2 s_e(y)s_e(z)\,\delta y \delta z$$

If it is assumed that the flow between two pores in sequence is controlled by the smaller of the two, say with size y, then the rate of flow that takes place through a fraction of the cross-sectional area, occupied by the sequences of pores with size between $(y - \delta y/2)$ and

$(y + \delta y/2)$ that are in contact with the pores with size between $(z - \delta z/2)$ and $(z + \delta z/2)$, is

$$-\frac{\text{Ge } g}{v} \frac{\partial h}{\partial x_n} [\theta_0 (1 - S_r)]^2 s_e(y)s_e(z) \, y^2 \delta y \delta z$$

in which the subscript n indicates the direction normal to the section. The pores with a size between $(y - \delta y/2)$ and $(y + \delta y/2)$ of the first surface are in contact with pores of all possible sizes of the second surface. The discharge rate through these pores is

$$-\frac{\text{Ge } g}{v} \frac{\partial h}{\partial x_n} [\theta_0 (1 - S_r)]^2 \left[s_e(y) \int_0^y s_e(z)z^2 dz\delta y + s_e(y)y^2 \int_y^R s_e(z)dz\delta y \right]$$

where R is the size of the largest pores that are still available for flow at the given degree of saturation. The first term gives the flow rate from the pores of the first surface with size between $(y - \delta y/2)$ and $(y + \delta y/2)$ into all the pores of the second surface that are smaller than y; the second term gives the flow rate into the pores of the second surface that are larger than y. Integration over y yields finally the total discharge per unit cross-sectional area of porous medium. Hence the intrinsic permeability, defined in Equation (8.23), can be written as

$$k' = \text{Ge } [\theta_0(1 - S_r)]^2 \left[\int_0^R s_e(y) \int_0^y s_e(z) \, z^2 dz dy + \int_0^R s_e(y)y^2 \int_y^R s_e(z) \, dz dy \right]$$
(8.46)

where y and z are dummy variables representing R. This result can be applied to fully saturated media simply by putting $R = \infty$. It can also be expressed directly in terms of the soil water characteristic function, by means of (8.5), Laplace's equation for capillary rise,

$$k' = \text{Ge } [(2\sigma/\gamma)\theta_0(1 - S_r)]^2 \left[\int_0^{S_e} \int_0^x [H(y)]^{-2} \, dy dx + \int_0^{S_e} [H(x)]^{-2} \int_x^{S_e} dy \, dx \right]$$
(8.47)

where now x and y are the dummy variables representing S_e. One can show by integration by parts that the first double integral on the right is identical with the second; thus (8.47) can be expressed in a more condensed form as

$$k' = (2 \text{ Ge}) [(2\sigma/\gamma)\theta_0 (1 - S_r)]^2 \int_0^{S_e} (S_e - x)[H(x)]^{-2} dx$$
(8.48)

Equations (8.47) and (8.48) can now be applied immediately with suitable expressions for $S_e(R)$ or $S_e(H)$ (see Brutsaert, 1968a). Although they can be applied to fully saturated media to obtain k_0 by putting $R = \infty$ and $S_e = 1.0$, respectively, they have been applied mostly to obtain the relative permeability $\kappa_r = k'/k_0'$.

Example 8.6. Calculation with the power function

As before, the integration is especially simple with Equation (8.14) and it produces

$$k' = \frac{\text{Ge}[(2\sigma/\gamma)\theta_0(1 - S_r)b]^2}{(b + 1)(b + 2)H_b^2} S_e^{2+2/b}$$
(8.49)

For the relative permeability this results again in (8.36) with the value of the exponent

$$n = 2 + 2/b \qquad (8.50)$$

The original version of the series-parallel model, both in its finite difference forms and in the integral forms (8.47) and (8.48), has been tested with experimental data (see Childs and Collis-George, 1950; Marshall, 1958; Millington and Quirk, 1964; Nielsen et al., 1960; Jackson et al., 1965). Although there is a wide variation for different soils, it appears to produce reasonable results for unstructured soils (without macropores). However, it also tends to overestimate the relative permeability somewhat under drier conditions.

Several shortcomings of the model are obvious from the assumptions invoked in its derivation. These are that: (i) the sizes of pores in sequence are independent of each other; (ii) the flow rate in a pore sequence is controlled by the smallest diameter; and (iii) the pores in sequence are lined up perfectly, and they are straight without tortuosity. Assumptions (i) and (ii) will not cause any overestimate in the calculated result. As regards (i), this can be seen by considering that in the parallel models the sizes of the pores in sequence are assumed to be totally dependent on each other, since each flow channel is assumed to have a uniform cross section over its whole length; parallel models, without tortuosity correction, severely overestimate the permeabilty. Thus the assumption of any partial correlation would result in a further overestimate. Similarly, as regards (ii), inclusion of the larger pore size in the sequence into the expression for the fluid velocity would also result in a larger rate of flow. This means that the overestimate is not the result of assumptions (i) and (ii) but mainly of assumption (iii).

To compensate for these shortcomings, in several studies use was made of the concept of tortuosity. Although the concept is intuitively clear, there has been no unanimity in defining it conceptually or mathematically. Clearly, the drier the soil is, the less perfect the remaining pores with water line up, and the more tortuous the flow paths become. For this reason, a common way of implementing the tortuosity effect has consisted of assuming that it is directly proportional with some power of the size of the largest pores, that contain water, and thus of the water content of the soil (see Millington and Quirk, 1964; Mualem, 1976), say S_e^c, where c is an empirical constant. As noted, this assumption was already used earlier in several of the parallel models (Burdine, 1953; Brooks and Corey, 1966). Subsequently, however, it was observed (Brutsaert, 2000) that this assumption is incapable of producing agreement with experimental data. Rather, it was found necessary to assume that the tortuosity of a flow path through any given pore depends on the characteristic spatial scale of that specific pore, and not just on the scale of the largest pores. Actually, this assumption had already been used by Fatt and Dykstra (1951), in their parallel model, with the physical justification that liquid flowing through smaller pores travels a more tortuous path; accordingly, they assumed that the tortuosity is inversely proportional to a power of the pore size, say r^c, in which c is another constant to be determined experimentally.

It is straightforward to incorporate this assumption into the series-parallel model presented above, to adjust it for tortuosity and possibly other factors that may not be fully taken into account. Thus, in Equation (8.46) the power of z and y should be taken as $(2+c)$ instead of 2, so that instead of (8.48) one obtains

$$k' = (2\,\text{Ge})[(2\sigma/\gamma)\,\theta_0\,(1 - S_r)]^2 \int_0^{S_e} (S_e - x)\,[H(x)]^{-2-c}\,dx \qquad (8.51)$$

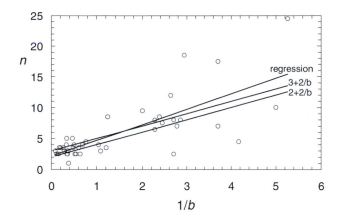

Fig. 8.31 **Dependence of n in Equation (8.36) on the exponent b in the power form of the soil-water retention relationship (8.14). The points are the experimental values from Mualem's (1978) data collection. The regression line is $n = 2.18 + 2.51/b$. Also shown are the lines obtained with Equations (8.44) and (8.49). (The parameter b tends to be smaller for finer-textured soils.)**

Example 8.7. Calculation with the power function

Again, this result can be readily integrated with (8.14) to yield

$$k' = \frac{2Ge[(2\sigma/\gamma)\theta_0(1 - S_r)b]^2}{(2b + c + 2)(b + c + 2)H_b^{2+c}} S_e^{2+(2+c)/b} \tag{8.52}$$

which is in the form of (8.36) with $n = 2 + (2 + c)/b$. A comparison with available experimental data for the relative permeability collected by Mualem (1978), revealed (Brutsaert, 2000) that a value of $c = 0.5$ produces good agreement, or

$$n = 2 + \frac{2.5}{b} \tag{8.53}$$

As illustrated in Figure 8.31, Equation (8.53) yields practically the same fit with the data as the regression relationship $n = 2.18 + 2.51/b$, with a correlation coefficient of $r = 0.75$.

8.4 FIELD EQUATIONS OF MASS AND MOMENTUM CONSERVATION

8.4.1 *Constant-density fluid in a rigid porous material*

Equation of continuity
In a porous medium, the infinitesimally small control volume, for which the continuity equation (1.8) is derived, consists of both pore space and solid matter. Therefore, the amount of fluid mass per unit volume is given by $(\rho_w\theta)$ in the case of water. Similarly, the mass flux per unit area of bulk porous material, comprising pores and solid matter, is given by $(\rho_w\mathbf{q})$, in which $\mathbf{q}$ (or q_i) is the specific flux as used in Darcy's law (8.19). Thus the equation of continuity (1.8), for a fluid with constant density but variable saturation,

becomes in the notation of porous media flow

$$\nabla \cdot \mathbf{q} = -\frac{\partial \theta}{\partial t} \qquad (8.54)$$

Conservation of mass and momentum: Richards's equation
Substitution of Darcy's law (8.19) in the equation of continuity (8.54) produces immediately

$$\nabla \cdot (k\nabla h) = \frac{\partial \theta}{\partial t} \qquad (8.55)$$

or, written out in full,

$$\frac{\partial}{\partial x}\left(k\frac{\partial h}{\partial x}\right) + \frac{\partial}{\partial y}\left(k\frac{\partial h}{\partial y}\right) + \frac{\partial}{\partial z}\left(k\frac{\partial h}{\partial z}\right) = \frac{\partial \theta}{\partial t} \qquad (8.56)$$

which is now usually referred to as the Richards (1931) equation. As such, Equation (8.55) is valid only for isotropic materials; it would be a straightforward exercise to extend the formulation to anisotropic materials. Under conditions of steady flow or under conditions of fully saturated flow, the right-hand side of (8.55) becomes zero. Under conditions of fully saturated flow in a uniform material, so that $\theta = \theta_0$ and $k = k_0 = $ constant, (8.55) reduces to the equation of Laplace, that is $\nabla^2 h = 0$, or written out in full

$$\frac{\partial^2 h}{\partial x^2} + \frac{\partial^2 h}{\partial y^2} + \frac{\partial^2 h}{\partial z^2} = 0 \qquad (8.57)$$

8.4.2 *General case of two immiscible fluids in an elastic porous material*

Biot (1941, 1955, 1956a, b) was probably the first to present a general theory of elasticity of a porous material saturated with an elastic fluid for the three-dimensional case with an arbitrary and variable load. This theory was subsequently (Brutsaert, 1964; Brutsaert and Luthin, 1964) extended to describe the elasticity of an unconsolidated granular material, containing two fluids in its interstices. Later Verruijt (1969) showed that Biot's theory for a saturated material can be simplified to describe groundwater movement in most cases of practical interest; he thus demonstrated that Biot's theory can often be reduced to Jacob's (1940) simple equation but also that in some cases the general theory is the only one that succeeds in explaining experimental results. In what follows, Verruijt's (1969) development is combined with Brutsaert's (1964) two-fluid extension of the theory to obtain a general formulation for unconfined and confined groundwater flow. Although the matter is straightforward, a careful exposition is desirable to bring out the significance of the underlying assumptions of the various more special groundwater equations used in the technical literature.

Strains

It is convenient to consider a fixed (Eulerian), infinitesimally small cubic element of a porous material, containing both water and air (or, more generally, a wetting fluid and a non-wetting fluid) in a Cartesian coordinate system. In this section the displacement vector of the solid part relative to its initial position is denoted by $\mathbf{u}(= u_x\mathbf{i} + u_y\mathbf{j} + u_z\mathbf{k})$. The corresponding displacement of the water $\mathbf{w}(= w_x\mathbf{i} + w_y\mathbf{j} + w_z\mathbf{k})$ and that of the air $\mathbf{v}(= v_x\mathbf{i} + v_y\mathbf{j} + v_z\mathbf{k})$

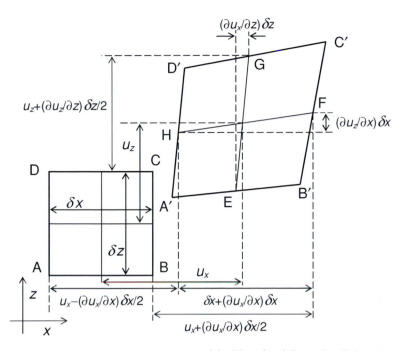

Fig. 8.32 Some of the displacements of a porous material subjected to deformation. Before the deformation the solid mass $\rho_s(1 - n_0)\delta\forall$ occupies the volume element $\delta\forall = (\delta x\,\delta y\,\delta z)$ at ABCD, and after the displacement this same solid mass has moved to A' B' C' D'. The center of the mass has moved from (x_0, y_0, z_0) to $(x_0 + u_x, y_0 + u_y, z_0 + u_z)$. The figure is shown in two-dimensions for clarity; the third coordinate y can be imagined as pointing out of the plane of the drawing.

are defined such that, when multiplied by the total or bulk cross-sectional area of porous material, they produce the displaced water and air volume, respectively.

The strain components of the solid are defined as $e_{xx} = \partial u_x/\partial x$, $e_{xy} = (\partial u_x/\partial y + \partial u_y/\partial x)/2$, etc. Their physical significance is illustrated in Figure 8.32. The position of the center of the cube shown in the figure is at (x_0, y_0, z_0) prior to the deformation, and the displacement components (u_x, u_y, u_z) refer to the displacement of the center of the cube. After the deformation the position of the point H is at

$$x_H = x_0 - \frac{\delta x}{2} + u_x - \frac{\partial u_x}{\partial x}\frac{\delta x}{2} + \frac{\partial^2 u_x}{\partial x^2}\left(\frac{\delta x}{2}\right)^2\frac{1}{2} - \cdots$$

and

$$z_H = z_0 + u_z - \frac{\partial u_z}{\partial x}\frac{\delta x}{2} + \frac{\partial^2 u_z}{\partial x^2}\left(\frac{\delta x}{2}\right)^2\frac{1}{2} - \cdots$$

The position of the point F is

$$x_F = x_0 + \frac{\delta x}{2} + u_x + \frac{\partial u_x}{\partial x}\frac{\delta x}{2} + \frac{\partial^2 u_x}{\partial x^2}\left(\frac{\delta x}{2}\right)^2\frac{1}{2} + \cdots$$

and

$$z_F = z_0 + u_z + \frac{\partial u_z}{\partial x}\frac{\delta x}{2} + \frac{\partial^2 u_z}{\partial x^2}\left(\frac{\delta x}{2}\right)^2\frac{1}{2} + \cdots$$

The normal strain is defined as the change in length of an element in a certain direction as a result of deformation, divided by the original length, in the limit for an infinitesimally small element. In the x-direction, the original length is δx and the deformed length $(x_F - x_H)$, which yields immediately $e_{xx} = \partial u_x / \partial x$; the same reasoning, *mutatis mutandis*, produces e_{zz} and e_{yy}. Note that the reasoning is similar to the derivation of Equations (1.5) and (1.6). The sum of the normal strains, which equals the fractional change in volume of the deformed cube of solid skeleton, is the volume strain, also called the dilatation, i.e.,

$$e = \nabla \cdot \mathbf{u} = e_{xx} + e_{yy} + e_{zz} \tag{8.58}$$

The shear strain is by definition one half the change in angle between two originally perpendicular elements, as deformation takes place, again in the limit for an infinitesimally small element. In the case of $A'B'C'D'$ the shear strain is one half the sum of the angle of HF with the x-axis and the angle of EG with the z-axis; the angle of HF with the x-axis is $(z_F - z_H)/\delta x$ and the angle of EG with the z-axis is $(x_G - x_E)/\delta z$, so that the (xz)-component of the shear strain is $e_{xz} = e_{zx} = (\partial u_x / \partial z + \partial u_z / \partial x)/2$; the two other shear strain components e_{xy} and e_{yz} are obtained in a similar way.

The relevant strains for the fluids are the changes in volume of fluid per unit bulk volume of porous material, that is, for the water,

$$e_w = \nabla \cdot \mathbf{w} = \frac{\partial w_x}{\partial x} + \frac{\partial w_y}{\partial y} + \frac{\partial w_z}{\partial z} \tag{8.59}$$

and similarly, the dilatation of the air,

$$e_a = \nabla \cdot \mathbf{v} = \frac{\partial v_x}{\partial x} + \frac{\partial v_y}{\partial y} + \frac{\partial v_z}{\partial z} \tag{8.60}$$

Because the fluid displacements represent fluid volume per unit bulk area of porous material, the corresponding changes in volume of fluid per unit volume of fluid are $e_w/(n_0 S)$ and $e_a/[n_0(1 - S)]$, respectively.

The displacements can readily be shown to satisfy the following equations of continuity, in accordance with Equation (1.8); namely, for the water:

$$\frac{\partial}{\partial t}(\rho_w n_0 S) = -\nabla \cdot \left(\rho_w \frac{\partial w}{\partial t} \right) \tag{8.61}$$

for the air,

$$\frac{\partial}{\partial t}[\rho_a n_0 (1 - S)] = -\nabla \cdot \left(\rho_a \frac{\partial v}{\partial t} \right) \tag{8.62}$$

and for the solid,

$$\frac{\partial n_0}{\partial t} = \nabla \cdot \left[(1 - n_0) \frac{\partial \mathbf{u}}{\partial t} \right] \tag{8.63}$$

in which ρ_w and ρ_a are the density of the water and of the air, respectively, n_0 is the porosity, and $S = \theta/n_0$ is the degree of saturation of the material with water, that is, the volume of water per unit volume of pore space and θ the volumetric water content. Observe that Equation (8.63) is based on the assumption that the density of the solid phase (namely, the grains but not the solid frame) is constant.

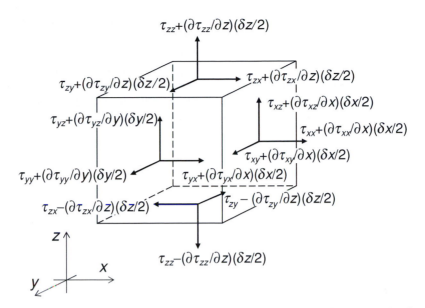

Fig. 8.33 Some of the effective or intergranular stress components acting on an elementary cube of porous material with dimensions $(\delta x \delta y \delta z)$, when the stresses at the center are $\tau_{xx}, \tau_{xy}, \tau_{xz}, \tau_{yx}, \tau_{yy}$, etc.; the first subscript indicates the direction of the surface on which the stress component acts, and the second subscript is the direction of the stress component itself.

Stresses

The total stress related to the aforementioned displacements, **u**, **w**, and **v**, consists of three parts, corresponding to the forces acting on each of three phases at a point in space. It can be written as (Biot, 1955; Brutsaert, 1964) $\tau_{xx} + \tau_{w} + \tau_{a}$, τ_{xy}, τ_{xz}, τ_{yx}, $\tau_{yy} + \tau_{w} + \tau_{a}$, τ_{yz}, τ_{zx}, τ_{zy}, and $\tau_{zz} + \tau_{w} + \tau_{a}$. The τ_{xx}, τ_{xy}, etc., components, namely those with two subscripts, represent the stress tensor acting on the solid part of the material, also known as the intergranular or effective stress. A few of the components of the effective stress tensor are illustrated in Figure 8.33, as they act on the sides of a small cubic element with volume $(\delta x \delta y \delta z)$. The first subscript of each effective stress component indicates the direction of the surface on which it acts, and the second subscript is the direction of the stress component itself. A component is taken as positive, when it acts in a positive direction on a positive surface, or in a negative direction on a negative surface. The part of the total stress tensor applied to the water is τ_{w}, given by

$$\tau_{w} = -\chi p_{w}' \tag{8.64}$$

where p_{w}' is the incremental fluid pressure over and above the initial hydrostatic pressure under equilibrium conditions prior to the displacement. This initial pressure is denoted by p_{wi} and the total pressure is $p_{w} = p_{wi} + p_{w}'$. The term χ is the effective stress function, a concept introduced for partly saturated soils by Bishop (1961) in 1955. It is generally accepted that χ equals zero when $S = 0$ and that it is close to unity when $S = 1$ at saturation, in accordance with Terzaghi's (1925, 1943) effective stress concept. It should be added that Terzaghi's concept (i.e. $\chi = 1$ for $S = 1$) is valid only for granular materials, if the grains are in point contact with each other, and if (see Bishop and Blight, 1963) the grains are incompressible. It would seem that besides the effect of partial occupancy of the pores by water, χ is also affected by the effective stress path, the pore geometries, the wetting angle, and probably other factors. The nature of χ has been the subject of uncertainty (see McMurdie and Day, 1960; Blight, 1967; Snyder and Miller, 1985). For the sake of simplicity, for small elastic displacements it has been assumed earlier (Brutsaert, 1964) that $\chi = S$. In general, it is probably safe to assume that χ is a unique function of S, i.e. $\chi = \chi(S)$. By applying Bishop's proposal to the air phase, one can write for the part of the total stress acting on the air

$$\tau_a = -(1 - \chi)p'_a \qquad (8.65)$$

where p'_a is the incremental air pressure over and above the initial pressure p_{ai}, existing prior to the displacement. Thus the total air pressure is $p_a = p_{ai} + p'_a$.

When the two immiscible fluids are in a "funicular" state, that is, consisting of connected filaments without isolated drops or occluded bubbles of entrapped gas, the total pressure p_w can be related to the total pressure p_a by the capillary pressure p_c, which is defined as follows:

$$p_w = p_a + p_c \qquad (8.66)$$

This capillary pressure equals the pressure decrease across the air–water interface considered in the derivation of Laplace's equation (8.3) via (8.4). Hysteresis (see Section 8.2.3) is always present. However, if the process in question involves only wetting, or only drying, so that hysteresis effects are avoided, $p_c = p_c(S, n_0)$ can be taken as a function only of the saturation and of the porosity.

The components of the stress tensor and the changes of the displacements of the three phases must satisfy the equations of motion (Brutsaert, 1964) or for slow displacements the simpler equilibrium equations (Verruijt, 1969). However, these are not needed in the present derivation.

Stress–strain relationship

If the solid strains and the changes in fluid content are small and if the processes involved are reversible, the stress components may in general be assumed to be linear functions of the strain components (Biot, 1941, 1955). This assumption yields a generalization of Hooke's law (Brutsaert, 1964) in the case of an isotropic porous material:

$$
\begin{aligned}
\tau_{xx} &= 2\mu e_{xx} + \lambda e + c_{sw}e_w + c_{sa}e_a \\
\tau_{yy} &= 2\mu e_{yy} + \lambda e + c_{sw}e_w + c_{sa}e_a \\
\tau_{zz} &= 2\mu e_{zz} + \lambda e + c_{sw}e_w + c_{sa}e_a \\
\tau_{xy} &= 2\mu e_{xy} \quad \tau_{xz} = 2\mu e_{xz} \quad \tau_{yz} = 2\mu e_{yz} \qquad (8.67) \\
\tau_w &= c_{sw}e + c_w e_w + c_{wa}e_a \\
\tau_a &= c_{sa}e + c_{wa}e_w + c_a e_a
\end{aligned}
$$

in which μ, λ, c_{sw}, c_{sa}, c_w, c_a, and c_{wa} are constants characterizing the elastic behavior of the material. These equations reduce to Biot's (1955) when the pores are saturated with one fluid and to Hooke's law for isotropic bodies if only the solid were present. Thus μ and λ represent the behavior of the solid. A fluid displacement does not result in a shear stress, only the rate of displacement does; therefore the fluid strains do not appear in the shear stress components. The coefficient c_w can be understood by considering the situation where wetting fluid is forced into the porous material, while the total volume of the other two is somehow kept constant, i.e. when e and e_a are kept zero. This means that owing to (8.67),

$$c_w = \chi K_w / n_0 S \tag{8.68}$$

where $K_w = \beta_w^{-1}$ is the bulk modulus of the wetting fluid and β_w its compressibility. Similarly, for the non-wetting fluid

$$c_a = (1 - \chi) K_a / n_0 (1 - S) \tag{8.69}$$

where K_a is the bulk modulus of the non-wetting fluid. The coefficients c_{sw}, c_{sa}, and c_{wa} indicate that, in principle at least, there should be a coupling between the volume changes of the three constituents as indicated by the subscripts. However, as is now shown, these coefficients are equal to zero if the density of the solid grains can be assumed to be constant. To see this, eliminate the fluid strains from Equation (8.67) to obtain an alternative form of the generalized Hooke law, namely,

$$
\begin{aligned}
\tau_{xx} &= 2\mu e_{xx} + (\lambda + c_1)e + c_2 \tau_w + c_3 \tau_a \\
\tau_{yy} &= 2\mu e_{yy} + (\lambda + c_1)e + c_2 \tau_w + c_3 \tau_a \\
\tau_{zz} &= 2\mu e_{zz} + (\lambda + c_1)e + c_2 \tau_w + c_3 \tau_a
\end{aligned} \tag{8.70}
$$

together with the fourth, fifth, and sixth of (8.67). The new constants are

$$c_1 = \left(2c_{sw}c_{sa}c_{wa} - c_{sw}^2 c_a - c_{sa}^2 c_w\right)c, \quad c_2 = (c_{sw}c_a - c_{sa}c_{wa})/c, \quad c_3 = (c_{sa}c_w - c_{sw}c_{wa})/c,$$
$$\text{and } c = \left(c_w c_a - c_{wa}^2\right)$$

Consider now a thought experiment in which τ_{xx}, the effective stress, is held constant but the fluid pressures are increased. In the case of a porous material containing only one fluid, this can be accomplished by placing an unjacketed saturated sample in the fluid and then increasing the fluid pressure. In the case of material containing air and water, this can be accomplished by increasing p_w and p_a in such a way that their difference p_c remains constant. It is clear that if the solid grain density is constant, this process does not increase the effective stresses, nor does it result in any solid displacement. Thus Equation (8.70) shows that $c_2 = c_3 = 0$. In other words, if the solid material (not the solid frame) is incompressible, one has also $c_{sw} = c_{sa} = 0$, which immediately results in $c_1 = 0$ as well. Moreover, if this is the case, one can relate μ and λ to the bulk modulus of the solid frame, as follows:

$$K_s = \frac{2}{3}\mu + \lambda \tag{8.71}$$

Stress versus rate of strain relationships for fluids

As noted in Section 8.3.1, Darcy's law represents the equation of creeping motion in porous material. When the motion takes place within the pores while the porous material itself is being subjected to deformation, the Darcy flux must be taken as the relative motion between the solid matrix and the fluids. Apparently (Verruijt, 1969), this concept was first proposed for liquid saturated media by Gersevanov around 1934. Biot did not use it in his original paper (Biot, 1941), but he introduced it in his generalized theory of elasticity (Biot, 1955). The displacement vector of the solid is an actual displacement length, whereas the displacement vectors of the two fluids are defined herein as volumes when multiplied by the total bulk cross-sectional area normal to them. Thus the relative velocity of the wetting fluid with respect to the solid is $\{[(\partial\mathbf{w}/\partial t)/(n_0 S)] - (\partial\mathbf{u}/\partial t)\}$ and an analogous expression for the nonwetting fluid, so that the respective Darcy fluxes are given by $[(\partial\mathbf{w}/\partial t) - n_0 S(\partial\mathbf{u}/\partial t)]$ and $[(\partial\mathbf{v}/\partial t) - n_0(1-S)(\partial\mathbf{u}/\partial t)]$.

When there are two fluids within the same pores, it is clear that the relative motion between the fluids may give rise to an additional "head" loss. This means that Darcy's law, which expresses a linear relationship between flux and the gradient of pressure and body force, can be written in a general form as follows:

$$-\nabla h_\mathrm{w} = \frac{\mu_\mathrm{w} n_0 S}{\gamma_\mathrm{w}} \left\{ \frac{1}{k_\mathrm{w}} \left[\frac{1}{n_0 S} \frac{\partial\mathbf{w}}{\partial t} - \frac{\partial\mathbf{u}}{\partial t} \right] + \frac{1}{k_\mathrm{wa}} \left[\frac{1}{n_0 S} \frac{\partial\mathbf{w}}{\partial t} - \frac{1}{n_0(1-S)} \frac{\partial\mathbf{v}}{\partial t} \right] \right\}$$

$$-\nabla h_\mathrm{a} = \frac{\mu_\mathrm{a} n_0(1-S)}{\gamma_\mathrm{a}} \left\{ \frac{1}{k_\mathrm{a}} \left[\frac{1}{n_0(1-S)} \frac{\partial\mathbf{v}}{\partial t} - \frac{\partial\mathbf{u}}{\partial t} \right] + \frac{1}{k_\mathrm{wa}} \left[\frac{1}{n_0(1-S)} \frac{\partial\mathbf{v}}{\partial t} - \frac{1}{n_0 S} \frac{\partial\mathbf{w}}{\partial t} \right] \right\} \tag{8.72}$$

where k is the (intrinsic) permeability and μ is the Newtonian viscosity. (Note that in what follows in this section for convenience of notation the prime symbol is omitted from the permeability terms k_w, k_a, and k_wa.) Recall that the total pressure is the sum of the initial and the incremental pressures, or $p_\mathrm{w} = p_\mathrm{wi} + p'_\mathrm{w}$. Since the initial pressure is hydrostatic, so that $\nabla z + (\nabla p_\mathrm{wi}/\gamma_\mathrm{w}) = 0$, it follows that

$$\nabla h_\mathrm{w} \left(\equiv \nabla z + \frac{1}{\gamma_\mathrm{w}} \nabla p_\mathrm{w} \right) = \frac{1}{\gamma_\mathrm{w}} \nabla p'_\mathrm{w} \tag{8.73}$$

where z is the vertical coordinate and $\gamma_\mathrm{w} = \rho_\mathrm{w} g$ the specific weight; the quantity h_w is defined in Equation (8.21). The subscript w refers to the wetting fluid; the same quantities, but with the subscript a, refer to the nonwetting phase. The cross-permeability term k_wa arises from the relative motion between the two fluids. For most practical problems, the effect of this relative motion is probably negligible. It is conceivable, however, that it becomes important under conditions of counterflow; this would be the case, for example, of water infiltration into a soil profile in which the displaced air is being prevented from escaping downward so that it bubbles upward while the wetting front moves down. The possibility of momentum exchange as a result of this relative motion has been considered already by Yuster (1951) and Scott and Rose (1953). Mainly because it is practically impossible to determine experimentally at present and probably small, it is omitted in what follows. Then Equations (8.72) become

$$\frac{\partial\mathbf{w}}{\partial t} - n_0 S \frac{\partial\mathbf{u}}{\partial t} = -\frac{k_\mathrm{w}}{\mu_\mathrm{w}} \nabla p'_\mathrm{w}$$

$$\frac{\partial\mathbf{v}}{\partial t} - n_0(1-S) \frac{\partial\mathbf{u}}{\partial t} = -\frac{k_\mathrm{a}}{\mu_\mathrm{a}} \nabla p'_\mathrm{a} \tag{8.74}$$

The equations presented so far form a complete system. They can be combined to eliminate certain less useful variables and to leave only those pertinent to most practical problems. One way of accomplishing this is to consider the solid displacements and the fluid pressures. Together with the porosity and the degree of saturation, these are seven variables: $\mathbf{u}$, p'_w, p'_a, n_0, and S. If it is assumed that the porous matrix is homogeneous and inert, so that μ and λ are constant and independent of S and thus of space and time (this would not be the case, for example, in a clay–water–air system), substitution of (8.67) or (8.70) into the well-known equilibrium equations (for incompressible grains) yields

$$\mu \nabla^2 \mathbf{u} + \nabla[(\mu + \lambda)e - \chi p'_\mathrm{w} - (1-\chi)p'_\mathrm{a}] = 0 \tag{8.75}$$

The fluid displacement terms in Darcy's law can be eliminated by multiplying each of (8.74) first by ρ_w and ρ_a, respectively, and then taking the divergence, after which substitution of the continuity equations (8.61) and (8.62) gives

$$\frac{\partial}{\partial t}\left[\rho_w n_0 S\right] + \nabla \cdot \left[\rho_w n_0 S \frac{\partial \mathbf{u}}{\partial t}\right] = \nabla \cdot \left(\frac{k_w \rho_w}{\mu_w} \nabla p'_w\right)$$

$$\frac{\partial}{\partial t}\left[\rho_a n_0 (1-S)\right] + \nabla \cdot \left[\rho_a n_0 (1-S) \frac{\partial \mathbf{u}}{\partial t}\right] = \nabla \cdot \left(\frac{k_a \rho_a}{\mu_a} \nabla p'_a\right)$$

(8.76)

Equations (8.75) and (8.76), together with solid continuity (8.63), form a closed system. Hence it should be possible to solve it for any consolidation or flow problem involving an elastic porous material occupied by two elastic fluids. Fortunately, for many problems this formulation is more general than necessary (see, however, Verruijt (1969)), and it is often possible to simplify it considerably as follows.

Simple case of constant vertical load

In soil mechanics and groundwater hydraulics the problem formulation is commonly simplified by two basic assumptions that were introduced by Terzaghi (1925, 1943) and Jacob (1940, 1950). First, compression is assumed to be strictly vertical without any horizontal solid displacements, and second, any changes in vertical compressive effective stress are balanced by equal and opposite changes in fluid stresses. These assumptions may be difficult to justify, but the resulting formulation has been used quite successfully in the solution of many problems in porous media saturated with one fluid. This suggests that the concept may also be valid in the simplification of certain problems involving two immiscible fluids. In the present notation the first assumption can be written as $u_x = u_y = 0$, so that $e_{zz} = e$ and the third of Equations (8.67) or (8.70) (for incompressible grains) yields

$$\tau_{zz} = (2\mu + \lambda)\,e$$

(8.77)

The second assumption can be written as $\tau_{zz} = -(\tau_w + \tau_a)$, which yields with (8.64) and (8.65)

$$\tau_{zz} = \chi p'_w + (1 - \chi)p'_a$$

(8.78)

Note as an aside, that Equation (8.78) agrees with the general observation that soils that are close to saturation do not easily disintegrate, but exhibit a certain degree of consistency and coherence. Indeed, if the effect of the air pressure can be neglected, in a soil that is close to saturation, the water pressure p'_w is negative and the effective stress factor χ is close to unity; hence the intergranular stress τ_{zz} is also negative, which means that the soil grains are drawn together and the soil exhibits a greater firmness. This effect can also be seen, for example, on a sandy beach just after the sea water has withdrawn during ebb tide; at that time the sand forms a harder surface than when it is submerged, with $p'_w > 0$, or than when it is totally dry.

Combining Equations (8.77) and (8.78) one obtains, instead of (8.75), simply

$$(2\mu + \lambda)\,e = \chi p'_w + (1 - \chi)\,p'_a$$

(8.79)

Substitution of (8.63) into the first of (8.76) yields

$$\rho_w S \frac{\partial e}{\partial t} + n_0 S \rho_w \beta_w \frac{\partial p'_w}{\partial t} + n_0 \rho_w \frac{\partial S}{\partial t} = \nabla \cdot \left(\frac{k_w \rho_w}{\mu_w} \nabla p'_w\right)$$

(8.80)

in which $\beta_w [= (\partial \rho_w/\partial t)/(\rho_w \partial p'_w/\partial t)]$ is assumed to be a constant, as a measure of the compressibility of the water, and in which $(\partial \mathbf{u}/\partial t) \cdot \nabla(\rho_w S)$ is assumed to be negligible on account of the small solid velocity. Similarly, the second of (8.76) reduces to

$$\rho_a (1 - S) \frac{\partial e}{\partial t} + n_0(1 - S)\rho_a \beta_a \frac{\partial p'_a}{\partial t} - n_0 \rho_a \frac{\partial S}{\partial t} = \nabla \cdot \left(\frac{k_a \rho_a}{\mu_a} \nabla p'_a \right) \tag{8.81}$$

in which now $(\partial \mathbf{u}/\partial t) \cdot \nabla [\rho_a(1 - S)]$ is assumed negligible. The combination of (8.79) with (8.80) and (8.81) yields the following diffusion-type equations:

$$\rho_w S\alpha \frac{\partial}{\partial t}(\chi p'_w) + \rho_w S n_0 \beta_w \frac{\partial p'_w}{\partial t} + \rho_w S\alpha \frac{\partial}{\partial t}[(1 - \chi)\, p'_a] + n_0 \rho_w \frac{\partial S}{\partial t}$$

$$= \nabla \cdot \left(\frac{k_w \rho_w}{\mu_w} \nabla p'_w \right)$$

$$\rho_a(1 - S)\alpha \frac{\partial}{\partial t}[(1 - \chi)\, p'_a] + \rho_a(1 - S)n_0 \beta_a \frac{\partial p'_a}{\partial t} + \rho_a(1 - S)\alpha \frac{\partial}{\partial t}(\chi p'_w) - n_0 \rho_a \frac{\partial S}{\partial t}$$

$$= \nabla \cdot \left(\frac{k_a \rho_a}{\mu_a} \nabla p'_a \right) \tag{8.82}$$

in which, on account of (8.77), α can be defined as the vertical compressibility of the solid frame

$$\alpha = (2\mu + \lambda)^{-1} \tag{8.83}$$

Note that this is different from the volumetric compressibility $K_s^{-1} [\equiv 3e/(\tau_{xx} + \tau_{yy} + \tau_{zz})]$, that is, the inverse of Equation (8.71).

The physical significance of the terms in Equations (8.82) can be explained as follows. The entire left-hand side of both equations represents the local rate of change of storage of the fluid in question at a point. In the case of the first of (8.82), which describes the flow of the water (i.e. the wetting fluid), the first term on the left is the rate of change of storage resulting from compression (or expansion) of the solid matrix caused by pressure changes in the water; the second term represents the storage rate of change caused by compression (or expansion) of the water; the third term is the rate of change of storage resulting from bulk volume changes of the solid matrix caused by pressure changes in the air. The fourth term shows the rate of change of water storage resulting from local changes of the degree of saturation. Finally, the right-hand side, which is a divergence of the Darcy flux, represents the storage rate of change as the difference between the inflow and outflow rate of water at the point in question. The different terms of the second of (8.82), which describes the flow of air, represent, *mutatis mutandis*, the same mechanisms as those in the first.

Before proceeding, for a better understanding of Equations (8.82) and their limitations, the basic assumptions may be briefly repeated.

1. The grains are incompressible.
2. The effective stress is obtainable by means of Bishop's parameter $\chi = \chi(S)$.
3. The solid displacements are sufficiently small, so that the solid frame is elastic within the range of $\mathbf{u}$; when this is not the case, this can sometimes be remedied. For example, as shown in Brutsaert and Corapcioglu (1976) the basic derivation leading to Equation (8.84) can be readily extended to flow in a viscoelastic aquifer.
4. The solid displacements are vertical only, with a constant total load. Verruijt (1969) has described situations of saturated flow when this assumption is not valid.

5.　The fluid compressibility β is defined in Equation (8.80) in terms of partial time derivatives.

6.　The terms $(\partial \mathbf{u}/\partial t) \cdot \nabla (\rho_w S)$ and $(\partial \mathbf{u}/\partial t) \cdot \nabla [\rho_a(1 - S)]$ are negligible.

Equations (8.82) are the main result of this section. In principle it should be possible to solve them for appropriate boundary conditions, provided the values of the parameters are known. However, for many common situations they are more general than necessary, and they can be simplified considerably.

Some special cases

(i)　Flow of one partially saturating, elastic fluid in an elastic porous material. Whenever the air pressure can be assumed to be constant, the second equation of (8.82) becomes irrelevant; the first equation of (8.82) can then be written as

$$\rho_w S\alpha \frac{\partial}{\partial t}(\chi_w p'_w) + \rho_w Sn_0\beta_w \frac{\partial p'_w}{\partial t} + n_0\rho_w \frac{\partial S}{\partial t} = \nabla \cdot \left(\frac{k_w \rho_w}{\mu_w} \nabla p'_w \right) \tag{8.84}$$

This equation was examined in Brutsaert and El-Kadi (1984) to study the relative effects of partial saturation and compressibility on the flow in unconfined systems. It may be noted that in the groundwater literature, various derivations have been presented; these have yielded results somewhat different from (8.84), mainly in the first term on the left. Some reasons for the discrepancies between these other equations and (8.84) stem from the neglect of the relative velocity in Darcy's law (8.72) and of the equation of continuity of the solid (8.63). The latter assumption is especially serious, since (8.63) involves the compression of the solid, which in turn gives rise to the compressibility α. Other differences result from the use of the total pressure p_w, rather than p'_w, as is done here, and also from the neglect of χ_w.

(ii)　Flow of one elastic fluid in an elastic porous material. This case is the one to which the theory of Biot (1941; 1955) is applicable. Because the pores are filled with one fluid only, one has $S = 1.0$, and $\chi_w = 1.0$; this reduces Equation (8.84) immediately to

$$\rho_w(\alpha + n_0\beta_w)\frac{\partial p'_w}{\partial t} = \nabla \cdot \left(\frac{\rho_w k'}{\mu_w} \nabla p'_w \right) \tag{8.85}$$

in which the symbol k_w has been replaced by the more common k' for the permeability. If now also the hydraulic conductivity $k = (\rho_w g k'/\mu_w)$ is assumed to be constant, (8.85) assumes the well-known linear form:

$$S_s \frac{\partial p'_w}{\partial t} = k\nabla^2 p'_w \tag{8.86}$$

where $S_s = \rho_w g[n_0\beta_w + (2\mu + \lambda)^{-1}]$. This form is the same as that of the equations describing heat conduction and diffusion (cf. also Equations (5.88) and (5.92)). Equation (8.86), but with various expressions for S_s, has been applied widely in the description of soil consolidation and of flow in confined aquifers. It was proposed for one-dimensional consolidation by Terzaghi (1925), who later (1943) extended it to three dimensions. Independently, Theis (1935) adopted the heat flow equation to analyze horizontal unsteady flow in an elastic artesian aquifer, but he justified it only on the basis of heuristic arguments concerning the analogy between Fourier's law and Darcy's law. But it was Jacob (1940; 1950) who derived this heat flow equation for confined aquifer flow in terms of the physical properties of the aquifer and the fluid. Subsequently, however, the exact derivation of Equation (8.86) remained the subject of controversy, until Verruijt (1969) showed how it can be reconciled with Biot's (1941; 1955) analysis.

In addition to the limitations inherent in Equation (8.84), the Terzaghi–Jacob equation (8.86) is also restricted by the assumption of a constant hydraulic conductivity. As shown earlier in this chapter, the permeability k' is dependent on the porosity n_0; for instance, in Equations (8.48) and (8.51) k' is proportional to n_0^2. Thus, in deriving (8.86) from (8.85), it is assumed that n_0 and $\rho_w = \rho_w(p_w)$ are constant, even though they are in fact unknown dependent variables. In other words, it is assumed that the fluid and the porous matrix are compressible on the left side of (8.85) but not on the right. This assumption may have its limitations whenever $\nabla p'_w$ is not small. Still, in spite of this inconsistency, for most problems the exact formulation of (8.86) in terms of physical properties of the porous material and the fluid is probably not very crucial, since (S_s/k) is usually determined from field experiments. Consequently, as irrotational or unidirectional displacement and the constancy of k' and ρ_w may be difficult to justify, the main problem is not how to express S_s in terms of n_0, β_w, μ, λ, etc., at the micro- or Darcy scale, but rather whether the heat flow equation (8.86) is adequate to solve the practical problem at hand at the larger scale of the aquifer.

(iii)   Flow of two immiscible fluids in an incompressible porous material. In this special case, the solid phase cannot move, so that the displacement $\mathbf{u}$ and the rate of displacement $(\partial \mathbf{u}/\partial t)$ of the solid aggregate are equal to zero; therefore Equations (8.76) can be written as:

$$n_0 \frac{\partial}{\partial t}(\rho_w S) = \nabla \cdot \left(\frac{\rho_w k_w}{\mu_w} \nabla p'_w \right) \tag{8.87}$$

and

$$n_0 \frac{\partial}{\partial t} (\rho_a (1 - S)) = \nabla \cdot \left(\frac{\rho_a k_a}{\mu_a} \nabla p'_a \right) \tag{8.88}$$

These equations are equivalent to those first proposed by Muskat and Meres (1936) but taking account of the solubility of the non-wetting fluid in the wetting fluid. Equations (8.88) have been used to study the effect of soil air movement on the infiltration of water (Le Van Phuc and Morel-Seytoux, 1972; Morel-Seytoux, 1973).

Whenever the effect of the non-wetting fluid is negligible, owing to small viscosity μ_a and small pressure changes $\nabla p'_a$, only the wetting fluid is of interest; if the density of this fluid ρ_w is constant, (8.87) reduces to

$$n_0 \frac{\partial S}{\partial t} = \nabla \cdot (k \nabla p'_w)/\gamma_w \tag{8.89}$$

which is equivalent to Equation (8.55), first derived by Richards (1931) for soil water movement.

REFERENCES

Arbhabhirama, A. and Dinoy, A. A. (1973). Friction factor and Reynolds number in porous media flow. *J. Hydraul. Div., Proc. ASCE*, **99**, 901–911.

Baker, F. G., Veneman, P. L. M. and Bouma, J. (1974). Limitations of the instantaneous profile method for field measurement of unsaturated hydraulic conductivity. *Soil Sci. Soc. Amer. Proc.*, **38**, 885–888.

Biot, M. A. (1941). General theory of three-dimensional consolidation. *J. Appl. Phys.*, **12**, 155–164.

(1955). Theory of elasticity and consolidation for a porous anisotropic solid. *J. Appl. Phys.*, **26**, 182–185.

(1956a). General solutions of the equations of elasticity and consolidation for a porous material. *J. Appl. Mech.*, **23**, 91–96.

(1956b). Theory of propagation of elastic waves in a fluid-saturated porous solid, I, Low-frequency range. *J. Acoust. Soc. Amer.*, **28**, 168–178.

Bishop, A. W. (1961). The measurement of pore pressure in the triaxial test. In *Pore Pressure and Suction in Soils*. London: Butterworths, pp. 38–46.

Bishop, A. W. and Blight, G. E. (1963). Some aspects of effective stress in saturated and partly saturated soils. *Geotechnique*, **13**, 177–197.

Blight, G. E. (1967). Effective stress evaluation for unsaturated soils. *J. Soil Mech. Found. Div., Proc. ASCE*, **93** (SM2), 125–148.

Boussinesq, J. (1868). Mémoire sur l'influence des frottements dans les mouvements réguliers des fluides. *J. de Mathématiques Pures et Appliquées, 2me Série*, **13**, 377–424.

(1914). Sur la vitesse moyenne ou le débit et la vitesse maximum ou axiale, dans un tube prismatique, a section régulière d'un nombre quelconque m de côtés. *Compt. Rend. Hebdomadaires des Séances de l'Académie des Sciences, Paris*, **158**, 1846–1850.

Braddock, R. D., Parlange, J.-Y. and Lee, H. (2001). Application of a soil water hysteresis model to simple water retention curves. *Transp. Porous Media*, **44**, 407–420.

Bradfield, R. and Jamison, V. C. (1938). Soil structure – attempts at its quantitative characterization. *Soil Sci. Soc. Amer. Proc.*, **3**, 70–76.

Brooks, R. H. and Corey, A. T. (1966). Properties of porous media affecting fluid flow. *J. Irrig. Drain. Div., Proc. ASCE*, **92**, 62–88.

Brutsaert, W. (1964). The propagation of elastic waves in unconsolidated unsaturated granular mediums. *J. Geophys. Res.*, **69**, 243–257.

(1966). Probability laws for pore-size distributions. *Soil Sci.*, **101**, 85–92.

(1967). Some methods of calculating unsaturated permeability. *Trans. Amer. Soc. Agr. Engrs.*, **10**, 400–404.

(1968a). The permeability of a porous medium determined from certain probability laws for pore size distribution. *Water Resour. Res.*, **4**, 425–434.

(1968b). A solution for vertical infiltration into a dry porous medium. *Water Resour. Res.*, **4**, 1031–1038.

(1979). Universal constants for scaling the exponential soil water diffusivity? *Water Resour. Res.*, **15**, 481–483.

(2000). A concise parameterization of the hydraulic conductivity of unsaturated soils. *Adv. Water Resour.*, **23**, 811–815.

Brutsaert, W. and Corapcioglu, M. Y. (1976). Pumping of aquifer with viscoelastic properties. *J. Hydraul. Div., Proc. ASCE*, **102** (HY11), 1663–1675.

Brutsaert, W. and El-Kadi, A. I. (1984). The relative importance of compressibility and partial saturation in unconfined groundwater flow. *Water Resour. Res.*, **20**, 400–408.

Brutsaert, W. and Lopez, J. P. (1998). Basin-scale geohydrologic drought flow features of riparian aquifers in the southern Great Plains. *Water Resour. Res.*, **34**, 233–240.

Brutsaert, W. and Luthin, J. N. (1964). The velocity of sound in soils near the surface, as a function of moisture content. *J. Geophys. Res.*, **69**, 643–652.

Buckingham, E. (1907). Studies on the movement of soil moisture, *Bureau of Soils, Bull. No. 38.* Washington, DC: US Department of Agriculture.

Burdine, N. T. (1953). Relative permeability calculations from pore-size distribution data. *Trans. Amer. Inst. Min. Engrs.*, **198**, 71–78.

Cahill, A. T. and Parlange, M. B. (1998). On water vapor transport in field soils. *Water Resour. Res.*, **34**, 731–739.

Carman, P. C. (1937). Fluid flow through granular beds. *Trans. Inst. Chem. Engrs. Lond.*, **15**, 150–166.
   (1956). *Flow of Gases Through Porous Media.* New York: Academic Press Inc.

Childs, E. C. (1942). Stability of clay soils. *Soil Sci.*, **53**, 79–92.

Childs, E. C. and Collis-George, N. (1950). The permeability of porous materials. *Proc. R. Soc. London*, A**201**, 392–405.

Clothier, B. E. and White, I. (1981). Measurement of sorptivity and soil water diffusivity in the field. *Soil Sci. Soc. Amer. J.*, **45**, 241–245.
   (1982). Water diffusivity of a field soil. *Soil Sci. Soc. Amer. J.*, **46**, 155–158.

Darcy, H. (1856). *Les fontaines publiques de la ville de Dijon.* Paris: Victor Dalmont.

Davidson, J. M., Stone, L. R., Nielsen, D. R. and LaRue, M. E. (1969). Field measurement and use of soil-water properties. *Water Resour. Res.*, **5**, 1312–1321.

DeVries, D. A. (1958). Simultaneous transfer of heat and moisture in porous media. *Trans. Amer. Geophys. Un.*, **39**, 909–916.

DeVries, D. A. and Philip, J. R. (1986). Soil heat flux, thermal conductivity and the null alignment method. *Soil Sci. Soc. Amer. J.*, **50**, 12–18.

Donat, J. (1937). Das Gefüge des Bodens und dessen Kennzeichung. *Trans. Intern. Congr. Soil Sci., 6th Congr., Paris, B*, pp. 423–439.

El-Kadi, A. I. and Brutsaert, W. (1985). Applicability of effective parameters for unsteady flow in nonuniform aquifers. *Water Resour. Res.*, **21**, 183–198.

Everett, D. H. (1954). A general approach to hysteresis, 3. *Trans. Faraday Soc.*, **50**, 1077–1096.
   (1955). A general approach to hysteresis, 4. *Trans. Faraday Soc.*, **51**, 1551–1557.

Fatt, I. and Dykstra, H. (1951). Relative permeability studies. *Trans. Amer. Inst. Min. Engrs.*, **192**, 249–255.

Feng, C. L. and Browning, G. M. (1946). Aggregate stability in relation to pore size distribution. *Soil Sci. Soc. Amer. Proc.*, **11**, 67–73.

Forchheimer, Ph. (1930). *Hydraulik*, 3. Aufl. Leipzig & Berlin: B. G. Teubner.

Gardner, W. R. (1958). Some steady-state solutions of the unsaturated moisture flow equation with application to evaporation from a water table. *Soil Sci.*, **85**, 228–232.

Gardner, W. R. and Mayhugh, M. S. (1958). Solutions and tests of the diffusion equation for the movement of water in soil. *Soil Sci. Soc. Amer. Proc.*, **22**, 197–201.

Gardner, W. R. and Miklich, F. J. (1962). Unsaturated conductivity and diffusivity measurements by a constant flux method. *Soil Sci.*, **93**, 271–274.

Gates, J. I. and Tempelaar Lietz, W. (1950). Relative permeabilities of California cores by the capillary-pressure method. *Drilling and Production Practice*, American Petroleum Institute, pp. 285–298.

Graetz, L. (1880). Über die Bewegung von Flüssigkeiten in Röhren. *Z. für Mathematik u. Physik*, **25**, 316–334.

Haines, W. B. (1930). Studies in the physical properties of soils: 5. The hysteresis effect in capillary properties and the modes of moisture distribution associated therewith. *J. Agr. Sci.*, **20**, 97–116.

Hoeksema, R. J. and Kitanidis, P. K. (1985). Analysis of spatial structure of properties of selected aquifers. *Water Resour. Res.*, **21**, 563–572.

Ibrahim, H. A. and Brutsaert, W. (1968). Intermittent infiltration into soils with hysteresis. *J. Hydraul. Div., Proc. ASCE*, **94**, 113–137.

Irmay, S. (1954). On the hydraulic conductivity of unsaturated soils. *Trans. Amer. Geophys. Un.*, **35**, 463–467.

Jackson, R. D., Reginato, R. J. and VanBavel, C. H. M. (1965). Comparison of measured and calculated hydraulic conductivities of unsaturated soils. *Water Resour. Res.*, **1**, 375–380.

Jackson, R. D., Reginato, R. J., Kimball, B. A. and Nakayama, F. S. (1974). Diurnal soil-water evaporation: comparison of measured and calculated soil-water fluxes. *Soil Sci. Soc. Amer. Proc.*, **38**, 861–866.

Jacob, C. E. (1940). The flow of water in an elastic artesian aquifer. *Eos Trans. AGU*, **21**, 574–586.
    (1950). Flow of ground water. In *Engineering Hydraulics*, ed. H. Rouse. New York: John Willey, pp. 321–386.

Katul, G. G., Wendroth, O., Parlange, M. B., Puente, C. E., Folegatti, M. V. and Nielsen, D. R. (1993). Estimation of in situ hydraulic conductivity function from nonlinear filtering theory. *Water Resour. Res.*, **29**, 1063–1070.

Kimball, B. A., Jackson, R. D., Nakayama, F. S. and Idso, S. B. (1976). Comparison of field-measured and calculated soil-heat fluxes. *Soil Sci. Soc. Amer. J.*, **40**, 18–25.

Klute, A. (1952). A numerical method for solving the flow equation for water in unsaturated materials. *Soil Sci.*, **73**, 105–116.

Kozeny, J. (1927). Über kapillare Leitung des Wassers im Boden, *Sitzungsberichte. Akad. der Wissensch. Wien, Math.-Naturw. Klass. Abt. 2a*, **136**, 271–306.

Lamb, H. (1932). *Hydrodynamics*, sixth edition. New York: Cambridge University Press.

Le Van Phuc and Morel-Seytoux, H. J. (1972). Effect of soil air movement and compressibility on infiltration rates. *Soil Soc. Amer. Proc.*, **36**, 237–241.

Leamer, R. W. and Lutz, J. F. (1940). Determination of pore-size distribution in soils. *Soil Sci.*, **49**, 347–360.

Libardi, P. L., Reichardt, K., Nielsen, D. R. and Biggar, J. W. (1980). Simple field methods for estimating soil hydraulic conductivity. *Soil Sci. Soc. Amer. J.*, **44**, 3–7.

Marshall, T. J. (1958). A relation between permeability and size distribution of pores. *J. Soil Sci.*, **9**, 1–8.

McMurdie, J. L. and Day, P. R. (1960). Slow tests under soil moisture suction. *Soil Sci. Soc. Amer. Proc.*, **24**, 441–444.

Miller, R. D. and Bresler, E. (1977). A quick method for estimating soil water diffusivity functions. *Soil Sci. Soc. Amer. Proc.*, **41**, 1021–1022.

Miller, R. J. and Low, P. F. (1963). Threshold gradient for water flow in clay systems. *Soil Sci. Soc. Amer. Proc.*, **27**, 605–609.

Millington, R. J. and Quirk, J. P. (1964). Formation factor and permeability equations. *Nature*, **202**, 143–145.

Milly, P. C. D. (1984). A simulation analysis of thermal effects on evaporation from soil. *Water Resour. Res.*, **20**, 1087–1098.

Morel-Seytoux, H. J. (1973). Two-phase flows in porous media. *Advan. Hydrosci.*, **9**, 119–202.

Mualem, Y. (1976). A new model for predicting the hydraulic conductivity of unsaturated porous media. *Water Resour. Res.*, **12**, 513–522.
    (1978). Hydraulic conductivity of unsaturated porous media: generalized macroscopic approach. *Water Resour. Res.*, **14**, 325–334.

Mualem, Y. and Miller, E. E. (1979). A hysteresis model based on an explicit domain-dependence function. *Soil Sci. Soc. Amer. J.*, **43**, 1067–1073.

Muskat, M. and Meres, M. W. (1936). The flow of heterogeneous fluids through porous media. *Physics*, **7**, 346–363.

Nakano, M. and Miyazaki, T. (1979). The diffusion and nonequilibrium thermodynamic equations of water vapor in soils under temperature gradient. *Soil Sci.*, **128**, 184–188.

Néel, L. (1942). Théorie des lois d'aimantation de Lord Rayleigh, 1. *Cah. Phys.*, **12**, 1–20.

(1943). Théorie des lois d'aimantation de Lord Rayleigh, 2. *Cah. Phys.*, **13**, 19–30.

Neuzil, C. E. (1986). Groundwater flow in low-permeability environments. *Water Resour. Res.*, **22**, 1163–1195.

Nielsen, D. R. and Biggar, J. W. (1959). *Measuring capillary conductivity*. Annual Report, Dept. Irrigation, University of California Davis. (See also (1961). *Soil Sci.*, **92**, 192–193.)

Nielsen, D. R., Biggar, J. W. and Erh, K. T. (1973). Spatial variability of field-measured soil–water properties. *Hilgardia*, **42**, 215–259.

Nielsen, D. R., Kirkham, D. and Perrier, E. R. (1960). Soil capillary conductivity: Comparison of measured and calculated values. *Soil Sci. Soc. Amer. Proc.*, **24**, 157–160.

Ogata, G. and Richards, L. A. (1957). Water content changes following irrigation of bare-field soil that is protected from evaporation. *Soil Sci. Soc. Amer. Proc.*, **21**, 355–356.

Olsen, H. W. (1966). Darcy's law in saturated kaolinite. *Water Resour. Res.*, **2**, 287–295.

Parlange, J.-Y. (1976). Capillary hysteresis and the relationship between drying and wetting curves. *Water Resour. Res.*, **12**, 224–228.

Philip, J. R. and DeVries, D. A. (1957). Moisture movement in porous materials under temperature gradients. *Trans. Amer. Geophys. Un.*, **38**, 222–232.

Polubarinova-Kochina, P. Ya. (1952). *Theory of Ground Water Movement* (translated from the Russian by J. M. R. DeWiest, 1962). Princeton: Princeton University Press.

Poulovassilis, A. (1962). Hystersis of pore water, an application of the concept of independent domains. *Soil Sci.*, **93**, 405–412.

Purcell, W. R. (1949). Capillary pressures – their measurement using mercury and the calculation of permeability therefrom. *Trans. Amer. Inst. Min. Met. Engrs., Petrol. Devel. Technol.*, **186**, 39–46.

Raats, P. A. C. (1975). Transformations of fluxes and forces describing the simultaneous transport of water and heat in unsaturated porous media. *Water Resour. Res.*, **11**, 938–942.

Reichardt, K., Nielsen, D. R. and Biggar, J. W. (1972). Scaling of horizontal infiltration into homogeneous soils. *Soil Sci. Soc. Amer. Proc.*, **36**, 241–245.

Richards, L. A. (1928). The usefulness of capillary potential to soil moisture and plant investigators. *J. Agric. Res.*, **37**, 719–742.

(1931). Capillary conduction of liquids through porous mediums. *Physics*, **1**, 318–333.

Rogowski, A. S. (1972). Watershed physics: soil variability criteria. *Water Resour. Res.*, **8**, 1015–1023.

Royer, J. M. and Vachaud, G. (1975). Field determination of hysteresis in soil-water characteristics. *Soil Sci. Soc. Amer. Proc.*, **39**, 221–223.

Russell, M. B. (1941). Pore-size distribution as a measure of soil structure. *Soil Sci. Soc. Amer. Proc.*, **6**, 108–112.

Schofield, R. K. (1938). Pore-size distribution as revealed by the dependence of suction (pF) on moisture content. *Trans. Intern. Congr. Soil Sci. 1st Congr. A*, pp. 38–45.

Scott, P. H. and Rose, W. (1953). An explanation of the Yuster effect. *J. Petrol. Technol.*, **5**, 19–20.

Snyder, V. and Miller, R. D. (1985). Tensile strength of unsaturated soils. *Soil Sci. Soc. Am. J.*, **49**, 58–65.

Swartzendruber, D. (1968). The applicability of Darcy's law. *Soil Sci. Soc. Amer. Proc.*, **32**, 12–18.

Talsma, T. (1970). Hysteresis in two sands and the independent domain model. *Water Resour. Res.*, **6**, 964–970.

Terzaghi, K. (1925). *Erdbaumechanik auf Bodenphysikalischer Grundlage*. Leipzig und Wien: Franz Deuticke.

(1943). *Theoretical Soil Mechanics*. New York: John Wiley.

Theis, C. V. (1935). The relation between the lowering of the piezometric surface and the rate and duration of discharge of a well using ground water storage. *Trans. Amer. Geophys. Un.*, **16**, 519–524.

Topp, G. C. (1971). Soil water hysteresis in silt loam and clay loam soils. *Water Resour. Res.*, **7**, 914–920.

Van Genuchten, M. T. (1980). A closed form equation for predicting the hydraulic conductivity of unsaturated soils. *Soil Sci. Soc. Amer. J.*, **44**, 892–898.

Verruijt, A. (1969). Elastic storage of aquifers. In *Flow Through Porous Media*, ed. R. J. M. DeWiest. New York: Academic, pp. 331–376.

Watson, K. K., Reginato, R. J. and Jackson, R. D. (1975). Soil water hysteresis in a field soil. *Soil Sci. Soc. Amer. Proc.*, **39**, 242–246.

White, I. and Sully, M. J. (1992). On the variability and use of the hydraulic conductivity alpha parameter in stochastic treatments of unsaturated flow. *Water Resour. Res.*, **28**, 209–213.

Willis, W. O. (1960). Evaporation from layered soils in the presence of a water table. *Soil Sci. Soc. Amer. Proc.*, **24**, 239–242.

Yuster, S. T. (1951). Theoretical considerations of multiphase flow in idealized capillary systems. *Proc. 3rd World Petrol. Congr.*, **2**, 437–445. Leiden, Netherlands: E. J. Brill.

PROBLEMS

8.1   Calculate the capillary rise between two parallel glass plates spaced a distance, d, apart. The fluid has a surface tension σ and a specific weight γ. Assume that the wetting angle is negligible.

8.2   Laboratory tests have revealed that, for a given sandy soil, the water content, θ, is related to the suction in the water ($H = -p_w/\gamma_w$) by the following empirical formula:

$$\theta = n_0 \left[\frac{1400}{1400 + 0.1H)^6} \right]$$

where n_0 is the porosity, and H is expressed in cm of water column. (a) Consider a field situation with a stationary (i.e., not moving) horizontal water table at a depth of 1.0 m below the soil surface. If the soil profile is in equilibrium (i.e., no flow) and evaporation is negligible, what is the water content at 0.5 m below the soil surface? (b) Consider (several months later), again, a horizontal water table at 1.0 m depth in that same soil profile. You know that the soil moisture profile was originally (say one day earlier) in equilibrium (with the water table at some unknown depth), but you suspect that the water table is now moving vertically. If the water content at the soil surface is 0.5 n_0, decide whether the water table is rising or falling. (There is no precipitation and evaporation at the surface.) Prove your answer.

8.3   The following table shows the F-distribution of the independent domain approach for Adelaide dune sand obtained by Talsma (1970). F is expressed as drainable porosity in percent per $(10\,\text{cm})^2$.

Hw (cm)	Hd=0	10	20	30	40	50	60	70	80	90	100
100									4.06		
90								4.94	0.48		
80							5.26	0.56	1.43		
70						6.85	0.96	1.67	1.75		
60					6.14	0.16	2.23	4.38	5.42		
50				6.61	1.43	1.35	2.47	5.18	4.22		
40			5.90	0.00	1.12	9.16	0.88	3.11	0.56		
30		1.59	−3.90	−1.12	4.78	0.16	3.75	0.80	0.32		
20	1.51	0.32	0.56	−0.96	0.00	0.00	−1.35	2.47	0.24		
10	1.43	0.48	0.24	0.40	0.00	0.00	0.00	0.00	0.00	0.00	

H_d (cm): 0   10   20   30   40   50   60   70   80   90   100

(a) Plot the wetting and drying boundary curves. (b) On the same graph, plot the following sequence: starting with a dry material at 100 cm suction, wet to 50 cm; drain again to 80 cm; wet again to 20 cm.

8.4   Use the same F-distribution as shown in the previous problem. (a) Plot the wetting and drying boundary curves. (b) On the same graph, plot the following sequence: starting with a fully saturated soil, drain to 70 cm; wet again to 30 cm; drain finally to 90 cm.

8.5   Consider the F-function used in Example 8.3 and shown in Table 8.1. (a) Calculate and plot the wetting and drying boundary curves. (b) On the same graph, plot the following sequence: starting with a dry soil, wet to 8 cm; then drain the soil to 20 cm; finally, wet again to 4 cm.

8.6   A two-dimensional flow is taking place with a pressure gradient, $\gamma_w^{-1}\nabla p_w = 0.02\mathbf{i} - 0.03\mathbf{j}$, at a point in a soil whose hydraulic conductivity tensor is:

$$\underline{\underline{k}} = \begin{pmatrix} k_{xx} & k_{xy} \\ k_{yx} & k_{yy} \end{pmatrix} = \begin{pmatrix} 1.2 & 0.003 \\ 0.003 & 0.2 \end{pmatrix}$$

The pressure is expressed as height of equivalent water column; the hydraulic conductivity is in cm h^{-1}; the x-axis is horizontal; and the y-axis is vertical. (a) What is the angle between the pressure gradient and the x-axis? (b) What is the angle between the hydraulic head gradient and the x-axis? (c) What is the specific flux (i.e., the rate of flow per unit bulk area normal to the direction of flow) vector? (d) What is the angle this flux vector makes with the x-axis?

8.7   You are given a two-dimensional flow net in a principal axes system describing the steady flow in a homogeneous, anisotropic soil. The equipotential lines (i.e., lines of constant h) are a set of equidistant straight lines making an angle of $-20°$ with the x-axis (or $+70°$ with the y-axis). The stream lines (i.e., lines tangent to the local direction of the flow) are also a set of equidistant straight lines, making an angle of $+40°$ with the x-axis (or $-50°$ with the y-axis). (a) Sketch this flow net. (b) If h increases from bottom to top in your sketch, and if $k_{xx} = 10^{-5}$ cm s^{-1}, determine k_{yy}.

8.8   Derive the expression (8.41) for the soil-water diffusivity from its component functions (8.36) and (8.15).

8.9   Derive the pore size probability density function, $s_e(R) = dS_e/dR$, implied by the soil-water characteristic functions (8.14) and (8.15). Plot these two functions for $H_b = 33$ cm and $b = 5.7$, and for $a = 0.03$ cm^{-1} and $b = 5.7$, respectively.

8.10   Extend the Richards equation (8.56) to the general case of an anisotropic material. Use Equation (8.54) as your starting point.

8.11   Derive Equation (8.45) from (8.44).

8.12   Derive Equation (8.49) from (8.47).

8.13   (a) Apply Equation (8.49) with Ge $= 1/8$ to calculate the intrinsic permeability at saturation (in cm^2) of the sandy soil, whose soil-water characteristic is shown in Figures 8.5 and 8.20. Assume that the parameters are $\theta_0 = 0.405$, $S_r = 0.1$, $b = 5.7$ and $H_b = 0.33$ m in Equation (8.14). (b) Calculate the saturated, hydraulic conductivity at 20 °C from the permeability obtained in part (a). (Note that, experimentally, it was measured to be $k_0 = 1$ cm min^{-1}, approximately.)

8.14   Prove Equation (8.70) starting from (8.67).

8.15   Derive the first of Equations (8.82) from the first of (8.74) (via the first of (8.76)).

8.16   Multiple choice. Indicate which of the following statements are correct. Hysteresis in the moisture content–suction curve of a soil (suction is negative pressure):
       (a)   is related to the geometry of the pores;
       (b)   can be determined by using only a tensiometer;
       (c)   is an important factor to be considered in the determination of the flow rate, when the soil is saturated;
       (d)   suggests that the flow takes place below the water table;
       (e)   must be considered in the analysis of problems involving alternate wetting and drying.

8.17   Multiple choice. Indicate which of the following statements are correct. Darcy's law is not applicable:
       (a)   to describe extremely unsteady phenomena;
       (b)   when the soil is non-homogeneous;
       (c)   when the hydraulic head becomes large;
       (d)   when the pressure is zero (i.e. atmospheric);

(e)   when the pore-filling fluid is non-Newtonian (i.e. when the shear stress is a nonlinear function of the rate of strain);

(f)   when the porous matrix is non-elastic in the sense of Hooke's law. Note: porous matrix refers to the bulk solid, but not to the material out of which the grains are made.

8.18   Multiple choice. Indicate which of the following statements are correct. A large, geological formation consisting of a uniform, partly saturated, anisotropic non-swelling porous sand has a hydraulic conductivity, that may be characterized as follows:

(a) $k = k(\theta, x, y, z)$ and independent of direction;

(b) $k_{ij} = k_{ij}(x, y, z)$ consisting of $2^2 = 4$ components;

(c) $k_{ij} = k_{ij}(x, y, z)$ consisting generally of 3 components regardless of the orientation of the axes system;

(d) $k_{ij} = k_{ij}(\theta)$ in which the form of the functional relationship with θ does not depend on x, y, z;

(e)   it is usually assumed that $k_{ij} = k_{ji}$.

8.19   Multiple choice. Indicate which of the following statements are correct. The hydraulic conductivity of a partly saturated clayey soil (i.e., the capillary conductivity):

(a)   is sharply reduced in the initial stages of reduction of water content;

(b)   as a function of water pressure usually displays marked hysteresis;

(c)   increases with increasing hydraulic gradient;

(d)   is temperature dependent;

(e)   may depend on the type of salt that is in solution in the water.

8.20   Multiple choice. Indicate which of the following statements are correct. It is known that Darcy's law fails to describe the flow in a porous medium when the Reynolds number increases into the range between 1 to 10 and up. This initial deviation is the result of:

(a)   the presence of electrically charged clay particles;

(b)   the fact that water is non-Newtonian at low velocities;

(c)   the effect of inertia (i.e. convective acceleration of fluid particles);

(d)   the onset of local turbulence;

(e)   flow instabilities in the pore necks so that the flow develops local separation eddies;

(f)   the fact that the flow is non-laminar but still creeping.

(8.21) Multiple choice. Indicate which of the following statements are correct. Assume that the flow of a wetting fluid in an elastic porous material can be described by the first of Equations (8.82). A simplified form of that equation, namely Laplace's equation, $\nabla^2 p_w = 0$, is valid to describe flow in a uniform (prior to flow), isotropic, fully saturated porous material whenever:

(a)   the flow is steady, the fluid compressible and the porous matrix compressible;

(b)   the flow is unsteady, the fluid incompressible and the porous matrix incompressible;

(c)   the flow is steady, the fluid compressible and the porous matrix incompressible;

(d)   the flow is unsteady, the fluid compressible and the porous matrix incompressible;

(e)   the flow is steady, the fluid is incompressible and the porous matrix compressible;

(f)   the flow is steady, the fluid and the porous matrix are both incompressible.

9 INFILTRATION AND RELATED UNSATURATED FLOWS

This chapter deals with the flow of water in the partly saturated zone of the near-surface soil, and with the transfer through the atmosphere–soil interface. At the local scale, as precipitated water reaches the ground surface, *infiltration* into the soil takes place. In between precipitation events, the atmosphere exerts its drying effect, and the water in the soil profile may move to the surface by *vapor diffusion* and by liquid *capillary rise*, where it evaporates.

As illustrated in Figure 8.23, a small decrease in water content below saturation can cause a significant decrease in conductivity, so that in most soils the difference in hydraulic conductivity above and below the water table can be large. At an interface between soils with different conductivities, the streamlines are known to exhibit a pronounced refraction; therefore, in many situations unsaturated flow above the water table can be assumed to be nearly vertical, whereas the saturated flow below the water table can be assumed to be more horizontal or parallel to underlying impervious layers. Accordingly, in this chapter, infiltration and related flow phenomena in the partly saturated zone of the soil are analyzed in a one-dimensional vertical framework. Similarly, in Chapter 10, it is shown how many saturated flow situations can be analyzed in the one-dimensional framework of hydraulic groundwater theory. Because the infiltration capacity constitutes an upper limit of maximal rate of entry into the soil, it is treated first in Sections 9.2 and 9.3; rain infiltration is treated in Section 9.4. Various parameterizations of the infiltration of precipitation and related processes at the catchment scale are covered in Section 9.5. Finally, Section 9.6 describes a few elementary mechanisms of capillary flow during inter-storm periods.

9.1 GENERAL FEATURES OF THE INFILTRATION PHENOMENON

Infiltration can be defined as the entry of water into the soil surface and its subsequent vertical motion through the soil profile. In most situations of practical interest, the soil profile is initially less than saturated. Therefore, if it can be assumed that the displaced air can escape freely, the flow of the infiltrating water in the soil is governed by the Richards equation (8.55). For vertical downward movement of water this can be written as

$$\frac{\partial \theta}{\partial t} = -\frac{\partial}{\partial z}\left(k\frac{\partial H}{\partial z}\right) - \frac{\partial k}{\partial z} \tag{9.1}$$

In Equation (9.1) $H = H(\theta)$ is the water suction or negative pressure expressed as equivalent water column, and z (contrary to its normal usage) denotes the depth, that is the vertical coordinate pointing down.

Infiltration can take place in one of two possible ways. When the surface water supply rate resulting from precipitation or other sources is intense enough, part of it remains ponded or runs off, and part of it infiltrates at the maximal rate; this maximal rate of infiltration is the *infiltration capacity*. When the intensity is low, all of the precipitated water seeps into the pores; this is *rainfall infiltration*.

9.1.1 Infiltration capacity

For the purpose of analysis, consider the simple case of a deep uniform soil profile. The soil surface is assumed to be covered by a layer of water, which is sufficiently thin, so that at the soil surface the water pressure is atmospheric and the soil saturated; also, the initial water content is assumed to be constant throughout the profile. The corresponding boundary conditions are then as follows

$$
\begin{aligned}
\theta = \theta_i \qquad & H = H_i \qquad z > 0 \qquad t = 0 \\
\theta = \theta_0 \qquad & H = 0 \qquad z = 0 \qquad t \geq 0
\end{aligned}
\qquad (9.2)
$$

The first of these two conditions represents the initial situation at $t = 0$ characterized by a constant water content θ_i and water pressure H_i, throughout the soil profile. The second represents the condition at the soil surface $z = 0$, maintained indefinitely after the onset of the infiltration. As an illustration of the kind of solution that can be expected, Figure 9.1 shows the water distribution during infiltration into a soil column, observed in a laboratory experiment. Without going into the details of the solution of (9.1) subject to (9.2), at this point it is useful to make some general observations regarding the short and long time nature of the flow.

Short-time behavior

In Equation (9.1), the first term on the right represents the flow caused by the pressure gradient owing to capillarity and the second term represents the flow caused by the Earth's gravity field. As the water starts to enter the relatively dry soil, the pressure differences in the water at the surface and in the soil are quite large and, as a result, the second term on the right is practically negligible compared to the first one. Therefore in its early phase, infiltration can be described by the following

$$
\frac{\partial \theta}{\partial t} = -\frac{\partial}{\partial z}\left(k\frac{\partial H}{\partial z} \right)
\qquad (9.3)
$$

When cast in the form of a diffusion equation by means of (8.32), (9.3) subject to (9.2) is referred to as a *sorption* problem. More detailed aspects of the short-time behavior of infiltration capacity are treated in Section 9.2.

Fig. 9.1 Measured (points) and calculated (lines) soil water content distribution during vertical infiltration into a vertical column of air-dry Columbia silt loam, with $\theta_0 = 0.45$. (From Davidson *et al.*, 1963.)

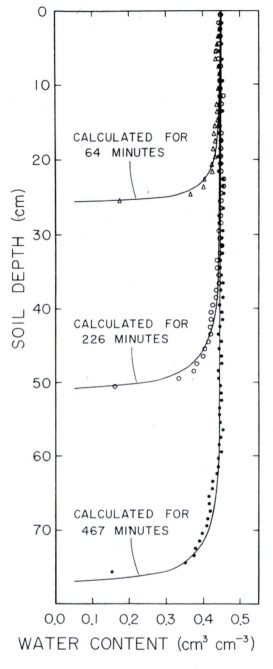

Long-time behavior

As illustrated in Figure 9.1, after longer times of infiltration, the water content profile near the surface gradually becomes more uniform and it eventually assumes the satiation value, or $\theta \to \theta_0$; similarly, the pressure in the upper layers of the soil becomes gradually atmospheric, or $H \to 0$. Hence, their vertical gradients $\partial\theta/\partial z$ and $\partial H/\partial z$ both approach zero. This means that $(\partial h / \partial z) \rightarrow -1$, so that from Darcy's law (8.19) it follows that the rate of infiltration f_c approaches the value of the saturated hydraulic conductivity, or

$$\lim_{t \to \infty} (f_c) = \lim_{\substack{z=0 \\ t \to \infty}} (q_z) = k_0 \qquad (9.4)$$

The problem of intermediate times of infiltration capacity is dealt with in Section 9.3.

9.1.2 *Rainfall infiltration*

Like the infiltration capacity, the infiltration of precipitation on the soil surface is governed by the Richards equation (9.1), but the boundary conditions are quite different from (9.2). Because in this case the water supply rate at the surface is smaller than the maximal rate the soil is capable of taking in, the surface water content θ_s is smaller than θ_0 by an unknown amount; the only known feature of the flow at the surface is the supply rate or precipitation intensity, and thus the flow rate into the soil. As long as the water supply rate is maintained, this unknown surface water content θ_s gradually increases. If the precipitation rate is large enough, eventually the surface water content reaches the maximum possible value θ_0; after this occurs, water starts to pond the surface. If t_p denotes the time from the start of the rainfall until the inception of ponding, the boundary conditions can be formulated as follows

$$
\begin{array}{llll}
\theta = \theta_i & H = H_i & z > 0 & t = 0 \\[2mm]
-D_w \dfrac{\partial \theta}{\partial z} + k = P & k \dfrac{\partial H}{\partial z} + k = P & z = 0 & 0 < t \le t_p \\[2mm]
\theta = \theta_0 & H = 0 & z = 0 & t \ge t_p
\end{array}
\qquad (9.5)
$$

where $P = P(t)$ is the intensity of the precipitation. The first condition represents again an initially constant water content θ_i throughout the soil profile. The second condition indicates that the flux is known at the surface; the third shows that, after ponding starts, the surface is satiated but the flux is no longer known. If the precipitation rate is small, however, the surface soil layer never becomes fully satiated; thus $t_p \rightarrow \infty$ and the third condition of Equation (9.5) is redundant. Further details on the solution of the rain infiltration problem are covered in Section 9.4.

9.2 INFILTRATION CAPACITY IN THE ABSENCE OF GRAVITY: SORPTION

Sorption, that is horizontal infiltration of water into a partly but uniformly saturated soil, when the movement of the displaced air is unimpeded, is a problem of long standing in the hydrologic literature. Although by itself this type of one-dimensional horizontal flow may not be very common in nature, the solution of this problem is of practical importance. First, it gives a good description of the initial stages of vertical infiltration of water ponded on the soil surface, that is the short-time behavior, while the effects of capillarity totally dominate those of gravity. Second, it has been useful as an essential part, or building block, of solutions for the later stages, obtained by various techniques.

9.2.1   *Diffusion formulation of this horizontal flow process*

Consider a uniform soil profile, which has an initial water content θ_i; the flow is allowed to start, when the water content at the soil surface is suddenly increased to $\theta_0(>\theta_i)$. The problem may be formulated by the one-dimensional form of Richards's (1931) equation without gravity term, that is Equation (9.3); to indicate that it describes horizontal flow, in what follows it is expressed in terms of the x-coordinate. Upon substitution of (8.32) this can be written in the form of a diffusion equation as

$$\frac{\partial \theta}{\partial t} = \frac{\partial}{\partial x}\left(D_{\mathrm{w}}\frac{\partial \theta}{\partial x}\right) \tag{9.6}$$

The boundary conditions are still (9.2); these may be recast in terms of the horizontal coordinate as

$$\begin{aligned}
\theta = \theta_i \quad &\quad x > 0 \quad &\quad t = 0 \\
\theta = \theta_0 \quad &\quad x = 0 \quad &\quad t \geq 0
\end{aligned} \tag{9.7}$$

in which θ_i is the initial water content and θ_0 is the water content maintained at the surface $x = 0$, where the water enters the soil. A simple experimental setup to study this problem is illustrated in Figure 9.2, and the experimental data that can be obtained with it are illustrated in Figure 9.3.

In the formulation of infiltration problems it is often convenient to normalize the water content, as follows

$$S_{\mathrm{n}} = \frac{\theta - \theta_i}{\theta_0 - \theta_i} \tag{9.8}$$

With this normalized water content the governing equation and boundary conditions become

$$\frac{\partial S_{\mathrm{n}}}{\partial t} = \frac{\partial}{\partial x}\left(D_{\mathrm{w}}\frac{\partial S_{\mathrm{n}}}{\partial x}\right) \tag{9.9}$$

and

$$\begin{aligned}
S_{\mathrm{n}} = 0 \quad &\quad x > 0 \quad &\quad t = 0 \\
S_{\mathrm{n}} = 1 \quad &\quad x = 0 \quad &\quad t \geq 0
\end{aligned} \tag{9.10}$$

Similarity approach
By the application of Boltzmann's (1894) transformation, which combines the space and time variables into one independent variable,

$$\phi = x\, t^{-1/2} \tag{9.11}$$

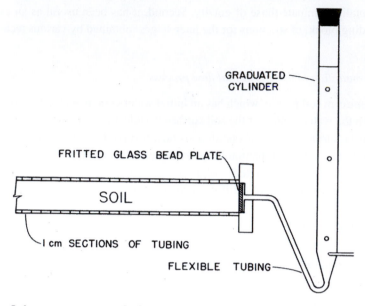

Fig. 9.2   Laboratory set-up to study the problem of sorption experimentally. The graduated cylinder maintains the water at the point of entry ($x = 0$), at constant pressure in the manner of a Mariotte bottle. At the start of the experiment ($t = 0$), the water supply through the flexible tubing is opened; after a certain time t, at the end of the experiment, the 1 cm sections of the horizontal soil column tubing can be rapidly taken apart, to determine their soil water content as $\theta = \theta(x)$. (From Nielsen *et al.*, 1962.)

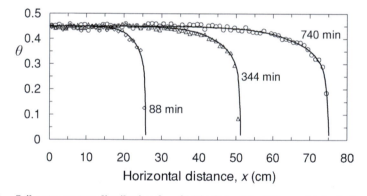

Fig. 9.3   Soil water content distribution $\theta = \theta(x)$ in Columbia silt loam, obtained after three different times of horizontal infiltration by means of the apparatus shown in Figure 9.2, with the pressure at the inlet maintained at -2 hPa (or roughly 2.04 cm of water column). The curve for 740 min is the best fit line through the data points (circles); the curves for 88 min and 344 min are calculated from the curve for 740 min on the basis of Boltzmann similarity, that is by multiplying it by $(88/740)^{1/2}$ and $(344/740)^{1/2}$, respectively. (After Nielsen *et al.*, 1962.)

Equation (9.6) can be simplified to an ordinary differential equation; this can be accomplished through the following steps

$$\frac{\partial \theta}{\partial t} = \frac{d\theta}{d\phi}\frac{\partial \phi}{\partial t} = -\frac{1}{2}x\,t^{-3/2}\frac{d\theta}{d\phi}$$

$$\frac{\partial \theta}{\partial x} = \frac{d\theta}{d\phi}\frac{\partial \phi}{\partial x} = t^{-1/2}\frac{d\theta}{d\phi}$$

$$\frac{\partial}{\partial x}\left(D_w\frac{\partial \theta}{\partial x}\right) = \frac{\partial}{\partial x}\left(D_w t^{-1/2}\frac{d\theta}{d\phi}\right) = t^{-1/2}\frac{d}{d\phi}\left(D_w\frac{d\theta}{d\phi}\right)\frac{\partial \phi}{\partial x}$$

$$= t^{-1}\frac{d}{d\phi}\left(D_w\frac{d\theta}{d\phi}\right)$$

(9.12)

Equating the first and third of (9.12), one obtains the desired ordinary differential equation

$$\frac{d}{d\phi}\left(D_w\frac{d\theta}{d\phi}\right) + \frac{\phi}{2}\frac{d\theta}{d\phi} = 0 \tag{9.13}$$

The boundary conditions (9.7) become now

$$\begin{aligned}
\theta &= \theta_i \ (\text{or } S_n = 0) & \phi &\to \infty \\
\theta &= \theta_0 \ (\text{or } S_n = 1) & \phi &= 0
\end{aligned} \tag{9.14}$$

The applicability of the Boltzmann transform indicates that the space variable x controls the flow in a similar manner as the time variable $t^{-1/2}$. Conversely, the use of this similarity in (9.11) requires a certain symmetry in the boundary conditions, to allow the combination of the two independent variables x and t into a single one ϕ. In the present case of (9.7), this is satisfied, in that θ assumes the same value for $t = 0$ as it does for $x \to \infty$, that is far away from the point of entry, where the water content will remain at θ_i; similarly, θ assumes the same value at $x = 0$ as it does for $t \to \infty$, that is after a long time, when the water content in entire soil profile will approach θ_0. Figure 9.3 illustrates the experimental validation of Equation (9.11); it confirms that if the experimental data were plotted as $\theta = \theta(\phi)$ (instead of as $\theta = \theta(x)$ for different values of time), all the points would collapse onto one single curve.

Infiltration
The solution of Equations (9.13) and (9.14) can be formulated as $\phi = \phi(\theta)$. Before actually knowing that solution, it is already possible to infer some essential features of the infiltration phenomenon. Indeed, without specifying the actual form of the solution, as illustrated in Figure 9.4 for the vertical case, the cumulative volume of infiltration can obtained by integration of the total water volume that has entered the soil; this can be an integral either of $(z\,d\theta)$, or of $(\theta\,dz)$. Hence, the cumulative volume of infiltration, with dimensions $[L^3\,L^{-2}]$, can be written in general as

$$F = \int_{\theta_i}^{\theta_0} x\,d\theta \tag{9.15}$$

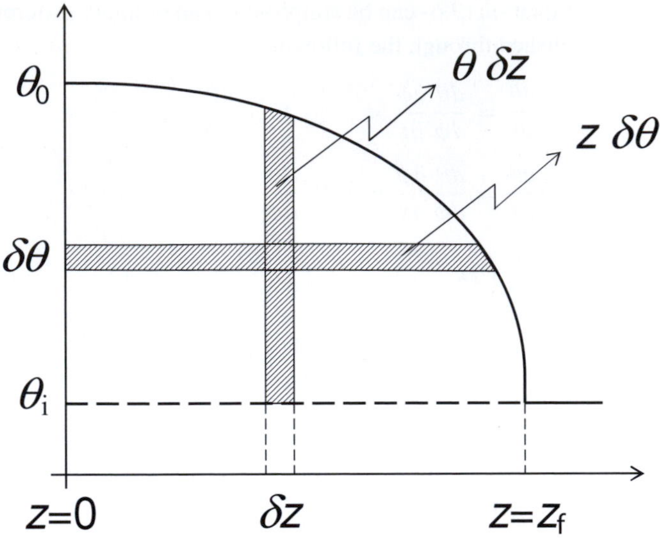

Fig. 9.4 Sketch illustrating the calculation of the infiltrated volume F as the area under the $\theta = \theta(z)$ curve, that is the water content profile in the soil at a given instant in time t. This can be done by integrating either the elemental area $(z\,d\theta)$ or the elemental area $(\theta\,dz)$. The coordinate z points down into the soil; $z = 0$ is where the water infiltrates, and $z = z_f$ is the position of the wetting front.

in which x is used instead of z to indicate the absence of gravity in the present case. With the solution in terms of the Boltzmann variable (9.11), this assumes the form

$$F = t^{1/2} \int_{\theta_i}^{\theta_0} \phi\,d\theta \tag{9.16}$$

The integral in this equation has constant limits and is therefore also a constant. Thus, for conciseness of notation it is often convenient to express horizontal infiltration in terms of the sorptivity, defined by Philip (1957a) as

$$A_0 = \int_{\theta_i}^{\theta_0} \phi\,d\theta \tag{9.17}$$

The cumulative infiltration (9.16) can now be written as

$$F = A_0 t^{1/2} \tag{9.18}$$

and the rate of infiltration $f = dF/dt$

$$f = \frac{1}{2} A_0 t^{-1/2} \tag{9.19}$$

The point here is that both equations indicate unequivocally how horizontal infiltration capacity proceeds in time, even though the solution is left unspecified so far.

Note that, because that solution can also be written as $\theta = \theta(\phi)$, the rate of infiltration can be expressed alternatively as a Darcy flux, or because $\partial\phi/\partial x = t^{-1/2}$,

$$f = -D_w \left.\frac{\partial\theta}{\partial x}\right|_{x=0} = -D_w \left.\frac{d\theta}{d\phi}\right|_{\phi=0} t^{-1/2} \tag{9.20}$$

which provides an alternative expression for the sorptivity. A second point of interest is that the form of Equations (9.17), (9.18) and (9.19) is already suggestive of a way to scale the variables that govern the sorption phenomenon. Indeed, because f has the same dimensions as the hydraulic conductivity, it is only natural to make it dimensionless with that variable; with (9.19) this produces then immediately a dimensionless time variable, as well. Thus one can construct the following dimensionless variables

$$f_+ = \frac{f}{k_0}, t_+ = \frac{k_0^2 t}{A_0^2}, \qquad \text{and} \qquad F_+ = \frac{k_0 F}{A_0^2} \tag{9.21}$$

Over the years, the sorptivity has come to be considered as one of the more fundamental flow properties of a soil whose relevance extends well beyond the phenomenon of sorption. As will become clear later in this chapter, the sorptivity also arises naturally in the formulation of vertical infiltration capacity and in the formulation of different facets of rainfall infiltration. Methods have been developed to measure the sorptivity in the field (see Talsma, 1969; Talsma and Parlange, 1972; Clothier and White, 1981; Cook and Broeren, 1994). It has also been used by White and Perroux (1987) to derive other field soil hydraulic properties such as the diffusivity $D_w(\theta)$, the hydraulic conductivity $k(\theta)$ and the soil water characteristic $H(\theta)$. As an illustration of the orders of magnitude of this quantity, the following values (in cm min$^{-1/2}$) were measured by Talsma and Parlange (1972) in the field: 0.97 (Bungendore sand), 0.08 (Pialligo sand), 0.17 (Barton clay loam); the respective satiated hydraulic conductivities, k_0 were 0.092, 1.08 and 0.093 cm min^{-1}.

Wetting front

In certain applications it is of interest to determine the position of the wetting front. The position of this front can be defined as the value of $x = x_f$, where the water content assumes a certain value $\theta = \theta_f(>\theta_i)$. Experimentally, this water content may be taken as the value at which the soil changes color as the water infiltrates. Mathematically, because the front can be quite sharp in many soils, it is often convenient to assume simply that it is located where the water content approaches $\theta = \theta_i$, or $S_n = 0$. Since ϕ is a function of θ, it is clear in light of Equation (9.11), that the position of the wetting front is directly proportional to $t^{1/2}$, or

$$x_f = \phi_f t^{1/2} \tag{9.22}$$

in which $\phi_f = \phi(\theta_f)$ is a constant for a given choice of θ_f. An experimental illustration of Equation (9.22) is shown in Figure 9.5.

9.2.2 *Some applications of the first integration*

Equation (9.13) can be integrated once, subject to the first of (9.14), to yield immediately

$$-2D_w \frac{d\theta}{d\phi} = \int_{\theta_i}^{\theta} \phi(y) \, dy \tag{9.23}$$

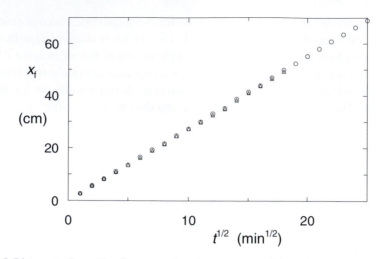

$t^{1/2}$ (min$^{1/2}$)

Fig. 9.5 Distance to the wetting front x_f against the square root of time t for the three experiments of horizontal infiltration into Columbia silt loam, already shown in Figure 9.3; the measurements of the front position were made by visual inspection of the change in color of the soil in the apparatus shown in Figure 9.2. The infiltration was allowed to proceed to distances of 25 (diamonds), 50 (triangles) and 75 cm (circles), respectively; the bulk densities were around 1.3 g cm^{-3}. (After Nielsen *et al.*, 1962.)

in which y is a dummy variable of integration representing θ. As probably first shown by Matano (1933), Equation (9.23) yields the following expression for the diffusivity,

$$D_w = -\frac{1}{2}\frac{d\phi}{d\theta}\int_{\theta_i}^{\theta}\phi(y)\,dy \tag{9.24}$$

This integral has been useful in several ways, most notably in the experimental determination of the soil water diffusivity, and also in the derivation of certain exact solutions for sorption and horizontal infiltration capacity.

Direct measurement of the soil water diffusivity

Equation (9.24) was the basis for the method of Bruce and Klute (1956) to determine the diffusivity $D_w = D_w(\theta)$ directly from a sorption experiment. Substitution of the Boltzmann transformation (9.11) into (9.24) produces the diffusivity in terms of the original variables x and t, as follows

$$D_w = -\frac{1}{2t}\left(\frac{dx}{d\theta}\right)\int_{\theta_i}^{\theta}x\,d\theta \tag{9.25}$$

This expression can be applied with a measured soil water content profile curve $\theta = \theta(x)$ obtained in a horizontal infiltration experiment of duration t, in a set-up like that shown in Figure 9.2. The diffusivity $D_w = D_w(\theta)$ can be readily calculated from any curve like those shown in Figure 9.3. As illustrated in Figure 9.6, this is done by estimating both the area under the curve of x vs θ and the slope of that same curve, at a series of values θ, for the given value of the elapsed time t. These numerical values of the integral and of the

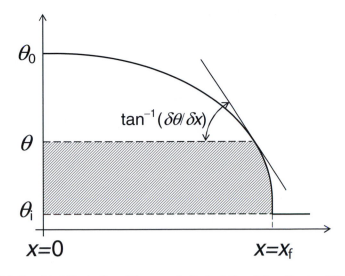

Fig. 9.6 Graphical illustration of the numerical calculation of the soil water diffusivity $D_w = D_w(\theta)$ by means of Equation (9.25) with the soil water profile $\theta = \theta(x)$ obtained from a horizontal infiltration experiment of duration t. In (9.25) the integral is the shaded area and $(dx/d\theta)$ is the inverse of the slope at θ.

derivative can then be used in Equation (9.25) to calculate the diffusivity at each of these values of θ. Another illustration of the method is the study by Clothier and White (1982); with the profile data shown in Figure 9.7 they applied it to compare the diffusivity function $D_w = D_w(\theta)$ in undisturbed and repacked soil columns, as shown in Figure 9.8.

An exact solution for soil water sorption

Equation (9.24) indicates that $D_w = D_w(\theta)$ can be determined when ϕ is known; in other words, (9.24) can also be applied in an inverse mode, to derive the form, which the diffusivity $D_w = D_w(\theta)$ must have, to produce any assumed functional form of the solution $\phi = \phi(\theta)$. This way Philip (1960) was able to list a large class of exact solutions of the nonlinear diffusion equation (9.6) subject to (9.7) for corresponding functional forms of the diffusivity. At the time, none of the obtained $\phi(\theta)$ functions seemed to be applicable to infiltration into soils, and they received relatively little attention in the hydrologic literature. However, it was subsequently shown that in certain cases by proper scaling, one such solution is adaptable to describe sorption in real soils, and that it can thus be made compatible with experimental data (Brutsaert, 1968; 1976). The simplest form of that solution is $\phi = (1 - S_n^m)$, which corresponds according to (9.24) with a diffusivity $D_w = m S_n^m [1 - S_n^m / (1 + m)]/2$. To obtain $D_w = D_{w0}$, that is the diffusivity at satiation for $S_n = 1$, this result must be scaled as follows

$$D_w = D_{w0}(1 + m) \left[S_n^m - S_n^{2m} / (1 + m) \right] / m \tag{9.26}$$

and the corresponding exact solution becomes

$$\phi = \left[2D_{w0} (1 + m) / m^2 \right]^{1/2} \left(1 - S_n^m \right) \tag{9.27}$$

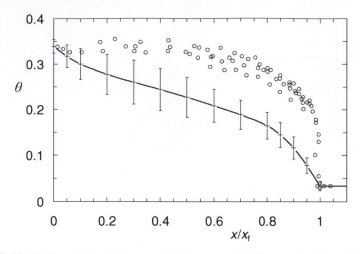

Fig. 9.7 Distribution of the water content θ during sorption in Bungendore fine sand, as a function of the scaled distance x/x_f (or ϕ/ϕ_f), in which x_f is the distance to the wetting front. The curve and cross points represent the mean (with the standard deviations) of the measurements in seven undisturbed field cores, and the circles are the data obtained in four repacked cores. The respective sorptivities for the two sets of experiments were on average $A_0 = 6.5 \times 10^{-4}$ m s$^{-1/2}$ and $A_0 = 1.26 \times 10^{-3}$ m s$^{-1/2}$, and the respective positions of the wetting front [see (9.22)] were on average $\phi_f = 3.43 \times 10^{-3}$ m s$^{-1/2}$ and $\phi_f = 4.45 \times 10^{-3}$ m s$^{-1/2}$. (After Clothier and White, 1982.)

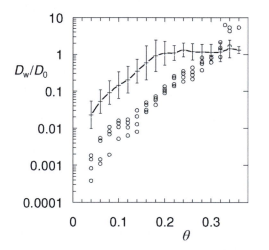

Fig. 9.8 The scaled soil water diffusivity D_w/D_0 as a function of the water content θ for Bungendore fine sand as obtained from the water content profiles shown in Figure 9.7 by means of Equation (9.25); in this case the normalizing diffusivity D_0 is the assumed constant value, which would produce the same sorptivity (see Equation (9.59)), and thus the same infiltration rate. The curve and cross points represent the log mean of the results (with their standard deviations) obtained from the θ profiles in the seven undisturbed field cores, and scaled with $D_0 = 4.09 \times 10^{-6}$ m^{2}s^{-1}; the circles represent the results obtained from the profiles in the four repacked cores and scaled with $D_0 = 1.34 \times 10^{-5}$ m^{2}s^{-1}. (After Clothier and White, 1982.)

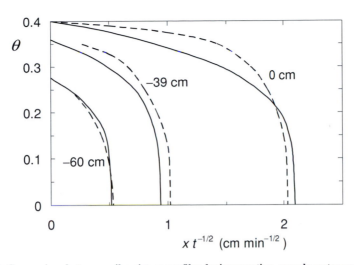

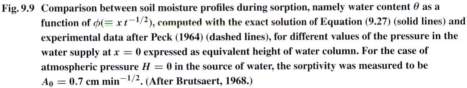

Fig. 9.9 **Comparison between soil moisture profiles during sorption, namely water content θ as a function of $\phi(\equiv x\,t^{-1/2})$, computed with the exact solution of Equation (9.27) (solid lines) and experimental data after Peck (1964) (dashed lines), for different values of the pressure in the water supply at $x = 0$ expressed as equivalent height of water column. For the case of atmospheric pressure $H = 0$ in the source of water, the sorptivity was measured to be $A_0 = 0.7$ cm min$^{-1/2}$. (After Brutsaert, 1968.)**

A comparison is shown in Figure 9.9 between soil water profiles calculated with this result and experimental data of Peck (1964). The curve for a water supply pressure of 0 cm at $x = 0$ is given by $\phi = 2.09(1 - S_n^4)$ cm min$^{-1/2}$ (Brutsaert, 1968); the curves for -39 cm and -60 cm (at $x = 0$) were obtained by renormalizing S_n with the water content at those pressures.

Although Equation (9.27) is an exact solution, its main shortcoming is that the required diffusivity (9.26) may not be flexible enough to provide a precise representation of the actual soil water diffusivity. The main advantage of (9.27) is that it can be used in testing the accuracy of other methods of solution. The sorptivity for this exact solution is readily found by means of (9.17),

$$A_0 = (\theta_0 - \theta_i)\,[2D_{w0}/\,(m+1)]^{1/2} \tag{9.28}$$

By combining (9.27) with (9.28) the water content profiles obtained with this exact solution can be expressed in dimensionless form, as shown in Figure 9.10. It can be seen that the wetting front becomes steeper with increasing values of m. The fractional values of m, namely $m = 0.25$ and 0.50 result in solutions with a shape, which is not very different from that of the linear case (9.56) (see below).

The position of the wetting front can be taken as the value of ϕ at $S_n = 0$; thus one obtains from (9.27)

$$\phi_f = [2D_{w0}(1 + m)/m^2]^{1/2} \tag{9.29}$$

Comparing this result with (9.28) one sees that the cumulative volume of infiltration is directly proportional with the position of the wetting front, or

$$F = (\theta_0 - \theta_i)\,\frac{m}{(m+1)}x_f \tag{9.30}$$

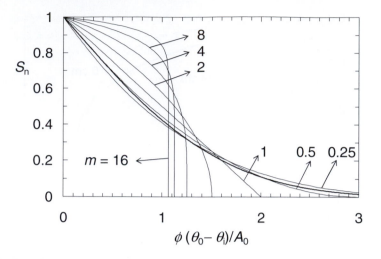

Fig. 9.10 The soil water content distribution obtained with the exact solution (9.27) in the scaled form $x(\theta_0 - \theta_i)/(A_0 t^{1/2}) = [(1+m)/m](1 - S_n^m)$ for different values of the parameter m. The heavy line represents the solution of the linear case (9.56).

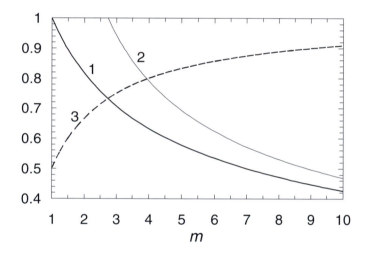

Fig. 9.11 The scaled sorptivity $A_0/[(\theta_0 - \theta_i)D_{w0}^{1/2}]$ (curve 1), the position of the wetting front $\phi_f/D_{w0}^{1/2}$ (curve 2), and the ratio of the infiltrated volume over the wetting front position $F/[(\theta_0 - \theta_i)x_f]$ (curve 3), obtained with the exact solution (9.27), as functions of the parameter m in the diffusivity function (9.26). D_{w0} is the diffusivity at satiation.

These expressions for the sorptivity A_0, the position of the wetting front ϕ_f, and the ratio F/x_f, obtained with the exact solution (9.27), are illustrated in Figure 9.11 as functions of the shape parameter m.

Integral constraint for approximate solutions

In some methods of solution the governing equation (9.13) is changed slightly to make it more amenable to mathematical analysis. Because physically (9.13) is based on the validity of Darcy's equation and the principle of mass conservation, any changes to it result in a violation of these physical principles. Therefore, in some cases it has been found useful to consider the integral of (9.13), with both of (9.14), namely,

$$2 \left(D_w \, d\theta / d\phi \right)_{\theta=\theta_0} + \int_{\theta_i}^{\theta_0} \phi \, d\theta = 0$$

or, in terms of S_n (9.31)

$$2 \left(D_w \, dS_n / d\phi \right)_{S_n=1} + \int_0^1 \phi \, dS_n = 0$$

and use it as a constraint on the approximate solution. An example of the application of this approach is found in the next section.

9.2.3 A nearly exact solution for strongly nonlinear soils

In the hydrologic literature many numerical solutions of (9.13) with (9.14) have been published starting with those of Klute (1952) and Philip (1955). Such solutions can be quite accurate, but their implementation in practical simulations, especially over larger areas, is still a cumbersome task. Therefore, it is often useful to describe the phenomenon with parameterizations that satisfy the dual requirement of physical realism and computational simplicity. Over the years a number of simple analytical solutions have been formulated, which satisfy this requirement. Although these solutions are not exact, their accuracy is still reasonable, and they involve much smaller mathematical error than those generated by the uncertainty of the parameter functions $k = k(\theta)$ and $H = H(\theta)$. Moreover, they are closed form and concise so that they are easy to apply. It can be shown (Brutsaert, 1976) that several of these solutions are special cases of the following general form

$$\phi = \left(2 \bigg/ \int_0^1 D_w \, S_n^a \, dS_n \right)^{1/2} \int_{S_n}^1 D_w(y) y^b dy$$ (9.32)

where a and b are constants. The specific values of a and b depend on the nature of the approximation used in the solution. For example, in the quasi-steady state solution (Landahl, 1953; Macey, 1959; Parlange, 1971), $a = 1$ and $b = 0$; in a second approximation of the quasi-steady state solution (Parlange, 1973), $a = 0$ and $b = -1$; in the sharp-front solution (Brutsaert, 1974), $a = b = -1$; finally, in a first weighting solution $a = -b = 1/2$ (Parlange, 1975), and in a second weighting solution $a = b = -3/2$. By comparing them with the exact solution (9.28), it was found (Brutsaert, 1976) that the error involved in all these solutions for ϕ is at most of the order of 3% or 4%; however, as will be shown next, the first weighting solution with $a = -b = 1/2$ is accurate within a few thousandths.

Derivation of general form

The general form of Equation (9.32) can be obtained by direct integration of (9.13), with the assumption that in its second term on the left, ϕ can be replaced by a power function of S_n. With this approximation (9.13) becomes

$$\frac{d}{d\phi} \left(D_w \frac{dS_n}{d\phi} \right) + \frac{d}{d\phi} \left(c S_n^{-b} \right) = 0$$ (9.33)

where c and b are constants. Integrating (9.33) with the first of conditions (9.14) (provided $b \leq 0$), one finds

$$D_w \frac{dS_n}{d\phi} + c\, S_n^{-b} = 0 \tag{9.34}$$

A second integration of (9.33) with the second of (9.14) yields

$$\phi = c^{-1} \int_{S_n}^{1} y^b D_w \, dy \tag{9.35}$$

The constants b and c remain to be determined. As shown elsewhere (Brutsaert, 1976), c can be determined in several ways. But the more accurate form, for the purpose of infiltration calculations, can be obtained by means of the integral condition (9.31). Substitution of Equation (9.34) for $S_n = 1$ into the first term and (9.35) into the second term of (9.31) produces upon integration by parts by means of Leibniz's rule (see Equation (A1))

$$-2c + c^{-1} \left[S_n \left(\int_{S_n}^{1} y^b D_w(y) dy \right) \right]_0^1 + c^{-1} \int_0^1 S_n^{1+b} D_w(S_n) dS_n = 0 \tag{9.36}$$

Because the second term of (9.36) is zero, this results in

$$c = \left[\int_0^1 S_n^{1+b} D_w(S_n) dS_n / 2 \right]^{1/2} \tag{9.37}$$

The solution (9.35) can therefore be written as

$$\phi = \left(2 \bigg/ \int_0^1 D_w S_n^{1+b} dS_n \right)^{1/2} \int_{S_n}^{1} D_w(y) y^b dy \tag{9.38}$$

Comparison with the more general form (9.32) shows that this method to determine c produces $a = 1 + b$.

   Some comments on this approximate method of solution are in order. Richards's equation in the form of Equation (9.13) is derived from the equation of continuity and Darcy's law, and thus it embodies the principles of mass and momentum conservation. Because (9.33) is only an approximation of (9.13), it may no longer satisfy these conservation principles. However, by constraining the solution of (9.33) with (9.31) in the determination of b and c, one ensures that this solution satisfies them at least in an integral or average sense.

Optimal value of the exponent b

   The procedure to derive an optimal value of b can be understood by recalling that Equation (9.27) represents an exact solution for a possibly approximate diffusivity function (9.26), whereas Equation (9.38) represents an approximate solution for an unspecified but presumably exact diffusivity function. Therefore, that value of b can be adopted, which makes (9.38) come closest to the exact result. Because infiltration is the phenomenon of interest, the sorptivity should be used for this purpose.

The sorptivity can be calculated by integration of (9.38) in accordance with (9.17); this readily produces

$$A_0 = (\theta_0 - \theta_i) \left(2 \int_0^1 D_w S_n^{1+b} d S_n \right)^{1/2} \tag{9.39}$$

The optimal value of b can now be estimated by solving (9.39) with the special diffusivity (9.26) and by comparing the result with the exact sorptivity (9.28); this yields

$$b = [(4m^2 + 8m + 5)^{1/2} - (2m + 3)]/2 \tag{9.40}$$

Because the numerical value 5 in Equation (9.40) may be approximated by 4, when m is not small, it is clear that b is close to -0.5. As mentioned, with this value of b it can be shown that the error in the sorptivity tends to be smaller than 1%.

Accordingly, in its most accurate form, i.e. with $b = -1/2$, the solution (9.38) becomes

$$\phi = \left(2 \bigg/ \int_0^1 D_w S_n^{1/2} d S_n \right)^{1/2} \int_{S_n}^1 D_w y^{-1/2} dy \tag{9.41}$$

Similarly, the sorptivity (9.39) becomes simply

$$A_0 = (\theta_0 - \theta_i) \left(2 \int_0^1 D_w S_n^{1/2} d S_n \right)^{1/2} \tag{9.42}$$

The position of the wetting front (9.22) follows directly from (9.41) with the lower limit of the integral taken as $S_n = 0$, or

$$\phi_f = \left(2 \bigg/ \int_0^1 D_w S_n^{1/2} d S_n \right)^{1/2} \int_0^1 D_w S_n^{-1/2} d S_n \tag{9.43}$$

Implementation with parametric diffusivity functions

Whenever it can be assumed that $\theta_i = \theta_r$ (or $S_e = S_n$), which is usually a good approximation when initially the soil is quite dry, the diffusivity functions (8.39) and (8.41) can be used to perform the above integrations of (9.32) and (9.39).

Example 9.1. Exponential diffusivity

In the case of the diffusivity (8.39), the sorptivity (9.42) can be shown to be (Brutsaert, 1976)

$$A_0 = D_{w0}^{1/2} (\theta_0 - \theta_i) C_1(\beta) \tag{9.44}$$

in which D_{w0} is the diffusivity at satiation, when $S_n = 1.0$. The term $C_1(\beta)$ depends on the value of the parameter β in (8.39); this can be calculated from the following

$$C_1(\beta) = \beta^{-1} ((2\beta - 1) + \exp(-\beta) M(-0.5, 0.5, \beta))^{1/2} \tag{9.45}$$

in which $M(a, b, z)$ is the confluent hypergeometric function, conveniently tabulated by Abramowitz and Stegun (1964, pp. 516−535). The dependency of $C_1(\beta)$ on β is illustrated

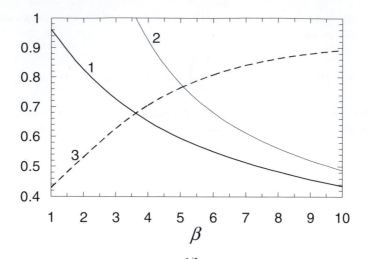

Fig. 9.12 The scaled sorptivity $A_0 / [(\theta_0 - \theta_i)D_{w0}^{1/2}]$ (curve 1), the position of the wetting front $\phi_f / D_{w0}^{1/2}$ (curve 2), and the ratio of the infiltrated volume over the wetting front position $F / [(\theta_0 - \theta_i)x_f]$ (curve 3), obtained with the solution (9.41), as functions of the parameter β in the exponential diffusivity function (8.39). D_{w0} is the diffusivity at satiation.

in Figure 9.12; typical values are $C_1(8) = 0.4828$ and $C_1(3) = 0.7256$. Equation (9.44) can provide a first estimate of A_0, when no other information is available. Because $D_{w0} = D_{wi}\exp(\beta)$, in which β is a constant, it follows that (9.44) also allows the formulation of the diffusivity function (8.39) in terms of the sorptivity, as shown in Brutsaert (1979). Similarly, substitution of the same diffusivity in Equation (9.43) produces immediately the position of the wetting front, namely

$$\phi_f = D_{w0}^{1/2} C_2(\beta) \tag{9.46}$$

in which

$$C_2(\beta) = 2\,(1 - \exp(-\beta)M(-0.5, 0.5, \beta) \cdot ((2\beta - 1)$$
$$+ \exp(-\beta)M(-0.5, 0.5, \beta))^{-1/2} \tag{9.47}$$

This result is illustrated in Figure 9.12. Again, (9.46) with (9.47) shows how the diffusivity function (8.39) for a given soil can be expressed in terms of the position of the wetting front, once β is known (Miller and Bresler, 1977; Brutsaert, 1979). Recalling the definitions of A_0 in (9.18) and of ϕ_f in (9.22), and also comparing (9.44) with (9.46), one can see that $A_0/\phi_f = (\theta_0 - \theta_i)C_1(\beta)/C_2(\beta)$; this means that when the soil water diffusivity is exponential, the cumulative horizontal infiltration F is related with the position of the wetting front x_f, as follows

$$F = C_3(\beta)(\theta_0 - \theta_i)x_f \tag{9.48}$$

where $C_3(\beta) = C_1(\beta)/C_2(\beta)$ is a number whose value depends on β. This result is also illustrated in Figure 9.12; typical values are $C_3(8) = 0.862$ and $C_3(3) = 0.626$.

Example 9.2. Power-form diffusivity

In the same way, in the case of the diffusivity (8.40), Equation (9.42) can be integrated readily to yield a sorptivity

$$A_0 = (2H_b(\theta_0 - \theta_i)k_0/b)^{1/2} (n - b^{-1} + 0.5)^{-1/2} \tag{9.49}$$

where H_b and b are the parameters of (8.14) and n is the power in (8.36).

With the diffusivity given by the somewhat more complex (8.41), the integral in (9.42) is a complete beta function, which can also be written as follows

$$A_0 = (2k_0/a)^{1/2} (\theta_0 - \theta_i)(\Gamma(n - b^{-1} + 0.5)\Gamma(b^{-1} + 1)/\Gamma(n + 0.5))^{1/2} \tag{9.50}$$

In Equation (9.50) a and b are the parameters of (8.15), n is the power in (8.36) and $\Gamma(\ )$ is the gamma function (see Abramowitz and Stegun, 1964). For most soils when n varies between 2 and 10 and b between 1 and 10, the square root term in (9.50) containing the gamma functions is likely to be of the order of unity. For instance, for a typical case of $n = b = 3$ it is equal to 0.7938. The position of the wetting front can be obtained from Equation (9.43) by a similar integration, to produce

$$\phi_f = (2k_0/[(\theta_0 - \theta_i)a])^{1/2} \left(\frac{\Gamma(n - b^{-1} + 0.5)\Gamma(b^{-1} + 1)}{\Gamma(n + 0.5)} \right)^{1/2} \frac{(n - 0.5)}{(n - b^{-1} - 0.5)} \tag{9.51}$$

Comparison of (9.50) with (9.51) indicates that with this diffusivity function, the infiltrated volume is proportional with the position of the wetting front as

$$F = C(n, b)(\theta_0 - \theta_i)x_f \tag{9.52}$$

in which the proportionality constant is given by $C(n, b) = (n - b^{-1} - 0.5)/(n - 0.5)$; for example, $C(3, 3) = 0.867$, which is similar to the result given by Equation (9.48).

9.2.4 A nearly exact solution for mildly nonlinear soils: linearization

For some soils the diffusivity can be assumed to be nearly independent of the water content θ; an example of this is shown in Figure 9.8. In such a case Equation (9.6) can be linearized and it reduces to the linear diffusion equation; accordingly, (9.13) can be written as

$$D_0 \frac{d^2\theta}{d\phi^2} + \frac{\phi}{2} \frac{d\theta}{d\phi} = 0 \tag{9.53}$$

where D_0 is the constant soil water diffusivity. By means of the ad-hoc substitution $p = d\theta/d\phi$, (9.53) can be integrated to yield

$$p = C_1 \exp(-\phi^2/4D_0) \tag{9.54}$$

A second integration yields

$$\theta = C_1 2D_0^{1/2} \int \exp(-y^2)dy + C_2 \tag{9.55}$$

where y is the dummy variable of integration representing $\phi/2D_0^{1/2}$ and C_1 and C_2 are constants to be determined from the boundary conditions (9.14). The integral term with limits between zero and infinity equals $(\pi^{1/2}/2)$; therefore, imposing these conditions, one obtains finally the solution

$$\theta = (\theta_0 - \theta_i)\, \text{erfc}\!\left(\phi/2D_0^{1/2}\right) + \theta_i \tag{9.56}$$

in which the complementary error function is, by definition,

$$\text{erfc}(y) = \frac{2}{\pi^{1/2}} \int_y^\infty \exp(-z^2)\,dz \tag{9.57}$$

The normalized water content $S_n = (\theta - \theta_i)/(\theta_0 - \theta_i)$ given by this solution is shown graphically in Figure 9.10. Application of (9.20) with Leibniz's rule (see Appendix) to (9.57), and comparison with (9.19) produce the following expression for the sorptivity

$$A_0 = 2(\theta_0 - \theta_i)(D_0/\pi)^{1/2} \tag{9.58}$$

Most natural soils have a soil water diffusivity, which is markedly dependent on the water content; therefore, the results obtained in this section with a linearized soil may appear suspect at first sight. However, as illustrated in Figure 9.8, this θ-dependency is not always strong, so that a linear model may still come close to describing the situation. Indeed, linearization tends to simplify the analysis considerably, and should therefore be of interest. The question remains what value should be assigned to the constant diffusivity D_0, to ensure that the linear model will reproduce the more important sorption features of the prototype. One possibility is to assign the value, which would reproduce the same infiltration rate and volume with the linearized soil as the nonlinear prototype soil. In this case one obtains immediately from Equation (9.58)

$$D_0 = \frac{\pi A_0^2}{4(\theta_0 - \theta_i)^2} \tag{9.59}$$

in which A_0, the sorptivity of the prototype, is to be determined independently. As another possibility, if no independent estimates of A_0 are available, one can use an empirical approximation proposed by Crank (1956, p. 256); from his calculations he had found that the weighted mean

$$D_0 = (5/3)(\theta_0 - \theta_i)^{-5/3} \int_{\theta_i}^{\theta_0} (\theta - \theta_i)^{2/3} D(\theta)\,d\theta \tag{9.60}$$

can yield initial rates with good accuracy for diffusivity functions $D = D(\theta)$ which increase with θ over several orders of magnitude.

9.3   INFILTRATION CAPACITY

The infiltration capacity, or the potential infiltration rate, was defined above as the maximal rate at which the soil surface can absorb water. Such conditions prevail when the soil at the surface is saturated, that is whenever its water pressure is at least atmospheric. The problem is usually analyzed by assuming that the surface is ponded with a very thin layer of water, so that the water pressure is essentially atmospheric.

9.3.1   *Diffusion formulation of vertical infiltration of ponded water*

The vertical downward movement of water into a dry soil, when the displaced air escapes freely, is described by Equation (9.1), the one-dimensional version of Richards's (1931) equation (8.55). Making use of Equation (8.32), one can express this as a diffusion equation, or

$$\frac{\partial \theta}{\partial t} = \frac{\partial}{\partial z}\left(D_w \frac{\partial \theta}{\partial z}\right) - \frac{\partial k}{\partial z} \tag{9.61}$$

To provide the maximal water supply rate at the point of entry $z = 0$, the soil surface is considered to be covered by a thin layer of water, so that at the surface the soil water pressure is atmospheric and the soil saturated. The initial water content is assumed to be uniform throughout the profile. This situation is described by the boundary conditions (9.2). Because the pressure H is eliminated in the diffusion formulation of the flow, these are simply

$$\begin{aligned} \theta &= \theta_i & z &> 0 & t &= 0 \\ \theta &= \theta_0 & z &= 0 & t &\geq 0 \end{aligned} \tag{9.62}$$

The solution of this problem is normally expressed as $\theta = \theta(z, t)$. Once this solution is known, it can be used in the form $z = z(\theta, t)$ to obtain the cumulative infiltration volume. As illustrated in Figure 9.4, this can be written as follows

$$F_c = \int_{\theta_i}^{\theta_0} z\, d\theta + k_i t \tag{9.63}$$

where k_i is the capillary conductivity at $\theta = \theta_i$; the symbol F_c is given the subscript c to indicate that it describes infiltration capacity. The second term on the right of Equation (9.63) represents the downward motion of the water initially present in the soil, under the influence of gravity; this is presumably negligibly small in most cases, if the soil is initially dry enough. The infiltration rate can then be immediately calculated as $f_c = dF_c/dt$. Alternatively, as before, the infiltration rate can also be determined as the Darcy flux at $z = 0$, that is in its diffusion form,

$$f_c = \left[-D_w \frac{\partial \theta}{\partial z} + k\right]_{z=0} \tag{9.64}$$

For this problem, that is (9.61) subject to (9.62), numerous numerical methods of solution have been presented in the literature. Again, however, in applications at the scale of a catchment and of a region, it is often desirable to describe the phenomenon by a concise, yet physically realistic, parameterization. Several such parameterizations are treated in what follows.

9.3.2   *Vertical infiltration as horizontal flow perturbed by gravity*

The time expansion by Philip (1957b; 1969) is a method of solution that has received much attention, as it was probably the first realistic attempt at solving the infiltration

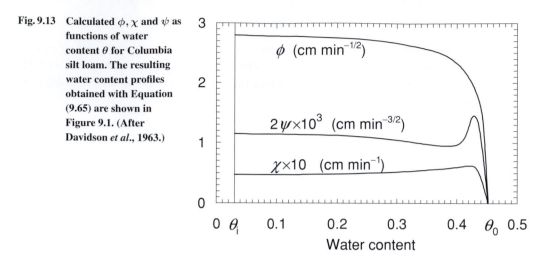

Fig. 9.13 Calculated ϕ, χ and ψ as functions of water content θ for Columbia silt loam. The resulting water content profiles obtained with Equation (9.65) are shown in Figure 9.1. (After Davidson *et al.*, 1963.)

problem; it has also stimulated subsequent advances in this field. This solution of (9.61), which is equivalent to a perturbation series around the solution of (9.6) for a deep homogeneous soil profile with the same boundary conditions, can be written as follows

$$z = \phi t^{1/2} + \chi t + \psi t^{3/2} + \omega t^2 + \cdots \tag{9.65}$$

where the functions $\phi = \phi(\theta)$ (see Equation (9.11)), $\chi = \chi(\theta)$, $\psi = \psi(\theta)$ and $\omega = \omega(\theta)$, etc., are each governed by a separate ordinary differential equation, one of which is (9.13) for ϕ; for each of these equations Philip (1957b) presented a numerical method of solution. The time expansion solution has been reported to produce good agreement with experimental data in the laboratory by Davidson *et al.* (1963) (see Figure 9.1). The functions ϕ, χ and ψ that were obtained in this calculation are shown in Figure 9.13.

Substitution of (9.65) into (9.63) produces the rate of infiltration $f_c = dF_c/dt$ as follows

$$f_c = \frac{1}{2} A_0 t^{-1/2} + (A_1 + k_i) + \frac{3}{2} A_2 t^{1/2} + 2A_3 t + \cdots \tag{9.66}$$

where A_0 is the sorptivity defined in (9.17), and $A_1 = \int_{\theta_i}^{\theta_0} \chi d\theta$, $A_2 = \int_{\theta_i}^{\theta_0} \psi d\theta$ and $A_3 = \int_{\theta_i}^{\theta_0} \omega d\theta$.

The main shortcoming of any series solution like (9.65) is that eventually, i.e. for large values of t, it fails to behave properly. This can be readily seen by comparing the rate of infiltration obtained by using Equation (9.66) with the actual rate given by (9.4). Although eventually the infiltration rate should approach the finite value k_0, the time expansions (9.65) and (9.66) fail to converge for large values of t; therefore, they can only be expected to be applicable for small and intermediate values of time.

9.3.3 A closed form of the series solution

Even though the series expansion (9.65) itself diverges, for some cases it can be shown (Brutsaert, 1977) to lead to a formulation for infiltration which is well behaved both for small and for large t. This formulation can be obtained by making use of an approximate – but quite accurate – method for solving the differential equations for ϕ, χ, ψ and ω. By using the power-form functions (8.36) and (8.41), for the asymptotic case of very large b (which in (8.15) represents soils with a narrow pore size distribution) one finds that the infiltration rate (9.66) can be approximated quite closely by

$$f_c = k_0 + \frac{1}{2}A_0 t^{-1/2}(1 - 2y + 3y^2 - 4y^3 \cdots) \tag{9.67}$$

where $y = k_0 t^{1/2}\beta_0/A_0$ and A_0 is the sorptivity, as before; β_0 is a constant that depends on the pore size distribution of the soil, which for most soils is of the order of 2/3. The main point of interest in Equation (9.67) is that for $y^2 < 1$ it can be expressed in closed form as a two-parameter algebraic infiltration equation, viz.

$$f_c = k_0 + \frac{1}{2}A_0 t^{-1/2}[1 + \beta_0(k_0 t^{1/2}/A_0)]^{-2} \tag{9.68}$$

which does not diverge for large t but instead tends to the proper limit $f_c = k_0$, as required by (9.4); also, for small t, Equation (9.68) approaches the proper limit, viz. $f_c = (1/2)A_0 t^{-1/2}$, as required by (9.19). This correct behavior at low and high values of t is also an indication that (9.68) is relatively insensitive to the exact value of β_0. The cumulative infiltration corresponding to (9.68) is

$$F_c = k_0 t + \frac{A_0^2}{\beta_0 k_0}\{1 - [1 + \beta_0(k_0 t^{1/2}/A_0)]^{-1}\} \tag{9.69}$$

For a more general comparison, it is again convenient to express these results in terms of dimensionless variables; Equation (9.68) confirms the scaling already formulated in (9.21) for horizontal infiltration; accordingly for infiltration capacity one has the following

$$t_+ = \frac{k_0^2 t}{A_0^2}, \quad f_{c+} = \frac{f_c}{k_0} \qquad \text{and} \qquad F_{c+} = \frac{k_0 F_c}{A_0^2} \tag{9.70}$$

Thus the scaled rate of infiltration can be written as

$$f_{c+} = 1 + \frac{1}{2}t_+^{-1/2}[1 + \beta_0 t_+^{1/2}]^{-2} \tag{9.71}$$

and the corresponding cumulative infiltration as

$$F_{c+} = t_+ + \beta_0^{-1}[1 - (1 + \beta_0 t_+^{1/2})^{-1}] \tag{9.72}$$

Equation (9.71) is illustrated in Figure 9.14, where it can also be compared with the time expansion expression (9.67), with the short time expression (9.19), i.e. $f_{c+} = t_+^{-1/2}/2$, and with the long time expression (9.4), i.e. $f_{c+} = 1$.

The convergence criterion for (9.67), that is $y < 1$ or $t_+ < \beta_0^{-2}$, suggests that the "small to intermediate values of time" as mentioned behind (9.66), for which (9.65) is valid, should satisfy at least

$$t < (1.5 A_0/k_0)^2 \tag{9.73}$$

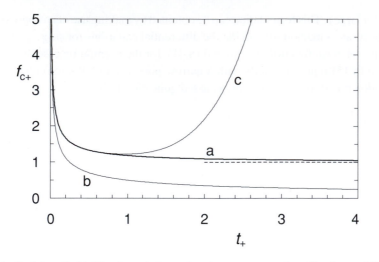

Fig. 9.14 Scaled vertical infiltration rate into a deep homogeneous soil profile, $f_{c+} = f_c / k_0$ as a function of scaled time $t_+ = k_0^2 t / A_0^2$ as given by Equation (9.71) (curve a). Also shown are the horizontal infiltration rate $f_{c+} = t_+^{-1/2}/2$ (curve b), and the time expansion (9.66) as given by (9.67) (curve c); the dashed line indicates the long-time asymptote $f_{c+} = 1$.

Major advantages of (9.68) and (9.69) are that they contain only the two parameters k_0 and A_0 and that a number of measurement and simple calculation (see Section 9.2.3) methods are available for these parameters.

9.3.4 Additional effects

The formulation of potential infiltration, as reviewed here, is for a rather idealized case of a deep homogeneous soil profile. In field situations there may be such important complications as the effect of air movement, areal variability and stratification of soil properties, an impermeable layer at shallow depth, the effect of crusting at the soil surface, fingering, non-uniform initial water content after redistribution during a period of drainage and evaporation, and others. Some are briefly touched upon in what follows.

Flow of air

Under certain conditions the movement of air can greatly reduce the movement of infiltrating water. For example, it is not unusual that, when there is a shallow impermeable layer or a shallow water table below a relatively flat surface ponded with water, bubbling of air can be observed; this is evidence of a counterflow of air, which undoubtedly (see Linden and Dixon, 1975) reduces the water intake rate. Several mathematical formulations have been developed to describe infiltration as a flow problem of two immiscible fluids. Examples of these can be found in the studies by McWhorter (1971) and Sonu and Morel-Seytoux (1976). When the objective is a rigorous physical description of the flow at the Darcy scale, infiltration should be considered as a two-phase flow phenomenon. However, when the objective is the derivation of a parametric equation to describe the phenomenon at the field scale, the one-phase flow assumption, with Richards's equation (8.46), is likely to be quite adequate as a first approximation. This is especially true when the soil profile is deep, without a shallow water table, or when, as is often the case in the field, there is some surface connected macroporosity (not accounted for by the soil moisture characteristic) as a result of shrinkage cracks, worm holes or root channels. Parlange and Hill (1979) have studied the air effect by comparing solutions in which the air movement is considered with that resulting from Richards's equation. For the case where the soil column is sealed at the bottom, the difference was found to be quite large; however, their results showed a difference of only 2% in water intake for the case where the air can move ahead of the wetting front without an appreciable pressure buildup. In experiments dealing with natural soils, a difference of 2% is very difficult to detect.

Soil variability

The spatial variability of soil properties has been studied with measurements in the field (see Nielsen *et al.*, 1973; Rogowski, 1972; Warrick *et al.*, 1977); more recently attempts have been made with remotely sensed data (Cosh and Brutsaert, 1999). However, it is still very difficult to use this type of information to determine infiltration over a larger area. The effects of stratification or layering of soil properties and of crusts at the surface have received considerable attention (Miller and Gardner, 1962; Philip, 1967; Bouwer, 1969; Hillel and Gardner, 1970; Ahuja and Swartzendruber, 1973; Bruce *et al.*, 1976). The details of instabilities at the wetting front during infiltration into stratified soils have also been investigated (White *et al.*, 1977; Selker *et al.*, 1992; Liu *et al.*, 1994a;b).

9.3.5   *Some other expressions for potential infiltration*

For most practical applications Equations (9.68) and (9.69) should be adequate as a parameterization of infiltration capacity. However, over the years, several other equations have been proposed and used in applied hydrology.

Truncated series expansion

A number of well-known equations can be considered truncated versions of Philip's time expansion series (9.66). Probably the oldest formulation was developed by Kozeny (1927), and can be written as

$$f_c = a t^b \tag{9.74}$$

where a and b are constants. Kozeny (1927) arrived at this form with $b = -1/2$, by making use of the analogy with flow into vertical capillary tubes, and by showing that this agrees with Wollny's (1884) experimental data. Equation (9.74) was later also proposed by Kostiakov (1932) and others (see, for example, Lewis, 1937) on empirical grounds. Theoretically, if (9.74) is considered as the first term of (9.66) the constants should be $a = A_0/2$ and $b = -1/2$, but with these values it would only be valid for short times. On the other hand, if (9.74) is to be used for large values of time, the constants should be $a = k_0$ and $b = 0$, in accordance with Equation (9.4). Equation (9.74) can be useful for certain purposes, but only over relatively limited time ranges, with values of a and b intermediate between these extremes and dependent on the range of interest.

Because the series (9.66) diverges for large values of time, Philip (1957a) proposed

$$f_c = a t^{-1/2} + b \tag{9.75}$$

where, a and b are constants. On the basis of the analysis leading to Equation (9.66), these constants can be estimated as $a = A_0/2$ and $b = (A_1 + k_i)$ at least for short to intermediate values of time. However, these values of a and b cannot serve to describe the phenomenon for large values of time; indeed, calculations of A_1 for different soils (Brutsaert, 1977; see also Equation (9.67)) show that it is usually of the order of $k_0/3$; thus with $b = (A_1 + k_i)$ in (9.75), f_c will also approach this value, rather than k_0, as required by Equation (9.4). This means that, strictly speaking, (9.75) can be applicable only over a limited time range, and that the values of a and b depend on that range. But in many situations of practical interest this should not be a serious obstacle, provided the constants a and b are considered curve-fitting parameters to suit the time range of interest.

Exponential decay equation

Horton (1939; 1940) proposed an empirical equation that has received wide attention in hydrology, in the form of an exponential decay function,

$$f_c = a + (b - a) e^{-ct} \tag{9.76}$$

where a, b and c are constants, which have to be estimated (Horton, 1942). Clearly, b is the initial infiltration rate and a should be equal to k_0. Although the exponential function is mathematically convenient in practical applications, this time dependency is hard to reconcile with the results of the theoretical analyses based on Richards's equation.

9.4   RAIN INFILTRATION

Observed rainfall rates in nature only rarely exceed the initial infiltration capacity of the soil. Therefore, in most situations, for a certain initial period at least, all the rainfall that reaches the ground surface without being intercepted infiltrates into the soil profile. During this initial phase, the surface water content gradually increases and the absorptive capacity of the soil decreases. There are two possible scenarios for what happens next, depending on the intensity of the precipitation (see Figure 9.15). Consider the simple case of a constant rate of precipitation on the surface of a deep homogeneous soil profile. If the rainfall intensity is smaller than the satiated hydraulic conductivity, i.e. $P \le k_0$, it will never exceed the ability of the soil to absorb the rainwater. Eventually, the surface water content will tend to reach a value θ_s, such that the hydraulic conductivity at that water content is equal to the precipitation rate, or $k(\theta_s) = P$. In the limiting case, if $P = k_0$, the soil surface will eventually, as $t \to \infty$, reach full satiation, or $\theta_s \to \theta_0$. The second scenario occurs when $P > k_0$. Although initially following the onset of precipitation all the rainwater infiltrates, after a finite period of time $t = t_p$ the soil surface becomes fully satiated, i.e. $\theta_s = \theta_0$. From that moment onward, conditions change markedly: as the surface soil is satiated and the rainfall intensity exceeds the infiltration capacity, ponding takes place and the excess precipitation may be evacuated by overland flow.

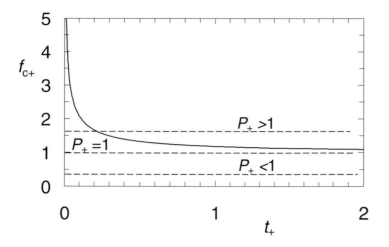

Fig. 9.15 Different precipitation intensities $P_+ (= P/k_0)$ (dashed lines) superimposed on the scaled infiltration capacity $f_{c+} = f_{c+}(t_+)$ of a homogeneous deep soil profile (solid curve). When $P > k_0$ the soil surface becomes satiated after a finite period of time $t = t_p$, and ponding starts. When $P = k_0$ the soil surface becomes satiated eventually as $t \to \infty$, but the rain can always seep into the soil and no ponding occurs. When $P < k_0$ the soil surface remains always unsaturated; eventually it reaches a water content $\theta = \theta_s$, such that the near-surface hydraulic conductivity satisfies $k(\theta_s) = P$, and ponding never occurs.

These changing conditions at the surface also have ramifications in the prediction, and possibly control, of erosion and related processes. Prior to ponding the surface soil is unsaturated, and therefore the soil water pressure is negative; this results in an effective stress (cf. Equations (8.64) and (8.78)), which provides some cohesion to the soil particles. Once the surface soil becomes saturated however, this effective stress vanishes, and the particles can be more easily carried away by the runoff. Thus the onset of ponding conditions may also mean the onset of erosion.

Another interesting feature of the solution of the rain infiltration problem is that it can provide estimates of some hydraulic properties of the soil by inverse calculations with measurements obtained in simple sprinkler irrigation experiments.

9.4.1   *Mathematical formulation*

During rainfall infiltration, the flow is again governed by Richards's Equation (9.1). Consider the case when $P > k_0$, which eventually results in a satiated soil surface. As already described briefly in Section 9.1.2, initially prior to ponding, the boundary condition at $z = 0$ is of the flux type; once the surface soil moisture becomes satiated and ponding starts, that boundary condition changes to a concentration-type condition. These conditions are formulated as the second and third of (9.5), respectively, and they indicate how the phenomenon takes place in two distinct phases. In principle, the complete solution of (9.1) subject to (9.5), should be of the form $\theta = \theta(z, t)$, which is the water content distribution in the soil profile.

Although numerical methods of solution can be obtained (Rubin, 1966; Smith, 1972), the problem remains a difficult one. A number of simpler approaches have provided ways of obtaining more concise parameterizations for certain aspects of the phenomenon. These have consisted of extensions of the approach of Green and Ampt (Mein and Larson, 1973; Swartzendruber and Hillel, 1975; Chu, 1978); empirical equations derived from the numerical solution of Equation (9.1) (Smith, 1972; Smith and Chery, 1973); and equations derived by the analytical solution of (9.1) on the basis of the quasi-steady state or other approximations (Parlange, 1972; Smith and Parlange, 1978; Broadbridge and White, 1987; White and Broadbridge, 1988; White et al., 1989).

Probably the most important part of any solution, for practical purposes, is the determination of the time to ponding t_p and the subsequent infiltration rate. In the following two sections, parameterizations are developed for this purpose.

9.4.2   Time to ponding

Consider for this analysis the simplest possible case of precipitation on the surface of a deep uniform soil profile. One of the oldest approximate methods for solving nonlinear diffusion equations like (9.1) consists of considering the problem as a succession of steady states. In groundwater theory it was used as early as 1886 by K. E. Lembke (Polubarinova-Kochina, 1962, p. 573) to approximate the Boussinesq equation (10.30), that is (9.6) with $D_w \sim \theta$, in the analysis of the drainage or desorption problem. Later, essentially the same method was applied by Landahl (1953) in the solution of the linear diffusion equation and was then generalized by Macey (1959) to the nonlinear diffusion equation (9.6) for sorption; Parlange (1971) applied the concept to derive a first estimate of the soil water profile $\phi = \phi(\theta)$ for sorption, that is (9.32) with $a = 1$ and $b = 0$. Parlange (1972) and Parlange and Smith (1976) then explored the same approach to study rainfall infiltration; this quasi-steady state approach is described next.

Sharp front approach

The approach is based on the assumption of a sharp wetting front, such that, once this front has passed a point, the water content θ is already so close to satiation that it does not change much from then on. Thus the term $(\partial\theta/\partial t)$ in the Richards equation (9.1) can be neglected, so that the term on the right-hand side becomes zero; this means that the specific flux is the same at all z, including at the surface, $z = 0$, where it is equal to the precipitation intensity P. Therefore, after one integration, (9.1) in its diffusion form (9.61) yields

$$P = -D_w \frac{\partial\theta}{\partial z} + k \qquad (9.77)$$

This is in accordance with the second of (9.5), and can be integrated a second time to yield

$$z = -\int_{\theta_s}^{\theta} \frac{D_w}{P - k} d\theta \qquad (9.78)$$

where θ_s is the water content at the soil surface $z = 0$, which changes with time as the precipitation proceeds. In Equations (9.15) and (9.63) the infiltrated volume F is obtained by integration of $(z\,d\theta)$; as illustrated in Figure 9.4, this can also be accomplished by integration of $(\theta\,dz)$, so that one can write instead

$$F = \int_{0}^{z_{\mathrm{f}}} (\theta - \theta_{\mathrm{i}})dz + k_{\mathrm{i}}\,t \tag{9.79}$$

where z_{f} is the position of the wetting front. In the present situation F is also the time integral of the precipitation rate, or

$$F = \int_{0}^{t} P\,dt \tag{9.80}$$

These two expressions for the cumulative infiltration, namely (9.79) and (9.80) can be combined to yield

$$\int_{0}^{t} P\,dt = -\int_{\theta_{\mathrm{i}}}^{\theta_{\mathrm{s}}} (\theta - \theta_{\mathrm{i}})(\partial z / \partial \theta)\,d\theta \tag{9.81}$$

in which it is assumed that k_{i}, the conductivity at the initial water content, is negligibly small, and the limits are θ_{s} at $z = 0$ and θ_{i} at $z = z_{\mathrm{f}}$. Substitution of (9.77) (or (9.78)) into (9.81) produces

$$\int_{0}^{t} P\,dt = \int_{\theta_{\mathrm{i}}}^{\theta_{\mathrm{s}}} \frac{(\theta - \theta_{\mathrm{i}})D_{\mathrm{w}}}{(P - k)}d\theta \tag{9.82}$$

This is an important result in that it represents a relationship between the water content at the soil surface and time, $\theta_{\mathrm{s}} = \theta_{\mathrm{s}}(t)$ prior to ponding.

The time to ponding $t = t_{\mathrm{p}}$ can be derived from Equation (9.82), by considering it as the time necessary for the surface soil moisture content to reach satiation, or $\theta_{\mathrm{s}} = \theta_{0}$. Thus one has

$$\int_{0}^{t_{\mathrm{p}}} P\,dt = \int_{\theta_{\mathrm{i}}}^{\theta_{0}} \frac{(\theta - \theta_{\mathrm{i}})D_{\mathrm{w}}}{(P - k)}d\theta \tag{9.83}$$

Both D_{w} and k are functions of the water content θ, so that, when they are known, it should in principle be possible to carry out the integration in Equation (9.83) for any arbitrary time distribution of the precipitation $P = P(t)$. The integration becomes especially simple if it can be assumed that D_{w} and k change rapidly in the vicinity of $\theta = \theta_{0}$. Take for example (8.36) for k and (8.40) for D_{w}; these are simple power functions, but the result would be the same with any other functions, such as for example exponential functions, which exhibit a similar behavior near $\theta = \theta_{0}$. After normalization of the water content with Equation (9.8) and substitution of these two functions, (9.83) becomes

$$t_{\mathrm{p}} = \frac{H_{\mathrm{b}}(\theta_{0} - \theta_{\mathrm{i}})k_{0}}{\langle P \rangle b} \int_{0}^{1} \frac{S_{\mathrm{n}}^{n-1/b}dS_{\mathrm{n}}}{(P - k_{0}S_{\mathrm{n}}^{n})} \tag{9.84}$$

in which $\langle P \rangle$ is the average precipitation rate during the event until the onset of ponding. After bringing the term $S_{\mathrm{n}}^{n-1/b}$ inside the differential, one recognizes immediately the presence of a number of terms which resemble the sorptivity (9.49) for the same power diffusivity (8.40), provided the exponent n is not small. Therefore to a good approximation; Equation (9.84) can be rewritten as

$$t_p = \frac{A_0^2}{2\langle P \rangle} \int_0^1 \frac{d S_n^{n-b-1+1}}{(P - k_0 S_n^n)} \tag{9.85}$$

Whenever $b = 1$ or both n and b are large, (9.85) can be integrated to yield the main result

$$t_p = \frac{A_0^2}{2\langle P \rangle k_0} \ln\left(\frac{P_p}{P_p - k_0}\right) \tag{9.86}$$

where P_p is the precipitation rate at the time of ponding, $t = t_p$. Equation (9.86) was first presented by Parlange and Smith (1976); they derived it from (9.83) in a somewhat different way, namely without the specific expressions for k and D_w used here. Indeed, it can be seen that this result is independent of the parameters in the two power expressions (8.36) for k and (8.40) for D_w.

Practical implementation

In many situations of practical interest, the rainfall intensity can be assumed to be constant during the storm or at least prior to the onset of ponding; in this case one has $P = \langle P \rangle = P_p = $ constant and Equation (9.86) can be written as

$$t_p = \frac{A_0^2}{2 P k_0} \ln\left(\frac{P}{P - k_0}\right) \tag{9.87}$$

As usual, Equation (9.87) can be made more general by expressing it in terms of dimensionless variables. Its form suggests immediately the same scaling of the time variable, already used in (9.21) and in (9.70); in addition, it suggests that the precipitation rate be scaled with the hydraulic conductivity, so that

$$t_{p+} = \frac{k_0^2 t_p}{A_0^2}, \quad \text{and} \quad P_+ = \frac{P}{k_0} \tag{9.88}$$

With these scaled variables (9.87) can be expressed as

$$t_{p+} = \frac{\alpha_p}{P_+} \ln\left(\frac{P_+}{P_+ - 1}\right) \tag{9.89}$$

where α_p is a constant equal to 0.5. As expected, both (9.87) and (9.89) show how the ponding is instantaneous at the beginning of the rainfall event, i.e. $t_p = 0$, when $P \gg k_0$, that is when the rainfall intensity is much larger than the satiated hydraulic conductivity of the soil. On the other hand, ponding never occurs, i.e. $t_p \to \infty$, when the hydraulic conductivity k_0 is equal to or larger than the precipitation rate P. These features are illustrated in Figures 9.15 and 9.16.

In the derivation of the results given in Equations (9.86), (9.87) and (9.89) it was pointed out that the same would be obtained with any other functions for k and for D_w in (9.83), as long as they change rapidly in the vicinity of $S_n = 1$. This kind of k and

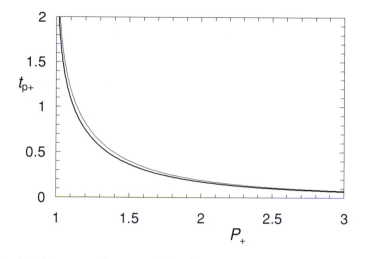

Fig. 9.16 Scaled time to ponding $t_{p+} = (k_0/A_0)^2 t_p$ against scaled rainfall rate $P_+ = (P/k_0)$; the rainfall rate is assumed to be constant during the rainfall event. The heavy line represents the solution obtained by the quasi-steady state approach, i.e. Equation (9.89) with $\alpha_p = 0.50$ (Parlange and Smith, 1976); the thin line is the relationship recommended by Broadbridge and White (1987) for field soils whose hydraulic properties are not well known, i.e. Equation (9.89) with $\alpha_p = 0.55$.

D_w behavior is typical for repacked soils in the laboratory and for undisturbed field soils at greater depth, which have little secondary structure. However, as illustrated in Figure 9.8, this is not always the case for the diffusivity D_w of undisturbed field soils near the surface, where secondary structure may affect the hydraulic characteristics. Broadbridge and White (1987) analyzed the performance of several approximate expressions for t_p by comparing them with the results of an exact solution for a special functional form of $k(\theta)$ and $D_w(\theta)$. They concluded that among them Equation (9.87) of Parlange and Smith (1976), or Equation (9.89) with $\alpha_p = 0.50$, is indeed quite accurate for soils with a structure similar to that of repacked soils; they also found that Equation (9.89) can be made to describe the time to ponding for most soils with a different structure by adjusting the constant α_p within a range between 0.50 and 0.66. For field soils, whose hydraulic properties are unknown, they recommended the use of (9.89), and they felt that $\alpha_p = 0.55$ would be a reasonable choice which would result in errors of at most $\pm 10\%$. Equation (9.89) with $\alpha_p = 0.55$ is also illustrated in Figure 9.16.

9.4.3 *Infiltration after the start of ponding: time compression approximation*

Prior to the time of ponding, the precipitated water can readily enter into the soil, and the rate of infiltration is equal to the precipitation rate, as indicated by the second of Equations (9.5). Once ponding starts, the soil surface is satiated, that is $\theta = \theta_0$, so that the boundary condition at $z = 0$ becomes the third of (9.5), which is the same as the second of (9.62) for the infiltration capacity problem. However, the initial condition at the start of ponding is quite different from the initial condition used to describe the infiltration capacity, i.e. the first of (9.62). In fact, the initial soil water content distribution at the start of ponding cannot be prescribed in general beforehand, because it will depend on the specifics of the duration and intensity of infiltration prior to ponding for each rainfall event. Because a detailed solution of Richards's equation for each rainfall occurrence is neither practical nor feasible, it is useful to explore further simplification of the problem.

Several parameterizations of rainfall infiltration that have been proposed in the past involve the concept of *time compression* (also called *time condensation*), or some assumption similar to it. Briefly, the underlying assumption is that the potential infiltration rate, at any given time after the onset of ponding within a rain storm period, depends only on the previous cumulative infiltration volume, regardless of the previous infiltration or rainfall history during that same storm. The time compression approximation (TCA) was introduced in the 1940s (see Sherman, 1943; Holtan, 1945) in the context of partitioning the rainfall on a watershed into runoff and infiltration, and was later applied in many other studies (see Reeves and Miller, 1975; Sivapalan and Milly, 1989; Salvucci and Entekhabi, 1994; Kim *et al.*, 1996).

Conceptually, TCA can be considered another instance of the application of the lumped kinematic approach, as formulated in Equation (1.10). Thus the soil profile is a one-dimensional control volume, the rate of infiltration f is the inflow rate Q_i, and the cumulative infiltration F is the storage S. After the inception of ponding, the inflow rate f is assumed to be a function of the storage F only, independent of the precipitation history.

General formulation
Let $f = f(t)$ and $F = F(t)$ denote the actual infiltration rate and actual cumulative infiltration, respectively; as these are functions of time, one also has the inverse functions $t = t(f)$ and $t = t(F)$. Similarly, $f_c = f_c(t)$ and $F_c = F_c(t)$ denote the same functions for the infiltration capacity, that is under potential conditions, as analyzed in Section 9.3. and subject to boundary conditions (9.2); their inverse can be written, respectively as $t = t(f_c)$ and $t = t(F_c)$. The basic assumption of TCA can be expressed as follows

$$
\begin{aligned}
f &= P && \text{for } t < t_p \\
f &= f_c\left(t(F_c = F)\right) && \text{for } t \geq t_p
\end{aligned}
\tag{9.90}
$$

With a constant (or average) rate of rainfall P, the cumulative infiltration at the time of incipient ponding is $(P\,t_p)$. Define now the (fictitious) compression reference time t_{cr} as the time period after the start of rainfall that would be required to produce the same infiltrated volume, but under potential conditions. Thus one has

$$
F(t_p) = P\,t_p = F_c(t_{cr})
\tag{9.91}
$$

from which t_{cr} or t_p can be estimated. Once t_{cr} and t_p are known, the cumulative infiltration is given by

$$
\begin{aligned}
F(t) &= Pt && \text{for } t < t_p \\
F(t) &= F_c(t - (t_p - t_{cr})) && \text{for } t \geq t_p
\end{aligned}
\tag{9.92}
$$

In light of (9.90), the rate of infiltration is

$$f(t) = P \qquad\qquad \text{for } t < t_p$$
$$f(t) = f_c(t - (t_p - t_{cr})) \quad \text{for } t \geq t_p \qquad\qquad (9.93)$$

In Equations (9.92) and (9.93) the time to ponding and the compression reference time remain to be determined. These two variables are related by (9.91); as this is one equation with two unknowns, additional information is needed to solve for t_p and t_{cr}. The time to ponding is a real physical quantity, whereas the compression reference time t_{cr} is essentially a parameter, arising in the TCA approximation. There are two possible procedures of estimating t_{cr}. In the first, it is obtained from the time to ponding; the latter is estimated independently, from measurements or with expressions like Equations (9.86) or (9.89). In the second procedure, use is made of the precipitation intensity to solve for both t_p and t_{cr} by means of the TCA approximation.

Estimation with the correct time to ponding
In the first procedure the value of t_p is determined independently and with a known or tolerable accuracy. The time to ponding can be measured directly, as in controlled situations during irrigation, or by appropriate observations; also, as reviewed in the previous section, there are reliable expressions available for this purpose, which are based on the solution of Richards's equation (see Parlange and Smith, 1976; Broadbridge and White, 1987). With t_p known, one obtains then as the inverse of Equation (9.91)

$$t_{cr1} = t(F_c = P\,t_p) \qquad\qquad (9.94)$$

in which the subscript 1 indicates that t_{cr} is obtained by the first alternative procedure. Note that with this procedure the infiltration rate has a discontinuity at $t = t_p$ (see Figure 9.17). This is unavoidable, and is a result of the approximate nature of TCA. However, the basic assumption of TCA, expressed in Equation (9.90), is satisfied.

Estimation from the precipitation intensity
In past applications of TCA, it has usually not been assumed that the time to ponding t_p can be determined independently. Rather, t_p has usually been estimated by assuming, that under potential conditions t_{cr} is the time after the start of rainfall that would be required to produce not only the same infiltrated volume, i.e. ($P\,t_p$) as at the time of ponding, but also the same infiltration rate, i.e. P. Thus one has the additional equation,

$$P = f_c(t_{cr2}) \qquad\qquad (9.95)$$

from which the compression reference time is obtained as the simple inverse function $t_{cr2} = t(f_c = P)$; the subscript 2 indicates that it is estimated by this second alternative procedure. The time to ponding can then be calculated by means of Equation (9.91) as

$$t_p = F_c(t_{cr2})/P \qquad\qquad (9.96)$$

This procedure is illustrated in Figure 9.18.

In the real world, it is certainly the case that at the time of ponding both $F = Pt_p$ and $f = P$ must be satisfied. But it should be kept in mind that the TCA concept is only

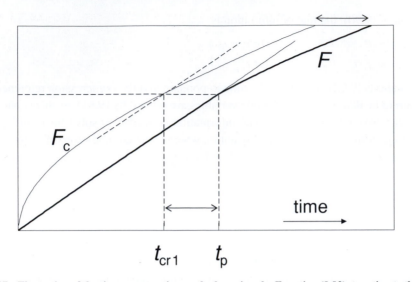

Fig. 9.17   Illustration of the time compression method, as given by Equation (9.92), to estimate the cumulative infiltration F (heavy line), resulting from precipitation rate P. In this version of the method the time to ponding is determined independently, and presumed to be correct. At the time of ponding $t = t_p$, the value of F is equal to that of the F_c curve at $t = t_{cr}$, but its slope is not equal to that of the F_c curve at $t = t_{cr}$; thus, the infiltration rate f has a discontinuity at $t = t_p$, but the infiltrated volume F satisfies the main premise of time compression.

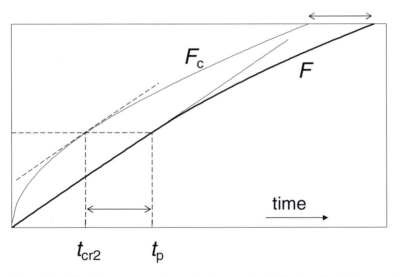

Fig. 9.18   Illustration of the time compression method as given by (9.92), to estimate the cumulative infiltration F (heavy line), resulting from precipitation rate P. In this version of the method, the time to ponding is calculated by assuming that, beside Equation (9.91), Equation (9.95) also holds. At the time $t = t_p$ both F and its slope, i.e. the infiltration rate f, are equal to the corresponding values of the F_c curve starting at $t = 0$. However, this procedure does not produce the correct value of t_p.

an approximation; therefore, it is impossible that both (9.91) and (9.95) are valid. This means that t_p, as obtained with Equation (9.96), is not valid either. The time to ponding is such a crucial parameter in describing rainfall infiltration, that the first procedure to determine the compression reference time should be used, if at all possible. Because TCA is mainly concerned with mass conservation and cumulative infiltration, and less so with infiltration rate, Equation (9.91) should have priority over (9.95). In other words, as pointed out by J.-Y. Parlange (Liu *et al.*, 1998), the first procedure with t_{cr1} is preferable over the second with t_{cr2}. The following example should give an idea of the practical implementation of the TCA concept and of the error involved in the estimation of t_p by means of the second procedure.

Example 9.3. Application with truncated time expansion

The above equations for rainfall infiltration can be applied with any of the available equations for infiltration capacity f_c or F_c discussed in Section 9.3. The calculations become especially simple when time t can be expressed explicitly as a function of infiltration capacity. Consider as an example the truncated time expansion (9.75) proposed by Philip (1957a). By using the procedure, with the correct value of t_p determined independently, one obtains from (9.94) for the compression reference time, implicitly

$$P\,t_p = 2a\,t_{cr1}^{1/2} + b\,t_{cr1}$$

and, after solution, explicitly

$$t_{cr1} = [-a + (a^2 + P\,bt_p)^{1/2}]^2/b^2 \tag{9.97}$$

in which a and b are the constants appearing in Equation (9.75).

With the second procedure, which is based on the infiltration rate constraint, by combining (9.95) with (9.75) one obtains the following

$$t_{cr2} = [a/(P - b)]^2 \tag{9.98}$$

The time to incipient ponding can be calculated by substitution of (9.98) into (9.91), which yields with the cumulative infiltration corresponding to (9.75),

$$t_p = \frac{a^2\,(2P - b)}{P\,(P - b)^2} \tag{9.99}$$

With the values of t_p and t_{cr} known, by either of the two procedures, the cumulative infiltration (9.92) can now be written, for this example with (9.75), as

$$F\,(t) = P\,t \qquad\qquad\qquad \text{for } t < t_p$$
$$F\,(t) = 2a[t - (t_p - t_{cr})]^{1/2} + b[t - (t_p - t_{cr})] \quad \text{for } t \geq t_p \tag{9.100}$$

and in a similar way, the rate of infiltration (9.93).

The error involved in the second TCA procedure can be determined by comparing the resulting expression for t_p, i.e. Equation (9.99), with the more accurate (9.89) derived in the previous section. As before, for the sake of generality, especially in this comparison with another expression, it is useful to express (9.99) in dimensionless form. It should be

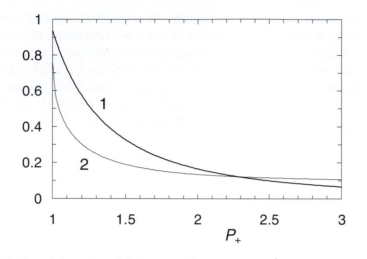

Fig. 9.19 Curve 1 shows the scaled time to ponding $t_{p+} = (k_0/A_0)^2 t_p$ against scaled rainfall rate $P_+ = (P/k_0)$, as calculated with the second procedure in the TCA approximation, i.e. Equation (9.101); the rainfall rate is assumed to be constant during the rainfall event. Curve 2 represents the negative error of this result relative to Equation (9.89) with $\alpha_p = 0.55$ (which is shown in Figure 9.16), and thus indicates that t_{p+} is underestimated with this procedure.

recalled that the constants in the infiltration equation (9.75), used in this example, can be taken as $a = A_0/2$ and $b = k_0/3$; with these values of the constants and in terms of the scaled variables shown in (9.88), Equation (9.99) can be written as

$$t_{p+} = \frac{0.5\,(P_+ - 1/6)}{P_+\,(P_+ - 1/3)^2} \tag{9.101}$$

This result is illustrated in Figure 9.19, and can be compared with the more accurate result shown in Figure 9.16. Also shown is the error inherent in Equation (9.101) relative to (9.89); it can be seen that the error in this second procedure can be considerable, and that it results in underestimates of the time to ponding, ranging between roughly 10% and 70%, depending on the rainfall intensity. As seen in Equations (9.92) and (9.100), an underestimated time to ponding t_p produces an underestimate of the infiltration F.

Accuracy of the TCA approximation

Several studies allow an assessment of the time compression approximation. One is the analysis of intermittent infiltration by Ibrahim and Brutsaert (1968), on the basis of the numerical solution of Equation (9.1) for conditions representing alternating potential infiltration and drainage (or redistribution) cycles. The hysteresis in the soil water characteristic was taken into account by means of the concept of independent domain. Inspection of the results shows that the cumulative infiltration, after restarting it following a drainage period of a given duration, can be obtained by merely time-shifting the initial (i.e. prior to the drainage) cumulative potential infiltration curve over a certain time period; however, the required time-shift period tends to be shorter than the drainage period, which is the time-shift assumed under TCA. This means that TCA usually underestimates infiltration. Similar results were obtained by Reeves and Miller (1975); although in some extreme cases the reported error was as large as 15% to 20%, in most cases it was considerably smaller.

Liu *et al.* (1998) estimated the error in the TCA solution by comparing it with the exact solution of rainfall infiltration for the special case of the linearized Richards equation without gravitational effect, that is Equation (9.6) with a constant $D_w = D_0$. The differences between the two solutions were very small, with the largest occurring near ponding. The maximal error in cumulative infiltration was found to be an underestimate of only 1.3% for the first TCA procedure with (9.94), and about 2.5% for the second procedure with (9.95) and (9.96); similarly small errors were obtained for the rates of infiltration. On the other hand the second procedure underestimated the time to ponding obtained with (9.96) by about 19%. This illustrates that t_p can be a sensitive parameter and that conversely, errors in t_p will cause much smaller error in the resulting F or f. The sensitivity issue was also dealt with by White *et al.* (1989).

9.5   CATCHMENT-SCALE INFILTRATION AND OTHER "LOSSES"

So far in this chapter, infiltration has been considered effectively as a point process. In applied hydrology it is usually necessary to estimate the process over larger areas, often with typical length scales on the order of kilometers. Over the years engineers, faced with the task of predicting storm runoff from precipitation, have developed various, mostly heuristic, approaches to deal with this problem. Some of these are reviewed in what follows.

9.5.1   Infiltration capacity methods

This approach consists of the simple extension of the available information on point infiltration over a larger area. It is currently implemented in many catchment water balance models, by subdividing the catchment area into appropriate subareas with assumed uniform infiltration characteristics; for each subarea an average or typical infiltration capacity relationship is adopted, which is then applied with the time compression approximation to deal with precipitation events. The main difficulties with this approach are the large spatial and temporal variability of soil properties and soil moisture content, that are normally encountered even in so-called homogeneous field situations. This means that it is never an easy matter to define an average $f_c(t)$ function for application over a larger area. As already pointed out in Section 9.3.4, this problem is still poorly understood and will have to be the subject of more research.

9.5.2   The loss rate concept

In many of the methods that have been in use to predict streamflow from rainfall observations for flood-control purposes (see Feldman, 1981), it is necessary to determine a rainfall *excess*, that is the part of the precipitation which generates the direct storm runoff. This is usually done by applying a loss rate to the observed precipitation intensity. Most of this "loss" is assumed to consist of infiltration; however, because it is difficult to consider them separately, other processes such as initial rainfall detention storage in depressions and rainfall interception, are usually included in the total loss. Because generally much

Fig. 9.20 A storm hyetograph showing the rainfall intensity P as a function of time, with an assumed initial loss and a subsequent constant loss rate. The remaining blank area is considered the excess precipitation, which produces the storm runoff.

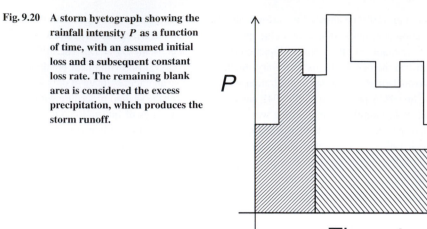

of the infiltrated water participates actively in the runoff generating process, the loss rate concept is of questionable validity and not soundly based. Nevertheless, as long as its limitations are kept in mind, the concept can be useful, especially in situations with limited data and for design purposes involving large flows and floods. There have been two major ways of applying the loss rate concept; in the first, which is used more commonly for large basins, it is assumed that the loss is independent of the rainfall, while in the second one, which is more widely used for smaller areas, the loss is assumed to be proportional to the rainfall.

Loss rate independent of rainfall
In this approach the loss rate is usually taken as a constant rate throughout the rainstorm event, and it is subtracted from the actual rainfall intensity to obtain the excess rainfall intensity. The underlying idea is that this loss rate represents mainly the space- and time-averaged behavior of the infiltration capacity which is controlled by the properties of the soil and which is independent of the rainfall intensities, as long as they are large enough. Several such indices have been proposed in the past, but Horton's (1937) method has probably been most widely used.

In brief, the loss rate is determined as the constant value that must be subtracted from the actual rainfall rate so that the resulting excess rainfall volume over the entire catchment is equal to the actual storm runoff volume; this storm runoff is derived from the streamflow hydrograph by subtracting the (assumed) baseflow from the total runoff. When rainfall is observed at several gages, their input must be properly weighted with their respective areas of influence (see Section 3.3.1) and several trial loss rates are required to obtain the solution.

This method is often applied in modified form, by the inclusion of an initial loss, which is a certain amount that is subtracted from the rainfall at the beginning of the storm event. The principle is illustrated in Figure 9.20. The initial loss can be defined as the loss that takes place before the onset of storm runoff in the stream, and it is usually envisioned to consist of interception storage, depression storage and initial high rate

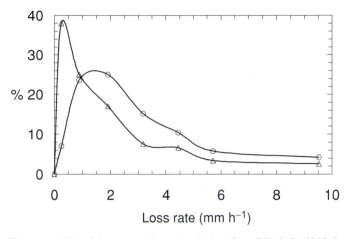

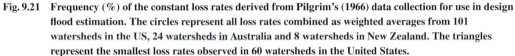

Fig. 9.21 Frequency (%) of the constant loss rates derived from Pilgrim's (1966) data collection for use in design flood estimation. The circles represent all loss rates combined as weighted averages from 101 watersheds in the US, 24 watersheds in Australia and 8 watersheds in New Zealand. The triangles represent the smallest loss rates observed in 60 watersheds in the United States.

infiltration. Various methods have been in use to determine the initial loss, but all of them have drawbacks. In the most obvious method the initial loss is taken as all the rain prior to the start of the rise in the stream flow; but this is not always applicable, because often the rain may be finished before the stream flow hydrograph shows any rise. This difficulty can be avoided by considering as initial loss the maximal isolated burst of rainfall observed in the record, which was not reflected in an obvious rise in the streamflow hydrograph. Another way is to make use of a "typical delay period," which can be derived from the record as the delay between short, intense storms and the subsequent start of the rise in the streamflow hydrograph; when the storm is sufficiently intense, the initial loss can be assumed to be negligible. This period can then serve to determine the start of the rainfall excess, i.e. the end of the initial loss period, in storms of longer duration. Examples of the application of the loss rate methodology can be found in the papers by Cook (1946) and Laurenson and Pilgrim (1963). Cordery (1970) has shown how the initial loss can be related to an antecedent precipitation index, which he used as a measure of the wetness of the catchment.

To give an idea of the values that can be expected for the constant loss rate, Figure 9.21 presents a summary of the data collection of Pilgrim (1966). The circles represent the loss rate frequency as a weighted average of 460 values from 101 watersheds in the United States, 150 values from 24 Australian watersheds and 116 values from 8 watersheds in New Zealand; the results for the three data sets were sufficiently similar so that they could be combined in one single curve. The triangles in Figure 9.21 represent the frequency of the smallest loss rates observed in 60 watersheds in the United States.

Loss rate proportional to rainfall intensity: the runoff coefficient
In the Rational Method, the peak runoff rate (Q_p/A) (expressed as volume rate of flow per unit catchment area) at the outlet of a catchment is assumed to be a fraction of the

rainfall intensity,

$$Q_p/A = CP \tag{9.102}$$

where A is the area of the catchment; C is a constant, also called the runoff coefficient, which ranges between 0 and 1 (see Table 12.2), depending on the nature of the surface. Although the basic approach was proposed some 150 years ago (Mulvany, 1850; Dooge, 1957), various versions of the Rational Method are still in common use in the design of road culverts and other structures draining small areas of a few square kilometers at most. Equation (9.102) suggests that the rainfall loss rate is simply proportional to the rainfall intensity, and equal to $[(1 - C)P]$. Physically, the assumption of proportional losses appears to be more compatible with the early stages of rainfall infiltration (see the second condition of Equations (9.5)) combined with interception losses (see Equations (3.14) and (3.19)) for short precipitation events. In contrast, the constant loss rate methods appear to reflect conditions during longer lasting events (see the third of Equations (9.5)) with eventually a near-constant infiltration capacity. It probably also explains the differences in the sizes of the catchments for which both methods have been applied in engineering. The Rational Method is treated in greater detail in Section 12.2.2.

9.6   Capillary rise and evaporation at the soil surface

The water that evaporates at the soil surface is transported to the surface through the underlying layers of the soil profile. This transport takes place both in the liquid and in the vapor phase; moreover, as evaporation is driven by radiation and other energy inputs, the transport involves not only water pressure gradients, but often also temperature gradients with a soil heat flux. However, as already discussed in Section 8.3.3, in many situations of hydrologic interest, some important features of the evaporation at the soil surface can be obtained on the basis of the isothermal flow equation, viz. Darcy's law (8.19). In particular, two flow problems have been the subject of previous research, that have practical relevance to soil surface evaporation; these are steady capillary rise from a water table to the surface, and unsteady desorption from a deep soil profile without a water table.

9.6.1 Steady capillary rise from a water table

This situation occurs when the water table is maintained at a constant level, from which water flows upward through the soil profile to the soil surface, where it is taken away by evaporation under constant atmospheric conditions. Under steady conditions in the soil profile $(\partial\theta/\partial t) = 0$, and the vertical flux is everywhere given by $q_z = E$. Thus, for a vertical coordinate system pointing upward with $z = 0$ at the water table where $p_w = 0$, one obtains from Equation (8.19)

$$z = -\frac{1}{\gamma_w} \int_0^{x=p_w} \frac{dx}{[1 + E/k(x)]} \tag{9.103}$$

in which x is a dummy variable representing the water pressure. This can be readily integrated for a uniform soil profile, provided the capillary conductivity $k = k(H)$ is known as a function of the soil water suction $H(= -p_w/\gamma_w)$. Gardner (1958) presented solutions of (9.103) with (8.37) for values of the parameter $c = 1, 3/2, 2, 3$ and 4.

Equation (9.103) produces the vertical pressure distribution of the soil water for any given rate of evaporation E. For relatively low values of E or for a soil profile with a water table at a shallow depth d_w below the surface, the value of H at the soil surface is relatively small, i.e. close to zero, and the soil surface is close to saturated. Hence, in such a case the rate of evaporation is governed by the prevailing atmospheric conditions, and not by the ability of the soil profile to transmit water. For a given depth of the water table d_w, as the drying power of the air is increased, the suction H at the soil surface will increase; with this increased gradient the rate, at which water moves upward and evaporates at the surface, will also increase. But eventually a limit is approached beyond which E cannot increase; in the limit E is totally controlled by the ability of the profile to transmit water, regardless of the drying power of the air. For most practical purposes it is probably sufficiently accurate to assume that the actual evaporation at any time is the lesser of the potential evaporation and of the limiting evaporation E_{lim}.

A satisfactory approximation of this limiting value E_{lim} can be obtained by assuming that the soil surface at $z = d_w$ is nearly dry or at field capacity, so that one can assume that $H \to \infty$ and $k \to 0$. Integration of Equation (9.103) with (8.37) then produces in general (Cisler, 1969) the following relationship between the limiting rate of evaporation and the depth of the water table,

$$d_w = \frac{\pi a}{c \sin(\pi/c)(a + bE_{lim})} \left(\frac{a + bE_{lim}}{E_{lim}} \right)^{1/c} \tag{9.104}$$

in which a, b and c are the parameters of Equation (8.37). Since in many cases $a > (bE_{lim})$, this is to a good approximation

$$E_{lim} = a \left(\frac{\pi}{c \sin(\pi/c)} \right)^c d_w^{-c} \tag{9.105}$$

The assumption of isothermal capillary flow is clearly an oversimplification. Especially near the soil surface, transport in the vapor phase is also likely to play a role, so that the limiting evaporation rate is probably larger than the predicted value. However, Gardner (1958) has estimated that this increase is not likely to exceed 20%. In any event, Equation (9.105) indicates that the limiting evaporation is proportional to d_w^{-c}. As shown in Figure 9.22, experimental results by Gardner and Fireman (1958) appear to confirm this. This gives some support to the isothermal flow assumption. (Note that in Figure 9.22 the curve is similar to, but not quite the same as (9.105), because the depth of the water table was simulated in the experiment by maintaining the bottom of the 1 m long column at a negative pressure rather than zero; however, the difference was shown to be small.) Equation (9.103) was used by Willis (1960) to study the steady flow from a water table in the case of a soil profile consisting of two layers of different texture. He concluded that the effect of stratification was pronounced for a system with the coarse-textured soil overlying the fine-textured soil, but not for the reversed condition.

9.6.2 *Unsteady drying of the soil profile and desorption-based parameterizations*

A high water table at a constant depth, as assumed in the previous section, is not a common occurrence; more often than not the water that evaporates from the soil surface is supplied by a release from storage in the soil profile. To facilitate the solution of this problem, it is instructive first to consider this drying process with constant atmospheric conditions.

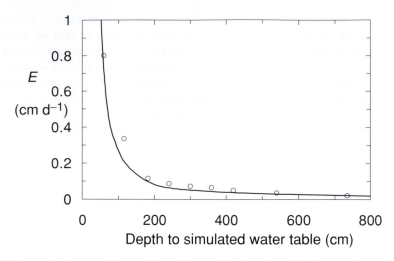

Fig. 9.22 Comparison between the experimental rate of steady evaporation from a column of clay soil and the curve calculated by means of Equation (9.103) with (8.37) in the form $k = 1100/[565 + (-p_w)^2] \, cm \, d^{-1}$, where p_w is in hPa. (After Gardner and Fireman, 1958.)

The first and second stages of evaporation from soil

Ever since the studies of Fisher (1923) and Sherwood and Comings (1933), among others, it has been customary to classify bare soil evaporation under constant external conditions in the laboratory into several stages of drying; from the hydrologic point of view, the first two of these stages are the more important ones. In the first stage, which prevails as long as the soil is sufficiently moist, the evaporation rate is primarily controlled by the atmospheric conditions. It is therefore best expressed in terms of measurements in the atmosphere. For a moist surface under natural conditions, several well-known formulations are available, which make use of atmospheric variables and which are treated in Chapter 4. Obviously, for constant atmospheric conditions, the rate of drying is constant. The duration of the first stage depends on the rate of evaporation and the ability of the soil profile to supply this rate.

As the soil near the surface dries out, the water supply to the surface eventually falls below that required by the atmospheric conditions. In this second or falling-rate stage, the rate of evaporation is mainly limited by the soil moisture conditions and the soil properties, and much less by the available energy. The transition from the first stage to the second stage may be quite abrupt at a given point on the surface; on a wider scale it is usually more gradual, because local transitions at different locations tend to occur at different times. It was noted by Jackson et al. (1976) that the transition from the first to the second stage can be observed visually by changes in color and in albedo.

In the second stage of drying, water moves also through the profile by diffusion of water vapor. And especially after the soil has become quite dry, the water transport through the profile is sensitive to the temperature gradients in the soil as well. Nevertheless, at least initially in the falling rate stage, it appears that the water moves primarily as a liquid. Although the matter is more complicated (Philip, 1957c; Cary, 1967), just like for the steady case, experimental evidence has shown that some of the more important features of the falling-rate stage can be obtained by means of the isothermal flow description with Richards's equation.

Isothermal flow in the absence of gravity: desorption

In a number of studies the formulation has been further simplified by considering the second stage of drying as a problem of desorption. This formulation, which was first used by Gardner (1959), is based on several additional assumptions, beside that of isothermal liquid flow. First, it is assumed that the effect of gravity is negligible. In other words, it is assumed that the drying rate of a vertical soil column is the same as that of a horizontal column, so that Equations (9.3) or (9.6) govern the flow. Second, the boundary conditions are taken to be same as given in (9.7), except that in desorption, instead of θ_0, let θ_s represent the water content at the dry soil surface, so that $\theta_i > \theta_s$; thus, when applied to this case, in the first of (9.7) it is assumed that the initial water content is uniform, and in the second that the water content at the surface is always very low. These conditions are equivalent with the assumption that the energy-limiting drying rate is so large, that the duration of the first stage of drying is negligibly short. Third, vapor transfer in the drier soil near the surface is neglected.

As was the case for the sorption problem, by applying the Boltzmann transform (9.11), Equation (9.6) can be reduced to the ordinary differential equation (9.13). Because $\theta_i > \theta_s$, in desorption the water content is normalized as

$$S_n = \frac{\theta - \theta_s}{\theta_i - \theta_s} \tag{9.106}$$

so that, instead of (9.14), the boundary conditions are

$$\begin{aligned} \theta = \theta_i \ (\text{or } S_n = 1) \qquad & \phi \to \infty \\ \theta = \theta_s \ (\text{or } S_n = 0) \qquad & \phi = 0 \end{aligned} \tag{9.107}$$

To date no general exact solution has been obtained for this problem, but only approximate solutions or exact solutions for certain types of diffusivity functions. Gardner (1959) made use of two solutions. One was the linearized solution obtained by means of a weighted mean diffusivity calculated by Crank's method; his second solution, which was presented graphically, was obtained by iteration for the exponential-type diffusivity (8.39). Parlange *et al.* (1985) proposed an approximate, but quite accurate method of solution for arbitrary diffusivity functions $D_w(\theta)$, in a manner similar to the techniques used for sorption. Finally, it has been shown by Brutsaert (1982) that there is a large class of $D_w(\theta)$ functions that admit exact solutions for desorption as formulated by Equations (9.13) and (9.107); one such function, in particular appeared to have practical relevance to flow in soils and other porous materials. A detailed discussion of these methods of solution is beyond the scope of the present treatment.

In the present context, however, the most interesting feature of any solution by means of the Boltzmann transform, regardless of the solution method and regardless of the assumed diffusivity function $D_w(\theta)$, is that the total water volume lost from the soil profile is proportional to the square root of time. Actually, by analogy with (9.19) and (9.17), for any solution $x = \phi(\theta)t^{1/2}$ it can readily be shown that the flux at the surface, that is the rate of evaporation can be written as

$$E = \frac{1}{2} De_0 t^{-1/2} \tag{9.108}$$

where De_0 can be referred to as the capillary desorptivity, and is defined by

$$De_0 = \int_{\theta_s}^{\theta_i} \phi \, d\theta \tag{9.109}$$

which is a constant for a given soil and given values of θ_i and θ_s.

Fig. 9.23 Cumulative evaporation from a bare soil surface as a function of the square root of time obtained in the laboratory with a 1.0 m long column of clay soil. (After Gardner, 1959.)

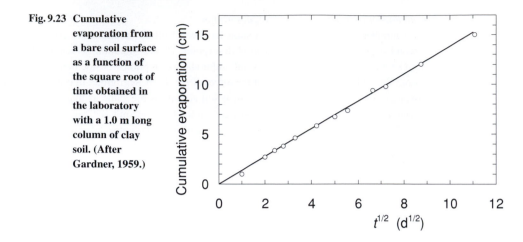

Good agreement was obtained by Gardner (1959) between (9.108) and the evaporation rate from an initially uniformly moist column of clay soil, 100 cm in length, subjected in the laboratory to a large potential evaporation rate of about 4 cm d^{-1}. These results are shown in Figure 9.23. Evidently, the column was long enough to be effectively semi-infinite for about 100 d. Similar results were presented by Gardner and Hillel (1962); in addition they observed that soon after the end of the first stage of drying, the rate of evaporation is independent of the initial drying rate, and that it depends primarily on the water content of the soil. These successful experimental tests of the desorption approach were obtained under constant atmospheric conditions in the laboratory, and should lend support to the underlying assumptions. Nevertheless, it can be expected that the neglect of gravity will somehow result in an overestimate of the evaporation rate, whereas the neglect of the vapor transfer must result in an underestimate of the evaporation. While these effects may be compensating each other to some extent, this will require more study.

9.6.3 Applications in the field

Under field conditions soil evaporation is more complicated than described in the previous section. Clearly, with a diurnal cycle of radiation, the surface boundary condition on θ is not simply a constant θ_0 as given in Equation (9.107), and the hourly changes of the surface water vapor flux even from dry soil are not given by (9.108) (see Jackson et al., 1973; Idso et al., 1974). Hourly values of near-surface θ and evaporation depend markedly on net radiation even after the soil surface has dried considerably. A simple explanation for this is that during the night in the absence of the driving solar radiation, the soil moisture is able to redistribute into some new equilibirum by early morning; this process involves hysteresis with drying from above during the day, and wetting from below during the night, resulting in the distinct diurnal pattern of the surface evaporation in the course of the following day. All this would appear to suggest that, under conditions of a diurnally varying evaporative demand, the two stages of bare soil evaporation and also the desorption approach may not be physically relevant; even relatively dry soil surfaces continue to change from an energy-limiting state to a soil-limiting state during the same day, and the transition is not instantaneous. Additional

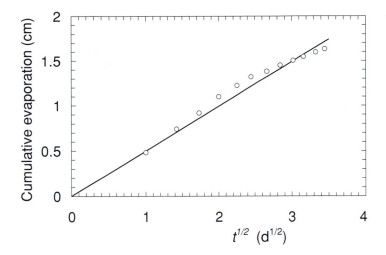

Fig. 9.24 **Cumulative evaporation from a bare sandy soil surface as a function of the square root of time in days, obtained in the field by means of a weighing lysimeter in Wisconsin. The straight line represents the integral of Equation (9.108) with a desorptivity $De_0 = 0.496$ cm d$^{-1/2}$. (Adapted from Black *et al.*, 1969.)**

effects come into play when the surface is not bare but, as is usually the case under natural conditions, covered with vegetation. Nevertheless, as will be shown below, under certain conditions the assumption that soil surface evaporation takes place in two stages, and that the second stage can be described by a desorption formulation, is still a useful construct to describe drying phenomena. But it should be kept in mind that, whenever (9.108) is applied under conditions other than those specified by (9.6) and (9.107), it does not really represent a capillary desorption phenomenon, in the strict sense of the word. Therefore, the value of De_0 obtained this way is probably better referred to as "effective" desorptivity.

Daily evaporation from bare soil

In several field studies it was found that $E = E(t)$ could be described reasonably well as a desorption phenomenon by Equation (9.108), provided this was done with daily time steps. In Wisconsin, Black *et al.* (1969) obtained good agreement between (9.108), in which t was set to zero after each heavy rainfall, and daily evaporation from a bare soil lysimeter measured during an entire summer. The measurements illustrated in Figure 9.24 suggested an effective desorptivity of around $De_0 = 0.496$ cm d$^{-1/2}$. Interestingly, this value was also not very different from the desorptivity $De_0 = 0.43$ cm d$^{-1/2}$, calculated by means of Crank's linearized solution. In light of the natural variability of the soil, and also of the likely errors stemming from the problem formulation and its linearization, the difference is small. Black *et al.* (1969) suspected that after a rainfall the evaporation would eventually depart from the $t^{-1/2}$ relationship, because of the finite depth of wetting. Still, it was possible to simulate an entire summer of evaporation this way. In a similar study in California, Parlange *et al.* (1992) assumed that daily evaporation from a bare soil lysimeter followed the pattern of (9.108) immediately after irrigation of the field; with this assumption their measurements suggested an effective desorptivity $De_0 = 0.58$ cm d$^{-1/2}$. In Arizona, Jackson *et al.* (1976) found that Equation (9.108) could be used to describe the second stage of drying on a daily basis from

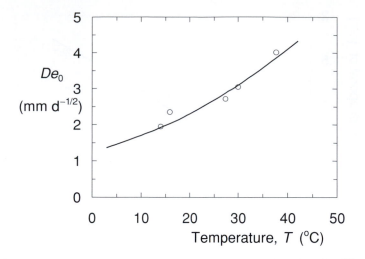

Fig. 9.25 The temperature dependency of the effective desorptivity De_0 in mm d$^{-1/2}$, as measured by a weighing lysimeter with loam in Arizona (open circles), and of the normalized water vapor diffusivity (curve). The indicated temperature is the average of the five daily averages for the first five days of the soil-limiting phase. (After Jackson *et al.*, 1976.)

those portions of a field whose surface moisture content was below a certain threshold value; however, they also found that De_0 exhibits a strong dependency on temperature, varying by a factor of about 2 from winter to summer. In Figure 9.25 these measured values of De_0 can be compared with the normalized temperature dependency of the water vapor diffusivity; the similarity of both dependencies suggests that vapor diffusion contributes substantially to the total water transport in the top layer of the soil.

In these three field studies with daily time steps, the transition between the first and second stage of drying appears to have been relatively abrupt. Jackson *et al.* (1976) concluded that at the field scale any gradual transition was mainly caused by the variability of the surface soil moisture, so that the field was partly in the first and in the second stage of drying. In the analyses of the bare-soil lysimeter measurements by Black *et al.* (1969) and by Parlange *et al.* (1992), the desorptive stage was assumed to have started immediately after the water application had ceased, and the first stage was dispensed with.

Time compression approximation

In the two bare-soil lysimeter studies just mentioned, the desorption formulation was implemented by simply plotting the cumulative evaporation $\sum E$ against the square root of time $t^{1/2}$ to fit a linear relation through the origin, and the end of the precipitation or the irrigation was taken as the starting point. Whenever the first stage is short or nonexistent, this procedure may be acceptable. However, because Equation (9.108) is a nonlinear function of t, the choice of the starting point of the second drying stage, i.e. $t = 0$, is critical for its proper performance. In many situations considerable evaporation can take place under first stage drying conditions and the transition can also be long; thus in general, this starting point is not known and a different approach is needed.

To obtain a more objective procedure, it was proposed by Brutsaert and Chen (1995) that Equation (9.108) should be recast in terms of a relative time $t_r = (t - t_0)$, in which t_0 is

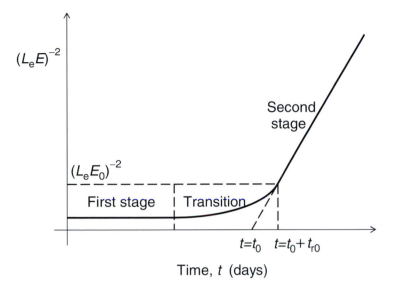

Fig. 9.26 Sketch of a method to determine the two stages of drying by means of Equation (9.110). (After Brutsaert and Chen, 1995.)

a time shift parameter. Then after fitting data of E^{-2} against t, in accordance with (9.108) or

$$E^{-2} = (2/De_0)^2 \, t_r \qquad (9.110)$$

one can derive De_0 and t_0 from the slope and intercept as sketched in Figure 9.26. In order to represent the cumulative evaporation as a function of $t_r^{1/2}$ in the range of validity of (9.108), the sketch illustrates that the starting point of the integration should be $t_r = t_{r0}$, i.e. $t = (t_0 + t_{r0})$, marking the end of the transition, rather than $t_r = 0$. At $t = (t_0 + t_{r0})$, the variable E^{-2} starts its linear relation with t_r. The cumulative evaporation after the onset of the desorptive regime is then

$$\sum E_2 = De_0(t_r^{1/2} - t_{r0}^{1/2}) \qquad (9.111)$$

in which the subscript 2 indicates the second stage of drying. The value of t_{r0}, marking the beginning of the second stage, can be related to the value of the cumulative evaporation at the end of the first stage, or of the transition if there is one, say $\sum E_1$, as follows

$$t_{r0} = \left(\sum E_1/De_0\right)^2 \qquad (9.112)$$

Thus (9.111) can also be expressed as

$$\sum E_2 = De_0 t_r^{1/2} - \sum E_1 \qquad (9.113)$$

As mentioned, (9.110) can be used as a regression equation with experimental data to estimate the values of the effective parameters De_0 and t_0; with a known or decided upon value of $\sum E_1$, Equation (9.113) allows the prediction of the cumulative evaporation after the onset of the second stage of drying.

Conceptually, it can be seen that this procedure is in fact based on a time compression assumption, which is analogous with that used to describe rainfall infiltration after the start of ponding, as described in Section 9.4.3. This means that the evaporation rate at the beginning of its desorptive phase is assumed to depend mostly on the water left in the soil profile and much less on the prior evaporation history. This assumption is consistent with the observations of Gardner and Hillel (1962) under constant laboratory conditions, mentioned in Section 9.6.2.

The effect of vegetation

The two-stage concept with the desorption approach for the second stage was originally developed for evaporation from bare soil. However, it was concluded by Brutsaert and Chen (1995) that the desorption formulation can also be put to use in the description of daily evaporation from grassland and other similar types of short vegetation. From their analysis of experimental data obtained during several drying episodes in a natural tallgrass prairie area in Kansas during FIFE, the First ISLSCP Field Experiment (see Sellers *et al.*, 1992), the following sequence of events appears to take place. Initially after rain, both soil surface and vegetation evaporate at a rate governed by the available supply only; this can be considered as a first stage of drying. (In this particular experimental setting, the first-stage behavior lasted as long as the volumetric soil moisture θ in the top 10 cm was in excess of 27%.) As the surface soil moisture becomes depleted below this first critical level, the water supply rate to the soil surface becomes a limiting factor, but at first the plant roots are still able to extract water from the soil at the energy-limiting rate. This may be referred to as the transition stage. In this stage the combined rate of evaporation from the surface and from the vegetation continues to decrease, until a second critical state of soil moisture is reached. (In the experiment this second critical state was reached when the moisture content went below about 17% and the vertical gradient started to exceed about 1.15% cm^{-1} at 5 cm.) At this point the vegetation becomes so stressed and relatively inactive that the drying takes place mainly from the soil surface; from that point on the daily evaporation proceeds like in a second stage of drying and it can be described simply by a desorption formulation, that is proportional to the square root of time. The transition period can last from several days to two weeks, depending on the soil moisture conditions and on the season. The longer transition periods were observed under conditions of lower net radiation and of higher soil moisture content at depths in excess of 50 cm. These results were in contrast with the observations of relatively abrupt transitions for bare soil.

Example 9.4. A monthlong drought period in tallgrass prairie

The longest documented drying episode during the FIFE experiment occurred in the fall of 1987. After a major rainfall event on day 253 (September 10) and minor rainfalls on days 258 and 259, no rain fell until day 288 (October 15). (See also Figures 2.22 and 4.12.) The recorded daily evaporation remained larger than the equilibrium evaporation E_e (see Equation (4.30)) until day 258, when the near-surface soil moisture content θ was 0.303. The evaporation rate became equal to E_e on day 259 and it dropped below it after that; therefore, day 260 was taken to be the start of the transition stage. In order to determine the end of the transition, the data were then analyzed as suggested by Equation (9.110), and linear regression yielded, as shown in Figure 9.27, $(L_e E)^{-2} = 2.0 (t - 271) \times 10^{-4}$, in which $(L_e E)$ is the average daily latent heat flux in W m^{-2} and t the time as day of the year. By choosing day 273 as the starting point for the integration, the cumulative evaporation after the onset of the second stage could be obtained by Equation (9.111) (or (9.113)), as shown in Figure 9.28. The fact that the data could be fitted to a straight line supports the

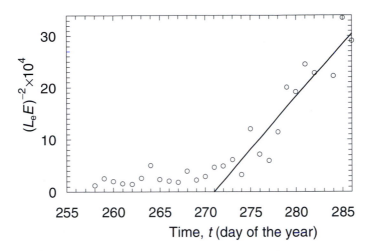

Fig. 9.27 Evolution of measurements of daily mean values of $(L_e E)^{-2}$ (with $L_e E$ in $W\,m^{-2}$) over a natural tallgrass prairie surface in eastern Kansas during a prolonged drying period in 1987. The straight line represents the relationship for the second stage of drying, namely $(L_e E)^{-2} = 2.0(t - 271) \times 10^{-4}$. The data were measured during the FIFE experiment. In the episode shown here, the last major rainfall had taken place on day 253 and minor rainfalls on days 258 and 259. (After Brutsaert and Chen, 1995.)

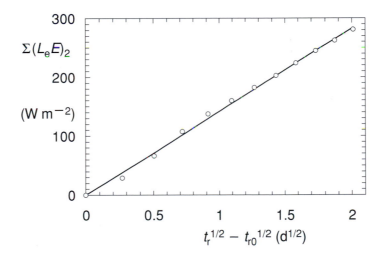

Fig. 9.28 Cumulative evaporation (with daily values in $W\,m^{-2}$) after the onset of the second stage of drying, $\sum(L_e E)_2$, as a function of the square root of time $(t_r^{1/2} - t_{r0}^{1/2})$ with the same data as the previous figure; the straight line represents (9.111) in the form $\sum(L_e E)_2 = 141.4(t_r^{1/2} - t_{r0}^{1/2})$ in which the summation starts on day 273, that is with $t_{r0} = 2$ days. (After Brutsaert and Chen, 1995.)

use of the desorption formulation in this case. The effective desorptivity derived from the slopes in Figures 9.27 and 9.28 is approximately $De_0 = 0.495$ cm $d^{-1/2}$.

Although this value of De_0 was obtained for a grassy surface, it is of the same order as values reported for other studies with bare soil. As mentioned, for a sandy soil Black *et al.* (1969) derived $De_0 = 0.496$ cm $d^{-1/2}$ from lysimeter measurements. Ritchie (1972)

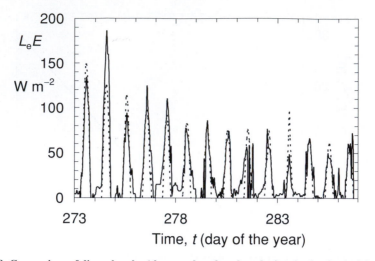

Fig. 9.29 Comparison of diurnal cycles (shown as hourly values during the daytime) of the calculated (by means of Equation (9.114)) and measured latent heat flux from natural tallgrass prairie in the FIFE experimental area in eastern Kansas during the later stages of a major drying episode in 1987; the time shift was taken as $t_0 = 271$. The solid lines represent the measurements, and the dashed lines represent the calculated values. (From Brutsaert and Chen, 1996.)

inferred the effective desorptivity for bare soil from several other studies; for a clay loam soil he reported 0.508 cm $d^{-1/2}$, for loam 0.404 cm $d^{-1/2}$, and for a black clay soil $De_0 = 0.350$ cm $d^{-1/2}$. In the study by Parlange *et al.* (1992) the reported value was $De_0 = 0.58$ cm $d^{-1/2}$. This similarity suggests that by the time evaporation from a grassy surface comes down to this stage, the vegetation becomes quite inactive, and most of the drying takes place from the surface, as if it is bare. The similarity for the different soil types also suggests, as already indicated by the results of Jackson *et al.* (1976) shown in Figure 9.25, that vapor diffusion plays an important role in the second stage of drying, in addition to the capillary rise of liquid water. Clearly, this phenomenon will require further study.

Diurnal variation by self-preservation approximation

In catchment hydrology the daily times scale is a common one; nevertheless, in many applications a daily time resolution is too coarse, and time steps of 30 min to 1 h are required. Further analysis of the same data observed over natural prairie, discussed in the previous section, also indicated (Brutsaert and Chen, 1996) that, while the total daily evaporation could be described with a $t^{-1/2}$ dependency, this day-to-day evolution is modulated during the day by the available energy at the surface, that is by the hourly radiation input. Moreover, during the daytime hours the surface energy budget often displays self-similarity or self-preservation, in the sense discussed in Section 4.3.4. This dual structure of the evaporative evolution during very dry conditions suggests that it can be described, by combining the desorption parameterization (9.108) for the total daily evaporation, or for any dimensionless counterpart (such as the evaporative fraction EF, the Priestley–Taylor α_e and possibly others), with the assumption of self-preservation as expressed in Equation (4.51). The combination of these two concepts yields the following evaporation rate at time $t = t_i$ of day t_d,

$$E_i = \frac{1}{2} De_{0d} (t_d - t_0)^{-1/2} F_d^{-1} F_i \qquad (9.114)$$

where t_0 is the time shift parameter, and the subscripts d and i refer to daily and instantaneous (say in practice, hourly or half-hourly) variables, respectively; F_d is the average flux on day t_d of some other flux term (beside the latent heat flux) in the surface energy budget, which can be assumed to exhibit a similar diurnal variation as the evaporation rate; F_i is that flux at time $t = t_i$ of the same day. Thus Equation (9.114) contains two time scales, t_d in units of days, and t_i in units of hours. As illustrated in Figure 9.29, by assuming $F = R_n - G$ this formulation was able to reproduce daytime hourly flux values over a period of 2 weeks during the second stage of drying already considered in Example 9.4. The approach was subsequently applied again and validated in a different experiment by Porté-Agel *et al.* (2000), who obtained similar results. Under the right circumstances, the approach based on Equation (9.114) may be useful in the disaggregation of daily, or even weekly, evaporation into hourly values, when more complete information is lacking.

REFERENCES

Abramowitz, M. and Stegun, I. A. (eds.) (1964). *Handbook of Mathematical Functions*, Appl. Math. Ser. 55. Washington, DC: National Bureau of Standards.

Ahuja, L. R. and Swartzendruber, D. (1973). Horizontal soil-water intake through a thin zone of reduced permeability. *J. Hydrol.*, **19**, 71–89.

Black, T. N., Gardner, W. R. and Thurtell, G. W. (1969). The prediction of evaporation, drainage, and soil water storage for a bare soil. *Soil Sci. Soc. Amer. Proc.*, **33**, 655–660.

Boltzmann, L. (1894). Zur Integration der Diffusionsgleichung bei variabeln Diffusionskoeffizienten. *Ann. Phys. (Leipzig)*, **53**, 959–964.

Bouwer, H. (1969). Infiltration of water into non-uniform soil. *J. Irrig. Drain. Div., Proc. ASCE*, **95**, 451–462.

Broadbridge, P. and White, I. (1987). Time to ponding: comparison of analytic, quasi-analytic and approximate predictions. *Water Resour. Res.*, **23**, 2302–2310.

Bruce, R. R. and Klute, A. (1956). The measurement of soil moisture diffusivity. *Soil Sci. Soc. Amer. Proc.*, **20**, 458–462.

Bruce, R. R., Thomas, A. W. and Whisler, F. D. (1976). Prediction of infiltration into layered field soils in relation to profile characteristics. *Trans. Amer. Soc. Agric. Eng.*, **19**, 693–698, 703.

Brutsaert, W. (1968). The adaptability of an exact solution to horizontal infiltration. *Water Resour. Res.*, **4**, 785–789.

(1974). More on an approximate solution for nonlinear diffusion. *Water Resour. Res.*, **10**, 1251–1252.

(1976). The concise formulation of diffusive sorption of water in a dry soil. *Water Resour. Res.*, **12**, 1118–1124.

(1977). Vertical infiltration in dry soil. *Water Resour. Res.*, **13**, 363–368.

(1979). Universal constants for scaling the exponential soil water diffusivity? *Water Resour. Res.*, **15**, 481–483.

(1982). Some exact solutions for nonlinear desorptive diffusion. *J. Appl. Math. and Phys. (ZAMP)*, **33**, 540–546.

Brutsaert, W. and Chen, D. (1995). Desorption and the two stages of drying of natural tallgrass prairie. *Water Resour. Res.*, **31**, 1305–1313.

(1996). Diurnal variation of surface fluxes during thorough drying (or severe drought) of natural prairie. *Water Resour. Res.*, **32**, 2013–2019.

Cary, J. W. (1967). The drying of soil: thermal regimes and ambient pressure. *Agric. Met.*, **4**, 357–365.

Chu, S. T. (1978). Infiltration during an unsteady rain. *Water Resour. Res.*, **14**, 461–466.

Cisler, J. (1969). The solution for maximum velocity of isothermal steady flow of water upward from water table to soil surface. *Soil Sci.*, **108**, 148.

Clothier, B. E. and White, I. (1981). Measurement of sorptivity and soil water diffusivity in the field. *Soil Sci. Soc. Amer. J.*, **45**, 241–245.

  (1982). Water diffusivity of a field soil. *Soil Sci. Soc. Amer. J.*, **46**, 155–158.

Cook, F. J. and Broeren, A. (1994). Six methods for determining sorptivity and hydraulic conductivity with disk permeameters. *Soil Sci.*, **157**, 211.

Cook, H. L. (1946). The infiltration approach to the calculation of surface runoff. *Trans. Amer. Geophys. Un.*, **27**, 726–747.

Cordery, I. (1970). Initial loss for flood estimation and forecasting. *J. Hydraul. Div., Proc. ASCE*, **96**, 2447–2466.

Cosh, M. H. and Brutsaert, W. (1999). Aspects of soil moisture variability in the Washita'92 study region. *J. Geophys. Res.*, **104** (D16), 19 751–19 757.

Crank, J. (1956). *The Mathematics of Diffusion*. Oxford: Clarendon.

Davidson, J. M., Nielsen, D. R. and Biggar, J. W. (1963). The measurement and description of water flow through Columbia silt loam and Hesperia sandy loam. *Hilgardia*, **34** (15), 601–617.

Dooge, J. C. E. (1957). The rational method for estimating flood peaks. *Engineering (London)*, **184**, 311–374.

Feldman, A. (1981). HEC models for water resources system simulation: theory and experience. *Adv. Hydroscience*, **12**, 297–423.

Fisher, E. A. (1923). Some factors affecting the evaporation of water from soil. *J. Agric. Sci.*, **13**, 121–143.

Gardner, W. R. (1958). Some steady-state solutions of the unsaturated moisture flow equation with application to evaporation from a water table. *Soil Sci.*, **85**, 228–232.

  (1959). Solution of the flow equation for the drying of soils and other porous media. *Soil Sci. Soc. Amer. Proc.*, **23**, 183–187.

Gardner, W. R. and Fireman, M. (1958). Laboratory studies of evaporation from soil columns in the presence of a water table. *Soil Sci.*, **85**, 244–249.

Gardner, W. R. and Hillel, D. I. (1962). The relation of external evaporative conditions to the drying of soils. *J. Geophys. Res.*, **67**, 4319–4325.

Hillel, D. and Gardner, W. R. (1970). Transient infiltration into crust-topped profiles. *Soil Sci.*, **109**, 69–76.

Holtan, H. N. (1945). Time condensation in hydrograph analysis. *Trans. Amer. Geophys. Un.*, **26**, 407–413.

Horton, R. E. (1937). Determination of infiltration capacity for large drainage basins. *Trans. Amer. Geophys. Un.*, **18**, 371–385.

  (1939). Analysis of runoff-plot experiments with varying infiltration capacity. *Trans. Amer. Geophys. Un.*, **20**, 693–711.

  (1940). An approach toward a physical interpretation of infiltration capacity. *Soil Sci. Soc. Amer. Proc.*, **5**, 399–417.

  (1942). A simplified method of determining the constants in the infiltration-capacity equation. *Trans. Amer. Geophys. Un.*, **23**, 575–577.

Ibrahim, H. A. and Brutsaert, W. (1968). Intermittent infiltration into soils with hysteresis. *J. Hydraul. Div., Proc. ASCE*, **94**(HY1), 113–137.

Idso, S. B., Reginato, R. J., Jackson, R. D., Kimball, B. A. and Nakayama, F. S. (1974). The three stages of drying of a field soil. *Soil Sci. Soc. Amer. Proc.*, **38**, 831–836.

Jackson, R. D., Kimball, B. A., Reginato, R. J. and Nakayama, F. S. (1973). Diurnal soil water evaporation: time-depth-flux patterns. *Soil Sci. Soc. Amer. Proc.*, **37**, 505–509.

Jackson, R. D., Idso, S. B. and Reginato, R. J. (1976). Calculation of evaporation rates during the transition from energy-limiting to soil-limiting phases using albedo data. *Water Resour. Res.*, **12**, 23–26.

Kim, C. P., Stricker, J. N. M. and Torfs, P. J. J. F. (1996). An analytical framework for the water budget of the unsaturated zone. *Water Resour. Res.*, **32**, 3475–3484.

Klute, A. (1952). A numerical method for solving the flow equation for water in unsaturated materials. *Soil Sci.*, **73**, 105–116.

Kostiakov, A. N. (1932). On the dynamics of the coefficient of water percolation in soils and on the necessity of studying it from a dynamic point of view for purposes of amelioration. 6th Comm. *Internatl Soc. for Soil Sci. Trans., Part A*, pp. 17–24.

Kozeny, J. (1927). Über kapillare Leitung des Wassers im Boden. *Sitzungsberichte, Akad. d. Wissensch., Vienna, Austria*, **136** (Part 2a), 271–306.

Landahl, H. D. (1953). An approximation method for the solution of diffusion and related problems. *Bull. Math. Biophys.*, **15**, 49–61.

Laurenson, E. M. and Pilgrim, D. H. (1963). Loss rates for Australian catchments and their significance. *J. Instn. Engrs., Australia*, **35**, 9–24.

Lewis, M. R. (1937). The rate of infiltration of water in irrigation practice. *Trans. Amer. Geophys. Un.*, **18**, 361–368.

Linden, D. R. and Dixon, R. M. (1975). Water table position as affected by soil air pressure. *Water Resour. Res.*, **11**, 139–143.

Liu, M.-C., Parlange, J.-Y., Sivapalan, M. and Brutsaert, W. (1998). A note on the time compression approximation. *Water Resour. Res.*, **34**, 3683–3686.

Liu, Y., Steenhuis, T. S. and Parlange, J.-Y. (1994a). Formation and persistence of fingered flow fields in coarse grained soils under different moisture contents. *J. Hydrol.*, **159**, 187–195.

(1994b). Closed-form solution for finger width in sandy soils at different water contents. *Water Resour. Res.*, **30**, 949–952.

Macey, R. I. (1959). A quasi-steady state approximation method for diffusion problems, 1. Concentration dependent diffusion coefficients. *Bull. Math. Biophys.*, **21**, 19–32.

Matano, C. (1933). On the relation between the diffusion-coefficients and the concentrations of solid metals (the nickel–copper system). *Japanese J. Phys.*, **8**, 109–113.

McWhorter, D. B. (1971). *Infiltration affected by flow of air.* Hydrology Paper No. 49, Colorado State University.

Mein, R. G. and Larson, C. L. (1973). Modeling infiltration during a steady rain. *Water Resour. Res.*, **9**, 384–394.

Miller, E. E. and Gardner, W. H. (1962). Water infiltration into stratified soil. *Soil Sci. Soc. Amer. Proc.*, **26**, 115–119.

Miller, R. D. and Bresler, E. (1977). A quick method for estimating soil water diffusivity functions. *Soil Sci. Soc. Amer. Proc.*, **41**, 1021–1022.

Mulvany, T. J. (1850). On the use of self registering rain and flood gauges. *Inst. Civil Engin. Proc. (Dublin)*, **4**, 1–8.

Nielsen, D. R., Biggar, J. W. and Erh, K. T. (1973). Spatial variability of field-measured soil-water properties. *Hilgardia*, **42**, 215–259.

Nielsen, D. R., Biggar, J. W. and Davidson, J. M. (1962). Experimental consideration of diffusion analysis in unsaturated flow problems. *Soil Sci. Soc. Amer. Proc.*, **26**, 107–111.

Parlange, J.-Y. (1971). Theory of water movement in soils: 1. One-dimensional absorption. *Soil Sci.*, **111**, 134–137.

(1972). Theory of water movement in soils: 8. One-dimensional infiltration with constant flux at the surface. *Soil Sci.*, **114**, 1–4.

(1973). Horizontal infiltration of water in soils: a theoretical interpretation of recent experiments. *Soil Sci. Soc. Amer. Proc.*, **37**, 329–330.

(1975). Comments on 'Determination of soil water diffusivity by sorptivity measurements' by C. Dirksen, *Soil Sci. Soc. Amer. Proc.*, **39**, 1011.

Parlange, J.-Y. and Hill, D. E. (1979). Air and water movement in porous media – compressibility effects. *Soil Sci.*, **127**, 257–263.

Parlange, J.-Y. and Smith, R. E. (1976). Ponding time for variable rainfall rates. *Can. J. Soil Sci.*, **56**, 121–123.

Parlange, J.-Y., Vauclin, M., Haverkamp, R. and Lisle, I. (1985). The relation between desorptivity and soil water diffusivity. *Soil Sci.*, **139**, 458–461.

Parlange, M. B., Katul, G. G., Cuenca, R. H., Kavvas, M. L., Nielsen, D. R. and Mata, M. (1992). Physical basis for a time series model of soil water content. *Water Resour. Res.*, **28**, 2437–2446.

Peck, A. J. (1964). The diffusivity of water in a porous material. *Australian J. Soil Res.*, **2**, 1–17.

Philip, J. R. (1955). Numerical solution of equations of the diffusion type with diffusivity concentration-dependent. *Trans. Faraday Soc.*, **51**, 885–892.

(1957a). The theory of infiltration, 4, Sorptivity and algebraic infiltration equations. *Soil Sci.*, **84**, 257–264.

(1957b). Numerical solution of equations of the diffusion type with diffusivity concentration-dependent, 2. *Austral. J. Phys.*, **10**, 29–42.

(1957c). Evaporation and moisture and heat fields in the soil. *J. Met.*, **14**, 354–366.

(1960). General method of exact solution of the concentration-dependent diffusion equation. *Austral. J. Phys.*, **13**, 1–12.

(1967). Sorption and infiltration in heterogeneous media. *Australian. J. Soil Res.*, **5**, 1–10.

(1969). Theory of infiltration. *Adv. Hydroscience*, **5**, 215–296.

Pilgrim, D. H. (1966). Storm loss rates for regions with limited data. *J. Hydraul. Div., Proc. ASCE*, **92**(HY2), 193–206.

Polubarinova-Kochina, P. Ya. (1952). *Theory of Ground Water Movement*, (translated from the Russian by J. M. R. DeWiest, 1962). Princeton: Princeton University Press.

Porté-Agel, F., Parlange, M. B., Cahill, A. T. and Gruber, A. (2000). Mixture of time scales in evaporation: desorption and self-similarity of energy fluxes. *Agron. J.*, **92**, 832–836.

Reeves, M. and Miller, E. E. (1975). Estimating infiltration for erratic rainfall. *Water Resour. Res.*, **11**, 102–110.

Richards, L. A. (1931). Capillary conduction of liquids through porous mediums. *Physics*, **1**, 318–333.

Ritchie, J. T. (1972). Model for predicting evaporation from a row crop with incomplete cover. *Water Resour. Res.*, **8**, 1204–1213.

Rogowski, A. S. (1972). Watershed physics: soil variability criteria. *Water Resour. Res.*, **8**, 1015–1023.

Rubin, J. (1966). Theory of rainfall uptake by soils initially drier than their field capacity and its applications. *Water Resour. Res.*, **2**, 739–749.

Salvucci, G. D. and Entekhabi, D. (1994). Equivalent steady soil moisture profile and the time compression approximation in water balance modeling. *Water Resour. Res.*, **30**, 2737–2749.

Selker, J., Parlange, J.-Y. and Steenhuis, T. (1992). Fingered flow in two dimensions, 2. Predicting finger moisture profile. *Water Resour. Res.*, **28**, 2523–2528.

Sellers, P. J., Hall, F. G., Asrar, G., Strebel, D. E. and Murphy, R. E. (1992). An overview of the First International Satellite Land Surface Climatology Project (ISLSCP) Field Experiment (FIFE). *J. Geophys. Res.*, **97**(D17), 18 345–18 371.

Sherman, L. K. (1943). Comparison of F-curves derived by the methods of Sharp and Holtan and of Sherman and Mayer. *Trans. Amer. Geophys. Un.*, **24**, 465–467.

Sherwood, T. K. and Comings, E. W. (1933). The drying of solids, V, Mechanism of drying of clays. *Ind. Engin. Chem.*, **25**, 311–316.

Sivapalan, M. and Milly, P. C. D. (1989). On the relationship between the time condensation approximation and the flux-concentration relation. *Jour. Hydrology*, **105**, 357–367.

Smith, R. E. (1972). The infiltration envelope: results from a theoretical infiltrometer. *J. Hydrol.*, **17**, 1–21.

Smith, R. E. and Chery, D. L. (1973). Rainfall excess model from soil water flow theory. *J. Hydraul. Div., Proc. ASCE*, **99**, 1337–1351.

Smith, R. E. and Parlange, J.-Y. (1978). A parameter-efficient hydrologic infiltration model. *Water Resour. Res.*, **14**, 533–538.

Sonu, J. and Morel-Seytoux, H. J. (1976). Water and air movement in a bounded deep homogeneous soil. *J. Hydrol.*, **29**, 23–42.

Swartzendruber, D. and Hillel, D. (1975). Infiltration and runoff for small field plots under constant intensity rainfall. *Water Resour. Res.*, **11**, 445–451.

Talsma, T. (1969). In situ measurement of sorptivity. *Aust. J. Soil Res.*, **7**, 269–276.

Talsma, T. and Parlange, J.-Y. (1972). One-dimensional vertical infiltration. *Austral. J. Soil Res.*, **10**, 143–150.

Warrick, A. W., Mullen, G. J. and Nielsen, D. R. (1977). Scaling field-measured soil hydraulic properties using a similar media concept. *Water Resour. Res.*, **13**, 355–362.

White, I. and Broadbridge, P. (1988). Constant rate rainfall infiltration: a versatile nonlinear model. *Water Resour. Res.*, **24**, 155–162.

White, I. and Perroux, K. M. (1987). Use of sorptivity to determine field soil hydraulic properties. *Soil Sci. Soc. Amer. J.*, **51**, 1093–1101.

White, I., Colombera, P. M. and Philip, J. R. (1977). Experimental studies of wetting front instability induced by gradual change of pressure gradient and by heterogeneous porous media. *Soil Sci. Soc. Amer. J.*, **41**, 483–489.

White, I., Sully, M. J. and Melville, M. D. (1989). Use and hydrological robustness of time-to-incipient-ponding. *Soil Sci. Soc. Amer. J.*, **53**, 1343–1346.

Willis, W. O. (1960). Evaporation from layered soils in the presence of a water table. *Soil Sci. Soc. Amer. Proc.*, **24**, 239–242.

Wollny, E. (1884). Untersuchungen über die kapillare Leitung des Wassers im Boden. *Forschungen auf dem Gebiete der Agrikulturphysik*, **7**, 269–308.

PROBLEMS

9.1   Suppose that an approximate (but sufficiently accurate) solution of the Richards equation (9.1), describing infiltration into an initially dry soil, with a thin layer of water ponded at the surface, can be written as follows: $z = 2.87 \, (1 - S^9) \, t^{1/2} + 0.04 \, (1 - S^{20})t$. The variable $S = \theta/\theta_0$ is the degree of saturation, z is the depth into the soil in cm, t is the time in min and θ_0 is the water content at satiation. Calculate the following. (a) The cumulative infiltration volume (per unit area of ground surface) as a function of time. (Specify the units of your result.) (b) The rate of infiltration as a function of time. (c) The depth of the wetting front after one minute. The wetting front is the depth below the surface, where the soil is just beginning to be wetted.

9.2   Calculate the same items, (a), (b) and (c), as in the previous problem; however, here the approximate solution of the Richards equation is as follows: $z = 2.90\,(1 - S^4)\,t^{1/2} + 0.05(1 - S^9)t$. The variable $S = \theta/\theta_0$ is the degree of saturation, z is the depth into the soil in cm, t is the time in min and θ_0 is the water content at satiation.

9.3   Derive Equation (9.71) from (9.68), by scaling the variables.

9.4   Derive an expression for the cumulative infiltration volume F_c from Horton's exponential equation (9.76).

9.5   Derive an expression for the cumulative infiltration volume from Horton's exponential equation (9.76). Try to give physical meaning to the parameters, b and c, by expressing them as functions of the sorptivity A_0 and of the hydraulic conductivity at saturation k_0. To accomplish this, compare F_c from the Horton equation with (9.69), such that both expressions produce the same infiltrated volume for very large values of t. Recall that $\beta_0 = 2/3$ in Equation (9.69) and that $a = k_0$ in Equation (9.76).

9.6   Assume that the infiltration capacity rate in a given soil can be described by $f_c = 0.5\,A_0\,t^{-1/2} + k_0/3$, in which A_0 is the sorptivity and k_0 is the hydraulic conductivity at satiation. Derive an expression for the cumulative infiltration capacity F_c.

9.7   A steady light rain $P = 0.45$ cm h^{-1} is infiltrating into a deep homogeneous soil, whose hydraulic conductivity (in cm d^{-1}) is given by Equation (8.37) (see also Figure 8.29), with $a = 170 \times 10^6$, $b = 2.5 \times 10^6$, $c = 4$, and H in cm. Two tensiometers measure the pressure at 0.5 m and at 1.0 m below the ground surface. If the manometers of these two tensiometers are located at 0.5 m above the ground, what is the pressure reading in each of these manometers? Express the result in cm of equivalent water column.

9.8   Consider the soil whose infiltration characteristics are given in Problem 9.6. (a) Calculate the time to ponding in terms of A_0 and k_0, for a rainfall intensity 1.3 times as large as the hydraulic conductivity, that is $P = 1.3\,k_0$. (b) Using this value of t_p, calculate the compression reference time t_{cr1} in terms of A_0 and k_0 by means of (9.91). (c) Write down an expression for the actual cumulative infiltration $F(t)$ making use of Equation (9.92). (d) Estimate the time to ponding for this case if the hydraulic conductivity is $k_0 = 0.08$ cm min^{-1} and the sorptivity is $A_0 = 1$ cm min$^{-1/2}$.

9.9   Multiple choice. Indicate which of the following statements are correct. The hydraulic conductivity of a partly saturated soil:
        (a)   becomes smaller when the soil becomes drier;
        (b)   is minimum near the wetting front during infiltration of ponded water (in contrast to near the surface);
        (c)   may increase with time during infiltration, as air, entrapped initially, goes into solution;
        (d)   is a function of the water content gradient;

(e)   is a function of the suction ($p < 0$) in the water in the soil;

(f)   the latter function exhibits hysteresis.

9.10   Multiple choice. Indicate which of the following statements are correct. The surface of a soil is kept ponded by a thin layer of water. The cumulative infiltration volume (not the rate) into a homogeneous, initially dry soil profile of infinite depth:

(a)   eventually (i.e. after a very long time;) becomes a linear function of time;

(b)   eventually becomes a constant and independent of time;

(c)   initially varies as $t^{-1/2}$ because mainly capillary forces are acting;

(d)   is initially equal to the hydraulic conductivity;

(e)   decreases as a smooth function with time.

9.11   Multiple choice. Indicate which of the following statements are correct. Infiltration capacity (which is the rate of vertical infiltration when the water supply at the soil surface is not limiting):

(a)   may vary considerably with time;

(b)   may depend on the rainfall rate (e.g. drizzle);

(c)   is a function of the permeability of the soil;

(d)   becomes, theoretically, equal to a constant after a long time of infiltration when the soil profile is very deep (i.e. without an impermeable layer at shallow depth) and uniform;

(e)   is largely independent of the vegetative cover of the surface or of the season of the year.

9.12   Assume that it is known that the similarity variable $\phi = x \, t^{-1/3}$ allows the reduction of the following partial differential equation:

$$2x\frac{\partial \theta}{\partial t} = \frac{\partial}{\partial x}\left(\theta^4 \frac{\partial \theta}{\partial x}\right)$$

to an ordinary differential equation, whose solution is $\theta = \theta(\phi)$. (a) Obtain that ordinary differential equation. (b) What are the restrictions on the problem geometry (time and space), as expressed in the boundary conditions, to permit this type of similarity variable (two to three sentences only)?

9.13   Consider the differential equation (9.13) and the boundary conditions (9.14). If the solution of this problem is $\phi = (1 - \theta)^n$ for $0 < \phi < 1$, and $\theta = 0$ for $\phi \geq 1$, in which n is a positive constant, what is the diffusivity, $D_w = D_w(\theta)$?

9.14   You are given the results of a horizontal infiltration experiment as shown in Figure 9.2. Initially, the soil is totally dry or $\theta_i = 0$ and its satiation water content is $\theta_0 = 0.4$. After $t = 100$ min, the following water content distribution was obtained.

x (cm)	0	5	10	15	17	19	20	20.5	21
θ	0.4	0.38	0.34	0.29	0.26	0.22	0.18	0.14	0

Calculate $D_w = D_w(\theta)$ (in cm^2 min^{-1}) for values of $\theta = 0.10, 0.25, 0.30, 0.35$ by solving Equation (9.25) graphically or numerically.

9.15   Same as previous problem for $\theta_i = 0.02$, $\theta_0 = 0.45$. After $t = 740$ min, the water content distribution was:

x (cm)	10	30	40	50	60	70	72	75	76
θ	0.45	0.45	0.45	0.44	0.42	0.36	0.33	0.20	0.10

Calculate $D_w = D_w(\theta)$ (in cm^2 min^{-1}) for $\theta = 0.2, 0.3, 0.4, 0.45$ by solving Equation (9.25) graphically or numerically.

9.16 Consider the horizontal infiltration experiment of Problem 9.15, which was allowed to run for $t = 740$ min. Tabulate the water content distribution $\theta = \theta(x)$, that would be observed, if the experiment were allowed to run for only $t = 370$ min.

9.17 Derive the expression for the sorptivity (9.28) from the exact solution (9.27). From this, derive an expression for the horizontal infiltration rate f.

9.18 Derive (9.39) from (9.38). Hint: use integration by parts, and follow up with Leibniz's rule (see Appendix).

9.19 A fairly accurate solution of the sorption problem is Equation (9.38) with $b = 0$. (It is not as accurate as (9.43), but it is easier to handle analytically.) Use this solution to calculate the sorptivity A_0 by means of (9.39) and also the position of the wetting front ϕ_f; give a simple relationship between F and x_f, as a function of β, if the diffusivity is given by (8.39). Compare with the more accurate result given in (9.48), and also presented in Figure 9.12, for the values $\beta = 3$ and 8.

9.20 Derive Equation (9.104) from (9.103).

9.21 Consider a homogeneous sandy soil profile, whose conductivity is given by (8.37) in cm d^{-1}, with $a = 170 \times 10^6$, $b = 2.5 \times 10^6$ and $c = 4$. Assume a potential evaporation of 0.4 cm d^{-1} from the bare soil surface; what is the smallest depth of the water table for which the soil (instead of the atmosphere) totally controls the evaporation?

9.22 Consider the Diablo loam, whose hydraulic conductivity is depicted in Figure 8.29. Use the values of the parameters a, b and c in Equation (8.37) from the figure, and calculate the maximal rate of evaporation (by steady capillary rise) from a bare soil surface for the following three cases; the water table is (a) at 0.5 m, (b) at 1.0 m and (c) at 1.5 m below the surface.

9.23 Multiple choice. Indicate which of the following statements are correct. The motion of a wetting fluid in two-phase immiscible flow problems is often described by Richards's equation. This formulation requires the following assumptions:
(a) Darcy's law is valid;
(b) conservation of thermodynamic energy;
(c) the non-wetting fluid is inviscid, so that it moves freely without pressure gradient;
(d) The effects of capillarity are negligible;
(e) The porous matrix is incompressible.

(9.24) Multiple choice. Indicate which of the following statements are correct. During vertical infiltration of ponded water (the water is maintained as a very thin layer of constant thickness) into a deep homogeneous dry soil:

(a)   the effect of gravity predominates initially, to become nearly negligible after a very long time;

(b)   the specific flux at the surface is equal to the infiltration rate;

(c)   the hydraulic conductivity is nearly constant at and on the wetting front;

(d)   the water pressure in the top layers of the soil becomes a constant with depth after a very long time;

(e)   initially, the infiltration volume changes linearly with time.

10 GROUNDWATER OUTFLOW AND BASE FLOW

A major portion of the precipitation that percolates into the soil profile eventually finds its way into creeks, rivers, lakes and other open water bodies. After the precipitation or other input has ceased for some time, the entire streamflow can be assumed to consist of the cumulative outflow from all upstream phreatic aquifers. The prediction of base flow is of some practical importance because it is the rate of flow, that a given river basin can sustain in the absence of precipitation and in the absence of artificial storage works. Accordingly, this type of flow is variously known as *base flow*, *drought flow*, *low flow* and *sustained* or *fair-weather runoff*. In engineering such flows have been studied in connection with problems of water supply and water quality in rivers during drought periods, and general basin and agricultural drainage.

During, and in response to, precipitation or snowmelt, the different pathways and the detailed mechanisms, by which water reaches the stream, are more complex than during drought flow episodes. Still, as will be seen in Chapter 11, it is generally agreed that also storm runoff from natural basins with vegetation is largely supplied into the streams by subsurface transport. Thus, subsurface drainage from the aquifers along the banks is one of the key elements in catchment hydrology, not only under drought conditions but also in response to precipitation. In this chapter, the subsurface outflow is first considered locally at the point where it enters the stream, by analysis of the groundwater flow process in the riparian unconfined aquifer; the first five sections describe the different formulations that are available for this purpose. In the last section of the chapter, the phenomenon is treated at the basin scale, by integration of these local groundwater outflows along the streams and channels in the basin.

10.1   FLOW IN AN UNCONFINED RIPARIAN AQUIFER

10.1.1   *General formulation*

A typical cross section of an unconfined aquifer, whose water flows into the adjoining stream, is sketched in Figure 10.1. As this flow system is usually relatively shallow and exposed to the atmosphere through a partly saturated soil moisture zone, the water pressures and effective stresses are rarely very large, so that the water and the solid matrix can be assumed to be incompressible (see Brutsaert and El-Kadi, 1984, 1986). Therefore, if the material can be assumed to be effectively isotropic, the flow, which involves combined saturated and partly saturated conditions, is in principle governed by

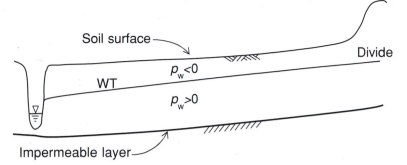

Fig. 10.1 Sketch of the cross section of an unconfined riparian aquifer. WT refers to the water table, which is the locus where the water pressure is atmospheric; above the WT the soil is partly saturated. The divide is assumed to be the catchment boundary.

the Richards equation (8.56) (or also (8.89)) rewritten here for easy reference,

$$\frac{\partial}{\partial x}\left(k\frac{\partial h}{\partial x}\right) + \frac{\partial}{\partial y}\left(k\frac{\partial h}{\partial y}\right) + \frac{\partial}{\partial z}\left(k\frac{\partial h}{\partial z}\right) = \frac{\partial \theta}{\partial t} \tag{10.1}$$

in which $h = (z + p_w/\gamma_w)$ is the hydraulic head, z the vertical coordinate, k the hydraulic conductivity and θ the water content of the aquifer. The boundary conditions can be prescribed in broad terms as follows. At the bedrock or impermeable bed underlying the aquifer and at the catchment divide, the flux is usually parallel to the boundary, or if n is the coordinate normal to the boundary, $\partial h/\partial n = 0$. At the ground surface, the specific flux **q** across the boundary can be given as the evaporation E, as the infiltration rate f, or as a combination of both; for simplicity, however, in the analysis of base flow it is often assumed to be zero, so that here also $\partial h/\partial n = 0$. At the stream channel boundary, the hydraulic head h is a constant that is equal to the height of the water surface in the stream above the reference level of the elevation, $z = 0$. Along the stream banks there is often a seepage surface, where the pressure is atmospheric, so that $h = z$. The initial conditions may vary, depending on the assumed initial moisture content distribution.

This problem is not easy to solve. Aquifer properties are generally not spatially uniform and may even change with time; thus beside $\theta = \theta(x, y, z, t)$ as the dependent variable, two additional nonlinear functions come into play; these are $k = k(x, y, z, t, \theta)$ and $H = H(x, y, z, t, \theta)$, in which, as before, $H = -p_w/\gamma_w$ is the suction expressed as equivalent water column. At present there are no methods available to determine the spatial variability of these parameter functions. Moreover, in geological deposits with an irregular geometry like that of the aquifer profile shown in Figure 10.1, the boundary conditions prevail underground, and they are invisible; they are therefore nearly impossible to validate or to formulate precisely. A general solution of this problem is obviously unattainable. Nevertheless, some crucial features of the flow phenomena can be brought out by the solution of special cases and simplified geometries.

One common simplification consists of the adoption of "effective" parameter functions; the basic concept was introduced in Section 1.4.3. In brief, it is based on the

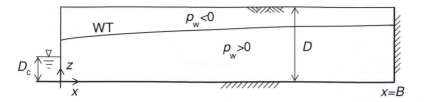

Fig. 10.2   Schematic representation of the cross section of an unconfined riparian aquifer, lying on a horizontal impermeable layer. The origin of the coordinates is taken at the stream ($x = 0$) and at the impermeable layer ($z = 0$); D_c is the water depth in the adjoining stream, D is the aquifer thickness, and B is the aquifer breadth, that is the distance from the stream to the divide. The water table (WT), is the locus where the water pressure p_w is atmospheric, or $p_w = 0$. Above the WT the soil is partly saturated, and the water pressure smaller than atmospheric; below the WT the water pressure is larger than atmospheric.

assumption that it is possible to define (or imagine) a spatially uniform model aquifer, with effective parameter functions $k = k(\theta)$ and $H = H(\theta)$ which, upon solution of Equation (10.1) with appropriate boundary conditions, produces the same flow characteristics of interest, as the real spatially variable prototype aquifer. A second simplification is based on the observation that the vertical dimensions of unconfined riparian aquifers tend to be much smaller than their horizontal extent. This has led to the assumption that the boundary conditions of real aquifers are rarely very different from those of a two-dimensional model aquifer of rectangular cross section; for purposes of flow analysis, the aquifer depicted in Figure 10.1 can thus be represented schematically as shown in Figure 10.2. These two simplifications have allowed some standardization of the groundwater outflow problem, while maintaining its main characteristics.

10.1.2   *Some common approximate formulations*

Even with the simplifications just mentioned, the governing Richards equation (10.1) remains highly nonlinear and most problems involving combined saturated and partially saturated flow must be analyzed by numerical methods. With the availability of high speed digital computers, at present there is no dearth of efficient numerical codes for this purpose, and rapid advances continue to be made in this field. One drawback of such exact solutions of (10.1), however, is that they cannot be easily parameterized in practical terms for incorporation in basin-scale analyses. Thus further simplifications, which may be valid under special conditions and for which solutions may be more readily obtained, are often called for.

In a first approximation the flow in the zone above the water table, where $p_w < 0$, is neglected, and the water table is treated as a true free surface; with the assumption of an effective hydraulic conductivity and porosity, the governing equation (10.1) reduces then to Laplace's. This case is discussed in Section 10.2. In a second approximate formulation, beside the free surface assumption, the distribution of the water pressure in the general direction normal to the flow is assumed to be hydrostatic; these two assumptions, also called the Dupuit assumptions, are the basis of the hydraulic groundwater theory, which is covered in Section 10.3. Linearization of the hydraulic groundwater approach constitutes a third approximate formulation, and this is covered in Section 10.4. Finally, the additional assumption that the hydraulic head gradient is equal to the slope of the land surface produces a kinematic wave formulation, which is the fourth approximation, treated in Section 10.5. But before looking more closely into these common simplified approaches in the remainder of this chapter, it is useful first to consider briefly some implications of solutions of Equation (10.1) itself.

10.1.3   A few features of combined saturated–unsaturated flow

Unsteady flow formulation

Some results of a numerical solution of Equation (10.1) were presented by Verma and Brutsaert (1970; 1971b) for the two-dimensional case of outflow from a horizontal unconfined aquifer with a rectangular cross section, after the cessation of recharge; this situation is illustrated in Figure 10.2. The soil water characteristic was assumed to be given by Equation (8.15) and the hydraulic conductivity by Equation (8.36). With the effective saturation S_e defined in Equation (8.6), the boundary conditions for this problem can be specified as follows

$$
\begin{array}{llll}
h = D_c & S_e = 1.0 & x = 0 & 0 \leq z \leq D_c \\[2pt]
h = z & S_e = 1.0 & x = 0 & D_c \leq z = h \\[2pt]
\dfrac{\partial h}{\partial x} = 0 & \dfrac{\partial S_e}{\partial x} = 0 & x = 0 & h \leq z \leq D \\[2pt]
\dfrac{\partial h}{\partial x} = 0 & \dfrac{\partial S_e}{\partial x} = 0 & x = B & 0 \leq z \leq D \\[2pt]
\dfrac{\partial h}{\partial z} = 0 & \dfrac{\partial S_e}{\partial z} = 0 & 0 \leq x \leq B & z = 0 \\[2pt]
\dfrac{\partial h}{\partial z} = 0 & \dfrac{\partial S_e}{\partial z} = 0 & 0 \leq x \leq B & z = D
\end{array}
\qquad (10.2)
$$

The first boundary condition is a result of the hydrostatic pressure distribution in the stream. At the seepage surface the pressure is zero (i.e. atmospheric) and the hydraulic head is equal to the height z, as indicated in the second condition. Above the seepage surface the water pressure is negative, and because no outflow is physically possible unless the pressure is at least atmospheric, this surface acts as an impermeable boundary as indicated in the third of (10.2). The boundary conditions given by the fourth and fifth of (10.2) express the no-flow or impermeable boundaries of the aquifer at the divide and at the underlying bed rock. As it is assumed that there is no evaporation or recharge, the ground surface acts like an impermeable boundary after the drainage starts, as indicated in the last condition; however, this condition at the ground surface can be readily replaced by the evaporative flux or recharge rate, if it is known.

The initial conditions, for $t = 0$, can be assumed to be those of a fully saturated aquifer, in which the water table coincides with the ground surface. This situation is

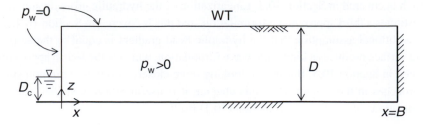

Fig. 10.3   Initial state for the unconfined riparian aquifer, shown in Figure 10.2. The water table (WT) is
assumed to be at the soil surface and the entire aquifer is saturated.

formulated as

$$\nabla^2 h = 0 \quad S_e = 1.0 \quad 0 \leq x \leq B \quad 0 \leq z \leq D$$

$$h = D \qquad\qquad\qquad 0 \leq x \leq B \quad z = D$$

(10.3)

Since $k = k_0$ at $S_e = 1.0$, which is assumed to be an effective hydraulic conductivity that
is constant over the whole flow domain, Equation (10.1) reduces to the Laplace equation,
as indicated in Equation (10.3). This initial state of the aquifer is shown in Figure 10.3.

Similarity criteria

By scaling the variables, it can readily be seen, that the only relevant (dimensionless)
parameters in this problem are: (i) those related to the soil, viz. n and b; (ii) those related to
the geometry of the flow, viz. $B_+ = B/D$ and $D_{c+} = D_c/D$; and (iii) one related to both soil
and geometry, viz. (aD). Because a^{-1} in Equation (8.15) can be considered a measure of
the thickness of the capillary fringe, $(aD)^{-1}$ expresses the relative importance of capillarity,
and thus of the partly saturated flow zone, with respect to the vertical dimensions of the
aquifer.

Results of some example calculations for $D_{c+} = 0$, $B_+ = 1.0$, with $n = 3$, $(aD)^{-1} =$
0.36, and $b = 1.5$ are shown in Figure 10.4, with $n = 3$, $(aD)^{-1} = 0.1$, and $b = 3$ in
Figure 10.5, and with $n = 8$, $(aD)^{-1} = 0.36$, and $b = 1.5$ in Figure 10.6. The values of
the parameters (mainly $(aD)^{-1}$ and b) for Figures 10.4 and 10.6 could represent, for example, a loam soil in an aquifer of approximately 3 m depth; those of Figure 10.5 a somewhat
coarser material in an aquifer of roughly the same depth. These calculations show that the
water table tends to fall faster for higher values of $(aD)^{-1}$, and for smaller values of b;
but the value of n does not appear to affect this very much. This illustrates that in real
situations where capillary effects are important, it may be deceiving to use the water table
as an upper boundary of the flow domain, because a large amount of water may be left in
the unsaturated zone above the water table. It can also be observed that for large values
of n and after large t the lines of equal hydraulic head are close to horizontal outside the
zone of saturation, whereas they are more nearly vertical within this zone. This refraction
phenomenon is mainly the result of the fact that the upper boundary of the saturated zone
constitutes a boundary between a zone of high and a zone of low conductivity.

The results of these numerical experiments can be used to derive similarity criteria to
determine different regimes of flow in unconfined aquifers by comparing them with the
results of special solutions for each of these regimes (Verma and Brutsaert, 1971a, b). In
principle, all three parameters n, b and $(aD)^{-1}$, which involve soil characteristics, should

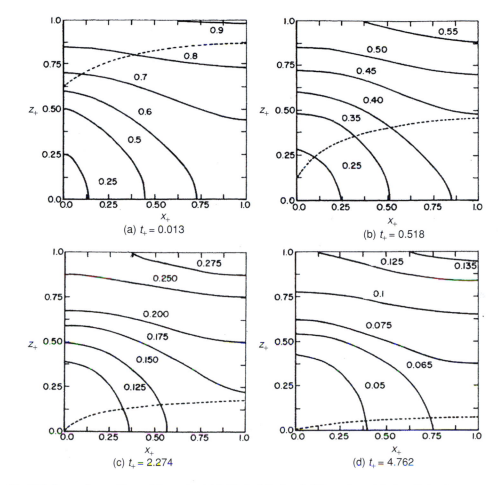

Fig. 10.4 Successive positions of the water table (dashed line) and of lines of equal hydraulic head
$h_+ = h/D$ **(solid lines), in an aquifer with** $D_{c+} = 0$, $B_+ = 1.0$, $(aD)^{-1} = 0.36$,
$n = 3$ **and** $b = 1.5$. **The indicated time values are scaled with** $[(\theta_0 - \theta_r)D]/k_0$. **(From Verma and Brutsaert, 1970.)**

represent some aspect of the transport in the partly saturated flow zone. However, the numerical calculations show that the effect of a change in b alone is very small, and also that the calculated outflow rates are relatively insensitive to changes in n of only a few units. Because n and b usually vary within a modest range for most soils, they are relatively unimportant as compared with $(aD)^{-1}$. This is illustrated in Figures 10.7, 10.8, 10.9 and 10.10. In Figures 10.9 and 10.10 the outflow rate is shown for an aquifer whose breadth B_+ has been increased from 1 to 5, and with the other parameters held constant except $(aD)^{-1}$, which is given the values 0.36 and 0.10. (The saturated two-dimensional case with a free surface (see Section 10.2 below) is shown for comparison.) Comparing these two figures, one sees that the difference between curve 1 and 2 decreases as B_+ increases. Thus the effect of decreasing $(aD)^{-1}$ from 0.36 to 0.10 becomes less pronounced as B_+ increases. This suggests that in nature, where the values of B_+ are usually much larger than the values

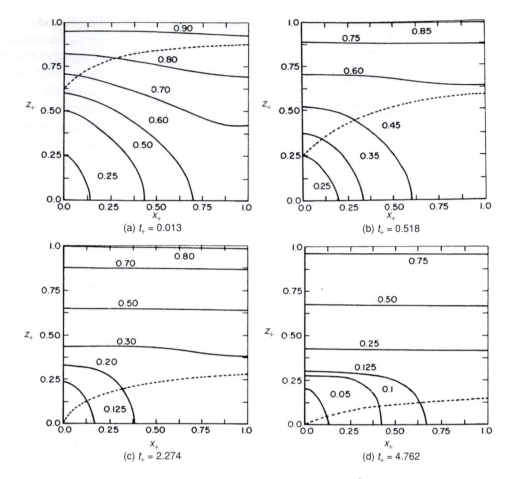

(a) $t_+ = 0.013$

(b) $t_+ = 0.518$

(c) $t_+ = 2.274$

(d) $t_+ = 4.762$

Fig. 10.5 Same as Figure 10.4 for an aquifer with $D_{c+} = 0$, $B_+ = 1.0$, $(aD)^{-1} = 0.10$, $n = 3$ and $b = 3$. (From Verma and Brutsaert, 1970.)

considered here, the effect of capillarity is likely to be even more pronounced than these numerical results indicate, and that probably a smaller value of $(aD)^{-1}$ is required before capillarity above the water table can be neglected in the calculation of outflow rates. In any event, these results show that it is mainly $(aD)^{-1}$ that can be used to determine whether or not the capillary flow above the water table is important.

Equation (8.15), from which the parameter a was obtained for these numerical experiments, may not always be the optimal way to parameterize the soil-water characteristic. Therefore, it is useful to broaden this criterion by defining a *capillary zone number*

$$Ca = \frac{H_c}{D} \tag{10.4}$$

in which H_c is a characteristic suction (negative pressure) required to reduce the degree of saturation of the soil to a certain fraction. This dimensionless quantity Ca can be implemented with the other expressions presented in Section 8.2.4 as well. For instance in the case of Equation (8.14) one can simply put $Ca = (H_b/D)$. It should be recalled that in Equation (8.15) the parameter a^{-1} represents the (negative) pressure head to reduce the

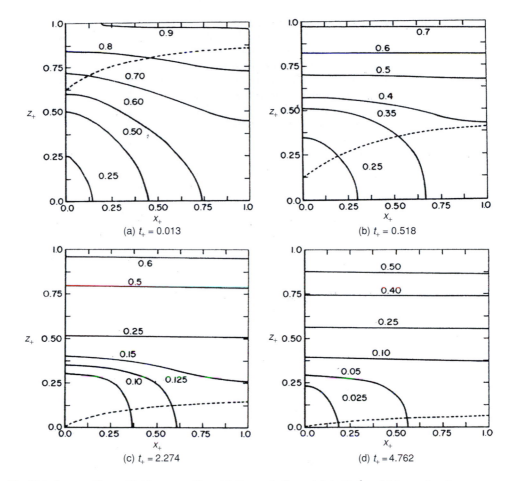

(a) $t_+ = 0.013$

(b) $t_+ = 0.518$

(c) $t_+ = 2.274$

(d) $t_+ = 4.762$

Fig. 10.6 **Same as Figure 10.4 for an aquifer with $D_{c+} = 0$, $B_+ = 1.0$, $(aD)^{-1} = 0.36$, $n = 8$ and $b = 1.5$. (From Verma and Brutsaert, 1970.)**

water content to $1/2$ of its value at saturation. In general, when Ca or $(aD)^{-1}$ is small, the partly saturated zone is relatively thin, and vice versa. As can be deduced from (8.5), the characteristic negative pressure H_c is larger for fine-textured materials, and smaller for coarse-textured soils. Put in practical terms, capillary flow effects are probably negligible in a 100 m deep sandy aquifer, but they are likely to be more important in a 2 m deep clayey soil profile.

The remaining two parameters, $B_+ = B/D$ and $D_{c+} = D_c/D$, can be used to satisfy the usual criteria for geometric similarity.

10.1.4 Initial state at the onset of drainage

The maximal outflow rate from an unconfined aquifer into an adjoining stream occurs when the aquifer is fully saturated. Such a situation can be assumed to exist at the end of heavy or prolonged precipitation, irrigation or snowmelt, and it can be taken to represent the initial state of the aquifer when drainage is about to start. As indicated in Equations (10.3), the

Fig. 10.7 Scaled outflow rate q_+ from an aquifer with rectangular cross section and with $D_{c+} = 0$, $B_+ = 1$, plotted against scaled time t_+. The rate of flow is scaled with $(D\,k_0)$ and the time variable with $[(\theta_0 - \theta_r)D]/k_0$. Curve 1 describes the outflow hydrograph for soil properties $(aD)^{-1} = 0.36$, $n = 3$ and $b = 2$; curve 2 for $(aD)^{-1} = 0.36$, $n = 3$ and $b = 6$; curve 3 represents the case in which the partly saturated zone above the water table is neglected (see Section 10.2). (After Verma and Brutsaert, 1971b.)

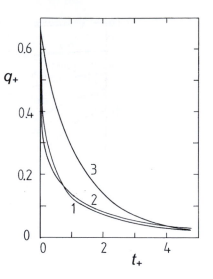

Fig. 10.8 Same as Figure 10.7. Curve 1 describes the outflow hydrograph for soil properties $(aD)^{-1} = 0.36$, $n = 3$ and $b = 1.5$; curve 2 for $(aD)^{-1} = 0.36$, $n = 8$ and $b = 1.5$; curve 3 represents the case in which the partly saturated zone above the water table is neglected (see Section 10.2). (After Verma and Brutsaert, 1971b.)

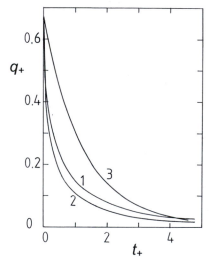

initial state of the unsteady drainage problem is thus taken to be a fully saturated aquifer, in which the water table is assumed to be at the ground surface; such a condition can be obtained by maintaining an infinitesimally thin layer of water on the surface at atmospheric pressure. The exact solution of this classical steady state problem was derived by Kirkham (1950).

Mathematical formulation

With the assumption of an effective hydraulic conductivity, the flow is governed by Laplace's equation (8.57), or for a two-dimensional cross section

$$\frac{\partial^2 h}{\partial x^2} + \frac{\partial^2 h}{\partial z^2} = 0 \qquad\qquad (10.5)$$

Fig. 10.9 Scaled outflow rate q_+ from an aquifer with rectangular cross section and with $D_{c+} = 0.5, B_+ = 1$, plotted against scaled time t_+. The rate of flow is scaled with $(D\,k_0)$ and the time variable with $[(\theta_0 - \theta_r)D]/k_0$. Curve 1 describes the outflow hydrograph for soil properties $(aD)^{-1} = 0.36, n = 3$ and $b = 1.5$; curve 2 for $(aD)^{-1} = 0.1, n = 3$ and $b = 3$; curve 3 represents the case in which the partly saturated zone above the water table is neglected (see Section 10.2). (After Verma and Brutsaert, 1971b.)

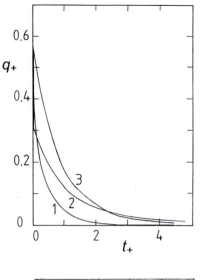

Fig. 10.10 Scaled outflow rate q_+ from an aquifer with rectangular cross section and with $D_{c+} = 0, B_+ = 5$, plotted against scaled time t_+. The rate of flow is scaled with $(D\,k_0)$ and the time variable with $[(\theta_0 - \theta_r)D]/k_0$. Curve 1 describes the outflow hydrograph for soil properties $(aD)^{-1} = 0.36, n = 3$ and $b = 1.5$; curve 2 for $(aD)^{-1} = 0.1, n = 3$ and $b = 3$; curve 3 represents the case in which the partly saturated zone above the water table is neglected (see Section 10.2). (After Verma and Brutsaert, 1971b.)

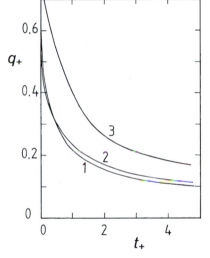

The boundary conditions (see Figure 10.3) are a combination of Equations (10.2) and (10.3), namely

$$
\begin{aligned}
h &= D_c & x &= 0 & 0 &\leq z \leq D_c \\
h &= z & x &= 0 & D_c &\leq z \leq D \\
\frac{\partial h}{\partial x} &= 0 & x &= B & 0 &\leq z \leq D \\
\frac{\partial h}{\partial z} &= 0 & 0 &\leq x \leq B & z &= 0 \\
h &= D & 0 &\leq x \leq B & z &= D
\end{aligned}
\tag{10.6}
$$

The solution can be obtained in several different ways. Kirkham (1950) derived it by generalizing an earlier solution in cylindrical coordinates for the problem of flow into an auger hole. It can also be derived by conformal mapping (Polubarinova-Kochina, 1952). However, separation of variables in the present coordinate system is probably the most straightforward. The result for the hydraulic head can be written as follows

$$h = D - \sum_{n=1,3,5,\dots}^{\infty} \frac{8D}{(n\pi)^2} \cos\left(\frac{n\pi D_c}{2D}\right) \cos\left(\frac{n\pi z}{2D}\right)$$

$$\times \cosh\left(\frac{n\pi(B-x)}{2D}\right) \Big/ \cosh\left(\frac{n\pi B}{2D}\right) \tag{10.7}$$

Outflow rate

The rate of flow into the open channel or water body at $x = 0$, expressed as volume of water per unit time and per unit length of channel (i.e. per unit width of aquifer normal to the main direction of the flow in the aquifer), can be derived by applying Darcy's law to the solution (10.7), to wit

$$q = -k_0 \int_0^D \left(\frac{\partial h}{\partial x}\right)_{x=0} dz \quad \text{or} \quad q = -k_0 \int_0^B \left(\frac{\partial h}{\partial z}\right)_{z=D} dx \tag{10.8}$$

which yield in either case

$$q = -k_0 \sum_{n=1,3,\dots}^{\infty} (-1)^{(n-1)/2} \frac{8D}{(n\pi)^2} \cos\left(\frac{n\pi D_c}{2D}\right) \tanh\left(\frac{n\pi B}{2D}\right) \tag{10.9}$$

The minus sign in front of the right-hand side indicates that the outflow is in the negative x direction. To allow comparison with other solutions and with experimental data, it is once again convenient to scale the variables and express the result in dimensionless terms. The form of Equation (10.9) suggests immediately the following

$$D_{c+} = D_c/D$$

$$B_+ = B/D \tag{10.10}$$

$$q_+ = q/(k_0 D)$$

Thus (10.9) assumes the form

$$q_+ = -\sum_{n=1,3,\dots}^{\infty} (-1)^{(n-1)/2} \frac{8}{(n\pi)^2} \cos\left(\frac{n\pi D_{c+}}{2}\right) \tanh\left(\frac{n\pi B_+}{2}\right) \tag{10.11}$$

In many situations of practical interest, the aquifer thickness D is much smaller than its horizontal dimensions, or $B_+ \to \infty$ and the water depth in the adjoining open channel is very small compared to the aquifer thickness, so that $D_{c+} \to 0$. These two conditions simplify (10.11) to the following

$$q_+ = \frac{-8}{\pi^2}\left(1 - \frac{1}{9} + \frac{1}{25} - \cdots\right) = -0.74245 \tag{10.12}$$

The significance of (10.12) is that it represents the maximal outflow rate from a fully saturated shallow extensive aquifer into an empty channel. As an aside, the sum inside the brackets is also known as Catalan's constant, which equals 0.915965.

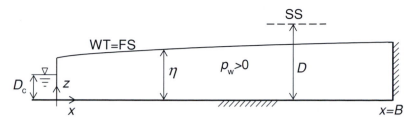

Fig. 10.11 Schematic representation of the flow domain some time after the onset of drainage in a two-dimensional unconfined riparian aquifer, lying on a horizontal impermeable layer. The flow above the water table (WT) is assumed negligible, so that the water table is a true free surface (FS). The effect of capillarity is parameterized by means of the drainable porosity (or specific yield). The initial position of the free surface is at the soil surface (SS), or $\eta = D$ as shown in Figures 10.2 and 10.3.

10.2   FREE SURFACE FLOW: A FIRST APPROXIMATION

Whenever the effects of capillarity can be assumed to be relatively unimportant, the flow in the partly saturated zone above the water table, in which $p_w < 0$, can be neglected. The moving water table can then be treated as a true free surface, which represents the upper boundary of the changing flow domain. As noted above, the dimensionless capillary zone number can provide an indication of the relative importance of the capillary effects in an unconfined aquifer. This capillary zone number is defined in Equation (10.4) as $Ca = H_c/D$, in which D is an average thickness of the unconfined aquifer under consideration, and H_c is a characteristic capillary rise above the water table in the aquifer, that is, a characteristic capillary suction, which reduces the degree of saturation of the soil to a certain fraction, say 50%. Thus, whenever Ca is small, the partly saturated zone above the water table can be eliminated from the flow domain, and the flow is assumed to take place only below the moving water table.

10.2.1   General formulation

Differential equation and boundary conditions
Because in this approximation the flow region below the free surface is fully saturated, the governing equation is again Laplace's Equation (10.5). For the simple two-dimensional case of an unconfined aquifer on a horizontal bed, which is initially fully saturated, the boundary conditions can be taken as (10.2) and (10.3) from which the partly saturated zone is eliminated (see Figure 10.11). They can be written as follows

$$
\begin{aligned}
h &= D_c & x &= 0 & 0 &\leq z \leq D_c & t &\geq 0 \\
h &= z & x &= 0 & D_c &\leq z = h & t &\geq 0 \\
\frac{\partial h}{\partial x} &= 0 & x &= B & 0 &\leq z \leq D & t &\geq 0 \\
\frac{\partial h}{\partial z} &= 0 & 0 &\leq x \leq B & z &= 0 & t &\geq 0 \\
h &= D & 0 &\leq x \leq B & z &= D & t &= 0
\end{aligned}
\tag{10.13}
$$

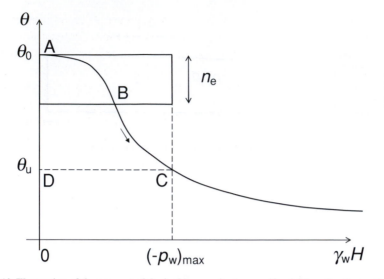

Fig. 10.12 Illustration of the concept of drainable porosity (or specific yield) n_e for the vertical soil column above a water table, in which the maximal water suction is $(-p_w)_{max}$ at the top of the column. The value of n_e can be defined by taking the area $[n_e \times (-p_w)_{max}]$ as the difference between the area $[(\theta_0 - \theta_u) \times (-p_w)_{max}]$ and the area ABCD under the soil water characteristic curve. The water drained from the soil at the top of the column is $(\theta_0 - \theta_u)$; thus n_e is the average amount drained per unit volume of soil in the column and $(\theta_0 - n_e)$ is the corresponding amount of water left in the soil.

Although this is a problem of unsteady flow, the time variable does not appear in the governing Laplace equation. As already hinted in the second of (10.13), which describes the seepage surface, the time variable enters the problem through the condition at the free surface, which constitutes the moving upper boundary. Before deriving this free surface condition, it is necessary to introduce the concept of the drainable porosity.

Drainable porosity
As the water table passes a point, say in the case of drainage when it is falling, the pores do not empty immediately, but the water is retained by capillarity and other mechanisms mentioned in Section 8.2.2; it is only when the water pressure decreases further, i.e. with increasing suction, as illustrated in Figures 8.3–8.6, that water drains from the pores. In the free surface approximation, the reality of this gradual transition is replaced by the assumption of the *drainable porosity*, n_e. This drainable porosity, which is also called the *effective porosity* or the *specific yield*, can be defined as the volume of water per unit volume of porous material, that is released or imbibed, as the free surface passes a given point. In general, the amount of water present in the pores at a point depends on the local water pressure. From this it follows that the drainable porosity must depend on the prevailing water pressure distribution above the water table and therefore on the nature of the flow situation. Figure 10.12 further illustrates how the concept can be interpreted with reference to the soil water characteristic. It can be seen that n_e depends on the value of the maximum suction in the soil column above the water table; because $(-p_w)_{max}$ changes with time during unsteady flow, in principle n_e may also be time dependent to some

extent. As shown, the soil water characteristic curve in Figure 10.12 describes drainage; obviously, consideration of repeated draining and wetting cycles with hysteresis would complicate the matter even more. This means that the drainable porosity n_e cannot be a unique physical property of a given porous material, and that it must be considered as a mere parameter that is to be adjusted and calibrated depending on the flow problem. This is its main limitation, which should be kept in mind in practical applications of the free surface approach in flow in porous media. But if this limitation is kept in mind, the concept can yield useful results in the parameterization of groundwater flow processes at the field and catchment scales.

Free surface condition

In principle, the condition at the free surface in a porous material is again Equation (5.1) as presented in Chapter 5. If the function $F = 0$ describing the free surface is taken as $F = F(x, z, t) = [\eta(x, t) - z] = 0$, in which η denotes the height of the free surface above the reference level $z = 0$, this can be written as

$$\overline{u}\frac{\partial \eta}{\partial x} - \overline{w} + \frac{\partial \eta}{\partial t} = 0 \qquad \text{at } z = \eta \tag{10.14}$$

in which, as before, $\overline{u}$ and $\overline{w}$ represent the x- and z-components of the true fluid velocity, which is also that of the free surface.

The velocity of a fluid is its volumetric rate of flow per unit cross-sectional area occupied by this fluid. Because the specific flux $\mathbf{q}$ in Darcy's equation is the volumetric flow rate per unit cross-sectional area of total or bulk porous material, it does not represent the true velocity of the fluid particles (cf. Section 8.3.1). Rather, with the assumption of a drainable porosity, the actual velocity of the fluid particles must be taken as $(\mathbf{q}/n_e)$. Therefore, with (q_x/n_e) and (q_z/n_e) as the x- and z-components of the velocity of the fluid and also of the free surface, Equation (10.14) becomes, in terms of the Darcy flux,

$$q_x\frac{\partial \eta}{\partial x} - q_z + n_e\frac{\partial \eta}{\partial t} = 0 \qquad \text{at } z = \eta \tag{10.15}$$

With Darcy's law one obtains finally

$$\frac{n_e}{k_0}\frac{\partial \eta}{\partial t} = \frac{\partial h}{\partial x}\frac{\partial \eta}{\partial x} - \frac{\partial h}{\partial z} \qquad \text{at } z = \eta = h \tag{10.16}$$

There is also a second way of implementing Equation (5.1) in a porous material to formulate the condition at the free surface. If the adopted free surface function is $F = F(x, z, t) = [h(x, z, t) - z] = 0$, one has instead of Equation (10.14)

$$\overline{u}\frac{\partial h}{\partial x} + \overline{w}\frac{\partial h}{\partial z} - \overline{w} + \frac{\partial h}{\partial t} = 0 \qquad \text{at } z = \eta = h \tag{10.17}$$

from which one obtains the free surface condition, as an alternative to Equation (10.16),

$$\frac{n_e}{k_0}\frac{\partial h}{\partial t} = \left(\frac{\partial h}{\partial x}\right)^2 + \left(\frac{\partial h}{\partial z}\right)^2 - \frac{\partial h}{\partial z} \qquad \text{at } z = \eta = h \tag{10.18}$$

Either (10.16) or (10.18) can be used in the solution of the problem; the choice depends usually on the specific mathematical aspects to be investigated.

Fig. 10.13 **Sketch illustrating the application of Equation (10.16) to calculate the rate of drawdown of the water table (WT), when treated as a free surface. The equipotentials are lines of constant hydraulic head h. (After Kirkham and Gaskell, 1951.)**

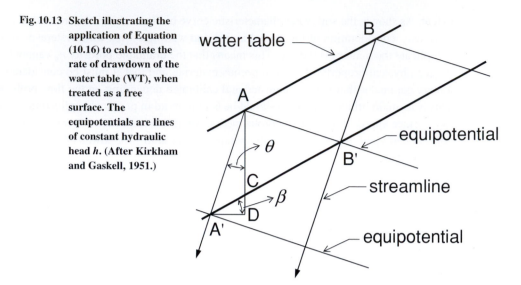

Example 10.1. Displacement of a free surface

The physical significance and the application of Equation (10.16) can be illustrated for the situation shown in Figure 10.13. An infinitesimally small portion of the water table AB moves along the streamlines AA′ and BB′ to a new position A′B′ during a short increment of time δt. If β is the slope of the water table and θ the angle between the streamlines and the vertical, then it can be seen that the vertical component of the distance of the water table fall AD is given by

$$AC = AA'(\cos\theta - \sin\theta \tan\beta) \tag{10.19}$$

According to Darcy's law the total distance traveled by the point A during δt is

$$AA' = -\frac{k_0}{n_e}\frac{\partial h}{\partial s}\delta t \tag{10.20}$$

where $\partial h/\partial s$ is the hydraulic gradient along AA′. Substitution of Equation (10.20) into (10.19) with the observation that

$$\frac{\partial h}{\partial s}\cos\theta = \frac{\partial h}{\partial z} \quad \text{and} \quad \frac{\partial h}{\partial s}\sin\theta = \frac{\partial h}{\partial x} \tag{10.21}$$

results in

$$AC = \frac{k_0}{n_e}\left(\frac{\partial h}{\partial x}\tan\beta - \frac{\partial h}{\partial z}\right)\delta t \tag{10.22}$$

This result, which was first derived by Kirkham and Gaskell (1951), is essentially a finite difference form of Equation (10.16).

10.2.2 Some features of free surface flow solutions

Probably the earliest solution of this type of problem was presented by Kirkham and Gaskell (1951) for the very similar flow situation of a falling water table during tile and ditch

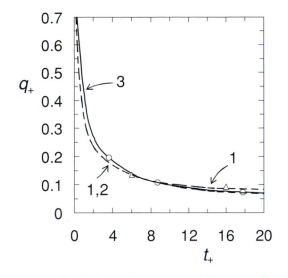

Fig. 10.14 Scaled outflow hydrographs $q_+ = q_+(t_+)$ from an aquifer with rectanglar cross section on a horizontal impermeable bed, calculated with the free surface water table assumption for $D_{c+} = 0$. The scaled rate of flow is defined as $q_+ = q/(D\,k_0)$ and the scaled time variable as $t_+ = k_0 t/(n_e D)$. Curve 3 describes the result of the two-dimensional analysis (see Section 10.2) with Laplace's equation and the free surface condition (10.16) or (10.22) for $B_+ = 6$. Curves 1 and 2 are the results obtained with the one-dimensional hydraulic approach (see Section 10.3) with Boussinesq's equation (10.30) for $B_+ = 6$ and 8, respectively. The circles and the triangles are experimental results for $B_+ = 6$ and 8, respectively. (After Verma and Brutsaert, 1971b.)

drainage. They derived the positions and shapes of a falling water table as a succession of steady flow conditions. Thus the distribution of the hydraulic head h for an initially known water table position was found by a numerical solution of Laplace's equation by means of a relaxation procedure. The next water table position, for a time δt later, was then determined by means of Equation (10.22), and so on. The method was extended in Brutsaert *et al.* (1961) by the inclusion of a partly saturated zone above the water table, and Laplace's equation was solved by an electrical network analog. Subsequently, other methods have been used to solve this and similar free surface flow problems in porous materials, namely perturbation techniques allowing linearization of Equation (10.16) (see Dagan, 1966; VandeGiesen *et al.*, 1994)), finite difference methods (see Verma and Brutsaert, 1971a) and boundary integral methods (Liggett and Liu, 1983).

The dimensionless number Ca can be used as the decisive criterion for the applicability of the free surface approach, as compared to the complete description of saturated–unsaturated flow. For example, Figures 10.7–10.10 illustrate how, unless $\mathrm{Ca} = (aD)^{-1}$ is small, the neglect of the capillary zone results in an overestimate of the outflow rate for small times, but in an underestimate for long times. As mentioned earlier, the other two parameters, $B_+ = B/D$ and $D_{c+} = D_c/D$, represent the criteria for geometric similarity. If the breadth of an unconfined aquifer B, that is the distance from stream to divide, is at least 10 times larger than the depth D, the outflow from a saturated aquifer can be satisfactorily reproduced by the application of hydraulic groundwater theory (see Section 10.3). Figure 10.14 illustrates this for an aquifer with $B_+ = (B/D) = 6$ and 8, in which the partly saturated zone above the water table is neglected; the experimental points were obtained with a Hele-Shaw viscous analog model (Ibrahim and Brutsaert, 1965). Figure 10.14 shows how, as time increases, the outflows obtained with the one-dimensional hydraulic approximation become practically the same as with the two-dimensional Laplace equation.

10.3 HYDRAULIC GROUNDWATER THEORY: A SECOND APPROXIMATION

Free surface representations of flow in unconfined aquifers, as outlined in Section 10.2, are usually easier to solve than those based on Richards's equation, which include also flow in the partly saturated zone above the water table. Nevertheless, the implementation of this simplification for problems in catchment hydrology is rarely straightforward and, even when obtainable, the resulting solutions can usually not readily be parameterized for this purpose. Therefore, further simplifications are called for. One very common approach is based on the observation that unconfined aquifers in natural catchments tend to be relatively thin compared to their horizontal extent. Thus beside the assumption that the water table is a true free surface, it is also assumed that under such conditions, the flow is essentially parallel to the ground surface and/or to the underlying impermeable bed. Specifically, the first assumption requires that the capillary zone number $\mathrm{Ca} = H_{\mathrm{c}}/D$ is small, whereas the second requires that the aquifer is shallow, so that $B_{+} = B/D$ is large. These two assumptions constitute the basis of the hydraulic groundwater theory. It will become clear below that the hydraulic approach is considerably simpler and more parsimonious than the more complete formulations described in Sections 10.1 and 10.2; moreover, in many situations it produces a solution which is quite close to that obtainable by a more complete formulation. Hence not surprisingly, this approach continues to be the method of choice in many investigations. The hydraulic approach is usually attributed to Dupuit (1863). It has also been referred to as the Dupuit–Forchheimer theory, to acknowledge the fact that Forchheimer (1930) applied it to many different problems.

10.3.1 *General formulation*

The governing differential equations for this approach can be derived by combining the continuity equation with Darcy's law adjusted for the hydraulic assumptions.

Adjustment of Darcy's law
The main assumption is essentially the same as that commonly made for open channel flow. It is that the curvature of the streamlines is very small, so that the pressure distribution is practically hydrostatic in the direction normal to the impermeable bed. For the two-dimensional cross section of the aquifer shown in Figure 10.15, Equation (5.5) is directly applicable and can be rewritten as

$$\frac{\partial p_{\mathrm{w}}}{\partial z} + \gamma_{\mathrm{w}} \cos \alpha = 0 \tag{10.23}$$

in which α is the slope angle of the underlying impermeable layer, and z is the coordinate normal to that layer. Observe that with a sloping bed x and z are related with the vertical

Fig. 10.15 Definition sketch for
two-dimensional
groundwater flow with a
water table (WT) above
a sloping impermeable
layer (IL). The WT is
assumed to be a true free
surface of the flowing
water, and flow above
the WT is neglected.

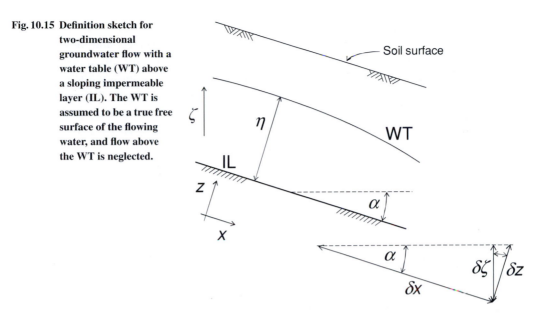

coordinate ζ by $\partial\zeta/\partial z = \cos\alpha$ and $\partial\zeta/\partial x = \sin\alpha$. Integration of Equation (10.23) yields

$$p_w = \gamma_w \cos\alpha\,(\eta - z) \tag{10.24}$$

where $\eta = \eta(x, t)$ is the height of the water table, again measured in the direction normal to the impermeable bed. From Equation (10.24) it follows that, for a constant bed slope α, the pressure gradient in the direction of flow x is given by

$$\frac{\partial p_w}{\partial x} = \gamma_w \cos\alpha \frac{\partial\eta}{\partial x} \tag{10.25}$$

This shows that this gradient depends only on the slope of the free surface, and is independent of z; put differently, $\partial p_w/\partial x$ is a constant along the direction normal to the impermeable bed. For a fluid of constant density the hydraulic head is Equation (8.18) or in the present notation $h = \zeta + p_w/\gamma_w$. With (10.25) the hydraulic gradient becomes

$$\frac{\partial h}{\partial x} = \cos\alpha \frac{\partial\eta}{\partial x} + \sin\alpha \tag{10.26}$$

Hence, under the assumption of hydraulic flow, Darcy's equation yields the specific flux

$$q_x = -k_0 \left(\cos\alpha \frac{\partial\eta}{\partial x} + \sin\alpha \right) \tag{10.27}$$

which, as observed below Equation (10.25), is a constant in any given cross section normal to the underlying bed at a distance x from the origin. A derivation of (10.27) was first presented by Boussinesq (1877), and was later clarified by Childs (1971); however, in both instances the approach differed somewhat from the one given here.

Equation of continuity

Because q_x is a constant along z, it is identical with the average, and the continuity equation (5.13) derived for free surface open channel flow is also directly applicable here. In (5.13) the term representing the displacement rate of the free surface $\partial h/\partial t$ becomes $\partial \eta/\partial t$ in the present notation. Because this is an actual velocity of the interface, the average velocity V in (5.13) must be replaced here by the "true" velocity in the porous material (q_x/n_e); similarly, the lateral inflow i must be replaced by a true recharge velocity (I/n_e), where I is the recharge rate, representing a source term as a volumetric flux per unit ground surface area of porous material. Thus one obtains

$$\frac{\partial \eta}{\partial t} + \frac{\partial}{\partial x}\left(\frac{q_x \eta}{n_e}\right) - \frac{I}{n_e} = 0 \tag{10.28}$$

With (10.27) this assumes the form

$$\frac{\partial \eta}{\partial t} = \frac{k_0}{n_e}\left[\cos\alpha\frac{\partial}{\partial x}\left(\eta\frac{\partial \eta}{\partial x}\right) + \sin\alpha\frac{\partial \eta}{\partial x}\right] + \frac{I}{n_e} \tag{10.29}$$

in which it is assumed, as is commonly done, that k_0, n_e and α are constant or can be treated as effective parameters. In the absence of lateral inflow and for a horizontal impermeable layer, Equation (10.29) becomes

$$\frac{\partial \eta}{\partial t} = \frac{k_0}{n_e}\frac{\partial}{\partial x}\left(\eta\frac{\partial \eta}{\partial x}\right) \tag{10.30}$$

Both (10.29) and (10.30) are forms of what is usually referred to as the Boussinesq equation. To repeat, the Boussinesq equation is based on the following assumptions. (i) The effect of unsaturated flow above the water table is negligible and it can be parameterized by an effective porosity or specific yield n_e; this is also the basis of the free surface approach (i.e. the first approximation). (ii) The pressure distribution in the direction normal to the bed is hydrostatic, which leads to (10.27), which is the basis of the hydraulic approach (i.e the second approximation).

The derivation of Equations (10.29) and (10.30) is presented here for a two-dimensional cross section of an unconfined aquifer. It is straightforward to consider the more general case of three-dimensional flow, with x as the coordinate pointing up the slope along the impermeable bed, and y as the horizontal lateral or span-wise coordinate, to obtain a more general form of the Boussinesq equation, namely

$$\frac{\partial \eta}{\partial t} = \frac{k_0}{n_e}\left[\cos\alpha\frac{\partial}{\partial x}\left(\eta\frac{\partial \eta}{\partial x}\right) + \sin\alpha\frac{\partial \eta}{\partial x} + \frac{\partial}{\partial y}\left(\eta\frac{\partial \eta}{\partial y}\right)\right] + \frac{I}{n_e} \tag{10.31}$$

Mathematically, Equation (10.31) can be characterized as a nonlinear advective diffusion equation, with a variable (i.e. a function of η) and anisotropic hydraulic (groundwater) diffusivity, whose two principal components are $D_{hx} = k_0\eta\cos\alpha/n_e$ and $D_{hy} = k_0\eta/n_e$, and with a hydraulic (groundwater) advectivity $c_h = -k_0\sin\alpha/n_e$.

A basic feature of the hydraulic approach is that two-dimensional flow is represented by a one-dimensional formulation in Equations (10.29) and (10.30); thus the unknown hydraulic head $h = h(x, z, t)$ is replaced by the unknown position of the water table

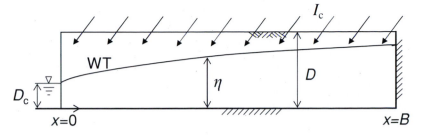

Fig. 10.16 Schematic representation of the cross section of an unconfined riparian aquifer, lying on a horizontal impermeable layer, under steady flow conditions. The position of the water table results from a steady and constant recharge rate I_c, and it has the shape of an ellipse without seepage surface at $x = 0$, when determined with the hydraulic approach.

$\eta = \eta(x, t)$. Similarly, three-dimensional flow is simplified to a two-dimensional problem in Equation (10.31), as the unknown hydraulic head $h = h(x, y, z, t)$ is replaced by the unknown height of the water table $\eta = \eta(x, y, t)$.

10.3.2   *Steady flow described with hydraulic theory*

Over the years hydraulic groundwater theory has been a powerful tool to solve a large number of important practical problems under steady state conditions. The main reason for its wide use is that under steady state conditions the Boussinesq equation becomes linear in η^2, which greatly simplifies the mathematical analysis. For instance, under conditions of steady flow over a horizontal bed and in the absence of lateral inflow, Equation (10.31) reduces to

$$\frac{\partial^2 \eta^2}{\partial x^2} + \frac{\partial^2 \eta^2}{\partial y^2} = 0 \qquad (10.32)$$

This is Laplace's equation in η^2, for which many solution methods are available. Moreover, because the problem is linear, known solutions for η^2 obtained for relatively simple boundary conditions can be extended to more complicated situations by the application of image methods and other methods of superposition. Two examples of steady aquifer outflow are presented in what follows.

Steady outflow resulting from a uniform precipitation
Under steady conditions and for an aquifer cross section with horizontal bed, as shown in Figure 10.16, one can write Equation (10.29) as

$$\frac{\partial}{\partial x}\left(\eta\frac{\partial \eta}{\partial x}\right) = -\frac{I_c}{k_0} \qquad (10.33)$$

where I_c is a constant recharge rate; this is usually taken as a climatological average rainfall for design purposes, but it may also represent irrigation or a negative rate of

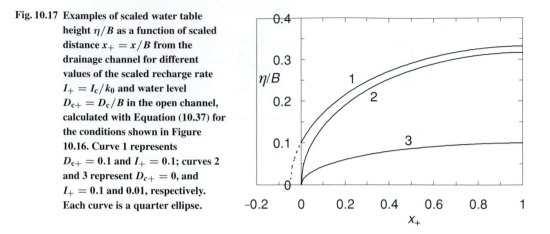

Fig. 10.17 Examples of scaled water table height η/B as a function of scaled distance $x_+ = x/B$ from the drainage channel for different values of the scaled recharge rate $I_+ = I_c/k_0$ and water level $D_{c+} = D_c/B$ in the open channel, calculated with Equation (10.37) for the conditions shown in Figure 10.16. Curve 1 represents $D_{c+} = 0.1$ and $I_+ = 0.1$; curves 2 and 3 represent $D_{c+} = 0$, and $I_+ = 0.1$ and 0.01, respectively. Each curve is a quarter ellipse.

evaporation from the water table or seepage through the bed, or any combination of several of these rates. The following boundary conditions

$$\eta = D_c \qquad x = 0$$
$$\frac{\partial \eta}{\partial x} = 0 \qquad x = B \tag{10.34}$$

indicate, as before in Equations (10.2) and (10.6), that the water level in the channel is D_c, and on account of (10.27) that the divide represents an impermeable boundary. Integrating (10.33) twice, one obtains for the height of the water table

$$\eta^2 = \frac{I_c}{k_0}(2Bx - x^2) + D_c^2 \tag{10.35}$$

Actually, this result can also be derived directly without formal recourse to the Boussinesq equation, by simply observing that at any point x according to (10.27) the rate of flow through the area η is $[-\eta k_0(\partial \eta/\partial x)]$; this equals the rate of recharge at the surface, which is given by $[-I_c(B - x)]$, and solution of this equality yields Equation (10.35).

Equation (10.35) can be generalized, to facilitate comparison with experimental results and other theoretical approaches, by scaling the variables as follows

$$\eta_+ = \frac{\eta}{B}, \quad x_+ = \frac{x}{B}, \quad D_{c+} = \frac{D_c}{B} \quad \text{and} \quad I_+ = \frac{I_c}{k_0} \tag{10.36}$$

This transforms (10.35) into

$$\eta_+ = \left[I_+\left(2x_+ - x_+^2\right) + D_{c+}^2 \right]^{1/2} \tag{10.37}$$

This result is illustrated in Figure 10.17 for a few examples.

Application of Equation (10.35) to $x = B$, where the water table has its maximal height $\eta = \eta_{max}$, yields immediately

$$B^2 = \frac{k_0}{I_c}\left(\eta_{max}^2 - D_c^2\right) \tag{10.38}$$

Equation (10.38) was originally intended as a design equation, to determine the spacing $(2B)$ of drainage ditches or underground pipe drainage systems of agricultural lands;

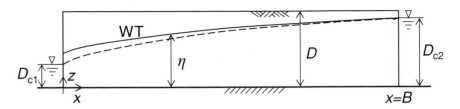

Fig. 10.18 Schematic representation of the cross section of an unconfined aquifer, lying on a horizontal impermeable layer between two open channels with constant water levels. If the water table (WT) is assumed to be a free surface, the resulting steady flow rate q between the two channels is given exactly by the Dupuit formula (10.43). The solid curve represents the true WT with a seepage surface and the dashed curve the WT obtained with the hydraulic approach.

in fact, with a number of subsequent improvements it still provides the basis for many of the soil drainage design procedures in use today. In order to apply it in its original form, the variables on the right-hand side of Equation (10.38) must be known or decided upon. Thus k_0 is the hydraulic conductivity of the soil, I_c is taken as the average rate of precipitation or other input during the period when drainage is needed most, D_c is the depth of the water in the drainage channel or, to a first approximation, the height of the drainage pipe above the impermeable layer, and η_{max} is the main design variable, namely the maximal allowable height of the water table above the impermeable layer.

Equation (10.38) has a long history. It is now often referred to as the ellipse equation, on account of the shape of the water table given by Equation (10.35) (see Figure 10.17). It was probably first derived by A. Colding in Denmark before 1872 for the case $D_c = 0$, after he became aware of earlier experimental results published in 1859 by S. C. Delacroix in France; interestingly, he also recommended a 10% reduction of any B value obtained with Equation (10.38), to make it agree better with these experimental data. Hooghoudt (1937), who knew indirectly of Colding's result through the work of others, was probably the first to derive (10.38) for arbitrary values of D_c; he later (Hooghoudt, 1940) adjusted it to make it more suitable for drainage with pipes. A detailed history of the equation and its more recent derivatives has been presented by VanderPloeg *et al.* (1999).

Steady flow between two parallel channels without precipitation
In this problem, as shown in Figure 10.18, the flow in the unconfined aquifer is described by the one-dimensional Laplace equation

$$\frac{\partial^2 \eta^2}{\partial x^2} = 0 \tag{10.39}$$

The boundary conditions are

$$\eta = D_{c1} \quad x = 0$$
$$\eta = D_{c2} \quad x = B \tag{10.40}$$

where D_{c1} and D_{c2} are the depths in the two channels. Integrating (10.39) once, one obtains

$$\eta \frac{\partial \eta}{\partial x} = C_1 \tag{10.41}$$

where C_1 is an integration constant. Comparison with Equation (10.27) for a horizontal bed shows immediately that $C_1 = -q/k_0$, in which $q = (\eta q_x)$. The variable q is the rate of flow in the aquifer between the two channels per unit length of channel $[L^2/T]$. Because the flow is steady, in the present situation q is a constant, that is, independent of x. A second integration, with the first of (10.40), yields the position of the free surface

$$\eta^2 = -\frac{2q}{k_0}x + D_{c1}^2 \tag{10.42}$$

Application of the second of (10.40) yields the rate of flow between the two channels, in terms of the hydraulic conductivity and the known water levels in the two channels, or

$$q = -\frac{k_0\left(D_{c2}^2 - D_{c1}^2\right)}{2B} \tag{10.43}$$

Again, the negative sign in Equation (10.43) merely indicates that the flow is taking place in the minus x direction. Equation (10.43) is known as the Dupuit formula (see also Dupuit, 1863, p. 236). This result is of considerable theoretical interest, because it can be shown to be exact. In other words, even though in the derivation of (10.43) use is made of the hydraulic assumptions, it has the same form as the solution for the same free surface problem, obtained when no use is made of the hydraulic assumptions. The fact that, in some cases, it produces the exact result, suggests that the hydraulic approach can be a powerful and reliable tool in the derivation of the groundwater flow rates. This has been confirmed in other instances as well. However, it is now also known that the hydraulic approach is not as accurate in the prediction of the geometry of the free surface. One obvious reason for this is that an inherently two-dimensional flow pattern is being described by a one-dimensional formulation. This precludes then, for example, the inclusion of a seepage surface in the boundary conditions, as was done in the second of (10.13). In hydraulic groundwater theory, there is no way to include the existence of a seepage surface and the first two of (10.13) must of necessity be combined into one condition, namely the first of (10.40) (or of (10.34)).

Exactness of the Dupuit formula

The proof proceeds as follows, for the situation sketched in Figure 10.18. Without consideration of the hydraulic approximation, the rate of flow through a vertical section at any point x between the two open channels is given by

$$q = -k_0 \int_0^\eta \frac{\partial h}{\partial x} dz \tag{10.44}$$

Recall that $h = h(x, z)$ and $\eta = \eta(x)$; application of Leibniz's formula (see Appendix, Equation (A2)) therefore allows (10.44) to be rewritten as

$$-q = k_0 \frac{d}{dx} \int_0^\eta h\, dz - k_0 h(x, \eta)\frac{d\eta}{dx} \tag{10.45}$$

Since $h(x, \eta) = \eta(x)$ defines the free surface, (10.45) can be integrated to yield

$$-qx = k_0 \int_0^\eta h(x, z)\, dz - k_0 \frac{\eta^2}{2} + C \tag{10.46}$$

The constant C can be determined by applying the boundary condition at $x = 0$, where $h = D_{c1}$ for $0 \leq z \leq D_{c1}$ and $h = z$ for $D_{c1} \leq z \leq \eta$. (The lower part of this boundary condition describes the hydrostatic conditions and constant hydraulic head in the canal, and the higher part the seepage surface; thus this condition differs from the first of (10.40), i.e. the corresponding condition of the hydraulic approach, which is incapable of incorporating a seepage surface.) After breaking up the integral in Equation (10.46) over the two parts of its range, one obtains

$$C = -k_0 \int_0^{D_{c1}} D_{c1}\, dz - k_0 \int_{D_{c1}}^{\eta_0} z\, dz + k_0 \frac{\eta_0^2}{2} \tag{10.47}$$

in which η_0 is the value of η at $x = 0$, or finally upon integration

$$C = -k_0 \frac{D_{c1}^2}{2} \tag{10.48}$$

Application of the second boundary condition, namely at $x = B$, where $h = \eta = D_{c2}$ over the whole range $0 \leq z \leq D_{c2}$, with insertion of (10.48), changes (10.46) into

$$-qB = k_0 \int_0^{D_{c2}} D_{c2}\, dz - k_0 \frac{D_{c2}^2}{2} - k_0 \frac{D_{c1}^2}{2} \tag{10.49}$$

Upon integration of (10.49), the desired result, namely the Dupuit formula (10.43), follows immediately. This confirms that, even though the Dupuit formula was originally obtained by means of the hydraulic approximation, it is in fact identical to the result obtainable with a more rigorous derivation. This fact should instill some confidence in the more general applicability of the hydraulic approach, as a very close approximation to describe flow rates in other situations as well.

The exact derivation of the Dupuit formula was probably first presented by I. A. Charnii, and it can also be found in the book by Polubarinova-Kochina (1952, p. 281). Later a similar proof was presented by Hantush (1962; 1963).

10.3.3   Unsteady flow described with standard hydraulic theory

It is again instructive to consider the phenomenon of outflow from an unconfined aquifer on a horizontal bed into a stream. This situation is still like the one described schematically in Figure 10.11; however, because now the hydraulic approximation is made use of, the two-dimensional problem has been reduced to a one-dimensional problem, and the z-coordinate is no longer part of the formulation.

Basic formulation
This flow problem was shown to be subject to boundary conditions (10.2) and (10.3) or, after a first approximation, to (10.13); when these conditions are translated to the

hydraulic approach with a horizontal aquifer, they become

$$\eta = D_c \quad x = 0 \qquad t \geq 0$$

$$\frac{\partial \eta}{\partial x} = 0 \quad x = B \qquad t \geq 0 \tag{10.50}$$

$$\eta = D \quad 0 \leq x \leq B \quad t = 0$$

The governing equation of this one-dimensional horizontal flow is the Boussinesq equation (10.30). However, even in this simple form, the equation is still nonlinear; this means that, in contrast to linear problems, there are no general solutions available, and that a specific ad hoc method must be devised for each new problem. Equation (10.30) subject to (10.50) can, of course, be solved numerically (Verma and Brutsaert, 1971a); in Figure 10.14 some results, obtained this way with (10.30) subject to (10.50), are compared with calculations with the complete free-surface formulation. There are, however, also two exact analytical solutions available for boundary conditions, that may be considered as short-time and long-time cases of Equations (10.50) with $D_c = 0$, and that are of interest in practical situations. These solutions are treated in the next two sections.

Outflow rate
Once the solution of this problem has been obtained as $\eta = \eta(x, t)$, the outflow rate q from the aquifer into the stream at $x = 0$ can be determined by applying the hydraulic form of Darcy's equation (10.27), as follows

$$q = -k_0 \left[\eta \frac{\partial \eta}{\partial x} \right]_{x=0} \tag{10.51}$$

Recall that in this chapter q denotes the volume of water per unit time and per unit length of channel (i.e. per unit span or per unit width of aquifer normal to the main direction of the flow in the aquifer), so that its dimensions are $[L^2 T^{-1}]$. In some cases it is more convenient to follow a procedure, analogous with that used to obtain Equation (9.15) for infiltration. Accordingly, as sketched in Figure 10.19, from continuity considerations the cumulative outflow volume per unit span (with dimensions $[L^2]$) from the aquifer at $x = 0$ can be written in general as

$$\forall = n_e \int_{D_c}^{D} x \, d\eta \tag{10.52}$$

which produces the outflow rate as $q = d\forall/dt$. Both (10.51) and (10.52) are used in the remainder of this chapter. The second approach to obtain the outflow is especially useful, whenever the method of solution is based on Boltzmann's transform, as was the case for sorption and infiltration in Chapter 9.

10.3.4   *Short-time outflow behavior*

As will become clear below, the short-time outflow behavior of an unconfined aquifer can be studied by analyzing the case of an infinitely wide aquifer, that is for $B \to \infty$

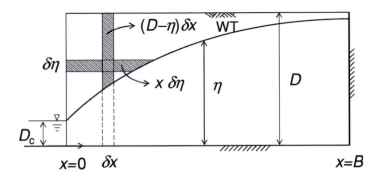

Fig. 10.19 Sketch illustrating the calculation of the drained volume as the area outside the $\eta = \eta(x)$ curve, that is the water table height at a given instant in time t. This can be done by integrating either the elemental area $(x\,d\eta)$ or the elemental area $(D - \eta)dx$. The point $x = 0$ is where the ground water exits the aquifer, and B is the breadth of the aquifer or the distance of the divide from the stream channel.

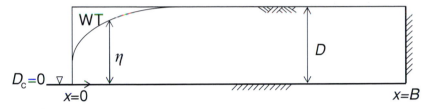

Fig. 10.20 Sketch illustrating the short-time water table configuration in an unconfined hydraulic aquifer, while the boundary condition at $x = B$ can be assumed to have no effect. The boundary conditions for this situation are Equations (10.53), and Boltzmann's transform is applicable.

(see Figure 10.20). If initially the aquifer is fully saturated, it can be assumed that as the outflow starts the flow condition at $x = 0$ is not "felt" further away from the channel, and that the flow proceeds as if the aquifer is infinitely wide. Subsequently, however, as drainage continues, the effect of this condition diffuses, not unlike a wave, away from $x = 0$, and as it approaches $x = B$, the short time solution gradually becomes invalid. The boundary conditions, describing flow from such an infinitely wide, initially saturated aquifer into an empty channel, can be formulated as

$$
\begin{aligned}
\eta = 0 \quad & x = 0 \quad t \geq 0 \\
\eta = D \quad & x > 0 \quad t = 0
\end{aligned}
\tag{10.53}
$$

Similarity considerations

Like in the sorption problem in Section 9.2, the nature of the semi-infinite flow domain and these boundary conditions imply a certain symmetry; because the aquifer must be empty after a very long time of drainage, the water level for $t \to \infty$, is the same as at $x = 0$ for all times; similarly, very far from the channel, i.e. for $x \to \infty$ the water will not "feel" the effect of the drainage and will remain at the original level it had at $t = 0$.

Hence, Boltzmann's transform, i.e.

$$\phi = xt^{-1/2} \tag{10.54}$$

can be used here as well. As will be shown next, this greatly simplifies the solution.

In the manner shown in Equations (9.12), the Boltzmann transformation (10.54) allows the reduction of the Boussinesq equation (10.30) to the following ordinary differential equation

$$\frac{k_0}{n_e} \frac{d}{d\phi} \left(\eta \frac{d\eta}{d\phi} \right) + \frac{\phi}{2} \frac{d\eta}{d\phi} = 0 \tag{10.55}$$

The boundary conditions (10.53) now become

$$\begin{aligned} \eta = 0 \quad &\phi = 0 \\ \eta = D \quad &\phi \to \infty \end{aligned} \tag{10.56}$$

Regardless of the method used, the solution of (10.55) with (10.56) is of the form $\eta = \eta(\phi)$ or $\phi = \phi(\eta)$. The cumulative outflow volume from the aquifer at $x = 0$ is given by Equation (10.52). Thus, once the solution $\phi = \phi(\eta)$ is known, in light of the Boltzmann transform, this outflow volume becomes

$$\forall = t^{1/2} n_e \int_0^D \phi(\eta) d\eta \tag{10.57}$$

Because the integral in (10.57) is a constant, for conciseness of notation, the outflow volume can be expressed in terms of the hydraulic desorptivity, defined as

$$De_h = n_e \int_0^D \phi(\eta) d\eta \tag{10.58}$$

The rate of outflow from the aquifer at $x = 0$, that is $q = -d\forall/dt$, can now be written as

$$q = -\frac{1}{2} De_h t^{-1/2} \tag{10.59}$$

This can probably serve as a more tangible and practical definition of the desorptivity than (10.58). Note that the outflow rate q can also be obtained by applying the hydraulic extension of Darcy's law at $x = 0$, namely Equation (10.51); naturally, for a known solution $\eta = \eta(x, t)$, the result should be the same as that obtained with (10.59) and (10.58).

> Before any solutions are discussed in detail, some interesting features of the hydraulic desorptivity De_h can be derived from additional similarity considerations. It stands to reason that the water table height η should be normalized with its initial value D; insertion of this normalized depth into Equation (10.55) reveals then immediately the dimensionless form of ϕ. Thus the problem can be cast in terms of the following scaled variables
>
> $$\begin{aligned} \eta_+ &= (\eta/D) \\ \phi_+ &= (n_e/k_0 D)^{1/2} \phi \end{aligned} \tag{10.60}$$

In terms of these dimensionless variables, the desorptivity (10.58) can now be written as

$$De_{\text{h}} = a\,(k_0 n_{\text{e}})^{1/2}\,D^{3/2} \tag{10.61}$$

In Equation (10.61) a represents the definite integral

$$a = \int_0^1 \phi_+ d\eta_+ \tag{10.62}$$

which is a dimensionless constant, and thus simply a number whose value depends on the solution.

To summarize, this brief analysis has shown how, except for a constant a, the exact functional form of the rate of outflow from the aquifer, that is (10.59) with (10.61), can be obtained by using similarity, that is almost by inspection, without actually deriving the solution. Two types of similarity were invoked here. The first, Boltzmann's transform, which results from the nature of the boundary conditions, involves a combination of the independent variables; it states that the dependency of η on x is similar to its dependency on $t^{-1/2}$. The second type of similarity involves the scaling of the variables to make the formulation dimensionless, and thus universally applicable to any aquifer, with any dimensions and consisting of any type of porous material.

Solutions

Several solutions of this problem have been derived. Polubarinova-Kochina (1952, p. 507) was able to obtain a solution, by transforming Boussinesq's Equation (10.30) to the Blasius equation for the viscous boundary layer. From her result it can be shown that

$$a = 0.664\,12 \tag{10.63}$$

but the details of the derivation are beyond the present scope. A similar but slightly more accurate procedure was later used by Hogarth and Parlange (1999). There are also several approximate solutions available, that while less accurate, still yield values of a close to (10.63). One such solution, based on an approximation of Equation (10.30) by successive steady states, was proposed in 1886 by K. E. Lembke (cited by Polubarinova-Kochina, 1952, p. 573); this assumption can be shown to lead to $a = (1/3)^{1/2}$, which is within 13% of Equation (10.63). Incidentally, the assumption of successive steady states is equivalent with the quasi-steady approach, which was used in the solution of the horizontal infiltration problem by Parlange (1971). A second approximate solution can be obtained by linearization; in 1947 J. H. Edelman (cited by Kraijenhoff, 1966) proposed its use to describe free surface groundwater flow; this solution yields $a = (4p/\pi)^{1/2}$, in which p is a parameter used to compensate for the approximation due to the linearization and discussed further in Section 10.4. Comparison with Equation (10.63) shows that the linear solution can produce the same result as the exact outflow rate, provided $p = 0.3465$.

Outflow rate

The rate of outflow from the aquifer into the adjacent stream or some other type of open water body is the main item of interest in catchment hydrology; on the basis of simple similarity, it was already shown to be given by Equation (10.59), in which De_{h} was defined as a constant but unspecified hydraulic desorptivity. By combining (10.59) with

Fig. 10.21 Successive scaled positions of the water table $\eta_+ = \eta/D$ in an unconfined riparian aquifer, as calculated with Boussinesq's solution $\eta_+ = F/(1 + at_+)$, for the indicated values of the scaled time $t_+ = [k_0 D/(n_e B^2)]t$. The variable $x_+ = x/B$ is the scaled distance from the stream and the function $F = F(x_+)$ is the curve shown for $t_+ = 0$.

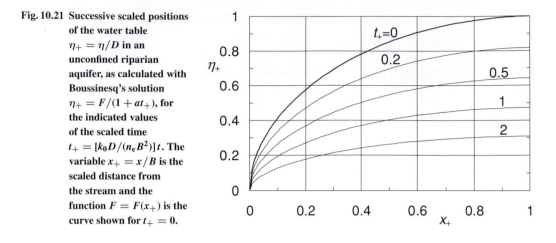

(10.61) and (10.63), for convenient reference the outflow into the stream can now be expressed in terms of the original values and parameters as

$$q = 0.332\,06(k_0 n_e D^3)^{1/2} t^{-1/2} \tag{10.64}$$

10.3.5   *Long-time outflow behavior*

As formulated in Equation (10.50), the third condition describes an initially saturated aquifer over the entire width of the aquifer, i.e. for $0 \le x \le B$. Boussinesq (1904) showed how, by relaxing this condition and by specifying the initial value of η only at $x = B$, it becomes possible to obtain an exact solution of (10.30). Thus instead of (10.50), the relaxed boundary conditions are the following

$$
\begin{aligned}
\eta &= 0 & x &= 0 & t &\ge 0 \\
\frac{\partial \eta}{\partial x} &= 0 & x &= B & t &\ge 0 \\
\eta &= D & x &= B & t &= 0
\end{aligned}
\tag{10.65}
$$

As will be shown below, Boussinesq's method of solution implicitly requires the assumption of self-preservation of the x-dependency of the water table height η, so that the shape of the water table remains the same with time. This water table shape is not arbitrary, but it results from the solution. As illustrated in Figure 10.21, that solution produces a water table that is curvilinear throughout the drainage process, from the beginning at $t = 0$ to the end when the outflow finally ceases. Normally, an initially saturated aquifer (cf. the third of (10.50)) will exhibit this kind of shape only after drainage has proceeded for some time; it is for this reason that the solution obtained in what follows is referred to as a long-time solution.

Similarity considerations

As was the case for the short-time solution, useful insight can be gained by making the formulation dimensionless. Again, it is reasonable to normalize the length variables

η and x with their respective maximal values D and B, as suggested by the boundary conditions (10.65). Insertion of these normalized variables in the governing differential equation (10.30) yields then the appropriate scaling of the time variable. Thus one ends up with the following scaled variables

$$x_+ = x/B$$

$$t_+ = [k_0 D/(n_e B^2)]t \tag{10.66}$$

$$\eta_+ = \eta/D$$

These dimensionless variables allow the Boussinesq equation (10.30) to be simplified as

$$\frac{\partial \eta_+}{\partial t_+} = \frac{\partial}{\partial x_+}\left(\eta_+ \frac{\partial \eta_+}{\partial x_+}\right) \tag{10.67}$$

and the boundary conditions (10.65) as

$$\eta_+ = 0 \qquad x_+ = 0 \qquad t_+ \geq 0$$

$$\frac{\partial \eta_+}{\partial x_+} = 0 \quad x_+ = 1 \qquad t_+ \geq 0 \tag{10.68}$$

$$\eta_+ = 1 \qquad x_+ = 1 \qquad t_+ = 0$$

This problem can be solved by separation of variables, that is, by assuming that the solution $\eta_+ = \eta_+(x_+, t_+)$ is a product of two functions, one dependent only on x_+ and one dependent only on t_+, or

$$\eta_+ = F_1(x_+)F_2(t_+) \tag{10.69}$$

Substitution of (10.69) into (10.67) produces

$$\frac{1}{F_2^2}\frac{dF_2}{dt_+} = \frac{1}{F_1}\frac{d}{dx_+}\left(F_1\frac{dF_1}{dx_+}\right) = C_1 \tag{10.70}$$

in which C_1 must be constant; since x and t are independent of each other, the F_1 and F_2 dependent parts of Equation (10.70) can only be equal to each other if they are constant. The solution of the differential equation for F_1 cannot be expressed in terms of common functions, but for the purpose of the present discussion it is not needed, so let it be assumed that it is known; it will be derived below. However, the solution of the differential equation for F_2 yields immediately

$$-F_2^{-1} = C_1 t_+ + C_2 \tag{10.71}$$

where C_2 is a second constant. Thus, putting $-F_1(x_+)/C_2 = F(x_+)$ and $a = (C_1/C_2)$, one can write the solution (10.69) in the following form

$$\eta_+ = \frac{F(x_+)}{1 + a t_+} \tag{10.72}$$

where a is a dimensionless constant, and $F(x_+)$ is a function of x_+ that satisfies the same differential equation as $F_1(x_+)$ and the conditions $F = 0$ for $x_+ = 0$, and $F' = 0$ and $F = 1$ for $x_+ = 1$, according to the boundary conditions (10.68). This solution indicates similarity, in that the initial shape of the water table is preserved throughout. In other words, once the water table has a certain shape, it keeps that shape; the height of the water table only becomes lower with time.

The outflow rate from the aquifer, Equation (10.51), can be written in terms of the dimensionless variables defined in (10.66) as

$$q = -\frac{k_0 D^2}{B} \left. \eta_+ \frac{\partial \eta_+}{\partial x_+} \right|_{x_+=0} \tag{10.73}$$

Observe that this suggests that the outflow rate be scaled with $(k_0 D^2/B)$, defining a dimensionless outflow rate

$$q_+ = \frac{Bq}{k_0 D^2} \tag{10.74}$$

With the solution (10.72) and putting

$$\left. F \frac{dF}{dx_+} \right|_{x_+=0} = b \tag{10.75}$$

one obtains in dimensionless form

$$q_+ = \frac{-b}{(1 + at_+)^2} \tag{10.76}$$

In (10.76) a and b are dimensionless constants, whose values depend on the solution of (10.70) for $F(x_+)$.

Again, as in the previous section, this brief derivation shows how, save for these two constants, it is possible to obtain the exact form of the outflow rate, without actually solving for $F(x_+)$, and mainly on the basis of two types of similarity considerations. The first type involves self-preservation of the shape of the water table; this follows from the fact that the solution can be obtained by separation of variables. The second type involves dimensional analysis to scale the variables.

Solution

As mentioned, Boussinesq (1904) presented the exact solution to this problem. This solution is greatly simplified by the use of the scaled variables and it proceeds as follows. The function $F_2(t_+)$ in Equation (10.70) has been solved for, and only $F_1(x_+)$ remains to be determined. With the transformations given below Equation (10.71) this requires the solution of the following ordinary differential equation

$$\frac{d^2}{dx_+^2}(F^2/2) = -aF \tag{10.77}$$

Putting $p = d(F^2/2)/dx$, one can readily check that the left-hand side of Equation (10.77) can be written as $p\, dp/d(F^2/2)$. A first integration of this result yields

$$\frac{p^2}{2} = -\frac{aF^3}{3} + C_3 \tag{10.78}$$

where C_3 is a third constant of integration. Performing a second integration one obtains from (10.78)

$$x_+ = \int_0^F \frac{y\, dy}{(2C_3 - (2a/3)y^3)^{1/2}} \tag{10.79}$$

in which use has been made of the first of the boundary conditions (10.68), namely $F = 0$ at $x_+ = 0$. The two constants a and C_3 will be determined next by imposing the remaining two boundary conditions (10.68). Imposition of the second of (10.68) requires the derivative of F. Thus, by means of Leibniz's rule (see Appendix) one obtains from (10.79)

$$1 = \frac{F}{(2C_3 - (2a/3)F^3)^{1/2}} \frac{dF}{dx_+} \tag{10.80}$$

At $x_+ = 1$ the derivative dF/dx_+ must equal zero, according to the second of (10.68); moreover, in light of (10.72) at $x_+ = 1$ the function F must be equal to one, according to the third of (10.68). Hence, for the left-hand side of (10.80) to be unity, it is necessary that $C_3 = (a/3)$. Finally, imposing the third of (10.68), i.e. $F = 1$ at $x_+ = 1$, also on (10.79), one obtains

$$1 = \left(\frac{3}{2a}\right)^{1/2} \int_0^1 y(1 - y^3)^{-1/2} dy \tag{10.81}$$

By putting $u = y^3$, it is easy to show that the integral in Equation (10.81) is equal to $B(2/3, 1/2)/3$, where the $B(\ )$ symbol denotes the complete beta function; hence (10.81) yields the following expression for the constant $a = [B(2/3, 1/2)]^2/6$. The value of this beta function can be readily calculated by expressing it in terms of gamma functions (see, for example, 6.2.2 in Abramowitz and Stegun, 1964), to obtain $B(2/3, 1/2) = \Gamma(1/2)\Gamma(2/3)/\Gamma(7/6) = 2.587\,11$, which produces immediately $a = 1.1155$. Substitution of these values of the constants a and C_3 into Equation (10.79) yields the x_+-dependent part of the solution as

$$x_+ = \frac{3}{B(2/3, 1/2)} \int_0^F y(1 - y^3)^{-1/2} dy \tag{10.82}$$

or, in a slightly different form,

$$x_+ = \frac{1}{B(2/3, 1/2)} \int_0^{F^3} u^{-1/3}(1 - u)^{-1/2} du \tag{10.83}$$

In the form of Equation (10.83) the solution is an incomplete beta function for the variable F^3. Numerical values of this solution for $F(x_+)$ have been presented by Aravin and Numerov (1953); they also indicated that in 1934 L. S. Leibenzon developed the approximation $F = (1.321x_+^{1/2} - 0.142x_+^{3/2} - 0.179x_+^{5/2})$; evidently, this expression involves a standard error of estimate for F of 10^{-3}. The function $F(x_+)$, as given by (10.82) or (10.83), is the curve for $t_+ = 0$ in Figure 10.21. Actually, Figure 10.21 shows the complete solution for the height of the water table η_+, as given by Equation (10.72), for different values of the time t_+. It can be seen that, indeed, with (10.72) the water table exhibits self-preservation and that it maintains the same curvilinear shape throughout the whole drainage process, from beginning to end. If the aquifer were initially saturated, as required in (10.50) and (10.53), the water table would only become curvilinear after enough water has drained; therefore, the solution of (10.30) with (10.65), is only be applicable to describe "long-time" outflow behavior.

Having obtained the solution, $F(x_+)$, one can now determine the value of the constant b, needed for the outflow rate in Equation (10.76). Again, to obtain b, as defined in

Equation (10.75), it is necessary to apply Leibniz's rule to (10.82), in the same manner as done earlier to obtain (10.80); this yields

$$1 = \frac{3}{B(2/3, 1/2)}(1 - F^3)^{-1/2} F \frac{dF}{dx_+} \tag{10.84}$$

Imposition of the boundary condition at $x_+ = 0$, i.e. $F = 0$ to this result, produces immediately $b = B(2/3, 1/2)/3 = 0.862\ 37$.

Outflow rate

With the two constants determined, for future reference it is convenient to rewrite the outflow rate (10.76) in terms of the original variables as

$$q = -\frac{bk_0 D^2}{B}\left(1 + \frac{ak_0 D}{n_e B^2}t\right)^{-2} \tag{10.85}$$

in which the two constants have the following values

$$\begin{aligned} a &= 1.115 \\ b &= 0.862 \end{aligned} \tag{10.86}$$

The applicability of this solution for large values of t has been confirmed experimentally by means of a Hele–Shaw model (Ibrahim and Brutsaert, 1965, 1966). Apparently, equations with the same and similar functional forms as (10.85) were first used by Maillet (1905; Boussinesq, 1904) in his analysis of drought flows of the Vanne River.

10.4 LINEARIZED HYDRAULIC GROUNDWATER THEORY: A THIRD APPROXIMATION

A major disadvantage of all formulations to describe unconfined groundwater flows, discussed so far in this chapter, is that they are nonlinear. Even the simplest, those based on the hydraulic approach, still suffer from this, and there is no general methodology available for their solution. It is understandable, therefore, that attempts have been made in the past to remedy this by linearization. Because the solution to a linear problem can be represented as a unit response function, it can be generalized and extended to many different boundary and initial conditions by simple superposition. Moreover, once obtained, such unit response functions can provide a direct link between the main underlying physical mechanisms captured by the Boussinesq equation, and the more abstract mathematical aspects of general linear systems (i.e. unit hydrograph) approaches at the catchment scale (see Chapter 12).

10.4.1 *General formulation*

The common way to linearize the Boussinesq equation, in the case of its simplest form, Equation (10.30), is to expand the second derivative on the right-hand side as follows

$$\frac{\partial \eta}{\partial t} = \frac{k_0}{n_e}\left[\left(\frac{\partial \eta}{\partial x}\right)^2 + \eta\frac{\partial^2 \eta}{\partial x^2}\right] \tag{10.87}$$

The basic assumption in the linearization is that position η of the free surface is never very different from an unperturbed average value, say η_0. Thus, because η remains close to constant, in (10.87) the first term on the right becomes negligible, and η in the second term can be replaced by η_0. Equation (10.87) then becomes

$$\frac{\partial \eta}{\partial t} = \frac{k_0 \eta_0}{n_e} \frac{\partial^2 \eta}{\partial x^2} \tag{10.88}$$

Equation (10.88) is in the form of the standard diffusion equation, with a constant hydraulic (groundwater) diffusivity

$$D_h = \frac{k_0 \eta_0}{n_e} \tag{10.89}$$

In a similar way, the more general form (Equation (10.31)) of the Boussinesq equation can be linearized to produce

$$\frac{\partial \eta}{\partial t} = \frac{k_0 \eta_0}{n_e} \left(\cos \alpha \frac{\partial^2 \eta}{\partial x^2} + \frac{\partial^2 \eta}{\partial y^2} \right) + \frac{k_0 \sin \alpha}{n_e} \frac{\partial \eta}{\partial x} + \frac{I}{n_e} \tag{10.90}$$

in which, as before, α is the slope of the underlying impermeable bed.

A second but less common way of linearizing the Boussinesq equation consists of multiplying both sides by η, and then bringing it inside the first derivatives or replacing it by η_0, whichever appears more appropriate. For instance, in the case of Equation (10.30) this yields

$$\frac{\partial \eta^2}{\partial t} = \frac{k_0 \eta_0}{n_e} \frac{\partial^2 \eta^2}{\partial x^2} \tag{10.91}$$

which is linear in η^2. This approach was probably first used by N. A. Bagrov and later by N. N. Verigin (Polubarinova-Kochina, 1952; Aravin and Numerov, 1953). A theoretical advantage of Equation (10.91) over (10.88) is that for steady conditions it reduces to (10.39) as it should; this means that it accords better with the hydraulic assumptions on which the Boussinesq equation is based. Nevertheless, the few studies on this have not been conclusive (Polubarinova-Kochina, 1952, p. 501; Brutsaert and Ibrahim, 1966) as to which of the two linearizations is preferable; but for some practical applications (see below) this may be immaterial.

A few comments are in order on the optimal value of η_0 to be used in the linearization. It stands to reason that the optimal value of η_0 should never be very different from the average water table height, namely

$$\langle \eta \rangle = \int_0^B \eta \, dx / B \tag{10.92}$$

The difficulty with this is that η is unknown. Nevertheless, two known solutions for special situations may give some indication. One occurs when D and D_c in the boundary conditions (10.50) have nearly the same value. Inspection of the Dupuit formula (10.43) for steady flow shows that it can be considered in some way as a finite difference form of

Darcy's law, with a hydraulic head gradient $(D_{c2} - D_{c1})/B$, and an average thickness of the flow zone given by $(D_{c2} + D_{c1})/2$; this suggests that when D and D_c have nearly the same value, a good approximation should be $\eta_0 = (D + D_c)/2$. The second case is encountered when the depth in the channel is negligible so that $D_c = 0$, and D is the only remaining parameter that can be used to characterize the average thickness of the flow zone. For such situations it is convenient to put

$$\eta_0 = pD \tag{10.93}$$

in which p is a constant adjustment parameter to compensate for the linearization. The linearized solution $a = (4\,p/\pi)^{1/2}$ for the short-time unsteady outflow rate by Edelman, mentioned earlier in Section 10.3.4, shows that it can simulate the exact result (10.63) provided $p = 0.3465$. This suggests that in the initial stages p probably lies in the vicinity of $1/3$. However, this value of $p = 0.3 \sim 0.4$ is applicable only for small to intermediate times, at most; for larger times, as the water table height η continues to decrease, the optimal value of η_0 is likely to become smaller as well.

10.4.2 Flow from a horizontal aquifer

Consider again the standard case of outflow from an initially saturated aquifer, after the cessation of rainfall or recharge, as described by boundary conditions (10.50). In the linearized system the governing differential equation is now (10.88).

Similarity considerations
As before in (10.66), it is convenient to scale the variables. The form of (10.88) and of (10.50) suggest that this be done as follows

$$x_+ = x/B$$
$$t_+ = [k_0\eta_0/(n_e B^2)]t \tag{10.94}$$
$$\eta_+ = (\eta - D_c)/(D - D_c)$$

These scaled variables allow the differential equation (10.88) to be written as

$$\frac{\partial \eta_+}{\partial t_+} = \frac{\partial^2 \eta_+}{\partial x_+^2} \tag{10.95}$$

Similarly, the boundary conditions (10.50) can be expressed in terms of the scaled variables as follows

$$\eta_+ = 0 \qquad x_+ = 0 \qquad t_+ \geq 0$$
$$\frac{\partial \eta_+}{\partial x_+} = 0 \qquad x_+ = 1 \qquad t_+ \geq 0 \tag{10.96}$$
$$\eta_+ = 1 \qquad 0 \leq x_+ \leq 1 \qquad t_+ = 0$$

Solution

Separation of variables in the form of a product solution like (10.69), and substitution in (10.95) produce in this case

$$\frac{1}{F_2}\frac{dF_2}{dt_+} = \frac{1}{F_1}\frac{d^2F_1}{dx_+^2} = -C_1 \tag{10.97}$$

where C_1 is a constant for the same reason as in (10.70), which is positive because η_+ and F_2 must always remain finite; this can be seen from the solution for the t_+-dependent part of (10.97), that is

$$F_2 = C_2\exp(-C_1t_+) \tag{10.98}$$

The solution of the differential equation for F_1 in (10.97) produces

$$F_1 = C_3\sin(\sqrt{C_1}x_+) + C_4\cos(\sqrt{C_1}x_+) \tag{10.99}$$

where C_3 and C_4 are constants. Application of the first of (10.96) indicates that $C_4 = 0$. Application of the second of (10.96) indicates that the constant C_1 must satisfy

$$\cos(\sqrt{C_1}) = 0 \tag{10.100}$$

This is the case, provided

$$\sqrt{C_1} = (2n-1)\pi/2 \tag{10.101}$$

in which n can assume any value $n = 1, 2, 3, \ldots, \infty$. Combining (10.98) and (10.99), with (10.101), one obtains

$$\eta_+ = C_n\sin((2n-1)\pi x_+/2)\,\exp(-(2n-1)^2\pi^2t_+/4) \tag{10.102}$$

where $C_n = (C_2C_3)$ is a constant which depends on the particular value of n used. Equation (10.102) satisfies the first two conditions of (10.96); the third remains to be satisfied. Inspection of (10.102) at $t_+ = 0$, shows that it is impossible to obtain $\eta_+ = 1$ with any single sine function, and that this is only possible if the solutions for the different values of n are summed as an harmonic series; the system is linear, and thus a sum of solutions is also a solution. Imposing the third of (10.96) on that series, one obtains

$$\sum_{n=1,2,\ldots}^{\infty} C_n\sin\left(\frac{(2n-1)\pi}{2}x_+\right) = 1 \tag{10.103}$$

The values of the constants C_n can be readily determined by the method of Fourier. This method consists of multiplying both sides of (10.103) by $\sin((2m-1)\pi x_+/2)\,dx_+$, and then integrating over the flow domain, which in this case covers the range $0 \le x_+ \le 1$. This produces

$$C_m = \frac{4}{\pi\,(2m-1)} \tag{10.104}$$

Finally, after insertion of (10.104) into the series of (10.102), the solution can be written as

$$\eta_+ = \sum_{n=1,2,\ldots}^{\infty} \frac{4}{\pi(2n-1)}\sin\left(\frac{(2n-1)\pi}{2}x_+\right)\exp\left(-\frac{(2n-1)^2\pi^2}{4}t_+\right) \tag{10.105}$$

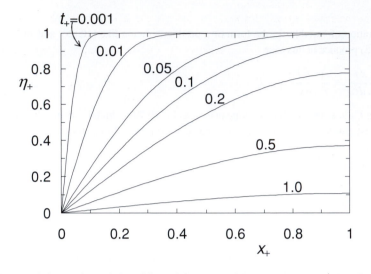

Fig. 10.22 Successive scaled positions of the water table $\eta_+ = (\eta - D_c)/(D - D_c)$ in an unconfined riparian aquifer, as calculated with the linear Boussinesq solution (10.105), for the indicated values of the scaled time $t_+ = [k_0\eta_0/(n_e B^2)]t$. The variable $x_+ = x/B$ is the scaled distance from the stream; for $t_+ > 0.2$, the solution is essentially reduced to the fundamental harmonic, (10.107), that is the first term in the series expansion.

Equation (10.105) is displayed for several values of the time t_+ in Figure 10.22. Reverting back to the original (dimensional) variables, by means of (10.94), one obtains from (10.105)

$$\eta = D_c + \frac{4(D - D_c)}{\pi} \sum_{n=1,2,\dots}^{\infty} \frac{1}{(2n-1)} \sin\left(\frac{(2n-1)\pi}{2B}x\right)$$

$$\times \exp\left(-\frac{(2n-1)^2\pi^2 k_0\eta_0}{4n_e B^2}t\right) \qquad (10.106)$$

This solution was already implicit in the work of Boussinesq (1903, 1904), who compared the problem to the "cooling of a prismatic impermeable rod, laterally impermeable, of length L, having its extremity $x = 0$ immersed in melting ice and its other extremity, $x = L$, impermeable to the heat just like the sides." But he felt that the solution would "reduce more or less rapidly to the simple fundamental solution of Fourier," that is, the first term in the series, so that the higher-order terms would be negligible. It was probably not until the work of Dumm (1954) and Kraijenhoff (1958) that the full series was used in hydrology.

The arguments of the exponential functions in the series in (10.105) (and in (10.106)) increase rapidly as $1, 9, 25, \dots$ Moreover, the amplitudes of the sine functions in the series decrease as $1, 1/3, 1/5, \dots$ These are the two features which made Boussinesq observe, that with time only the first term in the series survives. Thus the water table (10.105) gradually assumes the shape of the first sine function, and the long-time linear solution is the fundamental harmonic

$$\eta_+ = \frac{4}{\pi} \sin\left(\frac{\pi x_+}{2}\right) \exp\left(\frac{-\pi^2 t_+}{4}\right) \qquad (10.107)$$

Comparison between the first and second term in the series of (10.105) shows that this long-time solution can be assumed to be valid when $t_+ > 0.2$, where the error drops well below 1%. In terms of the original variables the long-time solution is from (10.107) and (10.94)

$$\eta = D_c + 4(D - D_c)\pi^{-1} \sin\left(\frac{\pi x}{2B}\right) \exp\left(\frac{-\pi^2 k_0 \eta_0 t}{4 n_e B^2}\right) \tag{10.108}$$

Outflow rate

After linearization, the outflow rate (10.51) from a hydraulic aquifer into the adjoining open water body becomes

$$q = -k_0 \eta_0 \left.\frac{\partial \eta}{\partial x}\right|_{x=0} \tag{10.109}$$

In terms of the scaled variables of (10.94) this can be written as

$$q_+ = -\left.\frac{\partial \eta_+}{\partial x_+}\right|_{x=0} \tag{10.110}$$

in which the rate of outflow is scaled with $k_0 \eta_0 (D - D_c)/B$, so that by definition

$$q_+ = \frac{Bq}{k_0 \eta_0 (D - D_c)} \tag{10.111}$$

Application of (10.110) with the general solution (10.105) yields

$$q_+ = -2 \sum_{n=1,2,\ldots}^{\infty} \exp\left(-\frac{(2n-1)^2 \pi^2}{4} t_+\right) \tag{10.112}$$

This result is illustrated in Figure 10.23. In terms of the original variables, after transformation with (10.94), the rate of outflow from the unconfined aquifer (10.112) can be written as

$$q = -2k_0 \eta_0 (D - D_c)B^{-1} \sum_{n=1,2,\ldots}^{\infty} \exp\left(\frac{-(2n-1)^2 \pi^2 k_0 \eta_0 t}{4 n_e B^2}\right) \tag{10.113}$$

As already noted, eventually with increasing time only the first term of the series remains, as the terms in $n = 2, 3, \ldots$ become negligible. Therefore, the long-time expression of the outflow rate is from (10.112)

$$q_+ = -2 \exp\left(-\frac{\pi^2}{4} t_+\right) \tag{10.114}$$

As shown in Figure 10.23, (10.114) becomes applicable for $t_+ > 0.2$, that is for $t > 0.2 n_e B^2/(k_0 \eta_0)$. Thus, when this criterion is satisfied, (10.113) yields the long-time outflow rate in terms of the original variables

$$q = -2k_0 \eta_0 (D - D_c)B^{-1} \exp\left(-\frac{\pi^2 k_0 \eta_0 t}{4 n_e B^2}\right) \tag{10.115}$$

If, as is often the case in small upland catchments, the water in the stream is shallow compared to the water table levels in the aquifer, it can be assumed that $D_c = 0$. With

Fig. 10.23 Scaled outflow hydrograph $q_+ = q_+(t_+)$ from a linearized hydraulic aquifer into an adjoining open water body (heavy line) as given by Equation (10.112); also shown is the long-time version (thin straight line), that is the first term in the series expansion, as given by Equation (10.114). The scaled variables are defined in Equations (10.94) and (10.111).

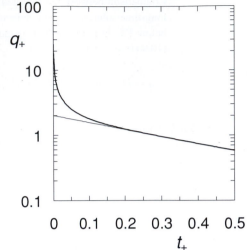

(10.93) this further simplifies (10.115) to

$$q = -2k_0 p D^2 B^{-1} \exp\left(-\frac{\pi^2 k_0 p D t}{4n_e B^2}\right) \qquad (10.116)$$

Unit response and response to an arbitrary input

Although (10.113) (or (10.116)) was obtained for rather specific boundary conditions, it is still broadly applicable. Indeed, as pointed out by Kraijenhoff (1958), it represents in fact the unit response, that is the Green's function for the linear hydraulic aquifer. The unit input applied over the whole aquifer (in two dimensions, that is per unit length of stream channel or per unit span), causing this response, is $[n_e(D - D_c)B]\delta(t)$; per unit area of ground surface this is $[n_e(D - D_c)]\delta(t)$. Therefore, the response of a linearized hydraulic aquifer to a delta function input, that is the unit response, from (10.113), is

$$u = -2k_0\eta_0(n_e B)^{-1} \sum_{n=1,2,\ldots}^{\infty} \exp\left(\frac{-(2n-1)^2\pi^2 k_0\eta_0 t}{4n_e B^2}\right) \qquad (10.117)$$

Arbitrary inputs into the aquifer by precipitation, snowmelt, etc., or negative inputs by evaporation at the surface, leakage through the impermeable bed, etc., can be dealt with by simple convolution (see Appendix). Thus if this input per unit horizontal area is $I = I(t)$, with dimensions [L/T], applied uniformly over the aquifer, i.e. independently of x, the resulting outflow rate at $x = 0$ is

$$q = \int_{t-t_m}^{t} I(\tau)u(t - \tau)d\tau \qquad (10.118)$$

where $u = u(t)$ is given by (10.117) and t_m is the memory inherent or assigned to the flow system. It should be noted that in a similar way, on account of the assumed linearity of the system, the solution for the position of the water table, namely (10.106), can in principle also serve as the basis for a unit response function resulting from a delta function input; this can then be used to predict the evolution of the water table resulting from arbitrary inputs (e.g. steady rainfall as in the following example) by convolution, i.e. with the analog of (10.118) applied to η instead of q. However, as noted earlier, the position of the water table in hydraulic groundwater theory is quite unreliable, and need therefore not be of further concern here.

Example 10.2

The application of (10.118) can be readily illustrated by considering the case of a constant input of unit intensity $I(t) = I_c$. The resulting outflow rate becomes steady when this input has been applied for a very long time. Thus one has from (10.118) with (10.117)

$$q = \frac{-2k_0\eta_0}{n_e B} \int_{-\infty}^{t} I_c \sum_{n=1,2,\ldots}^{\infty} \exp\left(\frac{-(2n-1)^2\pi^2 k_0\eta_0(t-\tau)}{4n_e B^2}\right) d\tau \qquad (10.119)$$

which upon integration becomes

$$q = \sum_{n=1,2}^{\infty} \frac{-8BI_c}{(2n-1)^2\pi^2} \exp\left(\frac{-(2n-1)^2\pi^2 k_0\eta_0 t}{4n_e B^2}\right) \left[\exp\left(\frac{(2n-1)^2\pi^2 k_0\eta_0 \tau}{4n_e B^2}\right)\right]_{-\infty}^{t} \qquad (10.120)$$

Applying the integration limits, and recalling that $1 + 1/9 + 1/25 + \cdots = \pi^2/8$, one obtains finally

$$q = -BI_c \qquad (10.121)$$

as expected. It goes without saying, that this case is the linearized version of the case already treated earlier, for which the position of the water table was shown to be given by Equations (10.35) and (10.37), illustrated in Figures 10.16 and 10.17.

Example 10.3

The next case to be considered is the outflow rate some time after a steady input I_c has ceased. This represents the outflow rate from the aquifer, with the initial shape of the water table given by (10.37) resulting from a steady infiltration, rather than by the third of (10.50) describing fully saturated conditions. This case is of practical interest, as the steady input may represent prolonged rainfall or irrigation. Indeed, the onset of drainage after prolonged rainfall or irrigation, which does not fully saturate the aquifer, is a common occurrence in humid regions. If $t = 0$ is the time when the steady input stops, the input $I = I(t)$ can be formulated as follows

$$\begin{aligned} I &= I_c && \text{for } -\infty < t < 0 \\ I &= 0 && \text{for } 0 \leq t \end{aligned} \qquad (10.122)$$

Thus (10.118) can be written as

$$q = \int_{-\infty}^{0} I_c u(t-\tau)d\tau + \int_{0}^{t} 0\, u(t-\tau)d\tau \qquad (10.123)$$

Fig. 10.24 Scaled outflow hydrograph $q_+ = q_+(t_+)$ from a linearized hydraulic aquifer into an adjoining open water body (heavy line) as given by (10.126), and the first term in the series expansion (thin straight line). Here the outflow rate is scaled with the initial outflow rate $q = BI_c$; this initial outflow rate results from a steady input I_c prior to $t_+ = 0$, as described in Example 10.3.

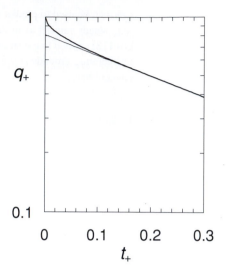

The second term on the right yields zero; therefore, upon integration of the first term on the right of (10.123) with (10.117), one can write

$$q = -\sum_{n=1,2}^{\infty} \frac{8BI_c}{(2n-1)^2\pi^2} \exp\left(\frac{-(2n-1)^2\pi^2 k_0\eta_0 t}{4n_e B^2}\right)\left[\exp\left(\frac{(2n-1)^2\pi^2 k_0\eta_0 \tau}{4n_e B^2}\right)\right]_{-\infty}^{0} \tag{10.124}$$

or, finally, after application of the limits of integration,

$$q = -\sum_{n=1,2}^{\infty} \frac{8BI_c}{(2n-1)^2\pi^2} \exp\left(\frac{-(2n-1)^2\pi^2 k_0\eta_0 t}{4n_e B^2}\right) \tag{10.125}$$

At $t = 0$, when the steady input ceases, this solution yields the initial condition (10.121), as it should. A feature worth noting in (10.125) is that it converges much more quickly to the first term in the series than (10.113), on which the unit response (10.117) is based. This rapid disappearance of the higher-order terms in (10.125) is best illustrated by expressing it in dimensionless form. By scaling the outflow rate with the initial flow rate, so that in the present case $q_+ = q/(BI_c)$, (10.125) becomes

$$q_+ = -\sum_{n=1,2}^{\infty} \frac{8}{(2n-1)^2\pi^2} \exp\left(\frac{-(2n-1)^2\pi^2 t_+}{4}\right) \tag{10.126}$$

where t_+ is as defined in (10.94). Equation (10.126) is illustrated in Figure 10.24, where it can be seen that only the first term survives, when $t_+ > 0.08$; this is in contrast with (10.112), illustrated in Figure 10.23 where the same can be seen to occur only for $t_+ > 0.2$. The difference between these two cases stems from the different initial condition. For (10.112) the aquifer is assumed to be initially fully saturated, whereas for (10.126), the water table is initially assumed to be described by the linearized version of (10.35).

Example 10.4

Consider now the outflow resulting from rainfall as an arbitrary input function of time $I = P(t)$. In principle there should be no problem in carrying out the convolution integration, analytically or numerically. However, rainfall data are usually recorded as bar graphs, that is as constant values over finite time intervals, hourly or daily. This simplifies the analysis somewhat, as it allows the formulation of the input function in the same way as in the previous example. In this example assume the following rainfall sequence

$$P = 0 \qquad \text{for } 0 \le t < t_1$$
$$P = 0.2P_c \qquad \text{for } t_1 \le t < t_2 \qquad\qquad (10.127)$$
$$P = 0.9P_c \qquad \text{for } t_2 \le t < t_3$$

in which P_c is a reference intensity rainfall (e.g. 5 mm h^{-1}).

Consider the case when the aquifer is dry at $t = 0$. Application of (10.118) with (10.127) and (10.117) produces different expressions $q(t)$ depending on the magnitude of t relative to t_1, t_2 and t_3. For instance, when $t_1 < t < t_2$ the outflow is

$$q = q(t) = \frac{-2k_0\eta_0 P_c}{n_e B} \int_{t_1}^{t} 0.2 \sum_{n=1,2,\ldots}^{\infty} \exp\left(\frac{-(2n-1)^2\pi^2 k_0\eta_0(t-\tau)}{4n_e B^2}\right) d\tau \qquad (10.128)$$

and after integration

$$q = -\sum_{n=1,2}^{\infty} \frac{8BP_c}{(2n-1)^2\pi^2} 0.2\left[1 - \exp\left(\frac{-(2n-1)^2\pi^2 k_0\eta_0(t-t_1)}{4n_e B^2}\right)\right] \qquad (10.129)$$

Similarly, for the case when $t > t_3$ one has

$$q = q(t) = \frac{-2k_0\eta_0 P_c}{n_e B}\left[\int_{t_1}^{t_2} 0.2 \sum_{n=1,2,\ldots}^{\infty} \exp\left(\frac{-(2n-1)^2\pi^2 k_0\eta_0(t-\tau)}{4n_e B^2}\right) d\tau\right.$$
$$\left. + \int_{t_2}^{t_3} 0.9 \sum_{n=1,2,\ldots}^{\infty} \exp\left(\frac{-(2n-1)^2\pi^2 k_0\eta_0(t-\tau)}{4n_e B^2}\right) d\tau\right] \qquad (10.130)$$

or, upon integration

$$q = -\sum_{n=1,2}^{\infty} \frac{8BP_c}{(2n-1)^2\pi^2}\left[0.2\left[\exp\left(\frac{-(2n-1)^2\pi^2 k_0\eta_0(t-t_2)}{4n_e B^2}\right)\right.\right.$$
$$\left. - \exp\left(\frac{-(2n-1)^2\pi^2 k_0\eta_0(t-t_1)}{4n_e B^2}\right)\right] + 0.9\left[\exp\left(\frac{-(2n-1)^2\pi^2 k_0\eta_0(t-t_3)}{4n_e B^2}\right)\right.$$
$$\left.\left. - \exp\left(\frac{-(2n-1)^2\pi^2 k_0\eta_0(t-t_2)}{4n_e B^2}\right)\right]\right]$$

$$(10.131)$$

The outflow rate produced by (10.127) is illustrated in Figure 10.25, for time values $t_1 = 2$ days, $t_2 - t_1 = t_3 - t_2 = 1$ day. In Figure 10.25, for conciseness of notation, q is scaled with BP_c and it is expressed in terms of scaled time t_+ defined in (10.94). Note that $t_+ = 0.1$ corresponds to roughly 1 day here; this conversion is obtained for typical (Brutsaert and Lopez, 1998; Eng and Brutsaert, 1999) field values for small catchments $k_0 = 0.001$ m s^{-1}, $n_e = 0.02$, $\eta_0 = 2$ m, and $B = 300$ m.

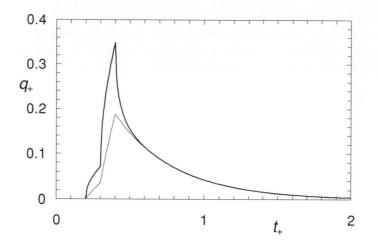

Fig. 10.25 Scaled outflow hydrograph $q_+ = q/(BP_c)$ from a linearized hydraulic aquifer into an adjoining open water body resulting from the precipitation event (10.127) given in Example 10.4. The full series solution is represented by the heavy line and the first term in the series expansion by the thin line. In this example $t_+ = 0.1$ is taken to be roughly equivalent with one day.

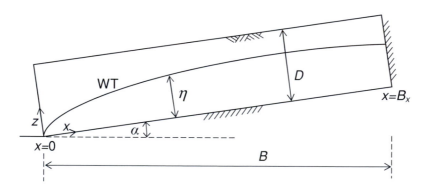

Fig. 10.26 Definition sketch of the cross section of an unconfined hillslope aquifer. The distance along the land surface from the stream to the divide is $B_x = B/\cos \alpha$.

10.4.3 Flow from a hillslope aquifer

Again, to derive the unit response function, it is convenient to consider the problem of outflow from an initially saturated aquifer (Figure 10.26). While the aquifer empties out, there is no surface recharge, so that the differential equation governing this phenomenon is (10.90) with $I = 0$. In the case of a sloping aquifer, the total flow rate at any distance x from the channel can be determined from (10.26); after linearization, this is

$$(q_x \eta) = -k_0 \left(\eta_0 \cos \alpha \frac{\partial \eta}{\partial x} + \sin \alpha \, \eta \right) \tag{10.132}$$

Therefore, the boundary condition at the divide, where $x = B_x$, is not simply the second of (10.50), but it must be adjusted to simulate the impermeable barrier in accordance with (10.132). Moreover, in hilly terrain, torrential streams tend to be shallow, and they usually have no effect on the groundwater flow in the adjoining hillslopes, so that it is safe to assume that $D_c = 0$. Hence, instead of (10.50), for an initially saturated sloping aquifer the boundary conditions can be formulated as follows

$$
\begin{aligned}
\eta &= 0 & x &= 0 & t &\geq 0 \\
\eta_0 \cos \alpha \frac{\partial \eta}{\partial x} + \sin \alpha \eta &= 0 & x &= B_x & t &\geq 0 \\
\eta &= D & 0 &\leq x \leq B_x & t &= 0
\end{aligned}
\tag{10.133}
$$

For the present two-dimensional case of flow down the slope, and in the absence of lateral inflow, with $I = 0$, the linearized form of the governing equation (10.31) (i.e. (10.90)) can be written as

$$
\frac{\partial \eta}{\partial t} = \frac{k_0 \eta_0 \cos \alpha}{n_e} \frac{\partial^2 \eta}{\partial x^2} + \frac{k_0 \sin \alpha}{n_e} \frac{\partial \eta}{\partial x}
\tag{10.134}
$$

Notice again that this equation is in the form of the linear advective diffusion equation, which was already encountered in the diffusion approach of open channel flow (cf. Equations (5.88) and (5.92)). In the present case of a sloping aquifer the hydraulic diffusivity is not simply (10.89), but it contains the slope effect, or

$$
D_h = \frac{k_0 \eta_0 \cos \alpha}{n_e}
\tag{10.135}
$$

In addition (10.134) contains a hydraulic (groundwater) advectivity

$$
c_h = -\frac{k_0 \sin \alpha}{n_e}
\tag{10.136}
$$

By analogy with flood wave propagation in open channels, a disturbance of the water table height η in a sloping aquifer can be visualized as undergoing two types of changes. The first is a deformation of its shape, which is governed by the diffusivity (10.135); the second is a displacement of this disturbance down the slope, whose rate of propagation is given by the advectivity (10.136).

Similarity considerations
The boundary conditions (10.133) and the form of the governing differential equation (10.134) suggest that the variables be scaled as follows (cf. Equation (10.94))

$$
\begin{aligned}
x_+ &= x/B_x \\
t_+ &= \left[k_0 \eta_0 \cos \alpha / \left(n_e B_x^2 \right) \right] t \\
\eta_+ &= \eta/D
\end{aligned}
\tag{10.137}
$$

Equation (10.134) becomes in terms of these variables

$$
\frac{\partial \eta_+}{\partial t_+} = \frac{\partial^2 \eta_+}{\partial x_+^2} + \mathrm{Hi} \frac{\partial \eta_+}{\partial x_+}
\tag{10.138}
$$

In (10.138) Hi can be called the groundwater hillslope flow number, which is defined as

$$\text{Hi} = \frac{B_x \tan\alpha}{\eta_0} \qquad (10.139)$$

Equation (10.138) is in the form of the advection diffusion equation; because it is dimensionless it has a unit diffusivity. From a comparison between Equations (10.95) and (10.138), it can be seen that the dimensionless parameter Hi, represents the relative magnitude of the slope term, that is the effect of gravity, versus the diffusion term. This ratio increases with the slope α and with the "shallowness" B_x/D of the aquifer. For large values of Hi the diffusion term, i.e. the first term on the right of Equation (10.138), is negligible, and the kinematic flow approximation (see Section 10.5) is valid. For small values of Hi (i.e. a small slope α and/or a relatively deep aquifer with large D/B_x)) the problem can be treated as one of horizontal flow, and the solution can be approximated by (10.106).

Solution

The boundary conditions (10.133) can be written in dimensionless form

$$\begin{array}{lll} \eta_+ = 0 & x_+ = 0 & t_+ \geq 0 \\[4pt] \dfrac{\partial\eta_+}{\partial x_+} + \text{Hi}\,\eta_+ = 0 & x_+ = 1 & t_+ \geq 0 \\[4pt] \eta_+ = 1 & 0 \leq x_+ \leq 1 & t_+ = 0 \end{array} \qquad (10.140)$$

The solution of (10.138), subject to (10.140) can be obtained by means of the Laplace transform. This can be shown (Brutsaert, 1994) to be

$$\eta_+ = \sum_{n=1,2,3\ldots}^{\infty} \frac{2z_n[\exp(-\text{Hi}/2) - 2\cos(z_n)]\sin(z_n x_+)\exp\left[-\left(z_n^2 + \text{Hi}^2/4\right)t_+ + (1 - x_+)\text{Hi}/2\right]}{\left(z_n^2 + \text{Hi}^2/4 + \text{Hi}/2\right)} \qquad (10.141)$$

In (10.141) z_n is the nth root of

$$\tan(z) = \frac{z}{-\text{Hi}/2} \qquad (10.142)$$

which has an infinity of roots, say $z_1, z_2, \ldots z_n, \ldots$; Equation (10.142) arises in many problems and its roots z_n have been tabulated (see Carslaw and Jaeger, 1959, p. 492; Abramowitz and Stegun, 1964, p. 224). Note that for very small values of Hi, one has $z_n = [(2n-1)\pi/2]$ which is in agreement with (10.101) for the horizontal case. For large values of Hi, the roots approach $z_n = (n\pi)$.

Outflow rate

The outflow rate can be obtained by applying (10.132) to (10.141) at $x = 0$. This yields

$$q_+ = -2 \sum_{n=1,2,3\ldots}^{\infty} \frac{z_n^2[1 - 2\cos(z_n)\exp(\text{Hi}/2)]\exp\left[-\left(z_n^2 + \text{Hi}^2/4\right)t_+\right]}{\left(z_n^2 + \text{Hi}^2/4 + \text{Hi}/2\right)} \qquad (10.143)$$

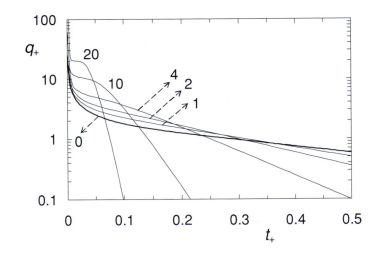

Fig. 10.27 Scaled outflow hydrograph $q_+ = q_+(t_+)$ from a linearized sloping hydraulic aquifer into an adjoining open channel as given by (10.143), for the values of the hillslope flow number Hi $= 0, 1, 2, 4, 10$ and 20; Hi $= 0$ represents the horizontal case. The scaled time variable is defined in (10.137) and the aquifer is initially fully saturated.

where the scaled outflow rate is defined as $q_+ = B_x q/(k_0\eta_0 D \cos\alpha)$ and the scaled time t_+ is defined in (10.137). The reader can verify that, when the hillslope flow number vanishes, or Hi $= 0$, this expression reduces immediately to the solution for the horizontal case (10.113) with $D_c = 0$.

The outflow hydrograph from a sloping aquifer (10.143) is illustrated in Figure 10.27 for different values of the hillslope flow number Hi. The long-time behavior of this outflow rate displays the typical exponential decay with time of linear systems, but the exponential function has two extinction coefficients; the first (z_n^2) reflects the diffusive character of the flow, and the second $(\text{Hi}^2/4)$ reflects its kinematic character, that is the effect of the steepness of the slope. As a result, the outflow rate displays two features which are worth noting. First, as is the case with (10.101), the values of z_n increase rapidly with n in the higher-order terms, so that these terms decay rapidly, regardless of the value of Hi. This means that for large values of t only the first term in the series survives, producing the straight lines in the semi-logarithmic plot of Figure 10.27; thus for larger values of t the rate of flow q decays exponentially in approximately the same way as in the horizontal case as a result of diffusion, but this rate is increased by the presence of the term containing Hi. The second feature is that, as the hillslope flow number Hi increases, the outflow hydrograph gradually displays a "hump," or rather a "pause" in its progress. Mathematically, this phenomenon results from the fact that, as Hi becomes larger, the relative importance of the diffusion term in Equation (10.138) (the first term on the right) decreases compared to the advective term (the second term on the right). This means that the nature of the flow in the aquifer becomes less diffusive and more kinematic, that is, increasingly driven by gravity on account of the slope. As will be discussed further in Section 10.5, kinematic motion is purely translatory without change in shape of the water table. In the present case described by Equation (10.138), in the early stages, while all the higher order terms in the series solution are still active, diffusion causes the water table near the outlet to spread out, a behavior not unlike that of flood waves in open channels, discussed in Chapter 7; but later on, as the kinematic effect takes over, the bulk of the water left in the aquifer tends to move down more as a translatory wave, causing the appearance of a "pause" in the hydrograph during the transition between the two regimes. The occurrence of this pause is related to the time when the height η_+ of the water table approaches zero at the divide, where $x_+ = 1$. The rate at which the water table height approaches zero at the divide, like the strength of the pause illustrated in Figure 10.27, depends mainly on the magnitude of the hillslope flow number Hi. Actually, it can be seen that Equation (10.141) predicts that at the divide η_+ can never become zero, and that with time it can only approach it exponentially. This is somewhat counterintuitive, because the hydraulic groundwater approach is based on the assumption that the water table is a true free surface and thus a sharp interface; therefore physically, there is no reason why such a sharp interface $\eta_+(x_+, t_+)$ would not be able to become 0 at $x_+ = 1$, and it would be expected that after this occurs, the point of $\eta_+ = 0$ would slide down along the bottom of the aquifer from $x_+ = 1$ in the direction of $x_+ = 0$. Originally, the fact that (10.141) does not predict such a sequence of events was thought to be caused by the shortcomings of the hydraulic approach with the Boussinesq equation; it was recently (Stagnitti *et al.*, 2004) shown, however, that this is not the case, and that it is in fact the result of the linearization of that equation. This should not be surprising, as it is well known that solutions of the linear diffusion equation usually do not exhibit sharp fronts, but rather long exponential tails; other examples of this or similar features can be found in applications of the linear diffusion equation in open channel flow (Equation (5.95) and Figure 5.9) and in infiltration (Equation (9.56) and Figure 9.10). Nevertheless, in the present context, the inability of Equation (10.141) to allow the water table height $\eta_+ = 0$ to move down along the bottom of the aquifer past $x_+ = 1$, may not be a crucial issue in hillslope hydrology. In real aquifers, the falling water table is not a sharp drying front, and the flow is more closely described by the Richards equation, than by the Boussinesq equation. This means that the solution of the linearized formulation, viz. Equation (10.141), with its asymptotic approach to zero, may not necessarily provide a worse approximation than the sharp interface description. Still, regardless of these shortcomings, the analysis of the linearized problem shows how with increasing Hi, the diffusive aspects of the phenomenon gradually become less important, and perhaps even irrelevant in the description of hillslope flows in actual catchments; this suggests that for large values of Hi, say in excess of 10, it may be preferable to use the considerably simpler kinematic approach outlined below in Section 10.5.

The short time limit of Equation (10.143) can be shown (Brutsaert, 1994) to be simply

$$q = - (k_0 \eta_0 n_e \cos \alpha / \pi)^{1/2} Dt^{-1/2} \tag{10.144}$$

This is Equation (10.59) with (10.61), as expected, with a value of the constant a given by

$$a = [4\eta_0 \cos \alpha / (\pi D)]^{1/2} \tag{10.145}$$

Actually, Equation (10.144) is the exact solution of (10.138) without the second term on the right, subject to (10.133), in which $B \to \infty$ and $\sin \alpha = 0$. This means that after a sudden cessation of the water supply (e.g. from rainfall) at the soil surface, the outflow proceeds initially as a diffusion phenomenon in an infinitely long aquifer. This is not unexpected. Recall that a similar type of behavior with initial $t^{-1/2}$ dependency occurs in other phenomena described by the advection diffusion equation as well; one example, covered in Chapter 9, is vertical infiltration of ponded water into a dry soil profile. Note also, that Equation (10.144) with (10.145) for $\alpha = 0$ is the solution for drainage from an infinitely long horizontal aquifer proposed by Edelman (cited by Kraijenhoff, 1966).

Unit response and response to an arbitrary input

Equation (10.143) represents the outflow rate following complete saturation of the hillslope aquifer. Therefore it is the unit response, that is Green's function, or the instantaneous unit hydrograph for a sloping aquifer. Again, the input applied over the whole aquifer (in two dimensions, that is per unit length of stream channel), causing the response (10.143), is $(n_e D B_x)\delta(t)$; per unit area of ground surface this is $(n_e D)\delta(t)$. Therefore, the response of a linearized hydraulic hillslope aquifer to a delta function input, that is the unit response, is

$$u(t) = \frac{-2k_0\eta_0 \cos\alpha}{(n_e B_x)} \sum_{n=1,2,3...}^{\infty} \frac{z_n^2[1 - 2\cos(z_n)\exp(\mathrm{Hi}/2)]\exp[-(z_n^2 + \mathrm{Hi}^2/4)t_+]}{(z_n^2 + \mathrm{Hi}^2/4 + \mathrm{Hi}/2)}$$

(10.146)

This means that, just like Equations (10.113)–(10.116) for the horizontal case, this solution can also accommodate arbitrary inputs such as infiltration of precipitation or snowmelt and leakage through the bottom bed, by convolution by means of (10.118). As before, this arbitrary input per unit ground surface area can be taken as $I = I(t)$, with dimensions [L/T], applied uniformly over the aquifer, i.e. independently of x. However, because of the slope some caution is called for; if I represents an input per unit horizontal area (as would be the case for rainfall), in (10.118) it should be replaced by $(I\cos\alpha)$.

The observations regarding the unit response function for the position of the water table, made below Equation (10.118) for horizontal aquifers, are equally applicable to the present case of sloping aquifers.

Example 10.5

Consider the same situation as described previously in Example 10.3. This problem concerns the formulation of the outflow rate from a hillslope aquifer after a long-duration steady precipitation event has ceased. The application of Equation (10.118) can follow along the same steps as outlined in (10.122) and (10.123). If the steady input rate per unit horizontal area is given by I_c, the input rate used in the convolution integral should be $(I_c \cos\alpha)$; with the unit response (10.146), this integral yields the following result

$$q(t) = -2B_x I_c \cos\alpha \sum_{n=1,2,3...}^{\infty} \frac{z_n^2[1 - 2\cos(z_n)\exp(\mathrm{Hi}/2)]\exp[-(z_n^2 + \mathrm{Hi}^2/4)t_+]}{(z_n^2 + \mathrm{Hi}^2/4 + \mathrm{Hi}/2)(z_n^2 + \mathrm{Hi}^2/4)}$$

(10.147)

Observe that Equation (10.125) represents the special case of (10.147) for a horizontal aquifer with $\alpha = 0$. Equation (10.147) is illustrated in Figure 10.28, for different values of Hi; to allow easy comparison, as before in (10.126) the flow rate is scaled with its initial value, namely as $q_+ = q/(I_c B_x \cos\alpha)$.

Example 10.6

Consider in this example the same input sequence as given in (10.127). By means of the unit response (10.146), the calculations can be carried out in the same way as in Example 10.4. As an illustration, the aquifer outflow rate for the case when $t > t_3$ can be written as

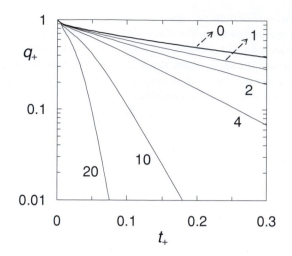

Fig. 10.28 Scaled outflow hydrograph $q_+ = q_+(t_+)$ from a linearized sloping hydraulic aquifer into an adjoining open channel as given by Equation (10.147), for the values of the hillslope flow number $\mathrm{Hi} = 0, 1, 2, 4, 10$ and 20. $\mathrm{Hi} = 0$ represents the horizontal case (see Example 10.3). The outflow rate is scaled with the initial outflow rate $q = (I_c B_x \cos \alpha)$, so that $q_+ = q/(I_c B_x \cos \alpha)$; this initial outflow rate results from a steady input I_c prior to $t_+ = 0$, as described in Example 10.5. The time is scaled as indicated in Equation (10.137).

[by analogy with (10.130)]

$$
q = q(t) = \frac{-2k_0\eta_0 P_c \cos^2 \alpha}{n_e B_x} \left\{ \int_{t_1}^{t_2} 0.2 \sum_{n=1,2,\dots}^{\infty} \frac{z_n^2[1 - 2\cos(z_n)\exp(\mathrm{Hi}/2)]}{(z_n^2 + \mathrm{Hi}^2/4 + \mathrm{Hi}/2)} \right.
$$

$$
\times \exp\left[\frac{-(z_n^2 + \mathrm{Hi}^2/4)\,[k_0\eta_0 \cos\alpha]\,(t - \tau)}{(n_e B_x^2)} \right] d\tau
$$

$$
+ \int_{t_2}^{t_3} 0.9 \sum_{n=1,2,\dots}^{\infty} \frac{z_n^2[1 - 2\cos(z_n)\exp(\mathrm{Hi}/2)]}{(z_n^2 + \mathrm{Hi}^2/4 + \mathrm{Hi}/2)}
$$

$$
\left. \times \exp\left[\frac{-(z_n^2 + \mathrm{Hi}^2/4)\,[k_0\eta_0 \cos\alpha]\,(t - \tau)}{(n_e B_x^2)} \right] d\tau \right\} \tag{10.148}
$$

This result can be readily integrated to yield (by analogy with Equation (10.131) for the horizontal case)

$$
q = -2B_x P_c \cos\alpha \sum_{n=1,2,\dots}^{\infty} \left[\frac{z_n^2[1 - 2\cos(z_n)\exp(\mathrm{Hi}/2)]}{(z_n^2 + \mathrm{Hi}^2/4)\,(z_n^2 + \mathrm{Hi}^2/4 + \mathrm{Hi}/2)} \right]
$$

$$
\times \left[0.2 \left[\exp\left(-(z_n^2 + \mathrm{Hi}^2/4)(t_+ - t_{+2})\right) - \exp\left(-(z_n^2 + \mathrm{Hi}^2/4)(t_+ - t_{+1})\right) \right] \right.
$$

$$
\left. + 0.9 \left[\exp\left(-(z_n^2 + \mathrm{Hi}^2/4)(t_+ - t_{+3})\right) - \exp\left(-(z_n^2 + \mathrm{Hi}^2/4)(t_+ - t_{+2})\right) \right] \right]
$$

$$
\tag{10.149}
$$

in which the scaled time variable is defined in Equation (10.137).

10.4.4 Incorporation of capillary flow zone

Attempts have been made to incorporate certain features of the partly saturated flow above the water table into the linearized hydraulic approach. In principle, such an approximation can be used for any type of free surface formulation, but until now mainly the linear Boussinesq equation has been considered. Examples of such studies are presented in the journal articles by Pikul *et al.* (1974) and Parlange and Brutsaert (1987).

10.5   KINEMATIC WAVE IN SLOPING AQUIFERS: A FOURTH APPROXIMATION

Equations (10.26) and (10.27) show how the flow is driven by a pressure gradient, as manifested by the inclination of the water table with respect to the underlying bed $\partial \eta / \partial x$, and also by gravity, as manifested by the magnitude of the bed slope $\sin \alpha$. The pressure gradient term results in diffusive transport, which appears as a second derivative in the Boussinesq equation; the bed slope term results in advective transport. For large values of the slope, and thus of the hillslope flow number Hi, the effect of advection overwhelms the diffusion. This can also be seen in Equations (10.134) and (10.138). In the kinematic wave approach, Hi is assumed to be sufficiently large that the pressure gradient term, leading to the diffusive term, can be simply neglected; thus the hydraulic gradient in (10.128) is assumed to be equal to the bed slope $\sin \alpha$, and (10.29) reduces to a first-order linear equation, as follows

$$\frac{\partial \eta}{\partial t} - \frac{k_0 \sin \alpha}{n_e} \frac{\partial \eta}{\partial x} = \frac{I}{n_e} \tag{10.150}$$

This approach was briefly introduced by Boussinesq (1877) for steep slopes; in the simple case of outflow without recharge I, he pointed out that, because (10.150) is in the form of a total derivative

$$\frac{d\eta}{dt} = \frac{\partial \eta}{\partial t} + \frac{\partial \eta}{\partial x} \frac{dx}{dt} = 0,$$

a water table height η travels down the slope at a speed

$$c_k = \frac{dx}{dt} = \frac{-k_0 \sin \alpha}{n_e} \tag{10.151}$$

Conversely, to an imaginary observer traveling down the slope at a speed given by Equation (10.151), it would appear that the height of the water table η does not change with time. This result is not unexpected, and it is analogous with open channel flow, in that the advectivity of the diffusion equation (10.136) and the celerity of the kinematic wave (10.151) are identical, or $c_k = c_h$. In contrast to the kinematic wave in open channel flow, Boussinesq's result (10.151) has the following two features. First, it can be seen that c_k is independent of η. This means that all values of η travel at the same speed, and the water table maintains its original shape as it moves downhill. For example, a rectangular input pulse of precipitation, which enters the aquifer instantaneously at the soil surface, will flow out from the aquifer into the stream channel as a time-delayed rectangular output pulse. Second, it could be argued that Equation (10.151) does not really describe a wave;

indeed, to the extent that the drainable porosity n_e represents the mobile water in the soil (so that $(\theta_0 - n_e)$ can be considered immobile water as indicated in Figure 10.12), Equation (10.151) is also the true velocity of the water in the aquifer. Therefore, it is perhaps preferable to refer to this phenomenon as kinematic flow, rather than kinematic wave. It can be seen that the reason for this equality of fluid and wave velocity is the inherent linearity of (10.150). On the other hand, it should also be remembered that, in spite of its name, the real physical significance of n_e is not obvious; it was introduced as a mere parameter to compensate for the neglect of the partly saturated flow above the water table in the soil by the free surface approximation. While the fraction of the soil volume below the water table occupied by flowing water is not as large as the total porosity n_0, it is probably larger than n_e, because not all the flowing water is removed by the drainage process. Therefore, the true velocity of the water in the aquifer is not likely to be as large as (10.151). This means that it can be expected that the water table motion down a steep slope may still have certain features of a wave; but on account of the obscure physical nature of n_e, these features are unclear and the phenomenon will require further study.

The applicability of the kinematic wave approach was studied by Henderson and Wooding (1964), by comparing their results presented in Section 6.2.2 applied to groundwater, i.e. for the case $a = 0$, with those obtainable with the full Boussinesq equation. They concluded that the differences can be significant for the decay phase, as obtainable, for example, from the groundwater analog of Equation (6.27). As mentioned, in the kinematic approach the hydraulic gradient is assumed to be equal to the bed slope; in practice, this is usually taken to be the same as the ground surface slope. The main practical drawback of this approach is that it is unsuitable when a wide range of slopes has to be considered, including very small ones in relatively flat terrain. However, for steep slopes or large values of the hillslope flow number Hi this approach can be a useful tool to describe the flow. Because the motion is strictly translatory, it also provides some justification for the application of the rational method (see Section 12.2.2) to describe subsurface storm runoff from very permeable hillslopes in hilly catchments for engineering design. The kinematic approach has been used in some catchment scale simulations of hillslope storm runoff (Beven, 1981).

10.6   CATCHMENT-SCALE BASE FLOW PARAMETERIZATIONS

10.6.1   General features

Base flow is the discharge rate in a river that results from the natural release of the water stored in the upstream river channels and adjoining riparian aquifers in the absence of precipitation, snowmelt, or other inputs. In general, this type of flow depends primarily on the physiographic characteristics of the basin, on the distribution of water storage in river channels and in groundwater aquifers, and possibly also on the evaporation from the basin. These physiographic characteristics are mainly the geomorphology of the landscape and of the stream network, and the configuration and nature of the riparian aquifers and near-surface soils; these characteristics reflect the geology and the climate

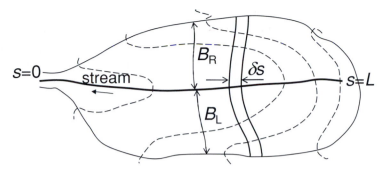

Fig. 10.29 Schematic plan view of a first-order catchment illustrating the integration of the local outflow rates from the riparian unconfined aquifers to derive the total outflow from the area according to Equation (10.152). The symbol s denotes the lineal coordinate along the river; $B_L = B_L(s)$ and $B_R = B_R(s)$ are the breadths of the aquifers on the left and right banks, respectively; the dashed lines indicate the height contour lines of the land surface.

of the basin, which can be considered as time invariant at the usual time scales of the major components of the hydrologic cycle. The effect of the channel storage under base flow conditions is normally quite small. Indeed, the water stored in river channels is usually several orders of magnitude smaller than that stored in active groundwater aquifers (see also Table 1.3). Moreover, typical travel times of water in a river tend to be orders of magnitude smaller than those in the adjoining groundwater aquifers. Under certain conditions, groundwater evaporation can have a seasonal effect; however, because groundwater evaporation can take place only from limited areas, usually near the river banks, where the water tables are close enough to the surface, this effect is often negligible (see Zecharias and Brutsaert, 1988b).

From these considerations, it is apparent that the base flow at any point in the river, $Q = Q(t)$, is mainly the result of groundwater drainage; thus it can usually be assumed to represent the instantaneous integral of all upstream local groundwater outflows, taken along the river channels all the way to the headwaters. With this assumption the flow rate in a river at the outlet of a catchment under base flow conditions can be formulated as

$$Q(t) = \int_0^L (|q_L| + |q_R|)\, ds \qquad (10.152)$$

where s is the upstream lineal coordinate along all river channels in the basin, L is the total length of these channels, and $q_L = q_L(s, t)$ and $q_R = q_R(s, t)$ the groundwater inflows from the left and the right bank, respectively (see Figure 10.29). Since the physiography of the basin does not change with time, the local groundwater outflow rates q_L and q_R are likely to be unique and time-invariant functions of the local groundwater storage. Because q_L and q_R depend on the coordinate s, Equation (10.152) indicates that the basin-scale base flow $Q(t)$ depends not only on the total water storage in the basin, but also on the areal distribution of that storage over the basin. However, the storage distribution does not always remain the same and it evolves, usually as a direct result of the spatial distribution of the antecedent precipitation. This means that $Q(t)$ need not be a unique function of time. Over smaller catchments, when the precipitation can be assumed to be sufficiently uniform, this may not be a problem. However, for larger basins it is not easy to define a unique base flow function from experimental data. This difficulty of non-uniqueness is aggravated by the fact that also the time reference, that is $t = 0$, is almost impossible to define. Indeed, in the case of a long-term streamflow record with base flow episodically interrupted by stormflow events resulting from precipitation, it is not a simple matter to identify the start of each base flow episode. In the past, this difficulty of non-uniqueness and uncertainty in time origin has been avoided mainly in two ways, namely by assuming that the low-flow recession hydrograph can be represented as an exponential decay or some other a priori adopted function, or by casting the recession hydrograph in differential form.

10.6.2 *Average base flow recession as an exponential decay process*

Whenever the time dependence of the flow rate in a river can be assumed to be an exponential decay process, this can be written in the following form

$$Q = Q_0 \exp(-t/K) \qquad\qquad (10.153)$$

where Q_0 is the flow rate at $t = 0$, and K is a constant, representing a characteristic storage delay in the watershed; both can be considered as parameters to be determined from observations. An important feature of Equation (10.153), and the main reason for its wide usage in practice is that, if (10.153) truly describes the flow, the value of K, as determined by regression or other techniques, should be totally insensitive to the choice of the time reference, $t = 0$. This means that in a semi-logarithmic plot of Q versus t, it should be possible to identify a base flow recession graphically, as the (straight) lower envelope of a number of tail end sections of low flow recession hydrographs, after shifting them horizontally until the best coincidence is obtained; the value of $-K^{-1}$ is obtained from the slope of that envelope.

Early representations of base flow by an exponential decay function have been reviewed by Hall (1968). In Barnes's (1939; 1959) approach, which is often quoted, the total recession hydrograph in a stream channel was assumed to be the sum of three exponential decay functions, namely the contribution by surface runoff, the contribution by interflow, and the contribution by the groundwater outflow from the watershed; eventually, after surface runoff and interflow are depleted, the recession consists only of ground water drainage. Some other examples of the wide practical application of Equation (10.153) in characterizing base flows can be found in the studies by Laurenson (1961), Feldman (1981) and Dias and Kan (1999), among many others.

Equation (10.153) is essentially in the form of (10.115), (10.116), and also the first term of (10.143), that is the first harmonic of the solution of the linearized Boussinesq equation. It is, of course, also the response of a lumped linear storage element, as used in the hydrologic systems approach (see Section 12.2.2). The fact that (10.153) has the same exponential form as these physically based expressions of Section 10.4 provides a strong indication that the storage delay constant K can be expected to depend on the soil

Fig. 10.30 **Daily discharge data in $m^3\, s^{-1}$ measured during stream flow recessions over a five-year period on Fall Creek, near Ithaca, NY. The rate of flow on the ith day is plotted against the rate of flow on the $(i+1)$th day. The area of this catchment is $326\ km^2$.**

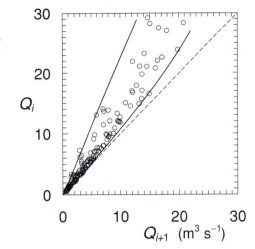

properties k_0 and n_e, the thickness D and the breadth B of the riparian aquifers, their slopes α, and possibly other basin characteristics. This will be further investigated in Section 10.6.3.

In some applications, the exponential outflow function is represented in a different form, as

$$Q_n = Q_0 K_r^n \tag{10.154}$$

in which $K_r = Q_i/Q_{i-1}$ is the depletion ratio, $n = (t/\Delta t)$ is the number of time intervals of duration Δt since the start of the recession, when t is taken as zero, and Q_i is the rate of flow at the ith time interval. Equation (10.154) is derived by assuming that the decrease in flow rate from any time t to time $(t + \Delta t)$ is constant, so that $Q_1 = Q_0 K_r$, $Q_2 = Q_1 K_r = Q_0 K_r^2$, and so on. This shows that (10.154) is just another form of (10.153) with $\ln(K_r) = -\Delta t/K$. It also means that, in light of the similarity of (10.153) with (10.115), (10.116) (and also the first term of (10.143)), K_r can be expected to depend on the same soil, aquifer, and basin characteristics as K.

The applicability of (10.153) or (10.154), that is the linearity of the basin for base flow, can be checked graphically by plotting Q_i versus Q_{i+1} for available recession flow data. An example of such data is shown in Figure 10.30. The upper envelope of these points describes the fastest rate of decrease in flow rate on record; it can therefore be assumed to be caused by the depletion of channel storage during a storm flow event and, as will be discussed below, depletion of the steeper hill slopes in the basin. On the other hand, the lower envelope, which describes the slowest rate of decrease on record, can be assumed to represent base flow recession, due to the depletion of groundwater storage in the riparian aquifers. If the base flow recession is truly linear and given by Equation (10.154), K_r must be constant; because K_r is the slope of the lower envelope, it should be a straight line. In the example shown in Figure 10.30, the points do not quite have a straight line lower envelope, and therefore for this basin the possibility of a nonlinear base flow regime cannot be excluded. This procedure of plotting Q_i versus

Q_{i+1} with the upper envelope was probably introduced by Langbein (1938), who applied it to characterize channel storage recession. It was later extended by Linsley et al. (1958) to characterize also the base flow recession by using the lower envelope.

10.6.3 Base flow decline rate: recession slope analysis

Under conditions when freezing, thawing and snowmelt do not play a role, long-term streamflow records normally consist of base flow episodes that alternate with episodes of storm flow resulting from precipitation. In general, except when it is an exponential function of time, the functional form of the base flow obtained from streamflow data is sensitive to the choice or definition of $t = 0$, that is the assumed start of each base flow episode. This uncertainty in the determination of a consistent time reference can be avoided by eliminating the time variable t from the analysis of the data, and by taking instead its differential dt. This can be done by considering not the hydrograph $Q(t)$ itself, but rather its slope as a function of Q, as follows

$$\frac{dQ}{dt} = f(Q) \tag{10.155}$$

where $f(\ )$ is a function that is characteristic for a given catchment. With actual stream-flow measurements Q_i versus Q_{i+1} at successive times Δt apart, this function can be approximated by

$$\frac{Q_{i+1} - Q_i}{\Delta t} = f\left(\frac{Q_{i+1} + Q_i}{2}\right) \tag{10.156}$$

The rate of decline of groundwater outflow is markedly slower than that of other streamflow input components, resulting from precipitation related events, such as over-land runoff or channel storage depletion. Therefore in the application of Equation (10.155) it can be assumed that base flows represent the smallest $|dQ/dt|$ for a given Q (or the largest rate of flow Q for a given $|dQ/dt|$). This means that in any graphical representation of $(Q_i - Q_{i+1})/\Delta t$ data versus $(Q_i + Q_{i+1})/2$ data points, for base flows the function $f(\ )$ in (10.155) can be taken as the lower envelope. The main objective of such a procedure is to capture some characteristics of the ensemble of many recessions, which cannot possibly be seen or detected by analyzing individual recessions. Indeed, in a natural catchment, hydrographs and their recessions come in many different shapes and they can vary greatly from one runoff event to the next. The shape of a hydrograph depends on many factors, such as the spatial distribution of the initial soil moisture content, the spatial distribution of the water table levels, and the spatial and temporal distribution of the prior precipitation events over the catchment. This infinity in possible outcomes and the large variability and non-uniqueness of shapes is illustrated by the fact that when one plots daily values of dQ/dt vs Q for a natural watershed one obtains a broad cloud of points. Figure 10.31 shows an example.

Thus the lower envelope is the locus of points for the slowest recession rate dQ/dt; conversely, it represents also the largest flow rate Q for any given recession rate dQ/dt. In principle, this largest flow rate is the one that would be observed (even though this may

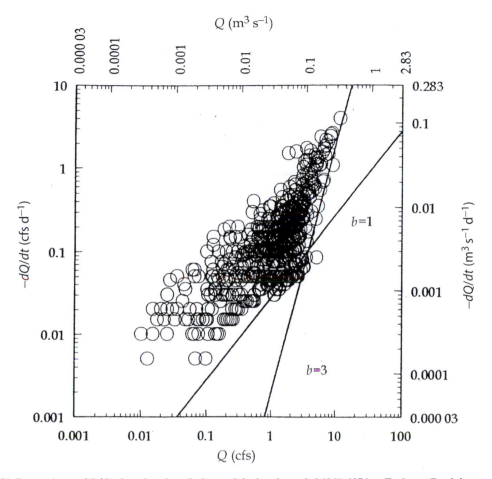

Fig. 10.31 Data points $-dQ/dt$ plotted against Q observed during the period 1961–1974 on Tonkawa Creek in Oklahoma, with the lower envelope lines with slopes 1 and 3, respectively, in accordance with Equation (10.157). The drainage area of this basin is $A = 67$ km^2, the total length of all stream channels is $L = 70$ km, and the estimated mean depth of the surface aquifers is $D = 1.6$ m. (From Brutsaert and Lopez, 1998.)

never occur) if the entire watershed were initially and uniformly saturated. In natural catchments, the exact form of the base flow function $Q = Q(t)$ is usually unknown, especially when the characteristics of the riparian aquifers deviate markedly from those of the idealized cases considered above in Sections 10.3, 10.4 and 10.5. Thus, beside the avoidance of the time origin problem, the procedure based on Equations (10.155) and (10.156) also has the advantage that it will give some insight into which of the above theoretical expressions may be applicable, if at all. Unfortunately, however, because it involves derivatives, this procedure is also sensitive to unavoidable inherent errors in the data. It is therefore advisable to constrain such envelopes somewhat on the basis of available theoretical considerations, as shown next.

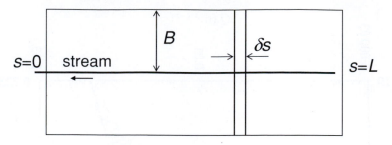

Fig. 10.32 Simplified schematic representation of the catchment shown in Figure 10.29, illustrating the use of the spatially constant effective parameters q and B to describe the catchment-scale groundwater outflow rate Q. Thus it is assumed that $Q = 2qL$ and $A = 2LB$.

Application with available groundwater outflow solutions

For several well-known solutions of the Boussinesq equation, describing groundwater outflow from an unconfined aquifer based on the hydraulic approach, it can be shown (Brutsaert and Nieber, 1977), that Equation (10.155) can be expressed as a power function,

$$\frac{dQ}{dt} = aQ^b \tag{10.157}$$

where a and b are constants.

Equation (10.157) is obtainable from each of these solutions by assuming geometric similarity of the drainage pattern and the channel network within the catchment. With this assumption, and by defining an equivalent or effective lateral inflow rate q into the stream, one can immediately integrate (10.152) as follows

$$Q = 2L\,|q| \tag{10.158}$$

As before, L is the total length of all tributary and main channel sections upstream from the gauging station where the stream flow is Q. Likewise one can define an effective aquifer breadth B, as the distance from channel to divide (see Figure 10.32) by

$$B = A/(2L) \tag{10.159}$$

in which A is the drainage area of the catchment, and (L/A) is known as the drainage density. Equation (10.159) is the same as the relationship proposed by Horton (1945) for the average overland flow distance in a catchment whose channel slope is much smaller than the land surface slopes.

The first of these solutions, that can be put in the form of (10.157), is the short-time outflow rate (10.64), which was obtained by Boltzmann similarity and which exhibits the characteristic $t^{-1/2}$ behavior. Upon substitution of (10.158), (10.64) yields the basin-scale outflow rate

$$Q = 0.664\,12\,\left(k_0 n_e D^3 L^2\right)^{1/2} t^{-1/2} \tag{10.160}$$

Operating on this result in the manner of (10.157), one obtains in this case for its constants

$$b_1 = 3$$
$$a_1 = -1.1336 \left(k_0 n_e D^3 L^2\right)^{-1}$$

(10.161)

in which the subscripts 1 indicate that it is the first solution which is considered here.

The second solution is the long-time outflow rate derived from the nonlinear Boussinesq equation, namely (10.76) or (10.85) with (10.86). After applying (10.158) and (10.159), this solution can be written in terms of basin-scale parameters as

$$Q = \frac{3.448 L^2 k_0 D^2}{A} \left(1 + \frac{4.46 L^2 k_0 D}{n_e A^2} t\right)^{-2}$$

(10.162)

With this expression the resulting constants for Equation (10.157) are

$$b_2 = 3/2$$
$$a_2 = -4.8038 k_0^{1/2} L (n_e A^{3/2})^{-1}$$

(10.163)

The third solution of interest is the long-time outflow rate (10.115) or (10.116) obtained from the fundamental harmonic of the linear solution. In the case of (10.116), substitution of (10.158) and (10.159) immediately produces the outflow rate in terms of catchment-scale parameters

$$Q = 8 k_0 p D^2 L^2 A^{-1} \exp\left(-\frac{\pi^2 k_0 p D L^2 t}{n_e A^2}\right)$$

(10.164)

Thus with this result Equation (10.157) has the constants

$$b_3 = 1$$
$$a_3 = -\pi^2 k_0 p D L^2 (n_e A^2)^{-1}$$

(10.165)

A fourth expression in the form of (10.157) can be obtained for a sloping aquifer, from the first term of (10.143). With (10.158) and (10.159) this can be written as

$$Q = \frac{8 k_0 p D^2 L^2 \cos \alpha}{A} \frac{z_1^2 [1 - 2\cos(z_1)\exp(\mathrm{Hi}/2)]}{(z_1^2 + \mathrm{Hi}^2/4 + \mathrm{Hi}/2)}$$

$$\times \exp\left[\frac{-(z_1^2 + \mathrm{Hi}^2/4) 4 k_0 p D L^2 \cos \alpha}{(n_e A^2)} t\right]$$

(10.166)

In this case the constants of (10.157) are

$$b_4 = 1$$
$$a_4 = \frac{-(z_1^2 + \mathrm{Hi}^2/4) 4 k_0 p D L^2 \cos \alpha}{(n_e A^2)}$$

(10.167)

The three solutions for horizontal aquifers can also be combined into a single expression, that can be applied with arbitrary values of b. This expression can be obtained by scaling Equation (10.157) with the dimensionless variables implicit in the Boussinesq equation and defined in (10.66) and (10.74). Application of (10.158) and (10.159) yields the following scaled time and the scaled outflow rate in terms of basin-scale parameters,

$$t_+ = 4k_0 DL^2 t/(n_e A^2)$$

$$Q_+ = AQ/(4k_0 D^2 L^2)$$

(10.168)

Thus (10.155) assumes the form

$$\frac{dQ_+}{dt_+} = a_+ Q_+^b$$

(10.169)

where a_+ is a (dimensionless) constant whose value depends only on b. As noted by Michel (1999), the numerical value of a_+ can be readily calculated for each of the theoretical values of $b = 1, 3/2$ and 3 from the respective expressions for a given in Equations (10.161), (10.163) and (10.165). The following interpolation formula (Brutsaert and Lopez, 1999) provides a close estimate of these theoretical values and may be useful for intermediate values of b over that range,

$$a_+ = 10.513 - 15.030b^{1/2} + 3.662b$$

(10.170)

Hydraulic aquifer characteristics at the basin scale

Equation (10.157) with (10.161)–(10.165) can be used to obtain an estimate of the effective hydraulic parameters of the riparian aquifers in the basin (see also Brutsaert and Nieber, 1977; Brutsaert and Lopez, 1998; Eng and Brutsaert, 1999). In the application of this approach a decision must first be made whether (10.161) or (10.163) is the more appropriate expression to describe the long-time outflow behavior of the basin. In past applications, this was done by inspection of the slope of the lower envelope of the low flows as they appear on a log–log plot of $|dQ/dt|$ vs Q. This is illustrated in Figure 10.31, in which the slope of the envelope happens to be close to one, or $b = 1$ in Equation (10.157). This has also been done by linear regression with all the data points of $\log(-dQ/dt)$ against $\log(Q)$. Neither procedure appears to be objective, and at present, it is still not clear how an appropriate a priori value of b can be determined, which describes the long-time behavior of a given basin. In the catchment studies by Brutsaert and Nieber (1977) and Troch *et al.* (1993), it was concluded that $b = 3/2$, whereas in Vogel and Kroll (1992), Brutsaert and Lopez (1998) and Eng and Brutsaert (1999) it was decided to be $b = 1$. This will require further study.

Once the appropriate long-time outflow expression and its value of b have been decided upon, the value of a_1 and a_3 (or a_2) can be determined from the lower envelopes with slopes 3 and 1 (or 3/2), respectively, on a log–log plot of the available $|dQ/dt|$ vs Q data. Examples of the procedure are shown in Figures 10.31 and 10.33. In what follows, the determination of the basin-scale aquifer parameters is outlined for the linear case with $b = 1$; however, the analogous analysis with $b = 3/2$ is straightforward, and can be left as an exercise for the reader.

The value of a_3 is related to the extinction coefficient of the exponential outflow equation (10.153) by

$$a_3 = -K^{-1}$$

(10.171)

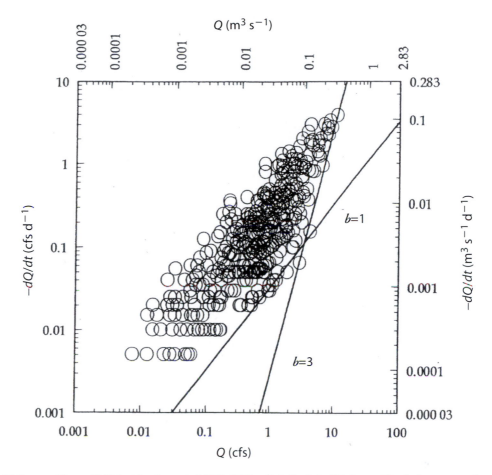

Fig. 10.33 Same as Figure 10.31, but for the period 1966–1977 on Salt Creek in Oklahoma. The drainage area of this basin is $A = 62\ \text{km}^2$, the total length of all stream channels is $L = 76\ \text{km}$, and the estimated mean depth of the surface aquifers is $D = 1.4\ \text{m}$. (From Brutsaert and Lopez, 1998.)

Thus, $-a_3^{-1}$ can be considered as a characteristic time scale for base flow drainage of a basin, from which also the storage half-life can be derived, as $-\ln(2)/a_3$. It is also related to the hydraulic diffusivity defined in Equation (10.89). Comparison between (10.165) and (10.89) with (10.93) produces immediately

$$D_{\text{h}} = -a_3 \pi^{-2} (A/L)^2 \tag{10.172}$$

In a similar way, the value of a_1 for the short-time envelope can be related to the hydraulic desorptivity defined in Equation (10.59). In this case, comparison between (10.161) and (10.61) with (10.63) shows that

$$De_{\text{h}} = (2a_1 L)^{-1/2} \tag{10.173}$$

Equations (10.165) (or (10.163)) and (10.161) can also be combined to determine the effective aquifer parameters from a_3 (or a_2) and a_1. However, because there are three

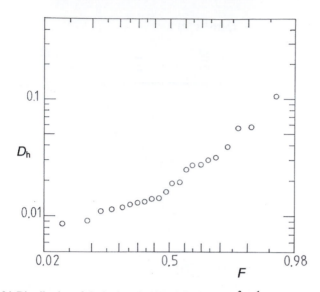

Fig. 10.34 Distribution of the hydraulic diffusivity D_h (in $m^2\,s^{-1}$) defined in Equation (10.89) and calculated with (10.172) from base flow measurements on 22 subbasins within the Washita River watershed complex in central Oklahoma. The scales of the coordinate axes are lognormal; the abscissa axis shows F as the probability of being smaller than or equal to, estimated by means of the Weibull plotting position $m/(n+1)$. (From Brutsaert and Lopez, 1998.)

parameters, namely k_0, n_e and D, with only two equations, one of the three must be known or must be estimated by some other independent method. For example, in case the mean aquifer thickness D can be assumed known (say, from soil maps or other surveys), the hydraulic conductivity and the drainable porosity can be immediately obtained by combining Equations (10.161) and (10.165), namely

$$k_0 = 0.5757\,(a_3/a_1)^{1/2}\,A\,(LD)^{-2}$$

and (10.174)

$$n_e = 1.9688\,(a_3 a_1)^{-1/2}\,(DA)^{-1}$$

Analogous equations can be derived by combining (10.161) and (10.163) for the non-linear case of the long-time outflow rate.

Example 10.7. Estimation of basin aquifer parameters

As an illustration of the results obtainable with this approach, Figure 10.34 displays the values of D_h derived with (10.172) from the a_3 values that were obtained in a study by Brutsaert and Lopez (1998); the study made use of streamflow data from 22 subbasins of the Washita River watershed in Oklahoma, with values of A ranging roughly between 1 and 500 km^2 and L between 2 and 670 km. These $-a_3$ values had a mean of 0.0316 d^{-1} (with a standard deviation of 0.0167 d^{-1}), which amounts to a mean storage half life of roughly 22 days. Figure 10.35 shows the values of the desorptivity De_h obtained with (10.173) from the a_1 values for the same 22 catchments.

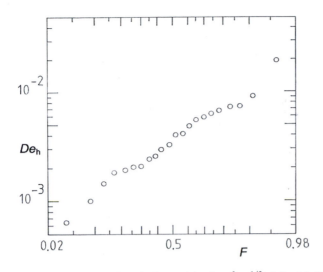

Fig. 10.35 Distribution of the hydraulic desorptivity (in $m^2\ s^{-1/2}$) defined in Equation (10.59) and calculated by means of (10.173) from the same river flow data and plotted as in Figure 10.34. (From Brutsaert and Lopez, 1998.)

Figures 10.36 and 10.37 show the ranges of values of k_0 and n_e, that were obtained by means of Equations (10.174) in the Washita River watershed. The values of these two parameters obtained with this method appear to lie well within the accepted ranges for field measurements in other studies. It is also not surprising that the hydraulic conductivities are several orders of magnitude larger than values to be expected on the basis of laboratory measurements or by "sandbox" standards. This is because of macropores and preferential flow paths that become operative at larger spatial scales. More importantly, however, the values shown in the figures are within a reasonable range from one another; this suggests that the arguably oversimplified hydraulic approach and the method of slope analysis can serve a useful purpose.

The same study also showed that the characteristic outflow time $(a_3)^{-1}\ (= K)$ is well correlated with spatial scale L, that is the length of all stream channels in the basin; the obtained correlation coefficient was $r = 0.66$. This is not unexpected in light of Equation (10.165), but as the drainage density L/A does not vary widely in this region, the predictive power of the relationship between a_3 and L was found to be weak. The short-time constant a_1 was found to be strongly related with stream length L; actually, the regression equation was calculated to be $a_1 = -5.46 \times 10^3 L^{-1.81}$ (in m) with $r = 0.91$, in good agreement with the $L-$ dependency in Equation (10.161). If the power of L is assumed to be exactly the same as in (10.161), the resulting median line through the data is

$$a_1 = -3.50 \times 10^4 L^{-2} \tag{10.175}$$

As illustrated in Figures 10.36 and 10.37, use of Equation (10.175) can constrain the erratic behavior of a_1 somewhat, and it results in considerably less variation (or scatter) in

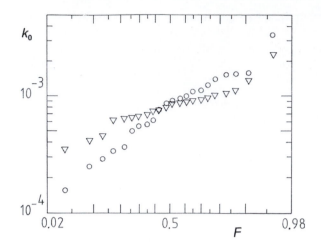

Fig. 10.36 Distribution of the hydraulic conductivity k_0 (in m s^{-1}) calculated by means of Equation (10.174) from the same river flow data and plotted in the same way as in Figure 10.34. The circles represent the k_0 values calculated with the individual a_1 values of each of the 22 catchments, and the triangles represent k_0 calculated with a_1 from Equation (10.175). (From Brutsaert and Lopez, 1998.)

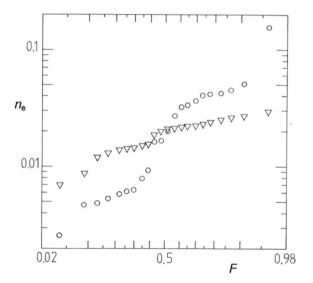

Fig. 10.37 Distribution of the drainable porosity (or specific yield) n_e calculated by means of Equation (10.174) from the same river flow data and plotted in the same way as in Figure 10.34. The circles represent the n_e values calculated with the individual a_1 values of each of the 22 catchments, and the triangles represent n_e calculated with a_1 from Equation (10.175). (From Brutsaert and Lopez, 1998.)

the estimated values of k_0 and n_e. Among the other parameters, the hydraulic conductivity was found to be only weakly (if at all) scale correlated in that range of scales of L ($> 2\,\mathrm{km}$). The hydraulic diffusivity D_h, the hydraulic desorptivity De_h, and the drainable porosity or specific yield n_e showed no evidence of any scale dependency in the same range of scales.

The effect of mean aquifer slope

From the analyses developed in this chapter it is clear that the slope of the riparian aquifers can be expected to exert a strong influence on the magnitude and evolution of the base flow from a basin. Cursory inspection of the unit response function of a hill slope (10.146) with the accompanying Figures 10.27 and 10.28 indicates that Hi can indeed play a major role. This was also brought out in the investigation of measured drought flow data from 19 watersheds in a mountainous section of the Appalachian Plateau by Zecharias and Brutsaert (1985; 1988a); the results of a factor analysis indicated that, among the geomorphic parameters that are related to groundwater outflow, total length of perennial streams, drainage density and average basin slope are most closely related to the process. Moreover, the influences of these three parameters on groundwater outflow behavior are independent of each other; thus, the inclusion of additional parameters would not necessarily yield a better relationship, and may result in redundancy. This empirical result is consistent with the linear basin-scale formulation of the phenomenon in Equations (10.166) and (10.157) with (10.167), which indicates that L, L/A and Hi are the only three geomorphic parameters which control the flow. For the present purpose, geomorphic parameters may be considered the ones that can be derived from topographic maps.

Unfortunately, in contrast to stream length and drainage density, until now attempts to include slope in basin-scale parameterizations have been less than successful. The problem was addressed in Zecharias and Brutsaert (1988b) in the context of the applicability of Equation (10.157) in hilly terrain. The same 19 representative catchments in the Allegheny Mountain section of the Appalachian Plateau, mentioned above, were analyzed on the basis of (10.157) with $b = 1$. The results showed that a, taken as the slope of the lower envelope of (linearly) plotted dQ/dt vs Q data, is dependent on drainage density (L/A) and on (k_0/n_e), in agreement with Equation (10.167), but surprisingly not on land surface slope. However, the results also showed, that in these same plots both the slope of the upper envelope and the mean slope through all the data points, decrease with time. For instance, in one watershed in the region, with flow values that occurred 2, 4, 6, and 7 days following a rainfall event, the slopes of the upper envelopes were observed to evolve as $a = 0.33$, 0.23, 019, and $0.15\ \mathrm{d}^{-1}$, respectively, whereas the lower envelopes remained at around $0.063\ \mathrm{d}^{-1}$. As illustrated in Figure 10.38, a similar evolution of a of the upper envelopes was observed in the Washita River Basin.

The value of the rate of flow Q of a receding hydrograph depends mainly on the storage of water in the watershed. But the upper envelope in graphical representations like Figure 10.38 provides information on the groundwater outflow regime in the early stages of a dry period when the rates of recession, i.e. $-dQ/dt$, are high. The successive values of a show that there are aquifers in the basin whose recession rates are initially large, but decrease sensibly as the rainless period continues. Advanced states of the outflow process, which are accompanied by small recession rates, are represented by the lower envelopes, whose a values in the successive scatter diagrams remain essentially the same. It is likely that the variation of the parameter a of the upper envelopes with time is largely the result of

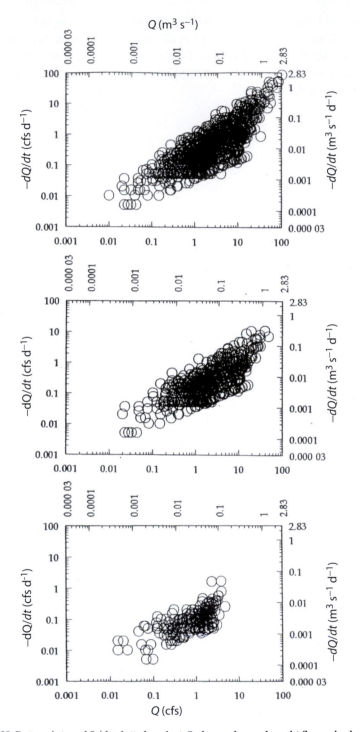

Fig. 10.38 Data points $-dQ/dt$ plotted against Q observed over drought flow episodes with the first day taken 1 (top), 3 (center) and 6 (bottom) days after rain. The flow measurements were made during the period 1962–1977 on West Bitter Creek within the Washita River basin in Oklahoma. The drainage area of this basin is $A = 154$ km^2, the total length of streams is $L = 161$ km, and the estimated mean depth of the surface aquifers is $D = 1.3$ m. (From Brutsaert and Lopez, 1998.)

the nonuniform distribution of the physical characteristics within a catchment. Equation (10.167) indicates that the steeper parts of a basin, where Hi is larger, must have faster depletion rates, and therefore larger values of a; such areas are usually located near the headwater sections of a basin. In contrast the downstream regions of a basin have smaller inclinations, hence relatively lower rates of depletion.

All this shows that, although the characterization of a basin as a single lumped unit with basin scale parameters is a useful paradigm, it has definite limitations. The total outflow rate is the sum of flow contributions from aquifer sections with unequal response characteristics. This total flow is initially dominated by the discharges of channel storage and of the steeper aquifers, which contribute a large fraction of the total flow during the first few hours or days of a recession period. As the recession progresses, however, these storage elements become rapidly depleted and the gentler parts of the aquifer, now being the main contributors, determine the outflow. This also means that the determination of the basin scale aquifer properties by means of an analysis of the lower envelopes is valid mainly in relatively flat and even terrain. In more rugged catchments the lower envelope tends to reflect the properties of the broader valley sections in the lower parts. Therefore, in practical applications for design purposes in hilly watersheds, it may be advisable to adopt an average value of a and b in Equation (10.157), say by regression through all the data points (rather than from the lower envelope alone), to describe representative basin scale parameters. These issues will require further study.

REFERENCES

Abramowitz, M. and Stegun, I. A. (Eds.) (1964). *Handbook of Mathematical Functions, Appl. Math. Ser.* 55. National Washington, DC: Bureau of Standards.

Aravin, V. I. and Numerov, S. N. (1953). *Theory of Fluid Flow in Undeformable Porous Media* (translated from the Russian by A. Moscona, 1965). Jerusalem: Israel Program for Scientific Translations.

Barnes, B. S. (1939). The structure of discharge recession curves. *Trans. AGU*, **20**, 721–725.

(1959). Consistency in unit graphs. *J. Hydraul. Div., Proc. ASCE*, **85** (HY8), 39–61.

Beven, K. (1981). Kinematic subsurface storm flow. *Water Resour. Res.*, **17**, 1419–1424.

Boussinesq, J. (1877). Essai sur la théorie des eaux courantes. *Mém. Acad. Sci. Inst. France*, **23** (1), footnote, pp. 252–260.

(1903). Sur le débit, en temps de sécheresse, d'une source alimentée par une nappe d'eaux d'infiltration. *C. R. Hebd. Séances Acad. Sci.*, **136**, 1511–1517.

(1904). Recherches théoriques sur l'écoulement des nappes d'eau infiltrées dans le sol et sur le débit des sources. *J. Math. Pures Appl., 5me sér.*, **10**, 5–78.

Brutsaert, W. (1994). The unit response of groundwater outflow from a hillslope. *Water Resour. Res.*, **30**, 2759–2763.

Brutsaert, W. and El-Kadi, A. (1984). The relative importance of compressibility and partial saturation in unconfined groundwater flow. *Water Resour. Res.*, **20**, 400–408.

(1986). Interpretation of an unconfined groundwater flow experiment. *Water Resour. Res.*, **22**, 419–422.

Brutsaert, W. and Ibrahim, H. A. (1966). On the first and second linearization of the Boussinesq equation. *Geophys. J. R. Astron. Soc.*, **11**, 549–554.

Brutsaert, W. and Lopez, J. P. (1998). Basin-scale geohydrologic drought flow features of riparian aquifers in the southern Great Plains. *Water Resour. Res.*, **34**, 233–240.

(1999). Reply. *Water Resour. Res.*, **35**, 911.

Brutsaert, W. and Nieber, J. L. (1977). Regionalized drought flow hydrographs from a mature glaciated plateau. *Water Resour. Res.*, **13**, 637–643.

Brutsaert, W., Taylor, G. S. and Luthin, J. N. (1961). Predicted and experimental water table drawdown during tile drainage. *Hilgardia*, **31**, 389–418.

Carslaw, H. S. and Jaeger, J. C. (1959). *Conduction of Heat in Solids*, second edition. Oxford: Clarendon Press.

Childs, E. C. (1971). Drainage of groundwater resting on a sloping bed. *Water Resour. Res.*, **7**, 1256–1263.

Dagan, G. (1966). The solution of the linearized equations of free-surface flow in porous media. *J. Mécan.*, **5**, 207–215.

Dias, N. L. and Kan, A. (1999). A hydrometeorological model for basin-wide seasonal evapotranspiration. *Water Resour. Res.*, **35**, 3409–3418.

Dumm, L. D. (1954). Drain spacing formula. *Agric. Eng.*, **35**, 726–730.

Dupuit, J. (1863). *Études théoriques et pratiques sur le mouvement des eaux dans les canaux découverts et à travers les terrains perméables*, 2me Ed. Paris: Dunod.

Eng, K. and Brutsaert, W. (1999). Generality of drought flow characteristics within the Arkansas River basin. *J. Geophys. Res.*, **104** (D16), 19 435–19 441.

Feldman, A. (1981). HEC models for water resources system simulation: theory and experience. *Adv. Hydroscience*, **12**, 297–423.

Forchheimer, P. (1930). *Hydraulik*, 3. Aufl. Leipzig/Berlin: B. G. Teubner.

Hall, F. R. (1968). Base flow recessions – a review. *Water Resour. Res.*, **4**, 973–983.

Hantush, M. S. (1962). On the validity of the Dupuit–Forchheimer well discharge formula. *J. Geophys. Res.*, **67**, 2417–2420.

(1963). Reply. *J. Geophys. Res.*, **68**, 594–595.

Henderson, F. M. and Wooding, R. A. (1964). Overland flow and groundwater flow from a steady rainfall of finite duration. *J. Geophys. Res.*, **69**, 1531–1540.

Hogarth, W. L. and Parlange, J. Y. (1999). Solving the Boussinesq equation using solutions of the Blasius equation. *Water Resour. Res.*, **35**, 885–887.

Hooghoudt, S. B. (1937). Bijdragen tot de kennis van eenige natuurkundige grootheden van den grond, 6. *Verslagen Landb. Onderzoek (Algemeene Landsdrukkerij, Den Haag)*, **43**, 461–676.

(1940). Bijdragen tot de kennis van eenige natuurkundige grootheden van den grond, 7. *Verslagen Landb. Onderzoek (Algemeene Landsdrukkerij, Den Haag)*, **46**, 515–707.

Horton, R. E. (1945). Erosional development of streams and their drainage basins: hydrological approach to quantitative morphology. *Geol. Soc. Amer. Bull.*, **56**, 275–370.

Ibrahim, H. A. and Brutsaert, W. (1965). Inflow hydrographs from large unconfined aquifers. *J. Irrig. Drain. Div., Proc. ASCE*, **91** (IR2), 21–38.

(1966). Discussion. *Jour. Irrig. Drain. Div., Proc. ASCE*, **92** (IR3), 68–69.

Kirkham, D. (1950). Seepage into ditches in the case of a plane water table and an impervious substratum. *Trans. Amer. Geophys. Un.*, **31**, 425–430.

Kirkham, D. and Gaskell, R. E. (1951). The falling water table in tile and ditch drainage. *Soil Sci. Soc. Amer. Proc.*, **15**, 37–42.

Kraijenhoff van de Leur, D. A. (1958). A study of non-steady groundwater flow, with special reference to a reservoir coefficient. *Ingenieur*, **70**, B87–B94.

(1966). Runoff models with linear elements. *Recent Trends in Hydrograph Synthesis, Comm. Hydrol. Onderzoek TNO, Versl. Mededel.*, **13**, 31–64.

Langbein, W. B. (1938). Some channel-storage studies and their application to the determination of infiltration. *Trans. Amer. Geophys. Un.*, **38**, 435–445.

Laurenson, E. M. (1961). A study of hydrograph recession curves of an experimental catchment. *J. Inst. Engineers, Australia*, **33**, 253–258.

Liggett, J. A. and Liu, P. L.-F. (1983). *The Boundary Integral Equation Method for Porous Media Flow*. London: Allen and Unwin.

Linsley, R. K., Kohler, M. A. and Paulhus, J. L. H. (1958). *Hydrology for Engineers*. New York: McGraw-Hill.

Maillet, Edmond (1905). *Mécanique et physique du globe, essais d'hydraulique souterraine et fluviale*. Paris: Librairie Sci., A. Hermann.

Michel, C. (1999). Comment on "Basin-scale geohydrologic drought flow features of riparian aquifers in the southern Great Plains" by W. Brutsaert and J. P. Lopez. *Water Resour. Res.*, **35**, 909–910.

Parlange, J.-Y. (1971). Theory of water movement in soils: 1. One-dimensional absorption. *Soil Sci.*, **111**, 134–137.

Parlange, J.-Y. and Brutsaert, W. (1987). A correction to free surface groundwater formulations due to capillarity. *Water Resour. Res.*, **23**, 805–808.

Pikul, M. F., Street, R. L. and Remson, I. (1974). A numerical model based on coupled one-dimensional Richards and Boussinesq equations. *Water Resour. Res.*, **10**, 295–302.

Polubarinova-Kochina, P. Ya. (1952). *Theory of Ground Water Movement* (translated from the Russian by J. M. R. DeWiest, 1962). Princeton: Princeton University Press.

Stagnitti, F., Li, L., Parlange, J.-Y., Brutsaert, W., Lockington, D. A., Steenhuis, T. S., Parlange, M. B., Barry, D. A. and Hogarth, W. L. (2004). Drying front in a sloping aquifer: Nonlinear effects. *Water Resour. Res.*, **40**, W04601, doi: 10.1029/2003WR 002255, 2004.

Troch, P. A., DeTroch, F. P. and Brutsaert, W. (1993). Effective water table depth to describe initial conditions prior to storm rainfall in humid regions. *Water Resour. Res.*, **29**, 427–434.

VandeGiesen, N. C., Parlange, J.-Y. and Steenhuis, T. S. (1994). Transient flow to open drains: comparison of linearized solutions with and without the Dupuit assumption. *Water Resour. Res.*, **30**, 3033–3039.

VanderPloeg, R. R., Kirkham, M. B. and Marquardt, M. (1999). The Colding equation for soil drainage: its origin, evolution and use. *Soil Sci. Soc. Amer. J.*, **63**, 33–39.

Verma, R. D. and Brutsaert, W. (1970). Unconfined aquifer seepage by capillary flow theory. *J. Hydraul. Div., Proc. ASCE*, **96** (HY6), 1331–1344.

(1971a). Unsteady free surface ground water seepage. *J. Hydraul. Div., Poc. ASCE*, **97** (HY8), 1213–1229.

(1971b). Similitude criteria for flow from unconfined aquifers. *J. Hydraul. Div., Proc. ASCE*, **97** (HY9), 1493–1509.

Vogel, R. M. and Kroll, C. N. (1992). Regional geohydrologic–geomorphic relationships for the estimation of low-flow statistics. *Water Resour. Res.*, **28**, 2451–2458.

Zecharias, Y. B. and Brutsaert, W. (1985). Ground-surface slope as a basin-scale parameter. *Water Resour. Res.*, **21**, 1895–1902.

(1988a). The influence of basin morphology on groundwater outflow. *Water Resour. Res.*, **24**, 1645–1650.

(1988b). Recession characteristics of groundwater outflow and baseflow from mountainous watersheds. *Water Resour. Res.*, **24**, 1651–1658.

PROBLEMS

10.1   Show that the two lines enveloping the data points plotted in Figure 10.30, must be straight, if the storm flow recession and the base flow recession in the basin are exponential functions of time.

10.2 Suppose that recession outflow data for a certain basin can be described by Equation (10.157) with $b = 2$. Derive the recession outflow hydrograph as a function of time, i.e. $Q = Q(t)$, for this case. Use two parameters in this function, namely a of (10.157) and Q_0, that is the flow rate at $t = 0$.

10.3 (a) Suppose that for a design project it is necessary to express the base flow recession as an exponential equation in the form of (10.153). Determine the value of K (in days) for Tonkawa Creek (see Figure 10.31) for which it is known that, for $b = 3$ in Equation (10.157), $a = -2.74 \times 10^{-5}$ s m^{-6}, and for $b = 1$ in Equation (10.157) $a = -3.24 \times 10^{-7}$ s^{-1}. (b) Determine the value of K_r in Equation (10.154) for this basin.

10.4 Solve the previous problem, 10.3 (a) and (b), for Salt Creek (see Figure 10.33) for which it is known that for $b = 3$ in Equation (10.157), $a = -3.90 \times 10^{-5}$ s m^{-6}, and for $b = 1$ in Equation (10.157) $a = -3.82 \times 10^{-7}$ s^{-1}.

10.5 (a) Derive Equation (10.160) from (10.64) by using the geomorphic relationship (10.158). (b) Derive from (10.160) the corresponding values of a_1 and b_1 for (10.157), as given by (10.161).

10.6 Derive Equation (10.162) from (10.85) with (10.86) using the geomorphic relations (10.158) and (10.159). (b) Then use (10.162) to derive (10.163) for a_2 and b_2.

10.7 (a) Derive Equation (10.164) from (10.116) by making use of the geomorphic relationships (10.158) and (10.159). (b) Then derive the corresponding values of a_3 and b_3 of (10.157), as given by (10.165).

10.8 Calculate the value of a_+ in Equation (10.169) for the case $b = 3$ by scaling (10.157) and (10.161) with (10.168). Compare this dimensionless number with that obtainable with the interpolation formula (10.170).

10.9 Combine Equations (10.165) and (10.161) to derive expressions for the effective hydraulic conductivity k_0 and for the effective unconfined aquifer thickness D in terms of a_1 and a_3. Assume that the drainable porosity n_e is known.

10.10 The recession flow data for Tonkawa Creek shown in Figure 10.31 yielded the parameters $a_3 = -3.24 \times 10^{-7}$ s^{-1} and $a_1 = -2.74 \times 10^{-5}$ s m^{-6}. This watershed has an area $A = 67.3$ km^2, total stream length $L = 70.1$ km and an average surface aquifer thickness $D = 1.6$ m. Calculate effective values of the hydraulic conductivity k_0 and of the drainable porosity n_e.

10.11 By combining (10.161) with (10.163), derive expressions for the regional values of k_0 and n_e, which can be used with the results of hydrograph analyses in which the slopes of the logs of dQ/dt versus those of Q, are $b = 3$ and $b = 3/2$ (cf. (10.174)). Assume that the mean near-surface aquifer thickness D is known.

10.12 Select a stream gauging station in your region of interest, preferably with a drainage area A smaller than 200 km^2. Obtain the daily flow data during periods of recession, for a number of years sufficient to produce an adequate data base. Plot these data, according to Equation (10.156) with logarithmic scales as illustrated in Figure 10.31. Estimate the length of all stream channels upstream from the station, from a map, if the drainage density is not known. Use Equations (10.174) to estimate effective regional values of k_0 and n_e. Test the sensitivity of this result to the assumed value of the effective regional aquifer depth D. (In the United States, such data records can be found on the web at http://waterdata.usgs.gov/usa/nwis/sw).

10.13   Multiple choice. Indicate which of the following statements are correct. The partly saturated zone above the water table during base flow, i.e. during drainage of an unconfined aquifer in the absence of rainfall:
   (a)   has a water pressure which is lower than atmospheric;
   (b)   obeys Laplace's equation $\nabla^2 h = 0$, when the aquifer is uniform and homogeneous;
   (c)   is sometimes described approximately by means of the hydraulic groundwater theory;
   (d)   is most important in shallow (instead of deep) aquifers consisting of fine textured (instead of coarse) materials;
   (e)   is thicker while flow is taking place, than after the complete cessation of base flow (assume that throughout the outflow process the water table is well below the ground surface);
   (f)   has an attenuating effect, such that the outflow from the aquifer is initially (shortly after the cessation of storm runoff) smaller than the value calculated by neglecting the partly saturated zone (and keeping the saturated flow parameters the same);
   (g)   is subject to the principle of continuity and Darcy's law;
   (h)   necessitates the consideration of hysteresis when the unconfined aquifer is initially fully saturated.

10.14   Multiple choice. Indicate which of the following statements are correct. Unconfined aquifers:
   (a)   prevent the deep seepage of water;
   (b)   containing water both at pressures larger and at pressures smaller than atmospheric, usually have a free surface, which is a sharp interface between a fully saturated and a completely dry region;
   (c)   can be an important source of base flow into rivers and lakes;
   (d)   can suffer water depletion due to evapotranspiration;
   (e)   always behave as linear reservoirs, whose outflow, in the absence of rain or other recharge, is given by an exponential decay function.

10.15   Multiple choice. Indicate which of the following statements are correct. Hydraulic groundwater theory requires that:
   (a)   Darcy's law is valid only in the partly saturated zone;
   (b)   the hydraulic head h does not vary along the impermeable layer;
   (c)   the impermeable layer is horizontal;
   (d)   the flow rate is proportional to the slope of the free surface;
   (e)   recharge from infiltration is to be neglected;
   (f)   the magnitude of the specific flux is constant (i.e. uniform) along the coordinate that is normal to the impermeable layer.

10.16   Multiple choice. Indicate which of the following statements are correct. The Boussinesq equation (10.30):
   (a)   requires the assumption that a practically saturated capillary fringe cannot be considered in the analysis;

(b)   in this form implies that the aquifer has a uniform drainable porosity that may however depend on t;

(c)   in this form implies that the aquifer has a constant and uniform hydraulic conductivity;

(d)   yields Laplace's equation under steady flow conditions;

(e)   is based on the assumption that the specific flux is not a function of x;

(f)   is applicable only when the horizontal length scale of the aquifer is much larger than the vertical scale.

10.17   Multiple choice. Indicate which of the following statements are correct. The recession curve of base flow, or drought flow, as a function of time observed at a streamflow gauging station:

(a)   is relatively (i.e. as compared with storm runoff) insensitive to the temporal storm and rainfall pattern over the basin;

(b)   depends primarily on the characteristics of the effluent ground water aquifers in the basin;

(c)   is often plotted as a straight line on log–log paper for engineering applications;

(d)   can sometimes be used to separate the amount of storm runoff due to a given storm, from the observed hydrograph;

(e)   may conceivably be affected by the rate of evapotranspiration from the basin.

10.18   Derive the solution (10.7) from Laplace's equation (10.5) for the boundary conditions (10.6). Hint. Use the method of separation of variables in a manner analogous to that leading to the solution (10.105).

10.19   Derive the expression (10.9) for the steady outflow rate from a saturated unconfined riparian aquifer, from the solution (10.7). Use either the first or the second of (10.8).

10.20   Show how Equation (10.72) is obtained from (10.70).

10.21   Determine the ratio of the second over the first term in the scaled outflow rate (10.112) obtained in the linear solution. For what value of the scaled time t_+, does it become smaller than 1%?

10.22   Consider an extensive unconfined aquifer on a horizontal impermeable layer bounded by a straight open channel (similar to Figure 10.20, with $B = \infty$); after the channel and the aquifer have both been dry (empty) for a very long time, this channel is suddenly (to alleviate flooding elsewhere) at $t = 0$ filled up to a level $D_c = 0.9D$. The flow in the aquifer is assumed to be governed by Boussinesq's equation (10.30). (a) State three boundary conditions for the Boussinesq equation (one of which is an initial condition), describing this situation. (b) Suggest a method of reducing the partial differential equation to an ordinary differential equation, which is permitted by these boundary conditions. (c) Give the functional relationship between x and t (except for one or more undetermined constants) for a given specific value of η. In other words, if the solution of the problem $\eta = \eta(x, t)$ is known and if η is given a certain value, what is the remaining relationship $x = x(t)$? Note. Do not try to find that solution; just assume that it is known and use it as such.

10.23   In Example 10.4, expressions are presented for $q = q(t)$, over the time intervals $t_1 < t < t_2$ and $t > t_3$ in Equation (10.129) and (10.131), respectively. Derive the expression for the time interval $t_2 < t < t_3$.

10.24   Reduce the linearized Boussinesq equation (10.88) to an ordinary differential equation, which can describe the short-time outflow behavior with boundary conditions (10.53). Hint. Follow the same steps as those used to obtain Equation (9.13).

10.25   Multiple choice. Indicate which of the following statements are correct. In the case of steady ground water flow above a horizontal impermeable layer, the hydraulic ground water theory implies that:
(a)   the streamlines are orthogonal to the free surface;
(b)   the pressure distribution is hydrostatic in the vertical;
(c)   the equipotentials (lines or surfaces of constant hydraulic head) are horizontal;
(d)   ground water recharge at the free surface cannot be taken into account;
(e)   potential flow theory with Laplace's equation is still applicable;
(f)   the horizontal scale of the problem is of the same order of, or smaller than, the vertical scale (e.g. the depth of the impermeable layer);
(g)   the flow region is fully saturated;
(h)   the water table is a true free surface;
(i)   effects of horizontal anisotropy can still be taken into account by adjusting the Boussinesq equation (10.31); in other words, if x and y are horizontal coordinates, k_{xx} need not be equal to k_{yy}, and they can be used to replace k_0 in the formulation;
(j)   however, flow in a soil profile with vertical anisotropy (i.e., when $k_{xx} = k_{yy} \neq k_{zz}$) cannot be described by hydraulic ground water theory.

IV FLOWS AT THE CATCHMENT SCALE IN RESPONSE TO PRECIPITATION

11 STREAMFLOW GENERATION: MECHANISMS AND PARAMETERIZATION

Streamflow is one of the main manifestations of the hydrologic cycle in nature. It is normally characterized by a hydrograph, that is the rate of flow in the stream channel as a function of time,

$$Q = Q(t) \tag{11.1}$$

A streamflow hydrograph at any point along a river is the integrated result of all flow processes upstream in the catchment, in response to precipitation, and possibly to snowmelt and other water inputs. Therefore, streamflow is not a local but a basin-scale phenomenon. In the previous chapters some of the more important transport phenomena and their mathematical formulations, that are amenable to analysis, have been considered. Most of these mechanisms are fairly well understood individually. However, at present there is still no unifying theory available that provides a coherent and satisfactory explanation for the integration of these different local mechanisms into the streamflow generation process. The main reason for this uncertainty is undoubtedly the large variation in drainage basins; each drainage basin behaves in many respects almost as if it were a law unto itself, and this has made it difficult to derive general relationships that are broadly applicable. But even for any given basin, it is often difficult to identify and quantify the different mechanisms that produce the observed $Q(t)$; the decomposition of an integral into its constituent parts, that is its inversion to obtain the integrands, like "unscrambling an omelet," is not a simple matter.

11.1 RIPARIAN AREAS AND HEADWATER BASINS

The transformation of precipitation, after it hits the land surface, into streamflow generally takes place over an area of land along the river channel that extends from the channel banks to the nearest divide. Thus each channel segment of a river system can be visualized as lying between two strips of riparian land on either side that feed water into it. While the mechanisms involved in the transformation from precipitation to stream flow depend on many factors, an important one to consider is the relative size of the river segment within the river and tributary system of the basin.

In geomorphology, it is customary to classify stream channels in a hierarchy of orders, in which the order of a stream depends on the number of upstream tributaries or bifurcations. Horton (1932; 1945) was probably the first to propose a downstream-moving ordering procedure. In this system, tributaries without branches are called first-order streams; the branches that receive only first-order streams are designated second-order streams, and those that receive one or more second-order and also

Fig. 11.1   **The order numbers of river channel segments in a natural basin drainage network according to the Horton–Strahler method.**

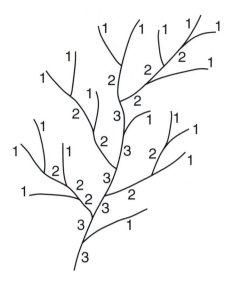

first-order streams are considered third-order streams, and so on. The definition of first- and second-order streams is clear and unambiguous in Horton's procedure, but the definition of third- and higher-order streams required some subjective decisions. To avoid these and to ensure that only one stream would bear the highest-order number in the basin, Strahler (1952) adjusted the procedure by stipulating that third-order streams can only be formed by the joining of any two second-order segments, and so on. The Horton–Strahler method, as it is now called, is illustrated in Figure 11.1; in this example, there are 18 first-order streams, five second-order streams, and one third-order stream.

Larger-order river channels usually do not receive much water locally from the riparian surfaces along their banks, but they receive most of their water from upstream through lower-order streams. The catchments that are drained by lower-order streams with no or very few tributaries can be called *headwater* basins, *source area* watersheds, or also *upland* watersheds. Because they feed into channels of progressively higher order, these lower-order catchments are crucial for a better understanding of runoff mechanisms in larger basins as well. An important feature for the analysis of runoff from such headwater catchments is that lower-order river channels tend to have relatively short residence times; thus any storm runoff hydrograph from a source area watershed is affected primarily by the nature of the soil mantle areas surrounding the stream and very little by the nature of the stream itself. Further downstream, however, as more and more tributaries join, the shape of the hydrograph evolves, and it will increasingly reflect the hydraulic characteristics of the channel network. The flow mechanisms in riparian areas and headwater basins, a topic often referred to also as *hillslope hydrology*, have been the subject of intense research in the past few decades. A knowledge of these mechanisms and of their interactions is not only essential to describe streamflow generation, but it is also the key to a better understanding of solute transport in the human environment and of the evolution of landforms and erosion.

Fig. 11.2 **Illustration of the overland flow (OF) mechanism as infiltration excess. The precipitation rate *P* exceeds infiltration capacity, and the water table is at the ground surface.**

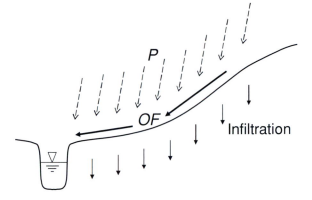

11.2   STORM RUNOFF MECHANISMS IN RIPARIAN AREAS

11.2.1   *Overland flow*

Infiltration excess overland flow

This type of flow occurs when the rainfall rate is larger than the infiltration capacity, so that there is an excess which runs off over the surface. Although this flow generation concept is sometimes associated with the name of Horton (1933), it goes back much earlier. It was already the basis of the well-known rational method, introduced 150 years ago by Mulvany (1850), and of the various runoff routing procedures subsequently derived from it by Hawken and Ross (1921) and others (see also Dooge, 1957; 1973). It is also implicit in the unit hydrograph, as originally proposed by Sherman (1932a; b). In these and other early studies concerned with maximal rates of runoff in problems of flooding and erosion, it was assumed that the infiltration rate is smaller than the precipitation rate over the entire catchment. In the rational method, the infiltration is taken as a fraction of the precipitation, whereas in the unit hydrograph approach and in Horton's work, the infiltration capacity or a related index is subtracted from the precipitation. Thus it was assumed that the infiltrated water is "lost" and that virtually all stormflow results from the overland flow of the precipitation excess (see Figure 11.2). In the prediction of extreme flows for design purposes in disaster situations, this assumption of overland flow was not unreasonable.

It is now understood that overland flow is not a universally occurring phenomenon, that in many situations it may not occur at all, and that its prevalence depends on the nature of the catchment and of the intensity of the precipitation. But it can be expected to be the main mechanism in catchments with relatively impermeable surfaces, and with only a thin soil layer; such surfaces cover mostly urban environments, factory and farm yards and other trampled soil areas, and rocky and stony areas with little or no soil or vegetation, as seen in arid and desert environments. Thus it occurs most frequently in areas where people live and work and in denuded arid regions. It can also occur on other more permeable surfaces, provided the rainfall is sufficiently intense. For instance, in a study of a 20 ha first-order agricultural catchment with steep slopes in semi-arid Shanxi (China), Zhu *et al.* (1997) reported that most storms generate no overland flow. However,

Fig. 11.3 Schematic illustration of the overland flow (OF) mechanism as saturation excess: (a) the position of the water table (WT) prior to the onset of precipitation and (b) during the precipitation event. The precipitation rate P is smaller than the infiltration capacity over the unsaturated portion of the land surface; overland flow takes place where the water table has risen to the ground surface.

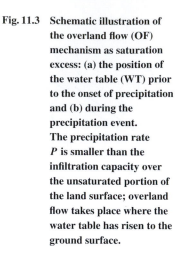
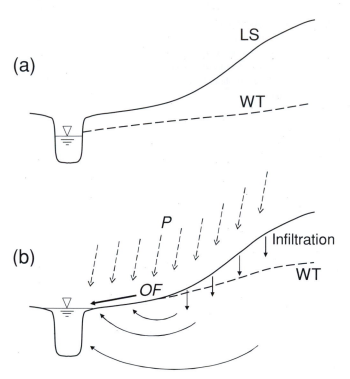

in 8% of the precipitation events infiltration excess overland flow was the predominant runoff process; rainfall intensity, rather than rainfall amount, was the decisive factor for its occurrence, although soil surface crusting also played a role. Occurrence and yield of overland flow varied spatially on account of the variability of the infiltration capacity.

In general, infiltration excess overland flow appears to be rare in natural basins covered with a thriving vegetation in more humid climates.

Saturation excess overland flow

This type of surface runoff occurs over land surfaces that are saturated by emerging subsurface outflow from below and perched water tables, regardless of the intensity of the rainfall (or snowmelt) (see Figure 11.3). It is a rapid and almost immediate transport mechanism to the stream channel, for the seepage outflow water and for the rainwater falling (or snow melting) on such areas. It usually takes place in conjunction with subsurface flow to the channel, but the relative magnitudes of surface and subsurface flows into the channel depend largely on the nature of the catchment and the precipitation. It is most often observed over limited areas in the immediate vicinity of the river channel where downslope subsurface flows emerge, and in wetlands, where the water table can rise rapidly to the surface; but it can also occur higher up in slope hollows, where elevation contours display strong curvature, thus forcing convergence of the flow paths. Outside of these saturated areas all the precipitation and other input can generally enter the soil surface.

Fig. 11.4 Schematic plan view of a second-order catchment illustrating the extent of the variable source areas (inside the dashed line) on which overland flow takes place: (a) under drought flow conditions; (b) and (c) after the onset of precipitation. The stream channels and the saturated areas near the stream channels expand as the precipitation continues.

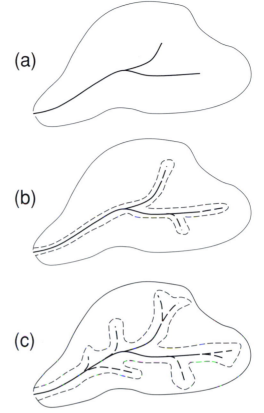

For instance, as early as 1961, US Forest Service hydrologists (Hewlett, 1974; Hewlett and Hibbert, 1967) reported that in forested hilly catchments in the Coweeta section in the southern Appalachians of North Carolina, the streamflow hydrograph rises as a result of precipitation on the channel itself and as a result of the expansion of these saturated areas in its immediate vicinity. The expanding and shrinking areas are often referred to as *variable source areas* (see Figure 11.4). On the basis of hill slope measurements in Vermont, Dunne and Black (1970a; b) also concluded that the stormflow originated from surface flow on limited areas along the stream channel. However, their interpretation of the mechanism was that this surface runoff was not fed significantly by subsurface outflow, but resulted mostly from rainfall on the expanding streamside areas; the role of the subsurface flow was mainly to control the expansion and subsequent contraction of the source areas.

But saturation excess overland flow does not always occur in the immediate vicinity of the stream. In a tropical rainforest in northeast Queensland, Bonnell and Gilmour (1978) and Elsenbeer *et al.* (1995a) observed that high intensity rainfalls generate widespread perched water table conditions close to the soil surface, which emerge easily; this results in saturation excess overland flow accompanied by subsurface flow within the top 20 cm. Evidence for this was taken to be the presence of pre-event water in the streamflow, that is water which was present in the soil profile prior to the rainfall event; if infiltration excess overland flow had been the only mechanism, all the storm runoff would have been event water, that is water furnished by the rainfall event. The ratio of event to pre-event water in the streamflow was found to depend on the rainfall duration and intensity. Because overland flow was so widespread, they concluded that in this type of tropical rainforest the variable source area concept does not apply. Elsenbeer (2001) subsequently surmised that overland flow may be a common flowpath in tropical rain forest catchments with "acrisol" profiles; these are soils, in which the clay content increases with depth, resulting in a decreasing hydraulic conductivity.

11.2.2   Subsurface stormflow

In many catchments under natural conditions infiltration is never exceeded, and the precipitation and other input can readily enter into the ground surface; thus the sub-sequent flow to the stream channel takes place below the surface, presumably through the soil mantle of the catchment. Lowdermilk (1934) and Hursh (1936) appear to have been among the first to propose subsurface flow as the main streamflow generation mechanism in forested hill slopes (see also Hewlett, 1974). It was later confirmed in several experimental investigations that subsurface flow can even be the only mechanism under certain conditions (see Roessel, 1950; Hewlett and Hibbert, 1963; Whipkey, 1965; Weyman, 1970).

The notion that subsurface flow is an important, and sometimes the only process of water transmission, was resisted by many on the grounds that porous media flow is generally much too slow compared with overland flow to be able to produce the observed streamflows. One early explanation of this paradox was suggested by Hursh (1944), who assumed that the transport takes place through secondary porosity of particle aggregates, forming a three-dimensional lattice pattern, and through hydraulic pathways consisting of dead root channels and animal burrows (see also Section 8.3.1). At the time, this possibility of macropore flow and piping seems to have been largely dismissed as unrealistic by experimentalists and mostly ignored by modelers. However, subsequent experimental work in the field, some of it with chemical and isotopic tracers, has produced ample and incontrovertible evidence not only for macropore flow and its importance, but for several other mechanisms enhancing subsurface flow as well. These are considered more closely in what follows.

Macropores and other preferential flow paths

The concept of preferential flow paths or macropores is an old one; "little channels" and "light soil, mixed with pebbles and roots of trees" were invoked as early as the 1680s by Mariotte to explain infiltration and to refute the claims of Seneca and Perrault that rain water cannot possibly penetrate the soil to be the source of springs. In general, macropores can be defined as secondary, often pipe-like structures of the soil matrix, that are the remains of purely physical processes, such as erosion initiated by desiccation cracking, and different forms of biological activity, such as decaying plant root channels

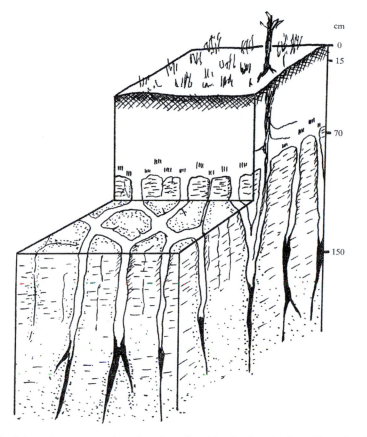

Fig. 11.5 **Schematic representation of a soil profile with a fragipan horizon. (From Smalley and Davin, 1982, after Van Vliet and Langohr, 1981.)**

and animal burrows of various sizes. Reviews on the subject have been presented by Jones (1971) and Beven and Germann (1982). Because soil drying and biological activity tend to take place near the ground surface, pipes and macropores are usually most abundant in the top soil layers and they tend to become less frequent with depth. Such structures are usually obvious features of soil profiles at banks and road cuts.

In addition to these macropores, different other types of preferential flow paths have been observed, which may also have ramifications for the relative transport of *pre-event* and *event* water to the stream channel. Recall that pre-event water, also called *old* water, is the water present in the soil mantle before the onset of the precipitation, whereas the event water, often called *new* water, is the water resulting from the precipitation. In one type of preferential flow, the paths can often be observed at the surface of clayey and loamy soils as cracks or fractures resulting from the shrinkage of the clay particles during drying episodes. At least during the initial stages of a precipitation event, before clay swelling closes them again, such cracks can facilitate the downward water movement in the profile. A somewhat related type of preferential flow has been observed in fissured fragipan horizons (Parlange *et al.*, 1989), as illustrated in Figure 11.5. A fragipan is typically a loamy clay layer with very low conductivity and higher bulk density than the overlying layers. However, in some cases during their evolution, fragipan horizons became fractured into a polygonal columnar structure with a network of interconnected vertical fissures, again, as a result of shrinkage of the clay particles; these cracks are then believed to have been filled with more permeable soil material from above, greatly facilitating water transport. The cracks are typically 10–20 cm wide. In another type of preferential flow, the paths are initially established as instabilities or *fingers* at infiltrating wetting fronts in coarse soils, when the infiltration rate is smaller than the saturated conductivity. A crucial point, however, is that, once established, these paths usually become permanent features of the profile, each time the soil is being rewetted (Glass *et al.*, 1989); exceptions may occur when the soil has undergone complete drying out or complete saturation, both of which are rare if not unlikely in nature. Figure 11.6 shows an example of the initial growth of fingers observed in the laboratory. Such fingers are not so obvious in the soil profile, but they become visible with dyes or other tracers. Other aspects of the nature and origin of this type of preferential paths have been clarified (see Selker *et al.*, 1992; Liu *et al.*, 1994a; b).

Although the existence of macropores has been known for a long time, the precise nature of their contribution to the streamflow generation processes has been emerging only gradually. A few examples follow of investigations in which macropores were observed to play a major role.

In a small (0.022 km^2) basin in east-central Honshu, Tanaka *et al.* (1981; 1988) observed that more than 90% of the storm runoff came from below the ground surface mainly through pipe flow; some saturation overland flow occurred over the gentler slopes ($S_0 \cong 0.12$) of the valley floor, when the rainfall exceeded 50 mm; the saturated area varied somewhat in location and extent from storm to storm, but it never occupied more than 4.5% of the total area (see Figures 11.7 and 11.8). No overland flow was ever observed on the steep ($S_0 \cong 0.50$) hillsides.

In a 0.47 ha forested catchment in Tennessee, Wilson *et al.* (1991) found that the initial subsurface stormflow water in moderate to high intensity events consisted mainly ($>70\%$) of new, i.e. event water; they concluded from this that it had bypassed the unsaturated soil matrix, in which the pre-event water was stored, via macropores without ever reaching the water table. Later on, however, as the flow continued, the fraction of old water gradually increased.

In a catchment under pasture in southern Australia, Smettem *et al.* (1991) and Leaney *et al.* (1993) observed that winter stormflow reaches the channel mainly through macropores, bypassing the soil matrix, and creating perched water table conditions immediately around these pores. In summer, however, overland flow was found to be dominant; they did not observe evidence of partial area sources, as only a negligible fraction of the catchment was occupied by wetland.

On a steep forested hillslope with cedar and cypress in Ibaraki in east-central Honshu, Tsuboyama *et al.* (1994) observed a dynamic system of macropores, which expanded and conducted increasing amounts of water as antecedent conditions became wetter. Continued studies on that same catchment (Noguchi *et al.* 1999; Sidle *et al.*, 2001) led

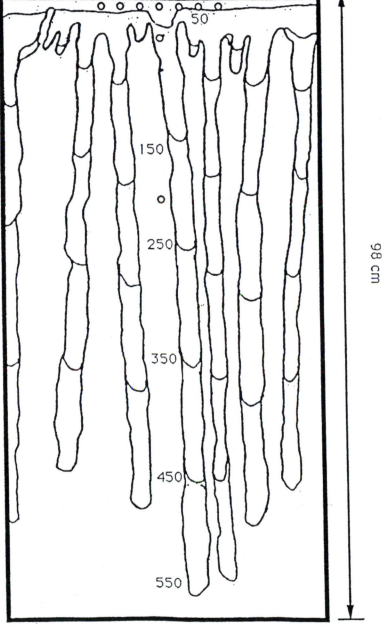

Fig. 11.6 Typical development of an unstable wetting front resulting in a persistent fingered flow pattern; the round holes indicate the positions of the tensiometers in the two-dimensional sand-filled chamber to monitor the water pressures during the experiment, and the numbers indicate time (s) after the start of the infiltration. (From Selker *et al.*, 1992.)

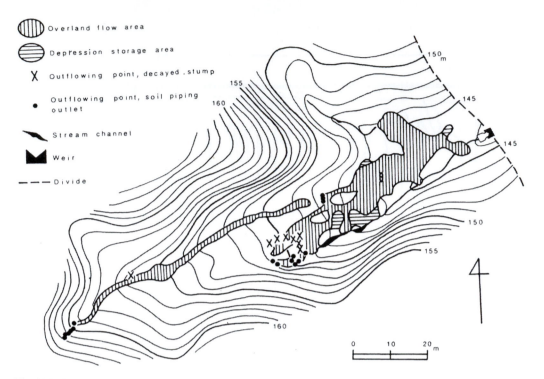

Fig. 11.7 Maximal extent of the saturated areas and distribution of subsurface outflow points at the peak discharge rate of a 195 mm storm in September 1980 on a steep 0.022 km² catchment within the source region of the Tama River. The saturated area occupied roughly 3.3% of the basin area and the area shown represents roughly one quarter of the basin area. (From Tanaka *et al.*, 1981.)

to the more specific view that, while individual macropore segments are usually shorter than 0.5 m, they tend to self-organize, as wetness increases, into larger flow systems with such preferential flow connections between them as buried pockets of organic material and loose soil, small depressions of bedrock substrate, and fractures in the weathered bedrock.

Chemical analysis of measurements on a 0.75 ha forested first-order catchment in the sub-Andean foreland basin of Peru by Elsenbeer *et al.* (1995b) indicated that the stormflow response is dominated by event water. This water traveled to the stream channel as a combination of overland flow and through pipes. Some pipe flow reached the stream directly, but some emerged to the surface before reaching the stream. The overland flow was thus generated by emerging pipe flow and directly by the rain. This made them observe that, from the perspective of the catchment, the distinction between pipe flow and overland flow is meaningless, as both mechanisms produced event water.

From observations in a semiarid pine forest in New Mexico, Newman *et al.* (1998) concluded that most of the lateral subsurface flow takes place in the B horizon through macropores. Thus throughout most of the year, the soil profile behaves like a two-domain

Fig. 11.8 The precipitation and runoff hydrograph of the 195 mm storm event on the catchment shown in Figure 11.7. (From Tanaka *et al.*, 1981.)

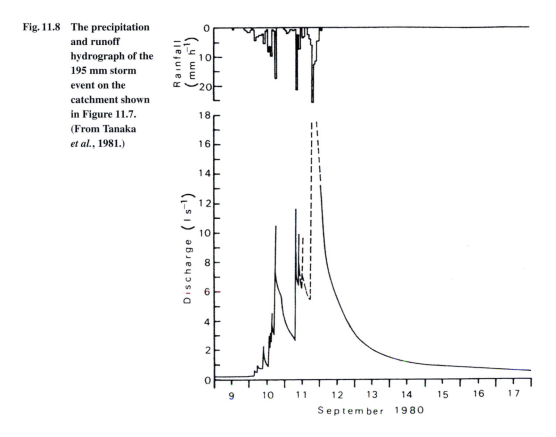

system; this consists of a macropore domain, which provides rapid subsurface flow that is not in equilibrium with the soil matrix, and of a matrix domain, in which the transport is very slow and in which evaporative processes cause major water losses and increased salinity (see Figure 11.9). Variations in the ratio of old to new water in the runoff were seen to depend mainly on the size of the precipitation event; macropores can conduct the flow directly or they may also feed shallow perched saturated zones overlying low permeability bedrock. Whenever the entire profile is fully saturated, as during snowmelt episodes, the two domains are connected, and large subsurface flow rates are produced.

In the above studies it was shown how subsurface flow through macropores and other preferential flow paths can play a major role in storm runoff generation. However, the specific interpretations of the measurements, especially on the relative roles of old and new water in this process, differed somewhat, and in some cases they were contradictory. Although this is largely the result of the wide variety in catchments that were being studied, it is no doubt also related to the differences in experimental techniques used in these studies. This was brought out, for example, in the long-term observations, carried out on the steep mixed evergreen forest catchments (1.63–8.26 ha) in a humid climate (2600 mm y^{-1}) at Maimai in New Zealand; a succession of detailed studies has

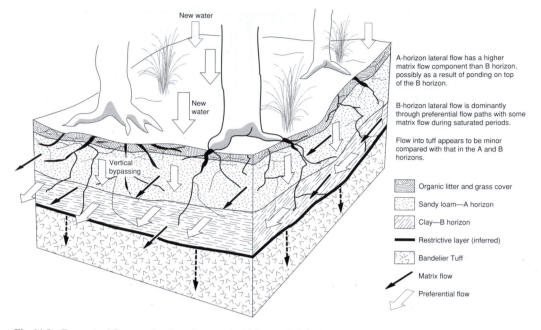

New water

New water

Vertical bypassing

A-horizon lateral flow has a higher matrix flow component than B horizon, possibly as a result of ponding on top of the B horizon.

B-horizon lateral flow is dominantly through preferential flow paths with some matrix flow during saturated periods.

Flow into tuff appears to be minor compared with that in the A and B horizons.

Organic litter and grass cover

Sandy loam—A horizon

Clay—B horizon

Restrictive layer (inferred)

Bandelier Tuff

Matrix flow

Preferential flow

Fig. 11.9   **Perceptual flow mechanisms in a semiarid forested slope in New Mexico. The lateral matrix flow in the A horizon is larger than that in the B horizon, possibly as a result of some ponding on top of the B horizon; in the B horizon the flow takes place mainly through preferential flow paths, with some matrix flow and leakage into the underlying tuff. (From Newman *et al.*, 1998.)**

illustrated how such interpretations can evolve over time, as more and better measurement techniques are brought to bear on the analysis (McGlynn *et al.*, 2002). In the early studies by Mosley (1979) it was concluded from local flow and dye tracer measurements in pits that macropore flow of mostly new water, in storms of moderate to large intensity, can bypass the soil matrix, where the pre-event water is normally stored, and is capable of generating the channel stormflow. On the basis of subsequent investigations with electrical conductivity and natural tracers, Pearce *et al.* (1986) and Sklash *et al.* (1986) arrived at a different conclusion; they deduced from the measurements that it was mainly old water throughflow that was responsible for hydrograph generation, and that the flow of new water above the ground surface or below it through the soil matrix or through macropores could not explain the streamflow response. To resolve these discrepancies, a third set of studies was carried out by McDonnell (1990; McDonnell *et al.*, 1991a) in which a chemical tracer analysis was supplemented with soil water pressure observations by means of tensiometers installed in near-stream, mid-hollow and upslope positions. It was observed that the soil water pressure response was dependent on storm magnitude, intensity and antecedent water content. For storm events producing peak runoff less than 2 mm h^{-1}, the water appeared to infiltrate downward as a wetting front in the soil matrix without appreciable macropore bypass flow; no water table developed along the slope and the streamflow consisted of old water issuing mainly from the near stream valley bottom groundwater. For events with peak storm runoff in excess of 2 mm h^{-1},

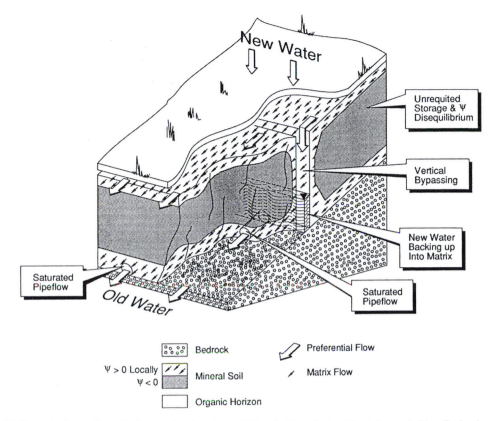

Fig. 11.10 Perceptual runoff production mechanisms in a midslope hollow of a humid catchment in New Zealand. As shown, the precipitation rate (P) exceeds the hydraulic conductivity (k_0) of the mineral soil, and moves down through vertical cracks. The invading new water perches at the soil–bedrock interface, and backs up into the newly saturated soil matrix, where it mixes with the much larger volume of stored old water. Once free water (with positive pore water pressures) exists, the larger pipes in the lower soil zones quickly dissipate transient water tables laterally downslope, producing a rapid throughflow response of well-mixed, albeit mainly pre-event water. (From McDonnell, 1990.)

the lower soil horizons along the slope responded almost instantaneously, indicating a rapid macropore flow, as Mosley (1979) had already surmised. The predominance of old water in the streamflow runoff was explained by McDonnell (1990) by the fact that the rapid flow of new rainwater through downward crack macropores backs up into the soil matrix at the soil–bedrock interface, which is still dry; this rapidly causes saturated conditions, and results in the emergence of well-mixed old water from the matrix into lateral pipe macropores and rapid downslope transport (see Figure 11.10). In a fourth set of experiments, Woods and Rowe (1996) dug a trench 60 m long and 1.5 m deep along the toe of a hillslope hollow, with 30 subsurface flow collection points along its length. The outflow from the hillslope was found to be very variable; this led to the conclusion that outflow data from single hillslope throughflow pits should not be extrapolated to an entire hillslope and further (Woods *et al.*, 1997) that this variability depends on wetness and surface topography. The latter conclusion was refuted by McDonnell and associates

Fig. 11.11 Schematic illustration of the rapid subsurface storm flow (SF) through various types of preferential flowpaths, pipes and macropores. The relative amounts of new (dashed arrows) and old water (solid arrows) in the mixing process depend mainly on the precipitation intensity and on the pre-storm soil moisture conditions.

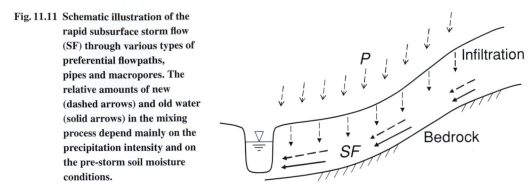

(McGlynn et al., 2002) on the basis of a fifth set of hillslope-scale tracer measurements with bromide at the same catchment. The main conclusion from that study was that it is not the surface topography, but rather the spatial pattern of the bedrock topography, with local preferential flow and mobile and immobile regions, conditioned by small local depressions in the bedrock, which controls the tracer outflow variability; tracer material and old water may remain trapped temporarily in such depressions and become mobilized only by a new storm event.

From measurements in a forested Canadian Shield basin in Ontario, Peters et al. (1995) concluded that preferential flow channels brought the water vertically down, after which it flowed laterally over the bedrock and that practically all the lateral flow occurred within a thin weathered zone near the soil–bedrock interface. The conductivity of this preferential flow layer appeared to be so large that some of the fast flows and peak runoff were suspected to be of the non-Darcy type. The storm runoff in the channel consisted of a mixture of event and pre-event water. This was interpreted to show that the fast infiltration of the event water caused saturated soil conditions above the bedrock, which in turn resulted in the downhill flow of both event and pre-event water; moreover, during the hillslope transport, there was ample opportunity for interaction between the event runoff water and the soil matrix.

In summary, the subsurface stormflow, observed in several of the hillslope experiments reviewed here, exhibited the common feature of unimpeded entry by new water from rainfall into the soil, followed immediately by rapid downslope flow through preferential paths, pipes and other macropores; this flow involved mixing with the old water already present in the soil profile (see Figure 11.11), to varying degrees depending on the intensity of the rain and on the initial moisture status of the soil mantle.

Throughflow in a shallow permeable layer
In many catchments covered with natural vegetation the soil mantle has a relatively permeable top layer consisting of organic debris and mineral soil with high organic content; typically, this layer has a thickness of only a few tens of centimeters and its bottom interface is characterized by an abrupt decline in hydraulic conductivity in the underlying mineral soil. Thus infiltrating rainwater tends to flow and build up along

this interface and develop a perched water table and fully saturated conditions, although deeper layers may remain partially saturated. In several experimental studies it has been observed that such layers can be effective enough to be a major, and sometimes even the main transport medium for stormflow. As noted above, this type of flow was observed to occur by Bonnell and Gilmour (1978), in conjunction with saturation overland flow, in a catchment in Queensland. The chemical signatures of hillslope waters in a catchment in Wales made also Chappell *et al.* (1990) conclude that this can indeed be the dominant mechanism for water and ion transport to lower near-stream riparian zones. Similarly, Jenkins *et al.* (1994) used natural tracers to characterize rain water, soil water and ground water in a moorland catchment in northeast Scotland. The interface between mineral soil layers and the upper organic layers of peaty podsol were identified as preferential pathways. Flow of water in this upper layer was observed to be triggered nearly instantaneously by the onset of the rain, and also to stop nearly as suddenly as the rain ceased; the water in this layer had a chemistry very similar to that of the rain. In the runoff hydrograph, the peak flow was found to be dominated by rain and soil water, whereas the recession part was dominated by pre-event groundwater.

Although they did not consider this type of flow as being representative of the entire catchment, McDonnell *et al.* (1991b) did observe it on small portions of the Maimai catchments in New Zealand. During a rain storm event of some 47 mm and with the soil water suctions initially ranging between $H = 60$ and 150 cm water column, most of the water was seen to flow out from the organic soil layer perched on the mineral soil profile; all the while, the lower soil profile remained only partly saturated. More recently, from an experimental study on seven nested (from 8 to 161 ha) forest catchments in the Catskill Mountains of New York, Brown *et al.* (1999) concluded that a large fraction of the rapid delivery to the stream took place through this same mechanism. Event water appeared to be most prevalent in the stormflow especially during dry conditions, with relative contributions between 50% and 62% near peak flow.

Wavelike mobilization of the water table
As illustrated in Figures 8.5, 8.6 and 8.7, for most soils within the nearly saturated capillary fringe, a small change in water content can result in a relatively large change in pore water pressure. This has led to the view that the addition of a very small amount of water to a relatively moist soil can raise the water table rapidly, almost as a pressure wave type of propagation, to produce a saturated soil profile. Wherever the profile becomes fully saturated this way, subsurface flow may emerge and saturation excess overland flow is also bound to occur. This type of water table rise may be especially fast in the lower parts of the hillslope and may result in the build up of an emerging groundwater mound, exhibiting greatly increased hydraulic gradients and groundwater discharge to the channel, and forming a partial or variable source area producing saturation excess overland flow as well. Thus the phenomenon is not unlike that depicted in Figure 11.3, except that here the water table rise is presumed to involve very little actual water movement.

Phenomena interpreted to be the result of this type of mechanism were observed, for example, in a swampy area by Novakowski and Gillham (1988) and in a grass-covered low relief basin by Abdul and Gillham (1989), both in Ontario. In these studies, the rise of the water table was most pronounced in the near-stream areas. The mechanism has also been inferred to occur in more rugged terrain. During a sprinkler irrigation experiment on a very steep (43%) forested hillslope near Coos Bay in Oregon, Torres *et al.* (1998) applied a sudden input spike, after the system had been driven to a steady state flow and the soil water pressures were mostly between 0 and −10 cm. They supposed that the timing and magnitude of the pore water pressure and of the discharge rate response to this sudden input were much faster than could be expected from advective water movement, and concluded that the fast response was triggered by a pressure wave moving undetected through the unsaturated zone; thus a small amount of rain on a wet soil profile can supposedly result in a rapid rise in the saturated zone, with a relatively slight increase in hydraulic gradient and a large increase in hydraulic conductivity. They also observed some preferential flow, but they felt that in this particular soil, this effect was minor compared to that of the soil water retention characteristics.

The concept that suction-saturated capillary fringe water can be easily converted into water below the water table, that is from a negative to a positive pressure, by a relatively small amount of rain, is undoubtedly realistic. Clearly, only a little additional water is required to mobilize the soil water, when the soil is already close to saturation. But the importance of this mechanism should be kept in perspective. For example, it can only be expected to be effective when the pore water pressure in the top layers of the soil is arrived at during a drainage phase and not during a wetting phase; as illustrated in Figures 8.14, 8.18 and 8.19, the capillary fringe is usually much smaller in the wetting cycle. Similarly, in the absence of any macropores or pipes, the water table (i.e. the locus of atmospheric- or zero-pore water pressure) can only be expected to move rapidly down a steep slope, if it is already close to the surface. As illustrated in Figure 10.12, the drainable porosity n_e is smaller when the water table is closer to the surface, that is when $(-p_w)_{max}$ is smaller. While not a perfect representation, hydraulic groundwater theory, as formulated by the Boussinesq equation (10.29) and its linearized form (10.134), is also fully consistent with this. This can be seen by considering the advectivity in Equation (10.29) (also (10.136)); rewritten here for convenience

$$c_h = -\frac{k_0 \sin \alpha}{n_e} \tag{11.2}$$

it describes the speed of propagation of a given water table height η (or of a disturbance of the water table resulting from rainfall) down the slope. This shows that, in the absence of preferential flow paths, large values of c_h can result only when the drainable porosity is small. As seen in Equation (10.151), this is equally consistent with the kinematic wave approximation.

Capillarity induced flow enhancement has also been linked to soil stratification. In situations where a fine-textured soil layer overlies a more coarse-grained material, the interface between the two layers can develop into a capillary barrier (Ross, 1990; Steenhuis *et al.*, 1991). On account of the different soil water characteristics in the two layers, for a given water pressure at the interface, at equilibrium the soil water content in the upper layer is normally larger than in the lower coarse-textured layer. As a result, the hydraulic conductivity in the upper layer may be considerably larger than in the lower layer; this is illustrated in Figure 8.26. In such a case infiltrating rain water will not readily enter into this lower layer but will tend to be diverted laterally and may cause a rapid rise in water table further down the slope if the water in the upper layer is already close to suction-saturated. Field observations within the source region of the Tama River in east-central Honshu by Marui (1991; Tanaka, 1996) on a hillslope unit, characterized by a 4 m thick fine-grained loam layer underlain by 15 m thick gravel layer, were consistent with this sequence of events. He observed a large-scale groundwater ridge along the steep hillslope. In addition, the air in the underlying partly saturated gravel seemed to be confined by the surrounding groundwater body, and by the saturated zone in the loam layer. In a separate study, Onodera (1991; Tanaka, 1996) inferred that the resulting air pressure increase may have led to increased groundwater outflow at the slope surface.

In conclusion, it stands to reason, that mechanisms related to capillarity can lead to so-called groundwater "ridging" not only in riparian areas, but also along hillslopes, wherever the capillary fringe is already close to the ground surface. However, until now no experiments have demonstrated that by itself this type of phenomenon is related to the hydrograph; thus, whether or not this mechanism can explain large subsurface stormflows remains to be answered.

11.3   SUMMARY OF MECHANISMS AND PARAMETERIZATION OPTIONS

11.3.1   *General considerations*

The brief review in Section 11.2 has shown that on the Earth's land surfaces one can encounter a bewildering range of hydrologic, climatic, topographic and soil conditions, which will favor widely different storm generation mechanisms. These mechanisms can be overland flow due to infiltration excess precipitation, or to saturation excess near the soil surface, resulting either from return outflow from the subsurface, or from rapidly mobilized capillary fringe water in the soil profile to full saturation. On steep slopes overland flow is more likely on converging sections in hollows. The mechanisms can also be subsurface flow of water in a number of different ways. Especially during large rainfall events, this can involve different types of macropores and preferential flow paths, namely as vertical bypass flow to some depth, and then as lateral flow through pipes or through a shallow porous soil layer with high organic content or at the soil bedrock interface. At the same time a slower and less localized throughflow takes place in the soil matrix. Several of these mechanisms have been found to be more than adequate to produce high-intensity runoff events. It is also striking that these mechanisms are not mutually exclusive and that in many situations they coexist and operate interactively in the production of streamflow; their relative importance then depends on the prevailing conditions, such as initial moisture conditions in the catchment and the magnitude of the precipitation.

In some cases the coexistence of different mechanisms can give rise to some unusual phenomena. For instance, under low initial conditions in a watershed in central Côte d'Ivoire, Masiyandima *et al.* (2003) observed double-peaked hydrographs resulting from the same rainfall burst; the first peak, which occurred while it rained, was produced by the rainfall on the saturated valley bottom; the second peak, which came minutes to hours after the first, resulted from the rain that had fallen on the area surrounding the valley bottom and that had traveled to the stream channel by subsurface flow.

All this underscores again the extreme complexity of the stream generation process. These observations suggest that a single unifying runoff model may not be possible nor even desirable, and they have profound implications for the development of modeling strategies for predictive purposes in applied hydrology.

Identification of major mechanisms

In order to keep the formulation sufficiently simple and parsimonious, it may be necessary to identify and include only the dominant mechanisms for any given set of conditions, and to accept some inevitable uncertainty resulting from the omission of the remaining minor mechanisms. On the basis of a knowledge of these local conditions, the analyst must then decide which mechanisms are the major ones that must be considered to represent a particular catchment. The insight gained by the recent field observations can also give some guidance in this. For instance, different kinds of subsurface flow can be assumed to dominate the runoff process in humid areas with an active vegetation. Well-developed mineral soils undoubtedly favor the development of preferential flow paths, whereas thin porous soils with organic litter probably lead to shallow lateral flow of the perched water above the less permeable soil or bedrock. Wetland areas near the stream may allow rapid mobilization or ridging of the water table, and the development of partial and variable source areas, on which saturation excess overland flow can take place. Infiltration excess overland flow will be prevalent during large precipitation events on unvegetated surfaces in arid regions and in areas subject to intense human activities.

Objectives of the analysis

This wide variety in possible mechanisms also means that, in the development of modeling strategies for engineering and other applied purposes, for a given catchment it may be advisable to adopt different formulations depending on the objectives of the analysis. For example, the prediction of disastrous flash flooding, under extreme precipitation conditions, may require an entirely different approach from those needed to describe solute transport and water quality in the environment, to analyze possible climate change scenarios under more normal flow regimes, or to assess the potential for erosion or landslides. For flood prediction, mainly the flows at a certain point along the river may be of interest; for climate change scenarios, surface–atmosphere interactions are of paramount importance; and for water quality purposes it may be crucial to know the pathways, in order to determine the fate and transport of admixtures and water pollutants; finally, erosion and landslide hazards tend to be related to the pore water pressure distribution and the local flow velocities.

Appropriate parameter values
But even for the same formulation, it may also be necessary to adopt different parameter values depending on the flow regime. The formulation of river flow usually requires different values of the roughness parameter in the Gauckler–Manning or Chézy equations, depending on whether it is low flow within the regular channel or high flow with flooding outside the banks. Similarly, in the description of hillslope outflow by some of the subsurface parameterizations of Chapter 10, the appropriate values of the effective hydraulic conductivity k_0 and of the thickness of the flow region η_0, used to represent stormflow conditions with active macropores, will be considerably larger than those appropriate for conditions of baseflow, after the water tables have subsided and many of the macropores in the upper soil layers have emptied and are no longer active. Actually, because of the high flow velocities, subsurface stormflows may not be of the Darcy type, and it may be necessary to use Forchheimer's Equation (8.34) with an additional transmission parameter beside the hydraulic conductivity.

Ultimately, the performance, in a general sense, of any kind of parameterization and of the resulting model, has to be judged on the basis of its ability to simulate or replicate observations of the variables of interest. As mentioned in Chapter 1, parsimony and robustness are important additional considerations. Different aspects of the modeling issue have been treated by Klemes (1986), Morton (1993), and Woolhiser (1996), among others.

11.3.2   *How to put it all together? Distributed versus lumped approach*

As already explained in Chapter 1, scale is the appropriate criterion to classify the different methodologies. Accordingly, one can distinguish two general classes of models that have been used in the past to simulate streamflow generation. In the *distributed* models, also called *runoff routing* models, the computational scales are much smaller than the flow domain characterizing the catchment, whereas in the *lumped* models the computational scale is essentially of the same order as that of the catchment.

The main feature of the distributed approach is that the basin outflow is obtained by tracking the water through its different transport phases in the basin interior. In brief, these phases are surface and subsurface transport into the stream channel network, in response to precipitation after it reaches the ground surface, and the subsequent open channel flow to the basin outlet; between precipitation episodes the basin outflow is dominated by baseflow and evaporation processes. The different mechanisms in each of these transport phases may be described by combining some of the formulations of the relevant processes, as presented in Chapters 2–10. These formulations invariably involve a number of assumptions neglecting certain aspects of the flow, which are considered to be less important; this means that they can be only simplified representations of reality. The distributed approach has been receiving increasing acceptance in recent years with the advent of digital computation and with the growing availability of higher-resolution data from digital terrain and other geographical information systems; rapid advances continue to be reported in the literature.

Among the main advantages of distributed models one can note that they allow the exploration of the consequences of various simplifying assumptions; as a result, they can lead to a better understanding of the various pathways and of the interplay between the main processes and related aspects of complex hydrologic systems in the real world. They can also be useful in the prediction of outflow from headwater catchments, provided their parameters can be determined. But this requirement subsumes also one of their main shortcomings. Ideally, the parameters should be determined a priori, that is independently from the model's performance. In many cases, however, this is impossible and the parameters must be estimated by calibration. But then, distributed models tend to contain so many parameters that it becomes practically impossible to estimate them all in objective and physically consistent ways. Another major drawback is that the underlying mathematical rigor of the parameterizations of the model components may instill in the practitioner a confidence and a sense of realism about their performance, that they do not deserve, on account of the many simplifications and uncertainties involved. As a result, the limitations of such models may not be fully understood by uninitiated users and they may be applied to situations for which they were not intended.

In contrast, the lumped models, whose computational scales are of the same order of magnitude as the catchment scales, rely on fewer parameters, which are generally easier to estimate from the available data. Therefore, they are easier to apply in basin outlet flow simulations for prediction and forecasting purposes. Unfortunately, as the computational scale increases, it becomes increasingly difficult to give a physical interpretation to these parameters, in the sense of the processes described in Chapters 2–10. This means that it is usually impossible to predict changes in these parameters, as the catchment undergoes physical changes, such as those resulting from an evolving land use or changing climate. Another drawback is that even when the catchment characteristics remain unchanged, catchment-scale parameters are incapable of accommodating spatial variability of the input (e.g. rainfall) and of the flow processes (e.g. infiltration and evaporation). Moreover, it is impossible to use this approach to describe the detailed flow paths required in the prediction of pollutant transport or erosion. In spite of all these shortcomings, the lumped approach continues to be useful in the prediction of streamflow for certain operational and design purposes. Specific implementations of this approach are further treated in detail in Chapter 12.

Again, in closing this review, it should be understood that, although a classification into distributed and lumped models is useful to bring some order in the multitude of possible approaches, it is also somewhat artificial. Comparison of the different methods treated in Chapters 5–10 has made it clear that the lumped kinematic approach is merely the simplest extreme in a continuous range of complexity levels, which can be applied in up-scaling the analysis from the finest resolution of the full space- and time-dependent conservation equations of momentum, energy and mass to the coarsest resolution, that is the scale of the catchment itself. However, the level of model complexity necessary for a specific application is still not well known; nor is it clear what scenarios warrant the use of more complex models or under what conditions a distributed model will consistently outperform lumped models. In other words, there is still no general consensus regarding the optimal simplifying assumptions that are most appropriate to describe streamflow generation under a given set of conditions. Although it could be argued that there never will be a consensus, this field is in an active state of development and rapid advances continue to be made.

REFERENCES

Abdul, A. S. and Gillham, R. W. (1989). Field studies of the effects of the capillary fringe on streamflow. *J. Hydrol.*, **112**, 1–18.

Beven, K. and Germann, P. (1982). Macropores and water flow in soils. *Water Resour. Res.*, **18**, 1311–1325.

Bonnell, M. and Gilmour, D. A. (1978). The development of overland flow in a tropical rainforest catchment. *J. Hydrol.*, **39**, 365–382.

Brown, V. A., McDonnell, J. J., Burns, D. A. and Kendall, C. (1999). The role of event water, a rapid shallow flow component, and catchment size in summer stormflow. *J. Hydrol.*, **217**, 171–190.

Chappell, N. A., Ternan, J. L., Williams, A. G. and Reynolds, B. (1990). Preliminary analysis of water and solute movement beneath a coniferous hillslope in mid-Wales, U.K. *J. Hydrol.*, **116**, 201–215.

Dooge, J. C. I. (1957). The rational method for estimating flood peaks. *Engineering (London)*, **184**, 311–313, 374–377.

(1973). *Linear theory of hydrologic systems*, Tech. Bull. 1468. Washington, DC: Agric. Res. Serv., US Department of Agriculture.

Dunne, T. and Black, R. D. (1970a). An experimental investigation of runoff production in permeable soils. *Water Resour. Res.*, **6**, 478–490.

(1970b). Partial area contributions to storm runoff in a small New England watershed. *Water Resour. Res.*, **6**, 1296–1311.

Elsenbeer, H. (2001). Hydrologic flowpaths in tropical rainforest soilscapes – a review. *Hydrological Processes*, **15**, 1751–1759.

Elsenbeer, H., Lorieri, D. and Bonnell, M. (1995a). Mixing model approaches to estimate stormflow sources in an overland flow-dominated tropical rainforest catchment. *Water Resour. Res.*, **31**, 2267–2278.

Elsenbeer, H., Lack, A. and Cassel, K. (1995b). Chemical fingerprints of hydrological compartments and flow paths at La Cuenca, western Amazonia. *Water Resour. Res.*, **31**, 3051–3058.

Glass, R. J., Steenhuis, T. S. and Parlange, J.-Y. (1989). Mechanism for finger persistence in homogeneous, unsaturated, porous media: theory and verification. *Soil Sci.*, **148**, 60–70.

Hawker, W. H. and Ross, C. N. (1921). The calculation of flood discharges by the use of a time contour plan. *Trans. Inst. Engrs Aust.*, **2**, 85–92.

Hewlett, J. D. (1974). Comments on letters relating to 'Role of subsurface flow in generating surface runoff, 2, Upstream source areas' by R. Allan Freeze. *Water Resour. Res.*, **10**, 605–607.

Hewlett, J. D. and Hibbert, A. R. (1963). Moisture and energy conditions within a sloping soil mass during drainage. *J. Geophys. Res.*, **68**, 1081–1087.

(1967). Factors affecting the response of small watersheds to precipitation in humid areas. In *Forest Hydrology*, ed. W. E. Sopper and H. W. Lull. New York: Pergamon Press, pp. 275–290.

Horton, R. E. (1932). Drainage-basin characteristics. *Trans. Amer. Geophys. Un.*, **13**, 350–361.

(1933). The role of infiltration in the hydrologic cycle. *Trans. Amer. Geophys. Un.*, **14**, 446–460.

(1945). Erosional development of streams and their drainage basins; hydrophysical approach to quantitative morphology. *Geol. Soc. Amer. Bull.*, **56**, 275–370.

Hursh, C. R. (1936). Storm water and absorption. *Trans. Amer. Geophys. Un.*, **17**, 301–302.

(1944). Subsurface-flow. *Trans. AGU*, **25**, 743–746.

Jenkins, A., Ferrier, R. C., Harriman, R. and Ogunkoya, Y. O. (1994). A case study in catchment hydrochemistry: conflicting interpretations from hydrological and chemical observations. *Hydrol. Processes*, **8**, 335–349.

Jones, A. (1971). Soil piping and stream channel initiation. *Water Resour. Res.*, **7**, 602–610.

Klemes, V. (1986). Operational testing of hydrological simulation models. *Hydrological Sciences J.*, **31**, 13–24.

Leaney, F. W., Smettem, K. R. J. and Chittleborough, D. J. (1993). Estimating the contribution of preferential flow to subsurface runoff from a hillslope using deuterium and chloride. *J. Hydrol.*, **147**, 83–103.

Liu, Y., Steenhuis, T. S. and Parlange, J.-Y. (1994a). Formation and persistence of fingered flow fields in coarse grained soils under different moisture contents. *J. Hydrol.*, **159**, 187–195.

(1994b). Closed-form solution for finger width in sandy soils at different water contents. *Water Resour. Res.*, **30**, 949–952.

Lowdermilk, W. C. (1934). The role of vegetation in erosion control and water conservation. *J. Forest.*, **32**, 529–536.

Marui, A. (1991). Rainfall-runoff process and function of subsurface water storage in a layered hillslope. *Geograph. Review Japan*, **64**, 145–166.

Masiyandima, M. C., VandeGiesen, N., Diatta, S., Windmeijer, P. N. and Steenhuis, T. S. (2003). The hydrology of inland valleys in the sub-humid zone of West Africa: rainfall-runoff processes in the M'be experimental watershed. *Hydrol. Processes*, **17**, 1213–1225.

McDonnell, J. J. (1990). A rationale for old water discharge through macropores in a steep, humid catchment. *Water Resour. Res.*, **26**, 2821–2832.

McDonnell, J. J., Stewart, M. K. and Owens, I. F. (1991a). Effect of catchment-scale subsurface mixing on stream isotopic response. *Water Resour. Res.*, **27**, 3065–3073.

McDonnell, J. J., Owens, I. F. and Stewart, M. K. (1991b). A case study of shallow flow paths in a steep zero-order basin. *Water Resour. Bull.*, **27**, 679–685.

McGlynn, B. L., McDonnell, J. J. and Brammer, D. D. (2002). A review of the evolving perceptual model of hillslope flowpaths at the Maimai catchments, New Zealand. *J. Hydrol.*, **257**, 1–26.

Morton, A. (1993). Mathematical models: questions and trustworthiness. *Brit. J. Philos. Sci.*, **44**, 659–674.

Mosley, M. P. (1979). Streamflow generation in a forested watershed, New Zealand. *Water Resour. Res.*, **15**, 795–806.

Mulvany, T. J. (1850). On the use of self registering rain and flood gauges. *Inst. Civ. Eng. Proc. (Dublin)*, **4**(2), 1–8.

Newman, B. D., Campbell, A. R. and Wilcox, B. P. (1998). Lateral subsurface flow pathways in a semiarid ponderosa pine hillslope. *Water Resour. Res.*, **34**, 3485–3496.

Noguchi, S., Tsuboyama, Y., Sidle, R. C. and Hosoda, I. (1999). Morphological characteristics of macropores and the distribution of preferential flow pathways in a forested slope segment. *Soil Sci. Soc. Amer. J.*, **63**, 1413–1423.

Novakowsky, K. W. and Gillham, R. W. (1988). Field investigations of the nature of water-table response to precipitation in shallow water-table environments. *J. Hydrol.*, **97**, 23–32.

Onodera, S. (1991). Subsurface water flow in the multi-layered hillslope. *Geograph. Review Japan*, **64** (Ser. A), 549–568.

Parlange, M. B., Steenhuis, T. S., Timlin, D. J., Stagnitti, F. and Bryant, R. B. (1989). Subsurface flow above a fragipan horizon. *Soil Sci.*, **148**, 77–86.

Pearce, A. J., Stewart, M. K. and Sklash, M. G. (1986). Storm runoff generation in humid headwater catchments, 1. Where does the water come from? *Water Resour. Res.*, **22**, 1263–1272.

Peters, D. L., Butte, J. M., Taylor, C. H. and LaZerte, B. D. (1995). Runoff production in a forested, shallow soil, Canadian Shield basin. *Water Resour. Res.*, **31**, 1291–1304.

Roessel, B. W. P. (1950). Hydrologic problems concerning the runoff in headwater regions. *Trans. Amer. Geophys. Un.*, **31**, 431–442.

Ross, B. (1990). The diversion capacity of capillary barriers. *Water Resour. Res.*, **26**, 2625–2629.

Selker, J., Parlange, J.-Y. and Steenhuis, T. (1992). Fingered flow in two dimensions, 2. Predicting finger moisture profile. *Water Resour. Res.*, **28**, 2523–2528.

Sherman, L. K. (1932a). Stream flow from rainfall by the unit-graph method. *Engin. News-Rec.*, **108**, 501–505.

(1932b). The relation of hydrographs of runoff to size and character of drainage basins. *Trans. Amer. Geophys. Un.*, **13**, 332–339.

Sidle, R. C., Noguchi, S., Tsuboyama, Y. and Laursen, K. (2001). A conceptual model of preferential flow systems in forested hillslopes: evidence of self-organization. *Hydrological Processes*, **15**, 1675–1692.

Sklash, M. G., Stewart, M. K. and Pearce, A. J. (1986). Storm runoff generation in humid headwater catchments, 2. A case study of hillslope and low-order stream response. *Water Resour. Res.*, **22**, 1273–1282.

Smalley, I. J. and Davin, J. E. (1982). *Fragipan horizons in soils: a bibliographic study and review of some of the hard layers in loess and other materials*, New Zealand Soil Bureau Bibliographic Report 30. New Zealand: Department of Scientific and Industrial Research.

Smettem, K. R. J., Chittleborough, D. J., Richards, B. G. and Leaney, F. W. (1991). The influence of macropores on runoff generation from a hillslope soil with a contrasting textural class. *J. Hydrol.*, **122**, 235–252.

Steenhuis, T. S., Parlange, J.-Y. and Kung, K.-J. S. (1991). Comment on "The diversion capacity of capillary barriers" by B. Ross. *Water Resour. Res.*, **27**, 2155–2156.

Strahler, A. N. (1952). Hypsometric (area-altitude) analysis of erosional topography. *Geol. Soc. Amer. Bull.*, **63**, 1117–1142.

Tanaka, T. (1996). Recent progress in Japanese studies on storm runoff processes. *Geograph. Review Japan*, **69** (Ser. B), 144–159.

Tanaka, T., Sakai, H. and Yasuhara, M. (1981). Detection of dynamic responses of subsurface water during a storm event with tensiometer and piezometer nests. *Hydrology (Japan Assoc. Hydrol. Sci.)*, **11**, 1–7.

Tanaka, T., Yasuhara, M., Sakai, H. and Marui, A. (1988). The Hachioji experimental basin study – storm runoff processes and the mechanism of its generation. *J. Hydrol.*, **102**, 139–164.

Torres, R., Dietrich, W. E., Montgomery, D. R., Anderson, S. P. and Loague, K. (1998). Unsaturated zone processes and the hydrologic response of a steep, unchanneled catchment. *Water Resour. Res.*, **34**, 1865–1879.

Tsuboyama, Y., Sidle, R. C., Noguchi, S. and Hosoda, I. (1994). Flow and solute transport through the soil matrix and macropores of a hillslope segment. *Water Resour. Res.*, **30**, 879–890.

Van Vliet, B. and Langohr, R. (1981). Correlation between fragipans and permafrost with special reference to silty Weichselian deposits in Belgium and northern France. *Catena*, **8**, 137–154.

Weyman, D. R. (1970). Throughflow on hillslopes and its relation to the stream hydrograph. *Bull. Intern. Assoc. Sci. Hydrology*, **15** (3), 25–26.

Whipkey, R. Z. (1965). Subsurface stormflow from forested slopes. *Bull.Intern. Assoc. Sci. Hydrology*, **10** (3), 74–85.

Wilson, G. V., Jardine, P. M., Luxmoore, R. J., Zelazny, L. W., Lietzke, D. A. and Todd, D. E. (1991). Hydrogeochemical processes controlling subsurface transport from an upper subcatchment of Walker Branch watershed during storm events, 1, Hydrologic transport processes. *J. Hydrol.*, **123**, 297–316.

Woods, R. A. and Rowe, L. K. (1996). The spatial variability of subsurface flow across a hillside. *J. Hydrol. (New Zealand)*, **35**, 51–86.

Woods, R. A., Sivapalan, M. and Robinson, J. S. (1997). Modeling the spatial variability of subsurface runoff using a topographic index. *Water Resour. Res.*, **33**, 1061–1073.

Woolhiser, D. A. (1996). Search for physically based runoff model – a hydrologic El Dorado? *J. Hydraul. Eng.*, **122** (No. 3), 122–129.

Zhu, T. X., Cai, Q. G. and Zeng, B. Q. (1997). Runoff generation on a semi-arid agricultural catchment: field and experimental studies. *J. Hydrol.*, **196**, 99–118.

12 STREAMFLOW RESPONSE AT THE CATCHMENT SCALE

In this chapter different formulations are considered that are available to transform the lumped water input by rainfall or melt events directly into streamflow output from the catchment area. The basic philosophy of this type of approach is that the physical processes are assumed to take place at the scale of the catchment, without consideration for the detailed subscale processes or for the intricacies of the flow paths inside the watershed. In a sense, this is analogous with the view espoused in continuum mechanics or thermodynamics (albeit in their own respective scale ranges), where only the "macro" or "everyday" properties of the fluids are taken into account, without consideration of their properties at the molecular or nuclear scales. Because by far most applications of the lumped approach in the past have been based on the assumptions of linearity and stationarity, these will be treated first and in greatest detail.

12.1 STATIONARY LINEAR RESPONSE: THE UNIT HYDROGRAPH

12.1.1 *Basic concept*

Definition

The unit hydrograph, or unit graph, can be defined as the hydrograph of unit volume of storm runoff produced by a unit volume of uniform (in space and in time) intensity excess rainfall over a unit period D_u, with the basic assumptions of *linearity* and *time invariance*. As defined here, the unit hydrograph is the response function of a linear system, which is treated in more general terms in the Appendix, and can be denoted by $u(D_u; t)$. Similarly, in what follows in this chapter, $y = y(t)$ and $x = x(t)$ will denote, respectively, the storm runoff per unit of catchment area and the intensity of excess rainfall or other input per unit catchment area, which directly produces this storm runoff; note that in some practical applications, the determination of y and x may require the abstraction of baseflow and of the interception or deep infiltration of the precipitation. According to the definition of Equation (A9), in the present context the assumption of linearity or superposition means that the hydrograph, resulting from any input pattern of rainfall or snowmelt, can be built up from separate unit hydrographs by superposition, that is, after scaling them in magnitude and after sequencing them in time. In light of the definition (A10), the assumption of invariance or stationarity means that the runoff from a given catchment due to a given input pattern is the same, regardless of the particular circumstances; thus there are no effects or feedbacks resulting from current conditions (e.g. season) or antecedent input during the event. The unit volume and unit period are arbitrary, but they can be taken, for example, as 1 cm over the entire area of the

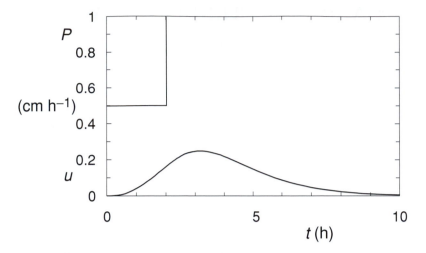

Fig. 12.1 Example of a unit hydrograph $u = u(D_u; t)$; its volume is 1 cm over the catchment area and it results from a unit volume precipitation excess with an intensity $P = 0.5$ cm h^{-1} lasting a unit duration $D_u = 2$ h.

catchment and 1 h or a multiple thereof, respectively. The concept of the unit hydrograph was introduced by Sherman (1932a,b) as a method of extending available data, in order to predict floods resulting from more complex and higher intensity storms, than those on record.

Example 12.1. Application of unit hydrograph

To illustrate this concept, Figure 12.1 shows an example of a 2 h unit hydrograph for a certain catchment; the numerical values are listed in Table 12.1. This hydrograph represents the storm runoff from the catchment, produced by a rainfall excess which has a unit volume, i.e. 1 cm, and a unit duration $D_u = 2$ h. To achieve this unit volume, the rainfall intensity is of necessity $x = 0.5$ cm h^{-1} over the 2 h period. This unit hydrograph can now be used to calculate the storm runoff produced by any pattern of spatially uniform excess rainfall. Consider the following sequence: $x = 1$ cm h^{-1} for $0 < t \leq 2$ h, $x = 2$ cm h^{-1} for $2 < t \leq 4$ h, and $x = 1.5$ cm h^{-1} for $4 < t \leq 6$ h. The first rainfall burst, i.e. 1 cm h^{-1}, has twice the intensity of the input that produces the unit hydrograph; therefore it produces a storm runoff hydrograph, whose magnitude is twice that of the unit hydrograph. The second burst, i.e. 2 cm h^{-1}, has four times the intensity of the input that produces the unit hydrograph, and so on. The ordinates of the resulting three hydrographs are then added to yield the storm hydrograph, as illustrated in Figure 12.2.

Practical limitations
The assumptions of linearity and invariance have their limitations, and the requirements of uniformity are rarely met. For example, the assumption of linearity implies that the

Table 12.1 Values of the 2 h unit hydrograph used in Example 12.1

Time (h)	Flow rate (cm h^{-1})	Time (h)	Flow rate (cm h^{-1})
0	0.0000	6.5	0.0654
0.5	0.0080	7	0.0477
1	0.0414	7.5	0.0342
1.5	0.0963	8	0.0242
2	0.1610	8.5	0.0169
2.5	0.2195	9	0.0118
3	0.2471	9.5	0.0081
3.5	0.2433	10	0.0055
4	0.2190	10.5	0.0037
4.5	0.1851	11	0.0025
5	0.1494	11.5	0.0017
5.5	0.1164	12	0.0011
6	0.0883	12.5	0.0008

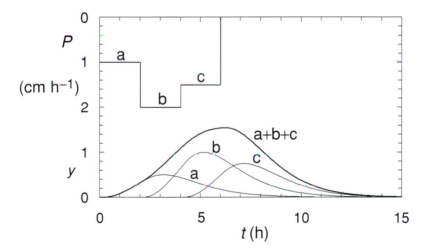

Fig. 12.2 Example of the storm runoff (heavy line) calculated by means of the 2 h unit hydrograph shown in Figure 12.1, resulting from a storm with excess precipitation rates of 1 cm h^{-1} for $0 < t \leq 2$ h, 2 cm h^{-1} for $2 < t \leq 4$ h, and 1.5 cm h^{-1} for $4 < t \leq 6$ h; these successive pulses and their responses are indicated by a, b and c. The respective volumes are 2, 4 and 3 cm over the catchment area.

time scales of runoff remain independent of the magnitude of the input. This assumption is acceptable as long as the flow rates do not deviate too much from some average or characteristic values. However, nonlinearities can be expected to show up when the flow rate magnitudes of interest cover a wide range; this holds especially true over smaller catchments. For instance, in the case of free surface flow, the Chézy and GM equations, (5.39) and (5.41), indicate that the velocity depends on the water depth. This means that the more water is flowing in the rills, gutters and creek channels of the basin, the

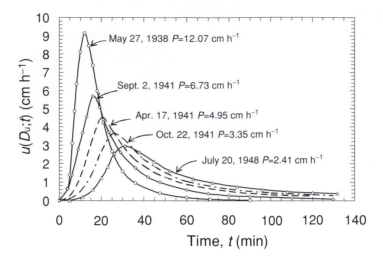

Fig. 12.3 **Illustration of the nonlinear features of the unit response $u(D_u;t)$ of a very small (0.11 km^2) agricultural catchment in Illinois, when the excess precipitation input rate x (denoted here as P) varies over a wide range between 2.4 and 12.1 cm h^{-1}. The unit duration D_u was nearly the same for all five cases and ranged between 10 and 14 min; the indicated time is from the start of the excess precipitation. (After Minshall, 1960.)**

shorter the time to peak of the outflow hydrograph will be, and thus also the higher the peak flow rate will tend to be. This type of nonlinearity is illustrated in Figure 12.3. The unit hydrographs were derived from field data in a study by Minshall (1960) on a small agricultural catchment of 0.11 km^2. The durations D_u were nearly the same for all five cases and ranged between 10 and 14 min, but the rainfall rates changed five-fold over a range between 24 and 121 mm h^{-1}. This type of response may be called *superlinear*. However, watersheds need not always behave this way. For instance, in the extreme case of a large flood, when the water spills over the banks of the channel onto the flood plain, it may happen that the flow is retarded by the larger roughness of the flood plain obstacles; the peak is then likely to arrive later than predicted by the unit hydrograph obtained from flows under more moderate flow conditions. This would be a case of *sublinear* response. The requirement of a spatially uniform rainfall input imposes an upper limit on the catchment area; for practical applications, an upper limit of the order of 1800 km^2 has been suggested by O'Kelly (1955).

12.1.2   *Extensions of original approach: alternative response functions*

S hydrograph

This type of response function results from a uniform input of unit intensity, which continues indefinitely. Thus with the same assumptions, as those used in the definition of the unit hydrograph, the S hydrograph can be obtained by superposing the unit hydrographs resulting from an uninterrupted sequence of unit rainfall volumes of unit duration. Observe that for a unit duration D_u the input intensity is of necessity $(1/D_u)$, if the total

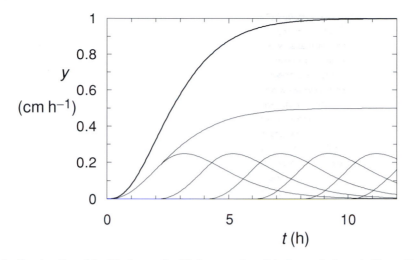

Fig. 12.4 **Construction of the S hydrograph with the example unit hydrograph shown in Figure 12.1. Several unit hydrographs are time-shifted by an amount $D_u = 2$ h and then summed; this result (thin line) simulates the outflow rate from the catchment, caused by a steady uniform input rate of $x = 1/D_u = 0.5$ cm h^{-1}. The S-hydrograph for a steady input rate of unit magnitude of 1 cm h^{-1} (heavy line) is obtained by multiplying this sum by 2 (i.e. D_u).**

input is to be a unit volume. Therefore, the superposed unit hydrographs must be scaled, that is divided by this intensity (or multiplied by D_u), to obtain the S hydrograph for a continuous input of unit intensity.

The main feature of the S hydrograph is that it allows the determination of the unit hydrograph for any other adopted unit period, say D'_u. This is accomplished graphically by time-shifting the S hydrograph D'_u time units to the right and by then subtracting the time-shifted from the original S hydrograph. Again, however, because the unit volume input pulse of duration D'_u must have an intensity equal to $(1/D'_u)$, the resulting difference must be multiplied by this amount to produce the output resulting from a unit volume input. Accordingly, the unit hydrograph for the new unit period is

$$u(D'_u; t) = \frac{1}{D'_u}[S_u(t) - S_u(t - D'_u)] \tag{12.1}$$

in which $S_u = S_u(t)$ is the S hydrograph.

Example 12.2. Scaling of the S and unit hydrographs

Consider again the 2 h unit hydrograph of Example 12.1, which is listed in Table 12.1 and shown in Figure 12.1. The S hydrograph can be derived as follows. First the outflow rates from several 2 h hydrographs, all of them time-shifted by 2 h, are summed; as illustrated in Figure 12.4, this simulates the outflow rate from the catchment caused by a steady rainfall excess rate of $x = (1/D_u) = 0.5$ cm h^{-1}. To obtain the outflow rate resulting from a steady uniform input rate of $x = 1$ cm h^{-1}, this summed hydrograph must be scaled by multiplying it by D_u, which is 2 in the case of the 2 h unit hydrograph. This

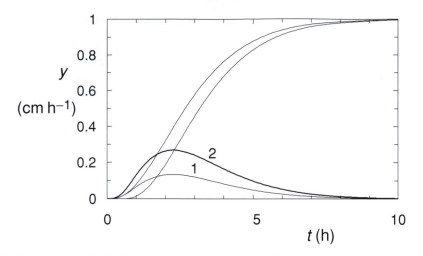

Fig. 12.5 **Construction of the half hour unit hydrograph from the S hydrograph example developed in Figure 12.4. Two S hydrographs are time-shifted by an amount $D'_u = 0.5$ h and then the difference between their ordinates is determined; this result (thin line 1) simulates the outflow rate from the catchment, caused by a uniform input rate of $x = 1.0$ cm h^{-1} lasting for 0.5 h. The unit hydrograph for a total input volume of unit magnitude of 1 cm (heavy line 2) is obtained by multiplying this difference by $1/D'_u = 2$.**

S hydrograph can be used to derive a unit hydrograph of any unit time period D'_u. If the unit hydrograph is sought for a unit period of, say, $D'_u = 0.5$ h, the operation described by Equation (12.1) is performed. As illustrated in Figure 12.5, first an S curve, time-shifted by half an hour, is subtracted from the original one. An input rate of $x = 1$ cm h^{-1} lasting for half an hour produces a total volume of 0.5 cm. Therefore this difference must be scaled by multiplying it by $1/D'_u = 2$, to obtain a unit runoff volume of 1 cm.

The instantaneous unit hydrograph

This is the outflow hydrograph resulting from an input of unit volume placed instantaneously and uniformly over the entire catchment surface, again under the assumptions of linearity and time-invariance. Clark (1945) was probably the first to apply this concept in runoff computations. As shown in the Appendix, an instantaneous input can be represented by a Dirac delta function. Hence the instantaneous unit hydrograph is in fact the impulse response or Green's function of the catchment. In the notation of the Appendix, it can be represented by $u = u(t)$. It also means that the outflow from the catchment, in response to a uniform rainfall input of intensity $x = x(t)$, can be obtained by means of the convolution integral, in any one of the forms (A11)–(A16), whichever is appropriate. For instance in the case of Equation (A14) this is

$$y(t) = \int_0^t x(\tau)\, u(t - \tau)d\tau \qquad (12.2)$$

in which $y = y(t)$ is the output, $x = x(t)$ the input, and $u = u(t)$ the unit response of the catchment. Recall that $y = y(t)$ is the storm runoff rate per unit area of catchment;

Fig. 12.6   Convolution operation with an instantaneous
unit hydrograph $u = u(t)$, as the analog of
the summation shown in Figure 12.2, in the
limit as $D_u \to dt \to 0$. (The values of y and u
are not drawn to scale.) (See also Figure A5.)

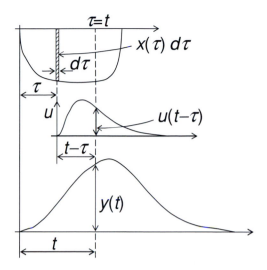

hence, for a streamflow rate $Q = Q(t)$ at the outlet of a catchment of area A, the function y represents Q/A, so that it has the dimensions [L/T], just like $x = x(t)$. This means that the dimensions of the unit response function in Equation (12.2) are $[u] = [T^{-1}]$, corresponding, for example, with the units of cm per hour of runoff per cm of rainfall input. In the operation represented by the convolution integral (12.2), $t = 0$ is defined as the start of the input rate $x = x(t)$. At any given value of time t, the total output rate y is the result of all past inputs from the start of the input at $\tau = 0$ until $\tau = t$, weighted at each instant τ with the unit response, as indicated in Figure 12.6, with the argument $(t - \tau)$. In the integration τ is the dummy time variable of integration, and t is treated as a constant. (A more mathematical illustration of this *convolution* or *folding* is given in Figure A5.)

Relationships between these different response functions
The instantaneous unit hydrograph $u(t)$ can be used to derive the finite duration unit hydrograph $u(D_u; t)$ by applying Equation (12.2) with the following input

$$
\begin{aligned}
x &= \frac{1}{D_u} \quad \text{for } 0 \le t \le D_u \\
x &= 0 \quad\;\; \text{for } t > D_u
\end{aligned}
\tag{12.3}
$$

This yields, say, for $t > D_u$

$$
u(D_u; t) = \int_0^{D_u} \frac{1}{D_u} u(t - \tau) d\tau
\tag{12.4}
$$

After putting $(t - \tau) = s$, so that $d\tau = -ds$, Equation (12.4) becomes

$$
u(D_u; t) = \frac{1}{D_u} \int_{t-D_u}^{t} u(s)\, ds
\tag{12.5}
$$

This indicates that the finite duration unit hydrograph at time t is the average of the instantaneous unit hydrograph taken over the period between $(t - D_u)$ and t. This is

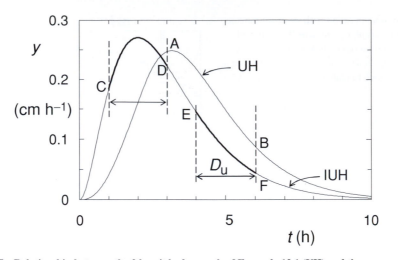

Fig. 12.7 Relationship between the 2 h unit hydrograph of Example 12.1 (UH) and the corresponding instantaneous unit hydrograph (IUH) for the same catchment. For example, the ordinates of the UH at points A and B are, respectively, the averages of the segments CD and EF on the IUH, in accordance with Equation (12.5).

illustrated in Figure 12.7. Because of this one-to-one relationship between the finite duration unit hydrograph and the instantaneous unit hydrograph, no distinction need be made between the two, and both can be referred to as unit response function.

The instantaneous unit hydrograph can also be used to derive the S hydrograph by simply applying Equation (12.2) with a constant unit input rate, $x(t) = 1.00$, starting at $t = 0$, that is a unit step function as defined in Equation (A8), or

$$S_u(t) = \int_0^t u(t - \tau)d\tau \qquad (12.6)$$

Note that $[x(t) = 1.00] = [L/T]$, so that also $[S_u(t)] = [L/T]$. Application of Leibniz's rule (A2) shows readily that the instantaneous unit hydrograph is the slope of the S hydrograph. Actually this also follows directly from Equation (12.1), as can be seen by letting the unit duration D_u approach zero, or in the limit as $D_u \to 0$

$$u(t) = \frac{d S_u(t)}{dt} \qquad (12.7)$$

12.2   IDENTIFICATION OF LINEAR RESPONSE FUNCTIONS

12.2.1   *From available data*

Simple storm events
In principle, it should be relatively straightforward to determine the unit hydrograph, whenever the rainfall occurs as a single burst of acceptably uniform intensity and of a suitable duration, and is accompanied by an easily identifiable streamflow hydrograph. This may be done by first subtracting the baseflow from the observed hydrograph and by

then scaling the remainder hydrograph to the desired unit volume, that is by dividing it by the observed excess rainfall depth. Trial and error may be required in the determination of the baseflow and of the precipitation losses, to ensure that the excess rainfall volume equals the storm runoff volume over the catchment area. Several unit hydrographs derived from different storm events of a similar duration can be averaged in order to obtain a more representative result. If necessary, a common duration can also be obtained by means of their respective S hydrographs.

Complex storm events

More often than not, precipitation events do not have a uniform intensity in time and the resulting runoff hydrograph may be quite irregular. The unit hydrograph must then be estimated by considering its mathematical operation in some detail. Precipitation and also streamflow data are usually given with discrete time steps. According to Equation (A18), the discrete analog of the convolution integral is

$$y(t) = \sum_{k=0}^{n} x(k\,\Delta\tau)\,u(\Delta\tau; t - k\,\Delta\tau)\Delta\tau \tag{12.8}$$

in which $\tau = n\Delta\tau (\leq t)$ is the time of the last input pulse prior to the designated response time $\tau = t$. If the unit period $\Delta\tau$ is literally taken as one, and the output and input times are discretized with the same resolution, this can be rewritten as

$$y_i = \sum_{k=1}^{i} x_k\, u_{i-k+1} \tag{12.9}$$

Note that by analogy with Equation (A16) this can also be written as

$$y_i = \sum_{k=1}^{i} x_{i-k+1}\, u_k \tag{12.10}$$

It should further be noted that y_i and u_i can designate either their respective average values over the ith time period, or their actual values at the end of that same period, depending on how the discretization is specified. Equation (12.9) (or (12.10)) leads to the following set of equations

$$y_1 = x_1 u_1$$
$$y_2 = x_1 u_2 + x_2 u_1$$
$$y_3 = x_1 u_3 + x_2 u_2 + x_3 u_1$$
$$\vdots$$
$$y_i = x_1 u_i + x_2 u_{i-1} + \cdots + x_i u_1$$
$$\vdots \tag{12.11}$$
$$y_{p-1} = x_1 u_{p-1} + x_2 u_{p-2} + \cdots + x_{p-1} u_1$$
$$y_p = x_1 u_p + x_2 u_{p-1} + \cdots + x_p u_1$$
$$\vdots$$
$$y_{m-1} = x_{p-1} u_n + x_p u_{n-1}$$
$$y_m = x_p u_n$$

Fig. 12.8 Example of convolution operation with discrete time variables according to Equation (12.13). The runoff values y_i are in cm h^{-1}.

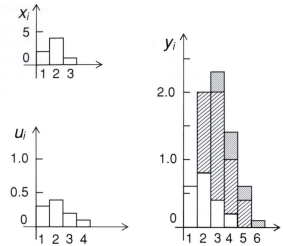

in which the subscript m denotes the number of discretized streamflow hydrograph ordinates, p the number of input pulses, and n the number of unit response function ordinates. Clearly, the number of ordinates of each of the functions must satisfy

$$m = p + n - 1 \tag{12.12}$$

which is illustrated in the following example.

Example 12.3. Numerical convolution

Consider a rainfall event consisting of $p = 3$ rainfall input pulses, and $m = 6$ measured output hydrograph ordinates; this implies a discretized response function with $n = 4$ ordinates. Equations (12.9) and (12.10) become for this case

$$y_1 = x_1 u_1$$
$$y_2 = x_1 u_2 + x_2 u_1$$
$$y_3 = x_1 u_3 + x_2 u_2 + x_3 u_1 \tag{12.13}$$
$$y_4 = x_1 u_4 + x_2 u_3 + x_3 u_2$$
$$y_5 = x_2 u_4 + x_3 u_3$$
$$y_6 = x_3 u_4$$

This is illustrated in Figure 12.8. for a storm with successive hourly input pulses $x_i = 2.0, 4.0, 1.0$ cm h^{-1}, on a catchment whose discretized unit response function has the successive ordinates $u_i = 0.3, 0.4, 0.2, 0.1$.

In the identification of the response characteristics of the system, as formulated in Equations (12.9)–(12.13), the ordinates $u_1, u_2, \ldots, u_n$ are the unknowns that need to be determined. Because there are m ($>n$) equations available, the system is over-determined. Nevertheless, one might be tempted to determine these

u-values algebraically by forward substitution, for example with Equation (12.11) as follows

$$u_1 = \frac{y_1}{x_1}$$

$$u_2 = \frac{y_2 - x_2 u_1}{x_1}$$

(12.14)

and so on for $u_3, u_4, \ldots, u_n$. This could also be done by backward substitution, starting with u_n, as follows

$$u_n = \frac{y_m}{x_p}$$

$$u_{n-1} = \frac{y_{m-1} - x_{p-1} u_n}{x_p}$$

(12.15)

and so on for the remaining values $u_{n-2}, u_{n-3}, \ldots, u_1$. This procedure would be fine if the data were accurate and if the real system in nature were indeed to perform as formulated in these equations. Unfortunately, hydrologic data are invariably subject to considerable error and natural catchments tend to exhibit some nonlinear and non-stationary features in their response characteristics. Therefore, an optimal solution must be sought, which makes use of all m available equations. Among the more common solution methods of a set of equations like (12.9) and (12.11) are those based on the least squares criterion (see Snyder, 1955) and on other mathematical programming techniques (see Deininger, 1969; Diskin and Boneh, 1973, Box *et al.*, 1994).

The method of least squares
The underlying criterion in this method consists of the minimization of the sum of the squares of the differences between the measured data y_i and the calculated values $\sum x_k u_{i-k+1}$ in Equation (12.9). These differences are called the residuals, say ε_i. A simple example will illustrate how their squares can be minimized.

Example 12.4. Application of least squares method

Consider again the simple case described by Equation (12.13) and illustrated in Figure 12.8. The sum of the squares of the residuals is

$$\sum_i \varepsilon_i^2 = (y_1 - x_1 u_1)^2 + (y_2 - x_1 u_2 - x_2 u_1)^2 + (y_3 - x_1 u_3 - x_2 u_2 - x_3 u_1)^2$$
$$+ (y_4 - x_1 u_4 - x_2 u_3 - x_3 u_2)^2 + (y_5 - x_2 u_4 - x_3 u_3)^2 + (y_6 - x_3 u_4)^2$$

(12.16)

This sum can be minimized by putting $\partial \sum \varepsilon_i^2 / \partial u_i = 0$ for each value of i. This yields respectively for $i = 1$ and 2

$$(y_1 - x_1 u_1)x_1 + (y_2 - x_1 u_2 - x_2 u_1)x_2 + (y_3 - x_1 u_3 - x_2 u_2 - x_3 u_1)x_3 = 0$$

and

$$(y_2 - x_1 u_2 - x_2 u_1)x_1 + (y_3 - x_1 u_3 - x_2 u_2 - x_3 u_1)x_2$$
$$+ (y_4 - x_1 u_4 - x_2 u_3 - x_3 u_2)x_3 = 0$$

(12.17)

and analogous equations for $i = 3$ and 4. These constitute a system of four linear equations for the four unknown values of u_i, which can readily be solved.

The technique may also be used in the solution of even more complex cases; various algorithms involving operations of matrix transposition and inversion are available and can be found in textbooks on numerical analysis.

Transform methods of identification
Beside the direct solution methods of Equations (12.9) and (12.11), numerous other methods are available to derive optimal u_i values. In several of these, by using a different formulation or a transformation of the original functions $y(t)$, $x(t)$, and $u(t)$, a simpler, usually algebraic, relationship between the three is obtained, which is more tractable for computation than the convolution integral (12.2) or (12.9). In the method of moments, the functions are characterized by their moments. The optimal u_i values, or the optimal constants in the function, that is used to describe $u(t)$, are determined such that the moments of the calculated output function are equal to the moments of the observed output function $y(t)$. In principle, this method is based on the application of the theorem of moments, as given by Equations (A22) and (A28). In harmonic analysis, the functions are described as Fourier series expansions. The optimal u_i values or the constants in $u(t)$, are determined such that the constants of the Fourier series expansion of the calculated output are exactly the same as those of the observed output. In Fourier and Laplace transform methods, the functions are formulated, respectively, in the frequency domain and in the s (i.e. the Laplace transform) domain.

A review of the application of many of the identification methods investigated in earlier years in catchment hydrology has been presented by Dooge (1973). In practice, however, some of these techniques of direct identification from available data result in response functions, which can be quite sensitive to small errors in the measurements, exhibiting such "non-physical" features as severe oscillations or negative values. Ways of coping with such problems have been discussed by, among others, Neuman and de Marsily (1976) and Singh (1976).

12.2.2 *More concise parameterizations by linear runoff routing*

The data needed to derive the unit hydrograph for a given basin are not always available. Therefore, it should be no surprise that over the years many attempts have been made to develop methods enabling the prediction of this unit response function from basin characteristics. The goal of these studies was to derive unit hydrographs for ungaged watersheds from maps and from some other readily available physical attributes. In one class of methods, empirical equations and empirical curves were used to describe the unit hydrograph, with parameters in terms of basin characteristics. However, because of their strictly empirical nature, their applicability tends to be limited to the region where they were developed, and they will not be considered any further here.

In another class of methods, which are of a more fundamental interest, various theoretical forms of the response function were proposed by postulating different combinations of model elements, which replicate the most important flow mechanisms to transform the precipitation into steamflow. In general, the response function is derived by routing a lumped rainfall excess input through a number of elements of storage and translation, which are patterned after the different processes as described in Chapters 2–10; therefore, such methods might, in a certain sense, be considered physically based. One of the main advantages of these methods is, that the resulting response functions usually require only few parameters; this makes them more general and also easier to calibrate. On the downside, however, since their computational scale is so large, the correspondence with the actual physical processes is not always clear. Indeed, as the computational scale becomes much larger than the variability scales of the basin, the parameters gradually lose their original physical meaning. Because of this ambiguity, these types of parameterizations have also been called conceptual models (see Dooge, 1973).

Linear translatory transport: the Rational Method
This is probably the earliest attempt to relate precipitation with the resulting runoff from a catchment. Evidently (Dooge, 1957), the method was pioneered already some 150 years ago by Mulvany (1850) in Ireland, but versions of it are still being used today in the design of small drainage structures. The underlying concept of the method is that each catchment has a (constant) *time of concentration*, t_c, which is the time needed for the water to flow from the most distant point of the catchment to the outlet. The peak discharge rate Q_p takes place when the entire catchment area A contributes to the outflow, and this occurs at the time $t = t_c$ after the onset of the rain at $t = 0$. Thus for a mean input intensity I (rainfall or snowmelt) over that period, the peak rate of flow is

$$Q_p = CIA \tag{12.18}$$

or, in input–output notation,

$$y_p = x \tag{12.19}$$

in which $y_p = Q_p/A$, such that $[y_p] = [L/T]$, and $x = CI$. The symbol C denotes the runoff coefficient, that is the fraction of the input resulting in direct storm runoff (see also Section 9.5.2). Note again that if Q_p is in $m^3 \, s^{-1}$, I in $(mm \, h^{-1})$ and A in km^2, Equation (12.18) should be written as

$$Q_p = 0.278 \, CIA \tag{12.20}$$

In principle, (12.18) (or (12.20)) should be applicable to drainage basins of any size, but in engineering practice its use is normally restricted to small catchments with $A \leq 15 \, km^2$.

   In its standard form the Rational Method can be applied as follows. The size of the drainage area A can be readily measured on topographic maps, after determining the ridge boundary line of the catchment. The value of C can be estimated from a knowledge of the surface conditions of the catchment by means of Table 12.2. For example, in urban areas a commonly used value is $C = 0.8$. The determination of the design input intensity, I, is probably the most difficult aspect in practice. Consider the case of rainfall input, so that $I = P$. First the duration of the design storm D is to be estimated; this is usually assumed to be equal to the time of concentration t_c, that is $D = t_c$. Several empirical

Table 12.2 Some values of C

Description of area	Runoff coefficients
Business	
Downtown	0.70 to 0.95
Neighborhood	0.50 to 0.70
Residential	
Single-family	0.30 to 0.50
Multi-units, detached	0.40 to 0.60
Multi-units, attached	0.60 to 0.75
Residential (suburban)	0.25 to 0.40
Apartment	0.50 to 0.70
Industrial	
Light	0.50 to 0.80
Heavy	0.60 to 0.90
Parks, cemeteries	0.10 to 0.25
Playgrounds	0.20 to 0.35
Railroad yard	0.20 to 0.35
Unimproved	0.10 to 0.30
Pavement	
Asphaltic and concrete	0.70 to 0.95
Brick	0.70 to 0.85
Roofs	0.75 to 0.95
Lawns, sandy soil	
Flat, 2%	0.05 to 0.10
Average, 2% to 7%	0.10 to 0.15
Steep, 7%	0.15 to 0.20
Lawns, heavy soil	
Flat, 2%	0.13 to 0.17
Average, 2% to 7%	0.18 to 0.22
Steep, 7%	0.25 to 0.35

(From ASCE and WPCF, 1982.)

equations are available for this purpose. The equation proposed by Kirpich (1940), on the basis of Ramser's (1927) data, is often quoted; this can be expressed as follows

$$t_c = 9.012 \ 10^{-5} \left(L/S_a^{1/2} \right)^{0.77} \tag{12.21}$$

where t_c is in hours, L is the length of the main channel from the furthest divide to the outlet in km, and S_a is the average (dimensionless) slope, that is the ratio of the fall of the main channel from the divide to the outlet and its length. As an alternative for very small catchments, the time of concentration can also be taken as the time to equilibrium (6.20) obtained analytically by means of the kinematic wave method. For turbulent flow, with the Gauckler–Manning formula (5.41), one has $a = 2/3$ and $K_r = S_0^{1/2} n^{-1}$, so that Equation (6.20) can be written as

$$t_c = (n^{0.60} P^{-0.40})(L/S_0^{1/2})^{0.60} \tag{12.22}$$

It is remarkable that the powers of L and S_0 are not very different from those in the strictly empirical formula (12.21). Note, however, the different conceptual origins of Equations (12.21) and (12.22); the former refers to the time required by the fluid particles to travel the length of the drainage area, whereas the latter refers to a wave motion, that is the time for the steady state signal to cover that same distance. Recall also that on the basis of experimental data on t_c reported in the literature, McCuen and Spiess (1995) recommended that Equation (12.22) should not be used when $(nL/\sqrt{S_0})$ exceeds 30 m.

Next, a decision must be made regarding the return period of the event T_r. This is usually taken as the expected lifetime of the structure. To give an idea, ASCE and WPCF (1982) suggest, depending on the economic justification, typical values of 5 y for storm sewers in residential areas, 20 y in commercial and high value districts, and 50 y or more for flood protection works.

Finally, with the duration of the rainfall event, i.e. D, and the return period of the design storm, i.e. T_r, both decided upon, the intensity P can be determined from the available intensity–frequency–duration data for the site. Figure 3.16 shows an example of such data. If deemed necessary, the rainfall intensity at the point can be converted to an area value by such means as illustrated in Figure 3.14.

The justification for equating the time of concentration with the duration of the selected design rainfall event is illustrated in Figure 12.9. This shows that, if it is assumed that $D < t_c$, only part of the drainage area can contribute to the outflow. On the other hand, if one assumes that $D > t_c$, the intensity P obtained for this longer duration, would be too small; indeed, as illustrated in Figure 3.16, for a given return period T_r, the rainfall rate P decreases with increasing duration D. Thus to allow the entire drainage area to contribute to the outflow rate, and to obtain the maximal intensity for the selected return period, it is reasonable to put $D = t_c$.

The Rational Method with Equation (12.21) (or (12.22)) and most of its subsequent variations in the engineering literature are based on the notion that storm runoff consists primarily of overland flow. As discussed in Chapter 11, on hillslopes with permeable soils, this is often a tenuous assumption, as most of the runoff takes place through subsurface flow paths. If for such situations Darcy's law is valid, the kinematic flow speed (10.151) yields the following time of concentration

$$t_c = \frac{n_e B_x}{k_0 \sin \alpha} \tag{12.23}$$

in which B_x is the hillslope length (see Figure 10.26), and n_e, k_0 and α are effective values of the drainable porosity, the hydraulic conductivity and the slope angle, respectively.

Fig. 12.9 **In the applicatication of the rational method to calculate the peak flow, the duration of the design storm is taken as the time of concentration, or $D = t_c$. If the duration were shorter, so that $D < t_c$, only part of the catchment would contribute to the runoff; on the other hand if a longer duration were assumed, so that $D > t_c$, the design rainfall rate, derived from available intensity–duration information (e.g. Figure 3.16), might be too small for the adopted design return period T_r.**

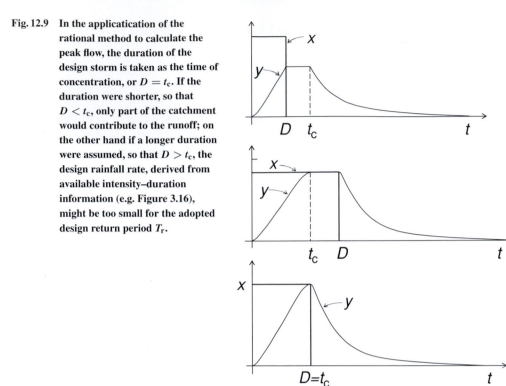

Linear translation with the time–area method

The rational method provides only the peak discharge rate, and over the years attempts have been made to broaden the approach, in order to allow a more complete description of the entire hydrograph including its rise and subsequent recession. This was done, for instance by Hawken and Ross (1921), who considered the effects of the drainage area shape and of the time variation of the storm rainfall. The effects of the shape of the drainage area and of the drainage net were accounted for by the introduction of the time–area(–concentration) function, or time–area diagram, which represents the distribution of the travel times in the basin to the outlet. This function is obtained by first establishing a travel time for each point in the basin, and by then sketching isochrones, which are lines connecting points of equal travel time. The time–area function $A_r = A_r(t)$ is a plot of the relative areas (as fractions of the total basin area) between different isochrones (equally spaced in time), against their respective travel times; thus it is the density function of the travel times to the outlet (Figure 12.10). In this approach, the time variation of the rainfall input was accounted for by a procedure, which, as pointed out by Nash (1958), was in fact a numerical convolution operation.

Just like in the standard version of the Rational Method, the basic assumption is that the entire catchment is equivalent with a plane on which the rainfall is transported to the outlet by translation. Because the system is linear, the transformation mechanism is that of a linear kinematic channel as formulated in Chapter 5. Since it is normalized,

Fig. 12.10 **Sketch of a time–area function $A_r = A_r(t)$, as an extension of the rational method. The dashed lines on the catchment map represent lines of equal travel time or isochrones.**

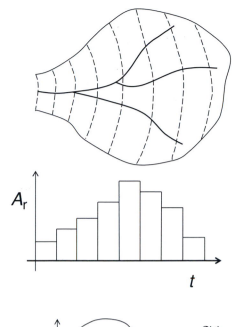

Fig. 12.11 **A linear translation element, as a mechanistic metaphor for the runoff derived by convolution (or routing) of the instantaneous input $\delta(t)$ through a time–area function.**

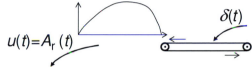

the time–area curve represents outflow resulting from an instantaneous unit input applied uniformly over the catchment area. Hence, the time–area function $A_r = A_r(t)$ is the unit response function of this type of catchment (see Figure 12.11), and the outflow resulting from a rainfall input $x(t)$ is given by Equation (12.2), or

$$y(t) = \int_0^t x(\tau)\, A_r(t - \tau)d\tau \qquad (12.24)$$

Note with Dooge (1973) that, while this approach made use of an instantaneous unit hydrograph in the form of the time–area function, it predated the formal invention of the unit hydrograph by Sherman (1932a,b) by about a decade. But the time–area approach never gained wide acceptance, probably because it takes insufficient account of storage mechanisms in the basin. In natural drainage basins precipitation cannot be immediately translated to the outlet, but a portion of it first must build up some water as storage on the vegetation and on the soil surface and in the pores of the soil profile, before any flow can take place. Therefore, it can be expected that, when the travel times in a basin are estimated on the basis of known velocities of overland flows and channel flows, the calculated peak outflow rates will tend to be severely overestimated. This realization led to increased efforts to include storage effects in subsequent developments.

In recent years the concept of translation, underlying the time–area–concentration function, has continued to be studied and used in a more formal way. This is done by

Fig. 12.12 **Width function $w = w(s_+)$ for the channel network shown in Figure 11.1. In this case, the variable s_+ is the topological distance from the outlet of the catchment, scaled with the longest (topological) stream length from the outlet.**

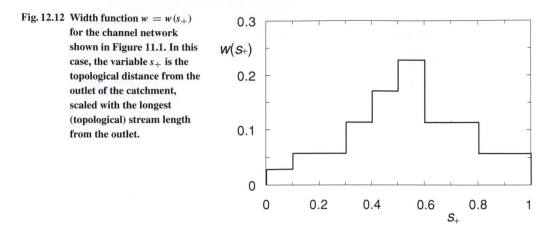

means of what is now usually called (see Kirkby, 1976) the *width function $w = w(s)$*; this function can be defined as the density function of the channel flow distances in the catchment from the outlet, and it describes the (normalized) number of links, or channel segments, as a function of distance s from the outlet. This distance variable s has variously been taken as the actual distance along the channels, as geometric distance, that is the piecewise straight line joining junctions, or as the number of links, that is the topological distance. (On average, these distances are related; apparently (Shreve, 1974), in large networks the longest topological distances differ only slightly from the longest geometrical distances.) The number of links is strongly related with basin area; moreover, average stream velocities are known (Wolman, 1955; Pilgrim, 1977; Rodriguez-Iturbe *et al.*, 1992) to remain relatively constant in the downstream direction, in spite of the decreasing slopes. Therefore, as the distances from the outlet can be taken to be roughly proportional to the travel times, the width function concept is essentially equivalent with the time–area function. However, because the width function is based on well-defined morphological characteristics of the river network, it can be determined more objectively and is therefore better suited for analysis. The correspondence between this concept and the time–area function as a unit response function was probably first pointed out by Surkan (1969), who used it to study the effect of stream channel pattern on the flow at the outlet of the basin. Subsequently, the width function has proved to be a useful tool for studying the stochastic properties of stream networks (see Kirkby, 1976; Veneziano *et al.*, 2000) and implicitly some of their translatory response characteristics (see Gupta and Waymire, 1983; Troutman and Karlinger, 1985; Rodriguez-Iturbe and Rinaldo, 1997).

Example 12.5. Construction of a width function

Consider again the hypothetical catchment shown in Figure 11.1. The number of channel links at topological distances 1, 2, 3, etc., from the outlet can readily be counted; they are respectively 1, 2, 2, 4, 6, 8, 4, 4, 2, 2. The density at each distance can be calculated by dividing the number of links by 35, that is the total number of channel links in this catchment. The results are shown in Figure 12.12.

Linear translation in series with one linear storage element

The transformation of a rainfall input hyetograph into a streamflow output hydrograph involves both a delay as a result of translatory effects and a deformation and attenuation as a result of storage effects. In a linear context, the simplest way of incorporating both effects is simply to add them. Thus it stands to reason that historically the next step in the development of linear runoff routing procedures consisted of the superposition of a linear time–area function, representing pure translation, with a linear reservoir, representing pure storage. In one of the better known implementations of this idea, Clark (1945) derived the instantaneous unit hydrograph from streamflow records by numerically routing the time–area concentration function of the basin through a single concentrated storage element by means of the Muskingum method with $X = 0$. Hence, in light of Equation (7.15), this type of storage element is characterized by

$$S = Ky \tag{12.25}$$

where y is the outflow rate, and S the storage, both per unit catchment area, so that $[y] = [L/T]$, $[S] = [L]$ and $[K] = [T]$; the parameter K is commonly referred to as the storage coefficient.

A similar approach was also applied successfully in the development of large-scale drainage schemes in a number of Irish catchments by O'Kelly (1955) and his fellow engineers at the Office of Public Works. However, in the early stages of this work it became clear that the routing through the concentrated storage element had such a smoothing effect on the time–area function, that the exact shape of the time–area function was not very critical, and that there was little loss in accuracy when it was replaced by an isosceles triangle. The main parameters in the applications of this concept were the time of concentration t_c, which is the time base needed to scale the triangular time–area function, and the storage coefficient K, or the delay, in the routing procedure by means of Equation (12.25). O'Kelly's report is noteworthy and it suggests that in many studies the importance of the time–area function, and of the width function, may have been exaggerated.

Various procedures have been used in the past to estimate the two parameters t_c and K. In Clark's (1945) application, it was assumed that the direct effect of the rainfall ceases at the inflection point of the recession limb of the outflow hydrograph, and that, from that time on, the outflow is merely a release from storage in the basin; accordingly he took t_c as the time between the end of rainfall and the inflection point of the falling leg of the hydrograph, and the storage coefficient from the recession after the inflection point with $K = -y/(dy/dt)$, which is obtained from Equation (1.10) (or (7.11)) with (12.25) and $x = 0$. In the procedure described by O'Kelly (1955) the shape of the instantaneous unit hydrograph is uniquely determined by the ratio K/t_c (see also Example 12.6); thus the general shape of the experimentally obtained unit hydrographs for the basin provided an estimate of this ratio, which allowed then in turn the estimation of the value of t_c (or K) by matching the peak outflow rates. Because the shapes could not always be fitted well, often various K/t_c ratios were tried yielding different t_c and K values. A review of some of the earlier methods to estimate these parameters has been presented by Dooge (1973, pp. 198–200). In several of these studies t_c and K were expressed in

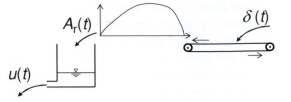

Fig. 12.13 A linear translation element placed in series with a linear tank element, as a mechanistic metaphor for the runoff derived by successive convolution (or routing) of the instantaneous input $\delta(t)$ through a time–area (or width) function and through a concentrated storage function.

terms of powers of basin characteristics, such as L, S_a and A. Note that both (12.21) and (12.22) suggest that t_c should be proportional to a power of the combined variable $(L/\sqrt{S_a})$.

To derive the unit response function of this model, it is necessary to consider first the unit response of a concentrated storage element. The flow through a linear storage element can be represented by the storage equation (1.10) (or (7.11)), or in the catchment-scale notation with an input $x(t)$, an output $y(t)$ and a storage per unit area $S(t)$, as

$$x - y = \frac{dS}{dt} \qquad (12.26)$$

After substitution of the concentrated storage function (12.25), this can be rearranged as

$$\left(\frac{d}{dt} + \frac{1}{K}\right) y = \frac{x}{K}$$

which, upon multiplication of both sides by $\exp(t/K)$, yields the solution

$$y = \frac{\exp(-t/K)}{K} \int x(t)\exp(t/K)\,dt + \text{constant} \qquad (12.27)$$

With a delta function input, that is $x(t) = \delta(t)$, the output of (12.27) is the unit response function for a single storage element; in light of (A7) this has the form

$$u(t) = \frac{\exp(-t/K)}{K} \qquad (12.28)$$

As could be expected, this is the same as the Muskingum response function (7.28) for $X = 0$.

The model of Clark (1945), O'Kelly (1955), and others (Dooge, 1973), in which the storm runoff is derived by successively routing the rainfall input through translation and storage, can thus be formulated by simply putting $A_r(t)$ in sequence with Equation (12.28). Hence, $A_r(t)$, which is the output from the translation operation, becomes the input into the storage element, whose unit response is (12.28). This is illustrated in Figure 12.13. The routing is accomplished by a convolution operation, which produces immediately the unit response function of this combined system,

$$u(t) = \int_0^t A_r(\tau)\exp[-(t-\tau)/K]\,d\tau/K \qquad (12.29)$$

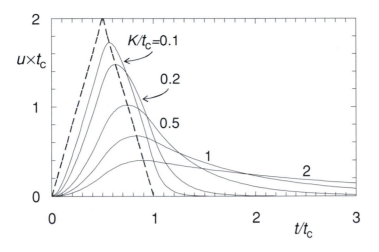

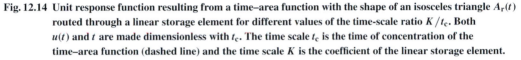

Fig. 12.14 Unit response function resulting from a time–area function with the shape of an isosceles triangle $A_r(t)$ routed through a linear storage element for different values of the time-scale ratio K/t_c. Both $u(t)$ and t are made dimensionless with t_c. The time scale t_c is the time of concentration of the time–area function (dashed line) and the time scale K is the coefficient of the linear storage element.

Example 12.6

Consider a hypothetical diamond-shaped catchment with the stream channel running along one of the diagonals; in the present context this produces a triangular time–area function (or width function), which can be formulated as follows

$$A_r = \frac{4t}{t_c^2} \qquad \text{for } 0 \le t \le t_c/2$$

$$A_r = \frac{-4t}{t_c^2} + \frac{4}{t_c} \qquad \text{for } t_c/2 < t \le t_c \tag{12.30}$$

$$A_r = 0 \qquad \text{for } t_c < t$$

where t_c is the time of concentration. Observe that the area under $A_r = A_r(t)$ equals unity, as it should. The unit response is calculated by applying (12.29) with (12.30). Thus, one has for $t \le t_c/2$

$$u(t) = \frac{4}{t_c^2 K} \int_0^t \tau e^{-(t-\tau)/K} d\tau \tag{12.31}$$

which upon integration results in

$$u(t) = \frac{4}{t_c^2}(t + K(e^{-t/K} - 1)) \tag{12.32}$$

Similarly for $t_c/2 < t \le t_c$, one can write

$$u(t) = \frac{4}{t_c^2 K} \int_0^{t_c/2} \tau e^{-(t-\tau)/K} d\tau - \frac{4}{t_c^2 K} \int_{t_c/2}^t \tau e^{-(t-\tau)/K} d\tau + \frac{4}{t_c K} \int_{t_c/2}^t e^{-(t-\tau)/K} d\tau \tag{12.33}$$

Fig. 12.15 Same as previous figure, for the case of a time–area function with the shape of a right-angled triangle (dashed line).

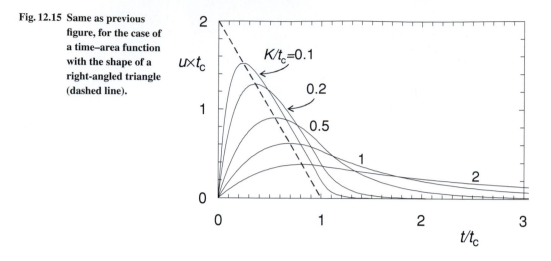

which yields

$$u(t) = -\frac{4}{t_c^2}(t - K - t_c) + \frac{4Ke^{-t/K}}{t_c^2}\left(1 - 2e^{t_c/(2K)}\right) \qquad (12.34)$$

It can be readily checked that Equations (12.32) and (12.34) yield the same value for u at $t = t_c/2$, as they should. For $t > t_c$, (12.33) must be integrated again, but with the upper limit at $\tau = t_c$ in the second and third integral, because A_r is zero beyond that point; this produces

$$u(t) = \frac{4Ke^{-t/K}}{t_c^2}\left(1 - 2e^{t_c/(2K)} + e^{t_c/K}\right) \qquad (12.35)$$

Again, it can be seen that both (12.34) and (12.35) produce the same result at $t = t_c$, as they should. The resulting instantaneous unit hydrograph, obtained by patching (12.32), (12.34) and (12.35) together over their respective time ranges, is shown in Figure 12.14. In principle, it should be possible to use this three-component unit response function with any input function $x(t)$ in the convolution integral (12.2), to calculate the actual outflow $y(t)$ analytically. However, because this $u(t)$ consists of three parts, this is rather involved, so that in practical applications it may be more convenient to convert $u(t)$ into tabular form and carry out the calculations numerically. An idea of the effect of the shape of the time–area function on the resulting unit response can be gained by comparing this result with the response function obtained with a right-angled triangle shown in Figure 12.15.

Combinations of linear storage elements

In yet another class of models, the basin outflow is derived by routing the rainfall input solely through a number of storage elements, without any formal or explicit translation in the formulation. Equation (12.28) is used as the response function of each of the storage elements. In what follows, a few examples are reviewed of this type of representation.

Fig. 12.16 A tank model representation of the Kitakami
River in northern Honshu by Sugawara and
Maruyama (1956). Both tanks are linear storage
elements with a unit response function given by
Equation (12.28). The bottom part of the fast
response tank represents 20 mm of initial loss.

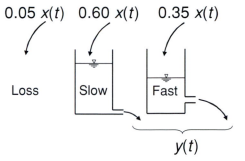

Fig. 12.17 A tank model representation of daily flows consisting of
baseflow (BF) and interflow (or subsurface stormflow SSF) by
Sugawara and Maruyama (1956). All three tanks are linear
storage elements with unit response functions given by
Equation (12.28) but different storage coefficients K.

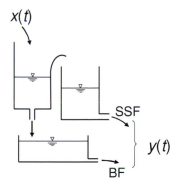

In the earliest description of this approach, now also known as the tank model, Sugawara and Maruyama (1956) and Sugawara (1961) gave a number of examples of combinations of linear storage elements, which had been used to describe basin outflows. For instance, Figure 12.16 shows the arrangement used to describe flood flows of the Kitakami River in northern Honshu; the basin was represented by two elements in parallel, one with $K = 33$ h that receives 60% of the input, and one with $K = 2.9$ h that receives 35% of the input; 5% of the total input and 20 mm of the initial input into the fast response tank were assumed to be "lost." Unlike in flood flows, in the description of daily flows, interflow and baseflow are more important; to simulate these, a different arrangement was used, which is illustrated in Figure 12.17. Initially after a drought period, precipitation flows out of the first tank into groundwater storage, from which the water flows out as baseflow. Only after the first tank has become full, does the overflow into the second tank result in subsurface stormflow runoff. Several arrangements were also proposed by Sugawara and Maruyama (1956) to accommodate spatial variation of the input characteristics of the basin. One of these is considered in the following example.

Example 12.7. Tank model allowing for spatial variability

Figure 12.18 illustrates an arrangement by which each storage element represents a subarea of the basin. Thus each tank receives as input the output from the upstream tank, in addition to the rainfall on the subarea it represents. Let α_1, α_2 and α_3 be the fractions of the total area represented by each tank. Then, for an input into the first tank given

Fig. 12.18 Example of a tank arrangement used by Sugawara and Maruyama (1956) to accommodate spatial variability of the input over the catchment area, to derive the unit response $u = u(t)$.

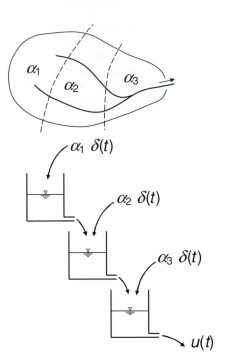

by $\alpha_1 \delta(t)$, one finds that the unit response of the first tank, $u_1(t)$, is Equation (12.28) multiplied by α_1. Similarly, the unit response from the second tank can be calculated by routing the output from the first tank, i.e. $u_1(t)$, plus the instantaneous rainfall on the second subarea, i.e. $\alpha_2 \delta(t)$, through the tank representing this second subarea,

$$u_2(t) = \int_0^t \left[\alpha_1 \frac{\exp(-\tau/K)}{K} + \alpha_2 \delta(\tau)\right] \frac{\exp[-(t-\tau)/K]}{K} d\tau$$

or   (12.36)

$$u_2(t) = \frac{\exp(-t/K)}{K} \left(\alpha_1 \frac{t}{K} + \alpha_2\right)$$

In the same way, one can show that the outflow from the third tank, resulting from an instantaneous input over the entire area, which is the unit response of the catchment, is given by

$$u(t) = u_3(t) = \frac{\exp(-t/K)}{K} \left[\frac{\alpha_1}{2}\left(\frac{t}{K}\right)^2 + \alpha_2 \frac{t}{K} + \alpha_3\right]$$ (12.37)

In a similar approach, Nash (1957) assumed that the transformation of catchment input into streamflow output is equivalent with a succession of routings through a series of n linear storage elements; thus, the input enters the first tank and is then successively routed through the second, the third, and so on (see Figure 12.19). The unit response of the Nash cascade, as it is sometimes called, can be derived as follows. The input of an instantaneous rainfall of unit volume produces an output given by Equation (12.28).

Fig. 12.19 The tank cascade, proposed by Nash (1957), consisting of n equal storage elements placed in series, as a metaphor for the response $u = u(t)$ of a catchment to an instantaneous input $x = \delta(t)$.

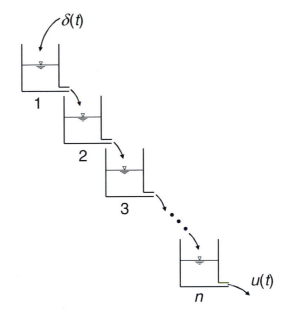

When this is taken as input into the second storage element, the output from that second storage is

$$u_2(t) = \int_0^t \frac{\exp(-\tau/K)}{K} \frac{\exp[-(t-\tau)/K]}{K} d\tau = \frac{t\exp(-t/K)}{K^2} \qquad (12.38)$$

This, in turn, is input into the third storage element and produces an output

$$u_3(t) = \int_0^t \frac{\tau\exp(-\tau/K)}{K^2} \frac{\exp[-(t-\tau)/K]}{K} d\tau = \frac{t^2\exp(-t/K)}{2K^3} \qquad (12.39)$$

The same process can be continued to obtain the outflow from the last storage element,

$$u_n(t) = \frac{(t/K)^{n-1}\exp(-t/K)}{(n-1)!K} \qquad (12.40)$$

which is the response function of the entire system. In order to allow the use of fractional values of n, the factorial can be replaced by the complete gamma function. Finally, the unit response of the entire catchment can be written as

$$u(t) = \frac{(t/K)^{n-1}\exp(-t/K)}{K\Gamma(n)} \qquad (12.41)$$

Equation (12.41) is known as the integrand of the incomplete gamma function or as the gamma density function. It has only two parameters, but it is quite flexible as it can accommodate a wide variety of hydrograph shapes; as illustrated in Figure 12.20, K can be considered a scale parameter and n a shape parameter. Equation (12.41) has been applied widely in watershed hydrology to parameterize unit hydrographs in terms of drainage basin characteristics.

Fig. 12.20 The unit response
function, obtained
with the gamma
density function
(12.40) or (12.41) as
$u = u(t/K)$, for
different values of the
parameter n.

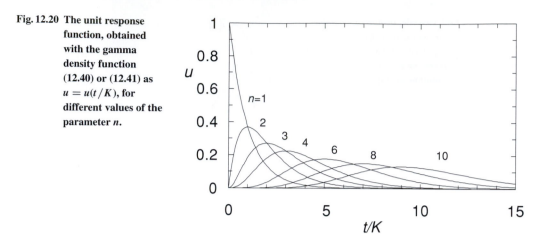

For example, Nash (1960) used Equation (12.41) to determine the instantaneous unit hydrographs for a number of British catchments by means of the method of moments. It can readily be shown (cf. Equation (13.9)) that its first moment about the origin is

$$m'_{u1} = nK \tag{12.42}$$

and that its second moment about the mean, or center of gravity (cf. Equations (13.10) and (13.12)), is

$$m_{u2} = nK^2 \tag{12.43}$$

Note that with $n = 1$ and $X = 0$ these are the same as Equations (7.31) and (7.34) for the Muskingum formulation. Because the moments of $u(t)$ can be calculated from available rainfall and streamflow records by means of the theorem of moments, as given by Equations (A22) and (A28), Nash (1959) was able to relate the parameters K and n directly with relevant basin characteristics; in this case, these were found to be drainage area, mean slope and length of the main channel. A similar study was carried out by Wu (1963) with catchments in Indiana. Actually, prior to its conceptual derivation by Nash, the incomplete gamma function had already been used by Edson (1951) on different grounds, to describe finite duration unit hydrographs. It was subsequently also used for this purpose by Gray (1961).

Several features of the tank cascade may help to explain, perhaps, why the integrand of the incomplete gamma function has been used so widely in applied hydrology. First, consider the case where n is allowed to increase indefinitely. As indicated by Equation (12.42), the center of gravity of the flood wave will then occur at a finite value of the time t, only if the storage coefficient of each tank in the cascade, K, is made to become very small. But if K is made to become very small, the second moment (12.43) indicates that the duration of the flood wave will become very short. At the same time, if the area under the wave curve is to maintain a magnitude of one, the magnitude of the peak must become very large. This is illustrated in Figure 12.21, for the case $nK = 1$. Thus, in the extreme case of $n \to \infty$, $K \to 0$, but finite nK, the unit response function (12.41)

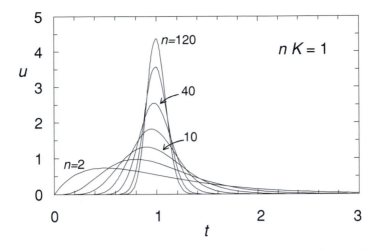

Fig. 12.21 **The unit response function (12.41) (i.e. the gamma density function) approaches a delta function as *n* becomes very large. The curves shown are for *n* = 2, 5, 10, 20, 40, 80 and 120. To allow easy comparison between the different curves as *n* increases, their centers of gravity, i.e. the first moments (12.42), are maintained at *t* = 1 by putting *K* = 1/n in Equation (12.41).**

assumes the characteristics of a delta function, as defined in Equations (A5)–(A7); it should be recalled that the delta function is also the unit response function of the linear kinematic channel, as formulated in Equation (5.124). This means that a tank cascade consisting of a finite number of storage elements is in fact intermediate between two extremes, namely "pure" storage action with $n = 1$, and "pure" translatory action with $n \to \infty$, and that Equation (12.41) provides some weighted average of both effects. Put differently, a finite set of storage elements is capable of accommodating not only storage effects but also the translatory effects of the hillslopes and of the channel network in the catchment.

A second feature is that, while Equation (12.41) is derived for a cascade consisting of storage elements with the same value of the storage coefficient K, this result is not very sensitive to this restriction. The following example illustrates this.

Example 12.8. Cascade with unequal storage elements

In the case of $n = 2$ tanks, each with a different value of the storage coefficient, namely K_1 and K_2, instead of Equation (12.38), the unit response function becomes

$$u_2(t) = \int_0^t \frac{\exp(-\tau/K_1)}{K_1} \frac{\exp[-(t - \tau)/K_2]}{K_2} d\tau$$

or, upon integration,

$$u_2(t) = \frac{\exp(-t/K_1) - \exp(-t/K_2)}{K_1 - K_2} \tag{12.44}$$

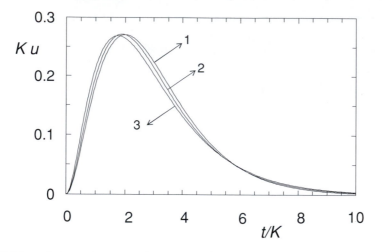

Fig. 12.22 Dimensionless unit response function $Ku(t/K)$ of a cascade of three storage elements (1) with equal storage coefficients K, as given by Equation (12.41) with $n = 3$ or Equation (12.39), (2) with unequal storage coefficients, namely Equation (12.45) with $K_1 = K_2 = 0.75K$ and $K_3 = 1.5K$, and (3) with unequal storage coefficients, namely Equation (12.45) with $K_1 = 0.4$, $K_2 = 1.0K$ and $K_3 = 1.6K$.

In the same way, for $n = 3$ tanks, with three different storage coefficients K_1, K_2 and K_3, instead of Equation (12.39), one obtains the following expression

$$u_3(t) = \frac{K_1[\exp(-t/K_1) - \exp(-t/K_3)]}{(K_1 - K_2)(K_1 - K_3)} - \frac{K_2[\exp(-t/K_2) - \exp(-t/K_3)]}{(K_1 - K_2)(K_2 - K_3)}$$

$$(12.45)$$

This process can be continued for any number of tanks. For $n = 3$ as an example, Figure 12.22 shows a comparison between Equations (12.39) and (12.45) with $K_1 = K_2 = 3K/4$, $K_3 = 3K/2$ and also with $K_1 = 0.4K$, $K_2 = K$, $K_3 = 1.6K$; the main point is that the difference between these response functions is not very large, because the total lag time, that is the first moment m'_{u1} has been kept the same in all three cases, such that $K_1 + K_2 + K_3 = 3K$ (cf. Equation (12.42)). The agreement would have been even better, if also the second moment had been kept the same in all three cases. This calculation illustrates why the passage of rain water through a succession of storage elements, such as interception and detention, soil moisture and groundwater, overland and channel flows, may still be described reasonably well by Equation (12.41), even though each one of these storage elements may have a different storage coefficient K.

Stochastic interpretations
Several of the above response functions, as combinations of storage tanks with or without linear channels, have been used with good results in the solution of a number of engineering problems. In some of these studies of specific catchment situations, ad hoc empirical relationships were derived for the parameters in these runoff representations.

However, because no direct link could be established between the parameters and the physical mechanisms in the catchment, like any other unit hydrograph they suffer from a lack of generality and they must always be calibrated to be of any use. For this reason the quest has continued for better formulations of the catchment scale processes involved in the transformation of precipitation into runoff.

One of the more active lines of endeavor has made use of stochastic concepts to describe the instantaneous unit hydrograph as the distribution of the arrival times of water at the catchment outlet. These approaches have typically consisted of linear routing of precipitation through topologically random channel networks with various probability distributions for the channel segments and with different assumptions regarding the holding time or travel time distributions of the water in the channel segments. As different concepts have matured, the width function has gradually emerged as the tool of choice to describe the structure of the channel network (Snell and Sivapalan, 1994; Marani et al., 1994; Veneziano et al., 2000), and has replaced earlier methods based on Horton–Strahler stream ordering (see Figure 11.1) and the resulting order ratios. Similarly, different attempts have been made to improve the formulation of the holding times from exponential distributions (Rodriguez-Iturbe and Valdes, 1979; Gupta et al., 1980) to more realistic response functions, such as obtained from the complete linear solution (5.72) (see Kirshen and Bras, 1983; Troutman and Karlinger, 1985), or from the diffusion approximation (5.95) (see Troutman and Karlinger, 1985; Rinaldo et al., 1991). The inclusion of hillslope outflows into the channel network has also been explored (see VanderTak and Bras, 1990; Robinson et al., 1995) with different hillslope response functions. Note that the assumption of an exponential distribution of residence times in a channel segment is equivalent with the assumption of a linear storage element as formulated by Equation (12.28). An overview of advances in this stochastic approach has been presented by Rodriguez-Iturbe and Rinaldo (1997).

With the growing complexity of such representations and the increasing number of the required parameters, these approaches are gradually evolving into direct simulation models; however, in the process the appeal of parsimony of the unit hydrograph is being lost, while its main limitations, namely linearity and time invariance are being kept. Also, although the description of the channel network is becoming increasingly realistic, the simulation of some critical processes at the catchment scale, involving the inclusion of hillslope mechanisms, with such thorny aspects as preferential flow and simultaneous transport of new and old water (see Chapter 11), has not received much attention so far; its inclusion in linear theory remains an elusive goal and will require more research.

12.3   STATIONARY NONLINEAR LUMPED RESPONSE

It is generally recognized that the transformation of precipitation and other inputs into streamflow can be quite nonlinear and non-stationary, so that the unit hydrograph is not always the proper method of approaching this problem. This is especially true in the case of extreme deviations, such as catastrophic floods, when rainfall–runoff systems tend to exhibit considerable nonlinearities, as manifested by the fact that runoff fails to be simply proportional to precipitation intensity. Over the past few decades various attempts have been made to incorporate nonlinearities in response formulations at the catchment scale. These can be subdivided into two broad categories, which are briefly considered in this section.

12.3.1 Functional analysis with nonlinear convolution

As outlined in the Appendix, a logical way to generalize the convolution operation to nonlinear systems is to make use of a Volterra integral series. In the case of a stationary, non-anticipatory system with no zero input response and with a finite memory m, this is Equation (A31), or

$$
y(t) = \int_0^m u_1(\tau)x(t-\tau)d\tau + \int_0^m \int_0^m u_2(\tau_1, \tau_2)x(t-\tau_1)x(t-\tau_2)d\tau_1 d\tau_2
$$

$$
+ \int_0^m \int_0^m \int_0^m u_3(\tau_1, \tau_2, \tau_3)x(t-\tau_1)x(t-\tau_2)x(t-\tau_3)d\tau_1 d\tau_2 d\tau_3 + \cdots
$$

$$\tag{12.46}$$

As before, the discrete analog of Equation (12.46) can be formulated by assuming that both rainfall and streamflow consist of piecewise constant values x_i and y_i, respectively, within the ith interval of time, where $(i-1)\Delta t \leq t \leq i\Delta t$. For the purpose of numerical analysis, Equation (12.46) can therefore be rewritten

$$
y_i = \sum_{j=1}^{m/\Delta t} u_{1,j} x_{i-j+1} + \sum_{j=1}^{m/\Delta t} \sum_{k=1}^{m/\Delta t} u_{2,jk} x_{i-j+1} x_{i-k+1}
$$

$$
+ \sum_{j=1}^{m/\Delta t} \sum_{k=1}^{m/\Delta t} \sum_{l=1}^{m/\Delta t} u_{3,jkl} x_{i-j+1} x_{i-k+1} x_{i-l+1} + \cdots
$$

$$\tag{12.47}$$

Different methods have been used in the past to apply the Volterra series formulation in the rainfall–runoff context. The main difficulty has invariably consisted of the identification of the response functions, $u_1, u_2, u_3, \ldots$, etc., in the case of Equation (12.46) or $u_{1,i}, u_{2,ij}, u_{3,ijk}, \ldots$, etc., in the case of Equation (12.47). Among the more practical examples of its application have been the studies by Amorocho and Brandstetter (1971), Bidwell (1971), Hino et al. (1971), Diskin and Boneh (1973), Liu and Brutsaert (1978) and Hino and Nadaoka (1979). Figure 12.23 shows an example of the results that were obtained with a two-term approximation of Equation (12.47) in the study by Diskin and Boneh (1973). In most of these studies one of the conclusions was that the nonlinear formulation is better able to simulate rainfall–runoff behavior of catchment areas than linear methods. This should not be surprising, as a representation with more adjustable parameters normally tends to produce a better fit. However, the numerical complexities of the computations are increased considerably.

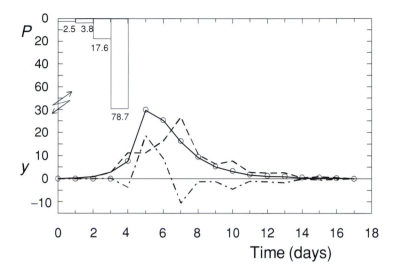

Fig. 12.23 Observed (circles) and simulated (solid line) flood hydrographs (in mm d^{-1}) on the Cache River at Forman, Illinois, resulting from a 103 mm storm over four days in March 1943. The simulation was carried out with the first two terms of the Volterra series (12.46); thus the total hydrograph is the sum of the first-order (linear) response (dashed line) and of the second-order response (dash-dotted line). (After Diskin and Boneh, 1973.)

12.3.2 Nonlinear runoff routing

This method of transforming rainfall into basin streamflow has mostly made use of concentrated storage elements, and is therefore also referred to as storage routing. In most cases the nonlinear storage function was assumed to be of the power type,

$$S = K_n y^m \tag{12.48}$$

where K_n and $m(\neq 1)$ are adjustable parameters; in this form the storage function can be considered a generalization of Equation (12.25). After substitution of (12.48) in the lumped equation of continuity (12.26) (or (1.10) or (7.11)) one obtains

$$x = y + K_n \frac{d(y^m)}{dt} \tag{12.49}$$

In what follows a few examples are presented of past attempts to include this type of nonlinearity in the catchment response behavior.

Horton (1941) was the first to use this approach; he proposed that flood hydrographs can be considered the result of a triangular "virtual channel inflow graph" produced by rainfall on the adjoining land, which is then routed through nonlinear channel storage by means of Equation (12.49). Horton (1936; 1937) estimated the parameters of the storage function (12.48) from quasi-steady open channel flow considerations. From analysis of a large number of flood events on different rivers, he showed that during a recession the channel storage behaves nearly the same as if the entire volume were concentrated in a single reservoir; but during rising stages it behaves as a reservoir of somewhat larger capacity. He reasoned by means of the GM equation (5.41) that m should lie around (3/5) for channels with a rectangular section, and around (3/4) for channels with a triangular section; although there was some scatter, an analysis of river recession flow data (cf. Equation (12.54)) confirmed that m was mostly between 0.6 and 0.8. Interestingly, although it was nonlinear, Horton's approach provided the impetus for the linear lag-and-route procedures of Clark (1945) and O'Kelly (1955), described in Section 12.2.2.

A second type of nonlinear runoff model, which has been used in a number of studies, consists of arrays of nonlinear storage elements, like Equation (12.48), in series and in parallel representing different components of the basin; the storage arrays are usually structured in the same arrangement as the actual stream channel network. Such arrays can be considered nonlinear analogs of the linear ones, examples of which are shown in Figures 12.16–12.19. One such routing procedure was described by Rockwood (1958), who used it to forecast streamflow in the entire Columbia River Basin on the basis of preceding streamflows and forecasts of basin inputs from snowmelt and rainfall. This large basin was assumed to consist of a number of subbasins, lakes and stream channels. Each subbasin was assumed to consist of two nonlinear storage elements in series representing surface runoff, which are placed in parallel with two storage elements in series representing subsurface runoff. Channel segments, mostly between 30 and 80 km long, were represented by three nonlinear storage elements in series. The routing procedure consisted essentially of the numerical solution of Equation (12.49) for each storage element, in which it was assumed that $m = 0.8$ and K_n was derived by trial routings.

Different arrays of nonlinear storage elements, each representing a subarea and each receiving the excess rainfall input on that subarea plus the outflow from the upstream storage elements, were devised by Laurenson (1964) and subsequently by Mein et al. (1974) to simulate storm flows from catchments in Australia. In this approach the catchment area is first subdivided in a number of approximately equal subareas along the major tributaries, and a nonlinear tank is then located at the center of gravity of each of the subareas and assigned a relative lag time of that location. At first this relative lag – or storage delay – time was assumed to be proportional to $\sum(L/S_0^{1/2})$, where L and S_0 are the length and slope of the reach through the subarea, and the summation was carried out from the location of the subarea to the outlet; however, it was subsequently found that putting it proportional to the distance from the outlet, i.e. $\sum L$, yielded the same results; this shows that the effects of slope, flow depth and surface roughness become irrelevant in this type of idealization. The parameter m was estimated by observing from comparison of (12.25) with (12.48) that

$$K = K_n y^{m-1} \tag{12.50}$$

Then in accordance with Equation (7.19), logarithmic regression was carried out for a number of storms between the time from the centroid of the excess rain to that of the storm runoff and the average storm runoff during the event, $\langle y \rangle$; the slopes of the regression lines were assumed to represent $(m - 1)$ of (12.50), and produced (see also Askew, 1970) a range of roughly $0.60 \leq m \leq 0.81$. These values are very similar to those of Horton (1941) and Rockwood (1958); Mein *et al.* (1974) concluded that $m = 0.71$ could be used as a typical value. As indicated in Equation (12.50), K_n is proportional to the time of travel; accordingly, this parameter was estimated by putting

$$K_n = C_1 K_1 \tag{12.51}$$

in which K_1 was taken to be the value of the relative travel time through the subarea represented by that particular storage tank, i.e. $(L/S_0^{1/2})$ or L mentioned earlier. With m and K_1 known, C_1 remained as the only unknown parameter of the model; this was estimated by trial-and-error routings with available data. A more optimal estimation of the parameters in this model was subsequently developed by Kuczera (1990) and Kuczera and Williams (1992). A similar nonlinear storage routing procedure was developed by Boyd *et al.* (1979). However, following Askew's (1970) findings, the storage coefficient (or lag) was made to depend also on the area A represented by the storage element, namely as $K = a A^b y^{m-1}$; in their case the constants were taken as $b = 0.57$ and the slightly different value $m = 0.77$.

Physical justification of nonlinear tanks

In the past, the nonlinear storage relationship (12.48) has been justified on physical grounds, mostly by considering open channel storage. The argument usually follows that originally developed by Horton (1936; 1937), based on a lumped kinematic analysis for quasi-steady, quasi-uniform flow. Thus the volume of water stored in a channel reach of length L is assumed to be given by $S_c = A_c L$, in which A_c is the average cross-sectional area in the reach. The channel is assumed to be wide enough, so that the hydraulic radius equals the mean water depth, or $R_h = h$, and the cross-sectional area of the channel equals the depth times the width, or $A_c = h B_c$. For steady uniform conditions, (5.39) (or (5.43)) produces then the outflow rate from the reach as $Q = C_r B_c S_0^b h^{a+1}$; the channel storage is in terms of the outflow from the reach

$$S_c = \left(\frac{B_c^a L^{a+1}}{C_r S_0^b} \right)^{1/(a+1)} Q^{1/(a+1)} \tag{12.52}$$

in which a and b are the parameters in the open channel equation (5.39). Hence, if it is assumed that all the storm runoff water in the catchment is stored in the stream channels, so that $S = S_c/A$, one has

$$S = \left(\frac{B_c^a L^{a+1}}{C_r S_0^b A^a} \right)^{1/(a+1)} y^{1/(a+1)} \tag{12.53}$$

in which, as before, $y = Q/A$, L is now the length of all stream channels in the catchment upstream from the point where Q is determined, and in which the other variables are assumed to be averages over the catchment area A. This result is in the form of (12.48), with $m = (a + 1)^{-1} = 0.60$ for the GM equation, and $m = 0.67$ for the Chézy equation.

But this derivation of Equation (12.53) is not wholly convincing, first, because obviously not all the storm runoff water is stored in channels, and second, because it is well known that a major part of most storm flows is generated by subsurface runoff, as explained in Chapter 11. Some estimate of groundwater storage can be obtained by considering unconfined flow under the hydraulic assumption. Under drainage conditions (i.e. without inflow), the outflow rate from a Dupuit–Boussinesq aquifer is given by Equations (10.85) with (10.86). These can be converted to outflow from a catchment area A with a stream channel length L by putting $Q = 2Lq$ and $B = A/2L$. With $y = Q/A$ as the outflow rate per unit area and $x = 0$, the lumped equation of continuity (12.26) (or (1.10) or (7.11)) produces the storage expressed as a layer of water of average thickness as follows

$$S = \int_t^\infty y(t)dt \tag{12.54}$$

Upon integration of (12.54) with (10.85) and (10.86), one obtains readily

$$S = \frac{0.416 n_e A}{L k_0^{1/2}} y^{1/2} \tag{12.55}$$

which is in the form of (12.48) with $m = 1/2$. In this result n_e is the effective drainable porosity, k_0 the effective hydraulic conductivity, A the drainage area of the catchment, and L the length of all the stream channels into which aquifer drainage takes place. If the system can be linearized, the solution is given by (10.113) and after longer times by (10.116). Integration of Equation (12.54) with the latter produces in the same way

$$S = \frac{n_e A^2}{\pi^2 k_0 L^2 p D} y \tag{12.56}$$

Again, this is in the form of (12.48), but now $m = 1$, in accordance with (12.25), as expected for a linear system.

In summary, most of the m values from field data reviewed here not only conform with the values expected for open channels, but they appear to be intermediate between the values for nonlinear and linear groundwater aquifers as well.

12.4   NON-STATIONARY LINEAR RESPONSE

In the definition of the unit hydrograph the two stipulated assumptions are linearity and time invariance. Until now, these two assumptions have mainly been studied separately, and their combined effect has not yet been fully explored. The incorporation of nonlinear effects into stationary systems, which is treated in Section 12.3, seems to have received more attention in the literature and relatively few studies have been devoted to nonstationary effects on linear catchment response. Yet, several experimental investigations reviewed in Chapter 11 have indicated that, for instance, the ratio of old and new water in the catchment outflow is affected not only by the intensity of the rain, but also by such factors as seasonal moisture status and the time since the start of the rain. Hence, as the catchment contains more or less water, different flow paths and mechanisms come into play in the production of the runoff, and this results in a non-stationary response.

In general, one can distinguish two ways of describing non-stationarity. One type of formulation makes use of a coarse time variable describing changes in catchment response at monthly and annual time scales; these changes could conceivably be cyclical, that is seasonal, or in the nature of a trend in the case of changes in land use or climate.

The second type of approach uses a finer time variable to describe response changes during the event itself, as a result of physical changes inside the catchment resulting from continued rain, snowmelt or flooding. These two types can be combined formally in a convolution integral, as an extension of Equation (12.2), namely

$$y(t) = \int_0^t x(\tau)\, u(\chi, \tau, t - \tau) d\tau \tag{12.57}$$

in which, as before, $y(t)$ is the output resulting from an input $x(t)$. In contrast to the stationary case of Equation (12.2), here the unit response $u(\chi, \tau, t - \tau)$ is a function of three time variables; the first variable χ is the coarse time scale in terms of months, seasons or years. The second, τ, is the dummy variable of integration, such that $0 \le \tau \le t$; however, as argument in the unit response, it denotes the time of the input $x(\tau)$, and thus specifies the response for the input at that time. The third, t, is the time for which the output is to be determined and $(t - \tau)$ is the time elapsed since the input $x(\tau)$. The convolution operation of Equation (12.57) describes superposition, which is the essence of linearity. However, as pointed out by Diskin and Boneh (1974), in contrast to the stationary case, here the convolution (or superposition) operation is generally not commutable. This means, for example, that two non-stationary systems connected in series, say A followed by B, will produce a different output when their order is reversed, B followed by A.

For the first type of non-stationarity the unit response in Equation (12.57) depends only on χ and $(t - \tau)$; thus it does not depend on the time τ, but only, say, on the season or the year. This means that during the input event the response can be considered stationary, and the concepts discussed in Sections 12.1 and 12.2 are applicable. Therefore, the second type is the one usually considered, when time variance effects are to be included. In this case, the unit response depends only on τ and $(t - \tau)$, and (12.57) can be simplified to

$$y(t) = \int_0^t x(\tau) u(\tau, t - \tau) d\tau \tag{12.58}$$

This is illustrated in Figure 12.24. In discrete form (cf. Equation (12.9)) this can be written as

$$y_i = \sum_{k=1}^{i} x_k u_{k, i-k+1} \tag{12.59}$$

or alternatively (cf. Equation (12.10)), as

$$y_i = \sum_{k=1}^{i} x_{i-k+1} u_{i-k+1, k} \tag{12.60}$$

where the first subscript of the response function refers to the time of the input, and the second indicates its role in the numerical convolution.

In several applications of the nonstationary linear approach in the past, the form of the unit response function has been assumed a priori. For example, Snyder *et al.* (1970) derived the catchment response by routing, what was essentially a time–area diagram through a linear storage element with time-dependent storage coefficient $K = K(\tau)$

Fig. 12.24 Illustration of the convolution operation $y(t) = \int_0^t x(\tau)u(\tau, t - \tau)d\tau$, with a nonstationary unit response $u = u(\tau, t)$. (The values of y and u are not drawn to scale.) The variable t denotes the time for which the runoff y is calculated and τ is the time since the start of the precipitation input x. The unit response, which is shown at only three instants of time τ, continually changes its shape as the precipitation continues. Compare this with the stationary case of Figure 12.6.

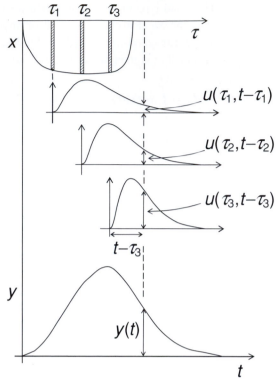

(see Equation (12.25)). In a similar vein, Mandeville and O'Donnell (1973) considered different combinations of time-variant linear channel and time-variant linear storage elements, one among them being a cascade of equal storage elements. Diskin and Boneh (1974) developed numerical least squares procedures to derive more general, i.e. not with a preconceived mathematical form, response functions $u_{i-k+1,k}$, as shown in Equation (12.60), from available rainfall–runoff data. Chiu and Bittler's (1969) study is probably the only one that considered both fine-scale and coarse-scale non-stationarity as formulated in Equation (12.57). The unit response function was obtained by routing the input through a single linear storage element, as given by (12.25) with a time-dependent storage coefficient $K = K(\tau)$ in the form of a power function

$$K = a\tau^{-b} \tag{12.61}$$

in which a and b were assumed to be functions of the coarse time variable χ. Rainfall–runoff data obtained in Pennsylvania indicated that a was a function of b and that b could be described well by a sine function of χ. Equation (12.61) indicates that in this study the storage coefficient K was observed to decrease as the precipitation continued; because K is the time of travel through the system (cf. Equations (7.19), (7.31) and (12.42)), this means that the unit response tended to become faster with storm duration. The approach was later extended by Chiu and Huang (1970) to include nonlinear effects by replacing (12.25) by its nonlinear analog (12.48).

REFERENCES

Amorocho, J. and Brandstetter, A. (1971). Determination of nonlinear functional response functions in rainfall–runoff processes. *Water Resour. Res.*, **7**, 1087–1101.

ASCE and WPCF (Joint Committee) (1982). Design and construction of sanitary and storm sewers, *ASCE (American Society of Civil Engineers) Manual on Engineering Practice* No. 37 and *WPCF (Water Pollution Control Federation) Manual of Practice* No. 9, 332 pp., 5th printing.

Askew, A. J. (1970). Derivation of formulae for variable lag time. *J. Hydrol.*, **10**, 225–242.

Bidwell, V. J. (1971). Regression analysis of nonlinear catchment systems. *Water Resour. Res.*, **7**, 1118–1126.

Box, G. E. P., Jenkins, G. M. and Reinsel, G. C. (1994). *Time Series Analysis, Forecasting and Control*, third edition. Englewood Cliffs, NJ: Prentice Hall.

Boyd, M. J., Pilgrim, D. H. and Cordery, I. (1979). A storage-routing model based on catchment geomorphology. *J. Hydrol.*, **42**, 209–230.

Chiu, C.-L. and Bittler, R. P. (1969). Linear time-varying model of rainfall-runoff relation. *Water Resour. Res.*, **5**, 426–437.

Chiu, C.-L. and Huang, J. T. (1970). Nonlinear time varying model of rainfall-runoff relation. *Water Resour. Res.*, **6**, 1277–1286.

Clark, C. O. (1945). Storage and the unit hydrograph. *Trans. ASCE*, **110**, 1419–1488.

Corps of Engineers (1963). *Unit hydrographs, Part I Principles and Determinations*, Civil Works Investigations – Project 152. Baltimore, MD: US Army Engineer District.

Deininger, R. A. (1969). Linear programming for hydrologic analyses. *Water Resour. Res.*, **5**, 1105–1109.

Diskin, M. H. and Boneh, A. (1973). Determination of optimal kernels for second-order stationary surface runoff systems. *Water Resour. Res.*, **9**, 311–325.

Diskin, M. H. and Boneh, A. (1974). The kernel function of linear nonstationary surface runoff systems. *Water Resour. Res.*, **10**, 753–761.

Dooge, J. C. I. (1957). The rational method for estimating flood peaks. *Engineering (London)*, **184**, 311–313, 374–377.

(1973). *Linear theory of hydrologic systems*, Tech. Bull. 1468, Agr. Res. Serv., US Department of Agriculture.

Edson, C. G. (1951). Parameters for relating unit hydrographs to watershed characteristics. *Trans. Amer. Geophys. Un.*, **32**, 591–596.

Gray, D. M. (1961). Synthetic unit hydrographs for small watersheds. *J. Hydraul. Div., Proc. ASCE*, **87**, 33–54.

Gupta, V. K. and Waymire, E. (1983). On the formulation of an analytical approach to hydrologic response and similarity at the basin scale. *J. Hydrol.*, **65**, 95–123.

Gupta, V. K., Waymire, E. and Wang, C. T. (1980). A representation of an instantaneous unit hydrograph from geomorphology. *Water Resour. Res.*, **16**, 855–862.

Hawken, W. H. and Ross, C. N. (1921). The calculation of flood discharges by the use of a time contour plan. *Trans. Inst. Engrs. Aust.*, **2**, 85–92.

Hino, M. and Nadaoka, K. (1979). Mathematical derivation of linear and nonlinear runoff kernels. *Water Resour. Res.*, **15**, 918–928.

Hino, M., Sukigara, T. and Kikkawa, H. (1971). Nonlinear runoff kernels of hydrologic systems. In *Systems Approach to Hydrology, Proc. First Bilateral U.S.-Japan Seminar in Hydrology*. Fort Collins, CO: Water Resource Publications, pp. 102–112.

Horton, R. E. (1936). Natural stream channel-storage. *Trans. Amer. Geophys. Un.*, **17**, 406–415.

(1937). Natural stream channel-storage (second paper). *Trans. Amer. Geophys. Un.*, **18**, 440–456.

(1941). Virtual channel-inflow graphs. *Trans. Amer. Geophys. Un.*, **22**, 811–820.

Kirkby, M. J. (1976). Tests of the random network model, and its application to basin hydrology. *Earth Surface Process.*, **1**, 197–212.

Kirpich, Z. P. (1940). Time of concentration of small agricultural watersheds. *Civil Eng. (N.Y.)*, **10**, 362.

Kirshen, D. M. and Bras, R. L. (1983). The linear channel and its effect on the geomorphologic IUH. *J. Hydrol.*, **65**, 175–208.

Kuczera, G. (1990). Estimation of runoff-routing model parameters using incompatible storm data. *J. Hydrol.*, **114**, 47–60.

Kuczera, G. and Williams, B. J. (1992). Effect of rainfall errors on accuracy of design flood estimates. *Water Resour. Res.*, **28**, 1145–1154.

Laurenson, E. M. (1964). A catchment storage model for runoff routing. *J. Hydrol.*, **2**, 141–163.

Liu, C.-K. and Brutsaert, W. (1978). A nonlinear analysis of the relationship between rainfall and runoff for extreme floods. *Water Resour. Res.*, **14**, 75–83.

Mandeville, A. N. and O'Donnell, T. (1973). Introduction of time variance to linear conceptual catchment models. *Water Resour. Res.*, **9**, 298–310.

Marani, M., Rinaldo, A., Rigon, R. and Rodriguez-Iturbe, I. (1994). Geomorphological width functions and the random cascade. *Geophys. Res. Lett.*, **21**, 2123–2126.

McCuen, R. H. and Spiess, J. M. (1995). Assessment of kinematic wave time of concentration. *J. Hydraul. Eng.*, **121**, 256–266.

Mein, R. G., Laurenson, E. M. and McMahon, T. A. (1974). Simple nonlinear model for flood estimation. *J. Hydraul. Div., Proc. ASCE*, **100**, 1507–1518.

Minshall, N. E. (1960). Predicting storm runoff on small experimental watersheds. *J. Hydraul. Div., Proc. ASCE*, **86**, 17–38.

Mulvany, T. J. (1850). On the use of self registering rain and flood gauges. *Inst. Civ. Eng. Proc. (Dublin)*, **4**(2), 1–8.

Nash, J. E. (1957). The form of the instantaneous unit hydrograph. *Comptes Rendus et Rapports, IASH General Assembly Toronto 1957, Int. Assoc. Sci. Hydrol. (Gentbrugge)*, Publ. No. 45, **3**, 114–121.

(1958). Determining run-off from rainfall. *Proc. Instn. Civ. Engrs., London*, **10**, 163–184.

(1959). Systematic determination of unit hydrograph parameters. *J. Geophys. Res.*, **64**, 111–115.

(1960). A unit hydrograph study, with particular reference to British catchments. *Proc. Instn Civ. Engrs., London*, **17**, 249–282.

Neuman, S. P. and de Marsily, G. (1976). Identification of linear systems response by parametric programming. *Water Resour. Res.*, **12**, 253–262.

O'Kelly, J. J. (1955). The employment of unit-hydrographs to determine the flows of Irish arterial drainage channels. *Proc. Instn. Civ. Engrs., London*, **4**, (III), 365–412, 428–436, 444–445.

Pilgrim, D. H. (1977). Isochrones of travel time and distribution of flood storage from a tracer study on a small watershed. *Water Resour. Res.*, **13**, 587–595.

Ramser, C. E. (1927). Run-off from small agricultural areas. *J. Agric. Res.*, **34**, 797–823.

Rinaldo, A., Marani, A. and Rigon, R. (1991). Geomorphological dispersion. *Water Resour. Res.*, **27**, 513–525.

Robinson, J. S., Sivapalan, M. and Snell, J. D. (1995). On the relative roles of hillslope processes, channel routing, and network geomorphology in the hydrologic response of natural catchments. *Water Resour. Res.*, **31**, 3089–3101.

Rockwood, D. M. (1958). Columbia basin streamflow routing by computer. *J. Waterw. Harbor Div., Proc. ASCE*, **84**, 1874.1–1874.15.

Rodriguez-Iturbe, I. and Rinaldo, A. (1997). *Fractal River Basins*. Cambridge: Cambridge University Press.

Rodriguez-Iturbe, I. and Valdes, J. B. (1979). The geomorphologic structure of hydrologic processes. *Water Resour. Res.*, **15**, 1409–1420.

Rodriguez-Iturbe, I., Rinaldo, A., Rigon, R., Bras, R. L., Marani, A. and Ijjasz-Vasquez, E. (1992). Energy dissipation, runoff production, and the three-dimensional structure of river basins. *Water Resour. Res.*, **28**, 1095–1103.

Sherman, L. K. (1932a). Streamflow from rainfall by the unit-graph method. *Eng. News-Record*, **108**, 501–505.

(1932b). The relation of hydrographs of runoff to size and character of drainage-basins. *Trans. Amer. Geophys. Un.*, **13**, 332–339.

Shreve, R. L. (1974). Variation of mainstream length with basin area in river networks. *Water Resour. Res.*, **10**, 1167–177.

Singh, K. P. (1976). Unit hydrographs – a comparative study. *Water Resour. Bull.*, **12**, 381–391.

Snell, J. D. and Sivapalan, M. (1994). On geomorphological dispersion in natural catchments and the geomorphological unit hydrograph. *Water Resour. Res.*, **30**, 2311–2323.

Snyder, W. M. (1955). Hydrograph analysis by the method of least squares. *J. Hydraul. Div., Proc. ASCE*, **81**, 1–25.

Snyder, W. M., Mills, W. C. and Stephens, J. C. (1970). A method of derivation of nonconstant watershed response functions. *Water Resour. Res.*, **6**, 261–274.

Sugawara, M. (1961). On the analysis of runoff structures about several Japanese rivers. *Jap. J. Geophys.*, **2**, 1–76.

Sugawara, M. and Maruyama, F. (1956). A method of prevision of the river discharge by means of a rainfall model. *Symposia Darcy (Dijon, 1956), Int. Assoc. Sci. Hydrol. (Gentbrugge)*, Publ. No. 42, **3**, 71–76.

Surkan, A. J. (1969). Synthetic hydrographs: Effect of network geometry. *Water Resour. Res.*, **5**, 112–128.

Troutman, B. M. and Karlinger, M. R. (1985). Unit hydrograph approximations assuming linear flow through topologically random channel networks. *Water Resour. Res.*, **21**, 743–754.

VanderTak, L. D. and Bras, R. L. (1990). Incorporating hillslope effects into geomorphologic instantaneous unit hydrograph. *Water Resour. Res.*, **26**, 2393–2400.

Veneziano, D., Moglen, G. E., Furcolo, P. and Iacobellis, V. (2000). Stochastic model of the width function. *Water Resour. Res.*, **36**, 1143–1157.

Wolman, M. G. (1955). *The natural channel of Brandywine Creek, Pennsylvania, US*. Geol. Survey Prof. Paper 271. Washington, DC: US Dept. of the Interior.

Wu, I.-P. (1963). Design hydrographs for small watersheds in Indiana. *J. Hydraul. Div., Proc. ASCE*, **89**, 35–66.

PROBLEMS

12.1   (a) Derive the 1 h unit hydrograph from the 2 h unit hydrograph given in Table 12.1. (b) Calculate the runoff in $cm\ h^{-1}$ resulting from the following pattern of excess rainfall: $x = 15\ mm\ h^{-1}$ for $0 < t < 1$ h; $x = 25\ mm\ h^{-1}$ for $1 < t < 2$ h; and $x = 39\ mm\ h^{-1}$ for $2 < t < 3$ h.

12.2   The table lists a storm runoff hydrograph resulting from a 4 h storm of presumably uniform (in space and time) but unknown intensity on a basin of 29.5 km^2. (a) Construct the S hydrograph, and determine the intensity of the storm rainfall (in $cm\ h^{-1}$) from the (smoothed) equilibrium flow rate of the S hydrograph. (b) Determine the 2 h unit hydrograph from the S hydrograph; be careful to scale the intensity over 2 h to ensure a volume of 1 cm. (c) Calculate the peak storm runoff (in $m^3 s^{-1}$) resulting from three successive 2 h periods of rainfall producing volumes of 0.4, 1.0 and 1.6 cm of runoff, respectively.

Time (h)	Storm runoff ($m^3 s^{-1}$)	Time (h)	Storm runoff ($m^3 s^{-1}$)
0	0	8	19.70
1	4.01	9	15.76
2	15.26	10	11.62
3	36.55	11	8.30
4	45.40	12	5.30
5	40.48	13	3.33
6	31.99	14	1.56
7	24.40	15	0

12.3   Below is a 6 h unit hydrograph (UH) from a drainage area of 875 km^2 on Goose Creek, near Leesburg, Virginia.

Time (h)	UH ($m^3 s^{-1}$)	Time (h)	UH ($m^3 s^{-1}$)	Time (h)	UH ($m^3 s^{-1}$)
0	0.00	28	27.34	54	1.00
2	0.13	30	13.34	56	0.87
4	0.40	32	10.00	58	0.67
6	1.80	34	7.34	60	0.60
8	7.34	36	5.34	62	0.53
10	17.34	38	5.27	64	0.47
12	29.34	40	4.20	66	0.40
14	44.68	42	3.47	68	0.33
16	66.69	44	2.80	70	0.27
18	104.03	46	2.20	72	0.20
20	113.37	48	1.80	74	0.13
22	112.04	50	1.47	76	0.07
24	81.36	52	1.20	78	0
26	59.35				

(a) Check what the unit volume is of this unit hydrograph from the (smoothed) steady equilibrium flow rate of the S hydrograph. (b) Find the peak flow (in $m^3 s^{-1}$) resulting from three successive 4 h periods of rainfall producing volumes of 5.1, 30.5, 16.5 mm of runoff, respectively (ignore base flow). (The above UH is derived from data in Corps of Engineers, (1963).)

12.4   Consider the following excess rainfall time sequence on a hypothetical catchment: $x_1 = 0.5$ cm h^{-1} from 1400 to 1500; $x_2 = 1.5$ cm h^{-1} from 1500 to 1600; $x_3 = 0.75$ cm h^{-1} from 1600 to 1700. This rainfall produced the following storm runoff hydrograph.

Time	Runoff (cm h^{-1}) at that time
1330	0
1430	0.250
1530	0.917
1630	0.958
1730	0.500
1830	0.125
1930	0

Calculate the 1 h unit hydrograph: in other words, determine the ordinates of the unit graph u_1, u_2, etc. for hour 1, hour 2, etc., respectively. (The calculation can be done exactly; assume that the system is perfectly linear, and that the data contain no errors.)

12.5   Multiple choice. Indicate which of the following statements are correct. The unit hydrograph:
   (a)   is the hydrograph of unit volume storm runoff generated by a storm of unit duration, under the assumption of linearity (or proportionality);
   (b)   method is also commonly applied in drought flow analysis;
   (c)   method always overpredicts the peak flow of very large events;
   (d)   tends to overpredict the time of the peak of very large events in hilly terrain;
   (e)   method can only be derived for a unit duration for which previous data happen to be available; in other words, to construct a 1 h unit hydrograph, one needs runoff data that were produced by 1 h storms;
   (f)   is sometimes derived synthetically from geomorphological data used in conjunction with empirical relationships;
   (g)   is based on the assumption that the time of peak depends on the rainfall intensity (i.e. proportionality);
   (h)   is based on the assumption that the baseflow mechanism is such that it gives a straight line on semi-log paper [i.e., $q = \beta \exp(-\alpha t)$ where α and β are constants];
   (i)   is not suitable for watersheds that are smaller than 4000 km^2;
   (j)   can, in principle, also be derived from a frequency analysis of stream flow data;
   (k)   can be derived from the S hydrograph that characterizes the watershed;
   (l)   for 2 h can be obtained by halving the unit hydrograph for 1 h;
   (m)   eventually (for large values of t) becomes equal to zero after the cessation of rainfall;
   (n)   curve is usually maximal at $t = 0$ (i.e. when rainfall starts) and decreases smoothly after that.

12.6   Assume that the S hydrograph for a continuous excess rainfall of constant intensity on a given watershed can be expressed by the following: $S_u(t) = 9000 \, t / (2 + t)$, where S_u is in m^3 h^{-1} and t in h. (a) Derive the instantaneous unit hydrograph for this watershed. Reduce to 1 cm, i.e. convert the outflow rate units into cm h^{-1}. (b) Given the following rainfall pattern $x = 5$ cm h^{-1} for $0 < t < 1$ h, and $x = 7.5$ cm h^{-1} for $1 < t < 2$ h; calculate (by means of the instantaneous unit hydrograph and convolution) the resulting discharge after 3 h.

12.7   Assume that the instantaneous unit hydrograph for a given watershed is $u(t) = (t + 1)^{-2}$ (its units are h^{-1}, if t is in h). (a) Derive an equation for the S hydrograph (specify the units) resulting from a uniform steady excess rainfall of 1 cm h^{-1} lasting indefinitely. (b) Calculate the runoff rate y (specify the units) after $t = 5$ h resulting from a uniform excess rainfall rate of 2 cm h^{-1} that lasts for 3 h (i.e. $x = 2$ for $0 < t \leq 3$ h). (c) Calculate the runoff rate (specify the units) after $t = 5$ h resulting from a uniform excess rainfall rate of 2 cm h^{-1} lasting for 3 h, that is followed by a rainfall rate of 1.8 cm h^{-1} lasting for 7 h (i.e. $x = 2$ for $0 < t \leq 3$ h) and $x = 1.8$ for $3 < t \leq 10$ h). Note: use the convolution integral and integrate to obtain the answers.

12.8   Multiple choice. Indicate which of the following statements are correct. The unit hydrograph method:
    (a)   is used mainly to calculate the return period of relatively rare events;
    (b)   results from the detailed physical analysis of watershed runoff phenomena;
    (c)   tends to yield more correct answers for very rare events, than it does for common events;
    (d)   in its usual form produces the total runoff, that is storm runoff together with long-term groundwater runoff;
    (e)   makes use of typical storm runoff hydrographs produced by uniform excess rainfall and reduced to a unit volume under the assumption of superposition and invariance;
    (f)   derives its practical appeal from the underlying principle of linearity;
    (g)   could, under certain conditions, also be applied to calculate groundwater outflow from a catchment, as explained in Chapter 10.

12.9   Consider the rainfall event of Example 12.3, in which $x_i = 2.0, 4.0, 1.0 \text{ cm h}^{-1}$; assume that the measured runoff is $y_i = 0.63, 1.97, 2.35, 1.38, 0.6, 0.12$. Calculate the unit hydrograph ordinates u_i by the method of least squares, with Equations (12.17) and their analogs for $i = 3$ and 4, and compare with the "exact" values given in Example 12.3.

12.10   Multiple choice. Indicate which of the following statements are correct. For a given location and a given time of year, maximum rainfall:
    (a)   intensity of a certain duration decreases with increasing exceedance interval (i.e. return period, T_r);
    (b)   intensity for a given return period decreases with increasing duration of the rain;
    (c)   volume for a given return period increases with increasing duration of the rain;
    (d)   intensity needed in the rational method is determined on the basis of a knowledge of a design return period and a duration equal to the time of concentration;
    (e)   can be determined for an ungauged site by areal interpolation of available data.

12.11   Multiple choice. Indicate which of the following statements are correct. The rational method in its classical form, $Q_p = CIA$:
    (a)   is normally used to calculate runoff resulting from infiltrated rainfall;
    (b)   will work better to calculate surface runoff from catchments larger than 100 km^2;
    (c)   is normally used to obtain a design peak runoff but not a design hydrograph;
    (d)   is based on the assumption that the infiltration rate is a constant loss rate, independent of rainfall intensity;
    (e)   is based on the simplest type of linear runoff model, namely a model consisting of pure storage.

12.12   Multiple choice. Indicate which of the following statements are correct. The rational method gives the flow rate in terms of the rainfall intensity.

(a) The basic formula, $Q_p = CIA$, can be used to determine storm outflow from confined aquifers.

(b) But its main application is in the design of highway culverts, sewers and other smaller structures.

(c) It is commonly used to compute the spillway design flood on larger basins in conjunction with the 1000 y rainfall.

(d) The runoff coefficient, C, tends to be smaller for rural areas than for urban areas.

(e) This method is more appropriate for flood prediction from areas larger than 10 km^2 than for smaller areas, because it is based on the assumption that watershed characteristics can be averaged.

(f) It is based on the implicit assumption (among some others) that the input (rain) is merely translated by means of a time lag to produce the output (runoff).

(g) It can also be justified on the basis of the kinematic wave equation with a linear rating curve, so that the velocity, V, is independent of water depth.

(h) It has the advantage of not requiring any information at all on the surface characteristics of the watershed.

(i) It is based on the assumption that the infiltration and other losses are a fraction of the rainfall.

12.13 In a manner similar to Example 12.6, derive the mathematical expression for the unit response function, $u(t)$, of a system consisting of a width function in the shape of a right-angled triangle (representing translation effects) placed in sequence with a linear storage element representing storage effects. (This case is shown graphically in Figure 12.15.) Thus, (a) give the equation describing $A_r(t)$, and (b) use this in the convolution integral with the unit response (12.28); the parameters of the system are K and t_c.

12.14 Derive the unit response function (12.37) of the system illustrated in Figure 12.18 by showing (and solving) the convolution operation needed to obtain (12.37) from the second of (12.36).

12.15 What would be the unit response function, $u(t)$, if the system shown in Figure 12.18 had four (instead of three) subareas, such that $\alpha_1 + \alpha_2 + \alpha_3 + \alpha_4 = 1$?

12.16 Show that the first moment about the origin of the unit response function (12.41) is (12.42), i.e., $m'_{u1} = nK$. (The moments are defined in Chapter 13.)

12.17 Show that the second moment about the mean of the unit response function (12.41) is (12.43), i.e. $m_{u2} = nK^2$. Make use of the fact that the second moments about the mean and about the origin are related as shown in Equation (13.12).

12.18 Determine the value of the power m in Equation (12.48) for storage in a reach of an open channel with triangular cross section. Hint. Follow the same reasoning as that leading to Equation (12.53).

12.19 Prove Equation (12.55) by carrying out the integration outlined in the text.

12.20 Prove Equation (12.56) by carrying out the integration outlined in the text.

12.21 In the analysis of the outflow record from the downstream end of a lake during a drought period (when the inflow Q_i is zero), it is found (surprisingly) that t versus $Q_e^{-1/3}$ plots as a straight line. (a) How would the relationship between dQ_e/dt and Q_e look like? In other words, derive a functional relationship, $dQ_e/dt = f(Q_e)$. (b) Determine the power m of the storage–discharge equation (12.48), namely $S = K_n Q_e^m$, which is to be used in the routing of floods through this lake by means of the equation $Q_i - Q_e = dS/dt$.

12.22   Multiple choice. Indicate which of the following statements are correct. The volume (not the rate) of direct storm runoff produced in a given watershed:

(a)   as a result of a given storm, is largely independent of the initial moisture conditions of the surface soils;

(b)   is always the same fraction of the rainfall volume;

(c)   as a result of a given storm, is totally independent of the baseflow that occurs at the time of the rainfall producing this event;

(d)   tends to become larger as the watershed becomes urbanized;

(e)   for engineering design purposes, is often estimated by subtracting the volume of baseflow from the total runoff volume.

13 ELEMENTS OF FREQUENCY ANALYSIS IN HYDROLOGY

One of the core questions in hydrologic data analysis is how to assign a probability of future occurrence or a risk estimate to an event of a given magnitude, on the basis of an available record of measurements. Over the years a number of concepts have been developed for this purpose, which are not usually part of standard treatments of elementary statistics. A few of these are reviewed in this chapter, together with some other indispensable fundamentals.

13.1 RANDOM VARIABLES AND PROBABILITY

In practical terms, a random variable may be defined as the magnitude of an event, which is the outcome of an experiment; the variable is called random because this magnitude cannot be predicted with certainty. Random variables can be classified as discrete or as continuous, or sometimes as a combination of the two.

When an event A occurs n_A times in an experiment, that is carried out n times, its relative frequency is the ratio (n_A/n). The probability of this event can then be defined as the limit of its relative frequency, when n is allowed to increase indefinitely, or

$$P(A) = \lim_{n \to \infty} \frac{n_A}{n} \qquad (13.1)$$

To be sure, this definition is intuitively appealing, as it gives a "feel" for the meaning of probability in everyday life. However, in any physical experiment the number n can only be finite, so that this limit is only a hypothesis. For this reason, the probability of an event is preferably defined axiomatically (Papoulis, 1965), as a number linked to the event, with the properties, that (i) the probability cannot be negative, or $P(A) \geq 0$; (ii) the probability of all possible outcomes, i.e. the probability of certainty, equals unity; (iii) if A and B are mutually exclusive events, the probability of A or B occurring equals the sum of their probabilities, or $P(A \text{ or } B) = P(A) + P(B)$. Clearly, the frequency definition (13.1) satisfies these axioms. In this sense, the word *relative frequency* usually refers to the empirical probability, that is the probability estimated from a finite sample, which is presumably drawn from an infinitely large population.

A random variable is called discrete when it can assume only certain values x_i, with $i = 1, 2, \ldots$; then the probability, that a discrete random variable X will assume a given value x_i, can be written as

$$p_i = P\{X = x_i\} \qquad (13.2)$$

Fig. 13.1 **Example of a rod graph, showing the relative frequency of a discrete random variable *X* for different values $x_i = \ldots, -2, -1, 0, 1,$ etc.**

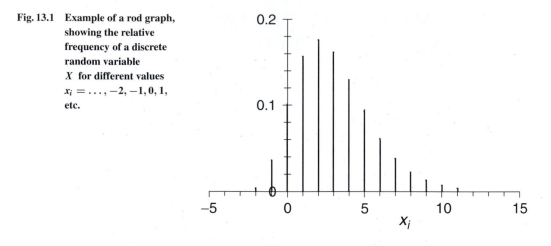

in which the symbol P denotes the probability of the event stated between the curly brackets. A rod graph is a natural way to represent the distribution of the probabilities as relative frequencies of a discrete variable (see Figure 13.1).

A random variable is called continuous when it can assume any value x in a certain range of real numbers, which may or may not be unbounded. In this case, since the variable X can assume an infinity of values, it is more appropriate to consider the probability that it is smaller than or equal to a given value x. This defines the (*probability*) *distribution function* as

$$F(x) = P\{X \leq x\} \tag{13.3}$$

where again $P\{ \}$ is the probability of the event between the curly brackets. For all values of x, where the distribution function is smooth, the (*probability*) *density function* can be defined by

$$f(x) = \frac{dF(x)}{dx} \tag{13.4}$$

Thus the probability that $X \leq x$ can also be written in terms of the density function as

$$P\{X \leq x\} = \int_{-\infty}^{x} f(y)dy \tag{13.5}$$

in which y is the dummy variable of integration. This is illustrated in Figure 13.2. In the same way the probability, that the random variable occur in a certain range $x_1 < X \leq x_2$, is given by

$$F(x_2) - F(x_1) = \int_{x_1}^{x_2} f(x)dx \tag{13.6}$$

Fig. 13.2 **Illustration of the relationship between the distribution function $F(x)$ and the density function $f(x)$. The density function is the slope of the distribution function; conversely, in accordance with Equation (13.5), $F(x_0)$ equals the area under the $f(x)$ curve to the left of x_0 shown in the lower figure.**

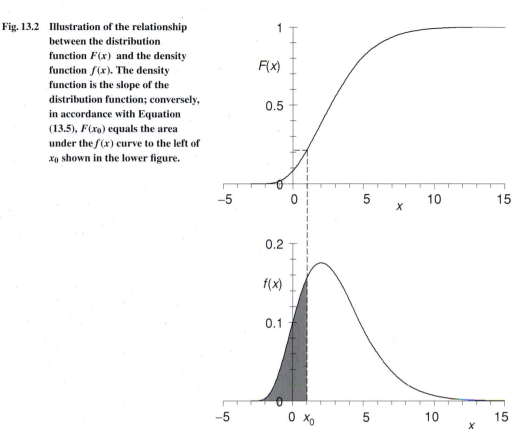

13.2 SUMMARY DESCRIPTORS OF A PROBABILITY DISTRIBUTION FUNCTION

13.2.1 *Moments*

In the case of a discrete variable X, the nth moment about the origin is defined as

$$m'_n = \sum_{\text{all } i} x_i^n p_i \tag{13.7}$$

whereas the nth moment about the mean μ is

$$m_n = \sum_{\text{all } i} (x_i - \mu)^n p_i \tag{13.8}$$

For a continuous type variable X, the nth moment about the origin, $x = 0$, can be defined as

$$m'_n = \int\limits_{-\infty}^{+\infty} x^n f(x)\,dx \tag{13.9}$$

and that about the mean $x = \mu$, as

$$m_n = \int_{-\infty}^{+\infty} (x - \mu)^n f(x)dx \tag{13.10}$$

Moments about the mean are also called central moments. In principle, the zeroth moments equal unity, or $m_0' = m_0 = 1$. The first moment about the origin is the *mean*, by definition, or $m_1' = \mu$; this is also called the *expected value* of the random variable, $E\{X\}$. From Equation (13.10) with $n = 1$, it follows that the first moment about the mean is zero, or $m_1 = 0$. The second moment about the origin is called the *mean square deviation*. The second moment about the mean is called the *variance*, and is usually denoted by $m_2 = \sigma^2$; its square root σ is called the standard deviation. When the standard deviation is made dimensionless with the mean, it is called the *coefficient of variation*, or $C_v = (\sigma/\mu)$. These parameters related to the second moment can be used to characterize the dispersion of the random variable. All odd central moments of distributions with a symmetrical distribution function are zero, or $m_3 = m_5 = \cdots = 0$. The lack of symmetry or skew is commonly expressed by the third moment about the mean, m_3. The coefficient of skew is defined by $C_s = (m_3/\sigma^3)$. A distribution function is symmetrical about the origin $x = 0$, if

$$F(x) = 1 - F(-x) \tag{13.11}$$

so that also $f(x) = f(-x)$. The central moments can be readily obtained from the moments about the origin. For the second and third central moments the relationships can be shown to be

$$m_2 = m_2' - {m_1'}^2$$
$$m_3 = m_3' - 3m_1'm_2' + 2{m_1'}^3 \tag{13.12}$$

The same relationships also hold when m_1', m_2' and m_3' denote the moments about any arbitrary reference, say $x = a$.

The moments of a distribution can be estimated directly from a set of n observations, X_i, with $i = 1, \cdots, n$. Sample estimators of the mean μ, variance σ^2, and skew coefficient C_s, are, respectively,

$$M = \overline{X_i} = \frac{\sum_{i=1}^{n} X_i}{n}$$

$$S^2 = \frac{\sum_{i=1}^{n} (X_i - M)^2}{n - 1} \tag{13.13}$$

$$g_s = \frac{n \sum_{i=1}^{n} (X_i - M)^3}{(n - 1)(n - 2)S^3}$$

The factors $(n - 1)$ for S, and $(n - 1)(n - 2)$ for g, rather than n and n^2, are introduced to reduce the bias in these estimators (Weatherburn, 1961).

13.2.2   Quantiles

By definition, the quantiles are the $(n - 1)$ values of the random variable, which partition the probability function domain (normally between 0 and 1), into n equal parts. Accordingly, the mth quantile, say x, is obtained by solving the following integral for x

$$\frac{m}{n} = \int_{-\infty}^{x} f(y)dy \quad \text{or} \quad \int_{-\infty}^{x} dF(y) \tag{13.14}$$

The median is the most widely used quantile; it is the value of the variable where the probability distribution equals $1/2$, and it is often used as another measure of the central tendency of the population, beside the mean. Thus, for a sample of n items, the sample median is the value of the item, which is found in the middle after all items have been ranked. The lower quartile of the sample is the value of x for which $m/n = 1/4$ in Equation (13.14), and the upper quartile that for which $m/n = 3/4$. When $n = 100$, quantiles (times 100) are also called percentiles.

13.2.3   Return period

The reciprocal of the probability that a certain value x will be exceeded is referred to as the *return period*, also called the recurrence interval or the exceedance interval, or

$$T_r(x) = \frac{1}{1 - F(x)} \tag{13.15}$$

This return period is the expected number of observations required until x is exceeded once.

This can readily be shown as follows. Let $p (= F(x))$ denote the probability that, at any trial in an experiment, the magnitude of the event will not be larger than x; then, the probability, that this magnitude will not be exceeded in the first $(k - 1)$ trials and finally will be exceeded in the last trial, is given by

$$P\{k \text{ trials until } X > x\} = p^{k-1}(1 - p) \tag{13.16}$$

This probability function is known as the geometric distribution (see Section 13.3.1 below). The average number of trials is the first moment, or from Equation (13.7),

$$\bar{k} = m'_1 = \sum_{k=1}^{\infty} kp^{k-1}(1 - p) \tag{13.17}$$

Except in case of certainty, one has $0 < p < 1$, so that $(1 + p + p^2 + \cdots) = (1 - p)^{-1}$; hence (13.17) reduces to $\bar{k} = (1 - p)^{-1}$, which proves the statement below Equation (13.15).

Equation (13.15) shows how the return period is uniquely related to $F(x)$. It can therefore be considered as an alternative to, and equivalent with, the probability distribution function. The term "return period" stems from the fact that, whenever the observations are made at regular intervals in time, the number of observations is a time expressed in the same units. It should be emphasized that the return period represents the average number of observations. This does not mean that the event will occur once every T_r number of observations. Thus a 100 y flood need not occur every 100 y. In fact, it may very well occur next year, or again, it may not occur at all for another 1000 y, although that is unlikely.

In practical applications, the return period as given by Equation (13.15) is normally used to characterize phenomena whose severity increases with their magnitude x. For example, a 500 y flood is more severe and causes more damage than a 100 y event. When, on the other hand, the severity of the event decreases with its numerical magnitude, the return period should be defined as

$$T_r(x) = \frac{1}{F(x)} \tag{13.18}$$

This will ensure that, for instance in the case of a drought, lower flows and lower rainfall amounts, as measures of drought severity, will be characterized by longer return periods. The theoretical significance of Equation (13.18) is, *mutatis mutandis*, the same as (13.15).

13.2.4   *Empirical probability plots*

It is often useful to plot the data from a hydrologic measurement record to gain a general idea of their statistical characteristics. The (empirical) probability plot, also known as the frequency plot or frequency curve, and sometimes as the quantile plot, is a common tool for this purpose. This plot is a graphical representation of the probability (of non-exceedance or exceedance) of the individual data in the record against their respective magnitudes, i.e. $F = F(x)$.

For a data record, consisting of n items, normally measured at regular intervals, the procedure is as follows. (i) Tabulate the n data, X_m, ranked in increasing magnitude, so that $X_1 \leq X_2 \leq \cdots \leq X_m \leq \cdots \leq X_n$; (ii) assign an order number to each item, in accordance with its respective subscript, namely $1, 2, \ldots, m, \ldots, n$; (iii) estimate the (empirical) probability P_m, that the magnitude of the item will not be exceeded, for each item by means of a suitable *plotting position* formula. This plotting position P_m is used as an estimate of the value of the unknown probability $F(x = X_m)$ for the observed event X_m. In the past numerous plotting position formulae have been suggested. Cunnane (1978) has presented a critical review of the history and properties of some of the more common ones.

Plotting position
Probably the oldest and intuitively the simplest is $P_m = m/n$. The problem with this expression is that the largest item on record is assigned a probability of one, that is certainty; in other words, with this formula it is assumed that the magnitude of the largest will never be exceeded in the future. Because this is impossible, the use of this plotting position is tantamount to discarding the largest item in the record. This difficulty may be

avoided by the formula $P_m = (m - 1)/n$, but this yields certainty for the smallest item in the record. Hazen (1930) proposed an intermediate position, namely $P_m = (m - 1/2)/n$. An obvious feature of Hazen's choice is that with Equation (13.15) it produces a return period $T_r = 2n$ for the largest item in the record; in other words, the resulting return period is twice as long as the period over which the data have been recorded. This is unacceptable, according to Gumbel (1958), because the estimated return period of the largest event should approach the length of the period of record, as n becomes large. These difficulties are not encountered in the Weibull formula

$$P_m = \frac{m}{n + 1} \tag{13.19}$$

Gumbel (1958), and many after him, have recommended Equation (13.19), by noting that beside the avoidance of difficulties for $m = 0$ or 1, so that all data can be plotted, it also has the following advantages: (i) it is independent of the distribution function $F(x)$; (ii) the return period of the largest (or smallest, as the case may be) observation approaches the number of observations n; (iii) all observations are equally spaced on the frequency scale, which means that the difference between the plotting positions of the mth and the $(m + 1)$th observation is a function only of n; (iv) it is intuitively simple and can be readily implemented. Its main advantage, however, is that it can be theoretically justified as the mean of the probability of the mth smallest observation; this can be proved as follows.

Derivation of Weibull plotting position

Consider again a sample of n observations, after they have been ranked in increasing magnitude, so that $X_1 < X_2 < \cdots < X_m < \cdots < X_n$. The probability distribution of the mth smallest observation by itself is given by Equation (13.3), or

$$F(X_m) = \int_{-\infty}^{X_m} f(x)dx \tag{13.20}$$

and the density of this observation by itself is

$$f(X_m) \tag{13.21}$$

However, X_m does not occur by itself, but in conjunction with $(n - 1)$ other observations; among these remaining $(n - 1)$ observations, $[(n - 1) - (m - 1)]$ observations exceed X_m and $(m - 1)$ do not. The probability of this occurrence is given by the binomial distribution (see Section 13.3.2 below), and equals

$$P\{(m - 1)\text{successes}, (n - m)\text{failures}\} = \frac{(n - 1)!}{(n - m)!(m - 1)!} F_m^{m-1}(1 - F_m)^{n-m} \tag{13.22}$$

in which the symbol $F_m = F(X_m)$ is introduced for conciseness of notation, and "success" refers to non-exceedance. The probabilities of each observation are independent. Moreover, the mth smallest observation, X_m, can occupy n different places in the sequence of the remaining $(n - 1)$ observations, namely in front of all of them and behind each one of them. Hence the density function of the mth smallest observation X_m, occurring in conjunction with the occurrence of the remaining $(n - 1)$ observations, is equal to n times the probability given by (13.22), multiplied by the density of X_m by itself (Equation (13.21)), or

$$\phi(X_m) = \frac{n!}{(n - m)!(m - 1)!} F_m^{m-1}(1 - F_m)^{n-m} f(X_m) \tag{13.23}$$

It is known (see Mood and Graybill, 1963) that, if y is a function of z, $y = y(z)$, and if y has a density function $f(y)$, then the density function of z is given by

$$g(z) = f[y(z)] \left| \frac{d[y(z)]}{dz} \right| \tag{13.24}$$

Applying this to the present case by putting $y = X_m$ and $z = F_m$, one obtains the density function of $F_m = F_m(X_m)$ from (13.23), namely

$$g(F_m) = \frac{n!}{(n - m)!(m - 1)!} F_m^{m-1}(1 - F_m)^{n-m} \tag{13.25}$$

The mean of F_m is the first moment, or, since $0 \le F_m \le 1$,

$$m_1' = \overline{F_m} = \int_0^1 F_m g(F_m) d F_m$$

or

$$\overline{F_m} = \frac{n!}{(n - m)!(m - 1)!} \int_0^1 y^m (1 - y)^{n-m} dy \tag{13.26}$$

in which y is a dummy variable of integration. The integral in this expression is a complete beta function, which can be expressed in terms of factorials (e.g. Abramowitz and Stegun, 1964, p. 258) as $m!(n - m)!/(n + 1)!$. This yields immediately $\overline{F_m} = m/(n + 1)$, which proves that the assumed plotting position of Equation (13.19) is in fact the mean of the probability of non-exceedance of the mth observation, that is $P_m = \overline{F_m}$.

Probability graph paper

An empirical probability plot on graph paper with linear scales usually results in an S-shaped curve, with considerable curvature. To facilitate interpretation of the plotted data and interpolation, it is desirable to eliminate or reduce this curvature by stretching or shrinking the scale in the appropriate range of values of P_m, so that the points plot more nearly along a straight line. A common way to accomplish this is the use of probability paper. Hazen (1914b) appears to have been the first to advocate the use of probability paper in hydrology.

Probability graph paper, for any given probability function $F(x)$, is designed in such a way that, when that function or the corresponding return period $T_r(x)$ is plotted against x, one obtains a straight line. The most common types of probability paper involve a probability function with two parameters, say a and b, whose values depend on the data under consideration; thus to make the probability paper generally applicable to any data set it is necessary to eliminate the dependency on these parameters. This can be done with a linear transformation of the type

$$y = a(x - b) \tag{13.27}$$

in which the nature of a and b depends on the particular function $F(x)$ that is being used. For instance, in the case of the normal distribution one has $b = \mu$ and $a = \sigma^{-1}$. In the case of the first asymptote for largest values, as will be seen below, b is the mode, that is the value of x where the density $f(x)$ is a maximum, and $a = (\pi/\sqrt{6})\sigma^{-1}$. This transformation results in a distribution function, say $F_Y(y)$, given by

$$F_Y(y) = F(b + y/a) \tag{13.28}$$

which is independent of the parameters. In principle, probability paper is constructed as a plot of x versus y, with the values of a and b left unspecified; when $F(x)$ is symmetrical, $y = 0$ is placed at the center of the scale. Parallel to the y-scale, a $F_Y(y)$-scale is plotted and in some types of probability paper a third scale is shown with $T_r(y)[= (1 - F_Y)^{-1}]$ values. However, most types of probability paper do not show the y-scale, and only one scale, either $F_Y(y)$ or $T_r(y)$, is displayed. While normal probability paper, which is based on the normal probability distribution, has been made available commercially in the past, nowadays normal scales can be generated by standard computer programs. For some applications, the x-variable is transformed logarithmically, so that it is $\log(x)$ which is plotted against $F_Y(y)$ or $T_r(y)$. Figures 10.34–10.37 illustrate applications of lognormal probability paper.

Example 13.1. Probability paper based on an extreme value distribution

The first asymptotic distribution for largest values, which is treated in detail below in Section 13.4.5, can be used here to illustrate the construction of probability paper. This function is given by $F(x) = \exp[-\exp(-y)]$ in which y is the linearly scaled or reduced variable defined in Equation (13.27). This distribution can be immediately inverted to yield $y = -\ln[\ln(F^{-1})]$ and $y = -\ln[\ln[T_r/(T_r - 1)]]$. One starts with graph paper in which one of the coordinate axes is designated as the y coordinate axis. Values of y can then be calculated for selected values of F covering the entire F range of interest, and marked with their F value on the y-axis, or (which is preferable to avoid clutter) on a separate axis parallel to y. The same is done for selected T_r values covering the entire T_r range of interest. As mentioned, on most types of probability paper not all three axes are shown, but only the resulting F- or T_r-axis. Figure 13.3 shows a lay-out with a y-axis and a T_r-axis.

13.2.5   *Theoretical probability distribution functions*

Countless mathematical functions are consistent with the definition of probability given above, which can be used to describe data sets. In frequency analysis, it is often useful

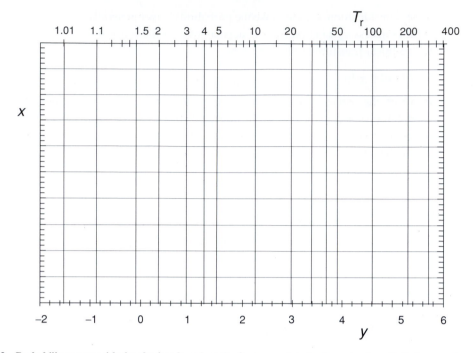

Fig. 13.3 **Probability paper with the abscissa based on the first asymptote for largest values. Both the y-scale and the T_r-scale are shown. (See Example 13.1.)**

to fit a mathematical probability distribution function to the available data; indeed such functions provide a smooth and succinct description of the data and they allow the formulation of objective confidence criteria. Moreover, although the procedure may be hazardous and should only be applied with great caution, a mathematical function may also allow some degree of extrapolation to estimate probabilities outside the range of the available data.

The application of most theoretical distribution functions can be justified on strict probabilistic considerations. Unfortunately, it is rare that such considerations are rigorously valid for data sets of hydrologic concern, and in most cases the actual mathematical form of the distribution function, that represents the population, is unknown. Thus the best that can be hoped for is that the selected distribution is simple enough and also physically plausible to be useful in practice.

Several procedures are available to determine the parameters in these distributions. Among them the method of maximum likelihood is commonly considered to be the best, in principle; however, several studies (see, for example, Stedinger, 1980; Martins and Stedinger, 2000) have revealed that this is not always the case, particularly in small samples. The present treatment concentrates mainly on the method of moments, which is usually simpler to apply. An additional feature of the method of moments is that it weights the larger observations more heavily, so that it may be more suitable in the analysis of large values. In the following two sections a few of the more common functions are considered which have been useful in the analysis of hydrologic events.

13.3   SOME PROBABILITY DISTRIBUTIONS FOR DISCRETE VARIABLES

13.3.1   *The geometric distribution*

This distribution was already discussed briefly in connection with the return period. For convenience, the reasoning can be briefly repeated here. Consider an experiment in which there are two possible outcomes, namely success with a probability p, and failure with a probability $q = (1 - p)$. If the trials are independent of one another, then it follows that the probability that $(k - 1)$ successes will be followed by one failure is given by Equation (13.16), or as formulated here

$$P\{k \text{ trials until first failure}\} = p^{k-1}(1 - p) \tag{13.29}$$

Equation (13.29) can be used next, for example, to calculate the probability that it will take k or fewer trials to incur failure. This is simply the sum of the probabilities, or if K denotes the number of trials needed to experience failure, as a random variable,

$$P\{K \leq k\} = \sum_{i=1}^{k} p^{i-1}(1 - p) \tag{13.30}$$

which yields immediately $1 - p^k$. It can also be seen that Equation (13.30) yields unity, that is certainty, when $k \to \infty$. Actually, (13.30) could also have been obtained by considering first the probability that failure would not occur for k trials; since the trials are independent, this is equal to p^k. Its complement is the probability that it will take k or fewer trials for failure to occur.

Example 13.2. Probability of exceedance of a flood of a given return period

Assume that "success" means that the annual maximum flow X in a river, also called the annual flood, does not exceed a given magnitude x, so that $p = F(x)$. As an example, consider the event for which $p = 0.98$; from Equation (13.15) it can be seen that this is a 50 y flood. Figure 13.4 shows the probability, that it will take exactly k years before that event is exceeded as a function of k, calculated with Equation (13.29). The probability that it will be exceeded the first year is 0.02. The probability that it will be exceeded after exactly 50 y is $0.98^{49} \times 0.02 = 0.007\,43$. Figure 13.5 shows the probability calculated with Equation (13.30), that it will take k or fewer years to exceed x, that is, the probability of the occurrence of a flood in excess of the same magnitude x. It can again be seen that the probability of this occurrence in the first year is 0.02; also, the probability that it will be exceeded before 50 y have passed is $1 - 0.98^{50} = 0.636$. The probability approaches one as k becomes large.

Example 13.3. Probability of sequences of dry or rainy days

The geometric distribution has been used by Hershfield (1970a, b; 1971) to study the probability of periods of a given length with or without precipitation. In this type of application p denotes the conditional probability of a success being followed by a success.

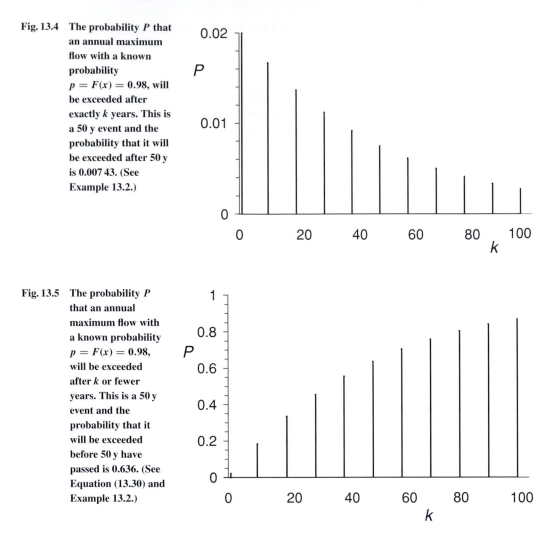

Fig. 13.4 The probability P that an annual maximum flow with a known probability $p = F(x) = 0.98$, will be exceeded after exactly k years. This is a 50 y event and the probability that it will be exceeded after 50 y is 0.007 43. (See Example 13.2.)

Fig. 13.5 The probability P that an annual maximum flow with a known probability $p = F(x) = 0.98$, will be exceeded after k or fewer years. This is a 50 y event and the probability that it will be exceeded before 50 y have passed is 0.636. (See Equation (13.30) and Example 13.2.)

In general, a success can refer to a day without rain, a day with rain, or some other desirable event, whatever the case may be. For the sake of illustration, let p denote the probability that a day without rain be followed by a day without rain. In this case, the form of Equation (13.29) arises again as the simple product of the probabilities of a dry day being succeeded by a dry day for a sequence of k days, multiplied by the probability of a dry kth day being succeeded by a rainy $(k + 1)$th day; in other words, (13.29) represents the probability of experiencing a sequence of exactly k dry days.

This product in (13.29) involves the assumption that the events are independent, or which is the same, that p remains constant and does not change with k, the duration of the period of dry days. Whether or not this is the case can be examined on the basis of the following considerations. In the present context, the probability that the dry period will last at most k days is the sum of the probabilities of all dry periods shorter than

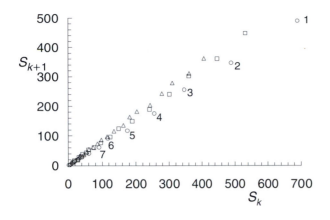

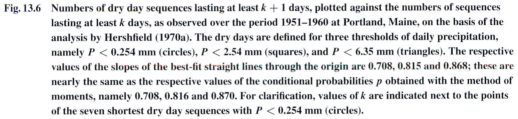

Fig. 13.6 Numbers of dry day sequences lasting at least $k + 1$ days, plotted against the numbers of sequences lasting at least k days, as observed over the period 1951–1960 at Portland, Maine, on the basis of the analysis by Hershfield (1970a). The dry days are defined for three thresholds of daily precipitation, namely $P < 0.254$ mm (circles), $P < 2.54$ mm (squares), and $P < 6.35$ mm (triangles). The respective values of the slopes of the best-fit straight lines through the origin are 0.708, 0.815 and 0.868; these are nearly the same as the respective values of the conditional probabilities p obtained with the method of moments, namely 0.708, 0.816 and 0.870. For clarification, values of k are indicated next to the points of the seven shortest dry day sequences with $P < 0.254$ mm (circles).

or equal to k days; this is given by Equation (13.30), which can readily be shown to yield $1 - p^k$. Conversely, the probability that a dry day sequence will last at least $k + 1$ days, that is, that it will be equal to or longer than $k + 1$ days, is the complement of (13.30), or p^k. Similarly, the probability that a dry day sequence will last at least k days is p^{k-1}. Hence, the validity of the assumption, that p is independent of k, can be readily examined empirically in any given situation by checking whether or not the ratio of the number of sequences, that last at least k days, over the number of those that last at least $(k - 1)$ days is a constant for all k, or

$$\frac{S_k}{S_{k-1}} = \text{const.} \tag{13.31}$$

where the constant should in principle be equal to p. As an illustration, Figure 13.6 shows a plot of the numbers of dry day sequences S_{k+1} against S_k observed at Portland, Maine, over the period 1951–1960, on the basis of precipitation data analyzed by Hershfield (1970a). A "dry" day or a day "without rain" was defined here as a day with precipitation less than a certain finite threshold; three such thresholds were considered, namely days with less than 0.254 mm, 2.54 mm and 6.35 mm. The slopes of the regression lines through the origin are 0.708, 0.815 and 0.868, respectively, which should be reasonable estimates of p for each of these three thresholds.

The more common way, however, to estimate the parameters is the method of moments. Because the geometric distribution (13.29) has only one parameter, viz. p, a knowledge of the mean of k will suffice. The mean duration of a dry period is given by

$$\bar{k} = \frac{N_{\text{DD}}}{N_{\text{DS}}} \tag{13.32}$$

in which N_{DD} is the total number of dry days in the data record under considera-tion, and N_{DS} is the number of dry periods in the same record. As can be seen from Equation (13.17), this mean duration is also given by $\overline{k} = (1 - p)^{-1}$, which shows how p can be determined immediately from the record by means of (13.32). The respective conditional probabilities of dry day sequences obtained this way with the data from Portland, Maine, on which Figure 13.6 is based, were found to be 0.708, 0.816 and 0.870 for the three thresholds; as was to be expected, these values are nearly the same as those obtained from the slopes of the lines through the points mentioned above.

This example was presented for sequences of dry days, or rather days with precipita-tion less than a certain finite threshold. The same reasoning also holds, *mutatis mutandis*, if p represents the probability of a rainy day (i.e. in excess of a certain threshold) being succeeded by a rainy day. Finally, it should be kept in mind that p, as used here in this example, is different from the probability of a dry day; the latter is defined as the ratio of the number of dry days over the total number of days (dry and rainy) of the record. However, Hershfield (1970b) has shown how the two may be related.

13.3.2   The binomial distribution

Consider again an experiment with two possible outcomes, namely success and failure, whose respective probabilities are p and $q = (1 - p)$. This distribution provides the answer to the question what the probability is of k successes and $(n - k)$ failures in n trials. This question is similar to the one leading to the geometric distribution. Indeed, if the sequence of independent successes and failures were made to occur in a certain specified order, the probability of the outcome would be given by $p^k q^{n-k}$; however, there are $n!/[k!(n - k)!]$ ways to select k items out of a total of n items. Therefore, if K denotes again the random variable for the number of successes, the total probability is the sum, or

$$P\{K = k\} = \frac{n!}{k!(n - k)!} p^k q^{n-k} \tag{13.33}$$

As before, it is assumed that the events are independent of one another.

Example 13.4. Probability of a year without freezing

An obvious application of the binomial distribution could be the following problem. From a 45 y temperature record at a certain location it is known that in 26 y (out of 45) the temperature did not go below $0\,°C$. Thus the probability of a frost-free year can be estimated to be $p = (26/45)$. Equation (13.33) then provides the probability of k frost-free years during a period of n years; in Figure 13.7 the results are illustrated for the different values of k during an 8 y period. A related problem is the determination of the probability of at least a certain number of years without freezing temperatures during a period of n years; for instance, what is the probability that during the next 5 y the site will experience a year with freezing temperatures not more than twice? The answer is

Fig. 13.7 The probability
$P = P\{K = k\}$ of k
frost-free years within
a period of $n = 8$
years at a location
where the probability
of a frost-free year is
estimated to be
$p = 26/45$. (See
Example 13.4.)

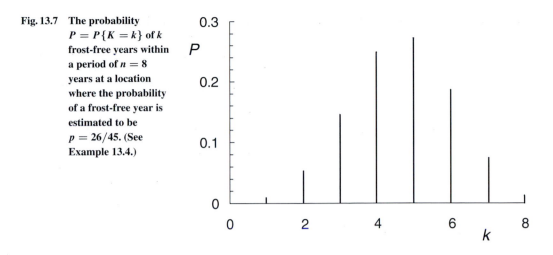

the probability of 3, 4, or 5 successes, that is the sum of the probabilities of respectively 3, 4, and 5 successes, or

$$P\{K \geq 3\} = \sum_{i=3}^{5} \frac{5!}{i!(5-i)!} \left(\frac{26}{45}\right)^i \left(\frac{19}{45}\right)^{5-i}$$

which equals 0.643.

13.4 SOME PROBABILITY DISTRIBUTIONS FOR CONTINUOUS VARIABLES

In flood and drought frequency analysis, the variable in question can usually assume any value, so that, within certain limits, it can be considered as being continuous. In this section some of the more common distribution functions are considered.

13.4.1 *The normal distribution*

This is a well-known function, but it is useful to consider it briefly, mainly as a reference or benchmark to compare other less common distribution functions with it later on. The normal or Gaussian density function has two parameters, namely the first two moments, and it can be written as follows

$$f(x) = \frac{1}{\sigma\sqrt{2\pi}} \exp\left[-\frac{1}{2}\left(\frac{x-\mu}{\sigma}\right)^2\right] \qquad -\infty < x \leq \infty \qquad (13.34)$$

in which μ denotes the mean, and σ is the standard deviation. This is a symmetrical bell-shaped curve that extends from minus to plus infinity on the x-scale; thus its median,

Table 13.1 Values of the normal distribution function, expressed as $[F(x) - 0.5]$ according to Equation (13.36)

$y = (x - \mu)/\sigma$	0	0.5	1.0	1.5	2.0	2.5	3.0
$0.5\,\mathrm{erf}(y/\sqrt{2})$	0	0.1915	0.3413	0.4332	0.4772	0.4938	0.4987

mode and mean values coincide, and its skew coefficient equals zero. The distribution function is the integral of Equation (13.34) or, according to (13.5),

$$F(x) = \frac{1}{\sigma\sqrt{2\pi}} \int_{-\infty}^{x} \exp\left[-\frac{1}{2}\left(\frac{y-\mu}{\sigma}\right)^2\right] dy \qquad (13.35)$$

This can also be written more concisely as

$$F(x) = \frac{1}{2} + \frac{1}{2}\mathrm{erf}[(x-\mu)/\sqrt{2}\sigma] \qquad (13.36)$$

in which the error function (cf. Equation (9.57)) is defined as

$$\mathrm{erf}(y) = \frac{2}{\sqrt{\pi}} \int_{0}^{y} \exp(-z^2)dz \qquad (13.37)$$

Note that $\mathrm{erf}(-y) = -\mathrm{erf}(y)$. The error function cannot be expressed in closed form; however, it has been tabulated and close approximations are available for computation (Abramowitz and Stegun, 1964, p. 299); it is also available in most computational software. For convenient reference a few values are presented in Table 13.1. As will be seen below, the values listed in Table 13.2 for zero skew (i.e. for $C_s = 0$) are the quantiles of $(x - \mu)/\sigma$ of the normal distribution for the indicated probabilities.

The use of this distribution can be justified by means of the Central Limit Theorem; this states that, if a random variable is the sum of n random, not necessarily independent, variables, each with its own, not necessarily normal, density function with a finite mean and variance, then the density of this random variable tends to the normal function (13.34) as n increases. In hydrology the normal distribution is commonly assumed to be applicable in the description of various types of central tendency observations, such as mean annual temperatures, annual river discharge rates, among others.

Example 13.5. Probability distribution of annual mean streamflows

In Figure 13.8 the annual mean values are plotted on normal probability paper for the Susquehanna River, as measured near Waverly, NY, for 57 years over the period 1938–1994, and for the Chemung River, measured at Chemung, NY, for 90 years during 1907–2000. Both stations are operated by the US Geological Survey. The Susquehanna River station is located in Bradford County, Pennsylvania, at approximately 41°59′05″ N, 76°30′05″ W, with the gage at 227 m above sea level, but the drainage area of 12 362 km² is in New York State; after correction (Korzoun et al., 1977) the long term

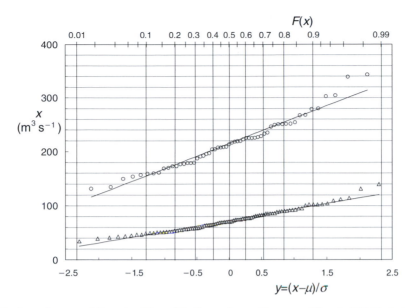

Fig. 13.8 **Annual means of the rate of flow of the Susquehanna River near Waverly, NY (circles), and of the Chemung River at Chemung, NY (triangles), plotted on normal probability paper. The theoretical straight line for the Susquehanna is calculated with a sample mean $M = 214.7$ m^3 s^{-1} and a sample standard deviation $S = 46.8$ m^3 s^{-1}, and for the Chemung with $M = 72.8$ m^3 s^{-1} and $S = 20.8$ m^3 s^{-1}. Both the y-scale and the $F(x)$-scale are shown. (See Example 13.5.)**

average annual precipitation over this basin appears to be of the order of 1200 mm. After sorting the streamflow data in ascending order, the non-exceedance probability $F(x)$ of each data point shown in the figure was estimated in accordance with Equation (13.19) as $P_m = m/58$. The sample mean and standard deviation of these data are respectively $M = 214.7$ m^3 s^{-1} and $S = 46.85$ m^3 s^{-1}, and these were used to calculate the theoretical distribution (13.35), which is also shown in Figure 13.8. The Chemung River station is located at approximately 42°00′08″ N, 76°38′06″ W, and 237 m above sea level; the contributing drainage area is 6491 km^2; over this second basin the average annual precipitation is probably closer to 1100 mm. For this basin each of the data points was assigned the non-exceedance probability $m/91$. The sample estimates of the moments of the Chemung record are $M = 72.79$ m^3 s^{-1} and $S = 20.81$ m^3 s^{-1}. The Chemung River is a major tributary of the Susquehanna, and the two rivers merge a short distance south of the two stations; thus while the two stations are located closely, the drainage areas do not overlap. The main point is that most of the data appear to obey the normal distribution fairly well, although it can also be seen that the wettest years plot above the best-fit line. As an aside, note that the largest mean flow rates at both stations were recorded for the year of Tropical Storm Agnes. This event took place in June of 1972; the total precipitation varied, with reported maxima of up to 450 mm, but over these two areas it was of the order of 150–250 mm over three days (Bailey et al., 1975). It caused the most severe flood ever experienced in the entire Susquehanna basin and resulted in the loss of some 120 lives; at the time, this flood was the most destructive and costly natural disaster in the United States. Tropical storms or hurricanes rarely penetrate this far north in full strength, but when they do they can materially affect the annual mean.

13.4.2 The lognormal distribution

Many natural phenomena, which have a lower bound and exhibit positive skew, cannot be described well by the normal distribution, but in some cases their logarithms can. According to the Central Limit Theorem, this would be the case when the random variable is the product of n variables, each with its own arbitrary density function with a finite mean and variance. Upon applying Equation (13.24) with $y = \ln z$ to (13.34), the lognormal density function results, and it can be written as

$$f(x) = \frac{1}{\sigma_n x \sqrt{2\pi}} \exp\left[-\frac{1}{2}\left(\frac{\ln(x) - \mu_n}{\sigma_n}\right)^2\right] \qquad -\infty < \ln(x) \leq \infty \qquad (13.38)$$

where μ_n and σ_n are the mean and the standard deviation of $\ln(x)$. When the logarithms of the data do not quite follow the normal distribution, introduction of a lower bound c, different from zero, may improve the fit. The density function becomes in this case

$$f(x) = \frac{1}{\sigma_{nc}(x - c)\sqrt{2\pi}} \exp\left[-\frac{1}{2}\left(\frac{y_n - \mu_{nc}}{\sigma_{nc}}\right)^2\right] \qquad -\infty < y_n \leq \infty \qquad (13.39)$$

in which $y_n = \ln(x - c)$, and μ_{nc} and σ_{nc} are the mean and standard deviation of y_n.

When c is known (e.g. from physical considerations about the lower bound), these two parameters can be estimatated by means of the first and second of (13.13), after replacing X_i by $\ln(X_i - c)$; alternatively, they can also be estimated directly from μ and σ, by inversion of the following two equations (see, for example, Chow, 1954)

$$\mu = c + \exp\left(\mu_{nc} + 0.5\sigma_{nc}^2\right)$$
$$\sigma^2 = (\mu - c)^2\left[\exp(\sigma_{nc}^2) - 1\right] \qquad (13.40)$$

However, because the moments of the X_i values are different from those of the $\ln(X_i - c)$ values, these two procedures yield different results; Stedinger (1980) has shown that for smaller samples the procedure with the moments of the logs yields better parameter estimates than that based on Equations (13.40).

The determination of c is not always easy. A rough idea of its value can be obtained graphically by plotting the data on log-normal probability paper, i.e. $\ln(X_i - c)$ versus $F_Y(y)$ (see Equation (13.28)) for different trial values of c; that value of c is selected which produces the best straight line. The value of c can in principle be obtained by the method of moments. Since the first two moments are already used in (13.40) to determine μ and σ, the third moment is needed. In the case of the lognormal distribution (see, for example, Chow, 1954), this is related to the second as follows

$$C_s = 3C_v + C_v^3 \qquad (13.41)$$

in which the coefficient of variation is given by $C_v = [\exp(\sigma_{nc}^2) - 1]^{1/2}$. The value of C_s can be calculated from the data by means of the third of (13.13), and with that value (13.41) can be solved for C_v as follows

$$C_v = \left[0.5\left(C_s + \left(C_s^2 + 4\right)^{1/2}\right)\right]^{1/3} + \left[0.5\left(C_s - \left(C_s^2 + 4\right)^{1/2}\right)\right]^{1/3} \qquad (13.42)$$

The value of c can then be obtained from the second of (13.40), by substitution of this value of C_v for $[\exp(\sigma_{nc}^2) - 1]^{1/2}$. It should be pointed out that this technique does not always produce good results; indeed, the third- and higher-order moments for most hydrologic data sets tend to be unreliable, so that some other method to estimate the parameters may be preferable. For instance, Stedinger (1980) has proposed a quantile method using the median and the smallest and largest observed values.

The idea of the lognormal distribution appears to have been introduced into hydrologic practice by Horton (1914) after a suggestion by his uncle George W. Rafter in the 1890s. But Hazen (1914a) was probably the first to state explicitly, ". . . that if the logarithms of the numbers representing the several floods are used instead of the numbers themselves, the agreement with the normal law of error is closer." For many years after that the lognormal distribution was the main function used in the United States to describe annual maximal river discharges for design purposes.

Example 13.6. Lognormal distribution applied to annual peak streamflows

The flow rate in the Cayuga Inlet near Ithaca, NY, has been measured since 1935 and the data have been published by the US Geological Survey (see also the web site http://waterdata.usgs.gov/nwis). The gaging station is located at 42°23′35″ N, 76°32′43″ W, at 133 m above sea level, and it has a drainage area of 91.2 km^2; after correction of the original precipitation data (Korzoun et al., 1977) the average annual precipitation was estimated to be of the order of 1100 mm. After sorting the annual peak flows of the 66 available years of record, each of them was assigned a probability $m/67$; the resulting data points are plotted with lognormal coordinates in Figure 13.9. The sample estimates of the moments of these flow rates were then calculated by means of (13.13) and found to be $M = 44.16\,\mathrm{m^3\,s^{-1}}$, $S = 33.51\,\mathrm{m^3\,s^{-1}}$ and $g_s = 1.969$. For the logarithms the values of these moments were found to be $M = 3.557$, $S = 0.6793$ and $g_s = 0.1281$. The small skew coefficient g_s of the logarithms suggests that the data are close to lognormally distributed and that c in (13.39) can probably be neglected. This is confirmed by the fact that the theoretical line, i.e. (13.35) or (13.36) applied with the logarithms of the data and with the values of M and S of the logarithms, and shown as the heavy straight line 1 in Figure 13.9, provides a good fit to the data. For comparison, some other theoretical curves are also shown in Figure 13.9. The generalized log gamma distribution (see Section 13.4.4) was calculated with the first three moments of the logarithms mentioned a few sentences earlier. The first asymptotic distribution (see Section 13.4.5 below) was calculated with the parameter values $\alpha_n = 0.038\,27$ and $u_n = 29.08\,\mathrm{m^3\,s^{-1}}$, and the generalized extreme value distribution (see Section 13.4.7 below) with the parameter values $a = -0.1057$, $b = 22.24$ and $c = 28.75\,\mathrm{m^3\,s^{-1}}$; the power distribution was applied with the parameters $a = 28.23$, and $b = 0.4745$.

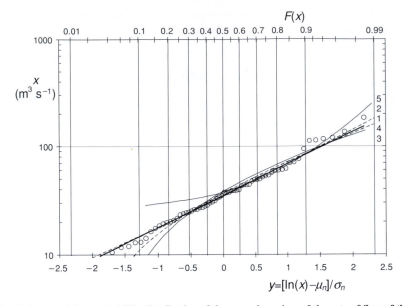

Fig. 13.9 Estimates of the probability distribution of the annual maxima of the rate of flow of the Cayuga Inlet near Ithaca, NY, plotted on lognormal probability paper. The heavy straight line 1 represents the lognormal distribution, which was calculated with the first two sample moments of the logarithms $M = 3.557$, and $S = 0.6793$. Also shown are the generalized log-gamma distribution (dashed line 2 curving upward), the first asymptote for largest values (thin solid line 3 curving downward), the generalized extreme value distribution (dashed line 4 curving downward) and the power distribution (solid curve 5). Both the y-scale and the $F(x)$-scale are shown. (See Example 13.6.)

13.4.3   *The generalized gamma distribution*

This distribution, which is often referred to as the Pearson Type III distribution, is a generalized form of the incomplete gamma function, by the inclusion of a lower bound c. Its density can be written as follows

$$f(x) = \frac{1}{b\Gamma(a)} \left(\frac{x-c}{b}\right)^{a-1} \exp\left(\frac{-(x-c)}{b}\right) \tag{13.43}$$

Except for the shifted origin, this has the same form as Equation (12.41). The three parameters can be related to the first three moments by $c = \mu - \sigma a^{0.5}$, $b = \sigma a^{-0.5}$, and $a = 4/C_s^2$, with $a > 0$, and $b > 0$ when $x > c$; if c is known, only the first two of these are needed (cf. Equations (12.42) and (12.43)). Once the parameters are known, the distribution function, which is the integral of (13.43), can be obtained from tables of the incomplete gamma function (Abramowitz and Stegun, 1964) applied to the variable $(x - c)$. An alternative method determines the probability function from its quantiles x_p, which in turn are obtained from the tabulated quantiles of the reduced variable $y_p = (x_p - \mu)/\sigma$; Table 13.2 shows these quantiles as functions of the skew coefficient, i.e. $y_p = y_p(C_s)$. Note that Table 13.2 yields the values of the normal distribution

Table 13.2 The value of the reduced quantile $y_p = (x_p - \mu)/\sigma$ as a function of skew coefficient C_s (or its sample estimate g_s) for the generalized gamma distribution

	Return period, T_r							
	1.25	2	5	10	25	50	100	200
	Probability of non-exceedance, $F(x_p)$							
	0.20	0.50	0.80	0.90	0.96	0.98	0.99	0.995
Skew C_s								
3.0	−0.636	−0.396	0.420	1.180	2.278	3.152	4.051	4.970
2.5	−0.711	−0.360	0.518	1.250	2.262	3.048	3.845	4.652
2.0	−0.777	−0.307	0.609	1.302	2.219	2.912	3.605	4.298
1.5	−0.825	−0.240	0.690	1.333	2.146	2.743	3.330	3.910
1.0	−0.852	−0.164	0.758	1.340	2.043	2.542	3.022	3.489
0.5	−0.856	−0.083	0.808	1.323	1.910	2.311	2.686	3.041
0	−0.842	0	0.842	1.282	1.751	2.054	2.326	2.576
−0.5	−0.808	0.083	0.856	1.216	1.567	1.777	1.955	2.108
−1.0	−0.758	0.164	0.852	1.128	1.366	1.492	1.588	1.664
−1.5	−0.690	0.240	0.825	1.018	1.157	1.217	1.256	1.282
−2.0	−0.609	0.307	0.777	0.895	0.959	0.980	0.990	0.995
−2.5	−0.518	0.360	0.711	0.771	0.793	0.798	0.799	0.800
−3.0	−0.420	0.396	0.636	0.660	0.666	0.666	0.667	0.667

when $C_s = 0$, i.e. when the skew is zero. For $0.01 \le F(x) \le 0.99$ and $C_s < 2$, the reduced quantiles can also be calculated with the Wilson–Hilferty approximation, namely

$$y_p = \frac{2}{C_s} \left[\left(1 - \frac{C_s^2}{36} + \frac{C_s y_{np}}{6} \right)^3 - 1 \right] \tag{13.44}$$

in which y_{np} is the corresponding quantile of the reduced or standardized normal variable; thus y_{np} values are the ones listed in Table 13.2 for $C_s = 0$, or may be obtained as the inverse of Equation (13.35) for $y_{np} = (x - \mu)/\sigma$. The accuracy of (13.44) and methods to improve it have been studied by Kirby (1972) and by Chowdhury and Stedinger (1991).

The application of the incomplete gamma distribution for flood frequency analysis, in the manner of Table 13.2, goes back to the work of Foster (1924). This distribution function has also been widely used for the same purpose in the Former Soviet Union (Sokolov, 1967). Matalas (1963) has observed that it can be used with success to represent low flow data, and that it performs equally well as the third asymptote for smallest values (see Section 13.4.6 below).

13.4.4 *The generalized log-gamma distribution*

A random variable is said to be described by the log-gamma distribution, also called the log-Pearson Type III distribution, when its logarithms obey a three-parameter gamma distribution. A common way (see Benson, 1968) of determining the probability with this distribution consists of the application of Table 13.2 with the first three moments of the logarithms of the data, as calculated with Equation (13.13).

In the late 1960s this distribution was recommended by a Federal interagency group for adoption by all government agencies in the United States for flood frequency analysis; this recommendation was arrived at (Benson, 1968; Thomas, 1985) to foster greater consistency and uniformity in planning for flood-plain management and water-resources development. It is now still widely used in the United States for this purpose. Further details and recommendations by this interagency group for the practical application of this distribution can be found in Bulletin 17B (Interagency Advisory Committee on Water Data, 1982). Beside the standard application of the method, Bulletin 17B also contains suggestions on plotting historical data (i.e. dating from prior to the period of record), and regionalization with data from hydrologically similar watersheds. A more comprehensive treatment of this distribution was presented by Bobée and Ashkar (1991).

As noted earlier, the skew coefficient tends to exhibit greater variability between samples than the mean and the variance. This may be overcome by regionalization (see Section 13.5.2 below). Several techniques have been used in the past to obtain a regional value, in place of the locally calculated value, if the data record is short. These include the construction of a map with iso-lines obtained by interpolation of the values computed at the existing gaging stations in the region; another possibility is the derivation of a regression relationship between the available skew values in the region and basin characteristics; finally, as a third possible approach, the skew may simply be taken as the average of all available skew values from the records in the region with long records; the average can also be weighted by multiplying each available value by the number of years of record at that gaging station divided by the average number of years of record of all stations in the region. Hardison (1974) has presented regional values of the skew coefficient for the annual peak flow rates in rivers in the United States, but with the availability of additional data since then, these results have gradually become obsolete. Tasker and Stedinger (1986) have further improved the estimation procedure of regional skew values.

Example 13.7. Log-gamma distribution applied to annual peak flows

In Council Creek, near Stillwater, Oklahoma, the flowrates have been measured from 1934 until 1993, but some peak flow information is also available for 1912; the data have been published by the US Geological Survey (see also http://waterdata.usgs.gov/nwis). The station is located at an elevation of 252 m above sea level, at $36°06'58''$ N and $96°52'03$ W in the Prairie region of North America and has a drainage area of 80.3 km^2; the corrected average annual precipitation was estimated (Korzoun *et al.*, 1977) to be

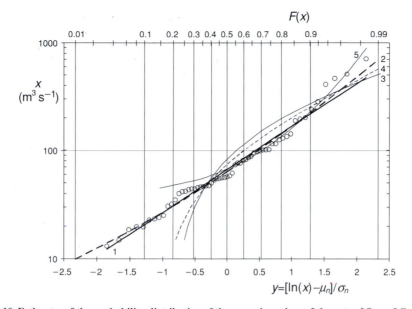

Fig. 13.10 **Estimates of the probability distribution of the annual maxima of the rate of flow of Council Creek near Stillwater, Oklahoma, plotted on lognormal probability paper. The heavy straight line 1 represents the lognormal distribution, which was calculated with the first two sample moments of the logarithms $M = 4.199$, and $S = 0.9145$. The generalized log-gamma distribution (dashed line 2 curving upward) was obtained with these same moments and with the sample skew $g_s = 0.3217$. Also shown are the first asymptote for largest values (solid line 3 curving downward), the generalized extreme value distribution (dashed line 4 curving downward) and the power distribution (curve 5). Both the y-scale and the $F(x)$-scale are shown. (See Example 13.7.)**

of the order of 1000 mm. The 61 available peak flow data points, which are displayed in Figure 13.10, have a sample mean $M = 104.5 \text{ m}^3 \text{ s}^{-1}$, a standard deviation $S = 127.9 \text{ m}^3 \text{ s}^{-1}$ and a skew coefficient $g_s = 2.964$; for the logarithms these same quantities are respectively 4.199, 0.9145, and 0.3217. The theoretical curve obtained with these moments of the logarithms by means of Equation (13.44) is shown as the upward curving dashed line 2 in Figure 13.10. Also shown in the figure are the theoretical curves for the lognormal distribution (with $c = 0$) (1), the first asymptote (3), the generalized extreme value distribution (4), and the power distribution (5). The parameters for the extreme value distributions obtained with the appropriate sample moments are respectively $\alpha_n = 0.01003 \text{ m}^{-3} \text{ s}$ and $u_n = 46.97 \text{ m}^3 \text{ s}^{-1}$, and $a = -0.1751$, $b = 74.01 \text{ m}^3 \text{ s}^{-1}$ and $c = 46.45 \text{ m}^3 \text{ s}^{-1}$; for the power distribution the parameters are $a = 39.91$ and $b = 0.7511$.

13.4.5   The first asymptotic distribution of extreme values

Extreme values and their initial distribution
When several samples consisting of, say, n items are taken from the same population, the mth smallest item in each sample is a random variable, which follows a certain distribution function. The form of this distribution function of the mth quantile depends

on m and n, and on the distribution function of the population from which the samples are drawn. When the distribution function of the population as a whole is mentioned in reference to the distribution function of the quantiles, it is often called the *initial distribution*. The extreme values of the samples are their smallest and their largest items. Obviously, the distribution functions of these extreme values of the samples depend only on n, the number of items in the samples, and on the initial distribution. When the size of the samples is very large, i.e. in the limit when $n \to \infty$, the distribution functions of the extreme values are called *asymptotic distribution functions* or *asymptotes*. Clearly, an extreme value asymptote no longer depends on m and n, but only on the nature of its initial distribution.

In the study of extreme values three general types of initial distribution $F(x)$ have been considered (Gumbel, 1954a; 1958). Each of these types results in a different functional form of the extreme value asymptotes. The first type, which is called the *exponential type*, comprises those distributions that for large x converge to unity at least as fast as the exponential function itself; all their moments exist. These types of distribution satisfy

$$\frac{f(x)}{1 - F(x)} = \frac{-f'(x)}{f(x)} \quad \text{and} \quad \frac{f(x)}{F(x)} = \frac{f'(x)}{f(x)} \tag{13.45}$$

for very large and for very small values of x, respectively. Since both numerator and denominator in these ratios are very small, this suggests the application of de L'Hospital's rule; hence one can continue the process, to obtain for very large x,

$$\frac{f(x)}{1 - F(x)} = \frac{-f'(x)}{f(x)} = \frac{-f''(x)}{f'(x)} = \cdots \tag{13.46}$$

and so on, and an analogous result for very small x. Examples of this type of distribution are the normal, the logistic, the gamma, and their logarithmically transformed distributions.

The distributions of the second type are also referred to as *Cauchy type* distributions; these are distributions, which do not have moments above a certain order. The *limited distributions* belong to the third type of initial distributions; these are distributions with an upper or lower bound or with both. In hydrology they are mainly of interest in the analysis of low flows and droughts. This brings up the point that the classification into these three separate types of initial distributions is not always rigid. For example, the lognormal distribution is of the exponential type at the upper end, since x can assume values all the way to infinity; however, it is of the limited type at the lower end of the distribution, because x cannot be smaller than zero or c, as can be seen in Equations (13.38) and (13.39).

The first asymptote for largest values
This distribution is also often referred to as the Gumbel distribution after the statistician who clarified and promoted its use (Gumbel, 1954a; 1958). Several derivations have been presented in the literature. One of the simpler derivations proceeds as follows. The starting point is the Taylor expansion of the initial distribution about some characteristic

large value of x, denoted by u_n, or

$$F(x) = F(u_n) + f(u_n)(x - u_n) + f'(u_n)\frac{(x - u_n)^2}{2!} + f''(u_n)\frac{(x - u_n)^3}{3!} + \cdots$$

$$(13.47)$$

From the definition of exponential type distributions (13.45) for large values it follows that

$$f'(u_n) = \frac{-[f(u_n)]^2}{1 - F(u_n)}; f''(u_n) = \frac{+[f(u_n)]^3}{[1 - F(u_n)]^2}; \text{ etc.} \qquad (13.48)$$

The value of u_n is fairly arbitrary, but if it is defined such that its probability is given by $F(u_n) = 1 - 1/n$, the derivation becomes especially straightforward. (Note that this is almost, but not quite, the average probability of the largest event in the sample, namely $1 - 1/(n + 1)$ shown in Equation (13.19)). After substitution of (13.48) and some algebra, (13.47) can be written as

$$F(x) = 1 - [1 - F(u_n)]\left[1 - \alpha_n(x - u_n) + \frac{\alpha_n^2(x - u_n)^2}{2!} - \frac{\alpha_n^3(x - u_n)^3}{3!} + \cdots\right]$$

in which, by definition $\alpha_n = f(u_n)/[1 - F(u_n)]$, which can be considered a constant parameter; thus, with the definition of u_n, this series becomes

$$F(x) = 1 - \frac{1}{n}\exp[-\alpha_n(x - u_n)] \qquad (13.49)$$

Recall that $F(x)$ is the initial distribution of the population from which the n items of the sample were taken, and that it indicates the probability that any item of the sample is smaller than or equal to x. The probability that all n items are smaller than or equal to x, is

$$G(x) = [F(x)]^n \qquad (13.50)$$

provided the items are independent of one another. Combination of Equations (13.49) and (13.50) produces

$$G(x) = \left(1 - \frac{1}{n}\exp[-\alpha_n(x - u_n)]\right)^n \qquad (13.51)$$

In the limit, as the size of the sample is allowed to increase indefinitely, so that $n \to \infty$, one obtains (Abramowitz and Stegun, 1964, 4.2.21) the asymptotic distribution function of the largest values

$$G(x) = \exp[-\exp(-y)] \qquad (13.52)$$

and the corresponding density function $g(x) = G'(x)$, or

$$g(x) = \alpha_n\exp[-y - \exp(-y)] \qquad (13.53)$$

where $y = \alpha_n(x - u_n)$ is the reduced largest value. The distribution and the density function of the extremes are denoted in this section by $G(x)$ and $g(x)$, merely to distinguish them from the initial distribution function and the initial density function, respectively.

Probability paper based on (13.52), often referred to as Gumbel paper, is illustrated in Figure 13.3.

In the derivation of Equation (13.52) its two parameters u_n and α_n were related to the properties of the initial distribution $F(x)$. In practice, however, the parameters are determined directly from the observed largest values. By means of the moment generating functions (Gumbel, 1958) the first two moments of (13.52) can be shown to be

$$\mu = u_n + \gamma/\alpha_n \quad \text{and} \quad \sigma = \pi/(\alpha_n \sqrt{6}) \tag{13.54}$$

where $\gamma = 0.577\,22$ is known as Euler's number. Thus the parameters u_n and α_n can be determined immediately from the first two moments of the largest values, as calculated with (13.13). Observe that, since the mean of the reduced variable is given by $\bar{y} = \gamma$ on account of the first of (13.54), it follows that $G(\mu) = \exp[-\exp(-0.577\,22)]=0.570$, so that $T_r(\mu) = 2.328$; in other words, the first asymptote (13.52) predicts that the return period of the mean is 2.33 time units (e.g. years in the case X, the random variable in question, represents the annual flood). Note also that by putting $\partial g(x)/\partial x = 0$ with (13.53), it is readily found that the mode is equal to u_n. Similarly, by putting $G(x) = 0.5$ with (13.52) one finds that the median is given by $(u_n + 0.36651/\alpha_n)$.

The first asymptote has been widely applied in the description of maxima. In hydrology, it has been especially useful in the analysis of annual floods, i.e. the yearly maximal discharges on record. It is useful to restate the assumptions on which the derivation is based, to gain a better understanding of its applicability. These are (i) the initial distribution is of the exponential type; (ii) the events, from among which the largest are considered, must be independent; (iii) the sample size n is infinitely large. In the case of the yearly floods, i.e. the maxima of the daily flows, these conditions are not really met. There is no doubt a strong correlation between successive daily flows; thus the truly independent events in one year are likely to last much longer than a single day; this greatly reduces their number from $n = 365$ to a value, which is probably not large enough to allow the use of an asymptote; this fact also obscures the nature of the initial distribution, which may not be exponential.

One of its practical disadvantages is that, since it has only two parameters, all moments above the second are related to the first two; this means, for example, that it has a third moment about the mean $m_3 = 2.404\,11/\alpha_n^3$, and therefore a constant skew coefficient, namely $C_s = m_3/\sigma^3 = 1.1395$. Only rarely does the estimated skew coefficient g_s of a record of annual floods have this value. Of course, other two-parameter distributions, such as the lognormal or gamma distributions, suffer the same drawback.

Another point of interest is the behavior of the first asymptote for extremely large events. Inversion of (13.52) with (13.15) yields the reduced variable as a function of the return period

$$y = -\ln\{\ln[1/(1 - 1/T_r)]\} \tag{13.55}$$

Since $\ln(1 + z) = z - z^2/2 + \cdots$ (provided $z \leq 1$ and $z \neq -1$)) (Abramowitz and Stegun, 1964, 4.1.24), it is readily shown that, for large T_r, Equation (13.55) can be written as

$$y = \ln(T_r) - \frac{1}{2T_r} \tag{13.56}$$

or, to a good approximation,

$$x = u_n + \alpha_n^{-1} \ln T_r \tag{13.57}$$

This shows that, if the largest events are plotted against T_r on semi-log graph paper, they should tend to a straight line in the range of very large values of T_r. This may be a useful procedure to apply, when no probability paper is available. It is remarkable also that Equation (13.57) is in the same form of, and thus provides a theoretical justification for, the equation proposed by Fuller (1914) to describe annual floods. Indeed, Fuller found empirically, ". . . by plottings of the existing data on American rivers" available to him, that the largest 24 h average rate of flow to be expected in T_r years is

$$Q = Q_{av}(1 + 0.8 \log T_r) \tag{13.58}$$

in which Q_{av} is the average annual flood and log denotes the decimal logarithm; Fuller (1914) also observed that Q_{av} is proportional to $A^{0.8}$, where A is the drainage area.

The first asymptote for smallest values
Whenever the initial distribution $F(x)$ is symmetrical about the origin, in accordance with Equation (13.11), the probability that an observation is larger than $-x$ is given by $[1 - F(-x)]$; hence the probability that the smallest in a sample of n independent observations is larger than $-x$ is

$$1 - {}_1G(-x) = [1 - F(-x)]^n \tag{13.59}$$

Proceeding in the same way as for the largest values, from (13.47) through (13.52), and making use of this symmetry, one obtains

$${}_1G(x) = 1 - \exp[-\exp(y)] \tag{13.60}$$

where as before $y = \alpha_n(x - u_n)$. Thus the first asymptote for smallest values can be obtained from that for the largest values by replacing x and u_n by $-x$ and $-u_n$, respectively. Most initial distributions are not symmetrical; but Gumbel (1958) has indicated how in the case of asymmetrical distributions the symmetry principle can be extended simply by adopting new parameters, say u_1 and α_1, instead of u_n and α_n. In other words, Equation (13.60) can still be applied but the reduced variable is $y = \alpha_1(x - u_1)$, and the parameters are derived from observations of the smallest values.

13.4.6 The third asymptotic distribution of extreme values

The third asymptote for largest values
This distribution is also known as the Weibull distribution for the Swedish engineer who first used it to analyze breaking strengths (Gumbel, 1954a; 1958). This third asymptote

is applicable to describe maxima when their initial distribution has an upper bound; if this upper bound is denoted by ω, it follows that the initial distribution is subject to

$$F(x) = 1 \quad \text{for } x = \omega$$

Several different derivations have been presented in the literature for this asymptote (see Gumbel, 1958, p. 273 ff.), which require certain assumptions regarding the manner in which $F(x)$ approaches unity. Probably the simplest, after Kimball (1942), consists of observing that the bounded variable $x \leq \omega$ can be transformed into an unbounded variable z as follows

$$\omega - x = a \exp[-b(z - c)] \tag{13.61}$$

where a, b and c are constants; this shows how $z \to \infty$ as $x \to \omega$. If it can be assumed that the resulting distribution function of z is of the exponential type, then according to (13.49), the initial distribution of x in the neighborhood of the upper bound $x = \omega$, can be described by

$$F(x) = 1 - \frac{1}{n}\left(\frac{\omega - x}{\omega - v}\right)^k \tag{13.62}$$

where the parameters in (13.61) have been changed to $a = \omega - v$, $b = \alpha_n/k$ and $c = u_n$. Proceeding as before, one finds immediately that the probability that all items in a very large sample are smaller than or equal to x, in the limit as $n \to \infty$, is

$$G_3(x) = \exp\left[-\left(\frac{\omega - x}{\omega - v}\right)^k\right] \tag{13.63}$$

The corresponding third asymptotic density function $g_3(x) = G_3'(x)$ is

$$g_3(x) = \frac{k}{\omega - v}\left(\frac{\omega - x}{\omega - v}\right)^{k-1} G_3(x) \tag{13.64}$$

The moments of the third asymptote are treated in detail below for the smallest values.

The third asymptote for smallest values

In hydrology it is mainly the third asymptotic distribution for smallest values that has been of interest. Indeed while, in principle at least, different types of common events, such as rainfall amounts, wind speeds or river flows, can often be assumed to be unlimited in magnitude, even the smallest of such events can never be smaller than zero. Thus the smallest values often have a definite lower limit below which they cannot go. As for the first asymptote, the symmetry principle (13.60) can be applied to derive the distribution of the smallest values from that of the largest values. The procedure consists of changing the sign of x, ω and v, and then assigning different values to the parameters, say ω_1 and v_1, to obtain

$$_1G_3(x) = 1 - \exp\left[-\left(\frac{x - \omega_1}{v_1 - \omega_1}\right)^k\right] \tag{13.65}$$

in which ω_1 is the lower limit, such that $x \geq \omega_1$ and $v_1 \geq \omega_1$ and also $_1G_3(\omega_1) = 0$. The corresponding density function is

$$_1g_3(x) = \frac{k}{v_1 - \omega_1} \left(\frac{x - \omega_1}{v_1 - \omega_1} \right)^k [1 - _1G_3(x)] \tag{13.66}$$

The nth moment about ω_1 can be determined as follows

$$m'_n = \int_0^\infty (x - \omega_1)^n {}_1g_3(x)dx$$
$$= -\int_0^\infty (x - \omega_1)^n d[1 - _1G_3(x)] \tag{13.67}$$

which yields, with (13.65),

$$m'_n = (v_1 - \omega_1)^n \Gamma(1 + n/k) \tag{13.68}$$

Hence the mean is

$$\mu = \omega_1 + (v_1 - \omega_1)\Gamma(1 + 1/k) \tag{13.69}$$

Similarly, since the second moment about ω_1 is $m'_2 = (v_1 - \omega_1)^2\Gamma(1 + 2/k)$; in light of (13.12), the variance is

$$\sigma^2 = (v_1 - \omega_1)^2 [\Gamma(1 + 2/k) - \Gamma^2(1 + 1/k)] \tag{13.70}$$

Higher-order moments can also readily be derived by means of Equation (13.68). In addition, it is easy to show that the median is given by $[\omega_1 + (v_1 - \omega_1)(\ln 2)^{1/k}]$ and the mode, which exists only for $k > 1$, by $[\omega_1 + (v_1 - \omega_1)(1 - 1/k)^{1/k}]$.

In practical applications, generally the only known fact is that the initial distribution is bounded on the left but that distribution itself is unknown. Thus, as was the case with the first asymptote for the largest values, the parameters can only be determined from the available smallest values. In case one out of the three parameters is known (usually the lowest value ω_1), the method of moments will require only the first two moments (13.69) and (13.70). If all three parameters have to be determined, the first three moments could be used, in principle. However, as noted earlier, the third moment is often unreliable, and a different approach may be desirable. Gumbel (1954a) has described a rapid method after a procedure developed by Weibull. First the value of v_1 is determined; since $_1G_3(v_1) = 1 - \exp(-1)$, one can take v_1 as the value of x which has an observed probability of 0.632 or, which is the same from Equation (13.18), a return period of $T_r = 1.58$ time units. For example, in the case of annual low flows or "droughts," v_1 can be taken as the magnitude of the 1.58 y event. The value of k can be determined from the probability of the mean $_1G_3(\mu)$, as observed from its plotting position. Thus k is the solution of the combination of (13.65) with (13.69), that is

$$_1G_3(\mu) = 1 - \exp[-\Gamma^k(1 + 1/k)] \tag{13.71}$$

The value of the lower limit ω_1 can then be determined from the variance, as given by Equation (13.70), in which v_1 and k are already known.

Finally, it should be noted that, just like the corresponding distributions for the largest values in (13.61), the first asymptote for smallest values is linked to the third by a logarithmic transformation, as follows

$$\ln\left(\frac{x - \omega_1}{v_1 - \omega_1}\right) = b(z - u_1) \tag{13.72}$$

where now $b = \alpha_1/k$. Hence, with (13.60) one obtains

$$_1G(z) = 1 - \exp\left[-\exp\left(\ln\left(\frac{x - \omega_1}{v_1 - \omega_1}\right)^k\right)\right] = {}_1G_3(x) \tag{13.73}$$

as given in (13.65). This shows how probability paper constructed for the first asymptote can be used for the third asymptote by plotting the logarithms of the magnitudes of the events, i.e. $\log x$ instead of x (see Example 13.1). Hence, the lay-out shown in Figure 13.3 can be used for this purpose, by changing the scale of the ordinate from linear to logarithmic.

Applications of this distribution to low flows in rivers have been presented by Gumbel (1954b) and Matalas (1963).

13.4.7 *The generalized extreme value distribution*

The first and third asymptotes were already shown in Equations (13.61) and (13.72) to be related by a logarithmic transformation. It should not be surprising, therefore, that the three asymptotes can be combined into a single expression. So far in hydrology, this idea has mostly been applied to the largest values. In this case the distribution function is usually written in the following form

$$F(x) = \exp[-(1 - a(x - c)/b)^{1/a}] \quad \text{for } a \neq 0 \tag{13.74}$$

in which a, b and c are constants. Clearly, when $a \to 0$, the term inside the square brackets approaches an exponential function, and Equation (13.74) reduces to the first asymptote (13.52). But in (13.74), a is not necessarily equal to zero; thus in this form the extreme value distribution has three parameters, and can therefore be considered more general. When $a > 0$, (13.74) is just another form of the third asymptote for largest values (13.63), with an upper bound at $x = c + b/a$; the parameters of the two forms are related by $k = a^{-1}$, $\omega = c + b/a$, and $v = c$. When $a < 0$, (13.74) has a lower bound at $x = c + b/a$ but it is unbounded for large x; therefore, this case is the one mostly used for the largest values. The density function corresponding to Equation (13.74) is

$$f(x) = \frac{1}{b}(1 - a(x - c)/b)^{-1+1/a} F(x) \quad \text{for } a \neq 0 \tag{13.75}$$

The central moments can be derived by first considering the nth moment about $x = x_1 = c + b/a$; in the case of $a < 0$, when x_1 is the lower bound, this is

$$m_n' = \int_{x_1}^{\infty} (x - c - b/a)^n dF(x) \tag{13.76}$$

Fig. 13.11 The parameter a in the generalized extreme value distribution (13.74), as a function of the skew coefficient C_s.

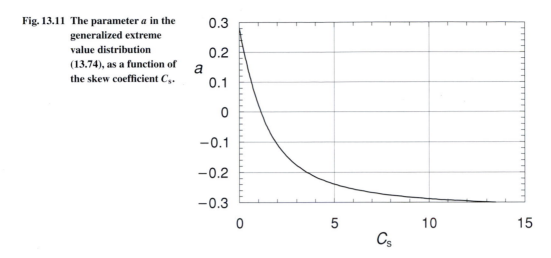

or

$$m'_n = \int_{x_1}^{\infty} (x - x_1)^n \exp[-(-a/b)^{1/a}(x - x_1)^{1/a}]d[-(-a/b)^{1/a}(x - x_1)^{1/a}]$$

(13.77)

This produces finally

$$m'_n = (-b/a)^n \Gamma(1 + an) \quad \text{with } a > -1/n$$

(13.78)

Hence the mean is $m'_1 + x_1$, or

$$\mu = c + (b/a)(1 - \Gamma(1 + a))$$

(13.79)

Similarly, on account of (13.12) the variance is

$$\sigma^2 = (b/a)^2[\Gamma(1 + 2a) - (\Gamma(1 + a))^2]$$

(13.80)

and the third central moment is

$$m_3 = -(b/a)^3[\Gamma(1 + 3a) - 3\Gamma(1 + a)\Gamma(1 + 2a) + 2(\Gamma(1 + a))^3]$$

(13.81)

As before, these first three moments can be used to estimate the parameters a, b and c. First, the parameter a can be determined by iteration from the sample skew coefficient g_s (see Equation (13.13)), expressed in terms of the ratio of Equations (13.81) and (13.80); a rough idea of the magnitude of a can be obtained from Figure 13.11. With this result b can be obtained from the sample variance S^2 and (13.80), and then c from the sample mean M with (13.79). If the data record is so short that the third moment must be considered unreliable, one can also apply the Weibull procedure, explained earlier for the third asymptote for smallest values. In brief, this consists of observing that Equation (13.74) produces $F(c) = \exp(-1)$; thus the parameter c can be estimated immediately from the available data as the value of x, which corresponds with a probability

$m/(n + 1) = 0.368$, or with a return period of $T_r = 1.58$ time units. The two remaining parameters a and b can then be determined from the first two moments (13.79) and (13.80).

The generalized extreme value distribution in the form of (13.74) was introduced in the environmental sciences by Jenkinson (1955), and it has subsequently found wide application in the prediction of various extreme phenomena, such as floods, rain events, wind speeds and wave heights; it has also come to be used in the estimation of regional flood frequencies (see, for example, Lettenmaier et al., 1987; Stedinger and Lu, 1995; Madsen et al., 1997; Martins and Stedinger, 2000). Its full potential continues to be explored (Katz et al., 2002).

Example 13.8. Extreme value distributions applied to annual peak flows

In this example a stream in a more arid climate is considered. At Palominas in Arizona the San Pedro River drains an area of some 1909 km^2, almost all in Sonora; the corrected average annual precipitation in this area was estimated (Korzoun et al., 1977) to be of the order of 400 mm. This gaging station is located at $31°22'48''$ N, $110°06'38''$ W, at an elevation of 1276 m above sea level. The 61 available annual peak flows measured from 1930 through 1999 are plotted against $T_r = 62/(62 - m)$ with first asymptotic coordinates in Figure 13.12. The first three moments of these data were estimated with Equation (13.13) as $M = 180.2 \, \text{m}^3 \, \text{s}^{-1}$, $S = 115.2 \, \text{m}^3 \, \text{s}^{-1}$ and $g_s = 1.436$; the corresponding moments of the logarithms were calculated to be, respectively, 5.000, 0.6466 and -0.2444. By means of Equation (13.54) the two parameters of the first asymptotic distribution for largest values were estimated as $\alpha_n = 0.011 \, 13 \, \text{sm}^{-3}$ and $u_n = 128.4 \, \text{m}^3 \, \text{s}^{-1}$; the curve calculated with (13.52) is shown in Figure 13.12 as the solid heavy straight line 3. Interestingly, it can be seen in the graph that the mean $M = 180.2 \, \text{m}^3 \, \text{s}^{-1}$ corresponds closely with a a value of the reduced variable $y = 0.58$, and with a return period $T_r = 2.33$ y; this is to be expected in light of the first of (13.54). The parameters of the generalized extreme value distribution were calculated with (13.79) through (13.81) as $a = -0.044 \, 92$, $b = 84.38 \, \text{m}^3 \, \text{s}^{-1}$ and $c = 127.6 \, \text{m}^3 \, \text{s}^{-1}$; the curve calculated with these parameters in (13.74) is shown as the heavy dashed line 4 in Figure 13.12). Again, it can be seen that, as expected from (13.74), the value of c corresponds closely with a return period of $T_r = 1.5$ y. For comparison, the curves based on the lognormal (with $c = 0$) (1), the generalized log-gamma (2), and the power distribution (5) are also shown in the figure. For the power distribution the parameters were taken as $a = 134.4$ and $b = 0.3854$.

13.4.8 Power law (or fractal) distribution

Many natural phenomena exhibit a type of self-similarity or scale invariance in their magnitudes, such that, for instance, the ratio of the event with return period $T_r = 100$ and that with $T_r = 10$, is equal to the ratio of those with $T_r = 1000$ and $T_r = 100$. Phenomena with this type of behavior are referred to as fractals (Turcotte, 1992). From this observation it follows that such phenomena obey a power law. Indeed, in this example

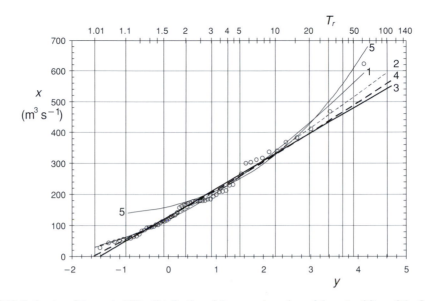

Fig. 13.12 Estimates of the probability distribution of the annual maxima of the rate of flow of the San Pedro River at Palominas, Arizona, plotted with first asymptotic coordinates. The heavy straight line (3) represents the first asymptotic distribution for largest values, which was calculated with the first two sample moments $M = 180.2$ m^3 s^{-1}, and $S = 115.2$ m^3 s^{-1}; the heavy dashed curve (4) represents the generalized extreme value distribution calculated with the same moments and with $g_s = 1.436$. Also shown are the lognormal distribution (thin solid curve 1), the generalized log-gamma distribution (dashed curve 2), and the power distribution (solid curve 5). Both the $y = \alpha_n(x - u_n)$ scale and the $T_r(x)$ scale (in years) are shown. (See Example 13.8.)

$x = x(T_r)$ satisfies

$$\frac{x(10)}{x(1)} = \frac{x(100)}{x(10)} = \cdots = \frac{x(10^n)}{x(10^{n-1})} = K_{10} \tag{13.82}$$

where K_{10} is a constant, in which the subscript indicates the ratio of the return periods. Thus, for the case of, say, a ratio of 2, the magnitude of an event with $T_r = 2^n$ is, by analogy with (13.82),

$$x(T_r) = K_2^n x(1) \tag{13.83}$$

Because $n = \ln T_r / \ln 2$, the logarithm of (13.83) can be rewritten as

$$\ln x(T_r) = (\ln K_2 / \ln 2) \ln T_r + \ln x(1)$$

which immediately results in a power law

$$x(T_r) = a T_r^b \tag{13.84}$$

with the constants $a = x(1)$ and $b = [\ln K_2 / \ln 2]$. Observe that the result obtained in (13.84) can be derived for any ratio of the return periods. Equation (13.84) with (13.15) yields the probability distribution function

$$F(x) = 1 - (x/a)^{-1/b} \tag{13.85}$$

which has a lower bound at $x = a$. The corresponding density function is

$$f(x) = \frac{a^{1/b}}{b} x^{-1-1/b} \tag{13.86}$$

In the practical application of this function, the parameters a and b can be derived simply by least squares linear regression of the logs of the observed values X against the logs of their return periods T_r, in accordance with Equation (13.84).

The power distribution has been found useful in the description of numerous phenomena, such as fragmentation, earthquakes, volcanic eruptions, mineral deposits, and land forms, among others. In hydrology, the power distribution probably found its earliest application in the description of rainfall intensities. Equation (3.3), whose origins go back at least to the work of Meyer (1917), is in the form of (13.84). (See also Figure 3.16.) A noteworthy feature in this particular application of Equation (13.84) is that its coefficient a is also a power function of the duration D of the rainfall event, for values of D in excess of 2 h. In a different context, namely in the description of capillary retention of water in soils, the form of Equations (8.14), (8.15) and (8.16), after substitution of (8.5), suggests a power distribution and fractal features of the smaller pores; this is illustrated for a sand in the example of Figure 8.20, indicating a straight line for large values of the capillary suction H.

More recently, the power distribution has also been used to describe flow maxima. Turcotte (1994) and Malamud et al. (1996) have presented cases where it provided a better fit with flood data than the generalized log-gamma distribution. However, the distribution appears to be more applicable to partial duration flow data than to annual flow maxima. Partial duration flow series contain all the data above a given pre-defined base, whereas annual flow series contain only the peak discharge rates observed during each year of the record. The main disadvantage of an annual series is that in some years a number of events may be larger than the annual event in other years. In the analysis of very large events this is rarely a problem, because the two types of data series tend to be nearly the same for events with return periods in excess of about three time units, years in this case. Hence, in the estimation of the parameters a and b in the power distribution for annual peak flows, it is advisable to use only the data whose return period is larger than 3 y, or whose probability of non-exceedance is larger than 0.67. This is illustrated below in Example 13.9. The performance of the power distribution, to describe annual peak flows in comparison with other distributions, is also illustrated Figures 13.9, 13.10, 13.12 and 13.14. It can be seen that it tends to produce smaller values of the non-exceedance probability and of the return period, and therefore will usually lead to more conservative design values.

Example 13.9. Power law distribution applied to annual peak flows

The Sheepscot River, at North Whitefield, Maine, drains a basin which is subject to strong maritime influence. The gaging station is located at 31 m above sea level at 44°13′23″ N and 69°35′38″ W; the upstream drainage area covers 376 km^2, and the corrected (Korzoun et al., 1977) long term average annual precipitation is around 1300 mm.

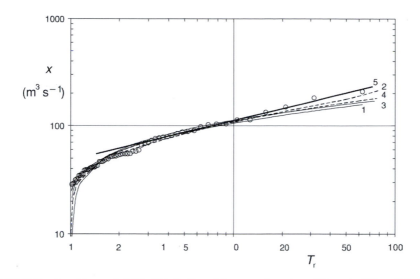

Fig. 13.13 Estimates of the probability distribution of the annual maxima of the rate of flow of the Sheepscot
River at North Whitefield, Maine, plotted with log–log coordinates. The heavy straight line
(5) represents the power distribution, which was calculated with the parameters $a = 48.10$, and
$b = 0.3657$. Also shown are the lognormal distribution (thin solid curve 1), the generalized log-gamma
distribution (dashed curve 2), the first asymptotic distribution for largest values (solid curve 3) and
the generalized extreme value distribution (dashed line 4). The $T_r(x)$ scale is shown in years. (See
Example 13.9.)

Measurements started in 1939, and the data have been published by the US Geological
Survey (see also http://waterdata.usgs.gov/). The 62 available values of the annual peak
flows are plotted against $T_r = 63/(63 - m)$ with log–log coordinates in Figure 13.13,
and with first asymptotic coordinates in Figure 13.14. It can be seen in Figure 13.13
that the data plot roughly along a straight line relationship for $T_r > 3$ y. Accordingly,
linear regression of the logs of the flow rates against the logs of the return periods in
excess of 3 y yielded the values of the parameters $a = 48.10$, and $b = 0.3657$. Equation
(13.84) with these parameters is plotted in Figures 13.13 and 13.14 as the heavy line 5.
Also shown in these two figures are the curves representing the lognormal distribution
(with $c = 0$), the generalized log-gamma distribution, the first asymptotic distribution
and the generalized extreme value distribution. The first three moments of the flow
rates used in the estimation of the parameters of these distributions were obtained with
Equation (13.13) as $M = 65.27 \, \text{m}^3 \, \text{s}^{-1}$, $S = 36.00 \, \text{m}^3 \, \text{s}^{-1}$ and $g_s = 1.926$, and the same
moments of the logarithms as 4.063, 0.4630 and 0.6788, respectively.

13.5   EXTENSION OF AVAILABLE RECORDS

13.5.1   *Historical information*

So far in this chapter the different methods of analysis have focused on data that are part
of a regular record of measurements carried out for a certain well defined purpose. In
many situations, however, additional information on the frequency of occurrence of the

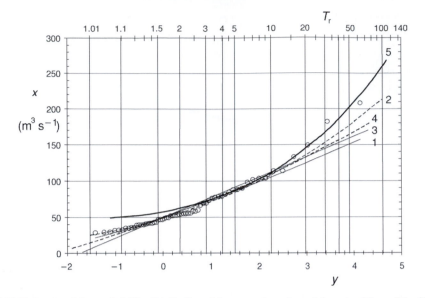

Fig. 13.14 **Estimates of the probability distribution of the annual maxima of the rate of flow of the Sheepscot River at North Whitefield, Maine, plotted with first asymptotic coordinates. The heavy solid line (5) represents the power distribution, which was calculated with the parameters $a = 48.10$, and $b = 0.3657$. Also shown are the lognormal distribution (thin solid curve 1), the generalized log-gamma distribution (dashed curve 2), the first asymptotic distribution for largest values (solid straight line 3) and the generalized extreme value distribution (dashed line 4). Both the $y = \alpha_n(x - u_n)$ scale and the $T_r(x)$ scale (in years) are shown. (See Example 13.9.)**

hydrologic phenomenon in question may be derived from a knowledge of events that date back to the time prior to the beginning of the systematic measurements. Because most hydrologic records are notoriously short, it is desirable to include such information whenever possible. Historical information may be derived from archived documents, or from evidence of a botanical or paleo-event (see Kochel and Baker, 1982; Stedinger and Baker, 1987) nature found in the natural environment.

Return periods

In practice, inclusion of information from the period prior to the start of the record is concerned with the assignment of a plotting position to each historical event, whose magnitude is known, in order to estimate the return period. As an illustration of possible scenarios for annual floods, consider the three cases discussed by Dalrymple (1960).

(i)   A single historical event, larger than any event during the regular period of record, is known to have occurred earlier. If N is the number of years since that historical event or, better, the time since the beginning of recorded historical information, the return period of the historical event can be taken as $(N + 1)$, and its probability as $[N/(N + 1)]$. The regular record of duration n is treated as usual.

(ii)   An historical event is known to have occurred and is the largest ever, until an even larger event occurs during the period of record. In this case the largest event during the period of record is assigned the return period $T_r = (N + 1)$, and the historical event $T_r = [(N + 1)/2]$. The remainder of the record is treated as usual with the second largest during the period of record assigned a value of $T_r = [(n + 1)/2]$, and so on for the third, etc.

(iii)   An historical record is available of all events above a certain base, such as for example "bankful stage," and it can be assumed that the distribution of the lesser events during the regular period of record is typical for that of the entire historical period. When the return periods or the corresponding plotting positions are obtained as outlined in the previous two cases (i) and (ii), there is a gap between data points of the regular record and those of the historical events, which causes some difficulty in deriving a best-fit curve. Such difficulty can be avoided, or at least alleviated, by means of Benson's (1950) procedure; this consists of weighting the lesser events (i.e. those below base) of the period of record more heavily by adjusting or "stretching" their order numbers, so that they cover the historical period. Consider H to denote the length of the historic period (e.g. the number of years since the first historical information became available until the present), Z the total number of events above base over that period, N the number of events below base during the period of record, and L the number of events that cannot be used (e.g. incomplete or missing records due to faulty equipment, etc.) during the period of record. Thus the weight assigned to each of the N lesser events is

$$W = \frac{(H - Z)}{(N + L)} \tag{13.87}$$

and their adjusted order number is

$$m' = Wm \tag{13.88}$$

For example, if the regular record consists of annual observations, by this procedure each data point below base is made to represent W years instead of 1 y. The plotting positions and return periods of the lesser events can now be determined as before in Section 13.2.4 but with the adjusted order number m'; for instance, with the Weibull plotting position, these are $P_m = m'/(H + 1)$ and $T_r = (H + 1)/(H + 1 - m')$. The larger events (i.e. those above base) are not weighted, but treated as usual, and their order numbers are not adjusted; thus they are in increasing magnitude $(H - Z + 1), (H - Z + 2), \dots, (H - 1), H$.

Estimation of moments

The same weighting method was also recommended in Bulletin 17B (Interagency Advisory Committee on Water Data, 1982) to adjust the moments for the parameter estimation of the generalized log-gamma distribution. From Equation (13.13) it follows immediately that, when the lesser observations are weighted in accordance with (13.87), the adjusted moments can be calculated from the data as follows

$$\widehat{M} = \frac{W \sum_{i=1}^{N} X_{Bi} + \sum_{i=1}^{Z} X_{Ai}}{H - WL}$$

$$\widehat{S}^2 = \frac{W \sum_{i=1}^{N} (X_{Bi} - \widehat{M})^2 + \sum_{i=1}^{Z} (X_{Ai} - \widehat{M})^2}{H - WL - 1}$$

$$\widehat{g}_s = \frac{(H - WL)}{(H - WL - 1)(H - WL - 2)\widehat{S}^3} \left[W \sum_{i=1}^{N} (X_{Bi} - \widehat{M})^3 + \sum_{i=1}^{Z} (X_{Ai} - \widehat{M})^3 \right]$$

(13.89)

where the circumflex denotes the adjusted moment, X_{Bi} is one of the N observations below base during the period of record, and X_{Ai} one of the Z observations above base during the entire historical period to the present.

Although commonly used in practice, the weighting method leading to Equation (13.88) for the plotting positions and to Equations (13.89) for the moments has its drawbacks (Hirsch and Stedinger, 1987). Better but more intricate methods have been proposed in the literature to accomplish the same objective. For instance, Cohn *et al.* (1997; 2001) have developed a procedure to estimate the parameters by means of the method of moments for the generalized log-gamma distribution, which was found to be more efficient than Equations (13.89), and which is nearly as efficient as the method of maximum likelihood.

13.5.2 *Regionalization*

Hydrologic data records are rarely available where they are needed. Moreover, even at the locations where a record is available, it is often too short to allow the reliable determination of the true distribution of the phenomenon of interest. Regional analysis, or regionalization, refers to the extension of available records in space. Its dual objective is to improve the record at regular measuring sites, and to provide estimates of frequency characteristics at sites, where no data are available. In what follows several methods are reviewed which have been found useful in the analysis of flood peaks.

Index-flood method

This method is probably the oldest and, as described by Dalrymple (1960), for many years it was the standard procedure used by the US Geological Survey. The underlying idea is that in a hydrologically homogeneous region the flood distribution functions for different streams are similar; in this case similarity means that, when the distribution functions are scaled with their respective index-flood, the resulting dimensionless distributions of all basins in the region can be assumed to have the same shape, which is independent of drainage area and of any other basin characteristics. Accordingly, the method comprises two components. The first component consists of a regional flood frequency curve. To derive this curve, first the flood distribution curve of each streamflow gaging site in the region is made dimensionless, that is normalized, by dividing the flow rates by the index-flood of the site; this index-flood is usually taken as the sample mean annual flood, but other measures, such as quantiles (Smith, 1989), have also been suggested. The regional flood frequency curve is then constructed as the average or the median curve of the available dimensionless curves. The second component of the method consists of a relationship between the magnitude of the index-floods and easily obtainable basin and climate characteristics. In principle, many different characteristics can be used for this purpose; however, in past practice usually only the drainage area has been considered as the significant characteristic. To summarize, the end products of the analysis of the available flow data are a dimensionless regional frequency curve, and a graph or a regression equation relating the index event with drainage area. These two relationships can then be used to predict the frequency curve for any ungaged catchment. In practical applications, the index event is first estimated from the drainage area of the ungaged catchment, and as mentioned possibly from other relevant characteristics; this index event is used in turn to dimensionalize the regional frequency curve. While the analysis to develop these two relationships is simple in principle, it also requires adjustment of all available records to a common base period, normally that of the station with the longest record. Examples of the application of this method can be found in Cruff and Rantz (1965) for coastal basins in California, and in Robison (1961) for New York State. In many studies based on this approach the mean floods, often taken as the events with $T_r = 2.33$ y (cf. Equation (13.54)), were found to be related to the drainage area by an equation of the power type

$$Q_{2.33} = aA^b \qquad (13.90)$$

where a and b are constants for a hydrologically homogeneous region. For most regions b was typically found to lie in the range between roughly 0.65 and 1.00; this is consistent with Fuller's (1914) earlier finding in relation to Equation (13.58).

The main difficulty experienced in applying the method is that, although tests have been proposed for this purpose, it is not immediately clear how a homogeneous region can be defined or delineated in terms of frequency curves with a similar shape and in terms of hydrologically relevant basin characteristics. A more serious problem is that the frequency is scaled with only one parameter, namely the index event, usually taken as the first moment. Thus it is implicitly assumed that higher moments have no effect, or that these higher moments (when made dimensionless as C_v and C_s) are constant within the region of hydrologic homogeneity. The limitations of this assumption have been studied (see Smith, 1992; Gupta et al., 1994; Stedinger and Lu, 1995; Robinson and Sivapalan, 1997a, b; Blöschl and Sivapalan, 1997). The method continues to be investigated (Hosking and Wallis, 1997).

Quantile estimation with multiple regression

In this approach, first the frequency curves are constructed for the stations for which data are available within the region of hydrologic homogeneity. On all these frequency curves the values of the quantiles Q_T are noted at several selected return periods, say, $T_r = 2$ (or 2.33), 5, 10, 20, 50, 100 and even 200 y. Each set of Q_T values is then related with relevant basin, climate or other characteristics, $B, C, D, \ldots$, as explanatory variables, by linear regression in a stepwise manner with an equation of the type

$$Q_T = a B^b C^c D^d \dots \tag{13.91}$$

in which $b, c, d, \dots$, are constants, whose values depend on the return period of the quantile. Characteristics to be considered may include drainage area, main channel slope, main channel length, mean annual precipitation, fraction of area with lakes and ponds, mean annual runoff, T_r y 24 h rainfall, mean basin altitude, fraction of basin area covered with forest, basin shape as ratio of main channel length and area, mean basin elevation, and possibly others. The final selection of the characteristics to be included can be made on the basis of their respective statistical significance and on the basis of the reduction of the standard error caused by their inclusion.

The basic idea and early applications of the method to the quantiles of annual floods were described by Benson (1962a, b) and Cruff and Rantz (1965). The method was also explored for other streamflow characteristics, beside the annual maxima, by Thomas and Benson (1970). The quantile regression procedure has subsequently been refined by means of a generalized least squares (GLS) procedure (Tasker and Stedinger, 1986; 1989) to take account of the fact that the stations may have records of unequal lengths and that concurrent observations at different stations may not be independent, but cross-correlated. With these improvements the method has become the main tool of the US Geological Survey to derive the frequency of flood flows for selected return periods T_r on a regional basis in different states. A nationwide summary of the information derived as of 2002 was compiled by Ries and Crouse (2002). However, as more information is becoming available and the streamflow records become longer, the regression equations are periodically being revised. For some examples of recent updates by state the reader is referred to the studies for Washington (Sumioka et al., 1998), Maine (Hodgkins, 1999), Colorado, (Vaill, 2000), West Virginia (Wiley et al., 2000) and North Carolina (Pope et al., 2001).

In these more recent studies the frequency relationships for each of the individual gaged sites were commonly derived on the basis of the generalized log-gamma distribution with a regionalized skew. In most cases the delineation of hydrologic regions within the state and the identification of the important explanatory variables for Equation (13.91) were next carried out in a stepwise manner by means of ordinary least squares regression. The regions were usually delineated by inspection of the statewide regression residuals. Once the explanatory variables were identified for each region, the final predictive equations for the different quantiles in terms of the basin characteristics were calculated by means of the GLS procedure, as outlined by Tasker and Stedinger (1989). The explanatory variables that were found to affect the flow peak quantiles varied greatly from one region to another, but the number of adopted variables was usually kept as small as possible and restricted to two or three at most. In all regions the size of the drainage was found to be the most important variable and key basin descriptor; for some regions it was actually concluded to be the only significant one (Pope et al., 2001; Wiley et al., 2000). It has also been suggested on the basis of scaling arguments (Gupta et al., 1994) that the sole dependency of Q_T on drainage area A, and on nothing else, can be used as the criterion to define a hydrologically homogeneous region. In regions with

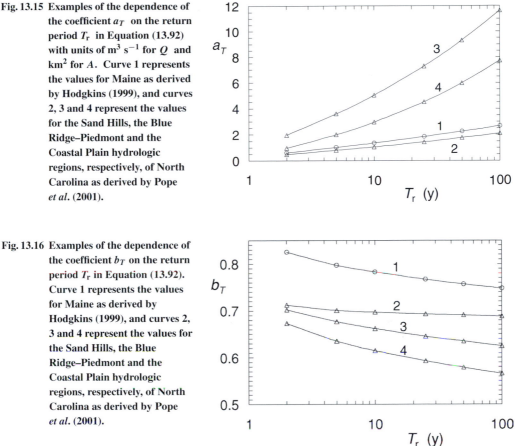

Fig. 13.15 Examples of the dependence of the coefficient a_T on the return period T_r in Equation (13.92) with units of m^3 s^{-1} for Q and km^2 for A. Curve 1 represents the values for Maine as derived by Hodgkins (1999), and curves 2, 3 and 4 represent the values for the Sand Hills, the Blue Ridge–Piedmont and the Coastal Plain hydrologic regions, respectively, of North Carolina as derived by Pope *et al.* (2001).

Fig. 13.16 Examples of the dependence of the coefficient b_T on the return period T_r in Equation (13.92). Curve 1 represents the values for Maine as derived by Hodgkins (1999), and curves 2, 3 and 4 represent the values for the Sand Hills, the Blue Ridge–Piedmont and the Coastal Plain hydrologic regions, respectively, of North Carolina as derived by Pope *et al.* (2001).

hilly and mountainous terrain the annual precipitation was often found to have some secondary relevance (Sumioka *et al.*, 1998) but sometimes it was the mean drainage basin slope which played this role (Vaill, 2000). In other regions the secondary variable was the fraction of the area occupied by lakes and wetlands (Hodgkins, 1999). In all past studies the quantiles were found to be related to drainage area A by a power function, as follows

$$Q_T = a_T A^{b_T} \qquad\qquad (13.92)$$

in which a_T and b_T are constants depending on the return period T_r of the quantile in question. The order of magnitude and tendency of these constants are illustrated for several regions in Figures 13.15 and 13.16. In most regions the constant b_T lies in the range between 0.5 and 0.9; again, this is consistent with Fuller's (1914) value of 0.8, and with the values reported for (13.90). However, b_T is usually not a constant, and typically it decreases with increasing T_r, as illustrated in Figure 13.16. This tendency indicates that the peak flow rate dependency on A decreases with increasing severity of the flood; put differently, it means that the runoff per unit area (Q_T/A) decreases more rapidly with increasing size of the drainage area as the severity of the flood increases.

As an aside, if Equation (13.92) is valid, it follows that, unless b_T is independent of T_r and a_T is a power function of T_r, the underlying assumption of the power distribution (13.82) cannot be generally valid. For example, substitution of (13.92) into (13.82), yields the following criterion for the validity of the power distribution

$$\frac{a_{100}}{a_{10}} A^{b_{100}-b_{10}} = \frac{a_{50}}{a_5} A^{b_{50}-b_5} \tag{13.93}$$

This shows that in this case the ratio K_{10} in (13.82) depends on the drainage area; in other words, the equality in (13.93) needed for the validity of the power distribution requires a certain size A of the basin, so that it is practically never satisfied. For instance, in the case of the typical values obtained for Maine (see Figure 13.16), for $A = 100 \, \text{km}^2$, the left-hand side of (13.93) is 1.67, and the right-hand side is 1.82; for $A = 1000 \, \text{km}^2$, the left-hand side of (13.93) is 1.54 and the right-hand side is 1.65.

Theoretical distribution functions with regionalized moments

The underlying assumption of this approach is that the moments in a hydrologically homogeneous region depend on known or measurable basin and climate characteristics. Thus, once the moments can be estimated for an ungaged basin within the region on the basis of these characteristics, it becomes possible to calculate the parameters of the selected probability distribution function. In principle, because several moments, namely the mean, the variance and the skew coefficient can be related to basin characteristics, the method is less restrictive than the index-flood method, which makes use of only the first moment. For two-parameter distributions the skew is not needed, and for three-parameter distributions, as it tends to be unreliable, a regional value can be assumed. The method has not been widely applied. For example, in their application to the Klamath Mountains in northern California, Cruff and Rantz (1965) found that the sample mean M and the sample standard deviation S are both related to the catchment area A and to the mean annual basinwide precipitation, respectively, by power functions similar to Equation (13.91).

REFERENCES

Abramowitz, M. and Stegun, I. A. (editors). (1964). *Handbook of Mathematical Functions, Appl. Math. Ser. 55*. Washington, DC: National Bureau of Standards.

Bailey, J. F., Patterson, J. L. and Paulhus, J. L. H. (1975). *Hurricane Agnes rainfall and floods, June–July 1972*. Geol. Survey Prof. Paper 924, Washington, DC: US Department of the Interior.

Benson, M. A. (1950). Use of historical data in flood-frequency analysis. *Trans. Amer. Geophys. Un.*, **31**, 419–424.

(1962a). *Evolution of methods for evaluating the occurrence of floods*, Geol. Survey Water-Supply Paper 1580-A. Washington, DC: US Department of the Interior.

(1962b). *Factors influencing the occurrence of floods in a humid region of diverse terrain*, Geol. Survey Water-Supply Paper 1580-B. Washington, DC: US Department of the Interior.

(1968). Uniform flood-frequency estimating methods for Federal agencies. *Water Resour. Res.*, **4**, 891–908.

Blöschl, G. and Sivapalan, M. (1997). Process controls on regional flood frequency: coefficient of variation and basin scale. *Water Resour. Res.*, **33**, 2967–2980.

Bobée, B. and Ashkar, F. (1991). *The Gamma Distribution and Derived Distributions Applied in Hydrology*. Littleton, CO: Water Resour. Press.

Chow, V. T. (1954). The log-probability law and its engineering applications. *Proc. Amer. Soc. Civil Eng., Hydraul. Div.*, **80**, 536.1–536.25.

Chowdhury, J. U. and Stedinger, J. R. (1991). Confidence intervals for design floods with estimated skew coefficient. *J. Hydraul. Eng., ASCE*, **117**, 811–831.

Cohn, T. A., Lane, W. L. and Baier, W. G. (1997). An algorithm for computing moments-based flood quantile estimates when historical flood information is available. *Water Resour. Res.*, **33**, 2089–2096.

Cohn, T. A., Lane, W. L. and Stedinger, J. R. (2001). Confidence intervals for Expected Moments Algorithm flood quantile estimates. *Water Resour. Res.*, **37**, 1695–1706.

Cruff, R. W. and Rantz, S. E. (1965). *A comparison of methods used in flood-frequency studies for coastal basins in California*, Geol. Survey Water-Supply Paper 1580-E. Washington, DC: US Department of the Interior.

Cunnane, C. (1978). Unbiased plotting positions – a review. *J. Hydrol.*, **37**, 205–222.

Dalrymple, T. (1960). *Flood-frequency analyses*, Geol. Survey Water-Supply Paper 1543-A. Washington, DC: US Department of the Interior.

Foster, H. A. (1924). Theoretical frequency curves. *Trans. Amer. Soc. Civil Eng.*, **89**, 142–203.

Fuller, W. E. (1914). *Flood flows. Trans. Amer. Soc. Civil Eng.*, **77**, 564–617, 676–694.

Gumbel, E. J. (1954a). *Statistical Theory of Extremes and Some Practical Applications*, Appl. Math. Ser. 33. Washington, DC: National Bureau of Standards.

   (1954b). Statistical theory of droughts. *Proc. Amer. Soc. Civil Engrs., Hydraulics Div.*, **80**, 439.1–439.19.

   (1958). *Statistics of Extremes*. New York: Columbia University Press.

Gupta, V. K., Mesa, O. J. and Dawdy, D. R. (1994). Multiscaling theory of flood peaks: Regional quantile analysis. *Water Resour. Res.*, **30**, 3405–3421.

Hardison, C. H. (1974). Generalized skew coefficients of annual floods in the United States and their application. *Water Resour. Res.*, **10**, 745–752.

Hazen, A. (1914a). Discussion on flood flows. *Trans. Amer. Soc. Civil Eng.*, **77**, 626–632.

   (1914b). The storage to be provided in impounding reservoirs for municipal water supply. *Trans. Amer. Soc. Civil Eng.*, **77**, 1539–1659.

   (1930). *Flood Flows, A Study of Frequencies and Magnitudes*. New York: John Wiley, Inc.

Hershfield, D. M. (1970a). Generalizing dry-day frequency data. *J. Amer. Water Works Assoc.*, **62**, 51–54.

   (1970b). A comparison of conditional and unconditional probabilities for wet- and dry-day sequences. *J. Appl. Meteor.*, **9**, 825–827.

   (1971). The frequency of dry periods in Maryland. *Chesapeake Sci.*, **12**, 72–84.

Hirsch, R. M. and Stedinger, J. R. (1987). Plotting positions for historical floods and their precision. *Water Resour. Res.*, **23**, 715–727.

Hodgkins, G. (1999). *Estimating the magnitude of peak flows for streams* in *Maine for selected recurrence intervals*, Water-Resour. Investig. Rept. 99-4008, Augusta, ME: US Department of the Interior, US Geol. Survey. (http://me.water.usgs.gov/99-4008.pdf)

Hosking, J. R. M. and Wallis, J. R. (1997). Regional Frequency Analysis: An Approach Based on L-Moments. Cambridge: Cambridge University Press.

Horton, R. E. (1914). Discussion on flood flows. *Trans. Amer. Soc. Civil Eng.*, **77**, 663–670.

Interagency Advisory Committee on Water Data (1982). *Guidelines for Determining Flood Flow Frequency*, Bulletin 17B. Reston, VA: US Department of the Interior, Geol. Survey, Office of Water Data Coordination.

Jenkinson, A. F. (1955). The frequency distribution of the annual maximum (or minimum) values of meteorological elements. *Quart. J. Roy. Meteor. Soc.*, **81**, 158–171.

Katz, R. W., Parlange, M. B. and Naveau, P. (2002). Statistics of extremes in hydrology. *Adv. Water Resour.*, **25**, 1287–1304.

Kimball, B. F. (1942). Limited type of primary probability distribution applied to annual maximum flows. *Ann. Math. Stat.*, **13**, 318–325.

Kirby, W. (1972). Computer-oriented Wilson–Hilferty transformation that preserves the first three moments and the lower bound of the Pearson Type 3 distribution. *Water Resour. Res.*, **8**, 1251–1254.

Kochel, R. C. and Baker, V. R. (1982). Paleoflood hydrology. *Science*, **215**(4531), 353–361.

Korzoun, V. I. *et al.* (editors) (1977). *Atlas of World Water Balance*, USSR National Committee for the International Hydrological Decade. Paris: UNESCO Press.

Lettenmaier, D. P., Wallis, J. R. and Wood, E. F. (1987). Effect of regional heterogeneity on flood frequency estimation. *Water Resour. Res.*, **23**, 313–323.

Madsen, H., Pearson, C. P. and Rosbjerg, D. (1997). Comparison of annual maximum series and partial duration series methods for modeling extreme hydrologic events. 2. Regional modeling. *Water Resour. Res.*, **33**, 759–770.

Malamud, B. D., Turcotte, D. L. and Barton, C. C. (1996). The 1993 Mississippi River flood: a one hundred or a one thousand year event? *Envir. & Eng. Geoscience*, **2**, 479–486.

Martins, E. S. and Stedinger, J. R. (2000). Generalized maximum-likelihood generalized extreme-value quantile estimators for hydrologic data. *Water Resour. Res.*, **36**, 737–744.

Matalas, N. C. (1963). *Probability distribution of low flows*. Geological Survey Prof. Paper 434-A. Washington, DC: US Department of the Interior.

Meyer, A. F. (1917). *The Elements of Hydrology*. New York: John Wiley & Sons, Inc.

Mood, A. M. and Graybill, F. A. (1963). *Introduction to the Theory of Statistics*, second edition. New York: McGraw-Hill Book Co.

Papoulis, A. (1965). *Probability, Random Variables, and Stochastic Processes*, New York: McGraw-Hill Book Co.

Pope, B. F., Tasker, G. D. and Robbins, J. C. (2001). *Estimating the magnitude and frequency of floods in rural basins of North Carolina – revised*, Water-Resour. Investigs. Rept. 01-4207. Raleigh, NC: US Department of the Interior, US. Geol. Survey. (http://nc.water.usgs.gov/reports/wri014207/pdf/report.pdf)

Ries, K. G., III, and Crouse, M. Y. (2002). *The National Flood Frequency Program, version 3: A computer program for estimating magnitude and frequency of floods for ungaged sites*, Water-Resour. Investigs. Rept. 02-4168. Reston, VA: US Department of the Interior, US Geol. Survey. (http://water.usgs.gov/pubs/wri/wri024168/#pdf)

Robison, F. L. (1961). *Floods in New York, magnitude and frequency*, Geological Survey Circular 454. Washington, DC: US Department of the Interior.

Robinson, J. S. and Sivapalan, M. (1997a). An investigation into the physical causes of scaling and heterogeneity of regional flood frequency. *Water Resour. Res.*, **33**, 1045–1059.

(1997b). Temporal scales and hydrological regimes: Implications for flood frequency scaling. *Water Resour. Res.*, **33**, 2981–2999.

Smith, J. A. (1989). Regional flood frequency analysis using extreme order statistics of the annual peak record. *Water Resour. Res.*, **25**, 311–317.

(1992). Representation of basin scale in flood peak distributions. *Water Resour. Res.*, **28**, 2993–2999.

Sokolov, A. A. (1967). Closing remarks. *Symposium on Floods and Their Computation*, Aug. 22. Leningrad, USSR: Unesco.

Stedinger, J. R. (1980). Fitting log normal distributions to hydrologic data. *Water Resour. Res.*, **16**, 481–490.

Stedinger, J. R. and Baker, V. R. (1987). Surface water hydrology: historical paleoflood information. *Rev. Geophysics*, **25**, 119–124.

Stedinger, J. R. and Lu, L.-H. (1995). Appraisal of regional and index flood quantile estimators. *Stochast. Hydrol. and Hydraulics*, **9**, 49–75.

Sumioka, S. S., Kresch, D. L. and Kasnick, K. D. (1998). *Magnitude and frequency of floods in Washington*, Water-Resour. Investigs. Rept. 97-4277. Tacoma, WA: US Department of the Interior, US Geol. Survey. (http://wa.water.usgs.gov/reports/flood-freq/tables.html)

Tasker, G. D. and Stedinger, J. R. (1986). Regional skew with weighted LS regression. *J. Water Resour. Plan. and Management, Proc. ASCE*, **112**, 225–237.

(1989). An operational GLS model for hydrologic regression. *J. Hydrol.*, **111**, 361–375.

Thomas, D. M. and Benson, M. A. (1970). *Generalization of streamflow characteristics from drainage-basin characteristics*. Geol. Survey Water-Supply Paper 1975. Washington, DC: US Department of the Interior.

Thomas, W. O. (1985). A uniform technique for flood frequency analysis. *J. Water Resour. Plann. Management Proc. ASCE*, **111**, 321–337.

Turcotte, D. L. (1992). *Fractals and Chaos in Geology and Geophysics*. Cambridge: Cambridge University Press.

(1994). Fractal theory and the estimation of extreme floods. *J. Res. Nat. Inst. Standards and Technology*, **99**, 377–389.

Vaill, J. E. (2000). *Analysis of the magnitude and frequency of floods in Colorado*, Water-Resour. Investigs. Rept. 99-4190. Denver, CO: US Department of the Interior, US Geol. Survey. (http://water.usgs.gov/pubs/wri/wri99-4190/pdf/wrir99-4190_V1.pdf)

Weatherburn, C. E. (1961). *A First Course in Mathematical Statistics*. Cambridge: Cambridge University Press.

Wiley, J. B., Atkins, Jr., J. T. and Tasker, G. D. (2000). *Estimating magnitude and frequency of peak discharges for rural, unregulated streams in West Virginia*, Water-Resour. Investigs. Rept. 00-4080. Charleston, WV: US Department of the Interior, US Geol. Survey. (http://water.usgs.gov/pubs/wri/wri004080/pdf/wri00-4080.pdf)

PROBLEMS

13.1   Prove both equations in (13.12).

13.2   Determine the second moment $m_2 = \sigma^2$ for the exponential distribution in terms of λ. The exponential distribution has a density $f(x) = \lambda e^{-\lambda x}$ for $x \geq 0$ and $f(x) = 0$ for $x < 0$.

13.3   Determine the fourth septile (i.e. $n = 7$) and the fifth octile (i.e. $n = 8$) for the exponential distribution as defined in the previous problem.

13.4   Calculate the mean μ and the variance σ^2 for the power distribution defined in Equations (13.85) and (13.86) in terms of a and b.

13.5   Calculate the 95th percentile for the power distribution defined in Equations (13.85) and (13.86) in terms of a and b.

13.6   What is the probability that a 100 y flood will be exceeded after exactly 100 y? What is the probability that it will be exceeded some time in the coming 100 y?

13.7   Multiple choice. Indicate which of the following statements are correct. When a flood of a certain magnitude is called the 50 y event, it means that:
    (a)   the probability that it will be exceeded in any given year is 98%;
    (b)   the probability that it will be exceeded once in any 3 y period is 5.8%;
    (c)   after it has been exceeded, it will take on the average 50 y before it will be exceeded again.
    (d)   it is the largest event that will occur during any period of record of 50 y;
    (e)   the probability that it will be exceeded during the course of a 1 y period is approximately 20%.

13.8   Multiple choice. Indicate which of the following statements are correct. Among the disadvantages of using $P_m = m/n$ (where $m = 1$ is the smallest and $m = n$ is the largest) to obtain the frequency of annual floods one has the following:
    (a)   the return period of the smallest event is 1 y;
    (b)   the return period of the largest event is equal to twice the period of record;
    (c)   the probability of the smallest event is equivalent to the assumption of certainty of a larger event;
    (d)   the probability of the largest event is equivalent to the assumption that a larger event cannot happen;
    (e)   it can be applied only to small samples.

13.9   Multiple choice. Indicate which of the following statements are correct. The Weibull plotting position $m/(n + 1)$ (in which m is the order number for the items ranked in increasing magnitude, and n is the sample size):
    (a)   is the mean of the probability ("that the event will be smaller than or equal to . . .") of the mth event of the sample;
    (b)   is exactly the probability that in any occurrence the magnitude of an event will be smaller than or equal to the mth event of the sample;
    (c)   is the probability of the mean of the mth event of the sample;
    (d)   can be used as an estimate of the probability that the mth event will not be exceeded;
    (e)   is applicable in the analysis only of the largest values but not of the smallest.

13.10   From a long-term rainfall record for a given location, we know that in summer on average 3 weeks out of 12 are without rain. What is the probability of having 6 weeks out of 12 without rain this coming summer? Assume, as a first approximation, that the likelihood of rainfall in summer is independent from one week to the next.

13.11   Multiple choice. Indicate which of the following statements are correct. For yearly events, the return period or recurrence interval T_r:
    (a)   in the case of floods, corresponds to the inverse of the probability that the event will be smaller than a given magnitude;
    (b)   signifies that, once the T_r year event has occurred, we are safe from any event exceeding it for the next T_r years;
    (c)   can be calculated from the probability that an event be smaller than, or equal to, a given magnitude;

(d)   can be used to calculate the probability of exceeding the T_r year flood each year during 5 subsequent years; that probability is $[(T_r - 1)/T_r]^5$;

(e)   the probability of not exceeding the T_r year event during the first 3 y of a 5 y period, and of exceeding that event in each of the remaining 2 y is $(1/T_r)^2 (1 - 1/T_r)^3$.

13.12   An annual flood record for a certain river is given below.

Year	Maximum flow rate $(m^3\ s^{-1})$	Year	Maximum flow rate $(m^3\ s^{-1})$
1991	269	1998	331
1992	374	1999	309
1993	207	2000	427
1994	241	2001	204
1995	393	2002	402
1996	289	2003	229
1997	535		

These data represent a sample from a population with some unknown probability distribution. Do not assume that the data obey some a priori distribution for parts (a), (b), (c) and (d). (a) Estimate the median flood from this sample. (b) Estimate the mean flood from this sample. (c) Estimate the 7 y flood from this sample. (d) Estimate, from this sample, the probability that next year the maximum flow rate will lie between 331 and 393 $m^3\ s^{-1}$. (e) Assume now that these data can be fitted by the exponential distribution. The density function is $f(x) = \lambda e^{-\lambda x}$ for $x \geq 0$, and $f(x) = 0$ otherwise. Estimate the parameter λ of this function from the available record by means of the method of moments.

13.13   What is the probability that a single observation will exceed the mean μ when the probability distribution function is the first asymptote (13.52)?

13.14   At a river gaging station, which has been operated for a very long time, it has been found that the probability distribution of the annual maximal flows can be described by $F(Q) = Q/(A + Q)$, in which Q is the magnitude of these annual events and A is a constant. Derive the probability distribution for decadal peak discharges (i.e. the maximal flows experienced in non-overlapping periods lasting 10 successive years) from the distribution of the annual peaks. Give the result in terms of A and Q.

13.15   It has been observed that the annual peak discharges Q (in $m^3\ s^{-1}$) for a given river can be described by Fuller's formula, as follows $Q = 294 (1 + 0.3 \ln T_r)$, where T_r is the recurrence interval (in years) of the peak discharge of magnitude Q. (a) Derive the probability distribution function $F = F(Q)$ from Fuller's formula. (b) What is the probability that 700 $m^3\ s^{-1}$ will be exceeded every single year of a given 4 y period? (c) What is the probability that the 700 $m^3\ s^{-1}$ flood will be exceeded only once, namely at the end of this 4 y period? In other words, what is the probability that this flood will not be exceeded during the first 3 y and then will be exceeded during the last year?

13.16   Select a river gaging station in your region of interest, preferably with a period of record in excess of 50 y. Tabulate the annual peak discharges for each water year on record. Calculate the first three moments of these discharges and of their logarithms. Then, carry out several or all of the following. (a) Determine for the generalized log-Pearson Type III distribution, the quantiles for the selected probabilities listed in Table 13.2 with these moments of the logarithms. (b) Repeat (a) with skew assumed to be zero. (c) Calculate the parameters α_n and u_n of the first asymptotic distribution for the largest values. (d) Calculate the parameters, a, b, and c, of the generalized extreme value distribution. (e) Plot the data and these four theoretical curves on log-normal probability paper. (f) Plot the data and these four theoretical curves on probability paper based on the first asymptote (see Figure 13.3). (In the United States, data records can be found on the web at http://waterdata.usgs.gov/usa/nwis/sw)

13.17   In the design of a bridge opening (i.e. clearance), it is necessary to determine the 40 y flood. Give the estimates according to the distributions determined in parts (a), (b), (c) and (d) of Problem 13.16.

13.18   Multiple choice. Indicate which of the following statements are correct. As defined in Equation (13.3), the theoretical distribution functions $F(x)$ that are used to describe the occurrence of hydrologic events:
(a)   have a magnitude, which ranges in general between $-\infty$ and $+\infty$;
(b)   have parameters that can be determined from observed data by the method of moments;
(c)   can never assume a value smaller than zero;
(d)   are symmetrical about the mean;
(e)   yield unity [i.e. $F(x) = 1$] when future events cannot be smaller than x.

14 AFTERWORD – A SHORT HISTORICAL SKETCH OF THEORIES ABOUT THE WATER CIRCULATION ON EARTH

14.1 EARLIEST CONCEPTS: THE ATMOSPHERIC WATER CYCLE

For as long as humans have been on Earth, they must have been acutely aware of their dependency on different forms of water in their environment. Water was literally vital for their health and sustenance but it could also be destructive and even lethal in severe weather, floods, and the other hazards they faced in their daily lives. Already in the earliest writings there are indications that among natural peoples in their primal stage it was a common notion that water in nature moves continually between different states in some repetitive, if not cyclical, fashion. Whatever is left of these early writings is not always easy to interpret, mainly because the meanings of even the most elementary concepts have evolved in the meantime. Nor is it always easy to distinguish profane views and naturalistic descriptions from the more sacred narratives and religious interpretations. Nevertheless, a cursory scan of some better known early writings yields several instances of water related imagery even in widely different cultural settings, in which the evidence is fairly clear, and which provide some idea on the thinking of early humans.

As early as the eighth century BCE in Greece, the poet Hesiod presented a remarkable description. In a passage with advice to farmers to get dressed warmly and to finish the work in time (Hésiode, 1928; also Hesiod, 1978; vv. 547–553), he wrote the following.

For the morning is cold when Boreas [the north wind] bears down; in the morning from the starry sky over the earth a fertilizing mist spreads over the cultivations of the fortunate; this [mist], drawn from ever flowing rivers, and lifted high above the earth by a storm wind, sometimes falls as rain toward evening, or sometimes blows as wind, while Thracian Boreas chases the heavy clouds.

This passage contains interesting features; it explains that mist is derived from river water, and that it may lead to rain; on the other hand, it implies that evaporation may be both a result and a cause of the wind. Apart from the reference to Boreas, the god of the north wind, Hesiod's passage appears quite naturalistic.

Several water cycle related passages appear in the Hebrew Bible. The oldest among them, written in the eighth century BCE, is probably (5,8) in the Book of Amos; it reads as follows (see, for example, *Oxford Study Edition*, 1976).

He . . . who turned darkness into morning and darkened day into night; who summoned the waters of the sea, and poured them over the earth; . . . he who does this, his name is the Lord.

Amos, a native of Judah, was by his own account originally a shepherd and a pruner of sycamore fig trees. From the context, that is from the first part of this quotation which refers to the cycle of day and night, it is possible that the second part refers to some kind of cyclical process as well; but if so, it is a cyclicity in the sense of periodicity and not in the sense of a water cycle. Here also, rain over the Earth results from evaporation from a water surface. A second, more recent, biblical passage of interest is (55,10–11) in the Book of Isaiah, namely, the following.

This is the very word of the Lord . . . and as the rain and the snow come down from heaven and do not return until they have watered the earth, making it blossom and bear fruit, and give seed for sowing and bread to eat, so shall the word which comes from my mouth prevail; it shall not return to me fruitless without accomplishing my purpose or succeeding in the task I gave it.

Isaiah also lived in the eighth century BCE but this chapter is now generally considered a later addition and attributed to an unknown prophet, who wrote in Babylon toward the end of the exile in the sixth century BCE. In this passage the physical phenomena serve mainly an allegorical purpose and their description is fairly naturalistic; they appear to occur on their own and not as a result of direct divine intervention. The description involves unambiguously some kind of cycle by which water returns to where it came from.

Notions on various cyclical processes were also held in ancient China. In a naturalist work "Chi Ni Tzu," probably of the late fourth century BCE (Needham, 1959, p. 467), atmospheric phenomena are described as follows.

Wind is the qi [or chhi, spirit, mind] of heaven, and the rain is the qi of earth. Wind blows according to the seasons and rain falls in response to wind. We can say that the qi of the heavens comes down and the qi of the earth goes upwards.

Because the rain is deemed to originate from the Earth even though it falls from above, the direct connection between evaporation and precipitation seems to be taken for granted here.

A passage in the *Chandogya Upanisad* (VI, 10), an important text in Hinduism, which was composed between 800 to 400 BCE, is less explicit; but it is suggestive of the same theme. The passage is an allegory to illustrate the essence of the Self or Being (see Anandatirtha, 1910, p. 458; Radhakrishnan, 1953, p. 460; Swahananda, 1965, p. 458) and can be translated as follows.

These rivers, my son, flow, the eastern toward the east, the western toward the west. They go from sea to sea. They become the sea itself, and while there, they do not know which river they are.

This text can be interpreted in different ways. The sentence "They go from sea to sea" could conceivably refer to sea currents, or to some underground seawater filtration as the origin of river springs, like that visualized by some in ancient Greece. Still, it is equally plausible that it refers to the evaporation of these waters from the sea and their subsequent precipitation back to the sea. The main point is that it implies a cyclical process.

The above descriptions are merely a few examples. A common feature of most of these early descriptions is that, wherever they imply a water cycling process, they refer to, or hint at, the atmospheric phase of the water cycle. Wherever evaporation is mentioned explicitly it is mostly, though not exclusively, assumed to take place from rivers and the sea. While some of the descriptions include flowing streams, they are silent on the origin of these streams or on whether or how the water returns to where the streams originated. The earliest speculations on this problem, which were not of an obvious mythical nature but based on observations, were probably those of the Greek natural philosophers.

14.2   Greek antiquity

The ancient Greeks are renowned for the large effort their natural philosophers made to arrive at a rational explanation of the world within that same world, without animistic or direct divine intervention. Inspection of their writings and other transmitted evidence indicates that water and various aspects of the water cycle played a central role in their cosmology. As seen in Hesiod's passage, the atmospheric phase of the hydrologic cycle was already a common concept among the Greeks in pre-philosophic times (see also Brutsaert, 1982). Therefore, it is mainly the evolution of their opinions on the origins of springs and rivers, that will be examined in what follows.

14.2.1 The Presocratics

The earliest Greek philosophers who were active in the sixth and fifth centuries BCE are customarily referred to as the Presocratics. Some of their writings were handed down to us in the form of fragments and some were paraphrased by later writers. Among these natural philosophers two competing theories prevailed on the origin of the water in springs, streams, and other fresh water bodies. These are the seawater filtration theory, which was probably the earlier of the two, and the rainfall percolation theory, which contains the essence of our present understanding.

Seawater filtration theory

The basic idea of this theory is that seawater seeps upward through the Earth, loses its salt by filtration and becomes the source of the springs and other surface waters (Figure 14.1). The written evidence points to Hippon as the earliest proponent of this view. Hippon of Rhegion, in what is now southern Italy, also called Hippon of Samos, was a contemporary of Pericles, so he must have flourished around the middle of the fifth century BCE. His opinion on the matter, in the only surviving fragment by him (Diels, 1961, p. 388) is formulated as follows.

Indeed all drinking waters originate from the sea; for the wells from which we drink are not deeper than the sea. So should the water not be from the sea, then from somewhere else. Now, the sea is deeper than the waters. Thus whatever waters are above the sea, all originate from it.

This fragment is rather terse and not very explicit. However, it should be seen in light of the fact that Hippon's other views were nearly identical with those of Thales, presented at least a century earlier. The following passage by Theophrastos in his *Physical Opinions* (Diels, 1879, p. 475) is revealing.

Of those who say that the original principle (arche) is one and movable, whom he (Aristotle) calls physicists, some contend that it is bounded; for instance, Thales of Miletos and Hippon, who appears even to (have) become an atheist, said that water is the first principle, being led to this by the observation of the phenomena; for heat thrives in moisture, dead matter dries out, the seeds of everything are moist, and all food is succulent; and naturally each thing is nourished by that from which it originates. Water is the principle of the moisture and bond of everything. Therefore, they maintained that water is the first principle of everything and that the earth evidently rests on water.

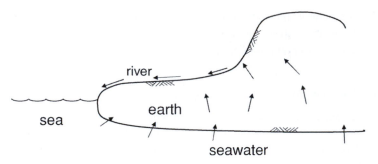

Fig. 14.1 **Sketch of the seawater filtration theory of the Presocratic philosophers in ancient Greece. The written evidence points to Hippon as the earliest proponent of this concept, but it was probably Thales with whom it originated.**

Thales of Miletos in Ionia flourished around 585 BCE, and he is generally considered to be the first Greek natural philosopher, with whom the formal inquiry started into the reality behind the changes in the Universe. He does not appear to have committed his ideas to writing, and no actual quotations of him have survived. While there is nothing on the origin of rivers or springs, the essence of Thales' views is well known and two of the most important ones are contained in the above passage; these are that the first principle of everything is water, and that the Earth rests on water. Hippon is mentioned here in the same breath as Thales, so it would be surprising if Hippon's opinion on the origin of rivers were very different from that of the old master. It is, therefore, difficult to disagree with Gilbert's (1907) opinion that Thales can reasonably be considered, as the actual originator of the seawater filtration theory, at least among the Greeks. But the roots of this theory may actually be much older. It is now known (see Eliade, 1978) that as early as the third millennium BCE, that is some 2000 years before Thales, in Sumer in lower Mesopotamia it was already a well established view that the Earth rests on the ocean.

Hippon's fragment does not mention the removal of the salt. But this aspect of the theory can be deduced from Aristotle's (1952, II 354 b,15) description, in his objections to this theory, some 200 years later.

It was this difficulty which led people to suppose that the sea was the primary source of moisture and of all water. So some say that rivers not only flow into it but out of it, and that the salt water becomes drinkable by being filtered.

This is a clear indication that the theory was around at the time of Aristotle and that it was taken seriously by many of his contemporaries.

Rainfall percolation theory

The earliest seeds of this second theory appear in the philosophical views of Anaximander of Miletos; Anaximander, a younger associate of Thales, was born around 610 BCE, and must have been in his prime around 565 BCE. While the issue of the origin of streams and rivers is not addressed directly, his main views can be deduced from the remaining evidence (see also Gilbert, 1907, p. 405). On the nature and the origin of the sea, Alexander of Aphrodisias, a well-known commentator, who flourished around 200 CE, summarized Anaximander's views as follows (Diels, 1879, p. 494).

Some of them (natural philosophers) say that the sea is the leftover of the original moisture. As the region around the earth was wet, the first of that moisture was evaporated by the sun and became the winds, and from it the turnings of the sun and of the moon as well; thus, as the turnings are caused by the same vapors and by their exhalations, it becomes then the provider of the same (moisture) for those revolving around them. The part of it (the moisture) that is left behind in the hollow places is the sea; therefore, it decreases, as it is continually evaporated by the sun, and eventually it will perhaps have to be dry. Anaximander and Diogenes (of Apollonia) arrived at this view, thus reports Theophrastos.

Anaximander's opinion on what happens to this continual evaporation was summarized by Hippolytus, a Christian writer of the early third century, who died in 235 CE; in his *Refutation* (see Mansfeld, 1992), Hippolytus described it as follows (Diels, 1879, p. 560, 6, 7; 1961, p. 84, 6, 7).

Winds are generated when the finest vapors of the air are separated off and whenever they are put into motion as they gather; rains are generated from the vapor that is released upward from the earth by the sun.

These two passages indicate that Anaximander considered the sea to be the remainder of the original water around the Earth; the evaporation from the sea is the cause, instead of the result, of the winds and also the cause of the rains. There is no specific mention of streams. Anaximander did not assume, as Thales did, that the Earth floats on water, which would then flow upward to the surface to feed springs and streams; instead, he is known to have posited that the Earth does not rest on anything and that it is suspended in the sky in some sort of equilibrium, because it is equidistant from everything on all sides. Therefore, it is unlikely that he would have assumed that the sea feeds the streams by some upward filtration, as asserted by Thales and Hippon. Rather, it would seem more natural in his scheme that it is a different source of water, perhaps rainwater, which is feeding the streams that flow into the sea. On the other hand, it is clear that he did not think that all the evaporated water ends up in streams and rivers, because the sea is gradually drying out; thus, he definitely did not propose a closed cycle. In any event, he seems to have started, or at least stimulated, a productive line of thought, as can be seen from the views of Xenophanes.

Xenophanes of Colophon (*c.* 570–460 BCE) was probably in his prime *c.* 530 BCE, which is roughly some 35 years after Anaximander. According to Aetius (in Diels, 1879, p. 371, 4; 1961, p. 125, III, 4, 4), a doxographer who probably lived in the first century CE (Mansfeld and Runia, 1997), Xenophanes said that

. . . what happens in the sky is caused by the heat of the sun; for, when the moisture is drawn out of the sea, the sweet part, which is distinguished by its fine texture, forms a cloud, and drips out as rain by compression like that of felt, and the winds vaporize it around. And he wrote emphatically

(an actual fragment follows in verse, Diels,1961, p. 136)

The sea is the source of the water, the source of the wind. For in the clouds, neither would the force of the wind, which blows outward, originate without the great sea, nor the flowing of the streams, nor the rainwater from the sky; but the great sea is the generator of the clouds, winds and streams. . . .

Regarding the saltiness of the sea, the opinion of Xenophanes is described by Hippolytus (Diels, 1879, p. 565, 14, 4; 1961, p. 122, 33, 14, 4) as follows.

The sea is salty, he says, because of the many admixtures which flow together into it.

All this indicates that Xenophanes had some idea of the hydrologic cycle, as we now know it. He not only includes streams in his description, but he specifies that together with the winds, with the rain and with the clouds, the streams are caused by the evaporation from the sea. The only possible interpretation is that this occurs indirectly through the rain on the land surface. This is further supported by his explanation that the saltiness of the sea is

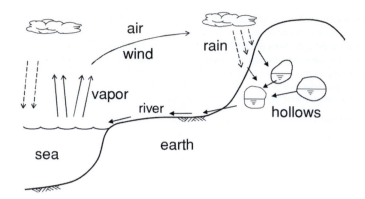

Fig. 14.2 Sketch of the rainfall percolation theory of the Presocratic philosophers in ancient Greece on the origin of rivers. The concept appears to have evolved from rough but seminal ideas by Anaximander, followed by more complete formulations by Xenophanes and Anaxagoras.

caused by the streams which flow into it carrying different salty admixtures picked up along the way. Clearly, the views of Xenophanes are a further development of Anaximander's.

Anaxagoras of Clazomenae (*c.* 500–428 BCE), who came some 70 years after Xenophanes, appears to have been even more explicit on the matter. It is again from Hippolytus (Diels, 1879, p. 562, 8, 4–5) that Anaxagoras is known to have said

... that the sea began to exist from the moist parts on the earth, that it originated this way as the waters in it were being evaporated or settled down, and also from the downflowing rivers; that the rivers take their substance from the rains and out of the waters that are in the earth; for this is hollow and that it has water in the caves.

But the most solid proof that Anaximander, Xenophanes, Anaxagoras and perhaps other Presocratics developed the notion, that the origin of streams and rivers can be accounted for by rain (see Figure 14.2), is found in its attempted refutation by Aristotle in his *Meteorologica*, some one to two centuries later. Evidently, at the time of Aristotle, the rainfall percolation theory was well enough established, that he considered it necessary to mount a head-on attack against it. Aristotle (1952, I 349 b,2) summarized the theory as follows.

Some people hold similar views about the origin of rivers. They suppose that the water drawn up by the sun when it falls again as rain is collected beneath the earth into a great hollow from which the rivers flow, either all from the same one or each from a different one: no additional water is formed in the process, and the rivers are supplied by the water collected during the winter in these reservoirs. This explains why rivers always run higher in winter than in summer, and why some are perennial, some are not. When the hollow is large and the amount of water collected therefore great enough to last out and not be exhausted before the return of the winter rains, then rivers are perennial and flow continuously: when the reservoirs are smaller, then, because the supply of water is small, rivers dry up before the rainy weather returns to replenish the empty container.

The statement, that "no additional water is formed in the process," is a clear indication that the rainfall percolation theory had eventually led to the concepts of a water cycle and of water mass conservation. From the space devoted to it in Aristotle's *Meteorologica*, the rainfall percolation theory was undoubtedly the more widely accepted at the time.

Both Anaxagoras and Aristotle refer to caves and hollows as the main underground storage spaces of water. This should not be surprising. About 65% of the terrain of Greece is limestone; this is easily eroded, resulting in a karst landscape (Higgins and Higgins,

Fig. 14.3 **Sketch of Aristotle's theory on the origin of rivers. While rain percolation provides a source of water, this is inadequate to supply the necessary amounts. Another important mechanism, not unlike the generation of rain above the Earth, is the formation of water resulting from cooling and condensation of rising vaporous air inside the Earth.**

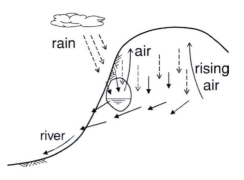

1996), with sinks, underground streams and caverns. Present day Greece is known to be among the regions of the world most endowed with caves, some seven thousand of them and of all kinds, large and small, vertical, horizontal, inland and along the coast.

14.2.2 Aristotle

It is generally agreed that Greek philosophy culminated with Aristotle (*c.* 384–322 BCE). Even though in antiquity he was acclaimed more as a logician than as a natural philosopher, his influence over the ensuing 18 centuries was to be so large, that it is necessary briefly to review his ideas here.

On the origin of rivers and springs

After having described the Presocratic rainfall percolation theory in the previous quotation, he immediately proceeds to present his own view (Aristotle, 1952, I 349 b,16).

> But it is evident that if anyone tries to compute the volume of water constantly flowing each day and then to visualize a reservoir for it, he will see that to contain the whole yearly flow of water it will have to be as large as the earth in size or at least not so much smaller.
>
> And though it is true that there are many such reservoirs in different parts of the earth, yet it is absurd for anyone not to suppose that the same cause operates to turn air into water below the earth as above it. If then cold condenses vaporous air into water above the earth, the cold beneath the earth must be presumed to produce the same effect. So not only does water form separately within the earth and flow from it, but the process is continuous.

Aristotle does not reject the rainfall percolation mechanism altogether; but he feels that the available underground storage and the amount of rain are inadequate to supply the observed river flows, so that there must be another important mechanism at work. That mechanism is the formation of water out of vaporous air beneath the Earth's surface (see Figure 14.3). Aristotle is correct in that water vapor does condense under the ground in caves; they are often wet and damp and water can be seen to drip from their walls and ceilings. It is now known, however, that the amounts produced this way are very small, and that regular precipitation exceeds by far any kind of condensation beneath the surface as the water supply for springs and streams. Compared to the rainfall theory of the Presocratics, Aristotle's explanation is definitely a step backward in the development of hydrologic theory.

Apparently, however, at this point Aristotle (1952, I 349 b,28) still does not feel that he has presented his argument strongly or clearly enough, because he continues as follows.

Besides, even if one leaves out of account water so produced and considers only the daily supply of water already existing, this does not act as a source of rivers by segregating into subterranean lakes, as it were, in the way some people maintain: the process is rather like that in which small drops form in the region above the earth, and these again join others, until rain water falls in some quantity; similarly inside the earth quantities of water, quite small at first, collect together and gush out of the earth, as it were, at a single point and form the sources of rivers. A practical proof of this is that when men make irrigation works they collect the water in pipes and channels, as though the higher parts of the earth were sweating it out. So we find that the sources of rivers flow from mountains, and that the largest and most numerous rivers flow from the highest mountains. Similarly the majority of springs are in the neighborhood of mountains and high places, and there are few sources of water in the plains except rivers. For mountains and high places act like a big sponge overhanging the earth and make the water drip through and run together in small quantities in many places. For they receive the great volume of rain water that falls (it makes no difference whether a receptacle of this sort is concave and turned up or convex and turned down: it will contain the same volume whichever it is); and they cool the vapor as it rises and condense it again to water.

Thus the argument is repeated and clarified by contrasting it with yet another theory which, as he explains, holds that rivers originate from preexisting or primal water stored in underground lakes. Reference is undoubtedly made here to the Tartarus theory of his teacher Plato (1975; 1993, 111 d, ff.), which Aristotle discusses and refutes more thoroughly later on (see 355 b,38). The passage is noteworthy in that it indicates that there were others who held this view. But this Tartarus, which also appears in Homer's poetry, is more a throwback to Greek mythology rather than natural philosophy and its discussion is beyond the present scope. Aristotle concludes the paragraph by summarizing once again his own opinion: springs and the sources of rivers result both from rainfall and from condensation inside the Earth.

On why the sea does not overflow

Beside the origin of rivers, Aristotle also concerned himself with the problem why the sea does not overflow, even though all rivers flow into it (Aristotle, 1952, II 355 b,15).

The place occupied by the sea is, as we say, the proper place of water, which is why all rivers and all the water there is run into it: for water flows to the deepest place, and the sea occupies the deepest place on earth. But one part of it is all quickly drawn up by the sun, while the other for the reasons given is left behind. The old difficulty why so great an amount of water disappears (for the sea becomes no larger even though innumerable rivers of immense size are flowing into it every day) is quite a natural one to ask, but not difficult to answer with a little thought. For the same amount of water does not take the same time to dry up if it is spread out as if it is concentrated in a small space: the difference is so great that in the one case it may remain for a whole day, in the other, if for instance one spills a cup of water over a large table, it will vanish as quick as thought. This is what happens with rivers: they go on flowing in a constricted space until they reach a place of vast area when they spread out and evaporate rapidly and imperceptively.

He calls it an "old difficulty," so it must have been a problem of long standing in Greek philosophy; indeed as seen earlier, Anaximander had already thought about it and had concluded that the sea may eventually dry up altogether. While Aristotle seems to have been the first on record to resolve the issue successfully by providing the correct explanation, it was considered elsewhere as well.

For instance, it appears to have been of concern in ancient China (see Lin, 1949). In the third century BCE during the Zhou (or Chou) dynasty, in the chapter "Autumn Floods", Zhuang Zi (or Chuangtse, d. 275 BCE), raised the issue, as follows.

There is no body of water beneath the canopy of heaven which is greater than the ocean. All streams pour into it without cease, yet it does not overflow. It is being continually drained off at the Tail-Gate, yet it is never empty. Spring and autumn bring no change; floods and droughts are equally unknown.

According to Lin (1949, p. 120), the editor of the treatise, this tail-gate (Wei-Lou or Wei Lu) is a mythical hole in the bottom or end of the ocean; this depletion mechanism to balance the river inflows is clearly different from the Hippon–Thales seawater filtration mechanism and from Aristotle's evaporation. The same issue was touched upon in the book *Lü Shi Qun Qiu* (or Lu-Shih-Chun-Chiu), written a few decades later during the Qin (or Chin) dynasty by a team of scholars under Prime Minister Lü Bu Wei (or Lu Buwei, d. 235 BCE) (P. K. Wang, 1996; personal communication, 2000), in the following passage (Needham, 1959, p. 467).

The waters flow eastwards from their sources, resting neither by day nor by night. Down they come inexhaustibly, yet the deeps are never full. The small (streams) become large and the heavy (waters in the sea) become light (and mount to the clouds). This is (part of) the Rotation of the Tao.

The terms within brackets probably represent the interpretation of the text by the translators; but this interpretation is not unreasonable and it would be difficult to come up with a different meaning. Thus here the invoked evaporation mechanism is the same as Aristotle's, and the authors clearly have some kind of hydrologic cycle in mind.

The problem was to continue to receive much attention throughout Western history, and this preoccupation stemmed directly from (1, 7) in Ecclesiastes (*Oxford Study Edition*, 1976) as follows

All streams run into the sea, yet the sea never overflows; back to the place from which the streams ran they return to run again.

Ecclesiastes dates from the third century, about a century after the death of Aristotle and of Alexander (the Great), when Hellenistic influences had been spreading like wildfire all over the Mediterranean world. The first part of this passage is so reminiscent of Aristotle's, that one has to wonder if the author of Ecclesiastes somehow had not been affected by Greek ideas. Ecclesiastes, like all the other Wisdom books, probably originated in the Jewish diaspora following the Babylonian exile, and possibly even in Alexandria, the very center of Hellenism. To be sure, the book is generally acknowledged to be quite different in literary style from the earlier books of the Hebrew Bible, and it has even been said that some ancient rabbis were distressed by its pessimism. On the other hand, however, the description in the second part is not quite the same as the explanation given by Aristotle. Aristotle unequivocally attributes the fact that the sea does not overflow to evaporation; in Ecclesiastes the way by which "they return" is not specified, but one cannot help inferring some kind of seawater filtration mechanism. At any rate, this passage shows that the "old difficulty" was of concern in Judaism. This preoccupation was also shared later by most Christian writers, and it was to endure well into the Middle Ages. But the theme kept recurring: Dobson (1777) contended that his data supported the wisdom in this biblical passage and, as recently as 1877, Huxley (1900, p. 74) used the passage in his description of the hydrologic cycle.

14.2.3 The Later Peripatetics

Upon Alexander's death in 323, Aristotle decided to leave Athens and he handed over the leadership of the Peripatetic School at the Lyceum to Theophrastos (*c.* 372–287 BCE). From the present vantage point, it would appear that Aristotle's *Meteorologica* continued to be held in high esteem because it was an essential part of the Aristotelian body of works, as it came to the Arab world and later to Western Europe in the thirteenth century. Evidently, however, not all the ideas of the old master were accepted uncritically later on by his successors, and some of them even seem to have been rejected outright. For instance, in the treatise *On Plants*, which is still formally attributed to Aristotle (1936; II 822b, 25) although it is known to be spurious, one reads the following.

> Rivers which arise under the ground from mountains behave in the same way. For the matter of which they are composed is rain; and when the water grows large in quantity and is forced into a narrow channel within, the excess of vapor rises from them, which cuts through the earth by pressure from within; and in this way springs and rivers make their appearance.

On Plants became associated with Aristotle's name probably because it was a product of the Lyceum and because it reflected the teaching at the school he had founded. But contrary to Aristotle's explanation in the *Meteorologica*, this passage unambiguously asserts that rivers are composed of rain, and there is no mention of underground condensation. Thus, among later generations at the Peripatetic School, it appears that it was the rainfall percolation theory which gained the upper hand, in spite of its original rejection by Aristotle.

To summarize at this point, Greek antiquity produced essentially four competing theories on the origin of rivers and springs, namely first and foremost, the rainfall percolation theory, which is the one still held today; in addition, there were the seawater filtration theory and the underground condensation theory. Finally, there was also the concept, quite likely based on early popular beliefs and mythology and seemingly less accepted by the philosophers, that rivers originate from underground reservoirs of primal water.

14.3   The Latin era

14.3.1 The Romans

The Romans are mainly praised for their engineering feats and their accomplishments in law and public administration. They are less known for their contributions to natural philosophy and as a result their writings often tend to be dismissed as mere reviews and commentaries on the Greeks. This may be true in general, but it is an oversimplification. With their practical orientation, the Romans usually relied more on observation than on speculation, arriving at interesting insights in some cases. Moreover, for several centuries their writings were the only source of ancient philosophy available in Western Europe; they are therefore an indispensable background to understand and trace the thought currents that brought about the scientific revolution.

The views of Lucretius (*c*. 99–55 BCE) in his work *On Nature* provide a revealing example of some aspects of natural philosophy in Rome. In the following passage Lucretius (1924, V, 261) deals with the problem of why the sea does not overflow and with the origin of springs.

> Moreover, there is no need to say how sea, rivers, and springs for ever well up in abundance with fresh waters and their streams flow unceasing: the great pouring down of waters from all sides makes it clear. But, bit by bit, whatever comes first of the water is taken off, and the result is that there is no excess of liquid in the sum total: partly because strong winds sweep the surface and diminish it, as does the sun on high unraveling it with his rays; partly because it is distributed abroad through all the earth underneath; for the pungency is strained off, and the substance of the water seeps back, and all meets at the sources of each river, whence it returns over the earth in a column of sweet water along the path which has once been cut for it in its liquid course.

A more elaborate but similar account is given in VI, 608–638. In contrast to Aristotle's explanation, evaporation is not the only reason why the sea does not overflow; seawater also flows back underground to feed the springs, in accordance with the original theory of Hippon and Thales. Also in contrast to Aristotle, who only considered the sun (Brutsaert, 1982),

Lucretius allows for the wind to be involved in the evaporation process. One of Lucretius' aims in writing his book was to promulgate the doctrines of Epikouros, whose natural philosophy, in turn, was derived from the atomic theory of Demokritos and Leukippos; this passage fully reflects this. The main principles of this theory are that nothing can be created out of nothing (or vice versa), which is equivalent with the principle of mass conservation, and that everything is made up of indivisible particles. This explains his view that on the whole there is no excess of water over the original amount, and that the winds are capable of sweeping water particles by evaporation. Unfortunately, beside the works of Lucretius and of Diogenes Laertius (1925) (third century CE), little is left that might give a better idea of what the Greek atomists themselves thought about these hydrologic phenomena.

A completely different example of Roman thought is the comprehensive treatise on architecture by Vitruvius (Marcus V. Pollio), a contemporary of Lucretius in the first century BCE. He composed it after having served as a military engineer under Julius Caesar in Gaul and in Spain. On the generation of spring water he wrote (see Vitruve, 1986, 8, 1) the following.

> We see, in fact, that the rain waters congregate in the hollows found at higher levels in the mountains, where the trees, which grow there in great number, keep the snow for a long time and where, as it melts little by little, it flows out imperceptibly through the veins of the earth; it is this water which, after it reaches the foot of the mountains, produces springs there.

Vitruvius is explicit and specific in attributing springs to rain water and snowmelt which, after infiltrating into the ground, flow out at lower levels. He undoubtedly gained this insight during his military campaigns up north in Gaul, where rainfall and all kinds of seepage outflow phenomena from hillsides are more obvious and more plentiful than in the more arid Mediterranean regions.

Similarly, the writings of Seneca (*c.* 4 BCE–65 CE), born in Cordoba, and teacher and later advisor of Emperor Nero, also give a good idea of the status of natural philosophy among educated Romans. In his work *Natural Questions* he quoted some 40 references, five among them Latin authors, but the remainder Greek. Book Three is devoted to the waters of the earth. He successively discusses five theories on ". . . how the earth supplies the continuous flow of rivers, and where such great quantities of water come from" (Seneca, 1971, III, 4–10.1). Before doing this, he also specifies "Whatever explanation we give of a river, the same will be so of streams and springs." In brief, these five theories are (i) the seawater enters the land by hidden paths (that is why the sea does not increase) and is filtered of its salinity while in transit; (ii) whatever rainfall the Earth receives is sent out again through the rivers; (iii) rivers are supplied by primal fresh water in vast underground reservoirs; (iv) within the deep cavities inside the Earth the stagnant cold air ceases to maintain itself and changes into water; (v) ". . . all elements come from all others: air from water, water from air, fire from air, air from fire . . . so why not water from earth?" Evidently, there are no precedents of this fifth theory, so this must be Seneca's own. The first two are, of course, the theories of the Presocratics, the third apparently a cleaned up version of Plato's Tartarus theory, and the fourth Aristotle's underground condensation theory. While Seneca seems to be willing to admit more than one theory, he is totally opposed to the rainfall percolation mechanism. Because Seneca was to exert such a profound influence on later thinkers, it is important to present his arguments in his own words (Seneca, 1971, III, 7).

> It is obvious that much can be said against this theory. First of all, as a diligent vine-gardener myself I assure you that no rainfall is so heavy it wets the ground to a depth beyond ten feet. All the moisture is absorbed in the outer surface and does not get down to the lower levels. How, then, is rain able to

supply an abundance to rivers since it only dampens the surface soil? The greater part of rain is carried off to sea through river-beds. The amount which the earth absorbs is scanty, and the earth does not retain that. For the ground is either dry and uses up what is poured into it or it is saturated and will pour off any excess that has fallen into it. For this reason rivers do not rise with the first rainfall because the thirsty ground absorbs all the water.

What about the fact that some rivers burst out of rocks and mountains? What will rains contribute to these rivers, rains which pour down over bare rock and have no ground in which to settle? Besides, in very dry localities wells are driven down to a depth beyond a distance of two hundred or three hundred feet and find copious veins of water at a level where rainwater does not penetrate. So you know that no water from the sky exists there nor any collection of moisture, but what is commonly called living water. The theory that all water comes from rain is disproved by another argument: the fact that certain springs well up on the high tops of mountains. It is obvious that they are forced up or are formed on the spot, since all rainwater runs down.

Seneca apparently admits that most rainwater makes its way to river channels, but he feels that this is a short-lived phenomenon and that these quantities are insufficient to maintain a continuous river flow. He bases this argument on observations in his vineyards, which are certainly perceptive, and similar to the findings of Perrault and de LaHire in the late seventeenth century, as will be seen below.

In the later stages of the Roman era Judaic and Christian views gradually gained in influence. In their writings the fathers, or early leaders, of the Christian church displayed a broad knowledge both of biblical accounts and of classical philosophy. But in their eclecticism among the different philosophical concepts they invariably accepted only those that could be reconciled with the biblical narrative. The set of homilies *On the Hexaemeron*, i.e. the six days (of creation), by Basileios of Cappadocia (*c.* 330–379 CE), is an example of this. Basileios had been educated in the classical tradition at Caesarea, Constantinople and Athens, and his writings generally reflect this background. In reference to Genesis (I,1,9) and Ecclesiastes (1, 7), he (Basil, 1963; 4,3) wrote the following.

For this reason, according to the saying of Ecclesiastes 'All the rivers run into the sea, yet the sea doth not overflow.' It is through the divine command that waters flow, and it is due to that first legislation, 'Let the waters be gathered into one place,' that the sea is enclosed within boundaries. Lest the flowing water, spreading beyond the beds which hold it, always passing on and filling up one place after another, should continuously flood all the lands, it was ordered to be gathered into one place.

Then, in (4, 6) he had this to say on the origin of rivers and springs.

In the first place, the water of the sea is the source of all the moisture of the earth. This water passing through unseen minute openings, as is proved by the spongy and cavernous parts of the mainland into which the swift sea flows in narrow channels, is received in the curved and sinuous paths and hurried on by the wind which sets it into motion. Then, it breaks through the surface and is carried outside; and, having eliminated its bitterness by percolation, it becomes drinkable.

Evidently, Basileios judged that among all available theories, the Hippon–Thales view was the main one in harmony with the creation events in Genesis and with the water cycle in Ecclesiastes. Similar views were promulgated some seventeen years later, around 389, by Ambrosius (*c.* 333–397) in his own *Hexameron*, which was partly inspired by that of Basileios. Ambrosius was then Bishop of Milan, but he had been converted to Christianity only at the age of 41, and his early education had been in the classical Latin tradition of the Roman upper class. His descriptions of the origin of rivers (Ambrose, 1961; 3, 2, 10; 3, 5, 22) are nearly the same as those of Basileios. The writings of Basileios and Ambrosius show how the fundamental concept of natural philosophy, as Thales had initiated it, was retained. Thus the Greek tradition of searching for an explanation of the physical world within that same world, without animistic or direct divine intervention, was continued. But the emphasis had shifted somewhat, since this knowledge had to serve as an aid for the transmission of the Christian doctrine and as an illustration of the wisdom of the Creator.

14.3.2 *The Early Middle Ages in the Latin West*

The *Book on Nature*, written around 613 by Isidorus Hispalensis of Sevilla (*c.* 560–636) for the benefit of Sisebut, king of Visigothic Spain at Toledo, illustrates how this interpretation and approach evolved and were transmitted into the early Middle Ages. Isidore (1960, 41,1) explains why the sea does not grow as follows.

Bishop Clemens says that it is because the naturally salty water consumes the flow of fresh water which it receives, in such a way that, however large the masses of water it receives, this salty element of the sea nevertheless absorbs them totally. Add to this what the winds take away, and what the evaporation and the heat of the sun absorbs. Finally, we see lakes and many ponds being consumed in a short time by the blowing of the wind and the glowing of the sun. And then Solomon says: the streams return to where they come from.

   From which it can be understood that the sea does not increase also because, after being returned to their sources through some conduits hidden in the deep, the waters flow back and run back along the usual course through their rivers. But the sea was made purposely so it would receive the runs of all rivers. While its depth is variable, the equality of its surface, however, cannot be discerned. As a result, it is believed that it is called a plain, because its surface is even. But the physicists say that the sea is higher than the land.

The title of Isidore's book is nearly the same as that of Lucretius; also, as noted by Fontaine (in Isidore, 1960) its outline is in many places similar to those of Aristotle, Lucretius, Pliny and Aetius. So to organize his subject matter, Isidore must have had some doxographic references at his disposal, or at least a monastery school manual of such material. But it is striking how in this particular instance, Isidore's treatment on the origin of streams comes closest to the opinion of Lucretius, quoted earlier. (Note that in the past Ecclesiastes has often, evidently mistakenly, been attributed to Solomon). Less than a decade later around 620, Isidore (Isidorus, 1911; 13, 14) again gave a similar account in his book *Etymologies*.

Therefore, the reason why the sea does not increase, although it receives all the streams and all the springs, is as follows: in part, because its own magnitude does not feel the inflowing streams; further, because the salty water consumes the fresh water flows; or because the clouds attract to themselves a large portion of the water; also partly because the winds sweep it up, or partly because the sun dries it up; finally, because after having percolated through some hidden openings of the earth and having been returned to the head of the streams and to the springs, it runs back.

   Isidore's writings rapidly spread all over Western Europe, and they had a huge impact. Bede (*c.* 673–735), a Benedictine monk at Jarrow in England, who lived some 100 years later, also wrote a book *On Nature*, which seems to be strongly inspired by Isidore's. His section 40 on why the sea does not increase (Beda, 1843) is an almost literal summary of Isidore's descriptions quoted above. Isidore's influence is also evident in the work of Hrabanus Maurus (*c.* 776–856) of Mainz. Entitled variously *On Nature* or *On the Universe*, it was written around 844, at the height of the Carolingian Renaissance. Intended as an aid for preparing sermons, the text is replete with biblical references and Christian allegories and Hrabanus comes across as a well-read author; however, for his explanation on why the sea does not increase and on the origin of streams and springs, his main source was clearly Isidore. His section on this topic (Rabanus Maurus, 1852; 11, 2) is taken nearly verbatim from Isidore's (13, 14) quoted above.

   These few examples show how by the end of the first millennium of the present era a number of concepts of Greek natural philosophy had been propagated in Western Europe through Isidore's writings. If Isidore deserves a place in this history, it is not on account of the originality or correctness – by today's standards – of his cosmological views. However, he was part of a tradition that has some scientific merit. To judge from his specification that the wind is a cause of evaporation, Isidore's hydrologic and meteorologic descriptions were inspired indirectly by those of Lucretius; they are thus related to the views of the earlier atomists Demokritos and Leukippos, rather than those of Aristotle.

14.3.3 The High Middle Ages and the Renaissance

These prevailing concepts in natural philosophy remained roughly the same until the beginning of the thirteenth century, when Aristotle's philosophical works began to draw greater attention in Western Europe. The Latin translations of these works were derived from Greek originals, as a result of intensified contacts with Constantinople during the crusades, and from Arabic translations mostly in Moorish Spain (see Jourdain, 1960; Peters, 1968). In contrast to Western Europe, where his theories had somehow been overlooked until then, possibly as a result of the emphasis on Epicureanism and Stoicism among the Romans, in the Arab world Aristotle had been held in high esteem once his works had become available in translation. This is witnessed by the fact (cf. Mieli, 1966; pp 95, 102) that the famous philosophers Al-Farabi (d. 950) from Turkestan, and the Iranian Ibn-Sina ("Avicenna," 980–1037) have also been called the second and the third master, respectively, after Aristotle. The history of Aristotle's theories in the Arabic world, their subsequent acceptance by the Latins, and their eventual penetration into the vernacular, make for some fascinating reading. In the case of the *Meteorologica*, the first three books were translated early on from a partly abbreviated and corrupted Arabic version by Gerardus Cremonensis (d. 1187), and the fourth book, which does however not deal with meteorologic phenomena, directly from Greek by Henricus Aristippus [d. 1162] (Grabmann, 1916). Roughly a century later, around 1260, a more faithful version of the first three books was produced from the original Greek by Guillelmus de Morbeka (Willem van Moerbeke, *c.* 1215–1286) (Brams and Vanhamel, 1989). As a result, in the course of the thirteenth century, copies of these Latin translations started to appear in Western Europe, and gradually made their influence felt. Also, not long after the Latin translation by Willem, toward the end of the thirteenth century a Norman cleric, Mahieu le Vilain, made a translation of the *Meteorologica* into the French vernacular. An indication of the tremendous influence Aristotle's works must have had, is the fact that for the period between 1200 and 1650 Lohr (1967–1973) lists more than 85 commentaries on the *Meteorologica*, some of them by famous scholars like Alfred of Sareshel (1988), Albertus Magnus, Thomas de Aquino, Johannes Buridanus, Nicholaus Oresme, Themo Judaei de Monasterio (Münster) and others (see also Thorndike, 1954; 1955; Ducos, 1998). Aristotle's influence continued for the next three centuries and at the height of the Renaissance European literature had become fully imbued with many of his physical theories. These theories served not merely as physical explanations, but they were also used as a rich source of metaphors and poetic imagery (Heninger, 1960).

But while Aristotle's ideas were ubiquitous and known by most scholars, they were far from universally accepted. The main effect of Aristotle's *Meteorologica*, like his other works, it seems, was that it generated a common vocabulary, within a coherent system of logic, which stimulated more thorough discussion and the formulation of new questions, but not necessarily the answers, about the nature of the Universe. Thus, contrary to what is usually assumed about medieval scholarship, Aristotle's theories were not always blindly accepted but they often provided the impetus for more correct interpretations. The writings of Buridanus (Jean Buridan, *c.* 1295–1358) from Béthune in Picardy, are a case in point. He is probably best known for his proverbial ass (asinus) and also for the fact that, in rebuttal to Aristotle and some 350 years before Newton, he had some idea of the principle of momentum conservation. In his book *Questions on the three books of Aristotle's Meteorologica* (Ducos, 1998, p. 82), Buridanus wrote the following.

For it is also said to be possible that the water of the sea is evaporated and that the vapor is changed into air, which is carried by the wind to a distant place, and descends there to the earth to replenish the pores to avoid a vacuum, and is there condensed and changed into water, which comes to the spring and then flows to the sea.

In this passage he seems to admit that Aristotle's mechanism of condensation inside the Earth may be possible, but then he continues in direct contradiction of Aristotle, pointing to the rain as the substantial source of the springs.

The waters of springs come from the rains in this manner, because there are in the earth large hollow spaces which receive much rain water in winter, which for some hollow spaces suffices to flow out through the year until the winter rains return, and thus they are perpetual springs, which flow from these hollow spaces. There are other smaller hollow spaces which cannot receive in themselves so much water, which would suffice to flow out through the whole year; therefore, the springs which flow from them dry up in summer.

In other words, if the condensed water could be a substantial source of spring water, springs would not dry up in summer. This shows that among some influential scholars at the University of Paris, rain was taken as the main, if perhaps not the sole, agent in the generation of springs.

Later examples, indicating that the rainfall percolation concept was not uncommon throughout this period, are the accounts by Bernard Palissy (1510–1589) (Palissy, 1888; 1957) and Guillaume de Salluste du Bartas (1544–1590) (du Bartas, 1988, p. 78). Both gave descriptions of the origin of springs and rivers that come generally quite close to the rainfall percolation mechanism as it is known today. It is worth noting that, just like Vitruvius, neither one was famous for his philosophical ideas; Palissy was known mostly for his practical and artistic talents as a ceramist, and Bartas, a soldier and diplomat, for his poetry. Although both were Huguenots, their specific ideas on the origin of springs do not appear to be literal biblical accounts.

But disagreement with Aristotle among some did not necessarily lead to improved concepts among others. For instance, Leonardo Da Vinci (1452–1519), in his notebooks (see MacCurdy, 1938, p. 22) first describes how heat raises water vapor to higher elevations, where it condenses and falls as rain and hail; he then explains how in a similar way the same heat also draws up water from the roots of the mountains, through channels inside these mountains like through the veins inside the human body, to their summits, where the water can flow out through cracks and crevices to create rivers. He also concludes ". . . that the water passes from the rivers to the sea, and from the sea to the rivers, ever making the self-same round . . ." thus implying the seawater filtration mechanism to arrive back at the roots of the mountains. Another example is the description given by Descartes (1596–1650) (1637, p. 179), which was also nearly the same as the seawater filtration theory of old. Hence, the fresh waters which flow into the sea, do not make it any larger because as many others leave it continuously. Some of these waters are raised in the air after being changed into vapors, and then proceed to fall back down as rain or snow on the earth; however, most of these waters penetrate through underground conduits to beneath the mountains; from there the heat, which is in the earth, raises them as vapor to the peaks, where they replenish fountains and rivers. Seawater moving through sand becomes fresh because the salty parts, which are larger, more rigid and interlaced, cannot follow the tortuous paths around the sand grains as easily as the more slippery and smaller fresh water parts, and they are left behind.

14.4 From philosophy to science by experimentation

In the course of the seventeenth century the general approach to science started to change, and gradually experimentation became an essential part of it. Pierre Perrault (1608–1680) and Edme Mariotte (1620–1684) were two central figures at this juncture of the history of hydrology. Their main merit was that, in contrast to the earlier writers on the subject, both relied on experiment and quantitative arguments. But to put their work in proper context, it is necessary to bear in mind the various opinions on the causes and mechanisms of river runoff, as they were then known to them.

14.4.1 The Common Opinion at the end of the seventeenth century

The book *On the Origin of Springs* by Perrault (1674) can provide some insight in this; the first half of it, covering 146 pages, is devoted to a thorough review of the better-known theories and explanations of the day. The authors discussed by Perrault are Plato, Aristotle, Epikouros, Vitruvius, Seneca, Pliny, Thomas de Aquino, Scaliger, Cardano, Agricola, Dobrzenski, Van Helmont, Lydiat, Davity, Descartes, Papin, Gassendi, Du Hamel, Schottus, Rohault, François and Palissy. For each of these authors Perrault first gives a brief description of the main features of the propounded theory, followed by his own critique and reasons for rejection. After completing the survey, he then singles out one of these theories and further specifies (p. 148) how those, who support this particular view,

. . . believe that the waters of the rains & of the melted snows, which fall on the earth, penetrate it until they encounter heavy (lit. greasy) soil or some other matter, which stops them; whereupon they flow to some opening on the slope of a mountain . . . They believe that the waters, which fall on the high plains, are the cause of the springs, by means of this penetration, which they assume (to take place). . . . They believe that the rains, which fall on the slope of hills, are lost & of no use for the springs, for the reason that from there they fall into the rivers which carry them to the sea . . . They also believe that it is the springs, which being joined together produce rivers, & that if there weren't any springs, there wouldn't be any rivers.

This description of the sequence of processes, which is elaborated on further on pp. 151–152, could have been written today, and it would not be out of place among the descriptions reviewed in Chapter 11. It is remarkable, therefore, that in 1674 Perrault calls this the "Opinion Commune" or "Common Opinion." But even more remarkable is the fact that he also points out that among his 22 "authors," by which he means the learned men and authorities on the subject, only four espoused this opinion, to wit Vitruvius, Gassendi, Palissy and François. In other words, although only a small minority among the expert natural philosophers held this view, he chooses to call it the Common Opinion. Could this mean that toward the end of the seventeenth century, almost everyone else, that is the person "in the street," was already of the opinion that springs and rivers are produced by rainfall percolation?

14.4.2 The first experimental analyses

Perrault's interpretations

Not much is known about the life of Pierre Perrault (1608–1680). He was born into a bourgeois family, had at least seven siblings and appears to have spent most of his life in Paris (see Hallays, 1926; Delorme, 1948; A. Picon in Perrault, 1993). Actually, more is known about several of his younger brothers: Claude (1613–1688), one of the original members of the Académie Royale des Sciences, was a physician, a naturalist and an architect; Nicolas (1623–1661) was a doctor in theology, who was expelled from the Sorbonne around 1655 for his Jansenism and known for his denunciation of the Jesuits; Charles (1628–1703) was controller of the King's buildings and author of the Mother Goose fairy tales. Like his father Pierre and his older brother Jean (1610–1669), Pierre Perrault was originally educated for the legal profession. With this background, he purchased the position of Receiver General of Finances for Paris. But because of some unexpected changes in the tax arrangements, around 1664 he came heavily into debt with the royal treasury and was subsequently forced to give up this post. At this point he was essentially broke and turned to hydrology and literature. It is unclear exactly why he set out to focus on the origin of springs. Was it a coincidence that around the same time his brother Claude translated the work of Vitruvius (Vitruve, 1986)? It should be recalled that Book 8 of that work is devoted to this very topic and that Pierre classified Vitruvius (correctly) as one of the proponents of the Common Opinion.

In any event, in the second half of his 1674 book he starts immediately (p. 148) by contrasting his own views with the Common Opinion, as quoted in the previous section, and then (p. 150) he states the two main difficulties with it, as he sees them.

The first is this supposed penetration of the earth by the waters of the rain, which to me does not seem possible in the manner they mean; the second is that I don't think that enough rain and snow water falls to soak the earth to the extent necessary, nor that there would still be enough left over to make the springs and rivers flow, which are produced by it, as they say, and in the manner they assume.

To support these two objections and to shed some light on the matter, Perrault proceeds to describe a soil water flow experiment he conducted. He took a 65 cm (2 pieds) long lead pipe with a diameter of 4.5 cm (20 lignes), closed off at the bottom with permeable cloth and filled with coarse river sand, and he inserted it about 1 cm (4 lignes) into the water contained in a wide shallow vessel (see Figure 14.4). (The stated dimensions are converted, here and in what follows, by assuming that 1 French inch or 1 pouce = 2.707 cm (Petit Larousse, 1964); also, 1 inch = 12 lines =1/12 foot.) After 24 h he observed that the water had risen and moistened the sand up to a level of 49 cm (18 pouces). To verify whether the risen water could flow out sideways to form springs, he made an opening in the pipe with a diameter of about 1.8 cm (7–8 lignes) at a height of about 5.4 cm (2 pouces) above the water surface, where he attached a small 5.4 cm long gutter, sloping down, in which he placed a strip of paper covered with a thin layer of sand in contact with that of the column. To his surprise, although the paper and the sand in the gutter became moist, never a single drop fell from this little gutter. To check further whether any water would ever flow out, he withdrew the sand column from the water and suspended it for half a day above an empty tray, but again no water flowed out of all that had earlier risen 49 cm. He then poured some water on the top of the column to soak the sand, but only three quarters of it came through at the bottom. The next day, after having poured on again the same amount, all the water passed through. Finally, the following day, he shook all the sand from the bottom of the pipe and observed that the soil which came out first was wet like mortar, whereas that which

Fig. 14.4 Reconstruction of the experimental set-up described by Perrault (1674) to measure the movement of water in a sandy soil. The soil was placed in a lead pipe with a length of 65 cm and a diameter of 4.5 cm; the bottom was closed off with permeable cloth. At a height of about 5.4 cm above the water surface an opening was made in the pipe to check whether any water, that had risen into the soil after the bottom had been inserted in the bath, would be able to flow out in the manner of a spring.

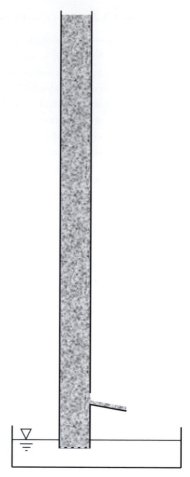

came out last was not so moist, even though he had twice poured water on the sand of the top, which came out last. He repeated the experiment with several other types of soil and set-ups, but the results were similar.

After drawing a number of general conclusions from this experiment, he returns to the two difficulties, which he raised earlier against the Common Opinion (p. 162).

As regards the first one, which is this penetration, which I don't think can take place, as they believe, I will say first, that if we are to believe Seneca and Lydiat. . . . the earth does not allow itself to be penetrated by the rain with such ease as is believed . . . but I add to this reasoning the everyday experiences one encounters with this penetration of the earth.

He further illustrates this inability of water to penetrate by describing the numerous drainage problems encountered by farmers and others dealing with soil water management. Following these general observations in the country side he turns once more to Seneca (p. 166).

The same Seneca asserts that the waters of the rain don't enter into the earth beyond ten feet, which he vouches for as a good wine-grower, which he says he is, who has often dug into the earth.

This shows again the profound influence Seneca's description of his vineyard experience (see Seneca, 1971) continued to have even after 17 centuries. Perrault then recounts how he himself conducted similar experiments and had pits dug in the earth on mountains, on hillslopes, in bottom lands, in cultivated gardens, after long and heavy rains, but he never found the earth moistened beyond a depth of 2 ft. Perrault next invokes the results from his own sand column experiments described above (p. 175).

> The second difficulty with this Common Opinion is, that I do not believe that the rains, which fall on the high plains, suffice to maintain the springs, not because of their smallness . . . but because of the waste & the loss of nearly everything which falls on these plains, without any of it benefiting the springs & live fountains . . . For before a certain quantity of water can traverse a certain quantity & thickness of earth, all the particles of this earth must be moistened, each one in particular & with all their surfaces; & this is a pure loss, for this water will only leave by evaporation, because of its adherent property, which causes it to attach itself to everything it touches, and to stay there suspended without moving downward, where its weight should normally attract it, as can be seen by our experiment.

With the Common Opinion disposed of, Perrault turns to the statement of Aristotle, quoted earlier, that the volume of the yearly flows of the rivers is ". . . as large as the earth in size, or at least not so much smaller." Thus he will allow the reader to judge

> . . . that these waters of the rivers will not equal the mass of the earth in one year, as he says, but even in a thousand years.

Follows now Perrault's celebrated analysis of the comparison of the flow in the head-waters of the Seine River in Burgundy with the rainfall on the upstream watershed. In brief, he estimated the distance between the source of the river and Ainay le Duc (now Aignay-le-Duc) as roughly 13.5 km (3 lieues) with an average distance to the divides on either side of roughly 4.5 km (1 lieue); with an average annual precipitation estimated at 51.96 cm (19 pouces, 2.333 lignes), this made him conclude that the total annual volume of precipitation over that area was of the order of 224 899 896 muids. (Units of length and volume were not always standardized and they tended to vary in different periods and in different regions; therefore it is not easy to check Perrault's calculations. However, since 1 muid equals 8 ft^3, adopting the conversion that 1 ft is equivalent to 32.484 cm, one finds that this volume is equivalent to roughly 6.167×10^7 m^3; to obtain this volume with the 51.96 cm of precipitation requires the magnitude of the lieue (i.e. the league) in this calculation to be about 4447.7 m. This result is remarkably accurate and shows that Perrault used the "lieue de terre" (land league), which according to the *Petit Larousse* (1964) has a formal length of 4445 m or 1/25 of a degree on a great circle.) He did not have any discharge measurements for the Seine at Ainay le Duc, but by comparing the flow situation to that of the Gobbelins River near Versailles, he guesses it to be about 36 453 600 muids per year, which is roughly equivalent with 1.0×10^7 m^3 per year or 8.42 cm of annual rainfall. This allows Perrault to conclude that

> . . . only one sixth of the water which falls as rain and snow on the upstream catchment is needed to make this river run continuously for an entire year . . .

and the remaining five sixths will serve to supply the losses, diminutions and wastes which one observes, as nourishment of vegetation, evaporation and useless outflows. The case of this one river also suggests that rain and snow should suffice for all the other rivers of the world as well, provided one takes the wastes into account.

After thus having shown that the Common Opinion cannot possibly be correct, also that the river flows are not as large as Aristotle had supposed and that the rains are more than adequate to feed the rivers, Perrault (p. 207 ff.) is ready to formulate his views on the origin of springs, the central topic of his treatise. In brief, water cannot penetrate the Earth directly to any appreciable depth. As a result, most of the rain and snow waters, which fall

Fig. 14.5 Symbolic representation of the origin of springs in Perrault's (1674) book, showing how nymphs carry water from the river to the mountain top where it can start to flow as a spring. (Courtesy Mandeville Collections Library, University of California, San Diego.)

on the mountains and hills, flow down from the slopes and end up in the rivers and in the creeks; under these rivers and the plains, which they drain, there are layers of clay and other impermeable material; therefore the river waters enter into the more permeable top layers of the plains, mostly laterally, often also by overflowing and flooding. Inside the Earth the water vaporizes by various mechanisms, namely by heat, by cold and by the movement of the air particles, whereupon this vapor rises inside the Earth to the summits of the mountains, where it condenses again to make springs. In support of this explanation, he also invokes the authority of several of the authors of his literature review, among whom Aristotle, Seneca, and Descartes, who had proposed similar mechanisms. His overall conclusion is that, while both springs and rivers are caused by precipitation, in the case of the springs the relationship is indirect, because the water must first enter into the rivers before it can produce springs. Hence springs are not the cause of rivers, but rivers are the cause of springs, so that if there were no rivers there would also be no springs. This imagined transport from rivers to springs was illustrated allegorically in Perrault's book, as reproduced in Figure 14.5.

Considering the state of measurement technology and of open channel hydraulics, Perrault's comparison between river runoff and precipitation was a remarkable feat. So, not surprisingly, in most reviews of the history of hydrology Perrault's work, with its emphasis on experimentation, is rightfully acclaimed as one of the significant landmarks of this science. On the other hand, however, it is usually overlooked, or not fully realized, that in fact one of the main objectives of Perrault's (1674) book, was to refute the largely correct Common Opinion. Thus in this sense, a large part of his work was also a major step backward. Perrault arrived at his erroneous notion mainly on the basis of a sand column experiment and of field observations similar to Seneca's. By today's standards and with present understanding of the underlying physics, his interpretation of these observations was wrong. The reason for this was Perrault's inability to grasp the effects of surface tension on the flow of water in a partly saturated soil. Clearly, the time was not ripe yet and a satisfactory explanation of his column experiment and his field observations would only be possible some 200 years later in the nineteenth century. In any event, whatever damage may have been caused by Perrault's book was soon undone by the more fundamental and perceptive work of Mariotte.

Mariotte's reaffirmation and proof of the Common Opinion

Few facts are known with certainty about the life of Edme Mariotte (Picolet, 1986). He was born around 1620 in Til-Châtel (or Tilchâtel) near Dijon in Burgundy and died in Paris in May 1684; he appears to have spent most of his early life in Burgundy, probably until 1666, when he was elected one of the original members of the newly founded Académie Royale des Sciences (de Condorcet, 1773) and he had to move to Paris. By 1634 he had received tonsure and was therefore a cleric, but there is no evidence that he received higher orders or was ever ordained into the priesthood. Perhaps as early as 1634 he was also appointed prior of St. Martin de Beaumont-sur-Vingeanne, which provided an annual income of some 300 pounds. But this did not involve major responsibilities and his life was essentially devoted to science. While he had many diverse interests (see Davies, 1974), he is now remembered mostly for the law of gases that bears his name, his discovery of the blind spot in the human eye, and his work on the laws of impact between bodies, among many other contributions. A fine example is the constant head device shown in Figure 9.2, which to this day is called a Mariotte flask. As member of the Académie he was also involved in the hydraulic works for the fountains at the king's new castle in Versailles. But it is his major work on this subject, namely *Treatise on the Movement of the Waters and of the Other Fluid Bodies* (Mariotte, 1686), published posthumously, which is of interest here. In the section "On the origin of springs," he first treats the formation of rains, and then unambiguously specifies what happens next (p. 19)

Having fallen, the rains penetrate the earth through little channels which they find there; thus, when one digs somewhat deeply into the earth, one usually encounters these little channels, whence the water, which gathers at the bottom of what one has excavated, makes the water of wells; but the water of the rains, which fall on the hills & on the mountains, after having penetrated the surface of the earth, mainly where it is light & mixed with pebbles & roots of trees, often encounters clayey soil or continuous rocky formation, which it cannot penetrate and along which it flows to the bottom of the mountain or some considerable distance from the summit, where it comes out again into the open, & forms the springs. This effect of nature is easy to prove, because firstly the water of the rains falls all year long in sufficiently large abundance to maintain the springs & the rivers, as we shall show later on by calculation; secondly, we observe every day that springs increase or decrease according to whether it rains or doesn't rain; & if two months go by without considerable rain, they decrease most of them by one half; & if the drought continues for another two or three months, most of them dry up & the others decrease down to one quarter. From this one may conclude that if there were a whole year without rain, there would be very few springs left, most of which would be very small, or that they would cease altogether.

With his own view clearly explained, Mariotte proceeds in detail to refute some of the mechanisms proposed by others and to provide proof of his own assertions. He first deals with those philosophers who assume that vapors rise from the depths of the Earth to condense into water inside the mountains when they encounter the upper vaults like in an alembic, whence the water flows out to form springs. Mariotte rejects this hypothesis by indicating,

Fig. 14.6 Sketch in support of Mariotte's assertion, that water condensed inside a mountain cannot possibly flow out as a spring. (From Mariotte, 1686; courtesy Division of Rare and Manuscript Collections, Cornell University Library.)

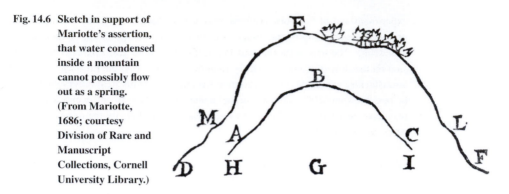

as illustrated in Figure 14.6, that if ABC is a vault in a mountain DEF, the water condensed on this concave surface ABC would fall down to HGI, instead of to L or M, so that it would be incapable of making a spring; he also rejects that there are many such caves. He counters the argument of some that there is earth beside or below ABC, by explaining that in this case the vapors will escape toward A and C, and will yield very little water; moreover, because there is always clayey soil where there are springs, it is unlikely that these condensed waters will be able to pass through from the inside of the mountains.

Next, without mentioning them, he deals with those, like Seneca and Perrault, who claimed that rain cannot penetrate into the soil.

Still others object that the summer rains, although very big, enter the earth only about half a foot, which one can observe in the gardens & in the tilled fields: I remain in agreement with the experiment. However, I maintain that in non cultivated soils & in the woods there are some little channels, which are quite close to the surface, in which rain water enters, & that these channels extend down to great depths, as one sees in deep dug wells, & that when it rains ten or twelve days in a row, at the end the top of the tilled soils becomes completely wet, & the remainder of the water passes in the little channels, which are below & which have not been broken by tillage.

He goes on to illustrate this with his own observations in the cellars of the Royal Observatory and inside several quarries. In these places water would drip down from the ceiling, but invariably this water could be seen to issue from small holes, crannies and cracks in the rocky vault, while the rest of the surface remained dry; also, this dripping was mostly in response to rain, and would cease during droughts, which suggests that springs are made in the same way. Among many other examples, he notes that during the dry summer of 1681 many wells and springs dried up, and that after a cold spell in the fall they continued to decrease; they would not have done this if the water had been formed by vapors raised from below and condensed by the cold of the surface. Furthermore springs, which are high up in the mountains, are always adjacent to even higher areas, and their flows are larger when these areas are larger; again, this indicates that they are produced by the rains which fall on these higher surfaces.

Finally (p. 30), he addresses the objection by some that the total yearly rain may not be able to supply enough to the great rivers which flow into the sea. He resolves the problem, like Perrault, by comparing river flow with the rainfall on the upstream watershed area; however, his watershed area is much larger and his estimation of the river discharge is also much more rigorous. From measurements over an eight-year period, he estimates the rainfall at Dijon to be about 46 cm (17 pouces), adding that a similar measurement by the "author of the book entitled 'On the Origin of Springs'" yielded a value of 51.96 cm (19 pouces, 2.33 lignes); but for the purpose of the exercise he decides to adopt a conservative value of 40.61 cm (15 pouces). (In his calculations Mariotte assumes that one lieue (league) contains 2300 toises (fathoms); as 1 toise equals 6 pieds, the length of his league is about 4482.8 m, which is slightly different from Perrault's assumed length.) With this value, and assuming that the Seine catchment upstream from Paris occupies roughly 60 286.27 km^2 (3000 square leagues), he figures that this catchment would receive roughly 24.479 km^3 (7.1415×10^{11} ft^3) of rain per year, on average. He estimates the average velocity of the Seine at the Pont Rouge in Paris from float velocity observations of around 1.35 m s^{-1} (250 ft min^{-1}), which he reduces to 0.54 m s^{-1} to account for the effect of bottom and side friction. With a cross-sectional area of the river of 211.04 m^2 (2000 ft^2) this velocity yields an average annual discharge of 3.6032 km^3 (1.0512×10^{11} ft^3); this is equivalent with about 6 cm of water over the whole catchment and is less than 1/6 of the annual rainfall. From this result Mariotte deduces that, even when evaporation, the moistening of surface soils and the replenishment of groundwater are taken into account, there is enough rainwater to produce springs and rivers.

Lest his readers not be convinced and still feel that this result applies only to rivers and not to fountains and springs, as Perrault had argued, Mariotte proceeds next (p. 34) to apply the same analysis to the great spring at Montmartre. He estimates its catchment area as 113 963 m^2 (30 000 square toises) and assumes a rainfall of 48.726 cm (18 pouces), which is equivalent to 55 529 m^3 per year or roughly 0.105 m^3 min^{-1} (107 pintes per minute; there are 35 pintes in a cubic foot). He then explains what happens in the field.

Now, the terrain of this mountain is sandy to a depth of 0.65 to 1.0 m (2 to 3 feet), & the bottom is clay soil; part of the water of the large rains first runs to the bottom of the mountain, part of the rest stays in the sand near the surface, and the rest flows between the sand and the clay; so, if we assume that it would be only the fourth part of the total, which is . . . 105 l/min (107 pintes per minute), that quarter would be around 26 l/min, which that spring should yield, & that's pretty close to what it yields, when it is running well.

Mariotte's work is without question one of the highlights in the history of hydrology. His treatment is clear and sound enough that it would not be out of place in present-day descriptions, like those reviewed in Chapter 11. His determination of the river discharge rate is based on solid reasoning, and therefore his comparison between precipitation and river flow is a marked improvement over Perrault's calculation a decade earlier. In addition, he shows cogently by different examples that rain water does penetrate the soil in sufficiently large quantities and to sufficiently large depths to be the only possible cause of springs. In this connection, his description of the "little channels or conduits" through which the water penetrates into saturated soil, should establish him as the originator of the concept of macropores. He further supports his ideas on the origin of springs by a mass balance comparison between rainfall and outflow rate from the spring at Montmartre. The reference to Perrault's rainfall measurements shows that Mariotte was familiar with Perrault's book; actually, it would be surprising if he had not been, because he had been working so closely with his brother Claude Perrault at the Académie. This probably also explains why he merely presented his own views, dispassionately, without criticizing or even mentioning Perrault's outlandish theory on the origin of springs.

14.4.3 Lingering doubts and slow acceptance of the Common Opinion . . .

It might be thought that, after the work of Mariotte had put the rainfall percolation theory for rivers and springs on a sufficiently firm foundation, the issue had been settled once and for all. On the other hand, while Mariotte's arguments were sound and indisputable, he had

Fig. 14.7 Reconstruction of the experimental set-up described by de La Hire (1703), which was intended to verify the downward movement of water in the soil profile at the lower terrace of the Observatory. A lead basin of 0.422 m², with 16 cm high sidewalls, was placed at 2.6 m below the surface; at one of its corners a 3.9 m long pipe was attached to permit outflow of captured water into an adjacent ditch.

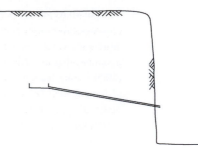

only addressed the issue of infiltration in the field, and had totally ignored the puzzling and paradoxical outcome of Perrault's column experiments (Figure 14.4). Because he put the emphasis on the role of macropores, perhaps he felt that the soil column set-up was irrelevant for field conditions.

This apparently did not escape de La Hire (1640–1718), who had in fact been the one to see to it that Mariotte's (1686) book was published posthumously. So, a few years later de La Hire (1703) published the results of another experiment, with a set-up that he specifically designed to check whether precipitation can penetrate the Earth until it would encounter some impermeable layer; he described it as follows (see Figure 14.7).

I chose a place on the lower terrace of the Observatory, and in 1688 I had a leaden basin with a surface area of 0.422 m² (4 feet) installed in the ground at a depth of 2.60 m (8 feet). This basin had sides ("rebords") of 16 cm (6 pouces) height, and it was slightly inclined toward one of its corners, where I had a 3.90 m (12 foot) long leaden tube soldered, which had a considerable slope and which entered in a small excavation at the other end. The basin was kept far from the wall of the excavation, in order that it would be surrounded by a greater quantity of soil similar to that which was on top, and that it would not dry out by the proximity of the wall.

From the present vantage point this set-up, which appears as a forerunner of the lysimeter, had serious shortcomings for its intended purpose; evidently, the basin side walls did not extend to the soil surface, so that percolating rainwater could move away laterally. With present day understanding of the flow in a partly saturated soil, it is no wonder that de La Hire had to report, that "not a single drop of water has come out through the tube in 15 years." He also conducted some experiments with a smaller basin at more shallow depths and under conditions of minimal evaporation, but here some water would only be collected after heavy rainfall and large snowmelt. From these percolation experiments he deduces that rainwater cannot penetrate the earth very deeply. He then proceeded to determine the evaporative loss from two individual fig leaves inserted in water, and this leads him to infer that rain alone is not sufficient to support vegetation in summer, let alone to feed the rivers. In the end de La Hire concludes that the rainfall percolation theory of Mariotte cannot be generally valid; rather, the explanation can only be that there are huge quantities of vapor inside cavities or hollows in the Earth in the form of an alembic, which rise from the waters at the level of the closest rivers or the sea through cracks in the rocks, and that these condense higher up, as a result of the cold at the surface of the Earth, and flow out as springs. Like Seneca's and Perrault's explanations before him, de La Hire's interpretation of his experiments was wide of the mark; indeed a correct explanation of his puzzling seepage phenomena would have to wait for the fundamental work of Laplace (1749–1827) in surface tension and its subsequent application by Buckingham (1907) in soil physics.

The works of Perrault and Mariotte promptly crossed the Channel and were deemed remarkable enough to be reported in the Philosophical Transactions of the Royal Society (Anonymous, 1675; 1686) immediately after their publication. But it is clear that not everybody accepted Mariotte's theory there. Edmond Halley's (1656–1741) reaction is a case in point. Without a doubt Halley was thoroughly familiar with developments in France. In 1681 he had already spent 6 months in Paris where he had become acquainted with several members of the Académie and other learned persons, and had purchased many books of interest to ship back to England (see Cook, 1998); in 1686, at the time of the publication of Mariotte's book, Halley was Clerk of the Royal Society and maintained an extensive international correspondence; he was also editor and publisher of the sixteenth volume of the *Philosophical Transactions*, which contained the review of Mariotte's book. All this makes him almost certainly the author of Anonymous (1686); it must also be his familiarity with this book, no doubt combined with his experiences at sea, which prompted Halley (1687) to engage in the study of evaporation, an aspect of the water cycle, which both Perrault and Mariotte had only dealt with obliquely in qualitative terms. From weight changes during evaporation of water from a small pan he deduces that, on warm days, evaporation amounted to approximately 2.5 mm (0.1 in) in 12 h; this was a reasonable result, as can be seen in Figure 4.16. Halley next uses Mariotte's method to determine the discharge rate of the Thames at Kingston Bridge; the determination of the flow rate this way was far from obvious at the time, as witnessed by the fact that some 15 years earlier Perrault had not quite known how to deal with this same problem. Estimating that the Mediterranean is fed by nine rivers, each of which is ten times larger than the Thames, he concludes that the total inflow into that sea amounts to hardly more than one third of the daily evaporation of 2.5 mm. At a first glance this conclusion is but a confirmation of Aristotle's (correct) explanation of why the sea does not overflow, some 20 centuries earlier. What was new, however, was that now an earnest attempt was made to base Aristotle's speculation on experimental evidence, and not just on everyday observation on a kitchen table. Although his pan evaporation measurements could provide only rough estimates of the actual values for the Mediterranean, Halley's study was probably the first in which evaporation was considered quantitatively in relation to streamflow.

What happens to this evaporated seawater in the global water cycle was the subject of a second paper (Halley, 1691). In brief, all of these vapors are eventually returned to the sea in various ways and this explains why the sea does not decrease even though the evaporation is so much larger than the river inflows. The greater part of these vapors is returned immediately to the sea as rains or dews without ever touching land. Part of the vapors, which are blown off the sea, falls on the lower lands where either it nourishes plants and is exhaled again, or it finds its way into the rivers, after the earth is saturated with moisture, to return to the sea. But most of these vapors are carried by the winds over the low lands to mountain ridges, where part of them precipitates ". . . gleeting down by the crannies of the stone . . .", and part enters the caverns of the hills, inside of which the vapors are collected ". . . as in an alembic, into basins of stone they find there . . ."; this condensed water then breaks out through the hillsides to form springs, which unite further down into rivulets, and eventually into rivers. (Halley's ideas on the origin of springs are also detailed in the *Journal Books of the Royal Society* (MacPike, 1932, pp. 217, 227).) Thus rain is not the only source of all springs. One may wonder why Halley rejected the explanation of Mariotte, whose book most likely had prompted his study in the first place, and why he was misled into invoking, beside rain, direct condensation on the ground and also Aristotle's underground vapor condensation and transport theory for the origin of springs. The explanation appears further down in the text, where he describes the earlier experience that had led him to this condensation theory. In 1677 he had been on an expedition to the Island of St Helena to chart the stars of the Southern Hemisphere; when he was carrying out nighttime celestial observations there on top of a hill some 800 m above sea level, the condensation was so heavy and fast that the droplets on his glasses had to be wiped off every 5–10 min, and the paper on which he recorded his observations became immediately so wet that it would not bear ink.

A more egregious example of reactionary science is Woodward's (1695) explicit reliance on the biblical Abyss or Plato's Tartarus as the ultimate water supply for the springs, rivers, vapors and rains of the earth. Woodward was a Fellow of the Royal Society, and also Professor of Physics at Gresham College in London; he was thus acquainted with Halley. Indeed in 1686 Halley had been elected Clerk of the Royal Society, which held its meetings at Gresham, and it was at that same college that he conducted his pan evaporation experiments (Halley, 1694). The learned men at Gresham College appear to have held various opinions on the origins of springs and rivers, but the Common Opinion was evidently not their favorite one.

Fortunately, the situation was not as dismal everywhere. One influential proponent of the Common Opinion in England was John Ray (1627–1705), naturalist and Cambridge professor until 1662, when he resigned out of religious principle (Raven, 1950). Early on, in fact one year before the publication of Perrault's book, he (Ray, 1673, pp. 296–300) expresses the view ". . . that all springs and running waters owe their rise and continuance to rain, seems to me more than probable . . ."; and he gives as specific reasons that he had never seen running waters breaking out near the top of hills unless there was enough earth above them to feed these springs, that springs generally abate in dry summers, that one seldom finds springs in clay grounds where water sinks in with difficulty, and that those, who would have fountains be fed by the sea, have still not given a satisfactory account of the ascent of water to the mountain tops and its efflux there; with filters and even pumps no such high ascents have ever been produced. He further argues that it is also unlikely that fountains can be attributed to ". . . watery vapors elevated by subterranean fires, or . . . diffused heat . . . , and condensed by the tops and sides of the mountains as by an Alembick head, and so distilling down and breaking out where they find issue", because the heat required to raise those vapors "through so thick a coat of earth" would be way too large. Finally, he also considers the general statement ". . . that rain sinks not above a foot or two deep into the earth . . ." as manifestly false; as evidence for his assertion he lists the internal flooding of coal mine pits and shafts during wet weather, the near complete absence of surface runoff on sandy and "heathy" grounds even during the heaviest rains, and the fact that the water outflows from caves in the sides of mountains generally increase in the rainy season and often stop completely in dry weather. In a later work Ray (1692; 1693) elaborates on this same theme but in more detail and with additional evidence. For instance, he mentions, without further specifics the "Ingenious French Author", who demonstrated in the Seine that rain may suffice to feed ordinary springs. It is unlikely that Ray had personally read Perrault's book. Rather, as a Fellow of the Royal Society since 1667, he was probably familiar only with the brief review by Anonymous (1675), which contains Perrault's comparison between rainfall and river flow in Burgundy, but nothing on Perrault's theory that springs originate from rivers, a view so at odds with his own. Ray also mentions his own observations on a little brook near his dwelling at Black Notley in Essex, which support his hypothesis that

". . . all its water owes its original to rain." In addition, he specifically addresses Halley's condensation theory. While he admits that Halley's condensation mechanisms may be partly valid in "fervid regions," he feels that they should be of little interest in the production of springs in more temperate countries. The Alps, which are above the fountains of four of the greatest rivers in Europe, are a case in point. Although the Alps are covered heavily with snow for six months of the year, and therefore cannot have any access to vapor, the rivers issuing from them continue to run, albeit low, all winter long without interruption; when the snow melts in spring, some of these Alpine rivers overflow their banks, although no rains fall; but later on after the snow has melted, the streams decay in spite of the vapors that condense on them, and in summer the streams flood again only when it rains; this proves that they are mainly fed by melted snow, as is also indicated by their "sea-green" color.

Ray's writings show that also in England, Halley's and Woodward's views notwithstanding, the Common Opinion was a well-established theory at the time. However, their main importance in the history of ideas stems from the fact that they are among the earliest and more articulate in the renewal of the long tradition, in which use is made of the hydrologic cycle as evidence for God's wisdom in the creation of the world. In this renewed form of the tradition, or "physical theology," which was to last nearly another 150 years, the hydrologic cycle served as a unifying and ordering concept to explain the wisdom behind a number of disparate phenomena on earth, such as mountains, floods and the size of the oceans, which might otherwise have appeared chaotic and paradoxical, in light of, and in contrast to, the obvious perfection of the new Newtonian mechanics. At the time, several others (see, for example, Bentley, 1693, pp. 31–32) were writing on the same theme; but Ray was by far the most popular and widely read author on the subject especially through his book *The Wisdom of God Manifested in the Works of Creation*. This book, first published in 1691, went through twelve editions (Ray, 1759) and continued to be issued until apparently as late as 1827. The underlying idea, namely that the ceaseless circulation of water on Earth is proof of a divine design, became almost a cliché and seems to have exerted a definite imprint on the thinking of intellectuals in England well into the nineteenth century; evidence for this fascination can be found (see Tuan, 1968) in the works of such well known intellectuals as, among others, W. Derham (1657–1735), A. Cooper (3rd Lord Shaftesbury) (1671–1713), J. Hutton (1726–1797), O. Goldsmith (1728–1774), J. Wesley (1703–1791), W. Paley (1743–1805), W. Buckland (1784–1856), J. Kidd (1775–1851), W. Whewell (1794–1866) and even the scientist John Dalton (1766–1844) (1793, p. 145). During the same period similar ideas in physical theology were popular also on the continent with, for instance, such authors as N.-A. Pluche (1688–1761), G. L. L. Buffon (1707–1788) in France, J. A. Fabricius (1668–1736) in Germany (who used the term "Hydrotheology"), and C. Linnaeus (1707–1778) in Sweden.

That the Common Opinion continued to deserve its name is also attested to in the books by the physicist Pieter Van Musschenbroek (1692–1761), the well-known inventor of the *Leyden Jar*, who held successive professorships at the Universities of Duisburg in Westphalia, and Utrecht and Leyden in Holland. In his description of water, Van Musschenbroek (1739, p. 417) asserts the following.

As the rain, the snow, hail and all the vapors fall on the earth, they penetrate it, & flow through the pores, the openings & the cracks, like through underground pipes to the lowest places. If these pipes or conduits are open at the top at one of their ends, fountains are formed thereof, from which the water gushes more or less high, depending on whether the opening in the earth is larger or narrower, or depending on whether the water in the underground conduits presses higher above this opening. But if the rain flows out on the surface of the earth into deep hollows, it forms the lakes & the swamps there, from which then rivers are born, which also owe their origin to the waters gushing from the fountains. Consequently, river water is either rain water, or fountain water, or both together.

In a later treatment of the same topic Van Musschenbroek (1769, p. 281) seems to have become aware that this rainfall penetration had presented some difficulties with others in the past. He now addresses the controversy, listing Seneca, Varin, de La Hire, and Buffon, as those who claim that rain cannot penetrate the earth beyond 4–10 ft; he then counters them with his own experience in Holland, as well as with that of Erndetl in Poland, and le Monnier in Auvergne, and repeats essentially his earlier description.

But in spite of the frequent appearance of the hydrologic cycle in physics and in physical theology alike, in none of the treatments reviewed here was there even a hint of calculations, of the kind made earlier by Perrault, Mariotte and Halley; in fact, during the century following their writings the basic notions on the origins of springs and streams did not undergo drastic changes, and many of the disagreements and uncertainties lingered on, it seems. This is brought out in Dalton's (1802a) paper, which he presented in 1799 before the Manchester Literary and Philosophical Society, and which he starts off with the following observation.

Naturalists, however, are not unanimous in their opinions whether the rain that falls is sufficient to supply the demands of springs and rivers, and to afford the earth besides such a large portion for evaporation as it is well known is raised daily.

This is followed by Dalton's rough estimates for all of England and Wales of average annual precipitation as $P = 787$ mm (31 in), on the basis of measurements at some 23 different sites; of annual dew, as 127 mm (5 in), on the basis of measurements by one Dr. Hales (probably Stephen Hales (1677–1761)); of river runoff, as $R = 330$ mm (13 in), by extending and correcting Halley's estimate for the Thames; and finally of evaporation, as $E = 635$ mm (25 in) per year, on the basis mainly of his own measurements with a simple lysimeter over a 3 y period at Manchester, which combined with the dew amounts to 762 mm annually "raised into the air." Combining these terms in a water budget (cf. Equation (1.1)), in which the dew fluxes cancel out, Dalton ends up with an annual deficit of $(330 + 635 - 787) = 178$ mm (7 in); he attributes this failure to close the budget to a possible underestimate of the average precipitation and, which he feels is more likely, to certain features of his lysimeter, which somehow lost water in heavy storms and which usually kept the soil surface more moist, and therefore must have evaporated more, than the earth around it. He summarizes this part of the paper.

Upon the whole then I think we may fairly conclude – that the rain and dew of this country are equivalent to the quantity of water carried off by evaporation and by the rivers. And as nature acts upon general laws, we ought to infer, that it must be the case in every other country, till the contrary is proved.

All this is fair enough, but evidently in Dalton's opinion, the closure of the water budget is a separate issue from the origin of springs and not a persuasive argument to prove that precipitation is their sole source. Thus he points out next (p. 367) that at the time

... There are three opinions respecting the origin of springs which it may be proper to notice.

1st.   That they are supplied entirely by rain and dew.
2d.   That they are principally supplied by large subterranean reservoirs of water.
3d.   That they derive their water originally from the sea, on the principle of filtration.

It is obvious that before we pay any attention to the latter two opinions, the causes assigned in the first ought to be proved insufficient by direct experiment. M. de la Hire is the only one who has attempted to do this ...

> It is remarkable that, at the dawn of the nineteenth century, these are still essentially the opinions which were being discussed among the Presocratics and Aristotle more than 2300 years earlier. Dalton proceeds then to show that the experimental disproof of the first opinion by de La Hire (1703), was in fact invalid and unwarranted, and this leads him finally to conclude (p. 371) as follows.

> The origin of springs may still therefore be attributed to rain, till some more decisive experiments appear to the contrary; and it becomes unnecessary to controvert the other two opinions respecting this subject.

14.5   CLOSING COMMENTS

This previous quotation was probably the last time that any other "opinions" on the origin of springs were brought up in the scientific literature. Still, although the debate on the main issue was closed, details of this "rainfall percolation" continue to be the subject of enquiry to this day, as seen in Chapter 11. In any event, Dalton's (1802a) analysis was a sign that the time was ripe for the rapid developments in the nineteenth century, that laid the foundations for the emergence of hydrologic science in its present form. For instance, it was Dalton (1802b) who introduced next several of the principles on which modern evaporation theory is based (see Brutsaert, 1982, p. 31). He proposed the law of partial pressures in gases and he determined the saturation vapor pressure of water as a function of temperature; he was then the first to express surface evaporation as a mass transfer equation nearly in its current form, and in recognition of this the mass transfer coefficient is still called the Dalton number. Fundamental developments by others followed in rapid succession throughout the nineteenth century, and several of the highlights are mentioned in the previous chapters of this book. But most of this is well-trodden terrain in the history of science, so there is no need to repeat the details.

Among the more striking facts of this historical sketch is that, while humans were able to grasp the essence and the significance of the atmospheric phase of the water cycle very early in prehistoric times, a full understanding of the origin of springs and streams took much longer.

The perceptions and opinions of those who commented on the movement of water in nature, were usually strongly affected by the specific hydrologic conditions in their immediate environment. Some of the early civilizations developed in rather arid and semi-arid climates, where rain, springs and streamflow were not always abundant, so that the linkages of the terrestrial water cycle were not very obvious. A case in point is the eastern Mediterranean region, where karst phenomena are ubiquitous and play a pronounced role. In this perspective many of the early concepts, such as the underground Tartaros or Abyss of Homer and Plato, and the caves of Anaxagoras and Aristotle, can be explained and are not as far-fetched as a superficial review might suggest. Similarly, to Thales or to the writer of Ecclesiastes, who must have known about underground seawater intrusion near the coast or in the Nile delta, the seawater filtration mechanism would not have been unreasonable.

The concept that finally survived, the rainfall percolation mechanism, is not a recent invention. In recorded history it can be followed as a thread running through the works of the pre-Socratics, the post-Aristotelian Peripatetics, Vitruvius in ancient Rome, Buridan and other medieval Schoolmen at the university of Paris, Bartas, Palissy and Gassendi in the Renaissance, and finally Mariotte, Ray and Van Musschenbroek, at the dawn of modern science. But all along it was only one of several competing theories. It is noteworthy that in many instances the rainfall percolation mechanism was advocated by active persons of a more practical inclination, rather than by philosophers. Also, its supporters often tended to have spent their formative years in the countryside in more humid climates with vegetation, and less in denuded arid regions or urban areas, where ubiquitous puddles and overland flow during rain indicate an almost total absence of infiltration. For example, Vitruvius, a rain and snow penetration advocate, had been a military engineer with Caesar's army in Gaul as a young man, before his career in architecture in Rome; during the Renaissance, Palissy was known mostly as a ceramic artist and du Bartas as a soldier and a diplomat. Both Perrault and Halley had grown up in urban environments, while Mariotte and Ray, who were proponents of the Common Opinion, had spent their youth in more rural settings. All this is consistent with the more recent findings on the occurrence of the different flow paths in the streamflow generation processes described in Chapter 11.

REFERENCES

Alfred of Sareshel, *Commentary on the Metheora of Aristotle*, Critical Edition; Introduction, and Notes by J. K. Otte. Leiden: E. J. Brill, 1988.

Ambrose, St, *Hexameron, Paradise, and Cain and Abel*, translated by J. J. Savage. New York: Fathers of the Church, Inc., 1961.

Anandatirtha (Madhvacharya) (1910). *Chandogya Upanisad* (Vol. 3, Part 2 of the Sacred Books of the Hindus). Allahabad: Panini Office, Bhuvaneswari Asrama, Bahadurganj. (Also New York: AMS Press, 1974.)

Anonymous (1675). A particular account given by an anonymous French Author in his book of the Origin of Fountains, printed 1674 at Paris; to shew that the Rain and Snow-waters are sufficient to make Fountains and Rivers run perpetually. *Phil. Trans. R. Soc. Lond.*, **10**, No. 119, 447–450.

Anonymous (1686). (Review of) Traité du mouvement des eaux et des autres corps fluides par feu Mr. Mariotte, A Paris An 1686, Octavo. *Phil. Trans. R. Soc. Lond.*, **16**, No. 181, 119–123.

Aristotle, *Meteorologica*, with an English translation by H. D. P. Lee. London: W. Heinemann, Ltd; Cambridge, MA: Harvard University Press, 1952.

Aristotle, *On Plants*. In *Minor Works*, with an English translation by W. S. Hett. London: W. Heinemann, Ltd; Cambridge, MA: Harvard University Press, pp. 141–233, 1936.

Basil, St., *Exegetic Homilies*, translated by A. C. Way, *The Fathers of the Church*, **46**. Washington, DC: The Catholic University of America Press, 1963.

Beda, Venerabilis, *De Natura Rerum*. In *The Complete Works of Venerable Bede in the Original Latin, Vol. VI*, ed. J. A. Giles. London: Whittaker and Co., 1843.

Bentley, R. (1693). *A Confutation of Atheism From the Origin and Frame of the World, A Sermon Preached at St. Mary-le-Bow, Dec. 5, 1692*. London: H. Mortlock.

Brams, J. and Vanhamel, W. (editors) (1989). *Guillaume de Moerbeke: recueil d'études à l'occasion du 700e anniversaire de sa mort (1286)*. Leuven: University Press.

Brutsaert, W. (1982). History of the theories of evaporation. In *Evaporation into the Atmosphere*, chapter 2. Dordrecht: D. Reidel Publ. Co. (Kluwer Academic), pp. 12–36.

Buckingham, E. (1907). *Studies on the movement of soil moisture*. Bureau of Soils, Bull. No. 38. Washington DC: US Department of Agriculture.

Cook, A. (1998). *Edmond Halley, Charting the Heavens and the Seas*. Oxford: Clarendon Press.

Dalton, J. (1793). On evaporation, rain, hail, snow, and dew. Sixth essay in *Meteorological Observations and Essays*. London: W. Richardson. (Also nearly verbatim second edition, 1843.)

    (1802a). Experiments and observations to determine whether the quantity of rain and dew is equal to the quantity of water carried off by the rivers and raised by evaporation; with an enquiry into the origin of springs. *Mem. Lit. Phil. Soc. Manchester*, **5** (part 2), 346–372.

    (1802b). Experimental essays on the constitution of mixed gases; on the force of steam or vapor from water and other liquids in different temperatures, both in a Torricellian vacuum and in air; on evaporation and on the expansion of gases by heat. *Mem. Lit. Phil. Soc. Manchester*, **5** (part 2), 535–602.

Davies, B. (1974). Edme Mariotte, 1620–1684. *Physics Education*, **9**, 275–278.

de Condorcet, Marquis (1773). *Éloges des académiciens de l'Académie Royale des Sciences morts depuis 1666, jusqu'en 1699*, Hotel de Thou, rue des Poitevins, Paris.

de La Hire, P. (1703). Sur l'eau de pluie, & sur l'origine des fontaines; avec quelques particularités sur la construction des citernes. *Histoire de l'Acad. Roy. des Sciences (Avec les Mémoires de Mathématique & de Physique pour la Même Année), Mém.*, 56–69 (see also Anonym., Sur l'origine des rivières, *Hist.*, 1–6). Paris: J. Boudet. (Republished in 1789 as second edition, pp. 72–90. Amsterdam: P. Mortier.)

Delorme, S. (1948). Pierre Perrault auteur d'un traité *De l'Origine des Fontaines* et d'une théorie de l'expérimentation. *Archives Internationales d'Histoire des Sciences*, **27**, 388–394.

Descartes, R. (1637). *Discours de la méthode, plus la dioptrique, les metéores et la géometrie, qui sont les essais de cette méthode*. Leyde [Leiden]: De l'imprimerie de Ian Maire.

Diels, H. (1879). *Doxographi Graeci*. Berlin: G. Reimer.

    (1961). *Die Fragmente der Vorsokratiker*, 10. Auflage herausgegeben von W. Kranz, 1. Band. Berlin: Weidmannsche Verlagsbuchhandlung.

Diogenes Laertius, *Lives of Eminent Philosophers*, Vol. II, with an English translation by R. D. Hicks. London: W. Heinemann; New York: G. P. Putnam's Sons, 1925.

Dobson (1777). Observations on the annual evaporation at Liverpool in Lancashire; and on evaporation considered as a test of the moisture or dryness of the atmosphere. *Phil. Trans. R. Soc. Lond.*, **67**, 244–259.

du Bartas, G. de Salluste, *La sepmaine ou création du monde*, texte préparé par V. Bol. Arles: Édns. Actes Sud, 1988. (First published in 1578.)

Ducos, J. (1998). *La Météorologie en français au moyen âge (XIIIe–XIVe siècles)*. Paris: Honoré Champion Éditeur.

Eliade, M. (1978). *A History of Religious Ideas, Vol. 1. From the Stone Age to the Eleusinian Mysteries* (translated from the French by W. R. Trask, 1962). Chicago: University of Chicago Press.

Gilbert, O. (1907). *Die Meteorologischen Theorien des Griechischen Altertums*. Leipzig: B. G. Teubner.

Grabmann, M. (1916). *Forschungen Über die Lateinischen Aristotelesübersetzungen des 13. Jahrhunderts, Beiträge zur Geschichte der Philosophie des Mittelalters*, **18** (5–6). Münster i. W: Aschendorffschen Verlagsbuchhandlung.

Hallays, A. (1926). *Les Perrault*. Paris: Perrin et Cie, Libraires-Éditeurs.

Halley, E. (1687). An estimate of the quantity of vapour raised out of the sea by the warmth of the sun. *Phil. Trans. R. Soc. London*, **16**, No. 189, 366–370.

(1691). An account of the circulation of the watery vapours of the sea, and of the cause of springs. *Phil. Trans. R. Soc. Lond.*, **16**, No. 192, 468–473.

(1694). An account of the evaporation of water, as it was experimented in Gresham Colledge in the year 1693. With some observations thereon. *Phil. Trans. R. Soc. Lond.*, **18**, No. 212, 183–190.

Heninger, S. K. Jr. (1960). *A Handbook of Renaissance Meteorology*. Durham, NC: Duke University Press.

Hesiod, *Works and Days*, ed. M. L. West. Oxford: Clarendon Press, 1978.

Hésiode, *Théogonie; les travaux et les jours; le bouclier*, texte établi et traduit par P. Mazon. Paris: Société d'Édition Les belles Lettres, 1928.

Higgins, M. D. and Higgins, R. (1996). *A Geological Companion to Greece and the Aegean*. Ithaca, NY: Cornell University Press.

Huxley, T. H. (1900). *Physiography*. London: MacMillan and Co. Ltd; New York: D. Appleton and Co. Ltd.

Isidore de Séville, *Traité de la nature (Liber de natura rerum)*, édité par J. Fontaine. Bordeaux: Féret et Fils, Éds., 1960.

Isidorus Hispalensis, E., *Etymologiarum Sive Originum Libri XX ("Etymologiae")*, Tomus II, Oxonii e Typographeo Clarendonio. Oxford: Oxford University Press, 1911.

Jourdain, A. (1960). *Recherches critiques sur l'âge et l'origine des traductions latines d'Aristote* (edition of 1843). New York: Burt Franklin.

Lin, Yutang (editor) (1949). *The Wisdom of China*. London: Michael Joseph Ltd., 1949.

Lohr, C. H. (1967). Medieval Latin Aristotle commentaries. *Traditio*, **23**, 313–413; (1968) **24**, 149–245; (1970) **26**, 135–216; (1971) **27**, 215–351; (1972) **28**, 281–396; (1973) **29**, 93–197.

Lucretius, *De Rerum Natura*, with an English translation by W. H. D. Rouse. London: W. Heinemann, Ltd.; Cambridge, MA: Harvard University Press, 1924.

MacCurdy, E. (1938). *The Notebooks of Leonardo da Vinci, Arranged, Rendered into English and Introduced*, Vol. II. New York: Reynal & Hitchcock.

MacPike, E. F. (editor) (1932). *Correspondence and Papers of Edmond Halley*. Oxford: Clarendon Pess.

Mansfeld, J. (1992). *Heresiography in Context: Hippolytus' Elenchos as a Source for Greek Philosophy*. Leiden: E. J. Brill.

Mansfeld, J. and Runia, D. T. (1997). *Aetiana: The Method and Intellectual Context of a Doxographer*. Leiden: E. J. Brill.

Mariotte, E. (feu) (1686). *Traité du mouvement des eaux et des autres corps fluides* (*mis en lumière par les soins de M. de la Hire, etc.*), Paris: Chez Estienne Michallet.

Mieli, A. (1966). *La science arabe*. Leiden: E. J. Brill.

Needham, J. (1959). *Science and Civilization in China, Vol. 3, "Mathematics and the sciences of the heavens and the earth"* (with collaboration of Wang Ling). Cambridge: Cambridge University Press.

Oxford Study Edition, The New English Bible, with Apocrypha. (1976). New York: Oxford University Press.

Palissy, M. B. (1888). *Discours admirables de la nature des eaux et fontaines, etc.* Paris: Martin Le Jeune (1580). (Reprinted in Fillon, B. (1888) *Les oeuvres de Maistre Bernard Palissy*, Vol. 2. Paris: L. Clouzot, Lib.

(1957). *The Admirable Discourses*, translated by A. La Rocque. Urbana: University of Illinois Press. (First published in 1580.)

Perrault, Ch. (1993). *Mémoires de ma vie*, précédé d'un essai d'Antoine Picon: *"Un moderne paradoxal"*. Paris: Macula.

Perrault, P. (anonymously) (1674). *De l'origine des fontaines.* Paris: Pierre Le Petit, Imprimeur & Libraire.

Peters, F. E. (1968). *Aristotle and the Arabs.* New York: New York University Press.

Petit Larousse, dictionnaire encyclopédique pour tous. (1964). Paris: Librairie Larousse.

Picolet, G. État des connaissances actuelles sur la biographie de Mariotte et premiers résultats d'une enquête nouvelle. In *Mariotte savant et philosophe (†1684): analyse d'une renommée*, Préface de P. Costabel. Centre A. Koyre. Paris: J. Vrin, pp. 245–276, 1968.

Plato, *Phaedo*, translated by D. Gallop. Oxford: Oxford University Press, 1975.

Plato, *Phaedo* (Greek text), edited by C. J. Rowe. Cambridge: Cambridge University Press, 1993.

Rabanus Maurus, *De Universo [De Rerum Naturis].* In *Patrologiae Cursus Completus*, Vol. 111. Paris: J.-P. Migne, pp. 10–614, 1852.

Radhakrishnan, S. (1953). *The Principal Upanisads.* London: George Allen & Unwin Ltd.

Raven, C. E. (1950). *John Ray, Naturalist, His Life and Works.* Cambridge: Cambridge University Press.

Ray, J. (1673). *Observations Topographical, Moral, & Physiological; Made in a Journey Through Part of the Low-Countries, Germany, Italy, and France: With a Catalogue of Plants . . .* London: J. Martyn Printer to the Royal Society.

  (1692). *Miscellaneous Discourses Concerning the Dissolution* and *Changes of the World.* London: S. Smith.

  (1693). *Three Physico-Theological Discourses.* London: S. Smith.

  (1759). *The Wisdom of God Manifested in the Works of Creation*, twelfth edition. London: J. Rivington.

Seneca, *Naturales Quaestiones*, with an English translation by T. H. Corcoran. London: W. Heinemann, Ltd; Cambridge, MA: Harvard University Press, Vol. I, 1971.

Swahananda, S. (1965). *The Chandogya Upanisad.* Mylapore, Madras: Sri Ramakrishna Math.

Thorndike, L. (1954). Oresme and fourteenth century commentaries on the Meteorologica. *Isis*, **45**, 145–152.

  (1955). More questions on the Meteorologica. *Isis*, **46**, 357–360.

Tuan, Y.-F. (1968). *The Hydrologic Cycle and the Wisdom of God: A Theme in Geoteleology.* Toronto: University of Toronto Press.

Van Musschenbroek, P. (1739). *Essai de physique*, traduit du hollandois par Mr. P. Massuet, Tome I. Leyden: S. Luchtmans.

  (1769). *Cours de physique expérimentale et mathématique*, traduit par M. S. de la Fond, Tome second. Paris: Bauche.

Vitruve, *Les dix livres d'architecture*, Traduction intégrale de Claude Perrault, 1673, revue et corrigée sur les textes latins et presentée par André Dalmas. Éditions Errance. Paris: A. Balland, 1986.

Wang, P. K. (1996). *Heaven and Earth.* Taipei: Newton Publ. Co., Ltd. (in Chinese).

Woodward, J. (1695). *An Essay Toward a Natural History of the Earth and Terrestrial Bodies, Especially Minerals: As Also of the Seas, Rivers and Springs.* London: Ric. Wilkin.

APPENDIX
SOME USEFUL MATHEMATICAL CONCEPTS

A1   DIFFERENTIATION OF AN INTEGRAL

Consider an integral, whose upper and lower limits $h = h(x)$ and $g = g(x)$ are differentiable functions of x, that is

$$F(x) = \int_{g(x)}^{h(x)} f(y, x)dy \tag{A1}$$

If $f(y, x)$ is continuous and smooth, the derivative of this integral can be written as

$$\frac{dF}{dx} = \int_{g(x)}^{h(x)} \frac{\partial f(y, x)}{\partial x}dy + f[h(x), x]\frac{\partial h(x)}{\partial x} - f[g(x), x]\frac{\partial g(x)}{\partial x} \tag{A2}$$

which is commonly known as Leibniz's formula. For a proof of the Leibniz formula the reader is referred to any good calculus textbook.

A2   THE GENERAL RESPONSE OF A LINEAR STATIONARY SYSTEM

The unit impulse
Physically, this can be described as an input or excitation of unit magnitude, imposed either suddenly and lasting a very short time over a wide area, or imposed locally and acting over a very small distance or area but in a steady fashion; of course, it can also be a combined spike action in both time and space. Typical examples are a hammer blow on a mechanical system, a more steady type of concentrated load, a voltage surge on an electrical system, a lightning stroke on a transmission line, or a sudden rainfall burst on a catchment. For example, as illustrated in Figure A1, one can construct a unit excitation applied over a finite interval Δt around t_0 as follows

$$E(t_0) = \begin{cases} 0 & \text{for } t < t_0 - \Delta t/2 \\ I/\Delta t & \text{for } t_0 - \Delta t/2 < t < t_0 + \Delta t/2 \\ 0 & \text{for } t > t_0 + \Delta t/2 \end{cases} \tag{A3}$$

Note that in this example I is taken as a constant; this need not be, and for the present purpose, the unit excitation can be equally well described by a function with a variable intensity $I = I(t)$ (such as, for example, an error function or a triangle) over the same interval $t_0 - \Delta t/2 < t < t_0 + \Delta t/2$ as in Equations (A3). Note further that, while t usually denotes time, in the present context it can also denote a spatial variable.

Fig. A1 **Example of an impulse function of magnitude**
I, **uniform intensity** $I/\Delta t$, **and duration** Δt.

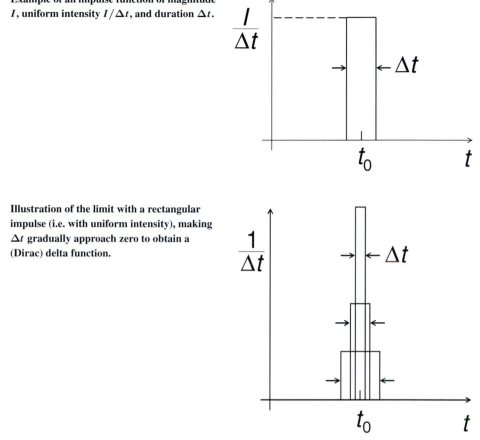

Fig. A2 **Illustration of the limit with a rectangular**
impulse (i.e. with uniform intensity), making
Δt **gradually approach zero to obtain a**
(Dirac) delta function.

The unit impulse or the (Dirac) delta function can be obtained from (A3), in the limit, as Δt is made to approach zero, as illustrated in Figure A2. Thus one has in this case

$$\delta(t - t_0) = \lim_{\Delta t \to 0} \begin{cases} 0 & \text{for } t < t_0 - \Delta t/2 \\ 1/\Delta t & \text{for } t_0 - \Delta t/2 < t < t_0 + \Delta t/2 \\ 0 & \text{for } t > t_0 + \Delta t/2 \end{cases} \tag{A4}$$

A similar procedure can be applied for other excitation functions as well; this is illustrated in Figure A3 for a triangular function.

In general, following these preliminaries, the delta function is often expressed a

$$\delta(t - t_0) = \begin{cases} 0 & \text{for } t \neq t_0 \\ \infty & \text{for } t = t_0 \end{cases} \tag{A5}$$

$$\int_{-\infty}^{+\infty} \delta(t - t_0)dt = 1 \tag{A6}$$

Fig. A3 **Illustration of the limit with a triangular impulse, making Δt gradually approach zero to obtain a (Dirac) delta function.**

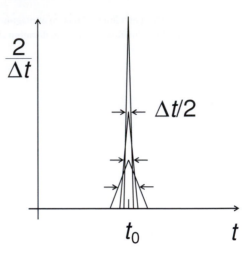

Equations (A4) and (A5) represent singular behavior, and they indicate that $\delta(t - t_0)$ is not continuous and not differentiable at $t = t_0$. Therefore, this definition cannot be taken literally, but it must be interpreted as suggestive of the limiting process involved. A better way to define the delta function is in the following integral form

$$\int_{-\infty}^{+\infty} \delta(t - t_0) f(t) dt = f(t_0) \tag{A7}$$

in which $f(t)$ is a continuous and smooth function. While the Dirac delta function is not a well-behaved function in the usual sense, it is classified as a *generalized function*. As explained in Greenberg (1971), one never talks about the values of a generalized function, but only about its action on the function $f(x)$, as indicated in Equation (A7).

The unit step function
A function closely related to the Dirac delta function is the Heaviside step function which can defined as follows

$$H(t - t_0) = \begin{cases} 0 & \text{for } t < t_0 \\ 1 & \text{for } t > t_0 \end{cases} \tag{A8}$$

and is illustrated in Figure A4. It is often convenient to think of the unit step function as the integral of the unit impulse function or, vice versa, of the impulse function as the derivative of the step function.

Unit response and actual response of a system
The transformation of input into output is called the response of a system. In general, a transformation or an operation **T** is said to be *linear* when the operation on the sum of two functions is the sum of the operations of each individual function; in mathematical terms, **T** is linear if the equality

$$\mathbf{T}[a\,x(t) + b\,y(t)] = a\,\mathbf{T}[x(t)] + b\,\mathbf{T}[y(t)] \tag{A9}$$

Fig. A4   **Illustration of the (Heaviside) unit step function**
$H = H(t - t_0)$.

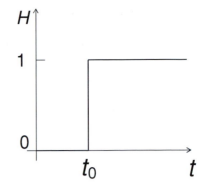

is valid for all constants a and b and for all x and y, such that $\mathbf{T}[x(t)]$ and $\mathbf{T}[y(t)]$ exist. A transformation or operation is said to be *stationary* or *invariant*, if for the same conditions at $t = 0$, it produces the same result under a coordinate translation; thus the transformation $z(t) = \mathbf{T}[x(t)]$ is stationary if

$$z(t - t_0) = \mathbf{T}[x(t - t_0)] \tag{A10}$$

remains valid for any value of t_0.

The *unit response* of a linear system $u = u(t)$ is its response to the unit impulse function $\delta(t)$. The unit response is also variously called the *impulse response*, the *Green's function*, and the *influence function* of the system. If these response characteristics are invariant in time or space (whatever domain t denotes), the response is $u(t_0 - t)$, when the input is $\delta(t_0 - t)$. Because the system is linear, it follows that the response becomes $x(t) u(t_0 - t)$ when the input is $x(t) \delta(t_0 - t)$. By the same token, upon multiplication of this response and of this input by dt and integration of both, comparison of the resulting input with Equation (A7) shows that when the input is $x(t)$, the response or output of the system is given by

$$y(t) = \int_{-\infty}^{+\infty} x(\tau)\, u(t - \tau)\, d\tau \tag{A11}$$

where τ is a dummy variable of integration. The operation shown in (A11) is called the *convolution integral*.

The upper and lower limits of Equation (A11) indicate that the output from the system is affected by input values of $x(t)$ for t all the way from minus to plus infinity. Time dependent hydrologic systems are causal and non-anticipatory; this means that they only depend on the values of present and past (but not future) values of the input function; such systems are also referred to as *hereditary* (Volterra, 1913). Thus whenever t denotes time, in hydrologic applications the upper limit of the integral in (A11) should be t, and one can write

$$y(t) = \int_{-\infty}^{t} x(\tau)\, u(t - \tau)\, d\tau \tag{A12}$$

Fig. A5   Illustration of the convolution or folding operation for a hereditary or causal system, with $t = 0$ defined as the start of the input rate $x = x(t)$. At any given value of time t, the total output rate y is the result (i.e. integral) of all past inputs from the start of the input until t, weighted at each instant with the unit response folded backwards. In this operation t is treated as a constant and τ is the dummy time variable of the integration.

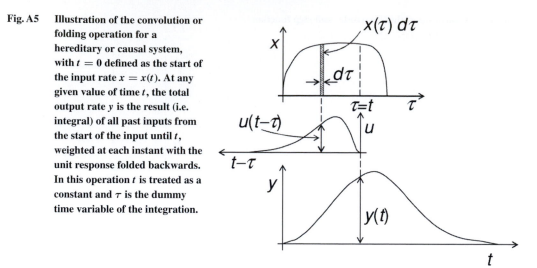

Equation (A12) describes the output from a system with a memory going back to $-\infty$. If the system only has a finite memory m, the lower limit of the integral can be changed to $(t - m)$, or

$$y(t) = \int_{t-m}^{t} x(\tau)\, u(t - \tau)\, d\tau \tag{A13}$$

Equation (A13) also describes the response of a system, in which the input starts m time units prior to t. Hence, if the input starts at $t = 0$, the convolution integral becomes

$$y(t) = \int_{0}^{t} x(\tau)\, u(t - \tau)\, d\tau \tag{A14}$$

In Equations (A12)–(A14), τ should be interpreted as the general time variable in the convolution operation, whereas t is the designated time at which the response is to be determined. The meaning of the name *convolution* or *folding* integral is illustrated for (A14) in Figure A5.

Because τ can be replaced by $(t - \tau)$ in Equations (A11)–(A14), each of these convolution integrals can be written in a form which is sometimes more convenient. For instance, in the case of Equation (A13) this is simply

$$y(t) = \int_{0}^{m} u(\tau) x(t - \tau)\, d\tau \tag{A15}$$

and in the case of Equation (A14) it is

$$y(t) = \int_{0}^{t} u(\tau) x(t - \tau)\, d\tau \tag{A16}$$

Fig. A6 **Approximation of a function $x = x(\tau)$ by a sequence of pulses of width $\Delta\tau$.**

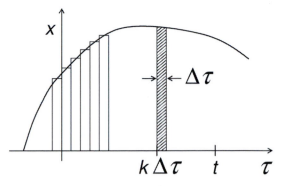

In numerical applications of the convolution integral, the input function $x(t)$ can be represented by a histogram consisting of pulses of width $\Delta\tau$, as illustrated in Figure A6. The contribution to the total response by the input pulse at $\tau = (k\ \Delta\tau)$ is given by

$$x(k\ \Delta\tau)\,u(\Delta\tau; t - k\ \Delta\tau)\Delta\tau$$

in which $u(\Delta\tau; t - k\ \Delta\tau)$ represents the response of the system to an input pulse of width $\Delta\tau$. The total response at t is the sum of the contributions by all input pulses, or

$$y(t) = \sum_{k=-\infty}^{+\infty} x(k\ \Delta\tau)\,u(\Delta\tau; t - k\ \Delta\tau)\Delta\tau \tag{A17}$$

which is the discrete analog of the general convolution integral (A11). Again, if t represents time, and the system is non-anticipatory with an input starting at $t = 0$, one obtains the discrete analog of (A14), or

$$y(t) = \sum_{k=0}^{n} x(k\ \Delta\tau)\,u(\Delta\tau; t - k\ \Delta\tau)\Delta\tau \tag{A18}$$

in which $\tau = n\ \Delta\tau (\leq t)$ is the time of the last input pulse prior to the designated response time $\tau = t$.

Relationships between the moments
The convolution integral provides also convenient relationships between the moments of the three functions involved, namely the input function $x(t)$, the output function $y(t)$ and the unit response function $u(t)$. Denote the mean values of t (or centers of area) of these three functions respectively as m'_{y1}, m'_{x1} and m'_{u1} and the nth moments about these means of the three functions, respectively as m_{yn}, m_{xn} and m_{un}. Assume for convenience that these functions are properly scaled, so that their zero-order moments $\int y\ dt, \int x\ dt$ and $\int u\ dt$ are equal to unity. The center of area of the output function, which is the first moment about the origin, can be calculated as

$$m'_{y1} = \int_{-\infty}^{\infty} t\ y(t)dt \tag{A19}$$

Making use of the convolution integral (A11) one obtains

$$m'_{y1} = \int_{-\infty}^{\infty} t \int_{-\infty}^{\infty} x(\tau)u(t-\tau)d\tau dt \qquad (A20)$$

Inversion of the order of integration, and substitution of $s = t - \tau$, so that $dt = ds$, changes Equation (A20) into

$$m'_{y1} = \int_{-\infty}^{\infty} x(\tau)d\tau \int_{-\infty}^{\infty} (\tau + s) \, u(s)ds$$

$$= \int_{-\infty}^{\infty} \tau x(\tau)d\tau \int_{-\infty}^{\infty} u(s)ds + \int_{-\infty}^{\infty} x(\tau)d\tau \int_{-\infty}^{\infty} s \, u(s)ds \qquad (A21)$$

Because the zeroth moments are equal to one, this yields the following relationship between the first moments about the origin

$$m'_{y1} = m'_{x1} + m'_{u1} \qquad (A22)$$

The nth moment of the output $y(t)$ about its center of area m'_{y1} can be written, after substitution of the convolution integral (A11) and of (A22), as follows

$$m_{yn} = \int_{-\infty}^{\infty} \int_{-\infty}^{\infty} (t - m'_{x1} - m'_{u1})^n x(\tau)u(t-\tau)d\tau dt \qquad (A23)$$

Again, inverting the order of integration, and putting $s = t - \tau$, so that $dt = ds$, one can rewrite (A23) as

$$m_{yn} = \int_{-\infty}^{\infty} x(\tau)d\tau \int_{-\infty}^{\infty} [(\tau - m'_{x1}) + (s - m'_{u1})]^n u(s)ds \qquad (A24)$$

The term with the square brackets can be expanded as follows

$$[(\tau - m'_{x1}) + (s - m'_{u1})]^n = (\tau - m'_{x1})^n + n(\tau - m'_{x1})^{n-1}(s - m'_{u1})$$

$$+ \frac{n(n-1)}{2!}(\tau - m'_{x1})^{n-2}(s - m'_{u1})^2 + \cdots + (s - m'_{u1})^n \qquad (A25)$$

This allows Equation (A24) to be written as

$$m_{yn} = \int_{-\infty}^{\infty} x(\tau)[(\tau - m'_{x1})^n + n(\tau - m'_{x1})^{n-1}m_{u1}$$

$$+ \frac{n(n-1)}{2!}(\tau - m'_{x1})^{n-2}m_{u2} + \cdots + m_{un}]d\tau \qquad (A26)$$

from which one obtains the main result

$$m_{y,n} = m_{x,n} + n\,m_{x,n-1}m_{u,1} + \frac{n(n-1)}{2!}m_{x,n-2}m_{u,2} + \cdots + m_{u,n} \tag{A27}$$

The commas have been introduced in the subscripts in this expression merely for clarity of notation. The first moments about the mean are zero; hence, for the smaller values of n, which are the ones of practical importance, one obtains from (A27)

$$m_{y2} = m_{x2} + m_{u2}$$

$$m_{y3} = m_{x3} + m_{u3} \tag{A28}$$

$$m_{y4} = m_{x4} + m_{u4} + 6m_{x2}m_{u2}$$

Equations (A22) and (A27) (or (A28)) are jointly sometimes referred to as the theorem of moments; it was introduced in the hydrologic literature by Nash (1959) for the purpose of deriving the moments of the unit response function from observed records of effective rainfall and storm runoff.

A3   THE GENERAL RESPONSE OF A NONLINEAR SYSTEM

The response of a nonlinear system can be described by a generalization of the convolution integral operation. This type of approach goes back to the work of Volterra (1913; 1959; also Barrett, 1963) who showed that a hereditary system can be described by a convergent series of integrals,

$$y(t) = F(x = 0) + \int_{-\infty}^{t} u_1(t, \tau)x(\tau)d\tau$$

$$+ \frac{1}{2!} \int_{-\infty}^{t} \int_{-\infty}^{t} u_2(t, \tau_1, \tau_2)x(\tau_1)x(\tau_2)d\tau_1 d\tau_2 + \cdots$$

$$\cdots + \frac{1}{n!} \int_{-\infty}^{t} \cdots \int_{-\infty}^{t} u_n(t, \tau_1, \ldots, \tau_n) \prod_{i=1}^{n} x(\tau_i)d\tau_i + \cdots \tag{A29}$$

in which the $u_i(\ )$ terms are the kernels of the integrals and the subscripts indicate their order. As before, $y(t)$ and $x(t)$ are the output and input of the system, as functions of time. If the system does not generate any output, when the input is zero, the first term on the right of (A29) can be omitted. Also, when the system is time invariant, it can be shown that the kernels must be of the form $u_1(t, \tau) = u_1(t - \tau)$, etc. The lower limit of (A29) indicates that the system has an infinite memory. Thus, assuming that the system has, first, no zero input response, second, time-invariant response characteristics, and third, a finite memory m, one can rewrite Equation (A29) (by analogy with (A13)) as follows

$$y(t) = \int_{t-m}^{t} x(\tau)\, u_1(t - \tau)d\tau$$

$$+ \frac{1}{2!} \int_{t-m}^{t} \int_{t-m}^{t} x(\tau_1)x(\tau_2)\, u_2(t - \tau_1, t - \tau_2)d\tau_1 d\tau_2$$

$$+ \frac{1}{3!} \int_{t-m}^{t} \int_{t-m}^{t} \int_{t-m}^{t} x(\tau_1)x(\tau_2)x(\tau_3)$$

$$\times u_3(t - \tau_1, t - \tau_2, t - \tau_3)d\tau_1 d\tau_2 d\tau_3 + \cdots \tag{A30}$$

To facilitate numerical computations, one can replace τ by $(t - \tau)$, etc., in (A30) to obtain, as in (A15) (after absorbing the factorials in the kernels),

$$y(t) = \int_{0}^{m} u_1(\tau)x(t - \tau)d\tau + \int_{0}^{m} \int_{0}^{m} u_2(\tau_1, \tau_2)x(t - \tau_1)x(t - \tau_2)d\tau_1 d\tau_2$$

$$+ \int_{0}^{m} \int_{0}^{m} \int_{0}^{m} u_3(\tau_1, \tau_2, \tau_3)x(t - \tau_1)x(t - \tau_2)x(t - \tau_3)d\tau_1 d\tau_2 d\tau_3 + \cdots \tag{A31}$$

REFERENCES

Barrett, J. F. (1963). The use of functionals in the analysis of non-linear physical systems. *J. Electron. Contr.*, **15**, 567–615.

Greenberg, M. D. (1971). *Applications of Green's Functions in Science and Engineering*. Englewood Cliffs, NJ: Prentice Hall.

Nash, J. E. (1959). Systematic determination of unit hydrograph parameters. *J. Geophys. Res.*, **64**, 111–115.

Volterra, V. (1913). *Leçons sur les équations intégrales et les équations intégro-différentielles*. Paris: Gauthier-Villars.

(1959). *Theory of Functionals and Integral and Integro-Differential Equations*. New York: Dover.

INDEX